Field Crop Diseases
Handbook

FIELD CROP DISEASES HANDBOOK

Second Edition

Robert F. Nyvall, Ph.D.

Professor, Plant Pathology
and
Superintendent, University of Minnesota
North Central Experiment Station
Grand Rapids, Minnesota

An **avi** Book
Published by Van Nostrand Reinhold
New York

An AVI Book
(AVI is an imprint of Van Nostrand Reinhold)

Copyright © 1989 by Van Nostrand Reinhold
Library of Congress Catalog Card Number 88-20776
ISBN 0-442-26722-3

Printed in the United States of America

Designed by Beehive Production Services

Van Nostrand Reinhold
115 Fifth Avenue
New York, New York 10003

Van Nostrand Reinhold (International) Limited
11 New Fetter Lane
London EC4P 4EE, England

Van Nostrand Reinhold
480 La Trobe Street
Melbourne, Victoria 3000, Australia

Macmillan of Canada
Division of Canada Publishing Corporation
164 Commander Boulevard
Agincourt, Ontario M1S 3C7, Canada

16 15 14 13 12 11 10 9 8 7 6 5 4 3 2 1

Library of Congress Cataloging in Publication Data
Nyvall, Robert F.
    Field crop diseases handbook / Robert F. Nyvall. — 2nd ed.
        p. cm.
    "An AVI book."
    Bibliography: p.
    Includes index.
    ISBN 0-442-26722-3
    1. Field crops—Diseases and pests—Handbooks, manuals, etc.
2. Plant diseases—Handbooks, manuals, etc. 3. Phytopathogenic
microorganisms—Control—Handbooks, manuals, etc.   I. Title.
SB731.N94   1989
633'.0893—dc19   88-20776
                                                                      CIP

To Martha, Nathan, and Sandra;
and to my parents,
Gertrude and Robert.

Contents

Preface

The purpose of *Field Crop Diseases Handbook* is to provide basic information in one volume on the diseases of many of the world's important field crops. The second edition contains considerably more information than the first edition. The first edition contained information on over 800 different diseases; the second edition includes information on approximately 1,200 diseases of 25 different plants. Investigations of the causal organisms continually contribute new knowledge or modify existing facts. Information on many of the diseases included in the first edition has either been expanded or modified in light of new research.

The purpose of the book, however, remains the same: to provide information on diseases of field crops in one volume to workers and students in all phases of agriculture. The organization of the book also remains basically the same. Diseases are grouped by causal organism and are then listed in each group alphabetically by common name. While some may criticize this method due to the variation in common names, it has proven to be a helpful way to categorize diseases, especially for nonplant pathologists. To facilitate finding a disease, other common names for the same disease are listed.

The writing of such a book, of necessity, entails some frustration. Taxonomically, genera and species of some organisms have undergone numerous changes. However, some of these changes may not be universally accepted. Therefore, the use of a latin binomial for a causal organism may not be standard among different scientists working on the same organism. Terms and descriptions may vary among plant pathologists and other scientists. Although, in my judgment, I have attempted to select the most common latin binomials and terms, I ask the indulgence of those who may not agree with my choices.

Most important, the value of a book such as this lies in its usefulness to its intended audience. If this book helps one student to identify the cause of a disease or one agricultural professional to understand how a disease happens, then its worth is vindicated.

ACKNOWLEDGMENTS

I would like to thank the following people: Mrs. Carolyn Frings for her help in the preparation of this manuscript, Erik Biever for his assistance in locating material and for the loan of the keys to the library, friends who have urged me on by asking "Is the book done yet?" and my mother for helping me to appreciate the value of plants.

Diseases of Alfalfa (*Medicago sativa* L.)

BACTERIAL CAUSES

Bacterial Leaf Spot

CAUSE: *Xanthomonas campestris* pv. *alfalfae* (Riker, Jones & Davis) Dye, (syn. *X. alfalfae* (Riker, Jones & Davis) Dows., and *Phytomonas alfalfae* Riker et al.) survives in infected residue in or on the soil, in hay and in debris associated with seed. During warm, wet weather the bacteria are splashed or blown onto alfalfa leaves and gain entrance into the plant through small wounds created by windblown soil particles or other means. Disease can also occur in hot, dry weather when soil particles wound the plant, thus aiding in inoculation. The bacteria multiply inside the plant and frequently ooze to the surface of the leaf where they may be splashed or rubbed onto healthy leaves.

DISTRIBUTION: Generally distributed wherever alfalfa is grown under warm, wet conditions.

SYMPTOMS: Infected seedlings are often killed or stunted during high-temperature conditions. Small, water-soaked spots occur in chlorotic areas on the leaves. These enlarge to lesions that are 2–3 mm in diameter with irregular shapes and chlorotic margins. Eventually lesions become light yellow to tan, often with a lighter center, a papery texture, and a translucent appearance. Lesions are more pronounced on the underside of leaflets. Lesions usually glisten in the light due to the dried bacterial exudate on their surface. Infected leaves usually drop prematurely. Stem lesions initially are water-soaked, then turn brown or black.

CONTROL:
1. Plant resistant cultivars. Plants within a cultivar differ in susceptibility.
2. Plant in the spring.

Bacterial Stem Blight

CAUSE: *Pseudomonas syringae pv. syringae* van Hall (syn. *P. medicaginis* (Sackett) E. F. Smith and *Phytomonas medicaginis* (Sackett) Bergey et al.) survives in infected plant residue in soil. The bacteria can only enter a plant through a wound in the epidermis that is most commonly caused by frost.

DISTRIBUTION: Western North America.

SYMPTOMS: The disease occurs in early spring and is associated with wet weather and low temperatures that cause injury or cracking of the epidermis of young stems. Normally, only the lower three to five internodes are affected. Disease normally occurs until the first harvest. Initially, lesions are water-soaked and yellow to green at the point of the leaf attachment to the stem. Lesions then extend down one side of the stem for one to three internodes. The lesions become light to dark brown or black with age, linear, with droplets of bacterial exudate on the surface. The lesions may extend into the crown and roots of older plants. Infected stems are shorter than normal, thin, and brittle. Leaves are water-soaked and chlorotic at the base, as are leaflet midribs and petioles. Leaves attached to infected areas on the stem become chlorotic and die.

CONTROL:
1. Hardy cultivars are less prone to injury; therefore, they will be less likely to become infected.
2. Harvest after all danger of frost is past. New growth will be healthy.

Bacterial Wilt

CAUSE: *Corynebacterium michiganense* pv. *insidiosum* (McCull.) Carlson & Vidaver (syn. *C. insidiosum* (McCull.) Jens. and *Aplanobacter insidiosum* McCull.) survives in diseased plant tissue and soil for several years. In the spring, bacteria enter roots through wounds made by a variety of causes, including winter injury and insects. Presence of the root knot nematode *Meloidogyne hapla* (Kofoid & White) Chitwood also increases the severity of bacterial wilt. During the summer, mowing machinery become contaminated by cutting diseased plants and spreading the bacteria to healthy plants. Bacteria are also spread from plant to plant by irrigation water or surface rain water. Long-distance spread is probably by hay and seed. The bacteria enter the water-conducting tissues of the stems, crowns, and roots, rapidly increase in number, and prevent water from moving up the plant. Seedlings are rarely infected the first year but normally become infected the second or third year. In advanced stages of the disease, bacteria multiply rapidly in diseased tissues that break down, releasing bacteria into surrounding soil water where they are redistributed to healthy plants.

The greatest incidence of bacterial wilt occurs in lower, poorly drained areas of fields. The disease can occur over a large area during periods of continuous wet weather.

DISTRIBUTION: Generally distributed wherever alfalfa is grown. However, bacterial wilt is considered to be most important in the United States.

SYMPTOMS: Bacterial wilt is first noticed as the dying of scattered plants in a field. Initially infected plants are stunted. Stems and leaves remain small after cutting, giving infected plants a dwarfed, bunchy appearance. Wilted plants at first turn yellowish or bleached with only the leaf margin initially affected. Eventually the whole leaf turns yellow. Such leaves are somewhat thickened, rounded at the tip, and tend to cup upward. Eventually the whole leaf becomes tan in color and dies, followed by death of the entire plant. Plants will wilt in periods of moisture stress during the day but recover their turgidity at night.

The crowns and taproots show a yellow to tan discoloration of the outer vascular area. This may appear as a ring of yellow to tan dots or continuous discolored tissue. The entire root then becomes rotted. Infected plants are more prone to winter kill than healthy plants.

CONTROL: Plant resistant cultivars.

Dwarf

CAUSE: Unidentified fastidious bacterium that also causes Pierce's disease of grape. At least 20 species of "sharpshooter" leafhoppers carry the bacterium to susceptible hosts and inject it during feeding.

DISTRIBUTION: California and Rhode Island.

SYMPTOMS: Affected plants are normally stunted with smaller than normal dark blue-green leaves. Xylem tissue of the taproot normally becomes brown when exposed to air.

CONTROL: Dwarf occurs sporadically and is now considered to be uncommon. However, a resistant variety may be grown.

Pseudomonas Viridiflava
Root and Crown Rot

CAUSE: *Pseudomonas viridiflava* (Burkholder) Dowson.

DISTRIBUTION: Not known.

SYMPTOMS: Symptoms in the field are a light brown dry rot of the crown and upper taproot. Light brown streaks extend beyond the rot into the vascular system for about one-third of the length of the root.

CONTROL: No control is reported.

FUNGAL CAUSES

Anthracnose

This disease is also called southern anthracnose.

CAUSE: *Colletotrichum trifolii* Bain & Essary survives as acervuli and mycelium in residue on machinery, on stems, and in crowns. Different races of *C. trifolii* occur. *Colletotrichum destructivum* O'Gara and *C. dematium* (Pers. ex Fr.) Grove f. *truncata* (Schw.) v. Arx. have also been reported as causal agents of anthracnose. During warm and moist or humid weather in the summer, spores are produced in acervuli and are airborne or splashed by water to healthy plants where new infections occur. In general, disease caused by races 1 and 2 is more severe as temperatures increase; however, light has no effect. Secondary inoculum is provided by spores produced in acervuli on new lesions that are splashed and blown to other plants.

DISTRIBUTION: Widely distributed in warm alfalfa-growing areas.

SYMPTOMS: Anthracnose is first detected in the summer or fall by the presence of wilted and dead yellow or light brown shoots scattered throughout a field. Stems and crowns are the plant parts that become infected. Large, diamond-shaped lesions with bleached or white centers are formed on the lower portions of the stems. Several small, black acervuli that have setae can be seen with magnification in the center of most lesions and appear hairy due to the numerous setae. Spores produced in the acervuli continue to produce new infections. Young dead shoots may droop to form shepherds crooks.

Colletotrichum trifolii grows from the stem into the crown, causing the infected tissue to appear blue-black or black. Diseased crowns are weakened, producing less vigorous growth, resulting in plants that are either predisposed to winter injury or are prematurely killed. Both *C. destructivum* and *C. dematium* f. *truncata* are mildly pathogenic on petioles and leaves but are primarily secondary invaders in stem lesions caused by *C. trifolii*.

CONTROL:
1. Plant resistant cultivars.
2. Cut alfalfa before losses become too severe.

Aphanomyces Seedling Blight

CAUSE: *Aphanomyces* sp. that has been reported is similar to *A. cochlioides* Drechs. in Ontario and to *A. euteiches* Drechs. in Ohio and Wisconsin. Infection is caused by zoospores that require wet soils. Infection occurs over a wide temperature range of 3–39°C; however, zoospores are motile in water at 3–5°C for up to 24 hours.

DISTRIBUTION: Ohio, Ontario, and Wisconsin.

SYMPTOMS: Seedlings damp off or are stunted.

CONTROL: No control is reported.

Brown Root Rot

This disease is also called Plenodomus root rot.

CAUSE: *Plenodomus meliloti* Dearn. & Sanford overwinters in residue in soil, possibly as pycnidia. Plant roots are infected when soils thaw out either in the spring or occasionally in the winter. Later in the growing season plants become resistant to infection.

DISTRIBUTION: Alaska, Canada, and Finland.

SYMPTOMS: Slightly sunken, brown lesions occur on both lateral roots and the taproot. Numerous dark pycnidia develop on or just below the surface of lesions. When rot proceeds to the crown, the plant dies. However, the plant may recover if rot stops and enough taproot remains to produce new branch roots below the crown.

CONTROL: Rotate alfalfa with resistant crops such as small grains.

Common Leaf Spot

This disease is also called Pseudopeziza leaf spot.

CAUSE: *Pseudopeziza trifolii* (Biv.-Bern.: Fries) Fckl. f. sp. *medicaginis sativae* (Schmiedeknecht) Schuepp (syn. *P. medicaginis* (Lib.) Sacc.) survives either as mycelium or apothecia in spots on diseased leaves that have fallen to the soil surface and are not decomposed. During periods of abundant moisture and moderate temperatures (15–24°C), ascospores are produced that are windborne to healthy leaves. Seedling stands that are grown under a thick cover crop such as oats can be severely diseased.

DISTRIBUTION: Generally distributed wherever alfalfa is grown.

SYMPTOMS: Plants are not normally killed by common leaf spot but defoliation reduces hay quality and yield. Small, circular, dark brown spots that are about 2 mm in diameter develop on leaflets. There tends to be a sharp line between spots and healthy-appearing tissue as the tissue surrounding a spot is usually not discolored. When spots are fully developed, their centers become thickened and a tiny, light brown, cup-shaped apothecium forms on the upper leaf surface. The apothecium can be observed through a hand lens and is a good identifying characteristic. Ascospores are produced in the apothecium, shot into the air, and carried by wind to healthy leaflets.

CONTROL:
1. Plant resistant cultivars.
2. Cut infected stands in prebloom or bud stage before leaves fall. This maintains hay quality and removes infected leaves that will be a source of infection later in the same growing season and for the next growing season.

Crown Rot

CAUSE: Several bacteria and fungi including *Phoma medicaginis* Malb. & Roum., *Fusarium solani* (Mart.) Appel. & Wr., *Rhizoctonia* *solani* Kuhn, *Pseudomonas marginalis* var. *alfalfae* (Brown) Stevens, *Erwinia amylovora* (Burrill) Winslow et al. var. *alfalfae*, *Serratia marcescens* Bizio, *F. oxysporum* Schlecht. ex. Fr., *F. acuminatum* Ell. & Ev. (syn. *F. roseum* (Lk.) emend. Snyd. & Hans. f. sp. *cerealis* (Cke.) Snyd. & Hans. 'Acuminatum'), *F. avenaceum* (Fr.) Sacc. (syn. *F. roseum* Lk. emend. Snyd. & Hans. f. sp. *cerealis* (Cke.) Snyd. & Hans. 'Avenaceum'), *F. tricinctum* (Corda) Sacc., *Cylindrocladium crotalariae* (Loos) Bell & Sobers, *Myrothecium* spp., and *Flexibacter* sp. These fungi and bacteria normally persist as saprophytes on dead plant material in the soil or as resting structures. Infections occur through wounds caused by machinery, frost, insects, and animals. The act of cutting alfalfa for hay or an injury by one of the former reasons causes a small portion of stem to be killed, allowing propagules of saprophytic fungi to begin growth on the dead tissue. Eventually this rot proceeds down the cut stem into the crown.

DISTRIBUTION: Generally distributed wherever alfalfa is grown.

SYMPTOMS: Crown rot can be found to some degree in almost every alfalfa plant that is more than one year old. Various amounts of the crown may be rotted with a brown to black discoloration. Rotted tissue may be interspersed with live tissue. The tissue that produces crown buds is often killed, resulting in a plant with few live stems. Crown rot progresses slowly for a number of years after the disease is initiated, resulting in a gradual thinning out of the stand from year to year.

CONTROL:
1. The last cutting of alfalfa should be made early enough to allow the root system to build up a supply of carbohydrates that will enable the plant to overwinter successfully.
2. Grow alfalfa cultivars adapted to an area.
3. Use good alfalfa management, fertilize properly, and strive to adjust soil pH between 6.2 and 7.0.
4. Avoid running livestock on field after the last cutting since hooves will break down crowns, resulting in wounds through which infection occurs.

Crown Wart

CAUSE: *Physoderma alfalfae* (Lager) Karling (syn. *Urophlyctis alfalfae* (Lagerh.) Magn. and *Cladochytrium alfalfae* (Lagerh.)) survives in the soil as resting spores. Spores germinate during wet soil conditions and liberate zoospores that swim to developing buds of the crown branches where they germinate.

DISTRIBUTION: Australia, Ecuador, Europe, New Zealand, and most commonly in the western United States. However, crown wart also occurs in the eastern and northeastern United States.

SYMPTOMS: Crown wart occurs in localized, warm, wet areas in the spring, being most common on heavy soils or in low-lying areas. Crown wart is most common in fields under irrigation. Irregularly shaped, white galls are formed on the crown near the soil surface. As the galls mature they become gray to brown in color and 3 mm to over 5 cm in size. The galls contain masses of resting spores that are released into the soil upon decomposition of the gall later in the growing season. When galls are cut open they are a mottled brown. Young leaves will rarely develop small galls.

CONTROL: No control measures are practiced since the disease is not considered serious.

Cylindrocarpon Root Rot

CAUSE: *Cylindrocarpon ehrenbergii* Wr., teleomorph *Neonectria caespitosa* (Fckl.) Wr., probably survives as sclerotia-like stromata in residue in soil. Infection normally occurs through a wound at the base of branch roots during the end of winter dormancy. Plants become relatively resistant later in the growing season.

DISTRIBUTION: Canada and Minnesota.

SYMPTOMS: Infection usually occurs below the crown. Initally there is a water-soaked area that eventually turns a dark brown. Depending on the severity of infection, either a small portion of root just below the crown or the entire root may be rotted. Stromata develop in cracks of the bark of the roots, giving it a dark and rough appearance.

CONTROL: Rotate alfalfa with resistant crops such as small grains.

Cylindrocladium Root and Crown Rot

CAUSE: *Cylindrocladium crotalariae* (Loos) Bell & Sobers, *C. clavatum* Hodges & Cardoso, and *C. scoparium* Morg.

DISTRIBUTION: Hawaii.

SYMPTOMS: Black and sunken lesions are present on roots and crowns. A brown root rot is sometimes present.

CONTROL: No control is reported.

Downy Mildew

CAUSE: *Peronospora trifoliorum* deBary (syn. *P. aestivalis* Syd.) is a fungus that overwinters as perennial mycelium in the crown buds and crown shoots. Oospores in residue probably provide another means of survival. In the spring, conidia are formed in darkness during periods of high relative humidity. They are airborne or splashed to healthy leaves. Germination occurs in free water and at an optimum temperature of 18°C.

DISTRIBUTION: Generally distributed wherever alfalfa is grown in the temperate areas of the world.

SYMPTOMS: Downy mildew is most damaging during the first year following seeding, especially during cool, wet springs. The disease disappears during warm, dry weather but may return during cool weather in the autumn.

The rate of plant growth and susceptibility determine the extent of invasion of the tissues. A rapidly growing plant shows only small areas of invasion on the leaves. Young leaflets, especially at the tips of stems, are dwarfed and twisted or rolled downward, with light green to yellow blotches. A grayish, cottony growth, which is the fungus mycelium, conidiophores, and conidia, is often visible on the underside of the leaflets especially during moist, cool weather or under conditions of high humidity.

When the entire stem is affected, all leaves and stem tissue are yellow. Stems are larger in diameter and much shorter than normal.

CONTROL:
1. Some cultivars tend to have a high percentage of resistant plants.
2. Cut alfalfa in prebloom stage to save foliage.
3. Spring planting reduces incidence of seedling infection.

Forked Foot Disease

CAUSE: *Pythium ultimum* Trow. is most frequently associated with this condition.

DISTRIBUTION: Likely wherever alfalfa is grown.

SYMPTOMS: This is a characteristic symptom associated with damping off of seedlings. Adventitious roots develop above lesions on the radicles of seedlings, giving the roots a forked appearance. The number of adventitious roots varies from one to six or more. Seedlings are dwarfed or stunted.

CONTROL: See Pythium Seed and Seedling Blight.

Fusarium Root Rot

CAUSE: *Fusarium* spp. including *F. roseum* (Lk.) emend. Snyd. & Hans. f. sp. *cerealis* (Cke.) Snyd. & Hans. 'Avenaceum,' and *F. roseum* (Lk.) emend. Snyd. & Hans. f. sp. *cerealis* survive in the soil as chlamydospores or as saprophytes on dead plant material. *Fusarium* spp. enter alfalfa roots by direct penetration or through wounds caused by feeding of nematodes, insects, or other agents. The primary role of wounding is not to breach the root surface but to alter the host pathogen interaction to favor fungal development in the root.

DISTRIBUTION: Generally distributed wherever alfalfa is grown.

SYMPTOMS: Seedlings may damp off either preemergence or postemergence, especially during warm, wet weather. Infected plants are usually stunted; leaves are yellow, have a bleached appearance, curl at the edges, and proceed to wilt.

The taproot when cut lengthwise is a light brown to black color. Necrotic areas often occur in the cortex of the branch roots and taproot, often in association with wounds.

CONTROL: Increased frequency of cutting, especially late in the growing season, increases the possibility of root rot occurring. Otherwise there is no practical means of control.

Fusarium Wilt

CAUSE: *Fusarium oxysporum* Schlecht. f. sp. *medicaginis* Weimer Snyd. & Hans. is the usual cause; however, *F. oxysporum* Schlecht f. sp. *vasinfectum* (Atk.) Snyd. & Hans. races 1 and 2 and *F. oxysporum* Schlecht f. sp. *cassia* Armst. & Armst. have been reported to cause similar symptoms. *Fusarium oxysporum* f. sp. *medicaginis* survives in soil for several years as chlamydospores and as mycelium in dead and live plant tissue. The fungus infects alfalfa by growing into small roots or wounds in the taproot and progresses up the xylem tissue, thereby causing a wilting. Disease is favored by high soil temperatures. Soil moisture does not affect disease severity.

DISTRIBUTION: Generally distributed wherever alfalfa is grown but is more severe in warm areas of the world.

SYMPTOMS: Fusarium wilt occurs in irregularly shaped areas in a field but a relatively small percentage of plants in the area show symptoms at one time. The greatest loss usually occurs in the first two years after planting, followed by death of scattered plants over the remaining years of the life of the stand.

The first evidence of Fusarium wilt is usually a rapid wilting of stems on one side of the plant. Leaves and stems appear light green in color or bleached, with leaves at the bottom of the plant often showing a slight pink color. The tips of stems may wilt during hot days or during periods when soil moisture is deficient but recover during cool nights or when soil moisture is replenished. The plant dies slowly, often requiring several months to succumb. Quite often leaves of affected plants become dry and brittle but retain their green color as in the case of freshly cut hay.

Plants that have been affected by Fusarium wilt for some time but have not died will be dwarfed or stunted. A longitudinal and cross section of a diseased taproot reveals a cinnamon brown to red discoloration, often apparent as streaks, only involving a portion of the woody area. As the disease progresses, the entire outer portion of the woody cylinder becomes discolored. This may also be evidenced as a light brown to red-brown dry rot in the crown extending down into the taproot.

CONTROL: There is no known satisfactory control. Apparently individual plants have been found to carry some resistance.

Leptosphaerulina Leaf Spot

This disease is also called lepto leaf spot, halo spot, pepper spot, brown leaf spot, and Pseudoplea leaf spot.

CAUSE: *Leptosphaerulina briosiana* (Poll.) Graham & Luttrel, (syn. *Pleosphaerulina briosiana* (Poll.), *Pseudoplea briosiana* (Poll.) Hoehn., and *Pseudoplea medicaginis* (Miles)) overwinters as mycelium or pseudothecia in infected leaves that remain on the soil surface. During cool, moist weather, spores are discharged and blown by wind to healthy, young leaves to begin new infections. Optimum germination occurs between 22°C and 25°C in the laboratory.

DISTRIBUTION: Generally distributed wherever alfalfa is grown.

SYMPTOMS: Leptosphaerulina leaf spot is most severe on young leaves that grow back after the first cutting and following a period of moist weather. In greenhouse tests, lesions were largest on the top leaves and decreased in size on successive older leaves. At first, numerous small, red-brown flecks appear on both surfaces of the leaf and on the petioles. As the flecks enlarge they acquire a tan center with an irregular brown border surrounded by a halo that causes the spots to merge. Large lesions occur under high light conditions. The infected leaves die and cling to the stem for a period of time. In older growth the upper leaves become infected and have typical symptoms but seldom die.

CONTROL: Effective control measures are not known; however, the following may reduce disease and severity.
1. Plant cultivars that are reported to have some disease resistance.

2. Plant certified, disease-free seed.
3. Cut plants in prebloom stage.
4. Rotate alfalfa with a nonsusceptible host such as soybeans, corn, or a small grain for at least two years.

Marasmius Root Rot

CAUSE: *Marasmius* sp.

DISTRIBUTION: Egypt

SYMPTOMS: In pathogenicity tests, roots are brown and water-soaked. The fungus grows in ropelike strands over the entire root system. Plants are killed in the field.

CONTROL: None reported.

Mycoleptodiscus Crown and Root Rot

CAUSE: *Mycoleptodiscus terrestris* (Gerd.) Ostazeski (syn. *Leptodiscus terrestris* Gerd.) overwinters as sclerotia in residue and soil. Conidia are probably formed in the summer and are disseminated by being forcibly ejected from the fruiting structure by setae that unfold as mucilage dries out.

DISTRIBUTION: Central and eastern United States.

SYMPTOMS: Disease is favored by warm, humid weather. Preemergence and postemergence damping off of seedlings occurs. The lateral roots, taproots, and crowns of older plants develop a brown to black rot. Crown rot probably develops from stem infections. The margin of decayed tissue is black while tissue behind the margin is normally lighter in color. Numerous sclerotia are usually present in decayed tissue. Red-brown lesions develop on stems and small spots develop on leaves.

CONTROL: None reported.

Myrothecium Root Rot

CAUSE: *Myrothecium roridum* Tode. ex Fr. and *M. verrucaria* (Alb. & Schw.) Ditm. ex Fr. Mycelium normally penetrates intact roots but will penetrate through wounds in some situations.

DISTRIBUTION: Wisconsin and Pennsylvania. However, Myrothecium root rot is probably more widely distributed.

SYMPTOMS: In laboratory tests using the slant-board evaluation, brown water-soaked rots with poorly delineated margins were initially evident on roots. Root growth in general was inhibited. Foliar symptoms were chlorosis, purpling of leaflet margins, leaf curling, mottling, and death of leaves and petioles. In greenhouse tests, rots were dark brown and occurred across entire roots with occasional streaks extending upward in vascular systems. However, rotted roots remained firm and were not water-soaked. Foliar symptoms were severe stunting in addition to other previously described symptoms.

CONTROL: None reported.

Phytophthora Root Rot

This disease is also called wet foot disease.

CAUSE: *Phytophthora megasperma* Drechs. formerly described as *P. cryptogea* Pethyb. & Laff. The causal organism is also reported in some literature as *P. megasperma* var. *medicaginis*. Survival is by oospores or mycelium in infected residue; however, only oospores free in soil or in root tissue are capable of long-term survival. Hyphae grows through soil to colonize root segments of various plants including black medic, birdsfoot trefoil, and corn. Colonization of organic matter may significantly increase its survival potential. These root segments then serve as sources of inoculum to infect alfalfa. Chlamydospores are also reported as survival structures. Chlamydospores germinate to form one or more germ tubes that penetrate tissue and eventually produce oospores, chlamydospores, and sporogenous hyphae. The latter emerge from tissue to bear sporangia that germinate directly or produce zoospores in presence of free water.

Sporangia are produced in flooded soil at optimum temperatures between 24°C and 27°C. Sporangia germinate either indirectly by releasing zoospores that swim to roots or directly to form mycelium or other sporangia in water.

The most frequent points of infection are tips of small roots and the base of fine lateral roots. Water saturation of soil predisposes roots to infection by increasing root damage and exudation of nutrient, which increases the chemolactic attraction of zoospores to the roots.

A high-temperature cultivar of *P. megasperma* from the desert areas of the southwestern United States has an optimum growth range between 27°C and 33°C. It has been designated as cultivar HTI (high-temperature isolate).

DISTRIBUTION: North central and eastern United States and Canada.

SYMPTOMS: The disease occurs in wet soils where drainage is poor during periods of excessive precipitation or irrigation. Leaves of infected plants become yellow or reddish and drop. Eventually the plants may be stunted, and they wilt and die.

Rootlets and taproots are rotted. The taproot may be rotted off at any depth and has a yellow-brown discoloration that later turns black. There is usually a sharp line between rotted and healthy tissue of the taproot that may occur at various depths below the soil line.

When the soil becomes drier the rot stops, and if enough taproot remains alive, side roots are produced and the plant remains alive. Such plants have a shallow root system and will produce abundant forage when surface moisture is plentiful. However, when surface moisture is depleted, little or no forage is produced.

Interaction of *P. megasperma* with either northern or southern root rot nematodes intensifies root rot.

CONTROL:
1. Plant resistant cultivars.
2. Improve water drainage in soils that tend to be poorly drained.

Pythium Seed and Seedling Blight

This disease is also called damping off.

CAUSE: *Pythium* spp. Species commonly associated with this disease include *P. debaryanum* Hesse, *P. irregulare* Buis., *P. splendens* Braun, and *P. ultimum* Trow. Other reported species include *P. aphanidermatum* (Edson) Fitzp., *P. myriotylum* Drechs., *P. pulchrum* Mind., and *P. rostratum* Butl. *Pythium paroecandrum* Drechs. was the Pythium species most commonly isolated from alfalfa roots in the Central Valley of California during the cooler months. Cold, wet weather in spring favors disease caused by most *Pythium* spp.; however, *P. aphanidermatum* & *P. myriotylum* are favored by high temperatures.

These fungi survive on crop residue primarily as oospores or as sporangia. Oospores and sporangia may germinate either directly by forming a germ tube or indirectly by producing zoospores that swim through soil water to a root or seed where they form a germ tube.

DISTRIBUTION: Generally distributed wherever alfalfa is grown.

SYMPTOMS: Preemergence damping off is inconspicuous and rarely observed by the grower since seeds and seedlings rapidly decompose. Infected seeds turn brown before germination. After germination, the radical and cotyledons turn brown and soft. Postemergence seedling blight is initially characterized by root tip necrosis with the hypocotyl and roots becoming light brown and water-soaked. Eventually most infected plants collapse and dry up. Surviving seedlings are stunted.

Pythium paroecandrum does not cause severe damping off but infects rootlets, thereby reducing shoot growth and development.

CONTROL:
1. A fungicide seed treatment is an aid to controlling the seed rot phase.
2. Apparently most seedlings become immune after a few days and plants older than two weeks seldom become infected.

Rhizoctonia Foliage Blight

CAUSE: *Rhizoctonia solani* Kuhn. is discussed under Rhizoctonia stem blight. Disease development occurs primarily during hot, humid weather in thick stands of alfalfa.

DISTRIBUTION: Generally distributed wherever alfalfa is grown in hot, humid weather; however, the disease is seldom severe.

SYMPTOMS: Rhizoctonia foliage blight is thought to be a progression of the disease Rhizoctonia stem blight. Lower leaves and stems are affected first but the disease may eventually progress halfway up the plant. Infected leaves appear watery and blue in color. Eventually leaves wilt, turn brown, then shrivel. A weblike growth of the fungus mycelium can be seen growing over the shriveled leaves and stems.

CONTROL: Increase air movement to allow drying of stems by proper grazing or cutting.

Rhizoctonia Root Canker

This disease is also called black root canker.

CAUSE: *Rhizoctonia solani* Kuhn. is discussed under Rhizoctonia stem blight. The fungus enters the taproot near the area where lateral roots emerge. Infection occurs during periods of high temperature and moisture. The fungus belongs in anastomosis group (AG) 4.

DISTRIBUTION: The irrigated areas of Arizona and California in the United States and in Australia and Iran.

SYMPTOMS: Affected areas in a field will vary in shape from circular to irregular and will be spotty with apparently healthy plants next to ones that are wilting. Taproots are covered by a large number of cankers that are oval to round in shape and 6–12 mm in diameter. Cankers vary in color from yellow to tan, often with a slightly darker border. When soil temperatures decrease during the winter, the affected areas heal and the lesions or cankers turn dark brown to black in color.

Cankers commonly occur where lateral roots emerge from the taproot and may often girdle it. Alfalfa plants are most damaged by cankers formed on the crowns.

CONTROL: No satisfactory control is known, but in irrigated areas of the southwestern United States, growers have reestablished stands by replanting infested fields in October or November.

Rhizoctonia Stem Blight

This disease is also called Rhizoctonia blight.

CAUSE: *Rhizoctonia solani* Kuhn. survives by forming sclerotia or by growing saprophytically on dead plant tissue. Under high-temperature and moisture conditions, mycelium that probably originated from a pre-colonized substrate will infect the crown or stem usually just below the soil line. Sclerotia or residue and soil containing propagules of *R. solani* may be splashed onto the plant foliage. Under flooding or irrigation, the area on a plant above the soil line may commonly be infected.

Rhizoctonia solani also causes a root canker and foliage blight that are discussed as separate diseases. Additionally, it causes a seedling blight.

DISTRIBUTION: Generally distributed wherever alfalfa is grown.

SYMPTOMS: Dead stems scattered through a stand may be an indication of the disease. Sunken cankers are usually found at the base of stems and vary in color from tan to red-brown to dark brown. The stems are usually girdled, causing leaves and tips of stems to yellow and wilt.

CONTROL:
1. Plant certified, high-quality seed in a well-prepared seed bed with good drainage.
2. Proper mowing or grazing is effective in partially controlling the disease since disease development is retarded when exposed to drying effects of direct sunlight and good air drainage.

Rust

CAUSE: *Uromyces striatus* Schroet. (syn. *U. striatus* var. *medicaginis* (Pass.) Arth.) survives in the milder climates of North America as urediospores and mycelium in infected alfalfa. The urediospores are then blown north in the summer and infect healthy plants during high humidity and temperatures between 21°C and 30°C. In North America the uredia and telial stages can be found on alfalfa. In Europe and Canada the rust is heteroecious with the presence of the aecial and pycnial stages forming in *Euphorbia* spp. Mycelium is perennial and systemic. Aeciospores infect alfalfa earlier in the season than infection by urediospores.

DISTRIBUTION: Generally distributed wherever alfalfa is grown under warm or temperate conditions.

SYMPTOMS: Rust is most prevalent later in the summer and fall. Red-brown pustules of urediospores rupture the epidermis of leaflets, petioles, and stems. Pustules are normally abundant on the undersides of leaflets. When pustules rupture, they appear as a powdery mass on the plant surface that easily rubs off and gives the fingers a reddish color. Sometimes a yellow halo may surround the spot. When the pustules are numerous, leaves may turn yellow, shrivel, and drop prematurely. Aecia develop on leaves and stems of *Euphorbia* spp. and may kill stem apices. Witches brooms and hypertrophy may occur on stems.

CONTROL: No satisfactory control known but some cultivars are more resistant than others.

Sclerotinia Crown and Stem Rot

CAUSE: *Sclerotinia sclerotiorum* (Lib.) deBary. Survival is by sclerotia in residue and surface soils. Sclerotia germinate during cool, wet weather in spring to produce either apothecia or mycelium. Ascospores from apothecia or mycelium can initiate infection. Sclerotia are produced on the surface of the host, in the pith, or under decaying plant parts on the soil surface. Disease is most severe in fields with plant residue remaining on the soil surface from the last harvest.

DISTRIBUTION: Eastern Washington.

SYMPTOMS: Similar to those caused by *S. trifoliorum* Eriks.

CONTROL: Similar to those recommended for *S. trifoliorum*.

Sclerotinia Crown and Stem Rot

CAUSE: *Sclerotinia trifoliorum* Eriks. (syn. *Pezziza ciborioides* Hoffm., *S. libertiana* Fckl., *S. ciborioides* Rehm., and *Whetzelinia sclerotiorum* (Lib.) Korf. & Dumont). The fungus survives during the winter and during the high summer temperatures as sclerotia that are on the soil surface, adhere to the plant surface, or are imbedded in infected stems and crowns. During cool, wet weather in the spring or autumn, the sclerotia germinate to form mycelial strands or apothecia in which asci and ascospores are produced. The ascospores are windborne and infect healthy plants. Mycelium grows only a short distance in soil and normally does not cause infection. Disease has been reported to be especially severe where alfalfa is planted into sod that has not been tilled. This is because a high number of plants grow in sod that are hosts for *S. trifoliorum*.

DISTRIBUTION: Cool, humid areas of Europe and North America.

SYMPTOMS: Sclerotinia crown and stem rot is most destructive on new seedlings and can be recognized by the small circular patches of dead or dying plants in the field. The first symptoms occur in the autumn as small brown spots on leaves and stems. Eventually the leaves and stems wilt and die; the fungus then spreads to the crown area. The disease is most destructive the following spring when the crown becomes soft with a gray-green discoloration. As the infected plant parts die, a white fluffy mass of mycelium grows over them. Sclerotia that appear as small black objects are later formed in the mycelial mass.

CONTROL:
1. Rotate alfalfa with a nonsusceptible host. Do not plant alfalfa for a minimum of 3 years.
2. Plow under residue to bury sclerotia.
3. Plant seed that is free of sclerotia.

Southern Blight

This disease is also called Sclerotium blight.

CAUSE: *Sclerotium rolfsii* Sacc. survives for several years in the soil or plant residue as small, brown sclerotia. Dissemination is by water and wind. The sclerotia germinate under hot, humid conditions to form mycelial strands that infect alfalfa plants. The teleomorph, *Pellicularia rolfsii* West (syn. *Corticium rolfsii* (Sacc.) Curzi), apparently is not an important source of infection. Southern blight is not an important disease of alfalfa except in some silty or sandy soils.

DISTRIBUTION: Southern Europe and the southern alfalfa-growing areas of the United States.

SYMPTOMS: The disease appears from a distance as a few scattered patches of dead plants in a field. Plants appear bleached to light tan in color. A white cottonlike mycelial growth is present on stems near the soil level. Numerous, small, tan to brown sclerotia that resemble seeds are formed in mycelium on the stems and crowns and appear as residue on the soil surface.

CONTROL:
1. Plow under infected residue.
2. Proper mowing or grazing aids in the control of the disease since the causal fungus does not grow well when exposed to direct sunlight and good air drainage.

Spring Black Stem

This disease is also called Ascochyta leaf spot and spring leaf spot.

CAUSE: *Phoma medicaginis* Malbr. & Roum. var. *medicaginis* Boerema (syn. *P. herbarum* West var. *medicaginis* Fckl. and *Ascochyta imperfecta* Pk.) overwinters as pycnidia in diseased tissue and may be seedborne. Spores are produced during cool (18–24°C), wet weather in the spring and are windborne to susceptible plants. Pycnidia are rarely produced during the growing season but occur later in the summer and autumn on lesions of diseased tissue. The fungus may be seedborne as mycelium on the seed coat.

DISTRIBUTION: Europe, North America, and South America.

SYMPTOMS: Spring black stem usually occurs in the spring but may be extensive during the fall; however, symptoms are usually less pronounced. Normally only the first cutting of hay is affected. Stem and petiole lesions at first are dark green and watery, then become dark brown to nearly black with a watery-appearing margin. Stem lesions enlarge and merge until most of the lower part of the stem is blackened. The fungus may grow into the crown and upper root. Young shoots are often girdled and killed.

Leaves have dark brown or black spots with irregular borders. The spots may enlarge and combine to cover most of the leaf area. As the disease progresses, leaves become yellow and drop. The loss caused by spring black stem is due mostly to defoliation. Although the most severe infections usually occur on lower leaves, leaves higher on the plant may also become infected. During wet, humid conditions seedpods may become discolored and shrivel. Spring black stem is normally not present on first-year alfalfa unless the fungus has been carried on seed or spores are airborne from nearby fields.

CONTROL:
1. Harvest hay before defoliation is severe. This also allows plants to dry out, making fungal sporulation and infection more difficult.
2. Plow under the plants in cases of severe infection.
3. Treat seed with a fungicide to prevent seedborne introduction of the disease. Use seed produced in arid areas.

4. Plant moderately resistant cultivars. No highly resistant cultivars are available.

Stagnospora Leaf Spot

CAUSE: *Stagnospora meliloti* (Lasch) Petr. is one of three phases of
the same fungus on alfalfa as well as other clovers. *Stagnospora meliloti*
is associated mainly with the leaf spot phase during the summer; *Phoma
meliloti* Allesch. is found mostly on stems as pycnidia in the autumn;
the perfect stage, *Leptosphaeria pratensis* Sacc. & Briard, is found as
perithecia on stems in late autumn and more commonly in spring. The
function of the Phoma stage is unknown.

The fungus overwinters as mycelium and pycnidia of *S. meliloti* and
perithecia of *L. pratensis* within the infected residue. However, little is
known of the role of the perithecial stage in infection. The fungi are
poor soil saprophytes and do not survive outside of infected residue.
During wet weather in the spring, conidia are extruded from pycnia and
splashed by rain water or carried by irrigation water to healthy plants.
Spores may infrequently be disseminated by wind. The leaf spot phase
caused by *S. meliloti* develops most commonly in the summer while the
Phoma stage apparently only occurs during low autumn temperatures.
The root rot phase evidently develops from stem and crown infections,
slowly developing for two to three years. It is thought to be caused
by *S. meliloti* and develops during moderate temperatures of 15–25°C.
Secondary spread is mostly by conidia and occurs under wet conditions.

DISTRIBUTION: Generally distributed but the disease is most common in warm, humid areas or where alfalfa is grown under irrigation.

SYMPTOMS: Spots on leaves and stems are similar in appearance.
They are 3–6 mm in diameter and circular to irregular in shape. The
center of the spot is pale buff to almost white in color with a light to
dark brown diffuse margin. In some instances the spots tend to show
faint concentric zones the same color as the margin. Infected leaves drop
soon after spots form. Pycnidia that are small and dark are found in
older spots scattered throughout the affected area.

CONTROL:
1. Rotate for two to three years with a nonsusceptible host.
2. Plant cultivars that are the most resistant.

Stagnospora Root and Crown Rot

CAUSE: *Stagnospora meliloti* (Lasch) Petr., teleomorph *Leptosphaeria
pratensis* Sacc. & Briard. *Phoma meliloti* Allesch. has also been reported

as a cause. Optimum growth is between 15°C and 25°C. The fungus enters the plant primarily through the stem tissue and progresses slowly down into the taproot. A further discussion of the fungus occurs under Stagnospora leaf spot.

DISTRIBUTION: Generally distributed wherever alfalfa is grown.

SYMPTOMS: A brown to black root rot with a necrosis and dry rot of the upper portion of the taproot and crown are common symptoms. Bark tissue of roots and crowns with lesions is cracked and irregular in shape. Lesions on the taproot occur 7.25–12.5 cm below the crown. Lengthwise sections through rotted areas show affected bark and xylem tissue with bright red-brown flecks. This latter symptom is a good diagnostic characteristic.

CONTROL:
1. Rotate for two to three years with a nonsusceptible host.
2. Plant cultivars that are the most resistant.

Stemphylium Leaf Spot

This disease is also called target spot.

CAUSE: *Stemphylium botryosum* Wallr., teleomorph *Pleospora herbarum* (Pers. ex Fr.) Rab. (syn. *P. vulgatissima* Speg.). Overwintering is by mycelium in infected crop residue or on seed and as perithecia. At any time of the growing season, spores are produced during moist weather and are windborne to healthy leaves. Temperature requirements depend on the biotype. Disease caused by the cool temperature biotype occurs at 8–16°C and for the warm temperature biotype at 23–27°C. The disease is more likely to be severe in the dense foliage of a heavy stand during late summer and autumn but it can be found throughout the growing season. The highest disease severity caused by the cool-temperature biotype occurs when plants are exposed to light before and after inoculation, followed by alternating dark/light periods and extended free moisture on leaves for several days. A Stemphylium-incited disease in Australia and South Africa has been attributed to a fungus that more closely resembles *S. vesicarium* (Wallr.) Simmons than *S. botryosum*. Disease caused by this Stemphylium is more common during cooler weather (see *Stemphylium vesicarium* leaf spot).

DISTRIBUTION: Asia, Australia, Europe, and the United States, particularly the southeastern United States.

SYMPTOMS: Spots are small, oval or irregular, dark brown with lighter centers, slightly sunken, and they appear on leaves. Normally they are surrounded by a pale yellow halo. The more or less circular spots enlarge and form concentric zones of light and dark brown tissues, resembling a target. A single large lesion can cause a leaf to turn yellow and drop prematurely. Older lesions may appear sooty from the abundant production of the large spores of *S. botryosum*, especially during wet weather. The spores may be seen with the aid of a hand lens.

Black areas may appear on the peduncles, petioles, and stems. Stems and petioles may be girdled in wet weather, causing the foliage to wilt and die. Severe defoliation may occur in certain years.

Different symptoms occur in the western United States, particularly California, due either to a different biological race of the pathogen or the environment. These symptoms are elongate lesions (3–4 mm) with irregular outlines, tan centers, and dark brown borders. Normally, defoliation does not occur. Symptoms are common in the spring in the interior valleys but occur throughout the year along the coast.

The Australian *Stemphylium* sp. causes circular to irregular leaf lesions with white to cream colored centers surrounded by a dark brown margin. On susceptible cultivars, the spots coalesce, causing a leaf blight. Only leaflets on actively growing shoots are affected.

CONTROL:
1. Plant cultivars that have a low to moderate resistance. No cultivar is highly resistant. In Australia the nondormant cultivars are most susceptible.
2. Harvest the stands early to save foliage and allow plants to dry out.

Stemphylium Vesicarium Leaf Spot

CAUSE: *Stemphylium vesicarium* (Wallr.) Simmons, teleomorph *Pleospora alii* (Ravenh.) Ces & DeNot. The causal fungus is seedborne. Seed becomes infected in wet or dry climates.

DISTRIBUTION: South Africa.

SYMPTOMS: Symptoms are of two types depending on the temperature of the area. In warm areas under overhead irrigation, such as are found in the central Transvaal, lesions are irregular in shape and circular to oval, 0.1–3.0 mm in diameter, light to dark brown, often lighter in the center, sometimes concentrically zonate, and presenting a target effect with diffuse margins frequently surrounded by lighter-colored halos.

In cooler areas, such as the southwest cape, lesions are circular to oval, 0.1–2.0 mm in diameter but uniformly bleached white or tan with a distinct black margin. Elongated black lesions up to 2 mm in length occur on petioles and stems.

Under both circumstances, lesions are usually distinct but may later coalesce. Defoliation may occur.

Symptoms on annual Medicago varied from black specks to spots 1.0–2.0 mm in diameter with tan centers and irregular brown margins.

CONTROL: None reported.

Other *Stemphylium* spp. Recorded on *Medicago* spp.

S. globuliferum (Vestergren) Simmons
S. sarcinaeforme (Cav.) Wiltshire
S. trifolii Graham

Summer Black Stem

This disease is also called Cercospora leaf spot and Cercospora black stem.

CAUSE: *Cercospora medicaginis* Ell. & Ev. overwinters as mycelium in infected plant residue. It may also be seedborne in warm humid areas but the incidence is apparently low under other conditions. During warm, moist weather later in the growing season, spores are produced that are either wind or rainborne to healthy plants. Secondary spread of conidia from infected stem and leaves occurs as plants grow taller and form a natural humidity chamber that prevents drying out.

DISTRIBUTION: Africa, Asia, Europe, South America, central and eastern United States.

SYMPTOMS: Summer black stem will generally first appear later in the growing season during hot, moist periods, usually after the first cutting of hay. Summer black stem is most damaging when harvesting is delayed. At first, the disease appears as small, brown spots on both leaf surfaces. The spots enlarge to form circular lesions, red-brown to smoky brown in color and 3–6 mm in diameter. During moist weather, the lesions will appear ash-gray due to the abundant production of spores on the lesion surface. In heavy infections, entire leaflets are killed and severe defoliation occurs. Frequently, only one or two lesions will develop on a leaflet.

The leaf spot phase is followed by the appearance of red to chocolate brown elliptical or linear lesions on petioles and stems. These lesions eventually enlarge and coalesce. Under moist conditions, the entire stem may be discolored. Smaller peduncles, petioles, and stems may be killed, resulting in further defoliation and loss of seed.

Symptoms on leaves may be confused with those of Stemphylium leaf spot but the latter are elliptical or irregular in shape and frequently have concentric rings. Symptoms on stems may be confused with those of spring black stem, but the lesions of summer black stem are a lighter brown, whereas those of spring black stem are a black-brown.

CONTROL:
1. Plant cultivars that have moderate resistance.
2. Harvest hay when plants are in early bloom stage to save leaves from falling. This also allows the foliage to dry out.
3. Do not allow cattle to graze on fields late in season as summer black stem is more severe in second year stands that do not make complete growth before winter dormancy.

Texas Root Rot

This disease is also called cotton root rot, Ozonium root rot, and Phymatotrichum root rot.

CAUSE: *Phymatotrichum omnivorum* (Shear) Duggar (syn. *Ozonium omnivorum* Shear) survives in soil to depths of 2 m as sclerotia and as brown sclerotial strands, primarily in alkaline soils. The sclerotia germinate in early summer during high soil temperatures and infect the roots. Low soil temperatures will kill the fungus. Conidia are produced on spore mats formed on the soil surface, particularly after a summer rain or during humid weather. However, they are not thought to be important either in infection or dissemination.

DISTRIBUTION: Mexico and southwestern United States.

SYMPTOMS: Infested areas may appear as circles or fairy rings of dead plants. Sometimes alfalfa in the center of the ring may not be affected. If plants become diseased in the center they are replaced with non-susceptible grasses. Lesions on the taproots are yellow to brown, sunken, and clearly defined although irregular in shape. Leaves of infected plants become yellow and bronze. Eventually the plant wilts and dies if lesions encircle the root or crown. Irregular-shaped spore mats, white to tan, sometimes appear on the soil surface.

CONTROL:
1. Plow under green manure crops or animal manure. This helps to suppress the disease.
2. Grow a nonsusceptible crop such as a grass for at least three years. However, if the sclerotia persist for a long time, this control measure is relatively ineffective.
3. Add sodium salts to soil. This helps to reduce disease severity.
4. Grow cultivars whose plants vigorously produce new roots. These varieties have some tolerance to the disease.

Verticillium Wilt

CAUSE: *Verticillium albo-atrum* Reinke & Berth. (dark mycelial species) overwinters as thick walled mycelium in soil, infected plants, and weed hosts. However, *albo-atrum* has limited saprophytic ability. Precolonized stems are effective as sources of inoculum until infected tissued decays. Persistence is adversely affected by moist soil. Conidia are produced during cool, moist weather and are disseminated by any means that moves infested plant residue. Secondary spread during the growing season is by harvest machinery, windborne conidia, and plant debris. Some evidence suggests that inoculum carried on the cutter bar is the most important means of dispersal within and between fields rather than airborne conidia. Several insects under laboratory conditions have been implicated in dissemination. Grasshoppers, *Melanoplus sanguinipes* (Fabricus) and *M. bivittatus* (Say); alfalfa weevils, *Hypera postica* (Gyllenhal); and wooly bears, *Apantesis blakei* Grote were fed infected alfalfa leaves. The fungus was able to safely pass through the digestive system and was detected in feces about one day after feeding. Fungus gnats, *Bradysia impatiens* Johannsen disseminate *V. albo-atrum* in the greenhouse and the leafcutter bee, *Megachile rotundata* (Fabricus), disseminate the fungus in the field. The pea aphid (*Acyrthosiphon pisum* (Harris)) is a vector by carrying spores as a surface contaminant on legs and antennae. *Verticillium albo-atrum* has been reported to pass through the digestive tract of sheep, thereby affording another means of spread. The fungus is also commonly carried directly on plant residue and is seedborne. Pollen can be infected in vitro. Under humid conditions, the fungus will colonize the pod and seed coat; however, the fungus reportedly does not survive for very long on the seed.

Plants are infected through the roots and mycelium grows into the xylem tissues. Conidia are then produced within the xylem and are the internal means of spread.

While *V. albo-atrum* infects a wide range of hosts, the strain on alfalfa is only known to infect alfalfa. *Verticillium dahliae* Kleb. (microscleortial species) also causes similar symptoms on alfalfa but is less virulent and only occasionally found in England.

DISTRIBUTION: Canada, Europe, and the United States.

SYMPTOMS: The first symptom is a flagging or wilting of the upper leaves on warm days. Initially, V-shaped, pink-orange to brown areas occur on leaflets. Leaflets on severely affected shoots are usually necrotic and twisted, forming spirals. The lower leaves eventually wilt, become yellow to white, and die. New shoots develop from the crown, but they may also become infected and die. Entire plants become stunted with dead, yellow stems.

Frequently stems remain green but attached leaves are bleached and dead. The base of infected stems may become covered with conidiophores, giving a grayish appearance. When the stem dies the infected area may become blackened. Flowering may become suppressed.

In advanced stages of the disease, plants are stunted, with yellow and dessicated shoots and leaves. Xylem tissue is brown. Often the discoloration can be traced from the taproot up into the stems. These symptoms can be confused with both bacterial and Fusarium wilt. A high percentage of plants in some resistant cultivars are symptomless carriers of the pathogen; however, plant height and dry weight are affected.

CONTROL:
1. Grow resistant cultivars.
2. Plant clean seed that is free of plant debris and treated with a fungicide seed protectant.
3. Harvest the youngest stands first.
4. Clean equipment before leaving an infested field.

Violet Root Rot

CAUSE: *Rhizoctonia crocorum* DC. ex Fr. (syn. *R violaceae* Tul.), teleomorph *Helicobasidium purpureum* Pat. survives in soil as sclerotia or as a saprophyte on plant residue. Infection apparently is more common when some moisture is present in low organic matter soils; however, the disease may also occur in loam soils.

DISTRIBUTION: Europe and infrequently in the United States.

SYMPTOMS: Violet root rot symptoms normally occur in mid-summer. Symptoms occur in more or less circular patterns in older stands and are associated with low areas that are subject to flooding and plants that have other root injuries. Plants on the edge of a diseased area are brown in contrast to surrounding green plants. Diseased roots are brown on the inside and are covered with thick mats of violet to cinnamon mycelium that may extend 20 cm or more below the soil line. Later the roots are rotted and shredded with a brown to dark violet discoloration. Tiny black sclerotia that are barely visible without magnification may be seen on infected roots.

CONTROL:
1. Plant alfalfa in well-drained soil.
2. Harvest hay in prebloom stage.
3. Rotate alfalfa with resistant crops such as small grains and corn.

Winter Crown Rot

This disease is also called snow mold.

CAUSE: An unidentified low-temperature basidiomycete (LTB) that survives as a soil inhabitant. However, *Coprinus urticicola* (Berk. & Br.) Buller is tentatively identified as being the causal organism. Optimum temperature for growth in culture is 12°C according to some reports and 13–15°C according to others. During favorable conditions in the fall, the fungus grows in close association with alfalfa crowns and produces HCN that is absorbed by crown tissue near 0°C. Mycelium invades crown buds in March and destroys tissue. When temperatures rise above freezing, alfalfa is no longer susceptible.

DISTRIBUTION: Alaska, Alberta, British Columbia, Manitoba, and Saskatchewan.

SYMPTOMS: Different-size areas of dead plants occur in an affected field. Dark brown rotted areas occur on crowns and infrequently on roots. The taproot may be unaffected even though the crown is dead. The taproot will eventually be rotted by secondary organisms. Often only the crown bud and underlying tissue is affected. Mycelium may grow on the soil surface in early spring.

CONTROL:
1. Plant cultivars that have *Medicago falcata* L. parentage.
2. Grow small grains instead of alfalfa in infested soil for a minimum of three years.
3. Application of borax to plants in the fall prevents HCN production; however, this treatment is sometimes phytotoxic.

Yellow Leaf Blotch

CAUSE: *Leptotrochila medicaginis* (Fckl.) Schuepp (syn. *Pseudopeziza jonesii* Nannf. and *Pyrenopeziza medicaginis* Fckl.) anamorph *Sporonema phacidiodes* Desm. overwinters as stromata in infected residue on the soil surface. During cool, wet weather in late spring, stromata give rise to apothecia in which ascospores are borne that constitute the primary inoculum. The spores are windborne to healthy plants.

DISTRIBUTION: Generally distributed wherever alfalfa is grown in temperate climates.

SYMPTOMS: Yellow leaf blotch is most conspicuous in rank, unusually tall stands. The lower leaves usually are the most affected. Young lesions appear as yellow stripes and blotches elongated parallel to the leaf veins. As the lesions enlarge they become fan-shaped or circular. The color changes to orange-yellow or brown. Small dark pycnidia develop mostly on the upper surface of the blotch. Eventually, pseudostromata and later stromata form in the center of the blotch, giving it a dark brown to black color. Stromata are uncommon on the undersurface of the leaf until after leaf drop. Dead leaves frequently remain attached to the stem for some time, curling downward as they dry. Similar yellow blotches that later turn dark brown may occur on the stems.

CONTROL:
1. Cut plants before leaf drop becomes severe.
2. Do not leave a high stubble or allow weeds to become a problem.

3. Rotate alfalfa for at least two years with a nonsusceptible host such as soybeans, corn, or a small grain.
4. Plant resistant cultivars.

Other Fungi Reported on Alfalfa

Fusarium chlamydosporum Wollenw. & Reinking (syn. *F. fusarioides* (Frag. & Cif.) Booth)
F. equiseti (Corda) Sacc.

MYCOPLASMAL CAUSES

Aster Yellows

CAUSE: Aster yellows is caused by a mycoplasma that is transmitted from diseased to healthy plants by leafhoppers.

DISTRIBUTION: Generally distributed wherever alfalfa is grown; however, the disease is usually not serious.

SYMPTOMS: Alfalfa plants are stunted and excessively branched. Flowers are chlorotic, sterile, and may develop leaf-like structures. Flowers will tend to remain on the plant rather than being shed.

CONTROL: No practical control is known.

Witches' Broom

CAUSE: Witches broom is caused by a mycoplasmalike organism that overwinters in several perennial plants such as Astragalus, Hedysarum, Lathyrus, Lotus, Medicago, Melilotus, and Trifolium. It is transmitted by grafting, dodder, and the leafhoppers *Scaphytopius acutus* (Say), *S. dubius* (Van Duzee), and *Orosius argentatus* (Evans).

DISTRIBUTION: Australia, Saudi Arabia, North America, USSR, and possibly other semiarid areas where alfalfa is grown.

SYMPTOMS: More than the normal number of crown buds develop that form thin, pale-green stems that are shorter than healthy plants, giving a broomlike appearance. The leaves are small and leaflets have a yellowing around their edge together with a crinkling. Flower buds develop slowly and may be green on some plants. Infected plants are relatively short-lived but will exhibit symptoms the following year if they survive the winter. Plants have a yellowish cast. Infected plants may appear to recover during cool weather in winter and spring but recurrent symptoms again develop during warm weather and moisture stress in summer.

CONTROL: No control known.

CAUSED BY
NEMATODES

Alfalfa Stem Nematode

CAUSE: *Ditylenchus dipsaci* (Kuhn) Filipjev. survives for years as larvae in infested stems in dry hay and in alfalfa crowns and soil. Larvae feed and reproduce primarily in the shoots near the crown, moving over the plant surface in a film of water and infecting plants through stomates. A female nematode lays between 75 to 100 eggs with optimum reproduction and infection during cool (15–21°C), wet weather. The nematode can also live unnoticed on a number of weed and crop hosts. It can be spread by farm machinery and free-flowing water. Infection occurs in heavy soils and areas of heavy rain and irrigation. Dissemination is by machinery, rain, and irrigation water.

DISTRIBUTION: Worldwide.

SYMPTOMS: Stem nematodes may ruin a stand in two to three years. Stems enlarge and become discolored, nodes swell, and internodes shorten. Plants are stunted in spots throughout a field. Plants with pure white shoots that are scattered through a field are also an indication of infestation. Infected plants have a bushy appearance due to swollen, short stems and abnormal proliferation of stems. Other indications are the presence of weedy areas in low-lying areas and increased severity of fungal and bacterial rots. Infested crown buds become swollen, spongy, and easily detached. Shoots arising from infected buds are severely dwarfed. Considerable yield loss may result with susceptible cultivars.

CONTROL:
1. Plant resistant cultivars.
2. Plow under infested stands.

3. Use nematicides where feasible.
4. Rotate alfalfa with other crops for two to three years.

Root Knot

CAUSE: Root knot nematodes *Meloidogyne hapla* (Kofoid & White) Chitwood, *M. arenaria* (Neal) Chitwood, *M. javanica* (Treub.) Chitwood, and *M. incognita* (Kofoid & White) Chitwood primarily infect alfalfa growing in sandy soils. The second stage or newly hatched larvae infect the root tip.

DISTRIBUTION: Generally distributed wherever alfalfa is grown.

SYMPTOMS: Top growth of infested plants is yellow and stunted. Roots branch excessively and have galls or knots that vary in size. The severity of bacterial wilt may also increase on plants infected by *M. hapla*.

CONTROL:
1. Use nematicides in areas of high nematode infestation.
2. Plant resistant cultivars.

VIRAL CAUSES

Alfalfa Enation

CAUSE: Alfalfa enation virus (AEV) is transmitted in the field by the cowpea aphid *Aphis craccivora* Koch. With the exception of graft transmission in the laboratory no other type of transmission is known to occur.

DISTRIBUTION: Eastern and southern Europe and Morocco.

SYMPTOMS: Leaflets are crinkled. Enations several millimeters long occur on the underside of the midvein of the crinkled leaflets. Affected plants may be normal size but are bushy.

CONTROL: No control measure is known.

Alfalfa Dwarf

CAUSE: Pierce's virus (PV). This virus infects several hosts besides alfalfa, including grasses, other legumes, shrubs, vines, and grapes. It is disseminated by several types of leafhoppers in the sharpshooter group. These leafhoppers are capable of retaining the virus for several months in their bodies.

DISTRIBUTION: Southeastern and southwestern United States.

SYMPTOMS: Infected plants decline in vigor over a period of several months and recover slowly after cutting. There are a large number of small, spindly stems with small leaves that are a darker green color than healthy leaves.

A brown to yellow discoloration occurs in the woody portion of the crown and roots similar to bacterial wilt; however, no discolored pockets of tissue extend into the bark.

CONTROL: Plant resistant cultivars.

Alfalfa Latent Virus

CAUSE: Alfalfa latent virus (ALV) is transmitted by the pea aphid, *Acyrthosiphon pisum* Harris, and sap inoculation.

DISTRIBUTION: Nebraska.

SYMPTOMS: Affected plants are apparently symptomless.

CONTROL: No control measure is known.

Alfalfa Mosaic

CAUSE: Alfalfa mosaic virus (AMV) overwinters in alfalfa and other perennial host plants. The virus is primarily disseminated from diseased to healthy plants by pea aphids, *Acyrthosiphon pisum* Harris, although other aphids may also be involved. The virus is also seedborne and is transmitted mostly through pollen and occasionally through the ovules. Infected seed is the most likely source of inoculum in new alfalfa-growing areas. The AMV complex is composed of many strains that differ in infectivity and other characteristics.

DISTRIBUTION: Occurs wherever alfalfa is grown.

SYMPTOMS: Older stands have a higher percentage of infected plants. Symptoms are best manifested during the cool weather of spring and fall. Many infected alfalfa plants may display no symptoms; however, the most common symptoms are yellow streaks parallel to the leaf veins and yellow or light green mottling often accompanied by distortion of the leaves. Stunting may often occur and infrequently plants may die.

CONTROL:
1. Plant virus-free seed.
2. Control aphids where possible.

Transient Streak

CAUSE: Lucerne transient streak virus (LTSV). The method of transmission in the field is not known; however, it is readily transmitted by sap in the laboratory.

DISTRIBUTION: Australia, Canada, and New Zealand.

SYMPTOMS: Symptoms are most obvious on newly expanded leaflets and fade as leaflets age. No symptoms occur during high temperature in the summer. Chlorotic streaks that vary in size from small spots to streaks 1–2 mm wide occur on main lateral veins of leaflets. Leaflets are often distorted around the streaks. Yield losses of up to 18 percent in the field have been demonstrated.

CONTROL: No control is reported.

—————BIBLIOGRAPHY—————

Allen, S. J.; Barnes, G. L.; and Caddel, J. L. 1982. A new race of *Colletotrichum trifolii* on alfalfa in Oklahoma. *Plant Disease* **66:**922–924.

Barnett, H. L. 1960. *Illustrated Genera of Imperfect Fungi.*
2nd ed. Burgess Publ., Minneapolis, MN.

Basu, P. K. 1981. Existence of chlamydospores as soil survival and primary infective propagules. *Phytopathology* **71:**202 (abstract).

Baxter, J. W. 1956. Cercospora black stem of alfalfa. *Phytopathology* **46:**398–400.

Blackstock, J. McK. 1978. Lucerne transient streak and lucerne latent, two new viruses of lucerne. *Aust. J. Agric. Res.* **29:**291–304.

Buchholtz, W. F. 1942. Influence of cultural factors on alfalfa seedling infection by *Pythium debaryanum* Hesse. *Iowa State College Agric. Exp. Sta. Res. Bull. 296.*

Carroll, R. B.; Jones, E. R.; and Swain, R. H. 1977. Winter survival of *Colletotrichum trifolii* in Delaware. *Plant Dis. Rptr.* **61:**12–15.

Chilton, S. J. P. et al. 1943. Fungi reported on species of Medicago, Melilotus, and Trifolium. *U.S. Dept. Agric. Misc. Pub. 499.*

Christen, A. A. 1982. Demonstrations of *Verticillium albo-atrum* within alfalfa seed. *Phytopathology* **72:**412–414.

Christen, A. A. 1983. Incidence of external seed-borne *Verticillium albo-atrum* in commercial seed lots of alfalfa. *Plant Disease* **67:**17–18.

Claflin, L. E., and Stuteville, D. L. 1973. Survival of *Xanthomonas alfalfae* in alfalfa debris and soil. *Plant Dis. Rptr.* **57:**52–53.

Claflin, L. E.; Stuteville, D. L.; and Armbrust, D. V. 1973. Wind-blown soil in the epidemiology of bacterial leaf spot of alfalfa and common blight of bean. *Phytopathology* **63:**1417–1419.

Cowling, W. A., and Gilchrist, D. G. 1981. Distinction between the 'Californian' and 'Eastern' forms of Stemphylium leafspot of alfalfa in North America. *Phytopathology* **71:**211 (abstract).

Cowling, W. A., and Gilchrist, D. G. 1982. Effect of light and moisture on severity of Stemphylium leaf spot of alfalfa. *Plant Disease* **66:**291–294.

Cowling, W. A., Gilchrist, D. G.; and Graham, J. H. 1981. Biotypes of *Stemphylium botryosum* on alfalfa in North America. *Phytopathology* **71:**679–684.

Devine, T. E.; Campbell, T. A.; and Hanson, C. H. 1975. Anthracnose disease ratings for alfalfa varieties and experimental strains. *USDA Agric. Res. Serv. Tech. Bull. No. 1507.*

Dickson, J. G. 1956. *Diseases of Field Crops.* 2nd ed. McGraw-Hill Book Co. Inc., New York.

Elliott, E. S.; Baldwin, R. E.; and Carroll, R. B. 1969. Root rots of alfalfa and red clover. *West Virginia Univ. Agric. Exp. Sta. Bull. 585T.*

Emberger, G., and Welty, R. E. 1981. Relationship of Fusarium wilt resistance and soil moisture to yield and persistence of alfalfa. *Phytopathology* **71:**766 (abstract).

Frosheiser, F. I.; Munson, R. D.; and Wilson, M. C. 1972. Alfalfa analyst. Printed by Certified Alfalfa Seed Council. *Iowa State Univ. Ext. Serv. Pm-537.*

Frosheiser, F. I., and Barnes, D. K. 1973. Field and greenhouse selection for Phytophthora root resistance in alfalfa. *Crop Sci.* **13:**735–738.

Gilbert, R. G. 1985. *Sclerotinia sclerotiorum* causing crown and stem rot of alfalfa. *Phytopathology* **75:**1333 (abstract).

Graham, J. H., coordinator. 1979. *A Compendium of Alfalfa Diseases.* American Phytopathological Society, St. Paul, MN, 65p.

Graham, J. H.; Devine, T. E.; and Hanson, C. H. 1976. Occurrence and interaction of three species of Colletotrichum on alfalfa in the mid-Atlantic United States. *Phytopathology* **66:**538–541.

Graham, J. H.; Kreitlow, K. W.; and Falkner, L. R. 1972. Diseases, pp. 497–526 in *Alfalfa Science and Technology,* C. H. Hanson, ed. American Society of Agronomy, Madison, WI.

Graham, J. H.; Peaden, R. N.; and Evans, D. W. 1977. Verticillium wilt of alfalfa found in the United States. *Plant Dis. Rptr.* **61**:337–340.

Griffin, G. D. 1968. The pathogenicity of *Ditylenchus dipsaci* to alfalfa and the relationship of temperature to plant infection and susceptibility. *Phytopathology* **58**:929-932.

Hancock, J. G. 1983. Seedling diseases of alfalfa in California. *Plant Disease* **67**:1203–1208.

Hancock, J. G. 1984. Prevalence and pathogenicity of *Pythium paroecandrum* on alfalfa in California. *Phytopathology* **74**:855. (Abstract).

Hemmati, K., and McLean, D. L. 1977. Gamete-seed transmission of alfalfa mosaic virus and its effect on seed germination and yield in alfalfa plants. *Phytopathology* **67**:576–579.

Horsfall, J. G. 1930. A study of meadow-crop diseases in New York. N. Y. (Cornell) Agric. Exp. Sta. Mem. 130.

Houston, B. R. et al. 1960. Alfalfa diseases in California. *Calif. Agric. Exp. Sta. & Ext. Serv. Circ. 485.*

Huang, H. C., and Hyvr, A. M. 1984. Transmission of Verticillium albo-atrum to alfalfa via feces of leaf-chewing insects. *Phytopathology* **74**:797. (Abstract).

Huang, H. C.; Hironaka, R.; and Howard, R. J. 1986. Survival of *Verticillium albo-atrum* in alfalfa tissue buried in manure or fed to sheep. *Plant Disease* **70**:218–221.

Hunt, O. J. et al. 1971. The effects of root knot nematodes on bacterial wilt in alfalfa. *Phytopathology* **61**:256–259.

Irwin, J. A. G. 1984. Etiology of a new Stemphylium-incited leaf disease of alfalfa in Australia. *Plant Disease* **68**:531–532.

Kreitlow, K. W.; Graham, J. H.; and Garber, R. J. 1953. Diseases of forage grasses and legumes in the northeastern states. *Pennsylvania Agric. Exp. Sta and Reg. Pasture Lab. USDA Bull.*

Kuan, T. L., and Erwin, D. C. 1980. Predisposition effect of water saturation of soil on Phytophthora root rot of alfalfa. *Phytopathology* **70**:981–986.

Leath, K. T., and Hill, R. R., Jr. 1974. *Leptosphaerulina briosiana* on alfalfa: Relation of lesion size to leaf age and light intensity. *Phytopathology* **64**:243–245.

Leath, K. T., and Kendall, W. A. 1983. *Myrothecium roridum* and *M. verrucaria* pathogenic to roots of red clover and alfalfa. *Plant Disease* **67**:1154–1155.

Lukezic, F. L. 1974. Dissemination and survival of *Colletotrichum trifolii* under field conditions. *Phytopathology* **64**:57–59.

Lukezic, F. L., and Leath, K. T. 1983. *Pseudomonas viridiflava* associated with root and crown rot of alfalfa and wilt of birdsfoot trefoil. *Plant Disease* **67**:808–811.

McVey, D. V., and Gerdemann, J. W. 1960. Host-parasite relations of *Leptodiscus terrestris* on alfalfa, red clover, and birdsfoot trefoil. *Phytopathology* **50**:416–421.

Marks, G. C., and Mitchell, J. E. 1970. Penetration and infection of alfalfa roots by *Phytophthora megasperma* and the pathological anatomy of infected roots. *Can. J. Bot.* **49**:63–67.

Ooka, J. J., and Vchida, J. Y. 1982. Cylindrocladum root and crown rot of alfalfa in Hawaii. *Phytopathology* **72**:955 (abstract).

Parmeter, J. R. 1970. *Rhizoctonia solani, Biology and Pathology.* University of California Press, Berkeley.

Pennypacker, B. W. 1983. Dispersal of *Verticillium albo-atrum* in the xylem of alfalfa. *Plant Disease* **67**:1226–1228.

Pennypacker, B. W.; Leath, K. T.; and Hill, R. R., Jr. 1984. Resistant alfalfa plants as symptomless carriers of *Verticillium albo-atrum*. *Phytopathology* **74**:855 (abstract).

Pennypacker, B. W.; Leath, K. T.; and Hill, R. R., Jr. 1985. Resistant alfalfa plants as symptomless carriers of *Verticillium albo-atrum*. *Plant Disease* **69**:510–511.

Pfender, W. F.; Hine, R. B.; and Stanghellini, M. E. 1977. Production of sporangia and release of zoospores by *Phytophthora megasperma* in soil. *Phytopathology* **67**:657–663.

Ribeiro, O. K.; Erwin, D. C.; and Khan, R. A. 1978. A new high-temperature Phytophthora pathogenic to roots of alfalfa. *Phytopathology* **68**:155–161.

Seif El-Nasr, H.I.; Abdel-Azim, O. F.; and Leath, K. T. 1983. Crown and root fungal disease of alfalfa in Egypt. *Plant Disease* **67**:509–511.

Seif El-Nasr, H. I.; Abdel-Azim, O. F.; and Leath, K. T. 1984. *Marasmius* root rot of alfalfa and Khella in Egypt. *Plant Disease* **68**:906–907.

Shyh-Jane Lu, N.; Barnes, D. K.; and Frosheiser, F. I. 1973. Inheritance of Phytophthora root rot resistance in alfalfa. *Crop Sci.* **13**:714–717.

Smith, K. M. 1972. *A Textbook of Plant Virus Diseases.* Academic Press, New York.

Smith, O. F. 1940. Stemphylium leaf spot of red clover and alfalfa. *J. Agric. Res.* **61**:831–846.

Stack, J. P., and Millar, R. L. 1985. Competitive colonization of organic matter in soil by *Phytophthora megasperma* f. sp. *medicaginis. Phytopathology* **75**:1020–1025.

Stack, J. P., and Millar, R. L. 1985. Relative survival potential of propagules of *Phytophthora megasperma* f. sp. *medicaginis. Phytopathology* **75**:1025–1031.

Streets, R. B., and Bloss, H. E. 1973. Phymatotrichum root rot. *Am. Phytopathol. Soc. Monograph No. 8*, 38p.

Stutz, J. C.; Leath, K. T.; and Kendall, W. A. 1985. Wound-related modifications of penetration, development, and root rot by *Fusarium roseum* in forage legumes. *Phytopathology* **75**:920–924.

Sundheim, L., and Wilcoxson, R. D. 1965. *Leptosphaerulina briosiana* on alfalfa: Infection and disease development, host-parasite relationships, acospore germination and dissemination. *Phytopathology* **55**:546–553.

Turner, V., and Van Allen, N. K. 1983. Crown rot of alfalfa in Utah. *Phytopathology* **73**:1333–1337.

Veerisetty, V., and Brakke, M. K. 1977. Alfalfa latent virus, a naturally occurring carlavirus in alfalfa. *Phytopathology* **67**:1202–1206.

Walker, J. 1956. Further diseases of clovers in New South Wales. *Agric. Gaz. N.S.W.* **67**:1–6.

Welty, R. E., and Rawlings, J. O. 1980. Effects of temperature and light on development of anthracnose on alfalfa. *Plant Disease* **64**:476–478.

Diseases of Barley (*Hordeum Vulgare* L. emend Bowden)

BACTERIAL CAUSES

Bacterial Leaf and Kernel Blight

CAUSE: *Pseudomonas syringae* pv. *syringae* van Hall may survive on plant residue and nonhost leaf surfaces. It may also be seedborne. The bacterium infects barley during cold, wet, and windy weather when it enters plants through stomates or wounds in the leaf.

DISTRIBUTION: North America.

SYMPTOMS: Normally only the top leaves, particularly the flag leaf, are affected. Initially, small, water-soaked lesions are formed that often coalesce to form larger lesions that involve the entire leaf. At first the lesions are gray-green but later turn tan or white. Lesions often form where the leaf bends. Eventually the whole leaf may die and curl inward from the edges.

Kernels may become infected with tan to dark brown spots with distinct margins. Lesions develop primarily on the lemma and average 2 mm in diameter. Kernel infection is most frequent before kernel development reaches the milky dough stage.

CONTROL: No control is known.

Bacterial Stripe Blight

This disease is also called bacterial stripe.

CAUSE: *Pseudomonas syringae* pv. *striafaciens* (Elliott) Young, Dye & Wilkie (syn. *P. striafaciens* (Elliott) Starr & Burk.) is seedborne and survives in infected crop refuse for at least two years. During cool, wet weather, bacteria are splashed or blown onto leaves. Warm, dry weather checks the spread of the bacteria.

DISTRIBUTION: Australia, Europe, North America, and South America.

SYMPTOMS: Initially, sunken, water-soaked dots appear on leaves. If the dots are abundant the leaf may die. The dots enlarge into water-soaked stripes or blotches that may extend the length of the blade. These stripes often have narrow, yellowish margins. As the stripes age, they become a translucent rusty brown. In moist weather bacteria exude in droplets from the stripes. Later they dry, forming white scales.

CONTROL:
1. Plow under infected residue.
2. Treat seed with a seed-protectant fungicide.
3. Plant resistant cultivars.

Basal Glume Rot

CAUSE: *Pseudomonas atrofaciens* (McCulloch) Stevens (syn. *Phytomonas atrofaciens* (McCulloch) Bergey et al.) is seedborne and probably overwinters in infected residue in the soil. Bacterial spores are wind disseminated on dust particles and become entrapped in water present in grooves or small spaces in the spikelets. Bacteria may also be disseminated by insects and probably by water splashing. The bacteria then multiply near glume joints when water is present but remain dormant when moisture is lacking.

DISTRIBUTION: Generally distributed wherever barley is grown.

SYMPTOMS: Symptoms are most likely to occur during wet weather particularly at heading time. The main symptom is a brown discolored area found at the base of the glumes covering a kernel. This discoloration is more evident on the inside than the outside of the glume. Usually only about the bottom one-third of the glume is discolored but sometimes the entire glume may be affected. Severely infected spikelets are slightly dwarfed and lighter in color than healthy ones. Sometimes the only sign of disease is a dark line at the attachment of the glume to the spike.

A diseased kernel would show a faint brown to black discoloration on its base. Infected leaves show small, dark water-soaked spots that will eventually elongate, turn yellow, then brown as the tissue dies. Seed filling is sometimes limited.

CONTROL:
1. Treat seed with a seed-protectant fungicide.
2. Seed should be thoroughly cleaned.
3. Rotate barley with resistant crops such as legumes.
4. Plow under infected residue.

Black Chaff

This disease is also called bacterial blight, bacterial leaf blight, bacterial leaf streak, and bacterial streak.

CAUSE: *Xanthomonas campestris* pv. *translucens* (Jones, Johnson & Reddy) Dye (syn. *X. translucens* (Jones, Johnson & Reddy) Dowson, *X. translucens* f. sp. *undulosa* (Smith, Jones & Ready) Hagb., and *Phytomonas translucens* (Jones, Johnson & Reddy) Bergey et al.) is seedborne and overwinters in infected crop residue. The initial infection probably comes from seedborne bacteria. Bacteria are disseminated by physical contact of leaves, splashing rain, and insects. The bacteria enter into the plant through natural openings such as stomata and wounds.

DISTRIBUTION: Generally distributed wherever barley is grown. However, black chaff is seldom a serious disease.

SYMPTOMS: Black chaff symptoms are confined mainly to leaves but chaff may be affected also. Symptoms usually occur after several days of damp or rainy weather. Infected plants grow slowly and may be stunted but usually the disease is not noticed until plants are about two-thirds grown. Small water-soaked spots occur on tender green leaves and sheaths of older plants and sometimes on seedlings. The spots enlarge and coalesce, becoming glossy, translucent stripes of various lengths that turn yellow or brown. Stripes may extend the length of a leaf and are usually narrow, limited by leaf veins. Occasionally a spot may become large and blotchy, causing the leaf to die, shrivel, and turn light brown. Severely diseased leaves die back from the tips. Under early morning humid conditions, droplets of milky bacterial exudate may be seen on the surface of diseased spots. The droplets dry into hard yellowish flakes that may be easily removed.

Water-soaked areas may occur on chaff of barley heads but bacterial exudate is not as evident as on leaves. Seed is not destroyed but it may be brown, shrunken, and carry bacteria that can potentially infect next year's crop. If a flag leaf is infected, the head may not emerge from the boot but break through the side of the sheath and be distorted and blighted.

If a newly developed stripe is cut crosswise, either piece will exude a bead of milky ooze at the cut edge.

CONTROL:
1. Plant disease-free seed.
2. Treat seed with a seed-protectant fungicide.
3. Rotate barley with other crops. The bacterium also infects wheat but not rye.

_____FUNGAL CAUSES_____

Anthracnose

CAUSE: *Colletotrichum graminicola* (Ces.)   G. W. Wilson (syn. *C. cereale* Manns and *Dicladium graminicolum* Ces.) is seedborne and over-winters as mycelium and conidia in infected residue, cereals, and wild grasses. Conidia are produced from mycelium during wet weather at an optimum temperature of 25°C and are then windborne to barley and other host plants.

DISTRIBUTION: Generally distributed wherever barley is grown. However, anthracnose is most severe when barley is grown on sandy soils that are low in fertility.

SYMPTOMS: Symptoms of anthracnose become apparent toward plant maturity. Premature ripening or whitening, a general reduction in plant vigor, or dying are gross symptoms. Specific symptoms gener-ally occur on the lower part of the plant. Elliptical-shaped lesions, 1–2 cm long, are produced. Initially the lesions are water-soaked but become bleached and necrotic. The entire crown and bases of stems also become bleached and later turn brown. Later, small, black raised spots, which are acervuli, develop on the surface of the lesions on lower leaf sheaths and culms. Acervuli may also develop on leaf blades of dead plants when moisture is plentiful. Under magnification, acervuli will appear as a clump of dark spines. Kernels may be shriveled if infection occurs early. A seedling and crown infection may occur under severe disease condi-tions.

CONTROL:
1. Rotate barley with a noncereal. Most cereals and grasses are susceptible to *C. graminicola*.

2. Improve soil fertility as anthracnose is most likely to develop on low fertility soils.
3. Treat seed with a fungicide.

Ascochyta Leaf Spot

CAUSE: *Ascochyta hordei* Hara var. *americana* Punith. is generally accepted as the cause of Ascochyta leaf spot. However, *A. graminea* (Sacc.) Sprague & Johnson, *A. sorghi* Sacc. (syn. *A. graminicola* Sacc.), and *A. tritici* Hori & Enjoji have also been reported on barley. *Ascochyta hordei* survives as mycelium and pycnidia in residue. Pycnidiospores are liberated during wet weather. Disease is associated with dense foliage, leaves in contact with soil, and high humidity.

DISTRIBUTION: Japan and the United States, primarily the north-western states, and New York.

SYMPTOMS: Spots occur primarily on lower leaves and are either round (4 mm in diameter), oval (2 by 4 mm wide to 8 by 12 mm long) or irregular, sometimes with several rings and located on the terminal third of the leaf blades. Initially, lesions are brown but fade to pale yellow, ash gray, or tan necrotic centers surrounded by a narrow brown band. Lesions at leaf tips often coalesce, causing dieback without the brown margin. The latter symptom is masked by frost injury. Pycnidia are present in infected tissue but are absent from frost-damaged tissue.

CONTROL: No control is necessary since the disease is considered to be of minor importance.

Barley Stripe

This disease is also called barley leaf stripe, barley stripe disease, Helminthosporium stripe, leaf stripe, pyrenophora leaf stripe, and stripe disease.

CAUSE: *Pyrenophora graminea* S. Ito & Kuribay, anamorph *Drechslera graminea* (Rab.) Shoem. (syn. *Helminthosporium gramineum* Rabh.). The fungus survives up to five years as mycelium in hull, pericarp, and seed coat. Conidia are produced in stripes of infected barley leaves under high-moisture conditions and are windborne to the heads of healthy plants. Spores that lodge near the tips of the glumes germinate under moist conditions and the mycelium grows between the hulls and kernels or into the seed coat of the grain. Infection occurs at any stage of development from preemergence of the head through the soft dough stage and over a wide range of temperature and moisture conditions. At seed germination, the fungus grows from the seed into the sheath that surrounds the leaves of the seedling, into the first leaf, and into succeeding leaves. Seed transmission is increased at soil temperatures of 12°C and below, and reduced or prevented above 15°C. Inoculum in the pericarp and the seed coat over the pericarp is most effective in causing the disease. The critical stage for infection of the germinating embryo is when the coleoptile reaches the apex of the seed until the seedling emerges from the soil. Eventually all leaves of the plant are infected. If the growing point becomes infected, the culm dies. Disease spread is directly related to moisture. Irrigation or high relative humidity near heading will increase disease incidence up to threefold compared to plants grown in a dry environment. Stripe disease is more common on winter than spring barley.

DISTRIBUTION: Generally occurs wherever winter barley is grown.

SYMPTOMS: The first symptoms of the disease usually occur at late tillering stage when small yellow spots that are easily overlooked appear on seedling leaves. Weeks before barley heads, the conspicuous and characteristic symptoms appear. These consist of one to several narrow, yellow to light tan stripes in the blade and sheath of a leaf. These stripes are parallel with each other and extend the entire length of the blade. The alternate yellow and green striping of a diseased leaf constrasts sharply with the uniform green of healthy leaves. The margins of the stripes turn red or dark brown with the center remaining tan or light brown. The diseased tissue dies with a split developing in the center and along the length of a stripe, causing the entire leaf to be shredded.

Infected plants are stunted and heads may not emerge while in others they emerge blighted, twisted, compressed, and brown and will stand erect in contrast to normal heads that turn down slightly when mature. Grain in infected heads is undeveloped or shriveled and brown. Infected plants become gray to olive gray in color due to *P. graminea* sporulating on diseased tissues. The spores are windborne to developing kernels of healthy plants where they germinate to continue the disease cycle when infected seed is planted the following growing season.

CONTROL:
1. Treat seed with a seed-protectant fungicide.
2. Plant resistant cultivars.

Cephalosporium Stripe

CAUSE: *Cephalosporium gramineum* Nis. & Ika. (syn. *Hymenula cerealis* Ell.& Ev.) survives as conidia and mycelium either in crop residue or soil for up to five years in the top 8 cm of the soil profile. Soilborne conidia serve as primary inoculum and infect roots during winter and early spring through mechanical injuries caused by soil heaving and insects.

Infection is most severe in wet, acid soils (pH 5.0) due to a combination of increased fungal sporulation and root growth. The subcrown internode can also be infected just as the seed is germinating. After infection, conidia enter xylem vessels and are carried upward in the plant where they lodge and multiply at nodes and leaves. The fungus prevents water movement up the plant and also produces metabolites that are harmful to the plant. At harvest time, *C. gramineum* is returned to the soil in infected residue where it is a successful saprophytic competitor with other soilborne microorganisms.

DISTRIBUTION: Great Britain, Japan, and in most winter-barley growing areas of North America.

SYMPTOMS: Dwarfed infected plants are scattered throughout a field, usually being more numerous in lower, wetter areas. During jointing and heading, distinct yellow stripes, one to four per leaf but usually one or two, develop on leaf blades, sheaths, and stems, usually occurring throughout the length of the plant. Thin brown lines consisting of infected veins usually occur in the middle of a stripe. The stripes eventually become brown, are highly visible on green leaves, but are still noticeable on yellow straw. Toward harvest, the culm at or below the node may become dark. Heads of infected plants are white and do not contain seed, or if seed is present, it is usually shriveled.

CONTROL:
1. Rotate barley for at least two years with a noncereal.
2. Plant winter barley later in the autumn or when the soil temperature at 10 cm is below 13°C. Such plants apparently have limited root growth, thus reducing the number of infection sites.
3. Infected residue should be plowed deeper than 8 cm where feasible.

Common Root Rot

CAUSE: Several fungi including *Cochliobolus sativus* (Ito & Kurib.) Drechs. ex Dast., anamorph *Bipolaris sorokiniana* (Sacc.) Shoemaker (syn. *Helminthosporium sativum* Pammel, King, & Bakke and *H. sorokinianum* Sacc. apud Sorokin.); *Fusarium culmorum* (W. G. Sm.) Sacc. (syn. *F. roseum* (Lk.) emend. Snyd. & Hans. f. sp. *cerealis* (Cke.) Snyd. & Hans. 'Culmorum'); and *F. graminearum* Schwabe (syn. *F. roseum* (Lk.) emend. Snyd. & Hans. f. sp. *cerealis* (Cke.) Snyd. & Hans. 'Graminearum'), teleomorph *Gibberella zeae* (Schw.) Petch (syn. *G. roseum* f. sp. *cerealis* 'Graminearum,' and *G. saubinettii* (Mont.) Sacc.). Conidia of *B. sorokiana* and chlamydospores of *Fusarium* spp. are capable of surviving for years in the soil. Additionally, these fungi are excellent saprophytes that easily colonize any plant material added to the soil.

Plants put under stress by drouth, warm temperatures, lack of nutrition, and Hessian fly injury are most subject to infection. Moisture is required for infection, but once initiated, disease development requires warm temperatures and moisture stress. Disease initiation by *B. sorokiniana* begins 25 to 30 days after planting and reaches maximum levels at about 60 days.

DISTRIBUTION: Generally distributed wherever barley is grown but it is more common where plants undergo moisture stress during the growing season.

SYMPTOMS: Common root rot is most likely to be noticed after heading as looking like drouth damage. Diseased plants are shorter and frequently show brown and dead lower leaves.

Early in the growing season, brown lesions may be noticed at the base of the leaf near the soil surface. Lesions may vary in size and extend partially or completely around the stem. Later in the growing season, the lower parts of leaves, tiller buds, crown roots, and internodes below and above crowns may show varying degrees of brownish discoloration indicative of rot. Such lesions are very distinct and rotted crown and crown roots can be readily seen if dead leaves and tillers are torn away from the crown. Lesions on the internode below the crown or between crown and seed are good indications of infection. At this stage plants will have many dead leaves starting with lower leaves dying first.

Survival of winter barley is reduced, particularly by infection caused by *B. sorokiniana*.

CONTROL: No controls are very satisfactory in controlling common root rot; however, the following may aid in control.
1. Rotate barley with noncereal crops.
2. Plow under infected residue where feasible.

Covered Smut

CAUSE: *Ustilago hordei* (Pers) Lagerh. survives as teliospores in soil and in crevices of seed. Smutted heads emerge at about the same time as those of healthy plants. A mass of smut spores that have replaced the kernels is covered by a somewhat persistent membrane that may remain intact until harvest time unless ruptured by wind, rain, or other means. The smut spores, referred to as either teliospores or chlamydospores, become windborne and land on or under the hulls of healthy barley kernels. The spores on the kernel surface usually remain dormant until seed is planted the following spring or autumn. Those spores deposited under the hulls may germinate with the resulting mycelium, growing only into the shallow layers of the seed just under the hull and remaining dormant until planting time.

When seed is planted, the resulting seedling becomes infected between germination and emergence by seedborne spores and mycelium or by soilborne spores. The mycelium of *U. hordei* grows within the plant tissues just behind the growing point. As the plant enters the boot stage, the smut fungus grows into the ovary of the flower and converts it into masses of teliospores before emergence. The smutted head then emerges from the boot in the same manner as a healthy head. The amount of infection of the new crop depends on the percentage of infested kernels that were planted. The highest percentage of smutted plants occurs in acid or neutral soils when the soil temperature is between 14°C and 25°C and soil moisture is high.

DISTRIBUTION: Generally distributed wherever barley is grown.

SYMPTOMS: The characteristic symptom is the somewhat persistent membrane enclosing the smut spores until plants are mature. Covered smut is first noticeable at heading time as black smutted heads that emerge from the boot in place of normal green heads. From a distance, smutted heads will be a grayish color. Black masses of smut spores, each enclosed by a grayish-white membrane, are found in place of healthy kernels in the affected heads. Each smutted head contains millions of spores. Physiologic races of *U. hordei* are present.

CONTROL:
1. Treat seed with a systemic seed-protectant fungicide.
2. Plant resistant cultivars.

Downy Mildew

CAUSE: *Sclerophthora macrospora* (Sacc.) Thirum., Shaw & Naras. (syn. *Sclerospora graminicola* (Sacc.) Schroet. and *S. macrospora* Sacc.) survives several years as oospores that are either in infected tissue or in soil when infected tissues eventually decay. It is an obligate parasite and cannot grow saprophytically on dead plant tissue. Oospores are seedborne and also disseminate in infected plant residue and soil that is either water or windborne. Oospores germinate in water-saturated soil to produce sporangia in which zoospores are produced and eventually released. The zoospores will swim through the water, settle on the developing seedling, and produce a germ tube that penetrates the plant. Oospores may survive for months in dry soil and germinate to form only a germ tube that directly penetrates a host.

Newly emerged seedlings are most susceptible. Plant parts, primarily leaves, that are in contact with infested water are subject to infection. Infection occurs over a wide temperature range of 7–31°C. Following infection, *S. macrospora* develops systematically within the plant, particularly in meristematic tissue.

DISTRIBUTION: Generally distributed wherever barley is grown. Downy mildew is found primarily on winter barley.

SYMPTOMS: Downy mildew occurs only in localized areas of fields where seedlings have been growing in flooded or waterlogged soil for 24 hours or longer. Diseased plants are scattered in or along standing water. Plants are dwarfed, deformed, and may tiller excessively. Infected plants have leathery, stiff, warty, and thickened leaves. Diseased plants may be flagged with prominently yellowed leaves. Affected heads are twisted in various ways and distorted. No seeds are formed in severely diseased plants. Plants may die before jointing.

In less severely diseased plants, dwarfing may be slight with one or more of the upper leaves stiff, upright, or variously curled and twisted. The heads and stems may not be deformed. Numerous round, yellow to brown oospores may be found in diseased tissue examined under magnification.

CONTROL:
1. Provide proper soil drainage where possible.
2. Control grassy weeds in a field that may serve as hosts.
3. Plant cleaned seed from disease-free plants to ensure no infested residue is disseminated with seed.

Dwarf Bunt

CAUSE: *Tilletia controversa* Kuehn (syn. *T. pancicii* Bub. & Ranjoj. and *T. hordei* Koern.). Isolates from barley can attack wheat and vice versa. The causal fungus survives as teliospores in the soil or on the surface of contaminated seed. When barley is planted, teliospores that are either on or close to the seed in the soil will germinate if moisture is present. A promycelium is formed on which 8 to 16 basidiospores are formed. The basidiospores then fuse in the middle with a compatible basidiospore to form an H-shaped structure. This structure then germinates to form yet another structure called a secondary sporidia. It is the secondary sporidia that finally germinate to produce mycelia that infect seedlings. The optimum temperature for germination is 1–5°C. The infection process takes from 35 to 105 days depending on temperature. These conditions generally occur under heavy snow cover over unfrozen ground. After infection, mycelium grows behind the growing point or meristematic tissue of the plant. The mycelium invades the developing head and displaces the grain. Eventually teliospores are formed within the seed. At harvest time, the infected seed is broken and teliospores are windborne to contaminate healthy kernels or the soil.

DISTRIBUTION: Utah, on winter barley.

SYMPTOMS: Dwarf bunt is restricted to areas where winter barley is grown under conditions of long continuous snow cover. Infected tillers are stunted to about one-fourth to one-half the normal height. Infected heads are compact and sori fill the seed coat to form a ball. These bunt balls are rounder and shorter than normal barley seed. Spores smell like rotten fish when bunt balls are broken over, particularly if teliospores are moist.

CONTROL: No control known.

Ergot

CAUSE: *Claviceps purpurea* (Fr.) Tul. overwinters as sclerotia in the soil and can survive for about one year. During the growing season, just before blossoming, a sclerotium germinates to produce a stromatic head in which perithecia are produced at the end of a stipe or stalk. Ascospores are formed in the perithecia and are primarily windborne and occasionally travel by splashing rain to young flowers. The ascospore germinates and mycelium grows into the young ovary. Conidia are produced in a sweet, sticky liquid called honeydew from mycelium growing in the ovary. Insects are attracted to honeydew and inadvertently disseminate conidia to healthy flowers where secondary infection occurs. Eventually ovaries enlarge and are converted into sclerotia progressively from the base to the tip of the kernel. Infection is favored by cool, wet weather, which prolongs flowering. Florets are most susceptible just before antithesis. Plants become resistant after pollination.

Alkaloid chemicals are produced in the sclerotium that can cause convulsions, gangrene, and death in animals and humans. Plants infected with barley stripe mosaic virus are more susceptible than healthy plants to infection. Several grasses and cereals are susceptible to *C. purpurea*.

DISTRIBUTION: Generally distributed wherever barley is grown; however, barley is usually not severely infected.

SYMPTOMS: The only part of the plant that is infected is the head. The first symptom is a yellowish, sticky substance (honeydew) that oozes from infected flowers and accumulates in droplets or adheres to the surface. Insects are attracted to honeydew and are found around infected flowers. Near maturity each infected kernel is replaced with the hard, black to blue-black sclerotium or ergot body composed of mycelium that protrudes from the glume. Sclerotia may be larger than seed.

CONTROL:
1. Deep plowing will place sclerotia in soil where they will not be able to release acospores into the air. Sclerotia are also decomposed in a year or so by soil microorganisms.
2. Control grassy weeds around fields as they may also be hosts for *C. purpurea* and increase inoculum in an area.
3. Rotate barley with resistant hosts.
4. Separate sclerotia from seed. Sclerotia may be harmful when fed to livestock.

Eyespot

This disease is also called Cercosporella footrot, footrot, strawbreaker, and culm rot.

CAUSE: *Pseudocercosporella herpotrichoides* (Fron) Deighton (syn. *Cercosporella herpotrichoides* Fron) survives in infected crop residue as mycelium. Conidia are produced during cool, damp weather either in autumn or spring and infect the crown and basal culm tissue. Conidia are disseminated primarily by splashing rain. Eyespot is favored by high soil moisture, dense crop canopy, and high humidity near the soil level. Plants may be predisposed to infection by spring frosts and excessive nitrogen fertilization. Roots are usually not infected. Winter barley is more likely to be infected than spring barley. *Pseudocercosporella herpotrichoides* has a wide host range and infects several other cereals.

DISTRIBUTION: Generally distributed wherever winter barley is grown.

SYMPTOMS: The disease is most conspicuous near the end of the growing season by the presence of lodged plants. The lodging causes straw to fall in all directions. Straw lodged by wind or rain falls in one direction. Whiteheads are also present at maturity.

Eye-shaped or ovate lesions with white to tan centers and brown margins develop first on the basal leaf sheath. Lesions darken with age. Similar spots form on the stem directly beneath those on the sheath and cause lodging. Roots are not infected but a necrosis occurs around the roots in the upper crown nodes. Under moist conditions, lesions enlarge and a black stroma-like mycelium develops over the surface of the crown and base of the culms, giving the tissues a charred appearance. The stems then shrivel and collapse. Plants also are yellowish or pale green with heads reduced in size and numbers.

CONTROL:
1. Rotate barley with legumes.
2. Plant spring barley or delay planting winter barley.
3. Plants grown in high-fertility soils are more likely to become diseased.

Gibberella Seedling Blight

CAUSE: *Gibberella zeae* (Schw.) Petch, anamorph *Fusarium graminearum* Schwabe (syn. *F. roseum* (Lk.) emend Snyd. & Hans. f. sp. *cerealis* (Cke.) Snyd. & Hans. 'Graminearum') overwinters as mycelium and spores on seed, and overwinters as mycelium and perithecia on barley, corn, rye, and wheat residue. Seedling infections are primarily from seedborne inoculum.

DISTRIBUTION: Generally distributed wherever barley is grown.

SYMPTOMS: Initially seedlings are stunted, then turn yellow and die. The root system shows a red-brown rot and may be covered with a white to pink mold.

CONTROL:
1. Treat seed with a fungicide-seed protectant.
2. Plant certified seed.
3. Plant when soil temperature is 15°C or below since warm, dry soil favors disease development.

Glume Blotch

CAUSE: *Septoria nodorum* Berk. teleomorph *Leptosphaeria nodorum* Müller, survives as mycelium in live plants and seed; and survives as pycnidia on infected crop refuse for up to three years. During warm (20–27°C), wet weather in the fall or spring, pycnidiospores are exuded from pycnidia within a gelatinous substance (cirrhi) that protects spores from drying out and radiation. Pycnidiospores are disseminated by splashing and blowing rain to lower leaves of healthy plants. Infection requires six hours or more of wetness. Spore germination and infection is optimum between 15°C and 25°C. The greatest losses occur when rainfall is excessive between flowering and grain harvest. New spores are produced in 10 to 20 days. Ascospores are produced in pseudothecia as barley matures in late summer or early autumn and are primarily windborne. Strains of *S. nodorum* that infect wheat will also infect barley.

DISTRIBUTION: Generally distributed wherever barley is grown.

SYMPTOMS: Glumes, culms, leaf sheaths, and leaves are infected; however, little damage usually occurs until the crop nears maturity. Small grayish or brown spots usually occur near the top third of a glume two to three weeks after the head emerges. The spots enlarge and become a chocolate brown color. Later the center of the spot becomes gray with black pycnidia that resemble tiny black dots scattered throughout.

The nodes of the stem are infected, turn brown, shrivel, and are also speckled with black pycnidia. Such infections often cause straw to bend over and lodge just above the nodes.

Infection of leaves causes light brown spots that may have a darker brown margin. Pycnidia will be present on both surfaces of a diseased leaf. If a flag leaf is infected, the head may be deformed. Infection of the leaf sheath causes a dark brown lesion that may include most of each leaf sheath. Severely infected plants are stunted.

Symptoms may be confused with black chaff or basal glume rot. Glume blotch spots do not form streaks and are not as sharply defined or as dark brown as those of black chaff. Glume blotch does not have the water-soaked appearance of basal glume rot.

A *Septoria* sp. that is different pathologically from *S. nodorum* has been reported to cause oval leaf spots with a buff center that contains numerous pycnidia. This *Septoria* sp. was also isolated from glumes.

CONTROL:
1. Plant certified, disease-free seed that has been cleaned and treated with a seed-protectant fungicide.
2. Plant the most tolerant barley cultivars if they are adapted to the area. Several winter barley cultivars have been reported to be resistant.
3. Plow under or burn infected residue where feasible.
4. Apply a foliar fungicide if conditions warrant it.
5. Rotate barley every three or four years with a nonsusceptible crop.
6. Plant winter barley after Hessian fly-free date.

Halo Spot

CAUSE: *Selenophoma donacis* (Pass.) Sprague & Johnson (syn. *Septoria donacis* (Pass.) and *Phyllosticta stomaticola* (Baeumler)). *Selenophoma donacis* var. *stomaticola* (Baeumler) Sprague & Johnson has been reported from Canada. The causal fungus overwinters as pycnidiospores, pycnidia, and mycelium in infected residue, seed, and overwintering cereals. During cool, moist weather, pycnidiospores are exuded from pycnidia and are splashed or windborne to healthy plants.

DISTRIBUTION: In cool, moist climates of Eastern Canada, Great Britain, northern Europe, and the United States. Halo spot rarely causes much damage.

SYMPTOMS: Numerous spots appear in the spring on leaves and sometimes culms of winter barley. Spots are less than 4 mm long, and elliptical, rectangular, or square-shaped. At first the spots have purple-brown margins that eventually fade as the spots become old. In time, the center of the spots becomes gray with small black pycnidia that are difficult to see. Sometimes spots may become so numerous that much of the leaf surface is destroyed.

CONTROL: Halo spot apparently does little damage to barley; therefore, controls are not necessary.

Leaf Rust

This disease is also called dwarf leaf rust and brown leaf rust.

CAUSE: *Puccinia hordei* Otth. (syn. *P. anomala* Rostr.) is a heteroecious rust fungus that apparently overwinters only as urediospores on volunteer and winter barley in the southern barley-growing areas of the United States. Urediospores are windborne northward in the spring to infect barley in the northern United States. The disease is most severe under warm and humid conditions.

Star-of-Bethlehem (*Ornithogalum* spp.) is the alternate host for P. *hordei* in Europe, Israel, and parts of Asia. In Israel, *Ornithogalum brachystachys* C. Koch, *O. trichophyllum* Boiss. et Heldr., *Dipcadi erythraeum* Webb et Bert., and *Leopoldia eburnea* Eig et Feinbr. have been reported to be alternate hosts of *P. hordei*. However, basidiospores do not infect nor have pycnia or aecia been found on the two species of Star-of-Bethlehem (*O. umbellatum* L. and *O. nutans* L.) found in the United States except under experimental conditions in the greenhouse.

The wheat leaf rust fungus, *P. recondita* Rob. ex Desm. f. sp. *tritici*, attacks barley in some areas but it is considered a weak parasite of barley.

DISTRIBUTION: Generally distributed wherever barley is grown. Leaf rust is most severe in areas where the crop matures late.

SYMPTOMS: Infection is usually late in the northern barley-growing areas, occurring just before heading. The lower leaves are attacked first with the disease progressing up the plant during warm, humid weather.

Uredia are small, oval, light yellow-brown pustules scattered irregularly on either side of the leaf. Heads are normally not infected except on highly susceptible cultivars. The symptoms are relatively inconspicuous until uredial development is quite abundant. Severely infected leaves turn yellow, making it difficult to see the tiny yellow-brown uredia. Teliospores are formed as plants near maturity, forming brown oblong to round telia that are usually covered by the epidermis. The telia tend to run together, forming gray patches on the leaves; however, they are normally less abundant than uredia.

CONTROL:
1. Plant resistant cultivars.
2. Apply foliar fungicides.

Leptosphaeria Leaf Spot

CAUSE: *Leptosphaeria herpotrichoides* de Notaris overwinters as mycelium in residue and ascospores in asci from pseudothecia on straw. Free water must occur on leaves for 48 hours or more for infection to occur.

DISTRIBUTION: Canada, Europe, United States, and barley-growing areas that have long periods of wet weather. It is considered a minor disease.

SYMPTOMS: Irregular, diffuse, yellow to tan spots occur on leaves. Under extended wet conditions, up to 50 percent of the leaf may be affected.

CONTROL: Some cultivars are more resistant than others.

Loose Smut

This disease is also called true loose smut, nuda loose smut, and brown loose smut.

CAUSE: *Ustilago nuda* (Jens.) Rostr. is seedborne, surviving as dormant mycelium within the embryo of barley seed. When infected barley seed germinates, the mycelium grows systemically within the plant toward the shoot apex and inhabits seed primordia, eventually producing the smutted heads filled with teliospores in place of healthy kernels. The smutted head emerges from the boot a day or two earlier than heads of healthy plants. Teliospores are windborne to flowers of healthy heads. During favorable weather with temperatures of 16–22°C and moisture provided by dews or light showers, the teliospores germinate. The germ tube penetrates the ovary of the flower and possibly the stigma with subsequent mycelial growth in the germ or embryo of the developing seed. As the grain matures, the loose smut fungus becomes dormant within the embryo of the seed until the following growing season. This fungus differs from other smut fungi on barley in that it infects the flowers. Losses are generally directly related to the percentage of infected heads. Thus, 10 percent of infected heads normally yields a 10 percent yield loss. Several physiologic races of *U. nuda* exist.

DISTRIBUTION: Generally distributed wherever barley is grown.

SYMPTOMS: Infected seed does not show any outward symptoms and germination is ordinarily not affected. Infected seed gives rise to smutted heads. Before heading, infected plants have dark green, erect leaves with chlorotic or yellowish streaks. The smutted heads emerge a couple of days earlier than healthy heads. The brown to dark brown spore mass is enclosed with a fragile, gray membrane that soon ruptures, releasing the spores to be windborne to healthy flowers. Mycelium that is intercellular and intracellular develops within gall tissues derived mostly from the rachilla cortex. The epidermis forms a membrane over the spore mass. Soon an erect, naked rachis that protrudes slightly above the healthy plants is all that remains of a smutted head. At maturity, the barren central axis of the smutted head remains erect among the reclining healthy ones.

Sori have been reported in flag leaves of greenhouse-grown 'Larker' barley. Sori appeared prior to or occasionally at the time of emergence of smutted or healthy heads from the leaf sheaths. Frequently no heads appeared and none were in the leaf sheath. Infrequently auricles, ligules, and sheaths were also involved.

CONTROL:
1. Treat seed with a systemic fungicide-seed protectant.
2. Grow resistant cultivars.

Net Blotch

CAUSE: *Pyrenophora teres* Drechs., anamorph *Drechslera teres* (Sacc.) Shoem. (syn. *Helminthosporium teres* Sacc.) overwinters either as seed-borne mycelium or as pseudothecia in infected barley residue. The causal fungus shows considerable variation; isolates have been found that produce spot-type lesions. Therefore, *P. teres* has been divided into two forms: *P. teres* f. *maculata* for the spot-causing isolates and *P. teres* f. *teres* for the net-causing isolates. During cool (10–15°C), humid weather, spores are produced on infected residue; however, it is not known if these spores are conidia or ascospores. The spores are windborne to healthy plants. Conidia produced on primary lesions serve as secondary inoculum. Sporulation occurs at a relative humidity near 100 percent. New infections occur as long as the weather remains cool and moist. Winter barley in Pennsylvania has the greatest severity in the earliest plantings; however, if spring conditions were warm and humid, previous differences in disease severity attributable to dates of planting were eliminated.

DISTRIBUTION: Net blotch is found wherever barley is grown.

SYMPTOMS: Barley is the only host of *P. teres*. Leaves, stems, and seeds are affected. Generally disease severity is greatest on the oldest leaves of winter barley. The first symptoms for the net symptom are brown spots or blotches that occur near the tips of the seedling leaves. Each young spot shows dark brown pigment as narrow lines that run both longitudinally and transversely within the lighter brown area of the spot. These lines give the spot a netted or crosshatched appearance. The netting of the spots is best seen when leaves are held up to a light source. The spots increase in length and form short, narrow streaks. Spots may coalesce and form long brown stripes with irregular margins. The netting or cross hatching is visible only at the margins of the older spots. At this stage of disease development, stripes do not extend into sheaths nor does the leaf tissue split along the stripes. Small brown streaks develop on the glumes, causing reduced yields and shriveled seed. Infected kernels have indistinct brown lesions at their bases. When a barley plant matures and the disease progresses, the net blotch fungus will eventually grow into the sheath and culm tissue where it produces perithecia. The stems at harvest are a dull brown and lack strength.

A spot form of net blotch has been found on leaves and leaf sheaths. These symptoms consist of dark brown, elliptical or fusiform lesions (3 mm by 6 mm), surrounded by a chlorotic zone of varying width. The entire leaf blade may be involved. Depending on the cultivar, some necrotic lesions may have little or no surrounding chlorosis. These symptoms may be confused with those caused by *Cochliobolus sativus* (Ito & Kurib.) Drechs. ex Dast.

CONTROL:
1. Treat seed with a fungicide seed protectant.
2. Rotate barley with resistant crops.

3. Plant resistant cultivars where practical.
4. Apply foliar fungicides when weather conditions favor disease development. A biological control consisting of a bacterium applied to leaves has shown promise in experiments.
5. Increasing in-row seed spacing from 2 cm to 4 cm between plants has been reported to decrease disease severity.
6. Increasing row spacing from 12 cm to 28 cm has been reported to decrease disease severity.
7. Application of nitrogen as ammonium nitrate at Zadok's growth stage 30 has been reported to decrease disease incidence.

Pink Snow Mold

CAUSE: *Fusarium nivale* (Fr.) Ces., teleomorph *Calonectria nivalis* Schaff. (syn. *C. graminicola* Berk. & Br., *Griphosphaeria nivale* (Schaff.) Med. & von Arx, and *Micronectriella nivalis* (Schaff.) Booth). Perithecia of *C. nivalis* develop in late spring and early summer during cool, wet weather. Winter barley is infected during cool, wet weather in the autumn by windborne ascospores, mycelium growing from residue, and conidia produced on residue. Infections occur on leaf sheaths and blades near the soil level. Mycelium will grow from primary infections when plants are covered by snow. Secondary infection is mainly by conidia produced in the spring and infrequently by ascospores.

DISTRIBUTION: Mountain valleys of the western United States, Japan, and Central Europe.

SYMPTOMS: Symptoms are most obvious after snow melt. The best symptom is the pink color of the mycelium and sporodochia that are visible at this time. Leaves are chlorotic and necrotic but remain dry and intact instead of crumbling. Perithecia develop in late spring and summer in lower leaf sheaths.

CONTROL: No known control other than to avoid barley where there is a persistent snow cover.

Powdery Mildew

CAUSE: *Erysiphe graminis* DC. ex Merat f. sp. *hordei* Em. Marchal overwinters as cleistothecia on plant residue. In areas where mild winters allow infected live leaf tissue to survive, the fungus may overwinter as mycelium and conidia. Ascospores serve as the primary inoculum during the spring in northern barley-growing areas. Most ascospores form within cleistothecia in the spring and are windborne to healthy plants.

Conidia are produced almost as soon as the mycelium becomes established on the leaf surface, especially during humid (100% RH optimum), cool (15–22°C optimum) weather but without the presence of free water. Conidia account for most of the secondary inoculum and spread of the disease during the growing season. Cleistothecia are formed on the leaf surface as the plant approaches maturity.

Barley is most susceptible during periods of rapid growth that results in tender, rank plants. Disease is favored by a dense stand, heavy nitrogen fertilizer, high humidity, cloudy weather, and cool temperatures.

DISTRIBUTION: Generally distributed wherever barley is grown.

SYMPTOMS: Superficial mycelium and conidia appear as light gray or white spots on the upper surface of leaf blades, sheaths, and floral bracts. Affected plant parts appear as if they have been dusted with a gray powder consisting of multiple conidia and conidiosphores of the fungus. Later the spots enlarge and darken as the plant matures. As the powdery-appearing areas enlarge, the leaf area that was first infected becomes yellow, then turns brown and dies. In some cultivars, leaf tissue adjacent to mycelium turns brown and becomes necrotic. Most growth occurs on the upper leaf surface and infrequently on the lower surface. Numerous small, round dark cleistothecia develop on the affected areas. These can be readily observed with a hand lens. Powdery mildew usually ceases to be a problem when the weather becomes dry and warm later in the growing season.

CONTROL:
1. Plant resistant cultivars.
2. Apply a foliar fungicide where economically feasible.

Pythium Root Rot

This disease is also called browning root rot.

CAUSE: *Pythium* spp., particularly *P. arrhenomanes* Drechs. (syn. *P. arrhenomanes* var. *canadense* Van. & Trusc), *P. graminicola* Subr., and *P. tardicrescens* Vanterpool. These fungi survive as oospores in infected residue and in soil for five years or more. In moist soil, the oospores germinate to form sporangia in which zoospores are produced. The zoospores swim to root tips where infection occurs within three to five hours. Infection normally occurs in autumn or spring. Optimum growth is in soils with a high water potential of −1 bar or less and at temperatures of 15–20°C. At warmer temperatures the oospore may germinate directly to form a germ tube that penetrates roots. Oospores are produced within infected tissue and are returned directly to the soil when plant residue decomposes or they remain in tissue that stays intact.

DISTRIBUTION: Wherever barley is grown but most common under wet soil conditions.

SYMPTOMS: Pythium root rot is an insidious disease that is active through the life of the plant. Early infections may kill plants before or after germination and before emergence leading to uneven stands. In later infections plant vigor is reduced with a reduction in tillering, stem elongation, green color, and head size. Additionally, heads on diseased plants are unable to emerge from the root.

   Crown or basal stem tissues are ordinarily not attacked. The lower leaves may dry up and turn brown due to poor root growth caused by fungi pruning the root tips. Rootlets may be destroyed with a slight yellowing to darker yellowing or browning that occurs mainly in the steele. Such roots have a soft and mushy texture. Brown lesions may be present on larger roots. If soils dry, new root growth may occur.

CONTROL:
1. Treat seed with a seed-protectant fungicide.
2. Drain areas in fields that tend to remain wet.
3. Apply either by broadcast or in a band the recommended amounts of phosphorous fertilizer at planting time.

Rhizoctonia Root Rot

This disease is also called bare patch, barley stunt disorder, purple patch, Rhizoctonia patch, and stunting disease.

CAUSE: *Rhizoctonia solani* Kuhn, teleomorph *Thanatephorus cucumeris* (Frank) Donk. Isolates of *R. solani* causing Rhizoctonia root rot may belong to anastomosis group (AG)3 whereas those causing sharp eyespot belong primarily to AG4 and a few belong to AG1. The causal fungus survives as sclerotia in soil or mycelium in infected residues of a large number of host plants. In Scotland the disease occurs on barley following heavily fertilized grass that has been grazed by cattle and is restricted to light, sandy soils within 40 km of Elgin, Scotland. In Australia the disease is found on calcareous mallee soils in regions around Salmon Gums. In the United States the disease occurs where barley was either directly drilled into stubble, planted with minimal prior tillage, or planted the same day the soil was tilled. The herbicide chlorsulfuron has been reported to increase root disease.

DISTRIBUTION: Australia, Canada, England, Oregon, South Africa, and the state of Washington in the United States.

SYMPTOMS: Symptoms have been reported to start six weeks after planting. Patches of plants with Rhizoctonia root rot are near circular with distinct borders and range in size from about 20 cm to 3 m in diameter. These patches tend to persist in the same position for up to four years. Plants inside the patch are stunted and chlorotic but those outside the patch are a normal green. Shoots tend to be spindly, stunted, and abnormally erect. Leaf development is retarded, and once formed, leaves senesce prematurely and become yellow, starting at the tip. Lesions on diseased roots are brown and often girdle the root. The cortices collapse and roots have pinched-off, pointed tips. Lesions are on primary and secondary roots as well as on subcrown internodes. Frequently roots are rotted back to brown stumps. Surviving roots often produce large numbers of lateral roots. This results in a heavily branched root system with a large amount of cortical tissues and root tips that have sloughed off. Basal leaf sheaths near the soil surface are often discolored brown. A number of plants die prematurely and those that survive are poorly developed and yield little grain.

CONTROL: The disease apparently does not occur in the northwestern United States if some time has elapsed between tillage and planting. Also the disease is more common in barley grown under conservation tillage. In Australia, cultivation reduces disease incidence compared to no-till. Under zero-tillage conditions, nitrogen in the form of ammonium sulfate, sodium nitrate, or urea reduces disease.

Scab

This disease is also called Fusarium head blight, Fusarium blight, and head blight.

CAUSE: *Gibberella zeae* (Schw.) Petch (syn. *G. roseum* f. sp. *cerealis* 'Graminearum,' and *G. saubinettii* (Mont.) Sacc.), anamorph *Fusarium graminearum* Schwabe (syn. *F. roseum* (Lk.) emend. Snyd. & Hans. f. sp. *cerealis* (Cke.) Snyd. & Hans. 'Graminearum'), *Fusarium avenaceum* (Fr.) Sacc. (syn. *F. roseum* (Lk.) emend. Snyd. & Hans. f. sp. *cerealis* (Cke.) Snyd. & Hans. 'Avenaceum'), *F. culmorum* (W. G. Sm.) Sacc. (syn. *F. roseum* (Lk.) emend. Snyd. & Hans. f. sp. *cerealis* (Cke.) Snyd. & Hans. 'Culmorum'), and *F. nivale* (Fr.) Ces., teleomorph *Calonectria nivalis* Schaff. are sometimes involved. The causal fungus overwinters as mycelium and spores on barley seed; as mycelium in crop residue; and as perithecia on barley, corn, wheat, and other crop residue infected the previous season. Ascospores are produced in perithecia and in conidia on mycelium during warm, moist weather, and are windborne to healthy plants. Secondary inoculum is provided mainly by conidia produced on infected heads. Disease severity is increased by wet, warm weather when grain is lodged or swathed and during formation and ripening of the kernels.

Scabby grain is toxic, particularly to swine. Several toxins are produced; zearalenone (F-2 toxin) is involved in causing hyperestrogenism and vomitoxin is involved in the refusal factor or vomiting syndrome.

Feed that contains 3 percent or more of scabby kernels may be poisonous to swine. Cattle and sheep normally do not react to scabby grain.

DISTRIBUTION: Generally distributed wherever barley is grown during warm, moist summer weather.

SYMPTOMS: One or more spikelets or the entire head may be infected. Infection begins in the flowers and spreads to other parts of the head, giving the appearance of premature ripening. Initially, the diseased spikelets, starting at their base, are water-soaked, then die and become light brown. Eventually the hulls change from a light to a dark brown color. When infection occurs late in the development of the grain, the hull may show the brown color only at its base. In severe cases, the entire kernels may become shrunken and brown. During wet weather a pink mold consisting of mycelium and conidia of the causal fungus can be seen on infected spikelets. Later, small black perithecia can be seen growing in the same diseased area.

An entire infected head is dwarfed with kernels a gray-brown color and light in weight. The interior of the kernels becomes floury and discolored. Infected grain may be harmful when fed to hogs, dogs, and humans. Sheep and cattle are not as badly affected.

CONTROL:
1. Do not plant barley following barley, corn, or wheat.
2. Plow under infected residue where feasible.
3. Do not spread manure that contains infected straw or corn stalks on soil in which barley is growing.
4. Early planting may allow grain to escape much of the warm, moist weather of summer.
5. Treat seed with a fungicide-seed protectant to control seedborne inoculum.
6. Scabby kernels may be separated from healthy kernels with cleaning equipment.

Scald

This disease is also called Rhynchosporium scald and barley scald.

CAUSE: *Rhynchosporium secalis* (Oud.) Davis (syn. *Marsonia secalis* Oud.) overwinters as stroma either in lesions on winter barley infected in the autumn or in older infected crop residue on the soil surface. During cool (10–18°C optimum), humid weather in the spring, conidia are produced on the stroma and are disseminated primarily by splashing water to healthy leaves. Conidia are disseminated a short distance by wind. Secondary inoculum consists of conidia produced in both new and old lesions during cool, humid weather.

Rhynchosporium secalis can exist as mycelium in the pericarp and hull of seed. Hyphae can invade the coleoptile as it emerges from the embryo at an optimum soil temperature of 16°C. However, the importance of seedborne inoculum is not well known.

DISTRIBUTION: Generally distributed wherever barley is grown. Scald is more severe under reduced tillage conditions and irrigation.

SYMPTOMS: Lesions occur on coleoptiles, leaves, leaf sheaths, glumes, floral bracts, and awns. Young lesions occur as oval to irregular blotches that are a dark or pale gray or blue-gray color. Later the lesions appear to be water-soaked in appearance. As the tissues die, the color of the lesions changes first to brown, then to light tan, bordered by a brown margin sometimes surrounded by a chlorotic area. Often lesions have a zonate appearance. Inconspicuous, brown spots may occur on the tips of glumes.

An unusual symptom developed on Ethiopian barley cultivars inoculated with a *R. secalis* isolate from Montana. Leaf blades were free of symptoms but brown discoloration and subsequent wilting of leaf sheaths occurred.

CONTROL:
1. Rotate barley with other crops. Barley is the only known host to *R. secalis*.
2. Plow under infected crop residues.
3. Plant resistant cultivars.

Sclerotinia Snow Mold

CAUSE: *Sclerotinia borealis* Bub. & Vleug. (syn. *S. graminearum* Elen.). Sclerotia in residue germinate in autumn to produce apothecia on which ascospores are produced. The ascospores are windborne to barley.

DISTRIBUTION: Mountain valleys of the western United States, central Europe, and Japan.

SYMPTOMS: Symptoms occur in isolated patches at snow melt. Dead leaves on plants are gray and thready. Numerous black sclerotia 2–4 mm long are produced in diseased tissue during late spring and summer.

CONTROL: No control other than to not plant barley in areas of prolonged snow cover.

Seed Discoloration

CAUSE: Some reported causes of seed discoloration are *Bipolaris sorokiniana* (Sacc.) Shoemaker, *Alternaria alternata* (Fr.) Keissler,

Fusarium culmorum (W. G. Sm.) Sacc., *F. graminearum* Schwabe, and *Cladosporium herbarium* Lk. ex Fr.

Discoloration is associated with the amount of rainfall between anthesis and harvest and is enhanced by late harvest.

Semiloose Smut

This disease is also called nigra loose smut, black smut, black loose smut, and intermediate loose smut.

CAUSE: *Ustilago nigra* Tapke (syn. *U. avenae* (Pers.) Rostr.) is seedborne, surviving as teliospores on the seed surface and in soil. Smutted heads emerge at about the same time as those of healthy plants. A mass of teliospores has replaced the kernels and is covered by a somewhat persistent membrane that may remain intact until harvest time unless ruptured by wind, rain, or other means. Teliospores become windborne and land on or under the hulls of healthy barley kernels. The spores on the kernel surface usually remain dormant until seed is planted the following spring or autumn. Those spores deposited under the hulls may germinate immediately. The mycelium grows only into the shallow layers of the seed just under the hull and remains dormant until planting time.

When seed is planted, the resulting seedling becomes infected between germination and emergence from the soil by seedborne spores and mycelium or by soilborne spores. The mycelium of *U. nigra* grows just behind the growing point. As the plant enters the boot stage, the smut fungus grows rapidly in the young flowers and converts them to masses of teliospores before emergence from the boot. The smutted head then emerges from the boot similar to a healthy head. Soil temperatures between 15°C and 21°C and a relatively dry soil are the most favorable conditions for infection. Physiologic races of *U. nigra* are present.

DISTRIBUTION: Africa, Asia, Australia, Europe, North America, Central America, and South America.

SYMPTOMS: *Ustilago nigra* infects barley similar to *U. hordei*, the causal pathogen of covered smut. The resulting symptoms generally vary between those of loose smut caused by *U. nuda* and covered smut, but more closely resemble those of loose smut. Semiloose smut can be distinguished from loose smut with certainty only by observing the smut spores and their germination under a microscope. *Ustilago nigra* chlamydospores are spherical to subspherical, 6.5–7.0 μ in diameter, dark brown to black with echinulations varying from slight to pronounced. The most reliable distinguishing feature is the sporidial germination type of *U. nigra* as opposed to the mycelial germination type of *U. nuda*. The chlamydospores of *U. nigra* germinate to form a three- or four-celled basidium bearing sporidia from each cell. *Ustilago nuda* chlamydospores are globose to subglobose or elongate, 5–7 μ in diameter, olivaceous-brown, lighter on one side, and minutely echinulate. The chlamydospores of *U. nigra* germinate to form a slender germ tube that in turn forms branches. The branches elongate and rebranch to form mycelium.

Smutted heads appear about the same time as healthy ones. The dark brown to black spore mass and variation in looseness of the mass are gross characteristics. The membrane around the spore mass varies from fragile, which breaks easily, to stronger and persistent. When membranes break, the spores are quickly washed or blown away, leaving a bare rachis. The more persistent membranes retain their spores until harvest.

CONTROL: Treat seed with a systemic seed-protectant fungicide.

Septoria Leaf Blotch

This disease is also called Septoria leaf spot, Septoria speckled leaf blotch, speckled blotch, and speckled leaf blotch. Speckled blotch is sometimes considered a separate disease caused by *Septoria passerinii*. However, it is also included in this discussion.

CAUSE: *Septoria avenae* Frank f. sp. *triticea* T. Johnson, (teleomorph *Leptosphaeria avenaria* Weber f. sp. *triticea* T. Johnson) and *S. passerinii* Sacc. survive as mycelium for two to three years in live plants and as pycnidia on infected crop refuse that is either buried or on the soil surface. During cool (15–25°C), wet, weather in the autumn or spring, pycnidiospores are exuded from pycnidia in a gelatinous drop or cirrhi that protects spores from drying out and radiation. Pycnidiospores are disseminated by splashing and blowing rain to lower leaves of healthy plants. Winter barley is infected in the fall and the mycelium will remain inactive during the winter but will resume growth during warm weather in the spring. Ascospores are produced in perithecia as barley matures in late summer or early autumn. They may infrequently serve as primary inoculum when they are disseminated by wind. Pycnidia are not common until late in the growing season. Therefore pycnidiospores produced on the previous year's residue probably serve as the most important source of inoculum thoughout the growing season. Barley becomes infected during cool weather when plant growth tends to be poor. Infection requires six or more hours of wetness.

DISTRIBUTION: Generally distributed wherever barley is produced but is most severe in cool, wet climates.

SYMPTOMS: Lesions are primarily on leaves and leaf sheaths and are yellowish to light brown and of various sizes. *Septoria passerinii* produces long linear lesions with definite margins parallel to the leaf veins. Those produced by *S. avenae* f. sp. *triticea* have indefinite margins as the yellowish brown area blends into the green of the leaf blade and sheath. Eventually the lesions merge and cover large areas of the leaf. The leaf margins often pinch or roll and become dry. Pycnidia that look like small, black specks are present in the diseased area and can be better viewed under magnification.

CONTROL:
1. Plant resistant cultivars.
2. Use foliar fungicides on susceptible cultivars when weather favors disease development.
3 Plow under or burn infected residue where feasible.
4. Rotate barley with nonsusceptible crops such as legumes.

Sharp Eyespot

This disease is also called Rhizoctonia root rot.

CAUSE: *Rhizoctonia solani* Kuhn and *R. cerealis* van der Hoeven. Most isolates of *R. solani* belong in the anastomosis group (AG)4 and a few belong in AG1. The causal fungi survive as sclerotia in soil or mycelium in infected residues of a large number of plant hosts. Barley may be infected any time during the growing season but it is damaged the most when seedlings are infected. Barley is not as susceptible as oats, rye, and wheat. Sclerotia germinate to form mycelium or mycelium grows from a precolonized substrate to infect roots and culms, particularly in dry (less than 20% moisture-holding capacity), cool soils.

DISTRIBUTION: Generally distributed wherever barley is grown.

SYMPTOMS: Serious losses are seldom caused. Diamond-shaped lesions that resemble those of eyespot occur on lower leaf sheaths. The lesions are more superficial than eyespot lesions. Lesions have light tan centers with dark brown margins and frequently have dark mycelium on them. Dark sclerotia may develop in lesions and between culms and leaf sheaths.
 Seedlings may be killed but plants may produce new roots to compensate for those rotted off. When roots are infected, plants may lodge and produce white heads. Diseased plants may appear stiff, have a grayish cast, and be delayed in maturity.

CONTROL: No control is effective. Vigorous plants growing in well-fertilized soil are not as likely to become severely diseased.

Speckled Snow Mold

CAUSE: *Typhula ishikariensis* Imai, *T. idahoensis* Remsb. (syn. *T. borealis* Ekstr.), and *T. incarnata* Lasch ex Fr. (syn. *T. itoma* Imai and *T. gramineum* Karst). *Typhula* spp. survive as parasites or sclerotia. Sclerotia germinate during wet weather in the autumn to produce basidiocarps or hyphae. Basidiospores are produced on the surface of the basidiocarp and are airborne to hyphae. Hyphae originating from sclerotia grow to seedlings.

DISTRIBUTION: Mountain valleys of the western United States, central Europe, and Japan. Speckled snow mold is most likely to be associated with snow cover.

SYMPTOMS: Numerous dark sclerotia 0.5–3 mm in diameter are produced under leaf sheaths, within plant tissues, or scattered throughout mycelium giving a plant a speckled appearance. Necrotic leaves are covered with a thick gray-white mycelium and are easily shattered when handled. These leaves commonly die, dry up, and disintegrate when snow cover is gone and temperatures rise. A few or many leaves may be infected. If the crown is infected the plant will die.

CONTROL: No control other than to not plant barley in areas of prolonged snow cover.

Spot Blotch and Associated
Seedling and Crown Rots

This disease is also called black point, head blight, Helminthosporium blight, Helminthosporium head blight, and kernel blight.

CAUSE: *Cochliobolus sativus* (Ito & Kurib.) Drechs. ex Dast., anamorph *Bipolaris sorokiniana* (Sacc.) Shoemaker (syn. *Helminthosporium sativum* Pammel, King, & Bakke, and *H. sorokinianum* Sacc. apud Sorokin.). The causal fungus is seedborne and overwinters as mycelium and conidia on plant residue and on seedling leaves of winter barley. *Cochliobolus sativus* may also overwinter as conidia in soil. In the spring, conidia are produced on either plant residue or wild grasses and are windborne to barley seedlings. Seedborne inoculum results in seedling blight, root, and crown rot in dry, warm soils. Leaf infections develop best under warm (20°C or higher), wet conditions. However, leaf spots will develop at low temperature with symptom development, especially leaf chlorosis, intensifying at higher temperatures (20°C) and high relative humidity. High light intensities support formation of leaf spots, while incubation under temporary low light conditions does not result in necrosis but in leaf chlorosis and accelerated senescence of leaves in a short time.

DISTRIBUTION: Generally distributed wherever barley is grown; however, spot blotch is most severe in areas of a warm, humid climate.

SYMPTOMS: The first symptoms are dark brown to black spots near the soil line or at the base of the sheaths that cover the seedling leaves. Infection may progress, turning seedlings yellow and killing seedlings either preemergence or, more frequently, postemergence. The seedling leaves of infected plants are dark green with dark brown lesions near the soil line that eventually extend into the leaf blade. The infected seedling is dwarfed and tillering is excessive. Crowns and roots may appear dark brown and rotted. With severe infections, heads may not emerge completely and kernels are poorly filled.

Leaf lesions of various sizes (1–5 mm wide by 2–25 mm long) and shapes appear on lower leaves after warm, moist weather. Individual lesions are round to oblong, dark brown, with a definite margin into the normal green of the leaf. The spots coalesce to form blotches that cover large areas of the leaf. Older lesions are olive due to sporulation of the fungus. Heavily infected leaves will dry up completely.

Lesions on floral bracts and kernels vary from small black spots to dark brown discoloration of the surface, resulting in a blackened end of the kernel. Browning or blackening of lemma and palea occurs with frequent discoloration of pericarp layer and darkening of underlying embryo region. Head blight occurring early causes sterility or killing of individual kernels soon after pollination. Infected kernels may result in seedling blight if planted. Additional discolored seed is discounted when sold and may be unacceptable for malting.

CONTROL:
1. Plant resistant cultivars.
2. Treat seed with a seed-protectant fungicide.
3. Apply a foliar fungicide if weather conditions favor spread of the disease.
4. Rotate with nonsusceptible crops such as legumes.

Stem Rust

CAUSE: *Puccinia graminis* Pers. f. sp. *tritici* Ericks. & E. Henn. is a heteroecious rust fungus that has barberries *Berberis vulgaris* L., *B. canadensis* Mill., *B. fendleri* A. Gray, and some species of Mahonia as the alternate host. Japanese barberry, *B. thunbergii* DC, is thought to be immune. The urediospore and teliospore stages are found on barley. The fungi overwinter as teliospores on barley. Basidiospores are formed from teliospores in the spring and are windborne to barberry. Pycniospores and aeciospores are formed on barberry leaves with the aeciospores being windborne to infect barley. Uredia are formed in aeciospore infections on barley. Urediospores provide secondary inoculum throughout the growing season. *Puccinia graminis* Pers. f. sp. *secalis* Ericks. & E. Henn. can also attack barley. For a more detailed discussion of stem rust see the chapter on wheat.

DISTRIBUTION: Generally distributed wherever barley is grown.

SYMPTOMS: Stem rust occurs on stems, leaf sheaths, leaf blades, glumes, and beards. Brick red pustules with ragged edges break through the epidermis. Teliospores form in old uredial pustules, producing black telia. With severe disease, the kernels are shriveled and rusted stems turn brown, become dry and brittle, and often break.

CONTROL:
1. Plant resistant cultivars.

2. Eliminate common barberry bushes from the vicinity of any barley fields.

Stripe Rust

This disease is also called barley stripe rust, barley yellow rust, glume rust, and yellow rust.

CAUSE: *Puccinia striiformis* f. sp. *hordei* West. (syn. *P. glumarum* (Schm.) Ericks. & E. Henn.) is a rust fungus whose alternate host is not known. The fungus oversummers as urediospores on residual green cereals, such as rye and wheat, as well as on barley and grasses. Urediospores are formed and are windborne to healthy plants. Disease development is optimum at temperatures between 10°C and 15°C with periodic rain or dew. Little or no infection occurs during warm weather in the summer. Teliospores are formed but are not known to function as overwintering spores.

DISTRIBUTION: In North America, stripe rust occurs at the higher elevations and cooler climates along the Pacific Coast and intermountain areas from Canada to Mexico. It also occurs under the same environmental circumstances in South America and in the mountainous areas of central Europe and Asia.

SYMPTOMS: Stripe rust symptoms usually occur only on leaves and heads. Symptoms vary but normally occur during cool weather early in the spring before the appearance of symptoms of other rust, especially in areas with mild winters. Yellow uredia appear on autumn foliage and new foliage early in the spring. Eventually uredia coalesce to produce long stripes between the veins on the leaf and sheath. Small linear lesions occur on floral bracts. Telia develop as narrow linear dark brown pustules covered by the epidermis.

CONTROL:
1. Plant resistant cultivars.
2. Apply foliar fungicides.

Take All

This disease is also called dead head and white head.

CAUSE: *Gaeumannomyces graminis* (Sacc.) von Arx & Oliv. var. *tritici* Walker was originally described as *Ophiobolus graminis* (Sacc.) Sacc. (syn. *Rhaphidophora graminis* Sacc.). The causal fungus survives as mycelium in infected live overwintering plants and mycelium and perithecia in residue for about ten months. It also lives parasitically on roots of a variety of grasses including corn. The fungus survives in soil outside for about five months of residue. Ascospores produced in perithecia are not considered important in the dissemination of the disease. Dissemination is by any means that will move soil. In the autumn or spring, seedlings probably become infected by the roots, growing into the vicinity of some crop residue precolonized by *G. graminis* and coming in contact with mycelium. Plant to plant spread of take all occurs by hyphae growing through the soil from an infected to a healthy plant or by a healthy root coming in contact with an infected one. Ascospores are produced in perithecia during wet weather but apparently are not disseminated a great distance by splashing water or wind. Disease is favored by high pH soils that remain cool and wet (-1.2 to -1.5 bars) for prolonged periods of time and are deficient in nitrogen and phosphorous.

DISTRIBUTION: Generally distributed wherever barley is grown.

SYMPTOMS: Take all is more severe on wheat than barley, and on barley than rye. Disease severity is related to time of infection with the earliest infections being the most severe. Take all normally occurs in patches or roughly circular areas in a field. The first symptom of take all is light brown to dark brown necrotic lesions on the roots. By the time an infected plant reaches the jointing stage, most of its roots are brown and dead. At this point many plants die, or if plants are still alive, they are stunted and the leaves are yellow.

The disease becomes most obvious as plants approach heading. The stand is uneven in height and plants appear to be in several stages of maturity. Infected plants have few tillers, giving the appearance of plants ripening prematurely. Heads are bleached and sterile. Such plants have white heads during grain filling. Roots are sparse, blackened, and brittle. Plants can be easily broken free of their crown when pulled from the soil. A very dark discoloration of the stem is visible just above the soil line, along with a mat of dark brown fungus mycelium under the lower sheath between the stem and the inner leaf sheaths. This dark discoloration is a good diagnostic characteristic.

CONTROL:
1. Rotate barley with a nongrass.
2. Plant winter barley as late as feasible.

Verticillium Disease

CAUSE: *Verticillium dahliae* Kleb.

DISTRIBUTION: Idaho.

SYMPTOMS: Leaves have longitudinal chlorotic stripes with brown vascular bundles. Symptoms are similar to those associated with Cephalosporium stripe.

CONTROL: None reported.

White Rot

CAUSE: *Sclerotium rolfsii* Sacc.

DISTRIBUTION: Brazil.

SYMPTOMS: White mycelium and sclerotia occur on necrotic areas of roots, crowns, and lower portions of stems. Infected plant will turn white and die.

CONTROL: None reported.

Yellow Leaf Spot

This disease is also called blight, leaf spot, and tan spot.

CAUSE: *Pyrenophora trichostoma* (Fr.) Fckl. (syn. *P. tritici-repentis* (Died.) Drechs.), anamorph *Helminthosporium tritici-repentis* Died. (syn. *H. tritici-vulgaris* Nisikado). The causal fungus overwinters as pseudothecia on infected residue. In the spring, during wet weather, ascospores are released to infect healthy plants. During wet weather in the growing season, conidia are produced in older lesions and are windborne to infect healthy tissue, thus serving as secondary inoculum. In the autumn, pseudothecia are produced on infected culms and leaf sheaths. Symptom development is favored by frequent rains and cool, cloudy, humid weather in the early growing season.

DISTRIBUTION: Barley is rarely affected; however, yellow leaf spot has been reported on wheat in many parts of the world.

SYMPTOMS: Symptoms first appear in spring on lower leaves and continue on to upper leaves into early summer. At first, tan flecks appear on both sides of the leaf. The flecks eventually become tan, diamond-shaped lesions up to 12 mm long, with a yellow border and a dark brown spot in the center that is due to *P. trichostoma* sporulating. Lesions may coalesce to cause large areas of the leaf to die, usually from the tip inward. Pseudothecia will eventually develop on crop debris as small, dark, raised bumps.

CONTROL:
1. Apply a foliar fungicide before the disease becomes severe and if weather conditions favor spread of the disease.
2. Plow under infected residue, if feasible, to reduce inoculum on the soil surface.
3. Most barley cultivars have a degree of resistance.

——————MYCOPLASMAL CAUSES——————

Aster Yellows

CAUSE: Aster yellows mycoplasma survives in several dicotyledonous plants and leafhoppers. Aster yellows mycoplasma is transmitted primarily by the aster leafhopper, *Macrosteles fascifrons* Stal., and less commonly by *Endria inimica* Say and *M. laevis* Rib. The leafhoppers first acquire the mycoplasma by feeding on infected plants, then fly to healthy plants and transmit the mycoplasma by again feeding on the healthy plants. Symptoms become most obvious between 25–32°C with the most severe symptoms occurring at the higher temperatures.

DISTRIBUTION: Generally distributed throughout eastern Europe, Japan, and North America. Aster yellows is rarely severe on barley.

SYMPTOMS: Seedlings will either die two to three weeks after infection, or if they survive, they will be stunted, leaves will generally yellow or have yellow blotches, and the heads will be sterile with distorted awns.

Infection of older plants causes leaves to become somewhat stiff and discolored from the tip or margin inward in shades of yellow, red, or purple. Root systems may not be well developed.

CONTROL: No control is reported.

——————CAUSED BY NEMATODES——————

Cereal Cyst Nematode

This nematode is also called the oat cyst nematode.

CAUSE: *Heterodera avenae* Woll. (syn. *H. major* (Schmidt) Franklin) survives as eggs and larvae within cysts in the soil for several years. Larvae emerge from eggs at optimum conditions of 10°C and moist soil in midwinter until spring and migrate to nearby roots. Larvae invade roots just behind the growing point at an optimum temperature of 20°C. As nematodes mature, they swell and erupt through root tissues. Males return to a vermiform shape but females become lemon-shaped as egg production begins. Within two months of infection the female dies and the body hardens into a cyst that protects eggs and larvae. One generation is completed each growing season. Dissemination is by any means that will move soil, especially machinery, water, and wind.

DISTRIBUTION: Australia, Middle East, northern Europe, Idaho, India, Japan, New Zealand, Ontario, Oregon, and Washington.

SYMPTOMS: The cereal cyst nematode is the most important nematode pathogen of barley with reported yield reduction up to 22 percent. A severe infestation is noticed by patches of permanently stunted plants that are overgrown by weeds. Symptoms are most pronounced on plants growing in light, sandy soils. Plants are stunted and leaves are discolored similar to nitrogen or phosphorous deficiency. Head emergence is delayed and number of spikelets is reduced.

Roots are stunted and knotted with pronounced proliferation. Immature white cysts appear on roots beginning approximately at heading. This is a good diagnostic symptom.

CONTROL:
1. Plant resistant cultivars.
2. Rotate barley with a nongrass crop.

Cereal Root-Knot Nematode

This nematode is also called the root-knot nematode.

CAUSE: *Meloidogyne naasi* Franklin is an endoparasitic nematode that overwinters as eggs in the soil. In the spring larvae hatch from eggs and infect roots. By the middle of the summer, females inside the root knot tissue release eggs into the soil.

DISTRIBUTION: Northern Europe, Iran, United States, and Yugoslavia.

SYMPTOMS: Stunted, yellow plants occur in patches that range in size from a few square meters to large areas. These plants produce short heads with no spikelets or no heads at all. Swellings or thickenings comprised of swollen cortical cells and bodies of nematodes containing egg masses can be found on roots especially near the root tips in the spring and summer. When the root knots are cut open, the egg masses will turn dark.

CONTROL:
1. Barley planted in the autumn does not become as severely diseased as barley planted in the spring.
2. Rotate barley with root crops such as sugar beets.

Root-Gall Nematode

This disease is known as "krok" in Scandinavia.

CAUSE: *Subanguina radicicola* (Grf.) Param. (syn. *Ditylenchus radicicola* (Grf.) Filip. and *Anguillulina radicicola* (Grf.) Gdy.) is a nematode that survives by continuous habitation in host roots. Larvae penetrate roots and develop in cortical tissue, forming a root gall in two weeks. The mature females begin egg production within a gall; eventually the galls weaken and release larvae that establish secondary infections. Each generation is completed in about 60 days.

DISTRIBUTION: Canada, Europe, the USSR, and the United States.

SYMPTOMS: Tips of outer leaves of seedlings are chlorotic and have reduced top growth. Roots are highly branched and have numerous galls that tend to be inconspicuous and vary in diameter from 0.5 to 6.0 mm. Roots may be twisted and bent at the gall site. At the center of larger galls is a cavity filled with nematode larvae.

CONTROL: Rotate barley with noncereal crops.

Root-Lesion Nematodes

CAUSE: *Pratylenchus crenatus* Loof and *P. thornei* Sher. & Allen overwinter as eggs, larvae, or adults in host tissue or soil. Both larvae and adults penetrate roots where they move through cortical cells, the females simultaneously depositing eggs as they migrate. Older roots are abandoned and new roots are sought as sites for penetration and feeding.

DISTRIBUTION: Generally distributed wherever barley is grown.

SYMPTOMS: Plants in areas of a field will appear stunted, chlorotic, and under moisture stress. Roots and crowns will be rotted when *Rhizoctonia solani* Kuhn infects through nematode wounds. New roots may become dark and stunted with resultant loss in yield.

CONTROL:
1. Plant in autumn when soil temperatures are below 13°C.
2. Soil nematicides could be used where high costs warrant them.

Stunt Nematode

CAUSE: *Merlinius brevidens* (Allen) Sidd. (syn. *Tylenchorhynchus brevidens* Allen) is an ectoparasite that feeds on external cells. Populations of the nematode are favored by wet soils.

DISTRIBUTION: Peru and the United States.

SYMPTOMS: Areas of affected plants are about 1 m in diameter. Damage has been reported only on winter barley. Plants are stunted and chlorotic with fewer tillers and kernels than healthy plants. Roots are short and stubby.

CONTROL: No control is reported.

VIRAL CAUSES

African Cereal Streak

CAUSE: African cereal streak virus (ACSV) is limited to the phloem. It is transmitted by the delphacid leafhopper *Toya catilina* Fennah. The natural virus reservoir is probably native grasses. It is not mechanically or seed transmitted. Disease development is aided by temperatures above 20°C.

DISTRIBUTION: East Africa.

SYMPTOMS: Initially faint broken chlorotic streaks begin near the leaf base and extend upward. Eventually definite alternate yellow and green streaks develop along the entire leaf blade. Later the leaves become almost completely yellow. New leaves tend to develop a shoe-string habit and die.

Young infected plants become chlorotic, severely stunted, and die. Older infected plants have yellow heads that are distorted. Seed yield is almost completely suppressed. Plants become soft, flaccid, and almost velvety to the touch. A phloem necrosis develops.

CONTROL: No control is reported.

American Wheat Striate Mosaic

CAUSE: American wheat striate mosaic virus (AWSMV) is transmitted by the leafhoppers *Endria inimica* Say and *Elymana virenscens* F.

DISTRIBUTION: Central United States and Canada.

SYMPTOMS: Plants are stunted with slight to moderate leaf striation consisting of yellow to white parallel streaks.

CONTROL: No control is reported.

Australian Wheat Striate

CAUSE: Australian wheat striate virus (AWSV), also known as Chloris striate mosaic virus (CSMV), is transmitted by the leafhoppers *Nesoclutha obscura* Evans and *N. pallida* (Evans) but is not sap transmissible.

DISTRIBUTION: Australia. The disease is of little importance in barley.

SYMPTOMS: Plants are dwarfed. Leaves have fine, broken yellow to gray streaks or stripes.

CONTROL: No control measures are necessary.

Barley Mosaic

CAUSE: Barley mosaic virus (BMV) is seedborne and is also transmitted by the corn leaf aphid, *Rhopalosiphum maidis* (Fitch), and mechanically. Infected seed has a lower germination rate than healthy seed. Oats and wheat are also hosts.

DISTRIBUTION: India.

SYMPTOMS: Plants are stunted and leaves are initially chlorotic. Later mosaic symptoms develop on leaves. Infected seeds are small and shriveled.

CONTROL: Plant resistant cultivars.

Barley Stripe Mosaic

This disease is also known as barley false stripe, false stripe, lantern head, and stripe mosaic.

CAUSE: Barley stripe mosaic virus (BSMV) can remain viable in seed for up to eight years. Optimum seed transmission occurs at 20–24°C. When infected seed germinates, the resulting seedling is infected. Wild oats (*Avena fatua* L.) is an overwintering host for BSMV and has the potential for contaminating previously virus-free crops. The virus is spread from infected to adjacent healthy plants when injured leaves rub against each other. It can also be transmitted by infected pollen but BSMV is not known to be transmitted by insects or other means. Greater spread of BSMV occurs in spring-planted rather than autumn-planted barley.

DISTRIBUTION: Southern Asia, Australia, Europe, Japan, western North America, and Russia.

SYMPTOMS: Symptoms will vary with the mode of infection and temperature. Optimum symptom development occurs between 20–24°C. Seedborne infection appears as yellow mottling or spots that are either narrow or wide, numerous or few, continuous or broken. The color of the spots may be light green, tan, yellowish, or bleached white while the rest of the leaf is green. Sometimes the leaf may be nearly white. Virulent strains of the BSMV cause brown stripes that are continuous or broken with irregular margins, often in a V-shape or chevron fashion on leaves. Sometimes plants may be severely stunted, florets may be sterile with poorly developed or no heads and kernels. Protein synthesis is affected.

CONTROL:
1. Plant resistant cultivars.
2. Plant disease-free seed.
3. Rotate barley with nongrass crops.

Barley Yellow Dwarf

This disease is also called yellow dwarf.

CAUSE: Barley yellow dwarf virus (BYDV) survives or is reservoired in small grains planted in the autumn and perennial grasses. Although perennial grasses serve as a large reservoir of BYDV, these may not serve as the most important source of inoculum for spread. However, in Canada, grasses, winter wheat and corn were considered to be of little importance as virus sources. Virus inoculum brought in by aphids from elsewhere was considered to be the main source of infection. In areas where oversummering is a factor, corn is a reservoir for both aphids and some isolates of the virus. The virus is transmitted by several species of aphids, including the corn leaf aphid, *Rhopalosiphum maidis* (Fitch); oat bird-cherry aphid, *R. padi* (Linnaeus); rice root aphid, *R. rufiabdominelis* (Sasaki); greenbug, *Schizaphis graminum* (Rondani); and English grain aphid, *Macrosiphum avenae* Fabr. The early instars of *S. graminum* are the most efficient vector for this insect species. Different types of BYDV are recognized based on transmission by different insects. MAV is transmitted specifically by *M. avenae*; RMV is transmitted specifically by *R. maidis*; RPV is transmitted specifically by *R. padi*; PAV is transmitted nonspecifically by *R. padi* and *M. avenae*; and SGV is transmitted specifically by the greenbug, *S. graminum*. An aphid, once it acquires BYDV, is capable of transmitting it for the rest of its life. The virus is not transmitted through eggs, newborn aphids, seed, soil, or by mechanical means. Disease development is most apt to occur under cool (10–18°C), moist conditions that favor grass and cereal growth and aphid reproduction and migration. Aphid flights may cover hundreds of kilometers when assisted by wind. Storm fronts likely aid flights of aphids from southern areas where most overwintering occurs to northern areas of the Midwest.

DISTRIBUTION: Africa, Asia, Australia, Europe, New Zealand, North America, and South America.

SYMPTOMS: Symptom expression is greatest at lower temperatures of around 16°C. A barley field will show very uneven growth due to stunting of plants that were infected early. This can be confused with nutrient deficiency and other unfavorable growing conditions. At first, faint yellow-green blotches occur near the leaf tip, margin, or well within the leaf blade. The blotchy discoloration within the leaf blade is a diagnostic characteristic for yellow dwarf and helps to distinguish it from other maladies. Discolored areas enlarge and coalesce toward the leaf base but tissue next to the midrib tends to remain green longer than the rest of the leaf. Eventually, infected plants are dwarfed and the color of the leaves becomes a bright yellow. The youngest leaves normally do not show any unusual discoloration. The bright yellow discoloration usually occurs on older leaves except in late infections when only the flag leaf may show symptoms. Sometimes discoloration may be a red or purple besides yellow.

In the laboratory, root dry weight, length, and numbers were reduced depending on stage of plant development at time of infection, that is, the earlier infection occurs in the development of the plant, the more severe the root symptom expression. Root development of infected plants is almost entirely halted at the time of symptom expression in the shoot. The degree of root supression is not necessarily correlated with BYDV resistance of barley.

CONTROL:
1. Plant resistant or tolerant cultivars.
2. Avoid planting in early autumn or late spring.

Barley Yellow Mosaic

CAUSE: Barley yellow mosaic virus (BYMV) survives in air-dried soil up to five years. *Polymyxa graminis* Led. is thought to be the vector. Dissemination is by any means that will move soil, particularly machinery, wind, and water. However, two types of BYMV have been reported from Germany. One is transmissible by soil and mechanically while the other is transmissible only by soil.

DISTRIBUTION: Belgium, England, France, Germany, and Japan.

SYMPTOMS: Symptoms normally do not occur above 18°C, are most prominent in the spring, and disappear as weather becomes warmer. Plants are stunted and yellow areas occur throughout a field. Initially leaves have yellow spots and short streaks. Later leaves turn completely yellow beginning at the leaf tip. Necrotic spots sometimes appear on older leaves. Some of these leaves may die prematurely.

CONTROL:
1. Plant resistant cultivars.
2. Plant barley in spring. No infections have been observed in spring-planted barley.

Barley Yellow Stripe

CAUSE: The causal agent has not been identified. It is transmitted by the leafhopper *Euscelis plebejus* Fn., but not by sap.

DISTRIBUTION: Turkey.

SYMPTOMS: Symptoms most commonly occur along field borders and other grassy areas that are infested by *E. plebejus*. Fine continuous stripes occur on leaves that are sometimes followed by leaf yellowing and death.

CONTROL: No control is reported.

Brome Mosaic

CAUSE: Brome mosaic virus (BMV) is transmitted to barley mainly by sap. Nematode, beetle, and transmission by nymphs and adults of red spider mites, *Tetranychus cinnabarinus* (Boisduvol) has also been reported.

DISTRIBUTION: North Central Europe, South Africa, the USSR, United States, and Yugoslavia.

SYMPTOMS: Infected plants are somewhat dwarfed and produce shriveled heads. Leaf symptoms are most obvious when plants are young but fade as plants mature and are often gone at heading. Initially light yellow or white spots and streaks occur on leaves. These spread rapidly over the leaf, eventually giving it a bright yellow mosaic pattern.

CONTROL: No control is considered necessary.

Cereal Chlorotic Mottle

CAUSE: Cereal chlorotic mottle virus (CCMV) is a rhabdovirus transmitted by cicadellids.

DISTRIBUTION: Australia and northern Africa.

SYMPTOMS: Severe necrotic and chlorotic streaks occur on leaves.

CONTROL: No control is reported.

Cereal Tillering

CAUSE: Cereal tillering virus (CTV) persists in planthopper vectors. The virus is transmitted to corn, oats, rye, wheat, and other grasses besides barley.

DISTRIBUTION: Finland, Italy, and Sweden.

SYMPTOMS: Plants are dwarfed with excessive tillering. Leaves are dark green and sometimes are malformed with serrated margin. Grain yields are reduced.

CONTROL: No control is reported.

Eastern Wheat Striate

CAUSE: Eastern wheat striate virus (EWSV) is transmitted by the planthopper *Cicadulina mbila* Naude but not by seed, soil, aphids, or mechanically. EWSV probably overseasons in the perennial Narenga grass.

DISTRIBUTION: India.

SYMPTOMS: Fine chlorotic stripes that become pronounced with plant age develop on leaves and leaf sheaths. Stripes will often become necrotic. Infected plants are stunted to some degree. Stunting and striate symptoms are more pronounced in barley than in wheat, and in plants infected when young rather than older plants. Plants that are infected while very young usually die.

Grain produced by infected plants is shriveled and of poor quality. Infected plants often produce only partially filled heads.

CONTROL: No control is reported.

Enanismo

This disease is also called cereal dwarf.

CAUSE: The causal agent has not been described but one or more viruses and a toxin produced by the leafhopper *Cicadulina pastusae* Rup. & DeLg. are likely responsible. Leafhopper adults and nymphs transmit the causal agent in a circulative fashion. Both sexes can transmit the virus but females are more efficient vectors.

DISTRIBUTION: Colombia and Equador. Oats and wheat are also affected.

SYMPTOMS: Seedlings are killed or stunted. Later infections cause less stunting but yellow leaf blotches with dark green stripes appear. Enations that look earlike or like galls occur on leaves. Galls develop on new leaves one to three weeks after insect feeding rather than developing on the actual leaves that were fed on. Barley yields may be reduced in localized areas. Plants infected near heading develop distorted heads with poorly filled kernels.

CONTROL: Plant later in the spring after vector activity declines.

Hordeum Mosaic

CAUSE: Hordeum mosaic virus (HMV). Rye, wheat, and other grasses are infected.

DISTRIBUTION: Alberta.

SYMPTOMS: Symptoms develop between 10°C and 30°C. A diffuse chlorotic mottle develops on leaves. Streaks develop only near the base of young leaves.

CONTROL: No control is reported.

Northern Cereal Mosaic

CAUSE: Northern cereal mosaic virus (NCMV) is transmitted by the planthopper, *Laodelphax striatellus* Fallen. Other insects that have been implicated as vectors are *Unkanodes sapporonus* (Mats.) and species of Delphacodes. Oats, rice, rye, wheat, and several other grass species are hosts.

DISTRIBUTION: Japan, Korea, and Liberia.

SYMPTOMS: Plants are stunted. Leaf symptoms are a chlorotic mosaic and leaf streak pattern. Initially, yellow-white or white-green specks occur on young and newly emerged leaves and leaf sheaths. Later yellow-white stripes occur parallel to leaf veins. Plants become stunted and tillering increases. Leaf blades are short, narrow, and have stripes along leaf veins. Few or no panicles are present. Panicles that are present have mostly unfilled grain.

CONTROL: No control is reported.

Oat Blue Dwarf

This disease is probably the same as moderate barley dwarf.

CAUSE: Oat blue dwarf virus (OBDV) is transmitted by the leafhopper, *Macrosteles fascifrons* Stal. It cannot be transmitted mechanically, by seed, or by soil. Immature leafhoppers may occasionally transmit OBDV. Symptom development is greater at temperatures between 16°C and 18°C than at higher temperatures.

DISTRIBUTION: Czechoslovakia, Finland, North America, and Sweden.

SYMPTOMS: Plants are stunted and tiller excessively. Leaves are shortened and stiff with enations along veins of the leaf and leaf sheath. There is a proliferation of lateral tillers and sometimes a blue-green color. Plants may fail to head and spikes produced on infected plants are usually sterile.

CONTROL: No control is reported.

Oat Pseudorosette

CAUSE: Oat pseudorosette virus (OPV) has the planthopper *Laodelphax striatellus* Fallen as the vector. Oats is most seriously affected and wheat is also susceptible.

DISTRIBUTION: USSR.

SYMPTOMS: Plants are stunted and tiller excessively.

CONTROL: Some cultivars are less susceptible than others.

Oat Sterile Dwarf

This disease is called oat base tillering disease in Finland and oat dwarf tillering disease in Sweden.

CAUSE: Oat sterile dwarf virus (OSDV) is transmitted by delphacid planthoppers, especially *Javesella pellucida* Fabr. and *Dicranotropis hamata* Boh. OSDV is transmitted after one month of incubation in its vectors, but it is rarely passed through eggs.

DISTRIBUTION: England, eastern and northern Europe.

SYMPTOMS: Plants are slightly stunted with few juvenile tillers produced. Leaves are a darker green color than normal.

CONTROL: Control grassy weeds and oat cover crops.

Rice Black-Streaked Dwarf

CAUSE: Rice black-streaked dwarf virus (RBSDV) is transmitted by the planthoppers *Laodelphax striatellus* Fallen, *Unkanodes sapproronus* (Mats.), and *U. albifascia* Mats. RBSDV is not seed or sap transmitted. The virus overwinters in winter wheat and is acquired within 30 minutes of feedings after an incubation period of 7–35 days. The virus is retained from season to season in all stages of planthoppers. Nymphs are more efficient vectors than adults but RBSDV is not transmitted through eggs.

DISTRIBUTION: Japan.

SYMPTOMS: Plants are severely stunted and have twisted leaves. Waxy swellings on veins are present on the under surface of leaves and on culms.

CONTROL:
1. Grow resistant rice cultivars to lessen chance of virus transmission to barley.
2. Do not grow barley in close proximity to rice.

Rice Stripe

CAUSE: Rice stripe virus (RSV) is transmitted primarily by the planthopper *Laodelphax striatellus* Fallen.

DISTRIBUTION: Japan, Korea, and Taiwan.

SYMPTOMS: Plants are dwarfed. Leaves have yellow-green to yellow-white stripes parallel to the midrib.

CONTROL: It is not necessary to control rice stripe in barley since the extensive use of resistant rice cultivars has reduced the vector population.

Russian Winter Wheat Mosaic

CAUSE: Winter wheat mosaic virus (WWMV) is disseminated by the leafhoppers *Psammotettix striatus* L. and *Macrosteles laevis* Rib. Oats, rye, and wheat are also infected.

DISTRIBUTION: Eastern Europe and the USSR.

SYMPTOMS: Seedlings are stunted with excessive tillering and necrosis. Surviving seedlings exist as rosettes without any leaf mosaic symptoms. Older infected plants have a typical mosaic and streak-mosaic pattern on leaves. Additionally, chlorotic dashes and streaks occur parallel to leaf veins. Older plants are slightly stunted.

CONTROL: No control is reported.

Soilborne Wheat Mosaic

This disease is also called wheat soilborne mosaic.

CAUSE: Soilborne wheat mosaic virus (SBWMV) survives in soil in *Polymyxa graminis* Led., a soilborne fungus that is an obligate parasite in roots of many higher plants. Soils may remain infected for years. The fungus enters root hairs and epidermal cells of roots as motile zoospores during cool (10–20°C), wet conditions in the autumn and infrequently in the spring. Once inside the plant, *P. graminis* replaces plant cell contents with plasmodial bodies that will either segment into additional zoospores or develop into thick-walled resting spores two to four weeks after infection. SBWMV is spread by any method such as tillage, water, or wind that will disseminate infested soil containing *P. graminis*.

The disease is most common in low areas of fields that tend to be wet. The disease is most severe on autumn-planted barley due to plant growth occurring during the same low temperatures that are conducive to dissemination and infection by *P. graminis*.

DISTRIBUTION: Argentina, Brazil, Egypt, Italy, Japan, eastern and central United States. It is not considered to be an important disease of barley.

SYMPTOMS: Symptoms are most prominent in the spring on the lower leaves and range from light green to yellow leaf mosaics. Plants may be severely or slightly stunted. Some strains cause rosetting. The youngest leaves and leaf sheaths are mottled and develop parallel spots or streaks. Warm weather prevents disease development, thus confining symptoms to lower leaves.

CONTROL:
1. Rotate barley with noncereal crops.
2. Plant barley later in the autumn.

Wheat Dwarf

CAUSE: Suspected to be a virus that is transmitted by several leafhoppers, primarily *Psammotettix alienus* Dahlb. and *Macrosteles laevis* Rib. Transmission does not occur through sap, seed, or soil.

DISTRIBUTION: Czechoslovakia, the USSR, and Sweden.

SYMPTOMS: Plants infected as seedlings are severely stunted and do not head out. Those infected later are less stunted. Leaves develop fine light green to yellow-brown spots and blotches over the surface. These may coalesce to cause yellowing and necrosis of the leaf.

CONTROL: No control is reported.

Wheat Streak Mosaic

CAUSE: Wheat streak mosaic virus (WSMV) survives in virus-infected plants and has a wide host range that includes corn, oats, rye, wheat, and several annual and perennial grasses. WSMV is mechanically transmitted but the normal means of spread from plant to plant is by wheat curl mite, *Aceria tulipae* Keifer. Mites persist in green wheat and overwinter on winter wheat. Once a mite has picked up WSMV from an infected plant, it is carried internally for several weeks. As winter barley plants mature, mites migrate to nearby volunteer cereals, grasses, or corn and infect them with the virus. However, some grasses are hosts for the mites but not the virus and vice versa; some are susceptible to both, whereas some are resistant to both.

During late summer or early autumn, the virus may be disseminated to volunteer cereals. In time, the virus may be disseminated from volunteer cereals to early-planted barley by mites. Mites may be windborne at least 2.4 km. However, wheat streak mosaic is a problem on barley only when mite populations build to extremely high levels in nearby fields.

Only the young mites acquire WSMV by feeding 15 or more minutes on infected plants. Neither the mite nor the virus can survive longer than one or two days in the absence of a living plant. No active virus has been detected in dead plants or in seed.

DISTRIBUTION: Eastern Europe, the USSR, western and central North America, particularly the Great Plains.

SYMPTOMS: Winter barley is commonly infected in the autumn, but symptoms ordinarily do not appear until the following spring. Infection depends on three factors: the population of wheat curl mites; nearness of virus-infected plants, especially volunteer cereals; and moisture to keep barley or other cereals vigorously growing where mites attain maximum reproduction. The greatest losses occur in early planted, autumn-infected plants.

The first symptoms consist of light green to light yellow blotches, dashes, or streaks in the leaves parallel to the veins. Plants become stunted, show a general yellow mottling, and develop an abnormally large number of tillers that may vary considerably in height. Stunted plants with sterile heads may remain standing after harvest at the same height or shorter than stubble. As infected plants mature, the yellow-striped leaves turn brown and die.

Heads may be sterile or partially sterile with shriveled kernels. In severe cases, plants may die before maturity. Synergistic effects are suspected between WSMV and other viruses such as barley yellow dwarf virus, making field identification difficult. Feeding mites often cause leaf edges to curl tightly in toward the upper midvein.

CONTROL:
1. Destroy all volunteer cereals and grasses two weeks before planting in adjoining fields and three to four weeks before planting in the same field as the weeds.
2. Plant barley as late as practical after Hessian fly-free date. If barley emerges in October or later it often escapes infection.

Wheat Yellow Leaf

CAUSE: Wheat yellow leaf virus (WYLV) is transmitted by the aphid *Rhopalosiphum maidis* (Fitch).

DISTRIBUTION: Japan.

SYMPTOMS: Leaves become yellow and blight. Plants die or ripen prematurely.

CONTROL: No control is reported.

──────────BIBLIOGRAPHY ──────────

Anikster, Y. 1981. Alternate hosts of *Puccinia hordei*. *Phytopathology* **72:**733–735.

Anonymous. 1961. Yellow dwarf. A problem disease of small grains. *USDA Agric. Res. Serv. ARS 22-70*

Bockelman, H. E.; Sharp, E. L.; and Bjarko, M. E. 1983. Isolates of *Pyrenosphora teres* from Montana and Mediterranean region that produce spot-type lesions on barley. *Plant Disease* **67:**696–697.

Boewe, G. H. 1960. Diseases of wheat, oats, barley and rye. *Illinois Nat. Hist. Surv. Circ. 48.*

Boosalis, M. G. 1952. The epidemiology of *Xanthomonas translucens* (J. J. & R.) Dowson on cereals and grasses. *Phytopathology* **42:**387–395.

Brooks, F. T. 1953. *Plant Diseases.* 2nd ed. Oxford University Press, New York.

Brown, J. K., and Wyatt, S. D. 1981. Corn as an oversummering host of barley yellow dwarf virus and aphid vector in eastern Washington. *Phytopathology* **71:**863 (abstr.).

Bruehl, G. W. 1953. Pythium root rot of barley and wheat. *USDA Tech. Bull. No. 1084*

Chico, A. W. 1983. Reciprocal contact transmission of barley stripe mosaic virus between wild oats and barley. *Plant Disease* **67:**207–208.

Cook, R., and Williams, T. D. 1972. Pathotypes of *Heterodera avenae*. *Ann. Appl. Biol.* **71:**267–271.

Crosier, W. F.; Nash, G. T.; and Crosier, D.C. 1970. *Ustilago nuda* on leaves of 'Larker' barley. *Plant Dis. Rptr.* **54:**927–929.

Cunfer, B. M. 1981. *Septoria* sp. on barley and *Hordeum pusillum*. *Phytopathology* **71:**869 (abstract).

Darlington, L. C.; Carroll, T. W.; and Mathre, D.E. 1976. Enhanced susceptibility of barley to ergot as a result of barley stripe mosaic virus infection. *Plant Dis. Rptr.* **60:**584–587.

Davies, K. A., and Fisher, J. M. 1976. Factors influencing the number of larvae of *Heterodera avenae* invading barley seedlings in vitro. *Nematologica* **22:**153–162.

Dehne, H. W., and Oerke, E. C. 1985. Investigations on the occurrence of *Cochliobolus sativus* on barley and wheat. 1. Influence of pathogen, host plant and environment on infection and damage. *Zeitschr. für Pflanzenkrankheiten und Pflanzenschutz* **92:**270–280.

Delserone, L. M., and Cole, H. JR. 1987. Effects of planting date on development of net blotch epidemics in winter barley in Pennsylvania. *Plant Disease* **71**:438–441.

Delserone, L. M.; Frank, J. A.; and Cole, H. Jr. 1983. Net blotch epidemics on winter barley in the fall as influenced by planting date. *Phytopathology* **73**:965 (abstract).

Dewey, W. G., and Hoffmann, J. A. 1975. Susceptibility of barley to *Tillertia controversa*. *Phytopathology* **65**:654–657.

Dubin, H. J., and Stubbs, R. W. 1986. Epidemic spread of barley stripe rust in South America. *Plant Disease* **70**:141–144.

Fargette, D.; Lister, R. M.; and Hood, E. L. 1982. Grasses as a reservoir of barley yellow dwarf virus in Indiana. *Plant Disease* **66**:1041–1045.

Fenne, S. B. 1956. Diseases of small grain. *Virginia Polytech. Inst. Agric. Ext. Serv. Bull. 151* (rev).

Fischer, G. W., and Hirschhorn, E. 1945. The Ustilaginales or "smuts" of Washington. *State College of Washington Agric. Exp. Sta. Bull. No. 459*.

Francki, R. I. B., and Hatta, T. 1980. Chloris striate mosaic virus. *Descriptions of Plant Viruses, No. 221*. Commonwealth Mycological Institute, Association of Applied Biology, Kew, Surrey, England.

Gill, C. C., and Westdal, P. H. 1966. Effect of temperature on symptom expression of barley infected with aster yellows or barley yellow dwarf viruses. *Phytopathology* **56**:369–370.

Harder, D. E., and Bakker, W. 1973. African cereal streak—a new disease of cereals in East Africa. *Phytopathology* **63**:1407–1411.

Hosford, R. M. Jr. 1978. Effects of wetting period on resistance to leaf spotting of wheat, barley, and rye by *Leptosphaeria herpotrichoides*. *Phytopathology* **68**:591–594.

Huftalen, C. S., and Bergstrom, G. C. 1986. First report of Ascochyta leaf spot caused by *Ascochyta hordei* var. *americana* on barley in New York. *Plant Disease* **70**:1074 (disease notes).

Hyung, L. S., and Skikata, E. 1977. Occurrence of northern cereal mosaic virus in Korea. *Korean J. Plant Prot.* **16**:87–92.

Inouye, T. 1976. Wheat yellow leaf virus. *Descriptions of Plant Viruses, No. 157*. Commonwealth Mycological Institute, Association of Applied Biology, Kew, Surrey, England.

Jenkins, J. E. E. 1974. New or uncommon plant diseases and pests. Botrytis diseases in barley. *Plant Pathology* **23**:83–84.

Johnson, D. A.; Tew, T. L.; and Banttari, E. E. 1977. Factors affecting symptoms in barley infected with the oat blue dwarf virus. *Plant Dis. Rptr.* **61**:280–283.

Kainz, M., and Hendrix, J. W. 1981. Response of cereal roots to barley yellow dwarf virus infection in a mist culture system. *Phytopathology* **71**:229 (abstr.)

Kelemu, S., and Sharp, E. L. 1986. Unique symptoms induced in Ethiopian barley cultivars by *Rhynchosporium secalis*. *Plant Disease* **70**:800 (disease notes).

Leukel, R. W., and Tapke, V. F. 1954. Cereal smuts and their control. *USDA Farmer's Bull. No. 2069*.

Leukel, R. W., and Tapke, V. F. 1955. Barley diseases and their control. *USDA Farmers's Bull. No. 2089*.

Lockhart, B. E. L. 1986. Occurrence of cereal chlorotic mottle virus in northern Africa. *Plant Disease* **70:**912–915.

Luzzardi, G. C.; Luz, W. C.; and Pierobom, C. R. 1983. Podridao branca dos cereais causada por *Sclerotium rolfsii* no Brazil. *Fitopathologia Brasileira* **8:**371–375 (in Portuguese).

Macnish, G. C. 1985. Methods of reducing Rhizoctonia patch of cereals in Western Australia. *Plant Pathol.* **34:**175–181.

Mains, E. B., and Jackson, H. S. 1924. Aecial stages of the leaf rusts of rye, *Puccinia dispersa* Erikss. and Henn., and of barley *P. anomala* Rostr. in the United States. *J. Agric. Res.* **28:**1119-1126.

Martin, R. A. 1985. Influence of seeding rate and nitrogen topdressing on net blotch development in barley. *Can. J. Plant Pathol.* **7:**446 (abstract).

Mathre, D. E. (ed.) 1982. *Compendium of Barley Diseases.* American Phytopathological Society. St. Paul, MN, 78p.

Mathre, D. E. 1986. Occurrence of *Verticillium dahliae* on barley. *Plant Disease* **70:**981 (disease notes).

McKay, R. 1957. *Cereal Diseases in Ireland.* Arthur Guiness, Son & Co., Ltd., Dublin.

Metz, S. A., and Scharen, A. L. 1979. Potential for the development of *Pyrenophora graminea* on barley in a semi-arid environment. *Plant Dis. Rptr.* **63:**671–675.

Murray, D. I. L., and Nicolson, T. H. 1979. Barley stunt disorder in Scotland. *Plant Pathol.* **28:**200–201.

Nagaich, B. B., and Sinha, R.C. 1974. Eastern wheat striate: A new viral disease. *Plant Dis. Rptr.* **58:**968-970.

Nutter, F. W., Jr.; Pederson, V. D.; and Timian, R. G. 1984. Relationship between seed infection by barley stripe mosaic virus and yield loss. *Phytopathology* **74:**363–366.

O'Brien, P. C., and Fisher, J. M. 1978. Factors influencing the number of larvae of *Heterodera avenae* within susceptible wheat and barley seedlings. *Nematologica* **24:**295–304.

Palival, Y. C. 1982. Role of perennial grasses, winter wheat, and aphid vectors in the disease cycle and epidemiology of barley yellow dwarf virus. *Can. J. Plant Pathol.* **4:**367–374.

Peters, R. A.; Timian, R. G.; and Wesenberg, D. 1983. A bacterial kernel spot of barley caused by *Pseudomonas syringae* pr. *syringae*. *Plant Disease* **67:**435–438.

Raemaekers, R. H., and Tinline, R. D. 1981. Epidemic of diseases caused by *Cochliobolus sativus* on rainfed wheat in Zambia. *Can. J. Plant Pathol.* **3:**211–214.

Richardson, M. J., and Noble, M. 1970. Septoria species on cereals—a note to aid their identification. *Plant Pathol.* **19:**159-163.

Roane, C. W.; Roane, M. K.; and Starling, T. M. 1974. Ascochyta species on barley and wheat in Virginia. *Plant Dis. Rptr.* **58:**455–456.

Rovira, A. D., and McDonald, H. J. 1986. Effects of the herbicide chlorsulfuron on Rhizoctonia bare patch and take all of barley and wheat. *Plant Disease* **70:**879–882.

Sampson, G., and Cloug, K. S. 1979. A Selenophoma leaf spot on cereals in the Maritimes. *Can. Plant Dis. Surv.* **59:**3.

Shands, H. L.; Shands, R. G.; Forsberg, R. A.; and Arawinko, Z. M. 1965. Barley. *Univ. Wisconsin Coll. Agric. Exp. Sta. Bull. 572.*

Simmonds, P. M. 1955. Root diseases of cereals. *Canada Dept. Agric. Pub. 952.*

Sinha, R. C., and Benki, R. M. 1972. American wheat striate mosaic virus. *Descriptions of Plant Viruses, Set 6, No. 99,* Commonwealth Mycological Institute, Association of Applied Biology, Kew, Surrey, England.

Slack, S. A.; Shepherd, R. J.; and Hall, D. H. 1975. Spread of seed-borne barley strip mosaic virus and effects of the virus on barley in California. *Phytopathology* **65:**1218–1223.

Tapke, V. F. 1943. Physiologic races of *Ustilago nigra. Phytopathology* **33:**324–327.

Tekauz, A., and Chico, A. W. 1980. Leaf stripe of barley caused by *Pyrenophora graminea:* Occurrence in Canada and comparison with barley stripe mosaic. *Can. J. Plant Pathol.* **2:**152–158.

Teviotdale, B. L., and Hall, D. H. 1976. Factors affecting inoculum development and seed transmission of *Helminthosporium gramineum. Phytopathology* **66:**295–301.

Thomas, P. L. 1981. Distinguishing between the loose smuts of barley. *Plant Disease* **65:**834.

Walker, J. 1975. Take-all disease of Graminae: A review of recent work. *Rev. Plant Pathol.* **54:**113–144.

Weller, D. M.; Cook, R. J.; MacNish, G.; Bassett, E. N.; Powelson, R. L.; and Peterson, R. R. 1986. Rhizoctonia root rot of small grains favored by reduced tillage in the Pacific Northwest. *Plant Disease* **70:**70–73.

Western, J. H. 1971. *Diseases of Crop Plants.* The Macmillan Press Ltd., London.

Diseases of Buckwheat (*Fagopyrum esculentum* Moench)

FUNGAL CAUSES

Blight

CAUSE: *Phytophthora parasitica* Dost.

DISTRIBUTION: USSR.

SYMPTOMS: Cotyledons are first attacked. Eventually the stem becomes infected, causing brown spots that spread in concentric circles. Infection results either in reduction of plant growth or death.

CONTROL: No control is reported.

Botrytis Rot

CAUSE: *Botrytis cinerea* Pers. ex Fr. Disease is favored by warm, wet weather. The fungus overwinters on residue and soil, and is externally seedborne.

DISTRIBUTION: USSR.

SYMPTOMS: Powdery spots form on leaves and stems. These spots are covered with a gray mold or a black film. Spots on the lower stem cause it to break off.

CONTROL: Apply a fungicide seed treatment.

Chlorotic Leaf Spot and Stipple Spot

CAUSE: Possibly *Bipolaris sorokiniana* (Sacc.) Shoemaker and *Alternaria alternata* (Fr.) Keissler. However, although *A. alternata* was isolated from lesions, it was not pathogenic when inoculated to buckwheat.

DISTRIBUTION: Manitoba.

SYMPTOMS: Chlorotic lesions are present on leaves mainly on the upper half of plants. Circular lesions, 12–26 mm in diameter, are randomly scattered over leaf blades. Lesions are of three types: spreading and uniformly chlorotic; spreading with concentric chlorotic bands alternating with normal dark green tissue; and small, restricted, tan lesions with sharply defined borders (stipple spot). Necrosis occurs in the older lesions and in the older chlorotic rings.

CONTROL: No control is reported.

Downy Mildew

CAUSE: *Peronospora ducometi* Siemaszko & Jankowska (syn. *P. fagopyri* Elen.). The fungus may possibly be seedborne and cause systemic infection.

DISTRIBUTION: Canada, Europe, and Japan.

SYMPTOMS: Large, circular, chlorotic lesions occur on leaves. Sometimes lesions will have concentric chlorotic bands alternating with normal dark green tissue. Some leaves that are badly infected have a mosaiclike appearance. The circular-type lesion appears first, generally on the leaves just below the top of the plant. As the disease progresses, most or all of the leaf becomes infected. Systemic infection occurs on the uppermost foliage, causing shortened internodes on upper stems of some plants, epinasty of leaves, and reduced number of seeds. Sparse conidia and conidiophores occur on lower leaf surface. Clumped conidia are purplish and can be seen with the unaided eye.

Another syndrome includes stunted seedlings and small stem diameter. Leaves of seedlings are rugose and mottled. These symptoms may be the result of seedborne infection.

CONTROL: No control is reported.

Fusarial Wilt

CAUSE: *Fusarium* sp. The causal fungus is reported to be externally seedborne and is also soilborne.

DISTRIBUTION: USSR.

SYMPTOMS: Wilting first occurs at the top of the plant, then proceeds to the entire plant. The lower stem becomes discolored, then necrotic, and eventually becomes covered with pinkish conidia, conidiophores, and mycelium. Reduced seed may be formed. Roots become necrotic and are easily pulled from the soil.

CONTROL: No control is reported.

Other Disease Reported on Buckwheat

Collar Rot—*Sclerotinia* sp.
Powdery Mildew—*Erysiphe commonis* Wallr. ex Fr. *fagopyri*
Ascochyta Leaf Blight—*Ascochyta fagopyri* Bres.
Leaf Spot—*Phyllosticta polygonorum* Sacc.
Ring Spot—*Cercospora fagopyri* Chupp & Muller

BIBLIOGRAPHY

Conners, I. L. 1967. An annotated index of plant diseases in Canada. *Canada Dept. Agric. Res. Branch. Pub. No. 1251.*

Savitskiy, K. A. 1970. *Gretchika* (Buckwheat). Kolos, Moscow, 312p.

Zimmer, R. C. 1974. Chlorotic leafspot and stipple spot, newly described diseases of buckwheat in Manitoba. *Canada Plant Dis. Surv.* **54:**55–56.

Zimmer, R. C. 1978. Downy mildew, a new disease of buckwheat (*Fagopyrum esculentum*) in Manitoba, Canada. *Plant Dis. Rptr.* **62:**471–473.

Zimmer, R. C. 1984. A possible new downy mildew syndrome on buckwheat seedlings. *Can. Plant Dis. Surv.* **64:**7–9.

4

Diseases of Corn
(*Zea mays* L.)

BACTERIAL CAUSES

Bacterial Leaf Blight

CAUSE: *Pseudomonas avenae* Manns. (syn. *P. alboprecipitans* Rosen) apparently does not survive well in infected residue or soil. Vasey grass, *Paspalum urvillei Steud.*, is considered a primary source of inoculum in Florida. Several other plants including oats and fantail are also attacked. In general, full-season field corn hybrids are more susceptible than short-season hybrids and sweet corn. Warm, rainy weather favors disease development but it will occur below 18°C.

DISTRIBUTION: Central and southern United States.

SYMPTOMS: Symptoms are first observed from late April to early June on leaves but no stalk rot, shank rot, or ear rot occurs. Infected plants may be stunted. Water-soaked lesions occur on leaves as they emerge from the whorl. Lesions stretch out and become elliptical until the leaf matures. Lesions are initially brown, then turn gray or white. They may coalesce, particularly during wet rainy weather, and form large necrotic areas. Leaves may shred during a wind storm.

CONTROL: Bacterial leaf blight is not considered a serious disease. Plant the most resistant hybrid.

Bacterial Stalk Rot

This disease is also called bacterial stalk and top rot.

CAUSE: *Erwinia chrysanthemi* pv. *zeae* (Sabet) Victoria, Arboleda & Munoz (syn. *E. carotovora* pv. *zeae* Sabet. Bacteria live as saprophytes on plant residue in the soil. The organism is also seedborne. During periods of high rainfall, overhead irrigation, or soils that remain wet for longer than usual amounts of time, bacteria are blown or splashed onto the plants. The bacteria then enter the plant through hydathodes, stomates, or wounds on leaves or stalks. Disease development is also aided by high temperatures (30–35°C), and poor air circulation.

DISTRIBUTION: Generally assumed to be distributed wherever corn is grown.

SYMPTOMS: About midseason, plants fall over as one to several internodes above the soil surface collapse and twist. Initially, water-soaked lesions occur in leaf sheath, extending into the leaf lamina as streaks. From these lesions, decay progresses into the pith. Affected node or nodes will be tan to dark brown, water-soaked, soft, and slimy with a disagreeable odor that resembles spoiled silage. Affected plants may remain green for several days because the vascular strands are not decomposed at first.

CONTROL:
1. Add chlorine to irrigation water.
2. Differences in resistance exist among corn lines.
3. Calcium hypochlorite was effective as a soil drench and calcium hydroxide and streptomycin were effective as sprays.

Bacterial Stripe

This disease is also called bacterial stripe and leaf spot.

CAUSE: *Pseudomonas andropogonis* (Smith) Stapp probably overwinters in infected residue. During extended warm, wet weather, bacteria enter the leaves through stomata. Some sorghum species including Sudangrass and Johnsongrass are also invaded.

DISTRIBUTION: Central and eastern United States.

SYMPTOMS: Leaves above the ears are seldom infected. Lesions are long and narrow with parallel sides, olive to amber in color, and water-soaked. Lesions are first evident on the lowest leaves and under favorable conditions spread up the plants. Lesions may eventually elongate and coalesce. In a few susceptible inbred lines, most of the leaves below the ear may be killed. Severely infected leaves may shred easily.

Susceptible inbred lines show pale yellow or green-white stripes. The stripes enlarge, coalesce, and continue to become lighter until the upper leaves are nearly completely white. These leaves are not infected but display a secondary effect of infection on lower leaves.

Other symptoms are described as circular to ellipsoidal, tan to brown spots with irregular margins. Spots are 1–4 mm in diameter with one or more darker brown rings within lesions. Some spots are surrounded by a chlorotic ring 1 mm wide. All spots tend to have a slightly sunken appearance. Sometimes spots coalesce into elongated blotches.

CONTROL: The disease is not economically important and has been severe on only a few susceptible inbred lines. Plant resistant hybrids and varieties.

Bacterial Top Rot

CAUSE: Unknown bacteria. The bacteria are probably splashed by water or windblown onto the plant and enter through hail wounds or other injuries.

DISTRIBUTION: Iowa; however, the disease probably occurs in other areas.

SYMPTOMS: Bacterial top rot occurs during hot, damp weather and has been observed to be associated with hail damage. Plant tops become gray to brown with shortened internodes. The outside of the stalk may be brown to black and appear water-soaked. The inside of the stalk is brown, slimy, and has a foul odor that somewhat resembles that of silage.

CONTROL: None known, but some hybrids appear to be less susceptible than others.

Chocolate Spot

CAUSE: *Pseudomonas syringae* pv. *coronafaciens* (Elliott) Young, Dye & Wilkie (syn. *P. coronafaciens* pv. *zeae* Ribeirs, Durbin, Arny & Uchytil.). Bacteria are disseminated and infection promoted by wind whipping leaves, thereby creating wounds. The disease occurs only in fields with potassium-deficient soils. The bacterium also infects oats and timothy.

DISTRIBUTION: Minnesota and Wisconsin.

SYMPTOMS: Lesions occur only on leaves. They are translucent, dark brown, elongated, nearly elliptical, up to 3 cm long but usually about 2 mm wide and 5 mm long, and surrounded by a broad yellow halo. Lesions are more numerous along leaf edges where potassium-deficiency symptoms first appear. Eventually, lesions coalesce, causing large areas of the leaf to die.

CONTROL:
1. Maintain proper potassium levels.
2. Hybrids differ in susceptibility.

Goss's Bacterial Wilt and Blight

This disease is also called bacterial wilt, blight, freckles, Goss's leaf freckles, leaf freckles and wilt, and Nebraska bacterial wilt.

CAUSE: *Clavibacter michiganense* subsp. nebraskense (syn. *Corynebacterium nebraskense* Vidaver & Mandel. The bacterium overwinters in infected corn residue consisting of leaves, stalks, cobs, and ears on or near the soil surface. It also overwinters in or on seed and survives readily in irrigation water. The bacterium is disseminated either externally or internally on seed and infested debris. Injury to plants aids in the infection process. Bacteria enter into plants in the presence of water from irrigation or rain. Sand blasting, hail, severe rain storms, and wind aid dissemination. Infection occurs during hot (over 27°C) and damp weather. Events that occur early in the growing season probably contribute more to the development of disease severity than those occurring later in the season. Different strains of the bacterium occur that have been classified into several groups.

DISTRIBUTION: Colorado, Iowa, Kansas, Nebraska, and South Dakota.

SYMPTOMS: Plants can be infected at all growth stages but seedlings are more susceptible than older plants. Symptoms tend to be similar to those of Stewart's wilt. Early infection causes seedlings to wilt, wither, and die. Later infection causes stunting, wilting, or various degrees of leaf blight. Yellow to gray-green streaks appear parallel to the veins. These are sometimes red on certain hybrids. As the streaks enlarge, the margins become wavy with dark green to black, water-soaked angular spots (freckles) occurring along veins. Droplets of bacterial exudate may appear on the surface of the diseased tissue. The droplets dry, leaving a crystalline substance that glistens in direct sunlight. The streaks may coalesce, forming large lesions and killing the leaf. Eventually the entire leaf dries up.

Systemically infected plants may have discolored water-conducting tissues. An orange bacterial exudate oozes from vascular bundles when a stalk is cut in cross section. A dry or water-soaked, slimy brown rot of roots and lower stalk may occur. Systemically infected plants may exhibit leaf symptoms resembling drouth stress rather than the previously described leaf symptoms. Seedlings as well as older plants can be killed following systemic infection.

CONTROL:
1. Plant resistant hybrids.
2. Rotate corn with soybeans, small grains, and alfalfa.
3. Deep plow corn residue in the autumn to reduce inoculum.

Holcus Spot

CAUSE: *Pseudomonas syringae* pv. *syringae* van Hall (syn. *P. holci* (Kendrick)) overwinters in infected residue. During warm, 25–30°C, wet, and windy weather early in the growing season the bacteria are splashed or blown onto the host where they invade the leaf through stomates. The disease is also present on a number of grasses including foxtail, millet, Sudangrass, Johnsongrass, and some cultivars of sorghum.

DISTRIBUTION: Prevalent in the eastern and midwestern United States. Holcus spot is probably present wherever corn is grown.

SYMPTOMS: Lesions on leaves are initially dark green and water-soaked. Later the lesions dry up and become tan to brown with a reddish to brown margin. Larger lesions may be surrounded by a yellowish halo. Lesions are round to elliptical and range in size from 2–10 mm in diameter.

CONTROL: The disease has not been of economic importance; therefore, control is normally not necessary.
1. Plant hybrids that are more resistant than others.
2. Rotate corn with resistant crops.

Stewart's Wilt

This disease is also called bacterial leaf blight, bacterial wilt, maize bacteriosis, Stewart's bacterial wilt, and Stewart's leaf blight.

CAUSE: *Erwinia stewartii* (Smith) Dye overwinters primarily within the body of the corn flea beetle, *Chaetocnema pulicaria* Melsh. Warm winter weather favors survival of the beetle and subsequent disease development the following season. If the sum of the mean monthly temperatures for December, January, and February total 37.8°C or more, Stewart's wilt may be potentially severe. Little or no disease is likely to occur when the sum of the mean temperatures is below 32.2°C. The bacterium is rarely spread through seed of infected dent corn but may be commonly spread by sweet corn seed.

Adult beetles begin to feed on corn seedlings in late spring and early summer. As the beetles feed, bacteria are placed by the beetle into the wound in the corn plant made by its feeding. Flea beetles will continue to spread bacteria throughout the growing season by feeding on infected plants, then flying to healthy ones. The twelve-spotted cucumber beetle, *Diabrotica undecimpunctata howartii* Bard; toothed flea beetle, *C. denticulata* Ill.; larvae of the seed corn maggot, *Hylemya cilicrura* Rond.; wheat wireworm, *Agriotes mancus* Say; and May beetle, *Phyllophaga* sp. may infrequently spread the bacteria but are not important in its overwintering.

High levels of ammonium and phosphorus increase susceptibility whereas high calcium and potassium tend to decrease susceptibility to infection by *E. stewartii*.

DISTRIBUTION: Central America, China, eastern and southern Europe, eastern United States, and USSR.

SYMPTOMS: Dent corn is not as susceptible as sweet corn except for possibly a few susceptible inbreds. In sweet corn, bacteria may be found in every part of the plant including roots.

The best symptom is the conspicuous streaks in the leaves that generally appear after plants have tasseled. At first the streaks are light green to yellow, later turning tan as the tissue dies. Individual streaks are long, with irregular or wavy margins, and tend to follow the veins of the leaf. Several streaks may coalesce and kill the entire leaf. Careful examination of the streaks, especially if the leaf is held up to the light, will show feeding scars made by the beetles. These appear as a tiny scratch, usually at a right angle to the streak.

Infected plants may produce bleached and dead tassels. Brown cavities may form in the stalk pith of severely infected plants near the soil line. Bacteria may spread through the vascular system and pass into the kernels. Masses of bacteria may ooze as yellow, moist beads from the cut ends of infected stalks or may stream from cut edges of infected leaf tissue.

CONTROL:
1. Plant resistant cultivars or hybrids.
2. Apply insecticides early to control corn flea beetles.

Yellow Leaf Blotch

CAUSE: *Pseudomonas* sp. Plants are infected in the seedling stage.

DISTRIBUTION: West Africa.

SYMPTOMS: Scattered, cream to yellow to light tan lesions occur on leaves. The lesions are nearly rectangular, sometimes with square ends, and vary in size from 14–45 mm long by 10–15 mm wide. There is a tendency for streaks or runners to follow the veins. On some plants, entire leaves become water-soaked and later wither and die. Young plants that are not killed may sometimes outgrow the disease and eventually produce an apparently healthy crop.

CONTROL: No control is necessary as the disease is not considered severe.

FUNGAL CAUSES

Alternaria Leaf Blight

CAUSE: *Alternaria alternata* (Fr.) Keissler (syn. *A. tenuis* Nees.) is a common saprophyte on plant residue. Infection occurs during long periods of heavy dew (16–20 hours) and after some form of injury has happened.

DISTRIBUTION: Midwestern United States.

SYMPTOMS: Initially chlorotic streaks that later become necrotic form on leaves of all ages. Leaves may be killed in one week if heavy and prolonged dews occur.

CONTROL: Some inbreds are moderately resistant.

Anthracnose Leaf Blight

CAUSE: *Colletotrichum graminicola* (Ces.) G. W. Wilson survives as mycelium and conidia in crop residue but will lyse in 16 and 14 days respectively when exposed directly to soil. The fungus is also seedborne as stroma or hyphae in the endosperm. During warm, wet weather conidia are produced in acervuli and are splashed or windborne to leaves.

Disease development is optimum at 30°C and periods of cloudy weather. Although seedlings may be infected, disease development, as indicated by lesion size, is best on the lowest leaves of older plants. Anthracnose leaf blight is more likely to occur earlier in the growing season when infested corn residue is left on the soil surface. Surface corn residue but not buried residue is an important source of inoculum. Disease is more severe in continuous corn. Disease is more severe when plants are also infected by the lesion nematode *Pratylenchus hexincisus* Taylor & Jenkins. *Colletotrichum graminicola* has a wide host range of grasses; however, isolates that infect small grains do not infect corn.

DISTRIBUTION: France, Germany, India, Philippines, Thailand, and the eastern United States.

SYMPTOMS: *Colletotrichum graminicola* causes leaf spots on young plants early in the growing season and stalk rot and leaf spots on maturing plants near the end of the growing season. Often the disease starts on the lower leaves early in the growing season with little disease development until late in the growing season. There is a decrease in susceptibility around pollination time followed by an increase toward plant maturity. Brown, spindle-shaped lesions 5–15 mm long, with yellow to red-brown borders occur on leaves. As lesions mature, their centers become gray and numerous acervuli that appear as black, spiny structures under magnification develop on the surface of dead tissue. Areas of yellow or yellow-orange often surround the spots. Concentric rings or zones are sometimes apparent within diseased areas. A rapid firing of infected leaves may occur under warm, wet conditions later in the season. This is due to a coalescence of lesions and is often referred to as top dieback.

CONTROL:
1. Plant resistant hybrids. However, hybrids susceptible to leaf blight may be quite resistant to the stalk rot phase and vice versa.
2. Rotate corn with nongrass crops.
3. Plow under infected residue where feasible.
4. Ensure soil has a balanced fertility.
5. Severity of stalk rot has been reported to decrease in artificially and naturally infected plants with increased nitrogen rates applied in either spring or autumn as anhydrous ammonia.

Anthracnose Stalk Rot

CAUSE: *Colletotrichum graminicola* (Ces.) G. W. Wilson. See Anthracnose leaf blight. However, primary inoculum for the stalk rot phase is not well understood. Conidia may wash behind leaf sheaths and invade the stalk. Inoculum may come from buried debris where resting cells can survive for several years. Disease is more severe in continuous corn. The fungus may infect corn roots and colonize lower internodes as normal senescence progresses. Disease is associated in New York with early or midseason infestations by European corn borer, *Ostrinia nubilalis* Huebner.

DISTRIBUTION: See Anthracnose Leaf Blight. However, stalk rot may occur where no leaf blight is present.

SYMPTOMS: Initially, narrow, oval, water-soaked lesions occur in rind after tasseling. These become tan to reddish and finally dark brown to black linear streaks. This usually occurs on the surface of the lower internodes late in the season before normal maturity. Frequently larger oval areas develop that involve most of the internodes, causing them to appear mostly dark brown or shiny black. When such stalks are split, the pith is a dark brown starting at the nodes. Some genotypes may have a soft and watery pith. Sometimes the internal discoloration may be present with little or no blackening of the surface. Severely affected stalks are likely to lodge.

However, symptoms may occur at different growth stages, depending on susceptibility of the plant. Very susceptible plants may be killed before pollination.

Sometimes portions of the stalk above the ear become a gray-green and die several weeks after pollination while the lower stalk remains green. The upper leaves may turn yellow or reddish, lodge, and drop off. Sometimes this occurs just before lower leaves begin to normally senesce. Top dieback may cause lodging but this is usually higher on the stalk than other stalk rot diseases.

CONTROL:
1. Plant resistant hybrids; however, resistance is not correlated with resistance to other stalk rot fungi.
2. Rotate corn with other crops.
3. Plow under infected residue where feasible.
4. Ensure soils have a balanced fertility.

Ascochyta Leaf and Sheath Spots

CAUSE: *Ascochyta zeae* Stout, *A. maydis* Stout, and *A. zeina* Sacc. probably survive as pycnidia in corn residue. During wet weather conidia are produced within pycnidia and are splashed or blown to corn.

DISTRIBUTION: United States.

SYMPTOMS: Lesions at first are ellipsoidal and red-purple to brown. Older lesions become elongated and irregular with brown margins. Tiny black dots, which are pycnidia, most of whose flasklike structure develops below the leaf surface, often develop in rows within the lesion.

CONTROL: The disease has not been considered important enough to warrant contol measures.

Aspergillus Ear Rot

CAUSE: Several species of Aspergillus, including *A. flavus* Lk. ex Fr., *A. glaucus* Lk. ex Fr., and *A. niger* Van Tiegh. The fungi enter kernels through growth cracks and injuries caused by corn earworms and other insects. *Aspergillus niger* has been observed most frequently in dry years, while *A. flavus* is favored by warm (32–38°C) temperatures and high humidity. *Aspergillus flavus* has been observed during dry years when kernels have been injured due to insects and drouth stress. Additionally, stress conditions that reduce yield may play a role in predisposing corn to infection by *A. flavus* or to increased aflatoxin productions once infection has occurred. Corn planted and harvested late and produced under nitrogen stress is a better substrate for preharvest aflatoxin production than corn grown under good management practices and supplied with adequate nitrogen.

Silk infection is favored by high day and night temperatures. *Aspergillus flavus* can also colonize silk tissue and invade developing kernels in the absence of insects or other injury.

Airborne spores of *A. flavus* may be an important source of inoculum. Open pollinated corn is more susceptible to infection than hybrid corn.

DISTRIBUTION: Generally distributed wherever corn is grown.

SYMPTOMS: *Aspergillus niger* appears as a black mold on infected kernels that are sometimes scattered over the whole ear. *Aspergillus flavus* appears as a green-yellow mold growing in the tracks of corn earworm or other insect damage and on injured kernels at the ear tip. This fungus has the potential to produce aflatoxin, a carcinogen or cancer-causing agent. *Aspergillus glaucus* is a greenish mold. All *Aspergillus* spp. will grow on corn kernels that are stored at a moisture content higher than 15 percent. Growth may occur at variable moisture contents.

CONTROL:
1. Dry corn to a moisture content of 15 percent or below as soon after harvest as possible.
2. Drouth conditions result in greater aflatoxin levels but either irrigation or tillage that reduces water stress reduces aflatoxin levels.
3. Differences in aflatoxin levels between genotypes exist.

Banded Leaf and Sheath Spot

CAUSE: *Rhizoctonia microsclerotia* Matz. (syn. *R. solani* f. sp. sasakii (Kuhn) Exner); teleomorph *Corticium solani* (Prill & Delacr.) Bourd. & Galz. (syn. *Thanatephorous cucumeris* (Frank) Donk. and *Pellicularia filamentosa* (Pat.) Rogers).

DISTRIBUTION: Africa, Asia, and Europe.

SYMPTOMS: Leaves and leaf sheaths have large gray, tan, or brown discolored areas that alternate with dark bands. Ears may have a brownish rot under warm, humid conditions.

CONTROL:
1. Plant resistant genotypes.
2. Apply foliar fungicides.

Black Bundle Disease

CAUSE: *Cephalosporium acremonium* Cda. (syn. *Acremonium strictum* Gams.) is seedborne and soilborne. Infection may be either directly or through wounds in the stalk.

DISTRIBUTION: Widely distributed.

SYMPTOMS: Initially leaves become purple or red at around the dough stage. Vascular bundles are blackened for a long length, normally through several internodes. Plants are barren or have nubbins for ears. Other symptoms are excessive tillering and multiple ears at one node.

CONTROL:
1. Some inbreds are resistant.
2. Apply seed treatment fungicides

Black Kernel Rot

CAUSE: *Botryodiplodia theobromae* Pat. is seedborne, soilborne, and windborne.

DISTRIBUTION: Africa, Asia, Central America, and South America.

SYMPTOMS: Initially kernels have a brown discoloration. Later kernels become black with small eruptions that may contain pycnidia. A dry rot develops.

CONTROL: No control is reported.

Borde Blanco

This disease is also called white border.

CAUSE: *Marasmiellus* sp. The disease occurs in humid lowlands and highlands where hanging mists occur during the rainy season. The fungus is soilborne.

DISTRIBUTION: Costa Rica, Mexico, and Nicaragua.

SYMPTOMS: The borde blanco fungus is primarily a pathogen of the leaf blade and sheath and is not associated with roots or the basal regions of the stalk. The leaf blight phase is characterized by blanched marginal lesions that may originate at any point along the leaf from the sheath-blade juncture to the apex but always beginning at the margin. Similarly, stalk lesions originate at the margin of the sheath, but the fungus penetrates through several sheath layers into the stalk.

Leaf lesions are 5–25 cm long and 1–3 cm wide. Lesions are characterized by concentric zones of slightly different discoloration. The white border lesions glisten in sunlight in severely infected fields. Lesions may be found on leaves of all ages.

Stalk lesions are numerous in fields showing a high percentage of infected leaves. Severe stalk rot has been observed in humid areas where foliage is dense.

CONTROL: No control is reported.

Brown Stripe Downy Mildew

CAUSE: *Sclerophthora rayssiae* Kenneth, Kaltin & Wahl var. *zeae* Payak & Renfro survives as oospores in infected residue in the soil and as mycelium in other grass hosts. The oospore germinates to form a sporangium in which zoospores are produced. Mycelium in other hosts gives direct rise to sporangia. Sometimes the sporangium may germinate directly to form a germ tube but most often zoospores are liberated from sporangia and germinate to infect corn leaves. Secondary spread is by the formation of sporangia. Sporangia are disseminated by wind and water.

Sporangia are produced at optimum temperatures of 20–22°C. Zoospore germination optimally occurs between 22°C and 25°C; thereafter, an optimum soil temperature between 28°C and 32°C favors disease development. A film of free water must be present in order for sporangial development and germination to occur. The weed *Digitaria sanguinalis* is also a host and may supply primary inoculum to early planted corn.

DISTRIBUTION: India, Nepal, Pakistan, and Thailand.

SYMPTOMS: Narrow (3–7 mm), yellow stripes with definite margins that are limited by veins develop on leaves. Later the stripes turn redpurple. Further development of lesions causes severe striping and blotching that may kill the plant and suppress seed development if symptoms occur prior to flowering. Sporangia develop on both sides of lesions and appear as a downy growth. Floral and vegetative parts are not malformed and leaves do not shred.

CONTROL:
1. Plant resistant cultivars or hybrids.
2. Plant before rainy season begins.
3. Apply foliar fungicides after symptoms appear or as a soil drench.

Charcoal Rot

CAUSE: *Macrophomina phaseolina* (Tassi) Goid. (syn. *M. phaseoli* (Maubl.) Ashby and *Botryodiplodia phaseoli* (Maub.) Thirum.). The anamorph *Sclerotium bataticola* Taub. is the sclerotial and mycelial stage of *M. phaseolina*. The causal fungus survives as sclerotia in residue and in soil. Strains of the fungus that infect other crops produce pycnidia but the strains infecting corn do not. Sclerotia are disseminated by any means that will move debris and soil. Sclerotia may germinate, cause a seedling blight, or infect roots of seedlings and young plants. Plants are frequently infected as they near maturity by sclerotia germinating to produce infection hyphae that penetrate roots and grow through cortical tissues into stalks. Disease development is most rapid in dry soils at temperatures near 37°C. A low soil temperature or high soil moisture will retard the disease. Tiny sclerotia are eventually produced on infected tissues.

DISTRIBUTION: Africa, Asia, Europe, North America, and South America.

SYMPTOMS: Plants nearing maturity are most likely to display symptoms. Brown, water-soaked lesions that later turn black appear on roots. As plants mature the fungus grows into the lower internodes, causing premature ripening and shredding, with the stalk breaking at the crown. The surface of rotted internodes may turn partly gray. Tiny, black spots, which are the sclerotia, show through the epidermis of the stalk and roots. The interior of the stalk is disintegrated with only the vascular bundles remaining intact and covered with sclerotia. The appearance of the numerous sclerotia gives the effect of charcoal, hence the name. Kernels are also infected and become completely black.

CONTROL:
1. Where irrigation is practiced, soils should be kept moist.
2. Rotate for two to three years with nonhosts.
3. Grow full-season hybrids that are adapted for the particular area.
4. Plow under infected residue where feasible.

Cladosporium Rot

CAUSE: *Cladosporium herbarium* Lk. ex Fr. and *C. cladosporioides* (Fres.) DeVries infects kernels damaged by early frost and have growth cracks.

DISTRIBUTION: Generally distributed wherever corn is grown.

SYMPTOMS: Symptoms are usually not obvious until corn has been harvested. Green-black streaks occur on scattered kernels. The discoloration starts where the kernels are attached to the cob and grows upward but seldom reaches the crown. After harvest, deterioration may occur if corn is stored at too high a moisture content.

CONTROL: No control is necessary.

Common Rust

This disease is also called common corn rust and common maize rust.

CAUSE: *Puccinia sorghi* Schw. is a heteroecious long-cycled rust fungus whose alternate host is wood sorrel, *Oxallis* spp. Urediospores, or the repeating stage, are windborne to corn where cool temperatures, 16–23°C, and high relative humidity favor disease development and spread. Teliospores replace urediospores in the pustule as the host nears maturity. In the spring teliospores germinate to form basidiospores that infect only *Oxalis* spp. The spermatial (pycnial) stage occurs on the upper leaf surface at the point of basidiospore infection. Pycniospores (spermatia) are formed in pycnia (spermagonia), and are extruded and disseminated by insects to receptive hyphae of an opposite mating type. Nuclei of pycniospores and receptive hyphae merge to initiate development of aecia on the lower leaf surface. Aeciospores from aecia are windborne and infect only corn. These infections give rise to urediospores. In warmer climates the urediospores overwinter and serve as primary inoculum, bypassing the alternate host. Aecial infection is then uncommon. Spores are airborne from tropical and semitropical areas into temperate corn-growing regions.

DISTRIBUTION: Uredial and telial stages are generally distributed wherever corn is grown. The aecial stage occurs infrequently in temperate areas of Europe, India, Mexico, Nepal, Russia, South Africa, and the United States.

SYMPTOMS: Rust symptoms appear soon after silking but older tissue is usually resistant. Sweet corn is generally more susceptible than field corn. Pustules may appear on any aboveground plant parts but are most abundant on leaves. Oval to elongate, golden-brown to red-brown pustules are sparsely scattered over both leaf surfaces. The urediospores break through the epidermis early in their development. As corn matures, pustules become brown-black with the formation of teliospores. The teliospores likewise break through the epidermis. The rupture of the epidermis by urediospores and teliospores differentiates common rust from southern corn rust.

During severe disease conditions, chlorosis and death of leaf sheaths and leaves may occur.

CONTROL:
1. Plant resistant hybrids.
2. Apply foliar fungicides when pustules first appear, particularly in seed-production fields.

Common Smut

This disease is also called boil smut.

CAUSE: *Ustilago maydis* (DC.) Cda. (syn. *U. zeae* Vng.) overwinters as teliospores (chlamydospores) in the soil. Teliospores germinate either directly or indirectly in water between 10°C and 35°C. Direct germination is by an infection hypha produced directly from a germinating teliospore. Indirect germination is by a promycelium first being formed by a teliospore and on which sporidia (basidiospores) are borne. Sporidia are airborne to leaves where mycelium from spores of opposite mating types fuse and form an infection hypha that infects young, developing corn tissues. Ears of corn are infected by infection hyphae growing through silk.

The parasitic mycelium stimulates host cells to increase in size and number, forming a gall. Eventually teliospores develop, causing the galls to be converted to a black powdery mass. Disease is favored by dry conditions and temperatures of 26-34°C. Dry weather, during which corn growth is retarded, favors smut. Abnormally cool, wet weather that retards growth of young corn plants also seems to be conducive to infection. Incidence of smut is higher among plants grown on soil high in nitrogen and particularly where heavy applications of barnyard manure have been made. Injuries from hail, cultivation, herbicide injury, and detasseling increase smut infection.

DISTRIBUTION: Worldwide; however, it rarely occurs in tropical areas. The disease is more prevalent in warm, drier areas.

SYMPTOMS: All aboveground parts of the corn plant are susceptible but embryonic tissue is particularly susceptible. Galls are first covered with a glistening white membrane. The interior of these galls soon becomes black and powdery with the membrane rupturing to expose the black sooty mass of spores. Galls on ears and stalks may reach 15 cm in diameter. Galls on leaves rarely develop larger than the size of peas (0.6–1.2 cm in diameter), become hard and dry, and contain few spores. Large galls on the ear and above the ear are more destructive than galls below the ear. Early infection may kill seedlings. Galls on the lower part of the stock may cause plants to be barren or produce several small ears. Such plants become reddish in the fall.

CONTROL:
1. Plant resistant hybrids.
2. Maintain a well-balanced soil fertility.
3. Avoid mechanical injuries to plants.
4. Remove and burn galls from infected plants before they rupture.

Corticium Ear Rot

CAUSE: *Corticium sasakki* (Shirai) Matsumoto. Ear rot occurs under extremely wet conditions.

DISTRIBUTION: India.

SYMPTOMS: Initially, irregular, water-soaked spots, that vary in size and shape, occur on leaf sheaths, and gradually enlarge in moist and warm weather. Profuse fungal growth sometimes appears on the surface of diseased spots. Brown mycelium progresses upward from the base of the ear and is commonly observed late in the season on husks, floral bracts, and pericarps of rotted kernels. Sclerotia are formed on infected ears in the later stages of disease development.

CONTROL: No control is reported.

Crazy Top

CAUSE: *Sclerophthora macrospora* (Sacc.) Thirum., Shaw & Naras. (syn. *Sclerospora macrospora* Sacc. and *Phytophthora macrospora* (Sacc.) Ito & Tanaka). The fungus survives in infected corn residue and wild grasses as oospores. The oospores probably germinate to form sporangia in which zoospores are produced. Zoospores are then liberated from sporangia in water-saturated soil. Infection occurs after planting and before plants are in the four- to five-leaf stage in soil saturated 24–48 hours. Germination and infection occur over a wide range of temperatures. Zoospores swim to the corn seedlings where they germinate and produce systemic mycelium.

Sporangia are rarely produced directly on sporangiophores projecting from stomates on leaves but numerous oospores are produced within infected leaves and leaf sheaths. Over 140 grass species are hosts of *S. macrospora*.

DISTRIBUTION: Africa, Asia, Australia, Canada, Europe, Mexico, and the United States. The disease occurs in localized wet areas in a field.

SYMPTOMS: The most conspicuous symptom is the partial to complete replacement of the normal tassel by a large, bushy mass of small leaves. No pollen is produced since normal flower parts in the tassel are completely deformed.

Normal ear formation may also stop causing ear shoots to be numerous, elongated, leafy, and barren. Affected plants vary greatly in height. Some may be severely stunted, with narrow yellow to brown-streaked leaves and numerous tillers. Others may be taller than average with an increased number of nodes and leaves above the ear and in the shank.

CONTROL:
1. Provide adequate soil drainage.
2. Control grassy weeds.
3. Do not plant in low, wet spots.

Curvularia Leaf Spot

CAUSE: *Curvularia pallenscens* Boed., *C. lunata* (Wakker) Boed. var. *aeria*, *C. inaequalis* (Shear) Boed., *C. maculans* (Bancroft) Boed., *C. intermedia* Boed., *C. tuberculata* Jain, *C. eragrostidis* (P. Henn) J. A. Meyer, *C. senegalensis* (Speg.) Subram., and *C. clavata* Jain. The fungi survive in soil and in residue.

DISTRIBUTION: Generally distributed in warmer or milder climates.

SYMPTOMS: Initially circular, small, 1–2 mm straw-colored lesions with reddish or dark brown margins develop on leaves. Lesions may coalesce to form necrotic areas up to 1 cm long. The centers of some lesions have circular grayish spots up to 1 mm in diameter. Chlorotic or yellowish halos may surround the lesion.

CONTROL: Foliar fungicides provide control, but they are not practical.

Didymella Leaf Spot

CAUSE: *Didymella exitialis* (Mor.) Muller.

DISTRIBUTION: India.

SYMPTOMS: Small, light-colored elongated to elliptical spots with brown margins form on leaves. Later the spots enlarge and coalesce, forming streaks or irregular areas in which perithecia may be formed.

CONTROL: No control is practiced.

Diplodia Ear Rot

This disease is also called cob rot.

CAUSE: *Stenocarpella maydis* (Earle) Sutton (syn. *Diplodia maydis* (Berk.) Sacc.) is discussed under Diplodia stalk rot. Ears are most susceptible from silking until about three weeks later. Commonly infection starts at the base of husks or an exposed ear tip but can also advance from the stalk through the shank. As the ear approaches maturity, the rot grows slower, ceasing completely when the kernels reach about 21 percent moisture content. As the ears grow in length, husk protection at the tip end becomes less and gradually the husks also become more loose, making it easier for spores to get to the ear or between the husks.

Pycnidia may be found on rotted ears, particularly on the inner husks. During wet weather conidia are exuded and become windborne. Dry weather early in the season followed by wet conditions just before and after silking favors ear infection.

DISTRIBUTION: Africa, Australia, Philippines, Romania, South America, and the United States.

SYMPTOMS: Infection usually begins at the base of the ear and pro-gresses toward the tip. Husks of ears infected early appear bleached in contrast to green healthy ears. When infection takes place within two weeks after silking, the entire ear becomes gray-brown, shrunk, and light in weight by harvest time. Such ears usually remain upright, with husks stuck tightly together owing to the growth of the fungus between them. Black pycnidia are often found at the base of the husks and on the faces of the kernels. Ears infected later in the growing season may show no external signs of the disease until the ear is broken or kernels are removed. White mold will be found growing between the kernels.

CONTROL:
1. Some inbred lines vary in resistance and tend to transmit this reaction to hybrid combinations. No inbred line or hybrid is completely resistant.
2. Harvest early to prevent grain from being weathered.
3. Dry corn to 15 percent moisture content and below to prevent further mold growth.

Diplodia Leaf Spot

This disease is also called Diplodia leaf streak and Stenocarpella leaf spot.

CAUSE: *Stenocarpella macrospora* (Earle) Sutton (syn. *Diplodia macrospora* Earle). The disease occurs under warm, humid (mean rel-ative humidity 50 percent day, 95 percent night) conditions in the semitropics. The pathogen survives in leaf tissues, infected ears, and leaf sheaths and as dormant mycelium in seed. Ears, leaves, and leaf sheaths are infected. However, it has been reported that all tissue can be attacked at all stages of growth.

DISTRIBUTION: Central America and North Carolina.

SYMPTOMS: Initially small yellowish spots occur on leaves. Later these become water-soaked lesions that turn into long, narrow chlorotic lesions or streaks up to 1.5 cm by 10 cm in size. Initially yellow margins along the lesions are evident. Pycnidia are scattered in the lesion. Frequently the fungus grows along the border of the leaf veins. Inoculated plants had lesions along veins the entire length of the leaf blade. Toxin production was indicated.

A leaf spot caused by *S. macrospora* has been described from North Carolina. However, symptoms differ somewhat from those described above. The lesions are described as circular to elongate and up to 25 cm long.

CONTROL: Difference in resistance exists between genotypes.

Diplodia Stalk Rot

CAUSE: *Stenocarpella maydis* (Earle) Sutton (syn. *Diplodia maydis* (Berk.) Sacc. and *D. zeae* (Schw.) Lev.) overwinters as conidia in pycnidia or mycelium in corn debris and as conidia and mycelium on seed. During warm, moist weather in late summer, conidia are exuded from pycnidia in long cirri and are disseminated by splashing rain, or after the cirri have dried up, by wind. Insects are also reported to disseminate spores. Most infection starts at any of the first three nodes aboveground but it may also start at higher and lower nodes. Infection develops at the place of origin of the leaf sheaths. Some infections develop at the junction of brace roots with the stalks and at the origin of roots at the soil level. Infection may also enter stalks by way of the mesocotyl from infected seed or infested soil. Growth of *S. maydis* then proceeds upward into the stalk and down into the roots. However, the entire plant is not invaded and infection does not commonly grow into the ear to cause a rot. Warm (28–30°C optimum), wet weather two to three weeks after silking favors spread and development of disease. Plants are predisposed to infection by high nitrogen and low potassium, higher than recommended plant population, and loss of leaf area. Early-maturing hybrids are usually more susceptible than longer-season hybrids.

All factors that predispose corn to stalk rot can possibly be unified in a hypothesis called the stress translocation balance of predisposition of corn to stalk rot. Predisposition is associated with carbohydrate shortage in root tissue, which is caused by the combination of the reduction of photosynthesis by the leaves and intraplant competition for carbohydrates by the developing kernels of grain. Consequently, root tissue has a weakened cellular defense system, allowing invasion and degeneration by soil microorganisms. Roots are then rotted, the plant wilts, and stalk rot develops.

DISTRIBUTION: Africa, Australia, Philippines, Romania, South America, and the United States.

SYMPTOMS: Symptoms commonly appear several weeks after silking. The first symptom before maturity is the presence of gray-green leaves. The leaves will then turn brown, seemingly in a day or two, giving the impression of an early frost. Dark brown lesions extend in either direction from the node.

When diseased stalks are split open, the pith is disintegrated and discolored with only the vascular bundles left intact. Stalks are weakened and break easily. Dark brown to black pycnidia may form in the autumn just beneath the epidermis and cluster around the lower internodes of the stalk. White fungal growth may also be present on the surface of the nodes.

CONTROL:
1. Plant resistant hybrids.
2. Balance soil fertility. Avoid high levels of nitrogen and low levels

of potassium. However, the severity of stalk rot in artificially inoculated plants has been reported to decrease with increasing nitrogen rates applied in either spring or autumn as anhydrous ammonia.

3. Plant the recommended plant population. Higher than recommended plant populations cause plants to be stressed and have less resistance to infection.

Ergot

This disease is also called Horse's tooth and diente de caballo.

CAUSE: *Claviceps gigantea* Fuentes, de la Isla, Ullstrup & Rodriguez, anamorph *Sphacelia* sp., survives as sclerotia on the soil surface or mixed with seed. In the spring, sclerotia germinate to produce stroma on the ends of stipes (stalks). The stromatic head contains perithecia in which ascospores are produced. Ascospores are windborne to flowers of corn. Eventually macroconidia and microconidia are produced in a sweet, sticky substance called honeydew, which is attractive to insects. Insects are then thought to inadvertently carry conidia to flowers of corn. Disease development is favored by high rainfall and optimum temperatures of 13–15°C. Eventually ergot bodies or sclerotia replace kernels.

DISTRIBUTION: Central Mexico. Ergot occurs at higher elevations.

SYMPTOMS: Ergot bodies or sclerotia replace kernels. At first they are cream-colored, later turning dark gray and comma-shaped (resembling a horse's tooth) and are up to 5 cm wide and 8 cm long. Honeydew may be dark due to dust and other material sticking to it. Insects are usually present on the honeydew.

CONTROL: There is no adequate control; however, some lines are apparently more susceptible than others.

Eyespot

This disease is also called brown spot.

CAUSE: *Kabatiella zeae* Narita & Hiratsuka, teleomorph, *Aureobasidium zeae* (Narita & Hiratsuka) Dingley, overwinters in corn residue, particularly residue on the soil surface. The fungus may also be seedborne. Conidia are produced on residue and are windborne to nearby plants.

Older leaves are more susceptible to infection than younger ones. Secondary infection is from conidia produced on new lesions that are windborne or splashed to plants. Disease development is favored by cool, humid weather. Irrigation is conducive to disease development.

DISTRIBUTION: North central United States, Pennsylvania, and Ontario. The disease also occurs in Argentina, Austria, Brazil, France, Germany, Japan, New Zealand, and Yugoslavia.

SYMPTOMS: The first symptoms are small, translucent or water-soaked ovoid to circular spots, 2–5 mm in diameter. These symptoms may be confused with genetic spotting or Curvularia spots. Later the tissue in the center of the spots dies and becomes tan surrounded by a narrow, brown to purple ring with a yellowish halo. Spots may coalesce to form large necrotic areas. The spots are still conspicuous even after the entire leaf dies and are found in greatest numbers on leaves of plants approaching maturity. Spots may also develop on leaf sheaths, outer husks, and kernels of severely infected plants.

CONTROL:
1. Plant resistant hybrids.
2. Plow under infected residue where feasible.
3. Rotate corn with other crops. Corn is the only plant that is known to be a host for *K. zeae*.

False Smut

CAUSE: *Ustilaginoidea virens* (Cke.) Tak. (syn. *U. oryzae* (Pat. Bref.)) is a fungus whose life cycle is similar to *Claviceps gigantea* Fuentes, de la Isle, Ullstrup & Rodriguez. Disease is favored by hot, wet weather. Rice and other Graminae are also hosts.

DISTRIBUTION: India, Central America, and the southern United States, but the incidence of the disease is rare.

SYMPTOMS: Sclerotia (galls or smut balls) usually replace only a few to several flowers on tassels. Sclerotia are 4–15 mm in diameter, olive green to black in color, with a velvety texture. Growth is irregular and subspherical. As sclerotia mature, the membrane ruptures, leaving the spores naked. Gradually the surface becomes rough. When the sclerotium is cut, the interior is composed of three layers: a greenish layer of mature spores; a light yellow-green mass composed of mycelium; and an inner yellowish layer consisting of host tissue, growing mycelium, and spores.

CONTROL: False smut is of no economic importance, therefore, no control is necessary.

Fusarium Kernel Rot

This disease is also called Fusarium ear rot.

CAUSE: *Fusarium moniliforme* Sheldon and *F. subglutinans* (Wollenw. & Reink.) Nelson, Toussoun & Marasas comb. nov. These fungi are essentially the same and are discussed under Fusarium stalk rot. Disease development is generally favored by dry, warm weather. Spores infect kernels through injury from corn earworms, corn borers, or other insects and cracks made by weather or other means. Additionally, rootworm beetles such as the western corn rootworm beetle (*Diabrotica virgifera* Lec.) may be important in the spread of fungi to corn ears and kernels. The presence of *F. moniliforme* increases as harvest is delayed beyond physiological maturity.

DISTRIBUTION: Generally distributed wherever corn is grown.

SYMPTOMS: Fusarium rot seldom involves the whole ear. Rot often may occur only at the tip of the ear or on kernels scattered over the ear. The color of infected kernels may vary from faint pink to red-brown. As the disease progresses, infected kernels show a powdery or cottony pink mold growth. Infection often becomes established around points where insects have entered the ear or channeled between the rows of kernels or where other damage may have occurred.

CONTROL:
1. Avoid planting hybrids that tend to be most susceptible. Hybrids with poor husk cover, or with weak seed coats in which kernels tend to pop, or silk cut are susceptible to infection. Differences in resistance are conditioned by the genotype of the pericarp. In mature stages, some high-lysine hybrids are quite susceptible to Fusarium ear rot as is sweet corn.
2. Harvest grain early to prevent grain from becoming further deteriorated due to weather.
3. Dry corn to 15 percent moisture content and below to prevent further mold growth.

Fusarium Leaf Spot

CAUSE: *Fusarium moniliforme* Sheldon is a fungus that causes kernel rot and stalk rot. Infection is favored by dry, warm weather.

DISTRIBUTION: Caribbean.

SYMPTOMS: Initially water-soaked spots appear on whorl leaves that later turn white and papery with brown borders. The whorl does not open and the leaf tips turn downward. Occasionally necrotic spots develop high on a stalk.

CONTROL: No control is reported.

Fusarium Stalk Rot

CAUSE: *Fusarium moniliforme* Sheldon., teleomorph *Gibberella fujikuroi* (Sawada) Wollenw., and *F. subglutinans* (Wollenw. & Reink.) Nelson, Toussoun & Marasas comb. nov. survives as chlamydospore-like structures and mycelium in corn residue and residue of other plants in the soil. Mycelium in residue may be present through precolonization or through colonization of residue placed in soil. It is also commonly seedborne and is thought to be systemic. *Fusarium moniliforme* may be primarily a surface contaminant occurring in cracks and natural openings in the seed coat.

Plants are infected earlier in the season than by Gibberella or Stenocarpella, usually before pollination. Conidia are produced on residue and are windborne to plants, or mycelium may grow through soil to corn roots from a previously infected substrate. *Fusarium moniliforme* commonly enters the plant belowground through various insect wounds or aboveground through corn borer wounds. It may also penetrate directly into roots and the stalk at the base of leaf sheaths.

Fusarium moniliforme is a pathogen of senescing tissues. Predisposition by a mild early season water deficit increases likelihood of chronic water stress during periods of high evaporative demand or limited soil water availability. This leads to earlier plant senscence and increased susceptibility to stalk rot pathogens. In Kansas, incidence of stalk rot was found to be highest under dryland conditions and lowest with a soil moisture level of 50 percent of field capacity.

In field plots, plants growing in reduced till had a higher population density in the rhizosphere than plants growing in chiesel till and conventional till plots. However, later in the season, stalks from all treatments were 100 percent infected. The teleomorph, *G. fujikuroi*, is rarely found in nature and is probably unimportant in the disease cycle.

Fusarium proliferatum (Matsushima) Nirenberg is also reported to be pathogenic to corn.

DISTRIBUTION: Generally distributed wherever corn is grown. It is usually more important in the drier, warmer areas.

SYMPTOMS: Symptoms are generally similar to Gibberella stalk rot. Rotting commonly affects roots and lower internodes. Leaves will turn brown, resembling an early frost. Dark lesions without definite margins may appear at nodes. Disintegration of the pith usually begins after pollination even though plants may be infected earlier in the season. When stalks are split, the pith tissue is shredded with the vascular bundles remaining intact. In damp weather, white to pale pink mycelium of the fungus will grow on the surface of the stalk at the nodes.

CONTROL:
1. Plant the most resistant hybrids. Hybrids are usually not completely resistant.
2. Balance soil fertility.
3. Plant the recommended plant population. Avoid overplanting.

Gibberella Ear Rot

This disease is also called red ear rot.

CAUSE: *Gibberella zeae* (Schw.) Petch is discussed under Gibberella stalk rot. Cool wet weather at silking favors development of the disease. Ears with loose, open husks are often more susceptible to infection than those with good husk coverage. Sap beetles (*Glischrochilus quadrisignatus* (Say)) are capable of transferring conidia and ascospores of *G. zeae*, both internally and externally, to corn ears and causing an increase in ear rot. Inbreds and hybrids differ in susceptibility.

DISTRIBUTION: Generally distributed wherever corn is grown.

SYMPTOMS: A reddish mold that usually starts at the tip of the ear is characteristic of the rot. Early infected ears may rot completely. Husks adhere tightly to the ear because of mycelium growth between husks and ears. Blue-black perithecia that can be easily scraped off are sometimes found on husks and ear shanks.

Corn affected with this disease is particularly toxic to hogs. The toxins deoxynivalenol (vomitoxin) and zearalenone may be produced. Zearalenone concentration may vary according to corn line and fungus isolate. However, it has been reported that the less virulent isolates produced relatively more zearalenone in culture and diseased ears. Hogs may sometimes refuse to eat infected ears due to production of vomitoxin. Vomitoxin causes vomiting and dizziness. In severe cases death may follow. Infected corn is also toxic to man, dogs, and animals with similar digestive systems. Zearalenone causes estrogenic disturbances in which young gilts come in heat and mammary glands of boar pigs enlarge.

CONTROL:
1. Harvest early to prevent continued mold growth in the field.
2. Dry corn to 15 percent moisture content and below to prevent further mold growth.
3. Several inbreds display resistance.

Gibberella Stalk Rot

CAUSE: *Gibberella zeae* (Schw.) Petch, anamorph *Fusarium graminearum* Schwabe (syn. *F. roseum* (Lk.) emend. Snyd. & Hans. f. sp. *cerealis* (Cke.) Snyd. & Hans. 'Graminearum'). The fungus overwinters as chlamydospores in the soil, perithecia and mycelium in corn debris, and rarely as mycelium and conidia on seed. Perithecia are produced on infected corn in autumn or spring. During warm, moist weather shortly after pollination, ascospores are exuded from perithecia and are windborne to infect nodes just above the soil. Also, conidia are produced from mycelium and are windborne to corn; however, this type of infection probably occurs earlier in the season. *Fusarium graminearum* apparently gains possession of pith tissue late in the growing season as stalks ripen and predominates as tissues senesce. In some areas Group II isolates dominate. Group II isolates are mostly airborne, cause diseases of aerial plant parts, and normally form perithecia. Infection develops at the place of origin of the leaf sheaths. An additional mode of infection that appears common in some areas but not in others is infection of roots by chlamydospores or mycelium. Roots are infected and eventually the fungus grows up into the crown and stalk. Seedborne infection is not thought to result in stalk rot.

DISTRIBUTION: Generally distributed where corn is grown. In the United States it is more important in the eastern and northern corn-growing areas.

SYMPTOMS: Symptoms generally appear several weeks after pollination. Leaves turn gray-green, then suddenly turn brown, resembling an early frost. Dark lesions appear at the nodes and extend into the internodes. Sometimes the lesions show concentric rings but this is relatively uncommon.

Stalks that are split open have a disintegrated pith with just the vascular bundles remaining. A considerable amount of pink discoloration is likely to be present. Small, round, black perithecia that are easy to scrape off may be found on diseased tissue.

Gibberella stalk rot is more prevalent in plants subjected to stresses that result in early senescence and a reduction of sugar to roots and stalks. Such tissue is more readily invaded by fungi. There are some indications that Gibberella stalk rot is more prevalent in plants growing under favorable conditions early in the season but subjected to unfavorable conditions after silking. *Gibberella zeae* also causes the disease scab on wheat, barley, oats, and rye.

CONTROL:
1. Plant the least susceptible hybrid.
2. Balance soil fertility. Avoid high levels of nitrogen and low levels of potassium. However, the severity of stalk rot in both artificially and naturally infected plants has been reported to

decrease with increasing nitrogen rates applied in either spring or autumn as anhydrous ammonia.

3. Plant the recommended plant population. Overplanting tends to create stress, making plants more susceptible to stalk rot.

Graminicola Downy Mildew

This disease also is called green ear.

CAUSE: *Sclerospora graminicola* (Sacc.) Schroet. overwinters in the soil or on seed as oospores. It is thought that primary infection occurs by oospores germinating to form germ tubes that infect corn leaves. Temperatures between 24°C and 32°C for two days after planting are conducive to heavy infection. Secondary spread is by formation of sporangia on infected leaves that are windborne to leaves. Sporangia infrequently germinate directly to form only a germ tube. More commonly, sporangia germinate indirectly to release three or more zoospores that eventually germinate to form a germ tube. Disease development is favored by high humidity, light rains or heavy dews, and an optimum temperature of 17°C. Oospores are formed in diseased tissue.

DISTRIBUTION: Israel and the United States. It occurs on various grasses throughout the world. However, it is not an economically important disease on corn.

SYMPTOMS: Symptoms normally appear about 10 days after the plumule emerges. Corn is stunted and chlorotic streaking occurs on leaves, together with gray blotching and mottling. Leaves may be thick, corrugated, and fragile. White downy growth occurs on discolored areas. Sometimes plants will outgrow the disease.

CONTROL: Since the disease is usually not serious on corn, no control is needed.

Gray Ear Rot

CAUSE: *Botryosphaeria zeae* (Stout) von Arx & Muller (syn. *Physalospora zeae* Stout), anamorph *Macrophoma zeae* Tehon & Daniels. The fungus overwinters as sclerotia in infected residue and as pycnidia and perithecia in infected leaves. The following spring ascospores from perithecia and conidia from pycnidia mature and are windborne or are spread by rain to infect leaves and ears. Occasionally infection takes place on the tassel neck or under the sheath of the uppermost leaf.

Perithecia and pycnidia are formed within the same large lesions on leaves but only sclerotia are formed in rotted ears and kernels. Long periods of warm to hot, wet weather for several weeks after silking favor disease development.

DISTRIBUTION: Africa, Europe, and the Eastern United States.

SYMPTOMS: The disease resembles diplodia ear rot in its early stages. A gray mycelial growth develops on and between kernels near the base of the ear. Early infection results in bleached husks that adhere tightly to the ear. At harvest, these ears are slate gray, lightweight, and remain upright. When an infected ear is broken, the cob will have sclerotia, evident as black specks scattered through it. Perithecia and pycnidia do not develop in the ears. Kernels may show black streaks beneath the seed coat.

CONTROL:
1. Some hybrids resistant to diplodia ear rot may also be resistant to gray ear rot.
2. Rotate corn with nonhosts.

Gray Leaf Spot

This disease is also called Cercospora leaf spot.

CAUSE: *Cercospora zeae-maydis* Tehon & Daniels and *C. sorghi* var. *maydis* Ell. & Ev. survive as mycelium in infected residue on the soil surface. A species of Mycosphaerella is thought to be the teleomorph of *C. zeae-maydis*. Conidia are produced from mycelium in residue during wet weather and are windborne to hosts. Plants are infected at all stages of growth but disease develops best later in the growing season due possibly to a thick plant canopy creating a favorable microclimate. Corn grown under conservation or no-tillage methods becomes more severely infected than corn grown under conventional tillage systems. Maize dwarf mosaic virus appears to predispose plants to gray leaf spot. Warm (22–30°C) weather favors disease development; however, long daily periods of high relative humidity and leaf wetness are the most important criteria for disease development. Barnyard grass, Johnsongrass, and Sorghum spp. may also be infected. Additionally *Zea diploperennis*, *Z. mays* ssp. *luxurians*, and *Z. mays* ssp. *mexicana* displayed typical eyespot lesions when inoculated. Genetic differences exist between isolates.

DISTRIBUTION: Africa, Central America, China, Europe, northern South America, India, Mexico, Philippines, southeast Asia, and the eastern United States.

SYMPTOMS: Symptoms appear on the lower leaves of mature plants at or near anthesis and progressively higher on plants later in the season. Initially, reddish water-soaked spots occur. These spots expand to long, narrow, tan or pale brown lesions that are up to 5 cm long. Lesions are rectangular and restricted by leaf veins. Lesions may have a grayish cast only when the causal fungi sporulate. Lesions may eventually coalesce, killing the leaf. Under severe disease conditions all the leaves may be killed. Severe stalk rot and stalk breakage may occur.

CONTROL:
1. Plant resistant hybrids.
2. Plow under infected residue to reduce inoculum where feasible.
3. Rotate corn with a nonhost.

Head Smut

CAUSE: *Sphacelotheca reiliana* (Kuhn) Clint. (syn. *Sorosporium reilianum* (Kuhn) McAlp. and *Ustilago reiliana* Kuhn) overwinters as teliospores in the soil and occasionally on seed. Survival is better in dry than in moist soil. Teliospores may germinate either one of two ways: directly by producing a germ tube or indirectly by producing a basidium on which sporidia (basidiospores) are formed. Sporidia of opposite mating types will then fuse, forming a germ tube or hyphae. In either example, seedlings are infected and the parasitic mycelium develops systemically, invading undifferentiated floral tissues. Part or all of these tissues develop into smut sori in which teliospores are produced. Disease is accentuated by nitrogen deficiency. Soil temperatures of 23–30°C and moderate to low soil moisture (-1.5 bar maximum) are optimum for seedling infection. Other hosts include pitscalegrass, sorghum, and sudangrass. Pathogenic specialization occurs with one race limited to corn and another to grain sorghum, forage-sorghum hybrids, and some sudangrass varieties. A hybrid of *S. reiliana* has been identified that infects both corn and sorghum.

DISTRIBUTION: Occurs in drier soils of Australia, India, western part of Mexico, New Zealand, South Africa, southeastern Europe, USSR, United States, and Yugoslavia.

SYMPTOMS: The first symptoms appear when tassels and ears are formed. However, chlorotic spots develop on the fourth or fifth emerged leaf of corn seedlings infected with *S. reiliana*. Such seedlings are likely to produce sori in inflorescences. Sori develop on tassels and infrequently on normal vegetative leaves. Galls are first covered with a membrane that soon breaks open to expose a powdery spore mass and the vascular bundles of the host. The appearance of the vascular bundles helps to distinguish head smut from common smut. If only parts of tassels and ears are converted to galls, leafy proliferations are found on these parts that are not found with common smut. Occasionally sori develop as long thin stripes in leaves. Smutted ears are rounded and lack silks. Plants with smutted tassels may be severely dwarfed and increased tillering may be common. When the tassel is not infected, most of the ears will be smutted except for a few nonsmutted nubbins. Sometimes multiple ears appear at nodes. Infected plants normally do not produce pollen.

CONTROL:
1. Plant resistant hybrids.
2. Rotate corn with other crops. The race of the fungus that infects corn usually does not infect sorghum and vice versa; however, a hybrid of *S. reiliana* has been identified that infects both corn and sorghum.
3. Fungicides applied to soil at or before planting may be useful in reducing inoculum.
4. Apply a fungicide-seed protectant.

Helminthosporium Leaf Disease

This disease is also called rostratum spot.

CAUSE: *Helminthosporium rostratum* Drechs. (syn. *Exserohilum halodes* (Drechs.) Leonard & Suggs., *E. rostratum* (Drechs.) Leonard & Suggs. emend Leonard, *Bipolaris rostrata* (Drechs.) Shoemaker, and *Drechslera rostrata* (Drechs.) Richardson & Fraser). The teleomorph is *Setosphaeria rostrata* Leonard. The causal fungus survives as mycelium and conidia in diseased tissue. Conidia are produced from mycelium and are windborne or splashed to corn where the older leaves are more likely to become infected than younger leaves. An optimum temperature of 30°C and a water-saturated atmosphere are optimum conditions for infection. Sudangrass, pearl millet, and several other grasses are also hosts.

DISTRIBUTION: India and the southern United States.

SYMPTOMS: Mature plants or older leaves are most likely to become infected. Initially, small pale yellow, elongated spots 1–2 mm by 2–5 mm occur on leaves. The spots gradually elongate to form longitudinal stripes between veins. In severe cases, lesions coalesce and may extend across the veins. Older lesions reach a size of about 2–3 mm by 2–40 mm and become light tan to cream with a light brown margin.

A stalk rot caused by *H. rostratum* has been reported from Florida. Typical stalk rot symptoms occurred in nodal and internodal tissue.

CONTROL: Plant resistant hybrids.

Helminthosporium Leaf Spot

CAUSE: *Helminthosporium carbonum* Ullstrup (syn. *Bipolaris zeicola* (Stout) Shoemaker and *Drechslera zeicola* (Stout) Subram. & Jain.). The fungus overwinters as mycelium and conidia in infected residue. The teleomorph *Cochliobolus carbonum* Nelson produces perithecia in pure culture but it has not been found in nature. Conidia are produced in spring from overwintered mycelium during moderate temperatures and high humidity and are windborne or splashed to lower leaves of corn. Wet weather favors the production and spread of spores and subsequent infection. At least three physiologic races of *H. carbonum* are known.

DISTRIBUTION: Eastern and midwestern United States.

SYMPTOMS: Races 1 and 2 are described here. Race 3 causes northern leaf spot. See Northern Leaf Spot for description of Race 3. Race 1 is highly virulent and infects only a few inbred lines. Race 2 is much less virulent and shows no distinct host specialization. The disease does not occur on commercial single- or double-cross hybrids.

Race 1 symptoms on the leaves are tan, oval to circular lesions that vary from small spots to 1 cm by 3 cm. A pattern of concentric zones is often evident in the lesions. Under optimum disease conditions, lesions may become numerous. Ears are also infected and have a black mold growing over the kernels, giving the ear a charred appearance.

Race 2 rarely infects leaves. When it does, oblong chocolate-colored spots are produced that range in size from small spots to 0.5 cm by 2.5 cm. The ear rot caused by Race 2 is indistinguishable from Race 1.

H. carbonum also causes a stalk rot characterized by a rapid and severe necrosis of the nodes and internode pith.

CONTROL: Plant resistant hybrids.

Hyalothyridium Leaf Spot

CAUSE: *Hyalothyridium* sp. causes disease under humid conditions.

DISTRIBUTION: Columbia, Costa Rica, and Mexico.

SYMPTOMS: Spots on leaves range from small, tan, elliptical spots with brown borders to large (1. 5 cm in diameter), tan, circular blotches that cover much of the leaf surface. Pycnidia are found in both types of lesions. In some areas only the smaller leaf spots are found.

CONTROL: No control is reported.

Java Downy Mildew

CAUSE: *Peronosclerospora maydis* (Racib.) C. G. Shaw (syn. *Sclerospora maydis* (Racib.) Butl.) overseasons in infected corn plants. Conidia are windborne to young plants and germinate under conditions of darkness, free moisture, and temperature below 24°C. Germ tubes of conidia enter plant through stomata. Conidia are produced on wet leaves in the dark. The fungus is seedborne as internal mycelium but this is not thought to be an important source of inoculum. Late planted corn is most vulnerable to infection early in the rainy season, particularly if it follows corn or sugarcane and is overfertilized with nitrogen.

DISTRIBUTION: Australia and Indonesia.

SYMPTOMS: Leaves of plants younger than four weeks old initially have a white to chlorotic streaking. Later the streaks become necrotic. Systemic infection results in chlorosis of upper leaves. Downy growth consisting of conidiophores and conidia in chlorotic streaks is common. Plants may be stunted, sterile, and lodge.

CONTROL:
1. Plant resistant hybrids and varieties.
2. Foliar fungicides were effective under greenhouse conditions.

Late Wilt

CAUSE: *Cephalosporium maydis* Samra, Sabet & Hingorani is generally soilborne and infects plants through roots and mesocotyl. The fungus initially develops slowly in the xylem tissue and rapidly when tassels emerge. The fungus is found in seed, but is evidently not transmitted by this mode. Disease development is favored by light sandy and heavy clay soils. Seedlings are more susceptible than older plants.

DISTRIBUTION: Egypt and India.

SYMPTOMS: Leaves wilt at tasseling, turning a dull green and then drying up. Vascular bundles are discolored. Lower portions of the stalk eventually dry up and become hollow and shrunken. Secondary organisms may cause a wet rot.

CONTROL:
1. Plant resistant hybrids.
2. Rotate corn with other crops.
3. Balance soil fertility.

Leptosphaeria Leaf Spot

CAUSE: *Leptosphaeria* sp.

DISTRIBUTION: High humid areas of the Himalayas and the United States.

SYMPTOMS: The disease is most obvious on lower leaves at silking and tasseling time. Small, light tan lesions form that later enlarge to become either concentric-shaped or streaked. The later symptoms bear little resemblance to each other and suggest symptoms of two separate diseases.

CONTROL: Differences in susceptibility exist among varieties.

Nigrospora Ear Rot

This disease is also called cob rot.

CAUSE: *Nigrospora oryzae* (Berk. & Br.) Petch is discussed under Nigrospora stalk rot. *Nigrospora oryzae* is a weak parasite. Corn that has been killed or in which normal development has been arrested because of stalk rot, leaf blights, cold weather, or root injury is susceptible to ear rot. Corn grown on poor soil appears more susceptible than corn grown in fertile soil. Infection usually starts at the butt end of the ear and sometimes at the tip.

DISTRIBUTION: Generally distributed wherever corn is grown.

SYMPTOMS: Infected ears are usually lightweight but do not show symptoms until ears are harvested. There is no visible mycelium growth between or over kernels. The cobs are usually soft and easily broken. Kernels are loose on the cob and are bleached or have whitish streaks starting at the tips and extending toward the crowns. The cob appears shredded, especially at the butt end, when ears are picked mechanically or later when ears are shelled.

Kernels may show gray mycelium and are covered with small, round, black dots. The dots are groups of spores, and when viewed under magnification, individual spores appear the size of pinpoints.

CONTROL:
1. Grow full-season, adapted hybrids that are resistant to stalk rots and leaf blights.
2. Soils must be well fertilized.
3. Apply fungicides as a seed treatment.

Nigrospora Stalk Rot

CAUSE: *Nigrospora oryzae* (Berk. & Br.) Petch (syn. *Basisporium gallarum* Moll); teleomorph is *Khuskia oryzae* Hudson. The fungus overwinters possibly as mycelium in plant debris in or on the soil. There is some evidence to suggest it is seedborne. Plants are infected that have been weakened or predisposed by other factors such as poor fertility, diseases, or injury. Conidia are formed on mycelium and are windborne to plants where they invade stalks through stomates. Plants nearing maturity are infected.

DISTRIBUTION: Generally distributed wherever corn is grown but it is not a serious disease.

SYMPTOMS: Superficial lesions form on lower internodes. Lesions are dark gray to black but have a slight bluish cast. Discoloration may appear as many separate, tiny blotches or larger, irregular blotches.

CONTROL: No control is necessary as healthy plants growing in well-fertilized soil are seldom infected.

Northern Corn Leaf Blight

This disease is also called Northern blight, Northern leaf blight, and Turcicum leaf blight.

CAUSE: *Exserohilum turcicum* (Pass.) Leonard & Suggs., (syn. *Bipolaris turcica* (Pass.) Shoemaker, *Drechslera turcica* (Pass.) Subram. & Jain, and *Helminthosporium turcicum* Pass.). Teleomorph is *Setosphaeria turcica* (Lutt.) Leonard & Suggs., (syn. *Trichometasphaeria turcica* Lutt.). The fungus overwinters as mycelium and conidia that have converted to chlamydospores in infected leaves, primarily the midrib and leaf sheath. In Israel *E. turcicum* also overwinters on sorghum plants. Successive nights with low temperatures of about 10°C induce formation of chlamydospores. The perfect stage, *T. turcica*, produces perithecia but it is rarely found in nature.

In the spring, conidia are formed from mycelium in residue and are windborne for long distances or splashed by rain to lower leaves of growing corn. Infection occurs when free water is present on leaf surfaces and the temperature is between 18°C and 27°C. Infection is retarded by dry weather. Successive spores are produced on leaf lesions and are spread to higher leaves. Sporulation is optimum during days following warm nights with 10–12 hours of relative humidity over 90 percent. Under ideal disease conditions, plants may be killed. Three physiologic races of the fungus are known to exist. Sorghum, Sudangrass, Johnsongrass, garnagrass, and teosinte are also infected.

DISTRIBUTION: Wherever corn is grown in a humid area. In the United States, the disease is frequently found in the eastern half of the corn belt extending eastward and southward.

SYMPTOMS: General leaf characteristic symptoms are elliptical, gray-green to tan lesions that are normally 5–10 cm long and 1.3 cm wide. Occasionally lesions may grow to 15 cm long and 3.8 cm wide. During damp weather, dark olive to black conidia are produced on surfaces of infected leaves, often forming in concentric zones, giving a targetlike appearance to the lesion.

When the HT 2 gene is present, the lesions will vary in appearance. On some hybrids a long, narrow, chlorotic streak extends the length of the leaf and may be confused with the symptoms of Stewart's wilt. Compared to a susceptible reaction, monogenic resistance is generally characterized by chlorotic lesions with little or no sporulation and polygenic resistance is characterized by fewer and smaller lesions. The lesions first appear on lower leaves, and as the season progresses, nearly all leaves of a susceptible plant are covered with lesions, giving the plant the appearance of having been injured by freezing temperatures. Husks can be infected but ears and kernels are not.

CONTROL:
1. Plant resistant hybrids. Resistance is of two types: monogenic and polygenic. Monogenic resistance genes are HT 1, HT 2, and HT 3. Another type of single-gene resistance is conditioned by the gene HtN, which mainly prevents formation of lesions on adult plants.
2. Apply a foliar fungicide before disease becomes severe.
3. Plow under infected residue to reduce inoculum.

Northern Leaf Spot

This disease is also called Helminthosporium leaf blight.

CAUSE: *Helminthosporium carbonum* Ullstrop (syn. *Bipolaris zeicola* (Stout) Shoemaker and *Drechslera zeicola* (Stout) Subram. & Jain.). The teleomorph is *Cochliobolus carbonum* Nelson. This disease is caused by Race 3 of *H. carbonum*. See Helminthosporium leaf spot.

DISTRIBUTION: North central United States.

SYMPTOMS: Lesions appear on leaf blades, sheaths, husks, and ears. Lesions are gray-tan and surrounded by a light to dark border. They are narrow and linear, ranging in size from 0.5–2 mm wide by 15–20 mm long. Ears are infected from the tip, butt, or through the sides of the husk. A black mold develops over the kernels.

CONTROL: Differences in resistance exist among inbreds. Some are either highly resistant or susceptible, with many inbreds being intermediate in reaction. Inbreds are usually more susceptible than hybrids.

Penicillium Rot

This disease is also called blue-eye.

CAUSE: *Penicillium oxalicum* Currie & Thom occurs on ears that have been injured by corn ear worms or other causes. Other species of Penicillium may also be involved. *Penicillium* spp. are soil-inhabiting fungi.

DISTRIBUTION: Generally distributed wherever corn is grown.

SYMPTOMS: Rot is most likely to occur on the tip ends but may be found on other parts of the ears. Kernels infected with *P. oxalicum* will have a grayish blue-green mold while other species of Penicillium may cause a brighter blue-green mold.

Penicillium spp. may also cause the storage rot known as blue-eye. The fungi may enter kernels through uninjured pericarps, even at a moisture content as low as 14 percent. The fungi grow in the germ and sporulate beneath the pericarp covering the germ area, giving this area a blue-green color.

Penicillium spp. may invade seed embryos and cause seedling blight.

CONTROL:
1. Grain should be dried to 14 percent moisture content or less after harvest.
2. Some genotypes have shown resistance.

Phaeocytostroma Stalk Infection

CAUSE: *Phaeocytostroma ambiguum* (Mont.) Petr. (syn. *Phaeocytosporella zeae* Stout) is a fungus that probably overwinters as pycnidia in infected residue.

DISTRIBUTION: France and Illinois.

SYMPTOMS: Light tan oblong blotches appear on maturing stalks near the soil line. Elongated pycnidia appear within lesions and have a short to long slit as an opening.

CONTROL: The disease is not considered important; however, fungicide seed treatments reduce infections.

Phaeosphaeria Leaf Spot

CAUSE: *Phaeosphaeria maydis* (P. Henn.) Rane, Payak & Renfro (syn. *Sphaerulina maydis* P. Henn.). The fungus survives in corn residue. Disease is favored by high rainfall and low night temperatures.

DISTRIBUTION: Brazil, Colombia, Costa Rica, Equador, India, and Mexico.

SYMPTOMS: Initially small (0.3–2.0 cm), round to oblong, light green or chlorotic lesions form on leaves. Later the lesions will be distributed randomly over a leaf and become white to light tan with dark brown margins. Spots may coalesce, forming larger irregular-shaped areas with perithecia, and infrequently, pycnidia develop in them.

CONTROL: No control is practiced.

Philippine Downy Mildew

CAUSE: *Peronosclerospora philippinensis* (Weston) C. G. Shaw (syn. *Sclerospora philippinensis* Weston) conidia are dispersed by wind and splashing water to young plants. Leaves are penetrated by germ tubes growing into stomata. The fungus grows into the stem where it becomes established in the shoot apex. The fungus is also seedborne. Night temperatures of 19–26°C and free moisture are essential for conidial production, germination, and infection. Other hosts include Johnsongrass, sorghum, sugarcane, and teosinte.

DISTRIBUTION: Africa, India, Indonesia, Nepal, Philippines, and Thailand.

SYMPTOMS: The first true leaf may show systemic symptoms of chlorosis or chlorotic stripes. Generally, long chlorotic streaks with downy growth consisting of conidia and conidiophores occur on leaves. Ears may abort, resulting in partial or complete sterility and tassels may be malformed, resulting in less pollen. Early infection results in stunted and dead plants.

CONTROL:
1. Plant resistant hybrids and varieties.
2. Rogue and destroy infected plants.
3. Apply systemic fungicide seed treatment and foliar spray.

Phomopsis Seed Rot

CAUSE: *Phomopsis* sp.

DISTRIBUTION: Kenya and the United States.

SYMPTOMS: Dark, ostiolate, globose, and erumpent pycnidia are present on seed. A thin, smooth, creamy white cirrhus is produced from the pycnidia. A loss in germination occurs.

CONTROL: No control is reported.

Physalospora Ear Rot

This disease is also called Botryosphaeria ear rot.

CAUSE: *Botryosphaeria festucae* Lib. von Arx & Muller (syn. *Physalospora zeicola* Ell. & Ev.); the teleomorph is *Diplodia frumenti* Ell. & Ev. Overseasoning likely occurs by perithecia and pycnidia in infested residue. Disease is favored by warm, humid weather.

DISTRIBUTION: Brazil and the United States.

SYMPTOMS: Partially infected ears have a few blackened kernels near the base. Severely infected ears have a dark brown to black, felt-like mold covering the whole ear.

CONTROL: No control is reported.

Physoderma Brown Spot

This disease is also called brown spot and Physoderma disease.

CAUSE: *Physoderma maydis* Miyabe (syn. *P. zeae-maydis* Shaw) is a fungus that overwinters as a thick-walled brown sporangia within corn debris or soil. Sporangia are released from infection pustules, disintegrating corn material, and soil and are disseminated by wind, water, insects, and machinery to susceptible plants. Corn becomes increasingly susceptible to *P. maydis* until about 45–50 days old and thereafter declines in susceptibility.

Moisture must be present in the whorl or behind the leaf sheaths together with relatively high temperatures, 22–32°C optimum, in order for *P. maydis* to germinate. A sporangium germinates by a lid or cap opening on one side to release a vesicle that ruptures to release 20–50 zoospores. The zoospores swim for one or two hours, then become amoebalike and penetrate young meristematic tissue with infection hyphae. The resulting mycelium enters mesophyl or parenchyma cells and forms vegetative structures from which sporangia develop in 16–20 days. Secondary spread may be by zoospores produced in thin-walled sporangia. Infection commonly occurs in a diurnal cycle, resulting in alternating bands of infected and healthy tissue as it emerges from the whorl. Disease development is more severe on plants planted in overwintered residue and is favored by warm, wet weather or areas of abundant rainfall and high mean temperatures.

DISTRIBUTION: Africa, Asia, Australia, Central America, and southeastern and midwestern United States.

SYMPTOMS: Initially, lesions occur mainly below the ear on the leaf blades, leaf sheaths, and stalks. Lesions first appear near the base of the leaf blade as small, round, yellowish spots. The lesions may occur in bands across the leaf blade with infected tissues turning chocolate brown to red-brown and merging to form large, irregular angular blotches. Cells of infected corn tissue disintegrate to expose dusty pustules or brown blisters containing large numbers of microscopic light to dark brown sporangia. Infection at nodes beneath the leaf sheaths and premature death of plants due to leaf blighting frequently cause stalk rot and lodging.

CONTROL:
1. Plant resistant hybrids.
2. Do not plant known susceptible hybrids in river bottom soils or other high-humidity locations.
3. Shred infected corn residue in the autumn and plow it under where feasible.
4. Apply foliar fungicides.

Physoderma Stalk Rot

CAUSE: *Physoderma maydis* Miyabe. See Physoderma brown spot.

DISTRIBUTION: See Physoderma brown spot.

SYMPTOMS: Water-soaked lesions are produced on stalk beneath the leaf sheath. Later lesions may coalesce to form brown blotches. Small pockets of brown sporangia may form in the stalk.

CONTROL: See Physoderma brown spot.

Pyrenochaeta Stalk Infection

CAUSE: *Pyrenochaeta terrestris* (Hans.) Gorenz, Walker & Larson, probably survives as pycnidia in infected residue in the soil. Plants are infected before pollination.

DISTRIBUTION: France and the United States; however, it is probably widely distributed.

SYMPTOMS: Only stalks below the soil surface are infected, where dark brown shallow blotches appear on stalks. Later a pink discoloration becomes evident in the outer rind that blends into the dark brown area.

CONTROL: No control is necessary since the disease is not important.

Pythium Seed and Seedling Blight

CAUSE: Several *Pythium* spp. including *P. debaryanum* Hesse, *P. irregulare* Buis., *P. paroecandrum* Drechs., *P. rostratum* Butl., *P. splendens* Braun, *P. ultimum* Trow., and *P. vexans* deBary. The fungi survive either as oospores, sporangia, or mycelium within infected residue.

During cool (10–13°C), wet soil conditions, sporangia will either germinate directly to form a germ tube or will germinate indirectly to form a vesicle in which zoospores are formed. Similarly, oospores will either germinate directly to form a germ tube or germinate indirectly to form a zoosporangium in which zoospores are formed. The zoospores swim through saturated soil by means of flagella to seeds or roots. Once in contact with the plant, the zoospores encyst and germinate to produce a germ tube that penetrates the plant. As mycelium grow through an infected plant, sporangia and oospores are produced on and within the plant.

DISTRIBUTION: Generally distributed wherever corn is grown.

SYMPTOMS: Disease is more severe on wet soils in low-lying areas in a field, soils high in clay and organic matter, and soils that have been compacted or remain wet for an extended period of time. Such soils also remain cool. Under these conditions germination is slow and seed is more susceptible.

Seeds may be rotted before germination. Seedlings may be infected either before or after emergence.

Seeds are brown, soft, usually very mushy and overgrown with mycelium. It is often difficult to find infected seeds in the soil because such seeds decompose very rapidly once infected. Rotted seed is usually covered with adhering soil that is bound by mycelium, making seed indistinguishable from surrounding soil.

Aboveground symptoms are a yellowing, wilting, and death of leaves. Lesions on the mesocotyl are initially brown and sunken. Eventually the mesocotyl becomes soft and water-soaked. Roots are water-soaked and brown with a sloughed appearance.

CONTROL:
1. Treat seed with a seed-protectant fungicide. However, fungicides are not effective after seed has germinated.
2. Seed stored two or more years at 15.5–35°C is more susceptible.
3. Mechanically damaged seed that has cracks or splits in the seed coat is more susceptible to damping off, since such cracks afford an entrance to fungi.
4. In general, sweet corn is more susceptible than dent corn. Popcorn is the most resistant.
5. Delaying planting until soil has become warm, above 13°C, minimizes hazards of poor stands. At warmer temperatures germi-

nation of the corn seed is rapid and the seedling becomes established, escaping infection by *Pythium* spp.

Pythium Root Rot

CAUSE: *Pythium graminicola* Subr. is the predominant causal fungus and probably survives in residue and soil as oospores and sporangia; however, some isolates apparently do not form oospores. Infection is either by direct germination of oospores and sporangia to form a germ tube or indirectly by zoospores formed in either structure and released into saturated soil. The zoospores swim to root surfaces where they encyst and germinate, forming a germ tube.

Pythium torulosum Coker & Patterson and *P. dissotocum* Drechs. are also associated to a lesser degree with Pythium root rot. Disease severity and recovery of fungi from diseased roots is greatest in poorly drained, wet soils. Pythium root rot is most prevalent during early spring and again in late summer.

DISTRIBUTION: Generally distributed wherever corn is grown.

SYMPTOMS: Secondary roots of young corn plants have scattered yellow-brown lesions. Primary roots become black and necrotic. Eventually the lesions coalesce and become dark brown to black, resulting in dark streaks and girdling of roots. Sometimes new roots form above a lesion. Similar lesions will be formed on new roots later in the growing season under wet soil conditions.

Stunting and yellowing of lower leaves occurs on young plants. Older plants may be generally stunted and also display yellowing, rolling, and wilting of leaves.

CONTROL:
1. Ensure soils are adequately drained.
2. Treat seed with a seed treatment fungicide.

Pythium Stalk Rot

CAUSE: *Pythium aphanidermatum* (Eds.) Fitz. (syn. *P. butleri* Subr.) survives as oospores, sporangia, or mycelium within infected residue in the soil. During hot (32°C or higher) and very humid weather, sporangia germinate either directly to form a germ tube or indirectly by releasing zoospores. Oospores likewise may germinate either directly to form a germ tube or indirectly to form a zoosporangium in which zoospores are formed. The zoospores swim through water and eventually form a germ tube that penetrates the plant.

DISTRIBUTION: Generally distributed wherever corn is grown under hot, humid conditions.

SYMPTOMS: Pythium stalk rot can occur any time during the growing season but is more likely to occur shortly before or after corn has tasseled. It is more likely to occur where air and soil drainage is poor. The disease is first recognized when plants fall over. The rot is usually confined to a single internode just above the soil line. The diseased area is brown, water-soaked, and soft. Stalks may be twisted and distorted but are not broken off. The plants remain green and turgid for several weeks because the vascular bundles remain intact. As the plants fall, they usually twist at the infected node.

CONTROL: Plant resistant hybrids.

Red Kernel Disease

CAUSE: *Epicoccum nigrum* Link. is a weak parasite that infects injured or weakened kernels.

DISTRIBUTION: Africa, Asia, Europe, and the United States.

SYMPTOMS: Red discoloration involving the endosperm of sweet corn.

CONTROL: No control is reported.

Rhizoctonia Ear Rot

CAUSE: *Rhizoctonia zeae* Voorhees overseasons as mycelium and sclerotia in kernels, soil, and on and in residue. Sclerotia are the main means of dissemination. Disease is favored by warm, humid weather.

DISTRIBUTION: Asia, Europe, North America, and South America.

SYMPTOMS: Initially a pink mold growth occurs on the ear. Later infected ears become dull gray. Numerous white to salmon to dark brown or black sclerotia develop on the outer husks.

CONTROL: No control is reported.

Rhizoctonia Root Rot

CAUSE: *Rhizoctonia solani* Kuhn and *R. zeae* Voorhees.

DISTRIBUTION: Widely distributed wherever corn is growing.

SYMPTOMS: Anastomosis group (AG) 2 of *R. solani* causes the most severe symptoms. *Rhizoctonia solani* causes postemergence damping-off, stunting or leaf necrosis, or chlorosis in seedlings. Isolates from other plants cause brown to black lesions on seminal, crown, and brace roots. Frequently roots decay and disintegrate 2–5 cm from the crown.

Rhizoctonia zeae is less virulent than *R. solani* and causes buff to tan lesions with dark brown borders on corn roots.

CONTROL: None practiced.

Sclerotium Ear Rot

CAUSE: *Sclerotium rolfsii* Sacc. Ear rot occurs under extremely wet conditions. The fungus survives in soil.

DISTRIBUTION: India and Pakistan.

SYMPTOMS: Symptoms are the same as for Corticium ear rot except mycelium on kernels is white instead of brown.

CONTROL: Fungicide seed treatments improved emergence in experiments.

Seed and Seedling Blight Other Than Pythium

CAUSE: *Stenocarpella maydis* (Earle) Sutton (syn. *Diplodia maydis* (Berk.) Sacc.), *Gibberella zeae* (Schw.) Petch, *Fusarium moniliforme* Sheldon, *Rhizoctonia solani* Kuhn, and possible other fungi. Most of these fungi are discussed under the stalk rot or ear and kernel blight caused by each. *Rhizoctonia solani* survives as mycelium in plant residue and as sclerotia in soil or residue. Seedling blight generally occurs under wet, cool, soil conditions that do not vary appreciably from those favoring Pythium blight.

DISTRIBUTION: Worldwide.

SYMPTOMS: Symptoms tend to be nondescript and may be lacking in many instances. Failure of seed to germinate, seedling to emerge, killing of seedling after emergence, or short, stunted plants are general symptoms. Seed may be rotted, overgrown with mycelium, and eventually disintegrate. The mesocotyl may be rotted and discolored, cutting off water and nutrients from the seed and primary root. If the adventitious roots from the first node above the mesocotyl are not well developed, the leaves wilt and the plants may die. Infected roots have brown to black, water-soaked flaccid lesions.

CONTROL: Treat seed with a fungicide-seed protectant. Once a seed has germinated, the fungicide does not afford further protection.

Selenophoma Leaf Spot

CAUSE: *Selenophoma* sp. Disease apparently is most likely to occur at lower temperatures (10–18°C).

DISTRIBUTION: Colombia, at an altitude of 2,700 m.

SYMPTOMS: Lesions on leaves are typical eyespots with concentric zonations bordered by a yellow halo. Under humid conditions (greater than 70% relative humidity), lesions extend to a length of 2 cm, eventually coalescing and causing severe leaf necrosis.

CONTROL: No control is reported.

Septoria Leaf Blotch

CAUSE: *Septoria* sp. causes disease under cool, humid conditions.

DISTRIBUTION: Not known for certain but may be common in cool, humid climates.

SYMPTOMS: Small, light green chlorotic or brown lesions are formed on leaves. Lesions coalesce to form large necrotic areas in which numerous pycnidia may be produced.

CONTROL: No control is practiced.

Sorghum Downy Mildew

CAUSE: *Peronosclerospora sorghi* (Weston & Uppal) C. G. Shaw (syn. *Sclerospora sorghi* Weston & Uppal and *S. graminicola* var. *andropogonis-sorghi* Kulk.). The fungus survives for several years as oospores either in infected crop residue or directly in the soil after residue has decomposed. Oospores germinate and infect young plants with mycelium growing systemically inside the plant. However, no infection by oospores occurs if seedlings emerge in cool soil. Conidia are also reported to provide primary inoculum. Conidia are produced on leaf surfaces and are windborne to infect leaves of plants three weeks old or younger. Conidia must germinate in high humidity and at optimum temperatures between 12°C and 20°C for some isolates and between 12°C and 32°C for others. Eventually oospores are formed within infected tissue and provide the means of survival. The fungus is not seedborne since *S. sorghi* can survive only in immature seeds planted soon after harvest. Johnsongrass, sorghum, and teosinte are also hosts.

DISTRIBUTION: Africa, Asia, Central America, Israel, Italy, South America, and the United States, primarily in the south central and midwestern states.

SYMPTOMS: Systemic infection occurs only in young plants. Initially young corn plants have chlorotic to white streaks on the leaves that run parallel to the midrib. Sometimes the entire half of a leaf on one side of the midrib will be chlorotic. The chlorotic area of a leaf includes the base of the leaf. Leaves are often more narrow and erect than those of healthy plants. Downy growth consisting of conidia and conidiophores may appear on both surfaces of infected leaves in chlorotic tissues. Severely infected plants are stunted, generally chlorotic, and may have phylliodied tassels in which normal floral parts are converted to small leaves, resulting in abnormal seed set. Secondary infection may take place on older plants and is indicated by small lesions scattered over the leaves with downy growth present in the lesions.

CONTROL:
1. Plant resistant hybrids.
2. Plow under infected residue.
3. Rogue out and destroy infected plants as they appear in the field.
4. Do not rotate corn with sorghum and vice versa.
5. Treat seed with a fungicide-seed treatment.
6. Plant in cool soil at the onset of the rainy season.

Southern Corn Leaf Blight

This disease is also called maydis leaf blight, southern blight, and southern leaf blight.

CAUSE: *Bipolaris maydis* (Nisik.) Shoemaker (syn. *Drechslera maydis* (Nisik.) Subram. & Jain. and *Helminthosporium maydis* Nisik.) overwinters as conidia and mycelium in plant residue. Race T also survives on kernels in storage. The teleomorph *Cochliobolus heterostrophus* Drechs. forms ascospores within perithecia but has not been found in nature. In the spring, conidia are windborne or splashed to growing plants. Race T produces five times more spores at 22.5°C and twice as many at 30°C as Race O. Infection is optimum when free moisture is present and the temperature is between 15.5°C and 26.5°C. A disease cycle may be completed in 60–72 hours. Production of new spores within lesions provides spores for secondary infection. Infection and subsequent losses may be severe if susceptible or moderately susceptible cultivars are grown in continuous corn culture with minimum tillage under overhead irrigation. Races O and T are probably the most common physiologic races; however, other races of the fungus are also present in some areas. Corn infected with maize dwarf mosaic virus is more susceptible to infection by *B. maydis* than noninfected corn.

DISTRIBUTION: The disease occurs wherever corn is grown; however, Race O is more likely to be present in the southeastern and midwestern United States.

SYMPTOMS: Race O infects only leaves, whereas Race T infects leaves, stalks, leaf sheaths, ear husks, and kernels. Lesions tend to form and expand more rapidly at 30°C than at lower temperatures.

Lesions caused by Race O on leaves are tan, 2–6 mm wide by 3–22 mm long with parallel sides and buff to brown borders. Race T lesions are larger (0.6–1.2 cm wide by 0.6–2.7 cm long), spindle shaped or elliptical, with yellow-green or chlorotic halos surrounding the lesions. Often lesions may have dark red-brown borders. Lesions may merge, blighting and killing the leaves.

Race T produces similar lesions on other plant parts. However, the lesions are usually very large, often being several centimeters in length. The lesions are usually tan with dark purple-brown borders. Mycelium may grow through the husks to the kernels where it will be a gray or a black color due to production of conidia. This gives ears and kernels a moldy or charcoal appearance. Seedlings from infected kernels may wilt and die within a few days after planting.

During warm (20–30°C), damp weather (100% relative humidity is optimum), Race T produces large numbers of spores giving lesions a black velvet appearance. Severely infected plants are predisposed to stalk rot.

CONTROL:
1. Plant resistant hybrids. Genetic and cytoplasmic sources resistant to Races O, T, and other races are available.
2. Apply foliar fungicides before disease becomes too severe and when weather conditions are conducive to disease development.
3. Plow under infected residue to reduce inoculum.

Southern Rust

This disease is also called Polysora rust.

CAUSE: *Puccinia polysora* Underw. is a rust fungus that produces the urediospore stage and rarely the teliospore stage. No alternate host is known. Urediospores constitute both primary and secondary inoculum. Disease development is favored by high temperatures (27°C) and relative humidity. Dew is always necessary, with 16 hours of dew at 26°C optimum for infection. Several races of *P. polysora* are known to exist.

DISTRIBUTION: Africa, southeast Asia, Central and South America, the West Indies, and the United States, primarily the southeastern states but occasionally the Midwest.

SYMPTOMS: Pustules occur on both leaf surfaces and resemble common rust pustules except they are smaller (0.2–2.0 mm long). Pustules are a light cinnamon brown, circular to oval in shape and densely scattered over the upper leaf surface. Pustule development on lower leaf surfaces is slower and less abundant than on upper leaf surfaces. The epidermis remains intact over the pustules for a longer time than common rust but eventually ruptures. Telia are chocolate brown to black, circular to elongate (0.2-0.5 mm in diameter), and often form a ring around the initial uredium. The epidermis over the telia will eventually rupture but it also remains intact longer than common rust. Severely infected leaves may become chlorotic and eventually dry up. Under severe conditions, pustules may occur on leaf sheaths and stalks.

CONTROL:
1. Plant resistant hybrids.
2. Apply foliar fungicides.

Spontaneum Downy Mildew

CAUSE: *Peronosclerospora spontanea* (Weston) C. G. Shaw (syn. *Sclerospora spontanea* Weston).

DISTRIBUTION: Philippines and Thailand.

SYMPTOMS: Symptoms are similar to those of Philippine downy mildew. Plants are generally stunted and chlorotic. During humid conditions at night, leaves are covered with a white fuzz consisting of conidia and conidiophores.

CONTROL: Similar to Philippine downy mildew.

Sugarcane Downy Mildew

CAUSE: *Peronosclerospora sacchari* (T. Miyake) C. G. Shaw (syn. *Sclerospora sacchari* T. Miyake) survives as mycelium in sugarcane.

Sporangia are produced at an optimum temperature between 20°C and 25°C at night in the presence of free water and are windborne to corn.

Sporangia germinate in the presence of free water and form mycelium that penetrate through stomata in plants less than one month old. Older plants are resistant. Mycelium then develops systemically. Sporangia are eventually produced and serve as secondary inoculum. Oospores are also produced but their function is not known. Other hosts are broomcorn, gamagrass, grain sorghum, sugarcane, and teosinte.

DISTRIBUTION: Australia and Southeast Asia.

SYMPTOMS: Initially small, round chlorotic spots appear on leaves. Eventually the systemic infection occurs as yellow to white stripes or streaks at base of third to sixth older leaves that may extend the length of the leaf. The streaks may be discontinuous in some hybrids or varieties. On later or mildly infected plants the streaks may disappear. Downy masses of sporangia appear at night on both surfaces of leaves, leaf sheaths, and husks during periods of high humidity and an optimum temperature of 25°C. Plants, especially tassels, are distorted with poorly filled ear and elongated ear shanks.

CONTROL:
1. Do not grow corn near sugarcane.
2. Plant resistant hybrids and varieties.

Tar Spot

CAUSE: *Phyllachora maydis* Maubl. is an obligate parasite. Dissemination is likely by windborne ascospores. Disease is favored by relatively cool, humid weather.

DISTRIBUTION: France, Central and South America.

SYMPTOMS: Disease becomes most severe after pollination and may cause premature dessication of the spike. Initially, light brown oval to circular lesions form (0.5–2.0 mm in diameter). These may be surrounded by a dark brown border. These may coalesce to form longer lesions up to 10 mm in length. Glossy, black, sunken spots frequently occur in the middle of a lesion but are also abundant outside a lesion, particularly when plants are heavily infected.

CONTROL: Differences in resistance exist between genotypes.

Tropical Rust

CAUSE: *Physopella zeae* (Mains) Cummins & Ramachar (syn. *Angiospora zeae* Mains) is a rust fungus that produces the urediospore and teliospore stages on corn. No alternate host is known. Urediospores constitute both primary and secondary inoculum. Disease development is favored by warm to hot and humid weather. At least two races of *P. zeae* are known. Teosinte is another host.

DISTRIBUTION: Central and South America and the Caribbean.

SYMPTOMS: Pustules occur in small groups generally 0.3–1.0 mm long on upper leaf surfaces. They are yellow or cream colored and covered by the epidermis except for a small pore or a split. Later the pustules develop into purple blotches that are circular to oblong (0.6 cm in diameter) with creamy centers. Dark brown to black telia develop in groups around uredia.

CONTROL: Plant resistant hybrids.

Unidentified Cercospora Leafspot

CAUSE: *Cercospora* sp. Possibly survives in residue.

DISTRIBUTION: Pennsylvania.

SYMPTOMS: Greenhouse inoculations cause pale tan lesions on leaves. Lesions are limited in lateral spread by major veins but ends are more irregular in shape than the rectangular lesions caused by *C. zeae-maydis*.

CONTROL: No control is reported.

Yellow Leaf Blight

This disease is also called Phyllosticta leaf spot.

CAUSE: *Phyllosticta maydis* Arny & Nelson, teleomorph *Mycosphaerella zeae-maydis* Mukunya & Boothroyd, overwinters as pycnidia or pseudothecia within infected corn or grass residue. Under cool, wet conditions, primarily in the spring, conidia ooze from pycnidia and are then windborne for a short distance or splashed to corn leaves; spread is also by infected leaves rubbing against healthy ones. Ascospores are produced that are also windborne to leaves. Young plants growing through residue on the soil surface are especially vulnerable. However, susceptible plants may be infected at any stage of growth. Conidia developing in pycnidia on large and older lesions provide secondary inoculum. Sudangrass and foxtails are also susceptible.

DISTRIBUTION: Africa, Argentina, Asia, Brazil, Canada, Romania, Taiwan, and the United States, primarily in northern corn-growing areas. However, the disease is also present in the Midwest and in southern states.

SYMPTOMS: Since *P. maydis* is not spread rapidly over a large area, and is usually restricted to isolated fields, a severely infected field may be found adjacent to one relatively free of the disease. Plants growing in fields containing residue from last year's crop may be most vulnerable to disease since residue contains the pathogen.

Gray to tan or brown lesions develop first on the lower leaves and average 0.3 cm wide by 1.3 cm long. The lesions are oval to elliptical in shape, often surrounded by a narrow red or purple margin with a wide yellow-green halo. The yellow discoloration of the leaf tissue surrounding the lesions gives the disease its name. Leaf symptoms may be difficult to distinguish from those of southern leaf blight caused by *Bipolaris maydis* (Nisik.) Shoemaker. Leaf sheaths and outer husks are also susceptible. Lesions on the older leaves may contain pycnidia that resemble tiny, brown-black specks. The pycnidia can be easily observed using magnification and will appear as small flask-shaped structures.

Severe infection results in yellowed leaves similar to those with nitrogen deficiency. Leaves eventually die and plants may be stunted. Infected plants may also get stalk rot.

CONTROL:
1. Plant the most resistant hybrids. Hybrids containing T cytoplasm are more susceptible than other cytoplasms.
2. Apply a foliar fungicide if economic and disease conditions warrant it.
3. Plow under infected residue.
4. Rotate corn with other crops.

Zonate Leaf Spot

CAUSE: *Gloeocercospora sorghi* D. Bain & Edg. survives as sclerotia and mycelium in infected residue. Conidia are produced from sporodochia in a slimy matrix and are primarily disseminated to plants by splashing rain. Sclerotia develop within diseased tissue. Bentgrass, Johnsongrass, sudangrass, and sugarcane are also hosts.

DISTRIBUTION: Africa, Central and South America, and the United States, mostly the southern corn-growing areas.

SYMPTOMS: Lesions on leaves are initially water-soaked and then become red-brown. They enlarge up to 5 cm in diameter forming a targetlike or zonate pattern. Small, black sclerotia 0.1–0.2 mm in diameter may develop within dead tissue.

CONTROL: The disease is not considered important enough to warrant control measures.

MYCOPLASMAL AND SPIROPLASMAL CAUSES

Corn Stunt

This disease is also called achaparramiento, maize stunt, and Rio Grande maize stunt.

CAUSE: Corn stunt is caused by a spiroplasma. It is not known how the organism survives in the absence of corn. In areas where corn is grown continuously, it is transmitted by the leafhoppers, *Dalbulus maidis* DeLong & Wolcott, *D. elimatus* Ball, *Graminella nigrifrons* (Forbes), and *Baldulus tripsaci* Kramer & Whitcomb, from maturing fields to newly planted fields. The organism is not seedborne or mechanically transmitted.

DISTRIBUTION: El Salvador, Jamaica, Mexico, Peru, United States, and Venezuela.

SYMPTOMS: Initially, a chlorosis occurs on the margin of the whorl leaf followed by a reddening of the tops of older leaves. Small, circular to elongated chlorotic spots then develop at the base of leaves of young plants. The spots often coalesce and become elongated stripes that may or may not have a distinct margin. Plants are stunted and bear numerous small ear shoots and suckers. In severe infections there is a proliferation of roots.

CONTROL: The disease is not currently considered important enough to warrant control measures.

Maize Bushy Stunt

Mesa Central corn (maize) stunt is also thought to be caused by the maize bushy stunt mycoplasmalike organism (MLO).

CAUSE: A MLO that is found only in the phloem of infected plants. It is not known how the MLO survives in the absence of corn. The MOL is transmitted by several leafhoppers but only *Dalbulus maidis* DeLong & Wolcott and *D. elimatus* Ball are efficient vectors. Corn and annual teosinte are the only hosts for the causal organisms.

DISTRIBUTION: Mexico, Central and South America, and the southern United States.

SYMPTOMS: Initially, a chlorosis of the whorl leaf margin occurs followed by a reddening of the tips of older leaves. The reddening may not occur with certain hybrids. Subsequent leaves develop a chlorosis of the margins, turn yellow or red, tear, twist, and are shortened. Plants have a bushy appearance due to numerous tillers that develop at the leaf axils and base. Plants are stunted and have numerous small ears.

CONTROL: Tetracycline inhibits symptom development.

Unnamed Disease

CAUSE: Spiroplasma. Disease occurs in late planted corn.

DISTRIBUTION: California.

SYMPTOMS: Bright red leaf blades at harvest time.

CONTROL: No control is reported.

CAUSED BY NEMATODES

Several kinds of nematodes attack corn; however, in most cases distribution is not well known. Additionally, controls tend to be general for all nematode-caused diseases. Following are practices that tend to reduce effects of nematodes:

1. Fertilize soils according to soil tests. Plants suffering from nutrient deficiency are more susceptible to injury.
2. Maintain good weed control. Weeds may act as reservoirs for the present crop or next year's crop.
3. Rotate perennial trouble spots to a crop other than corn. While crop rotation is usually advisable, little is known about the suceptibility of other hosts.
4. Apply nematicides to severely affected areas.

Awl Nematodes

CAUSE: *Dolichodorus* spp. are ectoparasites that are limited to wet soils in the southeastern United States. Some dissemination occurs by water. Feeding is at the root tip.

SYMPTOMS: The aboveground symptom is plant stunting. Belowground symptoms are stubby and thick roots with some lesions present.

Burrowing Nematodes

CAUSE: *Radopholus similis* Thorne is an endoparasite found in tropical and subtropical areas of the world. Survival does not occur for more than a few months outside of host tissue. Feeding occurs in cortex and reproduction occurs within roots. Some dissemination occurs by water.

SYMPTOMS: The aboveground symptom is slight stunting. Belowground symptoms are root lesions and decay.

Corn Cyst Nematode

CAUSE: *Heterodera zeae* n. sp.

DISTRIBUTION: Egypt, India, Pakistan, and Maryland in the United States.

SYMPTOMS: Aboveground symptom is stunted plants. Belowground symptom is white to tan cysts on roots.

Dagger Nematodes

CAUSE: *Xiphenema* spp. are ectoparasites. A species of *Xiphenema* reported to be pathogenic to corn is *X. mediterraneum* Martelli & Lamberti.

SYMPTOMS: Aboveground symptoms are slight to moderate stunting. Belowground symptoms are reduction in the feed roots accompanied by necrosis of a portion of root system.

Lance Nematodes

CAUSE: *Hoplolaimus* spp. are primarily endoparasites but sometimes feed as semiendoparasites and ectoparasites. A specific species reported to be pathogenic to corn is *H. galeatus* (Cobb) Thorne.

SYMPTOMS: Aboveground symptoms are stunting and chlorosis. Belowground symptoms are lesions that reduce root growth.

Needle Nematodes

CAUSE: *Longidorous* spp. are ectoparasites that reach their largest population primarily in sandy soils in the temperate regions. *Longidorous breviannulatus* n. sp. can potentially cause severe disease in Illinois, Indiana, and Iowa.

SYMPTOMS: Aboveground symptoms are thin stands of severely stunted and chlorotic plants. Occasionally plants may be killed in irregular patches during the first six to eight weeks after planting. Later damaged plants may become as tall as unaffected ones but stalks remain slender. If ears are present they are much reduced in size. Belowground symptoms include yellow discoloration, slightly swollen root tips, stubby roots, pruning of laterals, and scarcity of small feeder roots. When soil moisture is high, seminal roots are destroyed and bushlike crown roots proliferate near the soil surface. The prop root system is unaffected.

Ring Nematodes

CAUSE: *Criconemoides* spp. and *Macroposthonia* spp. are ectoparasites that are commonly found in the southeastern United States. A species of *Criconemoides* reported to be pathogenic to corn is *C. citui* Steiner.

SYMPTOMS: Aboveground symptoms are mild stunting. Below-ground symptoms are lesions on roots with some decay.

Root-knot Nematodes

CAUSE: *Meloidogyne* spp. are endoparasitic nematodes that cause the most damage to corn in the warmer regions of the world. A specific species reported to be pathogenic to corn is *M. incognita* (Kofoid & White) Chitwood. Populations of *M. incognita* were reported higher under minimum tillage in irrigated plots.

SYMPTOMS: Aboveground symptoms are stunting and chlorosis. Belowground symptoms include presence of galls on roots; roots are usually stunted and may be abnormally branched.

Root-lesion Nematodes

CAUSE: *Pratylenchus* spp. are endoparasites that feed and reproduce in the root cortex. Some species thrive in warm, sandy soils whereas others grow and reproduce better in heavy-textured soils in temperate climates. Corn is tolerant to populations of 500 nematodes per gram of dry root tissue if plant growth conditions are favorable. Some specific species reported to be pathogenic to corn are *P. brachyurus* (Godfrey) Filip. & Sch. Stek., *P. hexincisus* Taylor & Jenkins, *P. scribneri* Steiner, *P. thornei* Sher. & Allen, and *P. zeae* Graham.

SYMPTOMS: Aboveground symptoms include plant stunting. Leaves are chlorotic and purple to red discoloration may occur. Roots have reduced growth with few fibrous roots. Brown lesions may be present and a general brown to black root decay may occur.

Spiral Nematodes

CAUSE: *Helicotylenchus* spp. are either ectoparasites, semiendoparasites, or endoparasites that are generally found worldwide.

SYMPTOMS: Aboveground symptom is plant stunting. Belowground symptoms are a reduction in number of feeder roots with some root rot.

Sting Nematodes

CAUSE: *Belonolaimus* spp. are ectoparasites that are found only in sandy soils. Nematodes produce a phytotoxic enzyme. A specific species reported to be pathogenic to corn is *B. longicaudatus* Rau.

SYMPTOMS: Aboveground symptoms are stunting and chlorosis. Belowground symptoms are deep necrotic lesions on roots. Root tips are frequently destroyed, resulting in thick, stubby roots.

Stubby-root Nematodes

CAUSE: *Paratrichodorus* spp. and *Trichodorus* spp. are ectoparasites that are found primarily in sandy soils. Most nematode feeding is around the root tips. A species of Paratrichodorus reported to be pathogenic to corn is *P. minor* (Colbran) Siddgi. A species of Trichodorus reported to be pathogenic to corn is *T. christiei* Allen.

SYMPTOMS: Aboveground symptoms include severe stunting and chlorosis. Belowground symptom is a devitalizing of the root tips, causing stubby lateral roots. Roots sometimes appear thicker than normal.

Stunt Nematodes

CAUSE: *Tylenchorhynchus* spp. are ectoparasites that are generally found worldwide.

SYMPTOMS: Aboveground symptoms are moderate plant stunting and chlorosis. Belowground symptom is a reduction in root growth.

VIRAL CAUSES

American Wheat Striate Mosaic

This disease is also called wheat striate mosaic.

CAUSE: American wheat striate mosaic virus (AWSMV) is transmitted mainly by the painted leafhopper, *Endria inimica* Say, and also by *Elymana virenscens* F.

DISTRIBUTION: North central United States, Canada, and the USSR.

SYMPTOMS: Plants are stunted with slight to moderate leaf striation consisting of thin yellow to white parallel streaks.

CONTROL: No control is reported.

Barley Stripe Mosaic

CAUSE: Barley stripe mosaic virus (BSMV) is transmitted by seed. Corn is more susceptible to systemic infection at 25°C than at higher temperatures.

DISTRIBUTION: Worldwide.

SYMPTOMS: Chlorotic stripes on leaves.

CONTROL: None practiced on corn.

Barley Yellow Dwarf

CAUSE: Barley yellow dwarf virus (BYDV) is transmitted by several aphids. Irrigated corn is a reservoir of BYDV and its aphid vectors during the interim between summer harvest and fall planting of winter grains in eastern Washington. See BYDV in chapter two for further description.

DISTRIBUTION: Africa, Asia, Australia, Europe, New Zealand, and North America on several grass hosts.

SYMPTOMS: Leaves become chlorotic at the tips and margins. Later, beginning with the oldest, leaves become red to purple in color. Little or no stunting occurs.

CONTROL:
1. Avoid planting corn adjacent to or following infected grass hosts.
2. Differences in susceptibility have been reported among inbreds and hybrids.

Brome Mosaic

CAUSE: Brome mosaic virus (BMV) is transmitted by corn rootworms *Diabrotica undecimpunctata* Mannerheim and *D. virgifera* LeConte and nematodes *Longidorus macrosoma* Hooper, *Xiphinema coxi* Tarjan, and *X. diversicaudatum* (Micoletzky) Thorne. Several grasses and some dicotyledons are hosts.

DISTRIBUTION: Worldwide in other hosts but on corn in the United States.

SYMPTOMS: Chlorotic stripes of varying widths occur on young leaves. Occasionally young plants may wilt and die.

CONTROL: Avoid planting corn adjacent to or following infected grasses in a rotation.

Cereal Chlorotic Mottle

CAUSE: Cereal chlorotic mottle virus (CCMV) is transmitted by the insect *Nesoclutha pallida* (Evans). Several other grasses are hosts.

DISTRIBUTION: Queensland, Australia.

SYMPTOMS: Leaves initially have chlorotic stripes followed by a slight chlorotic mottling.

CONTROL: Some hybrids are resistant.

Cereal Tillering Disease

CAUSE: The causal virus is disseminated by *Laodelphax striatellus* Fallen and *Dicranotropis hamata* Boh. Barley, oats, and several grasses are also hosts.

DISTRIBUTION: Sweden.

SYMPTOMS: Infected plants are stunted with numerous shoots. Leaves have vein enations.

CONTROL: No control is reported.

Corn Chlorotic Vein Banding

This disease is also called Brazilian maize mosaic.

CAUSE: Corn chlorotic vein banding virus (CCVBV) is transmitted by the leafhopper *Peregrinus maidis* Ashm. CCVBV may be related to the maize mosaic virus.

DISTRIBUTION: Brazil.

SYMPTOMS: The disease occurs sporadically in a field. Chlorotic vein banding occurs along leaf veins. Plants are usually stunted. The disease is generally not considered serious and significant losses have not occurred.

CONTROL: No control is necessary.

Corn Lethal Necrosis

CAUSE: Corn lethal necrosis is caused by the synergistic action of maize chlorotic mottle virus (MCMV) in combination with either maize dwarf mosaic virus (MDMV) A and B or wheat streak mosaic virus (WSMV).

DISTRIBUTION: Kansas and Nebraska.

SYMPTOMS: It is assumed that corn plants are susceptible at all stages of development. Initially a bright yellow mottling develops in the leaves followed by a necrosis of the leaves inward from the margins and premature death of the plants. Normally, in maturing plants, necrosis begins at the tassel and progresses downward. Eventually mature plants die from the top down. Ears may be small, distorted, and have little or no kernel development. Kernels that are developed are often wrinkled and shriveled.

CONTROL: Grow a nonhost crop such as soybeans, sorghum, small grains, or alfalfa in fields where corn lethal necrosis has occurred. Tolerance has been identified in some hybrids and inbreds.

Cucumber Mosaic

CAUSE: Cucumber mosaic virus (CMV) is disseminated by several aphid species. Other hosts include numerous dicotyledons and monocotyledons.

DISTRIBUTION: CMV is distributed worldwide in several other hosts but on corn only in the United States.

SYMPTOMS: Mild symptoms start out as chlorotic, elliptical spots on leaves that eventually form stripes. Severe symptoms are death of seedlings or stunting with necrotic leaf spots.

CONTROL: No control is reported.

Cynodon Chlorotic Streak

CAUSE: Cynodon chlorotic streak virus (CCSV), a plant rhabdovirus, is transmitted by the planthopper *Toya propingua* (Fieber). The virus is not transmitted mechanically. It is endemic in Bermuda grass *Cynodon dactylon* L.

DISTRIBUTION: Morocco.

SYMPTOMS: Infected plants are stunted with narrow longitudinal chlorotic streaks on the leaves.

CONTROL: No control is reported.

Fiji Disease

CAUSE: The causal virus is disseminated by *Perkinsiella saccharicida* Kirkaldy and *P. vastatrix* Breddin. Other hosts include sorghum and sugarcane.

DISTRIBUTION: Java, New South Wales, New Guinea, and Philippines.

SYMPTOMS: Galls on leaf veins.

CONTROL: No control is reported.

Maize Chlorotic Dwarf

CAUSE: Maize chlorotic dwarf virus (MCDV) overwinters in Johnsongrass and is transmitted to healthy plants only by the leafhoppers, *Graminella nigrifrons* (Forbes) and *G. sonora* (Ball).

DISTRIBUTION: Southeastern United States.

SYMPTOMS: In combination with maize dwarf mosaic virus (MDMV) in corn, there is a more severe effect (chlorosis and severe stunting) on plants than single infections by either virus. Initially there is a chlorosis of young leaves in the whorl. As infected leaves unfurl, a fine chlorotic striping becomes associated with the smallest visible leaf veins, running for some length parallel to the veins. The early striping may become obscured as plants mature, with leaves becoming yellowish and reddish. The mosaic pattern caused by MDMV may obscure the fine chlorotic striping. Additional symptoms may be pronounced stunting and reddening, yellowing, and finally necrosis of leaf margins that may also be split horizontally. Diseased leaves are dull compared to the bright shiny appearance of healthy leaves.

CONTROL:
1. Plant tolerant hybrids.
2. Plant early in the growing season to avoid large leafhopper populations.
3. Control Johnsongrass to eliminate overwintering host.

Maize Chlorotic Mottle

CAUSE: Maize chlorotic mottle virus (MCMV) is transmitted by six species of beetles: the cereal leaf beetle, *Oulema melanopa* L.; the corn flea beetle, *Chaetocnema pulicaria* Mels.; the flea beetle, *Systena frontalis* (Fab.); the southern corn rootworm, *Diabrotica undecimpunctata* Mannerheim; the northern corn rootworm, *D. longicornis* (Say.); and the western corn rootworm, *D. virgifera* LeConte. MCMV is also mechanically transmitted. Continuous corn plantings have been reported to perpetuate MCMV, which survives in infected corn residue. It has been theorized that transmission of MCMV is also by soil. Newly hatched larvae of vectors in the absence of fresh corn roots forage on infected crop residues. MCMV would then be acquired and later transmitted by larvae feeding on developing corn roots. Virus survival in corn tissues and perhaps viability of beetle eggs in soil may occur only in soils of high water-holding capacities. Such soils can maintain infected crop residues in a proper state of hydration and preserve virus particles. Hosts include barley, Johnsongrass, rye, sorghum, and wheat.

DISTRIBUTION: Brazil, Kansas, Nebraska, Peru, and Texas.

SYMPTOMS: Fine chlorotic stripes that are parallel to the veins occur on the youngest leaves. The stripes coalesce to produce elongated chlorotic blotches that eventually become necrotic. Eventually there is a downward curling of the leaves followed by plant death. Growth is often stunted, tassels are distorted, and fewer ears are formed.

CONTROL: Crop rotation has been reported to reduce disease incidence in field plots.

Maize Dwarf Mosaic

CAUSE: Maize dwarf mosaic virus (MDMV) A through O. There are at least 13 and possibly more strains of the virus. Strains A through

M exclusive of B are believed to overwinter primarily in Johnsongrass, surviving in underground stems. Strain B would presently be considered as any strain of MDMV that infects corn and sugarcane but not Johnsongrass. However, several wild and cultivated grasses are susceptible to MDMV; in Mississippi, 70 percent of all grass species are susceptible in various degrees to MDMV-A and MDMV-B.

MDMV is closely related to sugarcane mosaic virus (SCMV). Most strains of MDMV may be considered as substrains of the Johnsongrass strain of SCMV (SCMV-Jg) Additionally isolates of MDMV-B could be synonymous with any of several strains of SCMV, which rarely or never infect Johnsongrass.

MDMV is transmitted mechanically and by carrying on the mouth parts of several different species of aphids. Disease appearance and spread are likely related to aphid numbers. MDMV is also seedborne in sweet corn. Young infected plants are more severely diseased than older plants; however, late plantings of sweet corn are more severely diseased than early plantings.

DISTRIBUTION: Strains A through O occur worldwide.

SYMPTOMS: Symptoms are extremely variable and are most severe on plants infected early; those infected at pollination time or later may appear normal. Initially plants may have a stippled mottle or mosaic of light and dark green on the youngest leaves that may develop into narrow streaks. Sometimes the mosaic appears as dark green islands on a chlorotic background. There may also be a shortening of upper internodes that imparts a feather duster appearance to the plants. Symptoms may occur on all leaves, leaf sheaths, and husks that develop after infection. As plants mature, the mosaic disappears and leaves become yellow-green, frequently showing blotches or streaks of red that are generally observed after periods of cool (15.5°C and below) night temperatures. Germ tube lengths of pollen tend to be shorter and virus may be present in silks of some sweet corn cultivars. Severely infected plants are barren. Infected plants are predisposed to root rot. Stalk strength is reduced due to stalks of diseased plants being smaller in diameter than healthy plants.

CONTROL:
1. Plant resistant hybrids.
2. Destroy Johnsongrass or other overwintering hosts.
3. Early planting of sweet corn may reduce losses.

Maize Leaf Fleck

CAUSE: Maize leaf fleck virus (MLFV) is disseminated by *Myzus persicae* Sulzer, *Rhopalosiphum padi* Linnaeus, and *R. maidis* (Fitch) in a persistent manner. Another host is harding grass *Phalaris stenoptera Hack*. MLFV has been reported seedborne in Canada and Europe.

DISTRIBUTION: California in the San Francisco Bay area.

SYMPTOMS: Older leaves and tips of other leaves first display symptoms. Small, round, yellow to orange spots occur together with tip and marginal burning on leaves. Eventually leaves become necrotic.

CONTROL: No control is reported.

Maize Line

CAUSE: Maize line virus (MLV) is disseminated by *Peregrinus maidis* Ashm.

DISTRIBUTION: East Africa.

SYMPTOMS: Wide chlorotic bands occur on leaves. Veins on leaf undersurface become thickened, causing a rough texture to the leaf surface. Plants are slightly stunted.

CONTROL: No control is reported.

Maize Mosaic

This disease is also called corn leaf stripe, corn stripe, corn mosaic, enanismo rayado, and sweet corn mosaic stripe. Additionally, maize mosaic has been confused with maize dwarf mosaic in some literature.

CAUSE: Maize mosaic virus (MMV) is transmitted by the leafhopper, *Peregrinus maidis* Ashm., and it may also be seedborne. MMV also infects sorghum and some wild grasses, which may provide a means of overseasoning.

DISTRIBUTION: Maize mosaic is a disease of the tropics and may be found in Africa, Australia, the Caribbean, Hawaii, India, and northern South America. It has also been reported in the United States.

SYMPTOMS: There may be considerable variation in symptoms. Initially, small, white flecks occur on one side of the midrib near the base of a young leaf. This is usually associated with whitening of leaf veins. The specks elongate to form fine discontinuous stripes that are parallel to the midrib. Sometimes the stripes coalesce, causing yellow bands to form on the leaves. In severe cases affected tissue becomes red to purple with a necrosis of the chlorotic areas. Stripes may also occur on sheaths, husks, and stalks. Moderate to severe dwarfing of new growth can occur.

Some symptoms occur only as broad chlorotic bands on leaves with both veins and interveinal tissue affected. The bands are usually not continuous but fade into spots or yellow areas at different lengths on the leaf. Bands may appear on one or both sides of the midrib.

CONTROL: Plant resistant hybrids and inbreds.

Maize Pellucid Ringspot

CAUSE: Maize pellucid ringspot virus (MPRV) is thought to be disseminated by an aphid.

DISTRIBUTION: Papua New Guinea and West Africa.

SYMPTOMS: Leaves have round, water-soaked spots.

CONTROL: No control is reported.

Maize Raya Gruesa

CAUSE: Maize raya gruesa virus (MRGV) is transmitted in a persistent manner by *Peregrinus maidis* Ashm. MRGV is acquired by a feeding acquisition period of 48–72 hours and an incubation period in the insect vector of 4–22 days.

DISTRIBUTION: Colombia.

SYMPTOMS: Chlorotic stripes (1 mm wide) occur on leaves.

CONTROL: No control is reported.

Maize Rayado Fino

This disease is also called fine striping disease and rayado fino.

CAUSE: Maize rayado fino virus (MRFV) is transmitted by nymphs and adults of the leafhoppers *Dalbulus maidis* Delong & Wolcott, *D. elimatus* Ball, *Graminella nigrifrons* (Forbes), *Baldulus tripsaci* Kramer

& Whitcomb, and *Stirellus bicolor* (Van Duzae). The incubation period is 7–21 days, with leafhoppers retaining the virus for prolonged periods, sometimes for life. Maize Colombian stripe (del rayado Colombiana del maize) and Brazilian corn streak are diseases of corn caused by strains of MRFV.

DISTRIBUTION: Central and South America, Florida, and Texas.

SYMPTOMS: Initially, a few small chlorotic dots or short stripes develop at the base and along the veins of young leaves. The dots become more numerous and begin to fuse; however, long continuous stripes are seldom found and a characteristic chlorotic stipple striping of the veins prevails. As plants become older, symptoms become less conspicuous and may disappear when the plant reaches maturity. Plants may be partially stunted and chlorotic.

CONTROL: No control is reported.

Maize Red Stripe

CAUSE: Maize red stripe virus (MRSV). It is not known how MRSV is spread.

DISTRIBUTION: Bulgaria.

SYMPTOMS: Red streaks occur on leaves. Plants are dwarfed with small ears.

CONTROL: No control is reported.

Maize Ring Mottle

CAUSE: Maize ring mottle virus (MRMV) is transmitted by seed.

DISTRIBUTION: Bulgaria.

SYMPTOMS: Chlorotic spots and rings occur on leaves.

CONTROL: No control is reported.

Maize Rio Cuarto Disease

CAUSE: A reolike virus.

DISTRIBUTION: Argentina.

SYMPTOMS: Plants are severely stunted, malformed, and dark green with stiff or brittle stalks and leaves. Enations on leaf veins may be present.

CONTROL: Genetic resistance is available.

Maize Rough Dwarf

This disease is also called nanismo ruvido.

CAUSE: Maize rough dwarf virus (MRDV) is disseminated only by the leafhopper *Laodelphax striatellus* Fallen.

DISTRIBUTION: China, Iran, Israel, and Europe.

SYMPTOMS: Early infection results in severely stunted plants with numerous enations on the leaf underside and sheath. Root systems are reduced in size and discolored. Ears are not formed, or if they are, they tend to be small and malformed.
   Later infection results in no noticeable stunting but enations are still formed.

CONTROL: Tolerant cultivars have been identified.

Maize Streak

This disease is also called maize streak disease. Bajra streak and maize mottle are caused by strains of the same virus.

CAUSE: Maize streak virus (MSV) is transmitted mainly by the leafhopper *Cicadulina mbila* Naude; however, other leafhoppers are also reported to transmit MSV. MSV is not seedborne or mechanically transmitted. Several specialized strains of the virus exist. Other hosts include African millet, barley, broomcorn, millet, oats, rice, rye, sugarcane, and wheat. Wild grasses are also infected. Fifty-four grass species are hosts.

DISTRIBUTION: Africa and Asia.

SYMPTOMS: Plants infected young are more severely diseased than older infected plants. Initially, circular (.5–2 mm in diameter), colorless spots occur on the lowest exposed portion of the youngest leaves. At first the spots are scattered, but as more leaf area becomes visible through growth, the spots become closer together and more numerous. Eventually narrow, broken, chlorotic stripes occur along the veins. Stripes may coalesce to form wider stripes. Stripes are evenly distributed on every leaf formed since the plant became infected. Early infection may cause dwarfing.

CONTROL:
1. Plant resistant hybrids.
2. Apply insecticides.
3. Do not plant corn adjacent to other hosts.

Maize Stripe

This disease is likely the same as maize hoja blanca (white leaf of corn) and Tsai's disease.

CAUSE: Maize stripe virus (MSV) is transmitted by the leafhopper *Peregrinus maidis* Ashm. An MSV isolate from Venezuela was mechanically transmitted to the sweet corn cultivar Iochief but not to field corn. Other hosts include barley, sorghum, and teosinte.

DISTRIBUTION: Worldwide in all tropical areas where *P. maidis* occurs.

SYMPTOMS: Initially, numerous chlorotic spots and streaks appear at base and extend outward on youngest leaves. As leaves expand, spots and stripes coalesce to form broad chlorotic bands. Later developing leaves become completely yellow. Infected plants may have acute bending at the apical tip. These plants may become necrotic and die. Severe infection results in reduced yields.

CONTROL: No control is reported.

Maize Tassel Abortion

CAUSE: Maize tassel abortion virus (MTAV) is disseminated by *Malaxodes farinosus* Fennah.

DISTRIBUTION: East Africa.

SYMPTOMS: The tassels are aborted. Small leaves grow at right angles to the stalk.

CONTROL: No control is reported.

Maize Vein Enation

CAUSE: Maize vein enation virus (MVEV) is disseminated by *Cicadulina mbila* Naude. Other hosts include numerous grasses, oats, rice, rye, sorghum, sugarcane, and wheat.

DISTRIBUTION: India.

SYMPTOMS: Galls form on lower leaf surfaces. Leaves become a darker green. Plants are stunted.

CONTROL: No control is reported.

Maize Wallaby Ear

CAUSE: Maize wallaby ear virus (MWEV) is transmitted by the leafhoppers *Cicadulina bipunctella bimaculata* Evans, *C. bimaculata* (Evans), and *Nesoclutha pallida* (Evans). Other hosts are rice, rye, sorghum, sugarcane, wheat, and several grasses.

DISTRIBUTION: Australia.

SYMPTOMS: Symptoms are of two types. The mild symptoms consist of galls and enations; however, plants normally recover. Severe symptoms consist of numerous galls on most veins, upright dark green leaves with edges that roll upward and inward and stand stiffly at right angles to the stalk. Plants may be severely stunted and yield is reduced.

CONTROL:
1. Adjust planting time.
2. Insecticides.

Maize White Line Mosaic

This disease is also called white line mosaic and stunt.

CAUSE: Maize white line mosaic virus (MWLMV). The virus is not mechanically transmitted. Infection is associated with time of season rather than plant age. Disease often occurs locally in poorly drained areas of a field and along edge rows. MWLMV has been transferred from diseased roots placed in sterile soil. Zoospores of an Olphidium-like fungus is considered the likely vector. A satellite virus, serologically related to a satellitelike particle associated with maize dwarf ringspot virus, is associated with MWLMV.

DISTRIBUTION: France, Italy, Michigan, New York, Ohio, Vermont, and Wisconsin.

SYMPTOMS: A mosaic pattern develops on leaves. This pattern is mostly interveinal and varies in intensity from a mild mottle to severe necrosis. Leaves develop chlorotic white lines 1–2 mm wide and 1–4 cm long near veins. Plants may be severely stunted with small or no ears. Some plants may develop a crozierlike hooking of the terminal portions. Plants may be barren or produce small ears with reduced kernel number. Infected plants may be symptomless.

CONTROL: Differences are reported among genotypes.

Millet Red Leaf

CAUSE: Millet red leaf virus (MRLV) is disseminated by *Macrosiphum granarium* (Kirby), *Rhopalosiphum maidis* (Fitch), and *Schizaphis graminum* (Rondani). Other hosts include Panicum and Setaria species.

DISTRIBUTION: China.

SYMPTOMS: Purple stem varieties have a reddening of leaf blades and sheaths; green stem varieties have a chlorosis.

CONTROL: No control is reported.

Northern Cereal Mosaic

CAUSE: Northern cereal mosaic virus (MCMV) is disseminated by *Delphacodes albifascia* (Mats.), *Laodelphax striatellus* Fallen, *Meuterianella fairmairei*, and *Unkanodes sapporonus* (Mats.). Other hosts include barley, grasses, oats, rye, and wheat.

DISTRIBUTION: Japan and Korea.

SYMPTOMS: Chlorotic striping occurs along one side of the leaf blades.

CONTROL: No control is reported.

Oat Pseudorosette

This disease is also known as zakuklivanie.

CAUSE: The oat pseudorosette virus is disseminated by *Laodelphax striatellus* Fallen. Other hosts include barley, oats, rice, wheat, and several species of grassy weeds.

DISTRIBUTION: Western Siberia.

SYMPTOMS: Infected plants have numerous tillers with malformed tassels. Leaves have a reddish discoloration.

CONTROL: No control is reported.

Oat Sterile Dwarf

CAUSE: The oat sterile dwarf virus (OSDV) is disseminated by *Dicranotropis hamata* Boh., *Javesella discolor* (Boheman), *J. obscurella* (Boheman), and *J. pellucida* Fabr.

DISTRIBUTION: Sweden.

SYMPTOMS: Infected plants have narrow leaves with enations that last a short time.

CONTROL: No control is reported.

Rice Black-streaked Dwarf

CAUSE: The rice black-streaked dwarf virus (RBSDV) is disseminated by *Laodelphax striatellus* Fallen, *Ribautodelphax albifascia* (Mats.), and *Unkanodes sapporonus* (Mats.). Other hosts include barley, several grasses, rice, and wheat.

DISTRIBUTION: Japan.

SYMPTOMS: Plants are stunted with chlorotic streaks on leaves.

CONTROL: No control is reported.

Rice Stripe

CAUSE: The rice stripe virus (RSV) is disseminated by *Laodelphax striatellus* Fallen, *Ribautodelphax albifascia* (Mats.), and *Unkan-*

odes sapporonus (Mats.). Other hosts include barley, millets, rice, and wheat.

DISTRIBUTION: Japan.

SYMPTOMS: Plants are stunted with chlorotic streaks on the leaves.

CONTROL: No control is reported.

Sugarcane Mosaic

CAUSE: Sugarcane mosaic virus (SCMV) overseasons in several grass hosts including Johnsongrass (SCMV-Jg); however, none has been significant as a reservoir of the virus for corn with the possible exception of Johnsongrass. SCMV is transmitted by several aphid species in a nonpersistent manner and by sap inoculation. It is not known to be seedborne in corn; however, SCMV is transmitted by seed at a low rate in sweet corn.

SCMV is closely related to maize dwarf mosaic virus (MDMV). Most strains of MDMV may be considered as substrains of SCMV-Jg; additionally strain B of MDMV (MDMV-B) could be synonymous with any of several strains of SCMV that rarely or never infect Johnsongrass.

DISTRIBUTION: Worldwide in tropical and subtropical areas; however, SCMV is normally not severe except occasionally in sweet corn.

SYMPTOMS: Initially a mild mottle occurs in the interveinal areas of the leaves. Chlorotic patterns soon occur and elongate to produce streaks or stripes with irregular margins. Plants may become severely stunted, are lighter in appearance than healthy plants, and produce ears with few or no kernels.

CONTROL:
1. Do not plant near sugarcane.
2. Control grassy weeds in and near fields.
3. Plant the most resistant hybrids especially for sweet corn.

Vein Enation

This disease is also called leaf gall.

CAUSE: Unknown virus that is possibly spread by the leafhopper *Cicadulina bipunctella* (Matsumara).

DISTRIBUTION: Philippines.

SYMPTOMS: Plants are stunted. Leaf size is reduced in length and width. Vein enation or galling makes leaves appear corrugated or creased. Often the young leaves enveloping the tassel will fail to unfurl.

CONTROL:
1. Rogue infected plants.
2. Spray insecticide.

Wheat Spot Mosaic

CAUSE: Wheat spot mosaic virus (WSMV) is disseminated by *Eriophyes tulipae* (Keifer). Other hosts include barley, numerous grasses, rye, and wheat.

DISTRIBUTION: Canada and Ohio.

SYMPTOMS: Chlorotic spots, streaks, and mottling occur on leaves.

CONTROL: No control is reported.

Wheat Streak Mosaic

CAUSE: Wheat streak mosaic virus (WSMV) overwinters in winter wheat and other winter small grains as well as a number of wild grasses. It is transmitted in the field by the wheat curl mite *Eriophyes tulipae* (Keifer). It can also be mechanically transmitted and is seedborne although transmission is at a low rate. Some corn lines are more susceptible to infection at 35°C than lower temperatures; however, many lines are susceptible regardless of temperature. Other hosts include cereals and numerous grasses.

DISTRIBUTION: Europe, Northern Africa, and North America.

SYMPTOMS: Younger plants are more susceptible than older plants. Small oval to elliptical spots at tips of younger leaves that elongate and develop parallel to the veins. Older leaves may become chlorotic at the tips. Ears may be poorly developed, with a general yellowing and stunting.

CONTROL: No control measures are necessary but differences in resistance exist among inbreds.

Wheat Striate Mosaic

CAUSE: American wheat striate mosaic virus (AWS + MV). See American Wheat Striate Mosaic on wheat.

DISTRIBUTION: North central United States, south central Canada, and the USSR.

SYMPTOMS: Upper leaves have very distinct, long, thin, white, chlorotic streaks.

CONTROL: No control is reported.

—————————BIBLIOGRAPHY—————————

Arny, D. C. et al. 1971. Eyespot of maize, a disease new to North America. *Phytopathology* **61:**54–57.

Attwater, W. A., and Busch, L. V. 1982. The role of sap beetles (*Glischrochilus quadrisignatus* (Say)) in the epidemiology of Gibberella corn ear rot. *Can. J. Plant Pathol.* **4:**303 (abstract).

Bains, S. S.; Jhooty, J. S.; Sokhi, S. S.; and Rewal, H. S. 1978. Role of *Digitaria sanguinalis* in outbreaks of brown stripe downy mildew of maize. *Plant Dis. Rptr.* **62:**143.

Bonde, M. R.; Peterson, G. L.; and Duck, N. B. 1985. Effects of temperature on sporulation, conidial germination, and infection of maize by *Peronosclerospora sorghi* from different geographical areas. *Phytopathology* **75:**122–126.

Boosalis, M. G.; Sumner, D. R.; and Rao, A. S. 1967. Overwintering of conidia of *Helminthosporium turcicum* on corn residue and in soil in Nebraska. *Phytopathology* **57:**990–996.

Boothroyd, C. W. 1981. Virus diseases of sweet corn. Pages 103–109 in D. T. Gordon, J. A. Knoke, and G. E. Scott, eds. Virus and viruslike diseases of maize in the United States. *Southern Cooperative Series Bull. 247,* 218 p.

Boothroyd, C. W., and Israel, H. W. 1980. A new mosaic disease of corn. *Plant Disease* **64:**218–219.

Bowden, R. L., and Stromberg, E. L. 1982. Chocolate spot of corn in Minnesota. *Plant Disease* **66:**744.

Bradfute, O. E.; Teyssandier, E.; Marino, E.; and Dodd, J. L. 1981. Reolike virus associated with maize rio cuarto disease in Argentina. *Phytopathology* **71**:205 (abstract).

Burns, E. E., and Shurtleff, M. C. 1973. Observations of *Physoderma maydis* in Illinois: Effects of tillerage practices in field corn. *Plant Dis. Rptr.* **57**:630–633.

Castillo, J., and Herbert, T. T. 1974. Nueva enfermedat virosa afectano al maiz en al Peru (A new virus disease of maiz in Peru). *Fitopatologia* **9**:79–84.

Christensen, J. J., and Wilcoxson, R. D. 1966. *Stalk Rot of Corn*. Monograph No. 3. American Phytopathological Society, St. Paul, MN.

Cohen, Y., and Sherman, Y. 1977. The role of airborne conidia in epiphytotics of *Sclerospora sorghi* on sweet corn. *Phytopathology* **67**:515–521.

Cook, G. E.; Boosalis, M. G.; Dunkle, L. D.; and Odvody, G. N. 1973. Survival of *Macrophomina phaseoli* in corn and sorghum stalk residue. *Plant Dis. Rptr.* **57**:873–875.

Cullen, D.; Caldwell, R. W.; and Smalley, E. B. 1983. Susceptibility of maize to *Gibberella zeae* ear rot: Relationship to host genotype, pathogen virulence, and zearalenone contamination. *Plant Disease* **67**:89–91.

De Agudelo, F. V., and Martinez-Lopez, G. 1983. Maize raya gruesa: A rhabdovirus transmitted by *Peregrinus maidis*. *Phytopathology* **73**:125 (abstract).

Dodd, J. L. 1980. Grain sink size and predisposition of *Zea mays* to stalk rot. *Phytopathology* **70**:534–535.

Doupnik, B., Jr., and Wysong, D. 1978. Nebraska corn variety tests for reactions to Goss's bacterial wilt and blight. *Univ. Nebraska Coop. Ext. Serv. and Agric. Expt. Sta. UNL-SCS 78-30.*

Gamez, R. 1969. A new leafhopper-borne virus of corn in Central America. *Plant Dis. Rptr.* **53**:929–932.

Gendloff, E. H.; Rossman, E. C.; Casale, W. L.; Isleib, T. G.; and Hart, L. P. 1986. Components of resistance to Fusarium ear rot in field corn. *Phytopathology* **76**:684–688.

Gilbertson, R. L.; Brown, W. M., Jr.; and Ruppel, E. G. 1985. Effect of tillage and herbicides on Fusarium stalk rot of corn. *Phytopathology* **75**:1296 (abstract).

Gilbertson, R. L.; Brown, W. M., Jr.; Ruppel, E. G.; and Capinera, J. L. 1986. Association of corn stalk rot *Fusarium* spp. and western corn rootworm beetles in Colorado. *Phytopathology* **76**:1309–1314.

Gingery, R. E.; Nault, L. R.; Tsai, J. H.; and Lastra, R. J. 1979. Occurrence of maize stripe virus in the United States and Venezuela. *Plant Dis. Rptr.* **63**:341–343.

Gordon, D. T.; Bradfute, O. E.; Gingery, R. E.; Knoke, J. K.; Louie, R.; Nault, L. R.; and Scott, G. E. 1981. Introduction: History, geographical distribution, pathogen characteristics, and economic importance. Pages 1–12 in D. D. Gordon, J. K. Knoke, and G. E. Scott, eds. Virus and viruslike diseases of maize in the United States. *Southern Cooperative Series Bull. 247*, 218 p.

Guthrie, E. J. 1978. Measurement of yield losses caused by maize streak disease. *Plant Dis. Rptr.* **62**:839–841.

Herold, F. 1972. Maize mosaic virus. *Descriptions of Plant Viruses.* Set 5. No. 94. Commonwealth Mycological Institute. Kew, England.

Hollier, C. A., and King, S. B. 1985. Effect of dew period and temperature on infection of seedling maize plants by *Puccinia polysora*. *Plant Disease* **69**:219–220.

Johnson, A. G.; Robert, A. L.; and Cash, L. 1945. Further studies on bacterial leaf blight and stalk rot of corn. *Phytopathology* **35**:486 (abstract).

Jones, R. K., and Duncan, H. E. 1981. Effect of nitrogen fertilizer, planting date, and harvest date on aflatoxin production in corn inoculated with *Aspergillus flavus*. *Plant Disease* **65**:741–744.

Jones, R. K.; Duncan, H. E.; Payne, G. A.; and Leonard, K. J. 1980. Factors influencing infection by *Aspergillus flavus* in silk-inoculated corn. *Plant Disease* **64**:859–863.

Jons, V. L.; Timian, R. G.; Gardner, W. S.; Stromberg, E. L.; and Berger, P. 1981. Wheat striate mosaic virus in the Dakotas and Minnesota. *Plant Disease* **65**:447–448.

Keller, N. P.; Bergstrom, G. C.; and Carruthers, R. I. 1986. Potential yield reductions in maize associated with an anthracnose/European corn borer pest complex in New York. *Phytopathology* **76**:586–589.

Kingsland, G. C. 1980. Effect of maize dwarf mosaic virus infection on yield and stalk strength of corn in the field in South Carolina. *Plant Disease* **64**:271–273.

Kitajima, E. W., and Costa, A. S. 1982. The ultrastructure of the corn chlorotic vein banding (Brazilian maize mosaic) virus-infected corn leaf tissues and viruliferous vector. *Fitopatologia Brasileria* **7**:247–259.

Kloepper, J. W.; Garrott, D. G.; and Kirkpatrick, B. C. 1982. Association of spiroplasmas with a new disease of corn. *Phytopathology* **72**:1004 (abstract).

Koehler, B. 1959. Corn ear rots in Illinois. *Univ. Illinois Agric. Expt. Sta. Bull. 639.*

Koehler, B. 1960. Cornstalk rots in Illinois. *Univ. Illinois Agric. Expt. Sta. Bull. 658.*

Kommedahl, T.; Sabet, K. K.; Burnes, P. M.; and Windels, C. E. 1987. Occurrence and pathogenicity of *Fusarium proliferatum* on corn in Minnesota. *Plant Disease* **71**:281 (disease notes).

Kucharek, T. A. 1973. Stalk rot of corn caused by *Helminthosporium rostratum*. *Phytopathology* **63**:1336–1338.

Lastra, R., and Carballo, O. 1985. Mechanical transmission, purification and properties of an isolate of maize stripe virus from Venezuela. *Phytopath. Zeit.* **114**:168–179.

Latterell, F. M., and Rossi, A. E. 1984. An unidentified species of Cercospora pathogenic to corn. *Phytopathology* **74**:852 (abstract).

Latterell, F. M., and Rossi, A. E. 1984. A Marasmiellus disease of maize in Latin America. *Plant Disease* **68**:728–731.

Latterell, F. M., and Rossi, A. E. 1983. *Stenocarpella macrospora (Diplodia macrospora)* and *S. maydis (D. maydis)* compared as pathogens of corn. *Plant Disease* **67**:725–729.

Latterell, F. M.; Rossi, A. E.; and Moreno, R. 1976. *Diplodia macrospora*: A potentially serious pathogen of corn in U.S. ? (abstr. no. 110) *Am. Phytopathol. Soc. Proc.* **3**:228.

Latterell, F. M.; Rossi, A. E.; and Trujillo, E. E. 1986. A previously undescribed Selenophoma leaf spot of maize in Colombia. *Plant Disease* **70**:472–474.

Leach, C. M.; Fullerton, R. A.; and Young, K. 1977. Northern leaf blight of maize in New Zealand: Relationship of *Drechslera turcica* airspora to factors influencing sporulation, conidium development, and chlamydospore formation. *Phytopathology* **67**:629–636.

Leonard, K. J., and Thompson, D. L. 1976. Effects of temperature and host maturity on lesion development of *Colletotrichum graminicola* on corn. *Phytopathology* **66**:635–639.

Levy, Y. 1984. The overwintering of *Exserohilum turcicum* in Israel. *Phytoparasitica* **12**:177–182.

Llano, A.; and Schieber, E. 1980. *Diplodia macrospora* of corn in Nicaragua. *Plant Disease* **64**:797.

Lockhart, B. E. L.; Khaless, N.; El Maataoui, M.; and Lastra, R. 1985. Cynodon chlorotic streak virus, a previously undescribed plant rhabdovirus infecting Bermuda grass and maize in the Mediterranean area. *Phytopathology* **75**:1094–1098.

Louie, R.; Gordon, D. T.; and Lipps, P. E. 1981. Transmission of maize white line mosaic virus. *Phytopathology* **71**:1116 (abstract).

Louie, R.; Gordon, D. T.; Madden, L. V.; and Knoke, J. K. 1983. Symptomless infection and incidence of maize white line mosaic. *Plant Disease* **67**:371–373.

McDaniel, L. L., and Gordon, D. T. 1985. Identification of a new strain of maize dwarf mosaic virus. *Plant Disease* **69**:602–607.

Maiti, S. 1978. Two new ear rots of maize from India. *Plant Dis. Rptr.* **62**:1074–1076.

Malek, R. B.; Norton, D. C.; Jacobsen, B. J.; and Acosta, N. 1980. A new corn disease caused by *Longidorus breviannulatus* in the Midwest. *Plant Disease* **64**:1110–1113.

Matyac, C. A., and Kommendahl, T. 1985. Occurrence of chlorotic spots on corn seedlings infected with *Sphacelotheca reiliana* and their use in evaluation of head smut resistance. *Plant Disease* **69**:251–254.

Matyac, C. A., and Kommendahl, T. 1985. Factors affecting the development of head smut caused by Sphacelotheca on corn. *Phytopathology* **75**:577–581.

Misra, A. P. 1959. Diseases of millets and maize. *Indian Agriculturist* **3**:75–89.

Mukunya, D. M., and Boothroyd, C. W. 1973. *Mycosphaerella zeae-maydis* sp. n., the sexual stage of *Phyllosticta maydis*. *Phytopathology* **63**:529–532.

Muller, G. J. et al. 1973. *Compendium of Corn Diseases.* The American Phytopathological Society. St. Paul, MN, 105 p.

Nault, L. R. et. al. 1978. Transmission of maize chlorotic mottle virus by chrysomelid beetles. *Phytopathology* **68**:1071–1074.

Niblett, C. L., and Claflin, L. E. 1978. Corn lethal necrosis—a new virus disease of corn in Kansas. *Plant Dis. Rptr.* **62**:15–19.

Nicholson, R. L.; Bergeson, G. B.; Degennaro, F. P.; and Viveiros, D. M. 1985. Single and combined effects of the lesion nematode and *Colletotrichum graminicola* on growth and anthracnose leaf blight of corn. *Phytopathology* **75**:654–661.

Norton, D. C., and De Agudelo, V. 1984. Plant-parasitic nematodes associated with maize in Cauca and Valle del Cauca, Colombia. *Plant Disease* **68**:950–952.

Nwigwe, C. 1974. Occurrence of Phomopsis on maize (*Zea mays*). *Plant Dis. Rptr.* **58**:416–417.

Ochor, T. E.; Trevathan, L. E.; and King, S. B. 1987. Relationship of harvest date and host genotype to infection of maize kernels by *Fusarium moniliforme*. *Plant Disease* **71**:311–313.

Payne, G. A.; Cassel, D. K.; and Adkins, C. R. 1985. Reduction of aflatoxin levels in maize due to irrigation and tillage. *Phytopathology* **75**:1283 (abstract).

Payne, G. A.; Cassel, D. K.; and Adkins, C. R. 1986. Reduction of aflatoxin contamination in corn by irrigation and tillage. *Phytopathology* **76**:679–684.

Payne, G. A.; Duncan, H. E.; and Adkins, C. R. 1987. Influence of tillage on development of gray leaf spot and number of airborne conidia of *Cercospora zeae-maydis*. *Plant Disease* **71**:329–332.

Payne, G. A., and Leonard, K. J. 1985. *Stenocarpella macrospora* on corn in North Carolina. *Plant Disease* **69**:613.

Payne, G. A., and Waldron, J. K. 1983. Overwintering and spore release of *Cercospora zeae-maydis* in corn debris in North Carolina. *Plant Disease* **67**:87–89.

Pedersen, W. L., and Brandenburg, L. J. 1986. Mating types, virulence, and cultural characteristics of *Exserohilum turcicum* race 2. *Plant Disease* **70**:290–292.

Pordesimo, A. N., and Aday, B. A. 1984. Vein enation or leaf gall of corn. *Philippines Phytopathol.* **20**:15 (abstract).

Rao, B.; Schmitthenner, A. F.; Caldwell, R.; and Ellett, C. W. 1978. Prevalence and virulence of Pythium species associated with root rot of corn in poorly drained soil. *Phytopathology* **68**:1557–1563.

Reifschneider, F. J. B., and Arny, D. C. 1980. Host range of *Kabatiella zeae*, causal agent of eyespot of maize. *Phytopathology* **70**:485–487.

Reifschneider, F. J. B., and Lopes, C. A. 1982. Bacterial top and stalk rot of maize in Brazil. *Plant Disease* **66**:519–520.

Ribeiro, R. De L. D.; Durbin, R. D.; Arny, D. C.; and Uchytil, T. F. 1977. Characterization of the bacterium inciting chocolate spot of corn. *Phytopathology* **67**:1427–1431.

Rich, J. R., and Schenck, N. C. 1981. Seasonal variations in populations of plant-parasitic nematodes and vesicular-arbuscular mycorrhizae in Florida field corn. *Plant Disease* **65**:804–807.

Roane, M. K., and Roane, C. W. 1983. New grass hosts of *Polymyxa graminis* in Virginia. *Phytopathology* **73**:968 (abstract).

Rupe, J. C.; Siegel, M. R.; and Hartman, J. R. 1982. Influence of environment and plant maturity on gray leaf spot of corn caused by *Cercospora zeae-maydis*. *Phytopathology* **72**:1587–1591.

Sardanelli, S.; Krusberg, L. R.; and Golden, A. M. 1981. Corn cyst nematode, *Heterodera zeae*, in the United States. *Plant Disease* **65**:622.

Schneider, R. W., and Pendery, W. E. 1983. Stalk rot of corn: Mechanism of predisposition by an early season water stress. *Phytopathology* **73**:863–871.

Schurtleff, M. C. (ed.) 1980. *Compendium of Corn Diseases.* American Phytopathological Society. St. Paul, MN, 105 p.

Sharma, H. S. S., and Verma, R. N. 1979. False smut of maize in India. *Plant Dis. Rptr.* **63**:996–997.

Sinha, R. C., and Benki, R. M. 1972. American wheat striate mosaic virus. *Descriptions of Plant Viruses.* Set 6, No. 99. Commonwealth Mycological Institute. Kew, England.

Smidt, M. L. , and Vidaver, A. K. 1986. Population dynamics of *Clavibacter michinagense* subsp. *nebraskense* in field-grown dent corn and popcorn. *Plant Disease* **70**:1031–1036.

Smidt, M. L., and Vidaver, A. K. 1987. Variation among strains of *Clavibacter michiganense* subsp. *nebraskense* isolated from a single popcorn field. *Phytopathology* **77**:388–392.

Stevens, C., and Gudauskas, R. T. 1982. Relation of maize dwarf mosaic virus infection to *Helminthosporium maydis* race O. *Phytopathology* **72**:1500–1502.

Sumner, D. R., and Bell, D. K. 1980. Root diseases of corn caused by *Rhizoctonia solani* and *Rhizoctoni zeae*. *Phytopathology* **70**:572 (abstract).

Sumner, D. R., and Bell, D. K. 1982. Root diseases induced in corn by *Rhizoctonia solani* and *Rhizoctonia zeae*. *Phytopathology* **72**:86–91.

Sumner, D. R., and Schaad, N. W. 1977. Epidemiology and control of bacterial leaf blight of corn. *Phytopathology* **67**:1113–1118.

Trujillo, G. E.; Acosta, J. M.; and Pinero, A. 1974. A new corn virus disease found in Venezuela. *Plant Dis. Rptr.* **58**:122–126.

Tsai, J. H. 1979. Occurrence of a corn disease in Florida transmitted by *Peregrinus maidus*. *Plant Dis. Rptr.* **59**:830–833.

Ullstrup, A. J. 1970. A comparison of monogenic and polygenic resistance to *Helminthosporium turcicum* in corn. *Phytopathology* **60**:1597–1599.

Ullstrup, A. J. 1978. Corn diseases in the United States and their control. *USDA Agric. Res. Serv. Agric. Hdbk. No. 199* (revised).

Uyemoto, J. K. 1983. Biology and control of maize chlorotic mottle virus. *Plant Disease* **67**:7–10.

Uyemoto, J. K.; Phillips, N. J.; and Wilson, D. L. 1981. Control of maize chlorotic mottle virus by crop rotation. *Phytopathology* **71**:910 (abstract).

Vakili, N. G., and Booth, G. D. 1981. *Helminthosporium carbonum*, a cause of stalk rot of corn in Iowa. *Phytopathology* **71**:910 (abstract).

Vidaver, A. K., and Carlson, R. R. 1978. Leaf spot of field corn caused by *Pseudomonas andropogonis*. *Plant Dis. Rptr.* **62**:213–216.

Vidaver, A. K.; Gross, D. C.; Wysong, D. S.; and Doupnik, B. L., Jr. 1981. Diversity of *Corynebacterium nebraskense* strains causing Goss's bacterial wilt and blight of corn. *Plant Disease* **65**:480–483.

Wallin, J. R. 1986. Production of aflatoxin in wounded and whole maize kernels by *Aspergillus flavus*. *Plant Disease* **70**:429–430.

Warren, H. L. 1975. Temperature effects on lesion development and sporulation after infection by races O and T of *Bipolaris maydis*. *Phytopathology* **65**:623–626.

Warren, H. L. 1977. Survival of *Colletotrichum graminicola* in corn kernels. *Phytopathology* **67**:160–162.

Waudo, S. W., and Norton, D. C. 1986. Pathogenic effects of *Pratylenchus scribneri* in maize inbreds and related cultivars. *Plant Disease* **70**:636–638.

Weston, W. J., Jr. 1921. Another conidial Sclerospora of Philippine maize. *Jour. Agric. Res.* **20**:669–685.

White, D. G.; Hoeft, R. G.; and Touchton, J. T. 1978. Effects of nitrogen and nitrapyrin on stalk rot, stalk diameter and yield of corn. *Phytopathology* **68**:811–814.

Windels, C. E., and Kommendahl, T. 1984. Late-season colonization and survival of *Fusarium graminearum* Group II in cornstalk in Minnesota. *Plant Disease* **68**:791–793.

Wright, W. R., and Billeter, B. A. 1974. "Red kernel" disease of sweet corn on the retail market. *Plant Dis. Rptr.* **58**:1065–1066.

Young, G. V.; LeFebvre, C. L.; and Johnson, A. G. 1947. *Helminthosporium rostratum* on corn, sorghum and pearl millet. *Phytopathology* **47**:180–183.

Zeyen, R. J., and Morrison, R. H. 1975. Rhabdoviruslike particles associated with stunting of maize in Alabama. *Plant Dis. Rptr.* **59**:169–171.

Zuber, M. S.; Darrah, L. L.; Lillehoj, E. B.; Josephson, L. M.; Manwiller, A.; Scott, G. E.; Gudauskas, R. T.; Horner, E. S.; Widstrom, N. W.; Thompson, D. L.; Bockholt, A. J.; and Brewbaker, J. L. 1983. Comparison of open-pollinated maize varieties and hybrids for preharvest aflatoxin contamination in the southern United States. *Plant Disease* **67**:185–187.

Zummo, N. 1976. Yellow leaf blotch: A new bacterial disease of sorghum, maize, and millet in West Africa. *Plant Dis. Rptr.* **60**:798–799.

5

Diseases of Cotton (*Gossypium barbadense* L. and *G. hirsutum* L.)

BACTERIAL CAUSES

Bacterial Blight

This disease is also called angular leaf spot and black arm.

CAUSE: *Xanthomonas campestris* pv. *malvacearum* (Smith) Dye (syn. *Bacterium malvacearum* Smith, *Phytomonas malvacearum* Smith, *Pseudomonas malvacearum* Smith, *X. campestris* (Pammel) Dowson and *X. malvacearum* Smith (Dowson)). The bacterium survives on lint of seed and probably within the seed itself. Overwintering also occurs in infested debris left in the field after harvest. In arid and semiarid areas, survival in soil usually occurs within residue consisting of stems and bolls. Greenhouse studies concluded that survival may also occur within buds of symptomless plants. The bacterium is disseminated on seed and by wind, water insects, animals, and machinery. Seed is especially important in long-distance spread.

Bacteria enter into plants through stomata and wounds especially during periods of high temperatures (30–36°C) and abundant moisture or high humidity (85% and above). Seed becomes infected if abundant moisture is present at harvest. Volunteer plants arising from infected seed often have infected cotyledons that may serve as a source of infection in the field. Different physiologic races of *X. campestris* pv. *malvacearum* are present in the field.

DISTRIBUTION: Generally distributed wherever cotton is grown.

SYMPTOMS: Initially, green, round to elongate, water-soaked translucent spots of different sizes appear on the lower surface of cotyledons. These spots appear 7–10 days after planting. Eventually infections penetrate to the upper side of leaves and appear as angular or irregular brown to black lesions between veins and along the main leaf vein. Infected areas dry up and become sunken and red-brown with a brownish to purplish margin. Lesions may also develop along leaf petioles. Leaf drop may occur. Lesions on hypocotyls of seedlings are black, elongated cankers that may girdle and kill the plant. This phase of the disease is called black arm. Similar lesions may also occur on the stem as the plant gets older. These lesions may also girdle a stem, causing death of the distal portion.

Bolls have round to irregular, or angular, black sunken spots. Boll-rotting fungi may invade bacterial lesions and discolor or destroy the boll. Boll rot frequently occurs during hot, humid weather when numerous insects may puncture or wound bolls.

A creamy bacterial exudate occurs on lesions of leaves and bolls during high humidity or after rain. Dried exudate will appear as a sugarlike coating or film on a dark lesion.

CONTROL:
1. Plant resistant cultivars.
2. Acid delint seed.
3. Rotate cotton for at least one year with a nonsusceptible crop.
4. Plow under infected residue immediately after harvest if erosion is not a problem.
5. Plant seed from disease-free plants.

Bacterial Lint Discoloration

CAUSE: *Pseudomonas sp.* likely survives in soil. It is probably introduced into bolls by insects.

DISTRIBUTION: United States.

SYMPTOMS: Fiber is discolored and fluoresces yellow to green under ultraviolet light (366 nm).

CONTROL: No control is reported.

Crown Gall

CAUSE: *Agrobacterium radiobacter* var. *tumefaciens* (Smith & Townsend) Keane & Kerr and new biotype 2. Infection is probably enhanced by root-knot nematodes (*Meloidogyne incognita* (Kofoid & White) Chitwood) but bacteria can penetrate root tissues without nematode involvement.

DISTRIBUTION: Israel and the United States.

SYMPTOMS: Plants are stunted. Galls are generally located on taproots, which often stop growing. Galls are 1–4 cm long, irregular in shape, and dark brown. Sometimes spherical-shaped galls, 1 cm in diameter, are found on lateral roots.

CONTROL: None practiced since crown gall on cotton is rare.

Erwinia Internal Necrosis

CAUSE: *Erwinia herbicola* (Lohnis) Dye. Necrosis depends on development of a secondary internal boll that splits the placentae and causes an opening to the outside of the boll and intertwining of the fibers of affected locules. This allows the bacteria to enter the boll where it spreads from the interplacentae space into locules. The bacterium may have some association with the stinkbug *Euschistus impictiventris* Stal., particularly in Imperial Valley of California.

DISTRIBUTION: California.

SYMPTOMS: Immature cotton bolls have necrotic tissue that is red-brown in color. Affected locules are soft and slimy. Affected seed coats are discolored and seed contents are completely decayed. Lint of infected mature open bolls are tan and locules are compact.

CONTROL: No control is reported.

FUNGAL CAUSES

Alternaria Leaf Spot

CAUSE: *Alternaria gossypina* (Thuem.) Hopkins survives as mycelium in infected residue. Conidia are produced from mycelium during moist conditions and are airborne to leaves. Secondary inoculum is from production of conidia within lesions. Infection may follow other diseases or occur in cotton under stress from low fertility or unfavorable environmental conditions.

DISTRIBUTION: Generally distributed wherever cotton is grown under warm, moist conditions.

SYMPTOMS: Symptoms may occur on young plant parts such as cotyledons, seedlings, petioles, and stems. They may also occur on mature leaves. Spots are roughly circular (13 mm in diameter), brown, with a papery consistency. Spots have concentric rings, giving them a targetlike appearance.

During moist weather, conidia are produced in spots, giving them a black appearance. Severely infected leaves may shrivel and shed prematurely, especially during moist weather late in the growing season.

CONTROL:
1. Treat seed with a fungicide seed-protectant.
2. Maintain proper soil fertility.
3. Avoid insect and mechanical injury.
4. Plow under infected residue where erosion is not a problem.
5. Early maturing cultivars that fruit heavily are more susceptible than later maturing cultivars that set bolls over a longer period of time.

Alternaria Stem Blight and Leaf Spot

CAUSE: *Alternaria macrospora* Zimm. survives in infected leaves that provide primary inoculum the following growing season.

DISTRIBUTION: Africa, China, South America, and southeastern United States.

SYMPTOMS: Initially, tiny dark brown, circular spots with deeply sunken centers appear on stems and leaf petioles of mature cotton. Spots develop into elliptical to oval cankers that cause stem and petioles to split longitudinally or crack into small pieces. Eventually the diseased stem or petiole breaks off at the canker.

Alternaria macrospora also infects bolls, bracts, and leaves of all but very young cotton plants under stress. At first lesions on leaves are small and brown or red-purple. Eventually the lesions enlarge to about 1 cm in diameter and may coalesce, causing irregular-shaped dead areas and defoliation. Mature spots have dead, gray centers in which cracks may develop, causing the center to fall out.

CONTROL:
1. Some cultivars are very susceptible.
2. Maintain cotton in vigorous and well-fertilized condition.

Anthracnose

CAUSE: *Colletotrichum gossypii* South., teleomorph *Glomerella gossypii* Edg., is seedborne as mycelium in or on the seed and conidia on the outside of the seed. It also survives as acervuli and perithecia in infected tissue. Infected seed may give rise to infected cotyledons on which conidia are produced. The conidia are airborne to healthy tissue. Primary inoculum may also arise from residue where perithecia produce acospores that become airborne to plants. Additionally, acervuli and mycelium on residue produce conidia that are either splashed or airborne to cotton. Disease is favored by moist or humid conditions. Seedling blight is most severe at temperatures of 20–26°C.

DISTRIBUTION: Generally distributed wherever cotton is grown; however, disease is most prevalent under humid conditions.

SYMPTOMS: Disease symptoms occur on all aboveground plant parts. The cotyledons of seedlings have small, reddish to light brown spots or a necrosis of the margin. Oblong, brown lesions occur on young hypocotyls and stems. During humid or wet weather, lesions may become pink due to formation of conidial masses. Severe seedling infection results in stem girdling, yellowing of the leaves, and lack of thriftiness. Hypocotyls have dark lesions, with the greatest injury occurring around the soil surface.

Mature or older plants usually become infected through the stem and infrequently through the leaves. Small, brown spots occur that are associated with injuries or angular leaf spots.

Infected bolls initially have sunken, small, reddish or red-brown spots. Eventually the spots become black to dark brown with a reddish margin. During damp weather the center of the spot becomes pink due to formation of slimy masses of conidia. Infection through a dead pistil results in internal rot of the boll.

Acervuli that look like pin cushions under magnification develop on lesions after tissue necrosis. Perithecia may also develop in dead tissue but are difficult to see and are evident as tiny dark bumps, which are the beaks of the perithecia. Most of the structure is submerged in dead tissue.

CONTROL:
1. Treat seed with a seed-protectant fungicide.
2. Plant recommended cultivars.
3. Rotate cotton with other crops.

4. Plow under infected residue where erosion is not a problem.
5. Control insects.
6. Fertilize properly.
7. Practice timely defoliation.
8. Skip-row planting provides for an open stand and prevents buildup of high humidity around lower bolls.

Ascochyta Blight

This disease is also called wet-weather blight and ashen spot.

CAUSE: *Ascochyta gossypii* Woron. overwinters as pycnidia and mycelium in infected crop residue in the soil and as spores and mycelium on seed. In the spring, during wet weather, spores are produced in pycnidia and are splashed or airborne to plants. Spores from seedborne inoculum also serve as an important source of primary inoculum. *Ascochyta gossypii* is most likely to infect young plants that are deficient in minerals, damaged by insects, hail, or other forms of mechanical wounding.

Pycnidia are formed in mature lesions and serve as a secondary source of inoculum. Disease is most severe under humid, moist conditions. Ascochyta blight may also be severe in cooler, humid cotton-growing areas.

DISTRIBUTION: Higher elevations in Mexico, South America, Africa, and Asia. In the United States, the disease is most prevalent along the northern fringe of the eastern and central cotton-growing areas.

SYMPTOMS: Seedling blight may occur either preemergence or postemergence. Symptoms occur on bolls, leaves, stems, and branches. Initially, small, round (2 mm in diameter), brown spots occur on leaves. These spots enlarge rapidly during humid weather and turn grayish, ash, or light brown, sometimes with red-brown borders. With continued lesion enlargement, the grayish centers become smaller. As tissues age, the center of the spot may fall out. Spots coalesce during wet, cool weather, resulting in defoliation and blighting of young plants.

Stems and branches have dark brown and elongated lesions that frequently occur at the base of leaf petioles. Lesions enlarge rapidly and become sunken and light brown in the center. Lesion centers decay early and tend to shred, break up, and fall out. The lesions may encircle the stem or branch and kill the plant part above the infection. Infection on bolls is similar to that on stems and branches.

CONTROL:
1. Treat seed with a fungicide-seed protectant.
2. Rotate cotton with nonsusceptible crops. Ascochyta blight is more severe when cotton follows cotton.
3. Plow under infected residue where erosion is not a problem.
4. Soil should have a well-balanced fertility.

Ashbya Fiber Stain

CAUSE: *Ashbya gossypii* Guill. likely survives as mycelium in infected residue. Asci, in which ascospores are produced, are formed within mycelial strands. The fungus enters the boll through feeding wounds of the cotton stainer bug.

DISTRIBUTION: Not reported.

SYMPTOMS: No external symptoms are visible on the boll. Fibers and seed have a tan to brown discoloration and eventually dry out.

CONTROL: No control is reported.

Ashy Stem

CAUSE: *Sclerotium bataticola* Taub. survives in soil and infected residue as sclerotia. The pycnicial stage is *Macrophomina phaseolina* (Tassi) Goid.; however, pycnidia are rarely formed in the stem. Sclerotia germinate on root surfaces and form several germ tubes that penetrate the root.

DISTRIBUTION: Generally distributed in the warmer cotton-growing areas of the world.

SYMPTOMS: Plants are infected from seedling to blossom stage. Leaves wilt and droop; eventually they become chlorotic and are prematurely shed. Plants may be stunted. Gray-white lesions develop on stems near the soil line and extend several centimeters up the stem. Small black sclerotia are scattered throughout the lesion and are embedded in the cortex and wood. Roots may be decayed. Plants may die but remain standing.

CONTROL:
1. Rotate cotton with other crops, preferably small grains.
2. Do not overplant. Crowded seedlings are more subject to infection.
3. Ensure that soils are well fertilized.
4. Plow under infected residue where erosion is not a problem.

Aspergillus Boll Rot

CAUSE: *Aspergillus flavus* Lk. ex Fr. survives in soil and residue as spores and mycelium. It enters bolls through wounds made by the pink bollworm, *Pectinophora gossypiella* Saunders, and other insects. Infection is more severe when carpels are infected and separated by *A. niger* Van Tiegh and *Rhizopus* sp. The fungus is also more pathogenic on senescent tissue at high humidity and a temperature of 32°C and above. Seed infection levels are affected by water potentials on the day of anthesis and inoculation, with the highest infection levels generally occurring in seed from plants with water potentials between −1.6 and −1.9 bar.

DISTRIBUTION: Generally distributed wherever cotton is grown.

SYMPTOMS: Lint is stained yellow and otherwise discolored. Lint fluoresces green-yellow under ultraviolet light.

CONTROL: Control insects on cotton.

Black Boll Rot

CAUSE: *Aspergillus niger* Van Tiegh. survives as a saprophyte on dead plant material of many kinds. Conidia are produced that are airborne and enter a boll primarily through wounds. Black boll rot is more severe in rank cotton growth and frequent rains during boll development and opening. In irrigated areas, boll rot is more severe where excessive moisture has occurred late in the season.

DISTRIBUTION: Generally distributed wherever cotton is grown.

SYMPTOMS: Initially a soft pinkish rot develops on the side or base of a boll. Eventually the rot changes from pink to brown, with abundant production of black spores in the affected area, giving a black sooty appearance.

CONTROL:
1. Avoid practices that promote rank growth, especially excessive nitrogen.
2. Use correct timing and control of irrigation.
3. Control insects after boll development.
4. Shield equipment to reduce mechanical injuries.
5. Use of bottom defoliation and two-stage harvesting will reduce exposure in field.
6. Keep cotton free of weeds.
7. Use skip-row plantings.
8. Rotate cotton with other crops.
9. Plow under infected residue so the old rotting bolls do not provide a source of inoculum to infect the current crop.

Black Root Rot

This disease is also called internal collar rot.

CAUSE: *Thielaviopsis basicola* (Berk. & Br.) Ferr. survives as chlamydospores on infected roots and in soil. Chlamydospores germinate to produce mycelium that infect roots through the root hairs and then grow to the endodermis. Disease is favored by temperatures of 15–20°C, wet and alkaline soils. Chlamydospores are produced as a black crust on roots.

DISTRIBUTION: Egypt, Peru, United States, and the USSR. The disease is uncommon.

SYMPTOMS: Seedling blight may occur in cool, wet weather. Seedlings are stunted with small, cupped, chlorotic leaves that have purplish borders with browning of the margins. Leaves are shed from severely diseased plants. The roots are brown and decayed with taproots smaller in diameter than those of healthy plants. The cortical tissue at the base of the stem may appear healthy but the vascular tissue becomes black or purple. The stem base and taproot may appear swollen, with a black crust consisting of chlamydospores on the roots. There is a sharp line between diseased and healthy tissue. Infected seedlings can be pulled from the soil easily. Within a few weeks diseased tissue is sloughed off and no symptoms remain.

An internal collar rot phase occurs on *Gossypium barbadense* from mid to late season. Leaves and stems wilt and collapse suddenly. Reddish lenticels and swelling and brown to purple-black discoloration of the stele occur in the crown.

CONTROL:
1. Plant in warm soils.
2. Rotate cotton with monocotyledons.

Cercospora Leaf Spot

CAUSE: *Cercospora gossypina* Cke. survives as conidia and mycelium on seed and in residue. The teleomorph, *Mycosphaerella gossypina* (Atk.) Earle may also survive as perithecia in infected residue. Primary infection may occur by conidia or ascospores being airborne to plants. Secondary inoculum is provided by conidia that are produced in leaf spots. Infection may follow other diseases or occur in cotton that is under stress from low fertility or unfavorable environmental conditions. Lack of soil moisture at time of greatest boll set and lack of proper amounts of nitrogen and potash cause the disease to be more severe. The disease is more severe on cotton grown on sandy soils.

DISTRIBUTION: Generally distributed in warm, humid areas.

SYMPTOMS: Symptoms first occur on older upper leaves and eventually on young leaves. Initially leaf spots are barely visible reddish points. Eventually leaf spots become about 2 cm in diameter, round to irregular in shape, with white centers surrounded by a purple margin. Masses of conidia may grow on infected areas, giving both lower and upper surface a dark appearance. As the lesion enlarges, the center may drop out giving a shot-hole effect.

Affected leaves turn brown and dry up, resulting in partial or complete defoliation. Young squares and bolls may fall from the plants. The main terminal and upper fruiting branches may die back under severe disease conditions. Branches on plants may be unaffected.

CONTROL:
1. Treat seed with a seed-protectant fungicide.
2. Maintain proper soil fertility.
3. Avoid insect and mechanical injury.
4. Plow under infected residue where erosion is not a problem.
5. Early-maturing cultivars that fruit heavily are more susceptible than later-maturing cultivars that set bolls over a longer period of time.

Botryodiplodia Seedling Blight and Boll Rot

CAUSE: *Botryodiplodia theobromae* Pat. is seedborne. It is considered a weak pathogen.

DISTRIBUTION: Brazil.

SYMPTOMS: Seedling death and boll rot.

CONTROL: No control is reported.

Charcoal Rot

CAUSE: *Macrophomina phaseolina* (Tassi) Goid. (syn. *M. phaseoli* (Maubl.) Ashby, *Macrophoma phaseolina* Tassi, *Rhizoctonia bataticola* (Taub.) Butler, and *Sclerotium bataticola* Taub.). The fungus survives as sclerotia in residue and soil and as saprophytic mycelium in the soil. Disease is most severe during high temperatures and periods of moisture stress.

DISTRIBUTION: Widely distributed throughout the world under hot, dry conditions.

SYMPTOMS: Infected plants appear to wilt and die rapidly. Affected plants appear to have a dry rot, with numerous tiny black sclerotia distributed throughout the infected stem, giving the tissue a grayish appearance similar to charcoal.

CONTROL: Management of water supply for irrigated cotton to reduce water stress during periods of high temperatures.

Choanephora Mold

CAUSE: *Choanephora cucurbitarum* (Berk. & Rav.) Thaxt. survives as mycelium, chlamydospores, and possibly spores in residue or soil. Blossoms become infected during wet, humid conditions.

DISTRIBUTION: United States.

SYMPTOMS: The corolla becomes covered with a white, fuzzy, fungal growth that consists of young mycelium and conidiophores. Later the fungal growth becomes sprinkled with black specks, which are the conidial heads borne on the end of the conidiophores.

CONTROL: No specific control but general controls for other boll rots are of benefit.

Colletotrichum Boll Rot

CAUSE: *Colletotrichum capsici* (Syd.) Butter & Bisby. Boll rot is reported to occur under exceptionally wet conditions.

DISTRIBUTION: Louisiana.

SYMPTOMS: Initially, a circular darkened lesion occurs that is often associated with a suture. Eventually the entire boll turns black. In advanced stages, the carpel wall is destroyed, leaving the skeletonized remains of the wall. Severely rotted bolls open often; however, the lint usually does not fluff, leaving a tight-locked boll. Lint from infected bolls is discolored and usually dislodges, falling to the ground.

CONTROL: No control is reported.

Corynespora Leaf Spot

CAUSE: *Corynespora cassiicola* (Burk. & Curt.) Wei overwinters in diseased cotton residue. It can also grow saprophytically on many different kinds of plant residue in the soil. Mycelium and chlamydospores in plant residue or chlamydospores in soil afford other means of survival. Conidia are splashed or blown to leaves and cause infection when free moisture is present or when the relative humidity is 80 percent or above.

DISTRIBUTION: The disease has been reported in Mississippi but is likely present in many cotton-growing areas in the world.

SYMPTOMS: Initially, pinpoint spots that are brick red develop on leaves. Later spots become roughly circular (2–6 mm in diameter), with a dark brown margin and a light brown center that may fall out. Mature spots may have alternate light and dark brown areas, giving a zonate appearance. On severely diseased leaves, spots coalesce, forming large, irregular necrotic areas. In cases of severe disease, leaves may yellow, wilt, and fall off the plant. Reddish spots also occur on petioles, frequently girdling them.

CONTROL: No control is usually necessary other than to maintain crop in vigorous growing condition.

Cotton Root Rot

This disease is also called Texas root rot and Phymatotrichum root rot.

CAUSE: *Phymatotrichum omnivorum* (Shear) Duggar survives in soil as sclerotia and mycelial strands. Survival could be facilitated by mycelial strands through yearly infection of susceptible dicotyledons in a rotation. However, mycelial strands are not pathogenic to tap roots unless attached to sclerotia. Mycelial strands may also form on monocotyledonous roots. Later sclerotia germinate to produce mycelium that infect roots. Sclerotia are abundantly produced on rotted roots. Conidia are produced on mycelial mats that form on the soil surface near dead plants when the soil is moist, but are not known to germinate. Almost 2,000 species of dicotyledonous plants may be infected by *P. omnivorum*.

Studies showed that no foliar symptoms develop until soil temperature at the depth at which inoculum occurs is above 22°C. Additionally soil moisture levels between −12 bars and −16 bars reduce the rate of disease development.

DISTRIBUTION. Southwestern North America, particularly Arizona, Texas, and Mexico. The disease is most common in soils with high pH.

SYMPTOMS: Expression of symptoms is associated with cortical senescence. *Phymatotrichum omnivorum* is capable of attacking roots 5 days after seedling emergence. Infected roots become water-soaked, discolored, and the cortex is sloughed 18–25 days after emergence, at which time foliar symptoms occur. Plant death occurs 27–50 days after emergence regardless of age that cotton plants are exposed to sclerotial inoculum of *P. omnivorum*.

When decay girdles the tap root, plants rapidly wilt and die. Initially leaves yellow and turn bronze, then wilt. Leaves dry up, turn brown, but remain attached to the plant. Bark and cambial tissues of the root turn brown with light tan strands of mycelium growing on the root surface. The bark easily peels off affected roots, revealing a reddish stain along the white woody tissue. Mycelial mats form on the soil surface near dead plants when soils are moist. At first the mats are white but later become tan and powdery due to production of conidia on the mat surface. Atypical symptoms have been observed only during periods of low soil moisture availability in Texas. The aboveground symptoms are initially a gradual wilting followed by leaf chlorosis and defoliation. Mature bolls and young leaves remain attached to the stem and the plants remain alive. Roots have discolored sunken lesions 10–20 cm below the soil surface.

CONTROL:
1. Plow under green manure crops. Crop residues can also be plowed under if nitrogen is applied to promote rapid decay.
2. Rotate cotton with cereals or grasses.
3. Plow deeply during hot, dry weather, thus exposing and killing the fungus.
4. Provide adequate fertility to ensure vigorous cotton growth.
5. Plant early and control insects and diseases to promote early maturation of bolls before *P. omnivorum* becomes fully active.

Cylindrocladium Black Rot

CAUSE: *Cylindrocladium crotalariae* (Loos) Bell & Sobers survives mainly as microsclerotia on infected residue. Mycelium may also serve as a means of survival in infected refuse. Microsclerotia are the main means of dissemination and germinate to form mycelium that penetrate roots.

DISTRIBUTION: Southeastern United States. This is a very minor problem and does not cause obvious symptoms.

SYMPTOMS: Ordinarily no aboveground symptoms are produced. Roots of susceptible cultivars are blackened and reduced.

CONTROL: No control is necessary. However, cotton may serve to maintain or increase levels of inoculum in the soil that may harm other crops such as peanuts.

Diplodia Boll Rot

CAUSE: *Diplodia gossypina* (Cke.) McGuire & Cooper survives as pycnidia in infected residue. During wet weather conidia are produced in pycnidia and are splashed or airborne to bolls where infection occurs through weevil punctures or wounds. Injured and noninjured bolls are infected at all stages of maturity but are most severely diseased when infected at ages 7 days or less or 40 days or older. Lower bolls that are in contact with soil are frequently infected. Disease development is optimum during moist conditions and a temperature of 3°C.

DISTRIBUTION: Generally distributed wherever cotton is grown.

SYMPTOMS: Initially, small, brown, water-soaked and sunken lesions occur on capsules or bracts. Pycnidia that appear as tiny black blisters appear on the boll surface. Black sooty masses of spores are released from pycnidia under high-humidity conditions and cover the boll surface. Lesions may enlarge and cover the entire boll. Bolls blacken, dry up, and split, exposing blackened and matted lint.

CONTROL: In general avoid practices that prevent plants from drying out.
1. Avoid practices that promote rank growth.
2. Use correct timing and control of irrigation.
3. Control insects after boll development.
4. Shield equipment to reduce mechanical injuries.
5. Keep cotton free of weeds.

Escobilla

This disease is also known as little broom, ramulose, superbrotamento, supersprouting, and witches' broom.

CAUSE: *Colletotrichum gossypii* South var. *cephalosporioides* Costa Fraga. Disease is most severe during periods of prolonged rainfall.

DISTRIBUTION: Brazil and Venezuela.

SYMPTOMS: Plants of all ages may be affected; however, those infected while young are most severely diseased. Infected plants have a dense shrublike appearance and have both healthy and diseased branches. Branches are swollen and twisted, with shortened internodes and swollen nodes resulting in an excessive number of short twigs and leaves.

Leaves on affected branches are small, dark green, wrinkled, and have reduced lobes and short petioles. Healthy-appearing leaves may occur on affected branches but have turned up margins and swollen petioles with internal browning. Small brown spots may occur on both leaves and shoots. Spots later turn gray with cracks developing in the center. Spots tend to elongate along leaf veins. Terminal buds are eventually killed, causing numerous lateral branching and tufts. Gray mycelium in which black dotlike acervuli appear eventually occurs on lesions.

Bolls on affected plants may remain green without opening. Seeds in unopened bolls may germinate abnormally.

CONTROL: No control is practiced.

Fusarium Boll Rot

CAUSE: *Fusarium moniliforme* Sheldon, *F. oxysporum* Schlecht. ex. fr., *F. roseum* Lk. emend. Snyd. & Hans., and *F. solani* (Mart.) Appel. & Wr. *Fusarium oxysporum* and *F. roseum* attack uninjured green bolls by initially infecting the bracts, then invading the capsule base through the receptacle. The most important factor limiting rot is boll age. Bolls 33–35 days old or older are more susceptible than younger fruits. Boll rot proceeds rapidly during moist conditions. Production of vast numbers of spores on plant debris such as flowers, bolls, and squares apparently provides the airborne inoculum necessary to initiate boll deterioration.

DISTRIBUTION: Southeastern United States.

SYMPTOMS: Initially, small, growing lesions occur along the bract margin, particularly at the apices of the toothed areas. Eventually the entire bract becomes necrotic. The capsule becomes invaded through the receptacle with tissues showing a dark blue-green discoloration. Decay is more rapid in dehiscing bolls.

CONTROL: No control is reported but development of resistant cultivars may be possible.

Fusarium Damping Off

CAUSE: *Fusarium* spp. survive saprophytically in infected residue and as chlamydospores in the soil. Seed and roots are infected under cool, wet soil conditions.

DISTRIBUTION: Generally distributed wherever cotton is grown.

SYMPTOMS: Seed may be rotted. Roots are tan to dark brown and rotted.

CONTROL: Control measures are basically the same as for pythium damping off.

Fusarium Wilt

CAUSE: *Fusarium oxysporum* Schlect. f. sp. *vasinfectum* (Atk.) Snyd. & Hans. survives as chlamydospores and sclerotia in the soil and as mycelium in infected tissue. The fungus has also been isolated from seed and gin trash. This may serve as a means of dissemination. Conidia, chlamydospores, or sclerotia germinate to infect roots. Usually the fungus enters roots only through wounds made by nematodes, including *Belonolaimus* spp., *Conemoides* spp., *Helicotylenchus* spp., *Meloidogyne* spp., *Pratylenchus* spp., *Rotylenchulus* spp., *Trichodorous* spp., *Tylenchorynchus* spp., and *Xiphenema* spp. Without the presence of nematodes, Fusarium wilt would not be a serious problem. Mycelium grows into xylem tissue and wilting occurs when water is prevented from moving up the plant. Sporulation occurs on the outside of dead plants and conidia are wind disseminated to other areas. Dissemination also occurs by any means that will move soil. Dispersal by furrow irrigation water has been demonstrated in Israel. Fusarium wilt is most serious in acid, sandy soil. At present, several races of the fungus are known to exist.

DISTRIBUTION: Generally distributed wherever cotton is grown.

SYMPTOMS: Fusarium wilt ordinarily appears about blossoming time, although seedlings may appear to damp off from infection by *F. oxysporum* f. sp. *vasinfectum*. The general appearance of a field will be spotty or uneven. Areas ranging in size from a few scattered plants to several acres will be affected with the most severe damage on the sandiest soils.

Initially, plants are stunted and unthrifty; eventually leaves wilt. Leaves turn yellow beginning at leaf margins, then between the veins, and shed from the bottom of the plant upward. The entire plant becomes bare of leaves, dies, and turns black, due in part to growth of saprophytic fungi.

If an infected plant does not die, it may remain stunted with lateral branches, sometimes outgrowing the main stalk. Frequently infected plants flower and mature earlier than healthy plants. A diagonal cut into the main stem will show a dark brown discoloration of the inner stem. Roots will also show symptoms of nematode injury.

CONTROL:
1. Plant resistant cultivars.
2. Rotate cotton with crops on which the cotton root knot nematode will not reproduce, including millet, peanuts, small grains, sorghum and Sudangrass.
3. Cotton stalks should be mowed and turned up with a bottom plow after harvest to reduce nematode populations. If erosion is a problem, plant a cover crop of rye. Most legumes are susceptible to root knot nematodes.
4. Leave soil in summer fallow and control weeds to deny nematodes hosts on which to reproduce.
5. Adequate potash fertilization sometimes reduces wilt severity.
6. Use a soil nematicide where cost can be warranted.
7. Avoid planting upland cotton in infested, sandy, acid soils if possible.

Gray Mildew

This disease is also called aerolate mildew, dahiya disease, false mildew, frosty mildew, moho blanco, and white mold.

CAUSE: *Ramularia areola* Atk. (syn. *Cercosporella gossypii* Speg.), survives as mycelium, conidia, or perithecia in infected residue. The teleomorph is *Mycosphaerella areola* Ehrlich & Wolf. The most rapid germination occurs at temperatures of 20–30°C and nightly wetting and daily drying of leaf tissues. Conidia are produced in lesions and are airborne to provide secondary inoculum. Disease is most severe under wet, humid conditions.

DISTRIBUTION: Generally distributed wherever cotton is grown, but gray mildew is most common in Africa and Asia.

SYMPTOMS: Symptoms normally appear toward the end of the growing season. Cotyledons have circular, water-soaked, dark green patches. Eventually cotyledons become chlorotic and wither with spots turning red-brown. Typical symptoms on leaves are small, light green to yellowish, angular spots on the upper leaf surface that are limited by the veins. During high humidity, the underside of the leaves is covered with white growth consisting of conidia and conidiophores. Severely infected leaves become chlorotic, dry, curly, red-brown, and drop prematurely. Bolls may open prematurely with loss of fiber quality.

Resistant cultivars have tiny, red-brown spots surrounded by a chlorotic area.

CONTROL:
1. Plant resistant cultivars.
2. Apply foliar fungicides.

Helminthosporium Leaf Spot

CAUSE: *Helminthosporium gossypii* Tucker overseasons in infected leaves.

DISTRIBUTION: Asia.

SYMPTOMS: Leaves, flower bracts, and bolls are infected. Spots on leaves and bracts are small, 1–8 mm in diameter, initially light red, but gradually turn dark purple with brown centers that fall out. Spots may be numerous, causing leaves to die and drop off prematurely. Spots on bolls are small and purplish but usually cause no damage to fiber.

CONTROL: No satisfactory control is known.

Nematospora Boll Stain

CAUSE: *Nematospora coryli* Pegl., *N. gossypii* Ashby & Nowell, and *N. nagpuri* Dastur & Singh survives in infected residue and possibly on nearby wild plants. Infection occurs through insect wounds.

DISTRIBUTION: *Nematospora coryli* is reported from California, *N. gossypii* in West India and Africa, and *N. nagpuri* in India.

SYMPTOMS: Initially, infected bolls display no external symptoms except tiny wounds or scars. Eventually lint becomes tannish and seed is shriveled and killed. Sheaths are discolored and affected bolls are shed.

CONTROL: The same control measures that are used to control black boll rot are followed.

Nigrospora Lint Rot

CAUSE: *Nigrospora oryzae* (Berk. & Br.) Petch is disseminated and introduced into the boll by the mite, *Siteroptes reniformis* (Krantz). A mutualistic form of symbiosis exists between the two organisms.

DISTRIBUTION: Africa, the USSR, and the United States.

SYMPTOMS: Lint fails to fluff and infected fibers are weakened. Lint is discolored by the presence of dark brown to black conidia on surface and in the lumen.

CONTROL: No control is reported.

Phomopsis Leaf Spot

CAUSE: *Phomopis* sp. The fungus probably survives as pycnidia in infected residue.

DISTRIBUTION: Louisiana.

SYMPTOMS: All aboveground parts of the plant may be attacked but the principal damage is to flower buds that may not open. Gray, water-soaked, slightly sunken areas may occur on leaves and stems. Later the infected areas are covered with pycnidia that resemble dark bumps.

CONTROL: No control is known.

Phytophthora Boll Rot

CAUSE: *Phytophthora parasitica* Dast. (*P. nicotianae* var. *parasitica* (Dast.) Waterhous). Infection possibly occurs in two ways: directly by the sporangia forming a germ tube or indirectly by formation of zoospores within sporangia. The zoospores are liberated and swim for a few minutes, become rounded, and germinate by means of a germ tube. Disease is most severe during conditions of high humidity. Other species of Phytophthora reported to cause a boll rot are *P. cactorum* (Lebert & Cohn) Schroter and *P. palmivora* (Butler) Butler.

DISTRIBUTION: Louisiana and probably other areas where cotton is grown under conditions of high humidity.

SYMPTOMS: Diseased bolls are blue-black to black but usually the entire surface becomes black in a couple of days. The boll surface becomes spongy and develops a soft watery rot. A white, mealy fungal growth consisting of sporangia and mycelium occurs on the boll surface. Bolls, particularly those approaching maturity, rot in two to three days.

CONTROL: No control is reported.

Powdery Mildew

This disease is also called manta blanca or white mantle when caused by *Salmonia malachrae*.

CAUSE: *Salmonia malachrae* (Seaver) Blumer & Muller (syn. *Erysiphe malachrae* Seaver) and *Leveillula taurica* (Lev.) Arn. (syn. *E. taurica* Les. and *Oidiopsis taurica* (Lev.) Salm.). The anamorph of *L. taurica* is *Oidiopsis gossypii* (Wakef.) Raychaudhuri (syn. *Ovulariopsis gossypii* Wakef.). The anamorph of *S. malachrae* is not known for certain. The fungi survive by continually infecting cotton or alternate hosts throughout the year. Infection is optimum at 85–100 percent relative humidity. *Salmonia malachrae* requires cooler temperatures than *L. taurica* (25–30°C) to cause disease.

DISTRIBUTION: *Leveillula taurica* is found in California, India, Peru, Sudan, the USSR, and the West Indies. *Salmonia malachrae* is found in the Antilles and South America.

SYMPTOMS: Symptoms caused by the two fungi differ. Initially *S. malachrae* normally appears only on the upper leaf surface as scattered circular areas of white mycelium. The areas coalesce to cover the entire leaf area. A severe infection late in the season gives the appearance of a field covered with snow. White to tan cleistothecia are scattered through mycelium. Infected leaves become chlorotic and curl. Defoliation may occur.

   Leveillula taurica causes initial symptoms of small (1.5–2.0 mm), powdery, white patches on the underside of the leaf. The patches may be angular or rounded depending on the cultivar. Sometimes no external symptoms occur. Later the patches will coalesce to cover the whole leaf surface; the leaf will become chlorotic, regardless of external symptoms, and fall. Cleistothecia are normally not present.

CONTROL: Control is normally not practiced.

Pythium Damping Off

CAUSE: *Pythium ultimum* Trow. is the most prevalent species found on cotton. Other *Pythium* spp. associated with seedling disease include *P. aphanidermatum* (Eds.) Fitz., *P. debaryanum* Hesse, *P. heterothallicum* Campbell & Hendrix, *P. irregulare* Buis., *P. polytylum* Drechs., *P. splendens* Braun, and *P. sylvaticum* Campbell & Hendrix. *Pythium* spp. generally survive in soil or residue as oospores and saprophytically as mycelium on plant residue. In cool (10–15°C), wet soil (17% or more), oospores germinate to form sporangium in which zoospores are formed. Zoospores swim in water to a plant surface where they encyst and germinate to form a germ tube. At higher temperatures, oospores may germinate directly and form a germ tube.

DISTRIBUTION: Generally distributed wherever cotton is grown.

SYMPTOMS: General symptoms are as follows. Seed may be rotted and soft. Initially hypocotyls and roots become water-soaked and pale green in color. Lesions on hypocotyls range from small discolored spots to large sunken, necrotic areas. Eventually the infected hypocotyls and roots become gray and dry up. The hypocotyl becomes constricted above the soil line, causing the seedling to fall over.

Root rot caused by *P. splendens* results in reduction of shoot growth. Severe root rot has not been observed under field conditions, although small, light brown lesions occur on tap and secondary roots.

CONTROL:
1. Plant only high-quality seed that has been treated with a seed-protectant fungicide.
2. Apply fungicide as a spray or dust into seed furrow.
3. If necessary, delay planting until soil temperature has warmed up to 20°C or more for at least three or four mornings before planting. This ensures rapid seed germination and vigorous seedling development.
4. Plant in bedded rows or slightly raised seed beds. Normally the soil temperature warms up faster than a furrow or level seed bed.

Rhizoctonia Leaf Spot

CAUSE: *Rhizoctonia solani* Kuhn. Leaf spots may result from airborne basidiospores. Disease is favored by high rainfall and warm temperatures particularly in thick foliage. See Soreshin.

DISTRIBUTION: El Salvador and Louisiana.

SYMPTOMS: Leaf spots are light brown, irregular in shape, and have dark purple borders. As spots enlarge, the center becomes chlorotic, then necrotic. Cracks develop that cause the center of older spots to drop out, giving a shot-hole and ragged appearance to infected leaves. During periods of high humidity, brownish growth of the fungus may appear on the underside of the leaf.

CONTROL: No control is necessary.

Rhizopus Boll Rot

CAUSE: *Rhizopus nigricans* Ehrenb. survives as mycelium and spores in residue and soil. Bolls become infected by spores under moist and warm weather conditions.

DISTRIBUTION: Generally distributed wherever cotton is grown.

SYMPTOMS: Infected areas of the boll initially are olive green. As the infected portion dries up, it becomes black. Fungal growth consisting of mycelium, sporangiophores, and sporangia, forms a dark gray mold over the boll.

CONTROL: The same control measures that are used to control black boll rot are followed.

Rust

CAUSE: *Cerotelium desmium* (Berkeley & Brooms) Arthur is a rust fungus where only the uredial stage is known.

DISTRIBUTION: India, South America, South Pacific, southeastern United States, and West Indies.

SYMPTOMS: The disease is confined mostly to leaves. The uredia are yellow-brown and initially start as small pustules that eventually become more powdery. The upper leaf surface is marked by small, purple-brown spots due to uredia being deeply immersed in host tissue. The spots coalesce into large patches, causing defoliation.

CONTROL: No control is reported.

Rust

CAUSE: *Puccinia schedonnardi* Kellerm. & Sw. is a heteroecious long-cycled rust whose alternate host consists of several grasses including *Muhlenbergia* spp. and *Sporobolus* spp. The pycnial and aecial stages occur on cotton; the uredial and telial stages occur on grasses.

DISTRIBUTION: Mexico and southwestern United States. The rust probably occurs in other cotton-growing areas also.

SYMPTOMS: Pycnia are yellow-orange and inconspicuous. Aecia develop around pycnia and are large, orange, raised spots on leaves, bolls, and bracts. Uredia are a light cinnamon-brown; telia are a dark brown.

CONTROL: No control is reported.

Soreshin

This disease is also called damping off, postemergence damping off, and Rhizoctonia seedling blight.

CAUSE: *Rhizoctonia solani* Kuhn survives as sclerotia and mycelium in soil or residue. The fungus also lives saprophytically on a wide range of dead plant tissue. Sclerotia germinate to produce myeclium that infects the hypocotyl near the soil line. The teleomorph, *Thanatephorus cucumeris* (Frank) Donk. produces basidiospores on a loose hyphal network. However, it is not certain what function the basidiospores perform in the life cycle of the fungus. Leaves may be infected by basidiospores or soilborne propagules other than basidiospores. Soreshin is most severe during periods of cool, wet weather.

DISTRIBUTION: Generally distributed wherever cotton is grown.

SYMPTOMS: Seedlings may damp off either preemergence or postemergence, usually during cool, wet weather. The hypocotyl is infected near the soil line and becomes necrotic, with the cotyledons drooping, then wilting. Under warm, dry soil conditions the hypocotyl lesions become red-brown, sunken cankers. Cankers may be either superficial linear lesions that grow slightly into cortical tissue or may completely girdle the stem near the soil line. If soil remains warm and dry, infected plants may partially recover by producing new roots, usually above the infected area of the stem or hypocotyl. Infrequently angular, brown leaf spots whose centers may fall out giving a shot-hole appearance will occur during moist weather; see Rhizoctonia Leaf Spot.

CONTROL:
1. Plant after soils have warmed up.
2. Rotate cotton with other crops; however, *R. solani* has a wide host range and can also exist saprophytically for several years in the absence of a host crop.
3. Plant on a bed because soil temperatures warm up faster than in a furrow or level soil.
4. Soil pH should be between 6.0 and 6.5.
5. Ensure that soil fertility is adequate.
6. Avoid chemical injury.
7. Plant only healthy, high-quality seed.
8. Treat seed with a seed-protectant fungicide.
9. Treat soil with an in-furrow granule or spray fungicide.

Southern Blight

This disease is also called Sclerotium rot and Sclerotium stem and root rot.

CAUSE: *Sclerotium rolfsii* Sacc., the binomials most commonly used at present for the teleomorph are *Corticium rolfsii* (Sacc.) Curzi and *Pellicularia rolfsii* West. However, *C. rolfsii* is sometimes considered to be a synonym for *P. rolfsii*. The causal fungus overwinters as sclerotia in the soil. It is also a saprophyte and colonizes dead organic matter in the soil. Sclerotia germinate under high soil temperatures and in relatively dry soil to form mycelium that infects seed or plants. Cool soil temperatures kill sclerotia and mycelium. *Sclerotium rolfsii* occurs most commonly in neutral to acid sandy soils.

DISTRIBUTION: Generally distributed wherever cotton is grown.

SYMPTOMS: Plants of all ages are infected. Initially, leaves wilt until an infected plant begins to lose leaves, blossoms, or bolls.

Eventually the plant dies. The wilting is caused by a canker that develops at or just below the soil surface. If soil moisture is present, a white, cottony, growth, which is the mycelium, will grow up the stem for a distance of 3 cm or more. Dark sclerotia will eventually develop in and on the mycelium. Dense white mycelium in which successive concentric rings of sclerotia are formed will grow out from the stem over moist soil for a distance of up to 10 cm. A swelling of the base of the stem and taproot has also been reported as a symptom.

CONTROL:
1. Plant the most resistant cultivar.
2. Rotate cotton with other crops such as small grains.
3. Plow infected cotton residue 12 cm deep.

Southwestern Cotton Rust

This disease is also called cotton rust.

CAUSE: *Puccinia cacabata* Arth. & Holw. (syn. *P. stakmanii* Presley) is a heteroecious long-cycle rust that has grama grass, *Bouteloua* spp., as the alternate host. The rust overwinters as teliospores that form on grama grass in late summer. During wet weather in the summer and night temperatures below 28.5°C, teliospores germinate to produce basidiospores that are airborne to cotton. Pycnia are formed, usually on the upper surface of cotton leaves. Aecia are produced on the underside of leaves, bracts, green bolls, and stems. Aeciospores are released from aecia, are airborne, and infect only grama grass in the presence of moisture. Uredia in which urediospores are produced are formed on grama grass. Urediospores are airborne and infect only grama grass. Telia, in which teliospores are produced, appear on grama grass in late summer as the grass matures. Disease is favored by high humidity and abundant rainfall.

DISTRIBUTION: Mexico and southwestern United States.

SYMPTOMS: Uredia are formed mostly on leaves of grama grass and appear as small, pale brown, raised areas, with the leaf epidermis turned back. Telia are formed either in the uredium or separately as dark brown to black raised pustules. Pycnia initially appear as small, somewhat inconspicuous yellowish spots, usually on the upper side of leaves; however, they may occur on any aboveground plant part. Eventually these develop into bright yellow pustules and later turn brown. Aecia are large and easily observed as orange-yellow, circular, and slightly raised lesions usually on the underside of cotton leaves. Later they fade to a light yellow. In severe rust infestations, leaves curl and plants may be defoliated. Stems and branches may become girdled and break. Bolls may be smaller in size.

CONTROL:
1. Plant resistant cultivars.
2. Apply a foliar fungicide spray before rust becomes severe.
3. Destroy grama grass in the vicinity of cotton fields; however, this is of limited value since spores may be airborne a considerable distance.

Stemphylium Leaf Spot

CAUSE: *Stemphylium* sp.

DISTRIBUTION: Generally distributed wherever cotton is grown. The disease is of minor importance.

SYMPTOMS: Small, roughly circular spots, with concentric rings giving a targetlike appearance. Spots tend to be discrete and usually do not spread or coalesce.

CONTROL: No control is necessary other than to keep plants in a thrifty, vigorous condition.

Tropical Rust

CAUSE: *Phakopsora gossypii* (Arth.) Hirat. f. Disease is favored by high relative humidity. Dissemination is by airborne urediospores.

DISTRIBUTION: Tropical areas around the world.

SYMPTOMS: Leaves are primarily affected but other aboveground plant parts may also become infected. Uredia are found on both leaf surfaces, are yellow-brown, 0.5–3.0 mm in diameter, and occur in purple lesions 1–5 mm in diameter.

Uredia are circular in shape on leaves but are elongated on pedicels and branches. Telia are found on lower leaf surfaces but are normally not present.

CONTROL:
1. Plant resistant cultivars.
2. Do not plant seed from diseased plants.
3. Removed infected residue.

Verticillium Wilt

CAUSE: *Verticillium dahliae* Kleb. However, no general agreement has been reached on the specific name, with many workers still preferring to call the causal organism *V. albo-atrum* Reinke & Berth. The causal fungus will be referred to as *V. dahliae* in this description because of the speciation criteria stated by Isaac. *Verticillium dahliae* forms microsclerotia whereas *V. albo-atrum* forms one- and two-celled conidia and sporophores with black bases.

The fungus survives as microsclerotia or dark resting hyphae in soil or infected residue. Some workers have reported formation of sclerotia by grouping together of several cells. Fungal dissemination is by contamination of seed lots by microsclerotia, wind, surface water, and equipment. Since *V. dahliae* has a large host range consisting of several weed species, inoculum may also originate from indigenous inoculum present on common weeds in an area.

Infectious propagules are thought to originate from several sources. Microsclerotia may produce multiple germ tubes and conidia. Conidia and hyphae are also found either free in the soil or in residue. Although both form hyphal germ tubes, those from hyphae infect roots and grow into the vascular system while it is not known what the role of conidia is in infection. Infectious hyphae grow through the root cortex into the xylem tissue. In the xylem, mycelium colonize vessels and produce conidia that move upward in the tissue. Propagules are stimulated to germinate by root exudates.

Verticillium wilt is most severe during cool, wet weather and ceases to be evident during warm, dry weather. However, *V. dahliae* can grow, sporulate, and form microsclerotia under relatively dry conditions. The only significant saprophytic growth in the soil occurs at the end of the growing season when infected moribund plants are returned to the soil. As a pioneer colonist, the fungus grows throughout infected debris and forms numerous microsclerotia. Foliar symptoms caused by strain SS-4 of verticillium wilt decline at about 25°C or 26°C and above, but vascular wilt symptoms continue to progress. Foliar symptoms caused by strain T-1 occur at temperatures above 26°C. Disease is favored by alkaline soils and cool temperatures.

Several strains of the fungus exist. The two most commonly found in the United States are SS-4 and T-1. The SS-4 isolate is a nondefoliating isolate while the T-1 isolate is considered the defoliating isolate.

DISTRIBUTION: In most cotton growing areas of the world except Egypt.

SYMPTOMS: The major effect of verticillium wilt is the reduction of growth and development of the plant, which results in reduced height, lateral branching, and dry matter accumulation in leaves, stems, roots, squares, and bolls. Cotton is susceptible at all stages of growth, but symptoms commonly appear about blossom time, following rainy weather, and vary according to fungus strain and temperature. Cotyledons of young plants infected with strain SS-4 become chlorotic and dry up. Chlorotic areas appear on leaf margins and between veins of lower leaves that eventually become necrotic. Chlorosis and necrosis of leaves proceed up the plant. Defoliation may occur.

Vascular discoloration is usually limited to the lower half of the plant. Normally the number of plants with foliar symptoms is less than plants with vascular discoloration. Plants may only be stunted with dark green leaves if infection occurs later in season. Epinasty may also occur but it is usually slight.

In susceptible plants infected by T-1 strain, symptoms in both young and old plants progress rapidly. Initially the terminal leaf curls downward, followed by severe epinasty and general chlorosis of upper leaves that soon defoliate. Terminal dieback normally occurs. At temperatures above 26°C, leaf symptoms similar to those caused by SS-4 occur except they are more severe and progress to the top of the plant. Fruiting branches and bolls may be dropped. A high percentage of bolls on the top of the plant do not mature and dry up. If plants are not killed, branches may form on lower plant portions to form bushy plants. Vascular discoloration may occur to the plant top.

CONTROL:
1. Plant tolerant cultivars. However, increases of inoculum of more aggressive strains occur when tolerant varieties are planted successive years in the same infested soil.
2. Rotate cotton with nonsusceptible crops. Do not grow cotton more than once every three years in the same soil. Most short-term rotations are of little value. Rotation with perennial rye grass has been shown to reduce the population of propagules.
3. Avoid deep cultivation.
4. Use skip-row planting where possible.
5. Plant populations should be planted at higher than normal populations.
6. Do not apply gin trash to field in areas where verticillium wilt is a problem.
7. Any cultural practice that increases soil temperature will partially aid in the control of the disease.

_____MYCOPLASMAL CAUSES_____

Leaf Crumple

This disease may be the same as Cotton Leaf Crumple.

CAUSE: Possibly a mycoplasm; however, some researchers still refer to the causal organism as a virus. Dissemination is by the whitefly *Bemisia tabaci* (Genn.) and by grafting techniques. The causal organism overwinters in cheeseweed, *Malva parviflora* L., and cultivated beans, *Phaseolus vulgaris* L., in Arizona.

DISTRIBUTION: Western United States.

SYMPTOMS: Symptoms are most severe in perennial cotton. Leaves curl downward due to hypertrophy of interveined tissue. There is frequent vein distortion and clearing and a mosaic appearance to leaf. Reddish spots sometimes occur in chlorotic interveinal tissue of senescent leaves. Floral parts tend to be irregularly shaped. Stunting commonly occurs in perennial cotton. Yield losses are most severe when plants are infested early in the growing season.

CONTROL: Tolerance and resistance occurs in some varieties; however, the following control measures have been suggested:
1. Control weed hosts.
2. Eliminate infected perennial cotton.
3. Plant clean seed.

Phyllody

CAUSE: A mycoplasm disseminated by a leafhopper, *Orosius* sp.

DISTRIBUTION: Ivory Coast, Mali, and Upper Volta.

SYMPTOMS: Leaves are string or straplike. Affected plants are sterile with virescence or green floral parts.

CONTROL: No control is reported.

CAUSED BY NEMATODES

Cotton Root Knot Nematode

CAUSE: *Meloidogyne incognita* (Kofoid & White) Chitwood. Infested soil contains eggs that hatch to produce larvae. The larvae move through the soil and penetrate the root tip; however, entry is not necessarily restricted to this area. The larvae migrate through the root to the vascular tissues, become sedentary and commence feeding on undifferentiated provascular tissue. Excretions of the larvae stimulate cell proliferation, resulting in the development of syncythia from which the larvae derive their food. Females lay 500–1,000 eggs in a gelatinous matrix that may be pushed out of the root into the soil.

DISTRIBUTION: Generally distributed where cotton is grown in sandy or sandy loam soil.

SYMPTOMS: Affected plants are less vigorous and shorter than healthy plants. Severely affected plants are lighter green and will wilt quickly when under moisture stress during the day but recover turgidity at night. Eventually wilting will become permanent.

The taproot is frequently missing or greatly reduced in size. Lateral roots contain numerous spindle-shaped galls or knots that become noticeable three to four weeks after infection. Galls may be found in lesser numbers on the taproot. Eventually the galls may attain a size of 6 mm in diameter.

CONTROL:
1. Rotate cotton with such crops as alfalfa, oats, barley, and sorghum for two or more years.
2. Keep field fallow and control weeds.
3. Fumigate soil.
4. Plant resistant cultivars.
5. Apply nematicide.

Lance Nematodes

CAUSE: *Hoplolaimus columbus* Sher. and *H. galeatus* (Cobb) Filipjev & Schuurmans-Stekhoven. *Hoplolaimus columbus* feeds primarily in cortex, but *H. galeatus* feeds as both an ectoparasite and endoparasite. Eggs are deposited in the cortex.

DISTRIBUTION: *Hoplolaimus columbus* is found mainly in South Carolina and Georgia and *H. galeatus* is found in the southeastern Coastal Plains.

SYMPTOMS: Both nematodes cause plants to become stunted and chlorotic. *Hoplolaimus columbus* has been reported to cause defoliation during moisture stress conditions.

CONTROL:
1. Rotate cotton with nonhost plants.
2. Apply nematicides.
3. Fumigate soil.

Miscellaneous Nematodes

CAUSE: Root lesion, *Pratylenchus* sp., and stubby root, *Trichodorus christiei* Allen.

DISTRIBUTION: Generally distributed where cotton is grown.

SYMPTOMS: Little injury generally occurs but some restricted root growth may occur. Roots may be coarse and stubby and some premature decay may occur.

CONTROL: No control is generally necessary.

Reniform Nematode

CAUSE: *Rotylenchulus reniformis* Lindford & Oliveira is a nematode where the female either is completely in the root or has just the posterior end protruding. Fifty to eighty eggs are laid outside the root in a gelatinous matrix that encompasses the exposed portion of the female body. Eggs are resistant to drying and hatch under favorable conditions. Larvae molt four times and develop into males and females; however, only females have been observed feeding. The life cycle is completed in 17–23 days.

DISTRIBUTION: Southeastern United States in localized areas in fields; however, it is a potential problem wherever cotton is grown in subtropical and tropical areas.

SYMPTOMS: Infected plants are severely stunted. Root growth is restricted and there are few large roots. Lateral roots may be coarse and stubby. Premature decay by seedling blight organisms may occur. Plants may be chlorotic and wilt. Weedy areas occur in infested fields.

CONTROL:
1. Rotate cotton with rice, sorghum, oats, mustard, turnip, corn, pepper, and grasses.
2. Fumigate soil.

Sting Nematode

CAUSE: *Belonolaimus longicaudatus* Rau feeds mostly on the outside of roots without penetrating them or becoming attached. The nematode feed at root tips and along the sides of succulent roots.

DISTRIBUTION: Sandy soils in coastal plains of the southeastern United States.

SYMPTOMS: Plants are stunted and chlorotic with frequent death of young plants. Roots may be stubby and coarse and have necrotic and discolored lesions, resulting in rotting. Roots may end in enlargements caused by repeated forming and killing of new branches. Nematode injury may provide an entrance wound for the Fusarium wilt fungus.

CONTROL:
1. Rotate cotton with tobacco, watermelons, and crotalaria.
2. Fumigate soil.
3. Apply nematicides.

VIRAL CAUSES

Abutilon Mosaic

CAUSE: Abutilon mosaic virus (AMV) survives in a wide number of host plants but *Sida* spp. are probably the most important. It is disseminated by the whitefly, *Bemisia tabaci* (Genn.). It is not seed transmitted but is graft transmissible.

DISTRIBUTION: Generally distributed wherever cotton is grown.

SYMPTOMS: Symptoms are most obvious on younger plants and become less pronounced in older ones. Infected plants have shortened internodes, making them dwarfed. Leaves are crinkled, blistered, malformed, and have a conspicuous mosaic of yellow and green areas. Chlorotic areas may be delimited by veins and become reddish, and disappear in older leaves. Younger infected leaves may be wrinkled, smaller, and have fewer lobes than healthy ones.

CONTROL: This is usually not a serious disease and control is not thought to be necessary.

Anthocyanosis

CAUSE: Anthocyanosis virus (AV) survives in *Gossypium barbadense*, *Sida* spp., and ratoon cotton. It has been artificially inoculated into and recovered from several other plants. AV is disseminated by *Aphis gossypii* Glover but not by seed or sap. A feeding period of 12 hours or longer is necessary for aphids to become moderately infective. AV is persistent within the aphid.

DISTRIBUTION: Brazil.

SYMPTOMS: Leaves initially develop chlorotic areas that turn red-purple in sunlight. Such purple areas are delimited by veins, but the entire leaf except for veins and a narrow band adjacent to the veins, may become purple. The disease is most severe on lower and middle leaves but may also affect the upper leaves of older plants.

CONTROL: Control insects on cotton throughout the growing season. It is not practical to eliminate the overwintering hosts.

Blue Disease

CAUSE: Unknown virus that is thought to be transmitted only by the aphid *Aphis gossypii* Glover. Plants of various ages are affected.

DISTRIBUTION: Africa.

SYMPTOMS: Plants are dwarfed with stems having a zigzag pattern of growth. Apical leaves initially bulge between the main veins and toward the base. Eventually the entire leaf is involved. At first the affected areas are a light green but later turn a blue-green.

CONTROL:
1. Plant resistant varieties.
2. Practice sanitation and destroy residue.

Cotton Leaf Crumple

This disease may be the same as Leaf Crumple.

CAUSE: Cotton leaf crumple virus (CLCV) is transmitted by the whitefly *Bemisia tabaci* (Genn.) at an optimal temperature of 32°C under experimental conditions. Bean (*Phaseolus vulgaris* L.), cheeseweed (*Malva parviflora* L.), and numerous plant species in the Malvaceae and Leguminosae are hosts. Perennial cotton serves as an inoculum reservoir. Yield reduction results from infection by CLCV at virtually any developmental stage.

DISTRIBUTION: India, northern Mexico, and southwestern United States.

SYMPTOMS: Leaves have mosaic pattern and together with floral parts are malformed and blistered. Plants are stunted and yields are affected.

CONTROL: Eliminate perennial cotton in vicinity of fields.

Cotton Leaf Curl

This disease is also called cotton leaf crinkle.

CAUSE: Cotton leaf curl virus (CLCV) survives in several different plant hosts from which it is transmitted to cotton at all stages of growth by the whitefly, *Bemisia tabaci* (Genn.). The virus is not transmitted by sap, seed, or soil but has been transmitted by grafting. Different strains of the virus may exist.

DISTRIBUTION: Africa.

SYMPTOMS: Only immature leaves usually develop symptoms. The lower surface of the smaller veins has intermittent thickening that eventually becomes continuous. When backlighted, the veins of a diseased leaf are darker green than the rest of the leaf.

The leaves formed after infection are small, crinkled, and curled either upward or downward at the edges. Enations are often formed on the thickened lower side of primary veins. Internodes are lengthened, twisted, and flattened. Often the entire plant is stunted. All parts of an infected plant are brittle. Yield may be greatly reduced.

CONTROL:
1. Plant resistant varieties.
2. Control whiteflies in the greenhouse.

Cotton Mosaic

CAUSE: Unknown virus transmitted by the whitefly, *Bemisia tabaci* (Genn.).

DISTRIBUTION: Africa.

SYMPTOMS: Leaf mosaic and mottling.

CONTROL: Plant tolerant varieties.

Cotton Small Leaf

This disease is also called cotton stenosis and smalling.

CAUSE: Cotton small leaf virus (CSLV). It is not known how the virus survives and is disseminated. The virus may be a mycoplasma.

DISTRIBUTION: India and Pakistan.

SYMPTOMS: The aerial portions of cotton are extremely stunted. Leaves develop in clusters, are malformed, variously lobed, and of different shapes and sizes. Leaves and epicalyx are usually mottled. Enations are produced on lower surface of veins. Flowers may remain small with bolls never forming.
  The taproot ends abruptly, giving rise to a large number of adventitious roots. Plants can be pulled easily out of the ground.

CONTROL: No control is known.

Cotton Terminal Stunt

CAUSE: Cotton terminal stunt virus (CTSV). The mode of survival of CTSV is unknown. Insects are suspected of being vectors and the virus is graft transmissable.

DISTRIBUTION: Southern Texas and adjacent areas in Mexico.

SYMPTOMS: Terminal growth of stems and branches is stunted. Young leaves tend to be small, misshapen, mottled, and cup either upward or downward. Tan to brown streaks occur in xylem of the main stem. Immature bolls have a dark internal discoloration. Commonly most bolls are shed as well as blooms and squares.

CONTROL: No control is reported.

Leaf Mottle

CAUSE: Leaf mottle virus (LMV). It is not known how LMV survives and is disseminated.

DISTRIBUTION: Sudan.

SYMPTOMS: Younger leaves are mottled, with the mottling most pronounced near the veins. Leaf lobes are distorted and elongated. The main stem was often stunted and flowering is reduced. Severely infected leaves are a light green.

CONTROL: No control is reported.

Leaf Roll

CAUSE: Leaf roll virus (LRV) is transmitted by different aphids including *Aphis gossypii* Glover, *A. laburni* Kaltenbach, *Myzus persicae* Sulzer, and *Epitetranychus althaeae*.

DISTRIBUTION: USSR.

SYMPTOMS: Plants are stunted, tend to droop, and become prostrate and spreading. Leaves have chlorotic margins, are shiny, brittle, and curl upward at the tip but downward at the edges. Stems and petioles become reddish and sticky. Fruiting is reduced and roots tend to be underdeveloped.

CONTROL: No control is reported. Some cultivars are less susceptible in certain areas than in others.

Tobacco Streak

CAUSE: Tobacco streak virus (TSV) survives in a wide number of host plants. It is not known how TSV is disseminated to cotton although insects are suspected.

DISTRIBUTION: The virus occurs on several other hosts around the world but is only known to infect cotton in Brazil.

SYMPTOMS: Infected plants are slightly stunted and have smaller than normal leaves that show a mosaic pattern. The mosaic pattern consists of light green areas between secondary veins and normal green areas, generally blistered, along the veins. Initial infection of cotton leaves results in necrotic rings or spots with some plants producing more axillary twigs than normal.

CONTROL: No control is warranted since the disease usually occurs too late in the season to do much damage.

Vein Clearing

CAUSE: Possibly a virus.

DISTRIBUTION: Texas.

SYMPTOMS: Initially vein clearing occurs followed by vein banding and eventually a mottle. Young leaves tend to cup downward. Plants tend to be stunted.

CONTROL: No control is reported.

Veinal Mosaic

CAUSE: Veinal mosaic virus (VMV). It is not known how VMV survives or is disseminated. It is easily graft transmissible.

DISTRIBUTION: Brazil.

SYMPTOMS: Plants have shortened internodes and are slightly stunted, giving them a compact and often bunchy top appearance. Leaves are a darker than normal green with veinal mosaic, roughening, and downward curling of the margins. Severely diseased leaves may be rolled. Necrotic lesions of the veins sometimes occur. Veinal mosaic may also occur on bracts.

CONTROL: No control is reported.

BIBLIOGRAPHY

Al-Beldawi, A. S., and Pinckard, J. A. 1970. Control of *Rhizoctonia solani* on cotton seedlings by means of benomyl. *Plant Dis. Rptr.* **54:**76–80.

Alderman, S. C., and Hine, R. B. 1981. Pathogenicity and occurrence of strands of *Phymatotrichum omnivorum*. *Pathology* **71:**198 (abstract).

Anon. 1966. *Cotton Diseases and Their Control*. Cotton Disease Council and National Cotton Council. Memphis, TN.

Arndt, C. H. 1944. Infection of cotton seedlings by *Colletotrichum gossypii* as affected by temperature. *Phytopathology* **34:**861–869.

Arndt, C. H. 1946. Effect of storage conditions on the survival of *Colletotrichum gossypii*. *Phytopathology* **36:**24–29.

Arndt, C. H. 1953. Survival of *Colletotrichum gossypii* on cotton seed in storage. *Phytopathology* **43:**220.

Ashworth, L. J., Jr.; Galanopoulos, N.; and Galanopoulou, S. 1984. Selection of pathogenic strains of *Verticillium dahliae* and their influence on the useful life of cotton cultivars in the field. *Phytopathology* **74:**1637–1639.

Ashworth, L. J., Jr.; Hildebrand, D. C.; and Schroth, M. N. 1970. Erwinia induced internal necrosis of immature cotton bolls. *Phytopathology* **60:**602–607.

Ashworth, L. J., Jr.; Rice, R. E.; McMeans, J. L.; and Brown, C. M. 1971. The relationship of insects to infection of cotton bolls by *Aspergillus flavus*. *Phytopathology* **61:**488–493.

Baehr, L. F., and Pinckard, J. A. 1970. Histological studies on the mode of penetration of boll rotting organisms into developing cotton bolls. *Phytopathology* **60:**581 (abstract).

Bagga, H. S. 1970. Fungi associated with cotton boll rot in the Yazoo-Mississippi Delta, 1966–1968. *Plant Dis. Rptr.* **54:**796–798.

Brown, J. K.; Mihail, J. D.; and Nelson, M. R. 1987. Effects of cotton leaf crumple virus on cotton innoculated at different growth stages. *Plant Disease* **71:**699–703.

Brown, J. K., and Nelson, M. R. 1986. Cotton leaf crumple virus transmitted from naturally infected bean from Mexico. *Plant Disease* **70:**981 (disease notes).

Brown, J. K., and Nelson, M. R. 1984. Geminate particles associated with cotton leaf crumple disease in Arizona. *Phytopathology* **74:**987–990.

Brown, J. K., and Nelson, M. R. 1987. Host range and vector relationships of cotton leaf crumple virus. *Plant Disease* **71:**522–524.

Correll, J. C. 1986. Powdery mildew of cotton caused by *Oidiopsis taurica* in California. *Plant Disease* **70:**259.

Devay, J. E.; Forester, L.; Garber, R. H.; and Butterfield, E. J. 1974. Characteristics and concentrations of propagules of *Verticillium dahliae* in air-dried field soils in relation to the prevalence of Verticillium wilt in cotton. *Phytopathology* **64:**22–29.

Evans, G.; Wilhelm, S.; and Snyder, W. C. 1966. Dissemination of the Verticillium wilt fungus with cotton seed. *Phytopathology* **56**:460–461.

Garber, R. H., and Presley, J. T. 1971. Relation of air temperature to development of Verticillium wilt on cotton in the field. *Phytopathology* **61**:204–207.

Gergon, E. B. 1982. Diplodia boll rot of cotton: Pathogenicity and histopathology. *Philippines Phytopathol.* **18**:6 (abstract).

Grinstein, A.; Fishler, G.; Katan, J.; and Hakohen, D. 1983. Dispersal of the Fusarium wilt pathogen in furrow-irrigated cotton in Israel. *Plant Disease* **67**:742–743.

Hancock, J. G. 1972. Root rot of cotton caused by *Pythium splendens*. *Plant Dis. Rptr.* **56**:973–975.

Hopkins, J. C. F. 1931. *Alternaria gossypina* (Thuem.) Comb., nov. causing a leaf spot and boll rot of cotton. *Trans. Brit. Mycol. Soc.* **16**:136–144.

Huisman, O. C., and Ashworth, L. J., Jr. 1976. Influence of crop rotation on survival of *Verticillium albo-atrum* in soils. *Phytopathology* **66**:978–981.

Isaac, I. 1967. Speciation in Verticillium. *Annu. Rev. Plant Pathol.* **5**:201–222.

Jeffers, D. P.; Smith, S. N.; Garber, R. H.; and DeVay, J. E. 1984. The potential spread of the cotton Fusarium wilt pathogen in gin trash and planting seed. *Phytopathology* **74**:1139 (abstract).

Joannou, N.; Schneider, R. W.; Grogan, R. G.; and Duniway, J. J. 1977. Effect of water potential and temperature on growth, sporulation, and production of microsclerotia by *Verticillium dahliae*. *Phytopathology* **67**:637–644.

Johnson, L. F., and Chambers, A. Y. 1973. Isolation and identity of three species of Pythium that cause cotton seedling blight. *Plant Dis. Rptr.* **57**:848–852.

Johnson, L. F.; Baird, D. D.; Chambers, A. Y.; and Shamiyeh, N. B. 1978. Fungi associated with postemergence seedling disease of cotton in three soils. *Phytopathology* **68**:917–920.

Johnson, W. M.; Johnson, E. K.; and Brinkerhoff, L. A. 1980. Symptomatology and formation of microsclerotia in weeds inoculated with *Verticillium dahliae* from cotton. *Phytopathology* **70**:31–35.

Jones, J. P. 1961. A leaf spot of cotton caused by *Corynespora cassiicola*. *Phytopathology* **51**:305–308.

King, C. J., and Presley, J. T. 1942. A root rot of cotton used by *Thielaviopsis basicola*. *Phytopathology* **32**:752–761.

Klich, M. A. 1987. Relation of plant water potential at flowering to subsequent cottonseed infection by *Aspergillus flavus*. *Phytopathology* **77**:739–741.

Ling, I., and Yang, J. Y. 1941. Stem blight of cotton caused by *Alternaria macrospora*. *Phytopathology* **32**:752–761.

McCarter, S. M. 1972. Effect of temperature and boll injuries on development of Diplodia boll rot of cotton. *Phytopathology* **62**:1223–1225.

Mathre, D. E.; Ravescroft, A. V.; and Garber, R. H. 1966. The role of *Thielaviopsis basicola* as a primary cause of yield reduction in cotton in California. *Phytopathology* **56**:1213–1216.

Pinckard, J. A., and Guidroz, G. F. 1973. A boll rot of cotton caused by *Phytophthora parasitica. Phytopathology* **63**:896–899.

Pizzinatto, M. A.; Soave, J.; and Cia, E. 1983. Patogenicidade de *Botryodiplodia theobromae* Pat. a plantas de diferentes idades e macao de algodoeiro (*Gossypium hirsutum* L.). *Fitopatologia Brasileira* **8**:223–228 (in Portuguese).

Pullman, G. S., and DeVay, J. E. 1982. Effect of soil flooding and paddy rice culture on the survival of *Verticillium dahliae* and incidence of Verticillium wilt in cotton. *Phytopathology* **72**:1285–1289.

Pullman, G. S., and DeVay, J. E. 1982. Epidemiology of Verticillium wilt of cotton: A relationship between inoculum density and disease progression. *Phytopathology* **72**:549–554.

Pullman, G. S., and DeVay, J. E. 1982. Epidemiology of Verticillium wilt of cotton: Effects of disease development of plant phenology and lint yield. *Phytopathology* **72**:554–559.

Rane, M. S., and Patel, M. K. 1956. Diseases of cotton in Bombay. I. Alternaria leaf spot. *Indian Phytopathol.* **9**:106–113.

Rane, M. S., and Patel, M. K. 1956. Diseases of cotton in Bombay. II. Helminthosporium leafspot. *Indian Phytopathol.* **9**:169–173.

Rathaiah, Y. 1976. Reaction of cotton species and cultivars to four isolates of *Ramularia areola. Phytopathology* **66**:1007–1009.

Rathaiah, Y. 1977. Spore germination and mode of cotton infection by *Ramularia areola. Phytopathology* **67**:351–357.

Rush, C. M.; Lyda, S. D.; and Gerik, T. J. 1984. The relationship between time of cortical senescence and foliar symptom development of Phymatotrichum root rot of cotton. *Phytopathology* **74**:1464–1466.

Rush, C. M.; Gerik, T. J.; and Kenerley, C. M. 1985. Atypical disease symptoms associated with Phymatotrichum root rot of cotton. *Plant Disease* **69**:534–537.

Rush, C. M., Gerik, T. J.; and Lyda, S. D. 1984. Factors affecting symptom appearance and development of Phymatotrichum root rot of cotton. *Phytopathology* **74**:1466–1469.

Schnathorst, W. C. 1973. Nomenclature and physiology of *verticillium* spp. with emphasis on the *V. albo-atrum* vs *V. dahliae* controversy. Pages 1–19 in *Verticillium Wilt of Cotton.* ARS-S19. Proceedings of a workshop conference held at the National Cotton Pathology Research Laboratory, College Station, Texas, 30 Aug. 1971–1 Sept. 1971. Published by U.S. Dept. Agric., Agric. Res. Serv., Publication Div., Beltsville, MD, 398p.

Sciumbato, G. L., and Pinckard, J. A. 1972. *Alternaria macrospora* leaf spot of cotton in Louisiana in 1974. *Plant Dis. Rptr.* **58**:201–202.

Sleeth, B. 1965. Terminal stunt, a virus disease of cotton. Texas A & M Univ., *Texas Agric. Exp. Sta. MP-766.*

Snow, J. P., and Mertley, J. C. 1979. A boll rot of cotton caused by *Colletotrichum capsici. Plant Dis. Rptr.* **63**:626–627.

Snow, J. P., and Sanders, D. E. 1979. Role of abscised cotton flowers, bolls and squares in production of inoculum by boll-rotting *Fusarium* spp. *Plant Dis. Rptr.* **63**:288–289.

Sobers, E. K., and Littrell, R. H. 1974. Pathogenicity of three species of Cylindrocladium to select hosts. *Plant Dis. Rptr.* **58**:1017–1019.

Sparnicht, R. H., and Roncadori, R. W. 1972. Fusarium boll rot of cotton: Pathogenicity and histopathology. *Phytopathology* **62**:1381–1386.

Tarr, S. A. J. 1964. *Virus Diseases of Cotton*. Commonwealth Mycological Institute. Misc. Pub. No. 18.

Watkins, G. M. (ed.). 1981. *Compendium of Cotton Diseases*. American Phytopathological Society, St. Paul, MN, 85p.

Weber, G. F. 1973. *Bacterial and Fungal Diseases of Plants in the Tropics*. University of Florida Press, Gainesville.

Wheeler, J. E., and Hine, R. B. 1972. Influence of soil temperature and moisture on survival and growth of strands of *Phymatotrichum omnivorum*. *Phytopathology* **62**:828–832.

Wilhelm, S.; Sagen, J. E.; and Tietz, H. 1974. Resistance to Verticillium wilt in cotton: Sources, techniques of identification, inheritance trends, and the resistance potential of multiline cultivars. *Phytopathology* **64**:924–931.

Wilhelm, S.; Sagen, J. E.; and Tietz, H. 1985. Phenotype modification in cotton for control of Verticillium wilt through dense plant population culture. *Plant Disease* **69**:283–288.

Wrather, J. A.; Sappenfield, W. P.; and Baldwin, C. H. 1986. Colonization of cotton buds by *Xanthomonas campestris* pv. *malvacearum*. *Plant Disease* **70**:551–552.

Zutra, D., and Orion, D. 1982. Crown gall bacteria. (*Agrobacterium radiobacter* var. *tumefaciens*) on cotton roots in Israel. *Plant Disease* **66**:1200–1201.

6

Diseases of Crambe (*Crambe abyssinica* Hochst. ex. R. E. Fries)

FUNGAL CAUSES

Alternaria Leaf Spot

CAUSE: *Alternaria brassicicola* (Schw.) Wiltsch. is seedborne.

DISTRIBUTION: Likely distributed wherever crambe is grown; however, this is not known for certain.

SYMPTOMS: Artificially inoculated plants have small, black, linear lesions that occur on petioles, stems, and veins of the lower leaf surfaces and on seed pods. Stem lesions are commonly 1–3 mm in length and frequently coalesce. A brown to black discoloration girdles tips of stems. Seed pods may be blackened and reduced in size, and seed maturation may be prevented.

Leaf spots are brown to black, oval, and 1–5 mm in length. Leaves quickly become chlorotic and drop off.

Seedlings may damp off postemergence. Stem tips and base of petioles become blackened, wilt, and dry up.

CONTROL:
1. There is a low level of resistance to *A. brassicicola* in some crambe cultivars, making it possible to screen for resistance.
2. Treat seed with a seed-protectant fungicide.

Alternaria Leaf Spot

CAUSE: *Alternaria circinans* (Berk. & Curt.) Bolle. The fungus is seedborne, which probably serves as one of the major sources of disease inoculum.

DISTRIBUTION: The disease is reported in the United States but may occur elsewhere.

SYMPTOMS: Black spots and streaks occur on stems, leaves, and fruits. Spots occur on seed hulls, with many hulls containing shriveled, nonviable seeds.

CONTROL: No control known for certain, but seed treatment with a fungicide would seem a viable option.

Aphanomyces Root Rot

CAUSE: *Aphanomyces raphani* Kendrick. It is not known for certain how *A. raphani* survives, but it is probably as oospores in infected residue or in the soil. Zoosporangia are formed in which zoospores are produced. Oospores are eventually produced in roots. The disease is most severe under wet soil conditions.

DISTRIBUTION: Not known. The disease has not yet been observed in the field and has been studied under greenhouse conditions.

SYMPTOMS: Seedlings are stunted, but artificially inoculated ones did not damp off. Roots and hypocotyls are blackened. The fungus moves up the hypocotyl into the cotyledon, causing chlorosis and blackening of the infected tissues. Oospores are evident under magnification in infected root cortex.

CONTROL: No control is necessary.

Fusarium Wilt

CAUSE: *Fusarium oxysporum* Schlecht. f. sp. *conglutinans* (Wr.) Snyd. & Hans. Race 2.

DISTRIBUTION: Not known.

SYMPTOMS: Seedlings may be susceptible to damping off until they reach a height of 6.5 cm. An early symptom is interveinal yellowing frequently followed by leaf abscission. Plants may also be stunted.

CONTROL: No control is reported.

Sclerotinia Stem Rot

CAUSE: *Sclerotinia sclerotiorum* (Lib.) deBary.

DISTRIBUTION: United States.

SYMPTOMS: Symptoms are similar to those observed on rapeseed and mustard. Sclerotia are found in hollow stems of infected plants.

CONTROL: No control is reported.

VIRAL CAUSES

Beet Western Yellows

CAUSE: Beet western yellows virus (BWYV). For further information on BWYV see sugar beets.

DISTRIBUTION: BWYV has been reported on sugarbeets from Asia, Europe, and North America. It is not known what the distribution of the virus on crambe is.

SYMPTOMS: Initially, artificially inoculated plants display a lighter green discoloration to the entire plant followed by an interveinal reddening of the lower leaves. The reddening intensifies and involves more tissue as the disease progresses. Older infected leaves become thickened, brittle, and red except for green areas adjacent to veins.

Young leaves inoculated with a severe virus strain develop black necrotic pinpoint spots. Necrotic areas develop later along stems. Severely infected plants may be killed. Yield is reduced on plants that are not killed.

CONTROL: No control is reported.

Turnip Mosaic

CAUSE: Turnip mosaic virus (TMV) is mechanically transmissible. It is also transmitted by the aphids *Myzus persicae* Sulzer and *Brevicoryne brassicae* Linne but it is not seedborne.

DISTRIBUTION: Turnip mosaic virus is distributed in Europe, Japan, North America, and South Africa on other hosts but the distribution on crambe is unknown.

SYMPTOMS: Artificially inoculated plants showed a systemic mottling that occurred in all leaves. The inflorescence is severely stunted and distorted. Necrosis occurs on stems and inflorescences. Seed pods do not mature on infected plants.

CONTROL: No control is reported.

BIBLIOGRAPHY

Armstrong, G. M., and Armstrong, J. K. 1974. Wilt of *Brassica carinata, Crambe abyssinica,* and *C. hispanica* caused by *Fusarium oxysporum* f. sp. *conglutinans* race 1 or 2. *Plant Dis. Rptr.* **58**:479–480.

Duffus, J. E. 1975. Effects of beet western yellows virus on crambe in the greenhouse. *Plant Dis. Rptr.* **59**:886–888.

Holcomb, G. E., and Newman, B. E. 1970. *Alternaria circinans* and other fungal pathogens on *Crambe abyssinica* in Louisiana. *Plant Dis. Rptr.* **54**:28.

Horvath, J. 1972. Reaction of crambe (family: Cruciferae) to certain plant viruses. *Plant Dis. Rptr.* **56**:665–666.

Humaydan, H. S., and Williams, P.H. 1975. Additional cruciferous hosts of *Aphanomyces raphani. Plant Dis. Rptr.* **59**:113–116.

Kilpatrick, R. A. 1976. Fungal flora of crambe seeds and virulence of *Alternaria brassicicola. Phytopathology* **66**:945–948.

Leppik, E. E. 1973. Diseases of crambe. *Plant Dis. Rptr.* **57**:704–708.

Thornberry, H. H., and Phillippe, M. R. 1965. Crambe: Susceptibility to some plant viruses. *Plant Dis. Rptr.* **49**:74–77.

White, G. A., and Higgins, J. J. 1966. Culture of crambe . . . a new industrial oilseed crop. *USDA Agric. Res. Serv. Production Res. Rept. No. 95.*

7

Diseases of Field Beans
(*Phaseolus vulgaris* L.)

BACTERIAL CAUSES

Bacterial Wilt

CAUSE: *Corynebacterium flaccumfaciens* (Hedges) Dows. is seed-borne. It is not known to overwinter in plant residue or soil. Bacteria are disseminated by driving rains and hail that cause mechanical injury to plants. Bacteria then enter into the plant through wounds.

DISTRIBUTION: Generally distributed wherever beans are grown.

SYMPTOMS: A few water-soaked lesions may occur on leaves but these are not as numerous as the lesions of bacterial blight or halo blight. Plants arising from infected seed are frequently killed while still small.
A general wilting of infected plants occurs. At first leaves wilt when temperatures are high and become turgid again when temperatures are reduced. Eventually as the disease progresses, leaves turn brown and drop from the plant. Sutures of the pods are discolored and the pods become flaccid. The seed may be invaded or a crust of bacteria may form externally on the seed. Infected white seed may appear yellowish due to bacteria that are visible through the seed coat.

CONTROL: Plant disease-free seed.

Bean Wildfire

CAUSE: *Pseudomonas syringae* pv. *tabaci* (Wolf & Foster) Young, Dye & Wilkie (syn. *P. tabaci* (Wolf & Foster) Stevens). This is a different strain from the *P. tabaci* that causes wildfire of tobacco.

DISTRIBUTION: Brazil.

SYMPTOMS: Symptoms of wildfire are confined to the leaves. Initially, lesions on leaves are water-soaked. These turn to light to dark brown lesions that are surrounded by a pronounced chlorotic halo. These symptoms closely resemble those of halo blight; however, pods and seeds are not infected. Chlorotic halos are produced at relatively high temperatures but halo blight bacterium induce the chlorotic effect at cooler temperature regimes only.

CONTROL: No control is reported.

Brown Spot

This disease is also called bacterial brown spot.

CAUSE: *Pseudomonas syringae* pv. *syringae* van Hall is seedborne and can overwinter in infected residue. Overwintering also occurs on hairy vetch, *Vicia villosa* Roth; common vetch, *V. sativa* L.; and kudzu, *Pueraria thunbergiana* (Sieb. & Zucc.) Benth. Disease severity increases when leaves are moist for extended periods of time. Higher populations of the bacterium occur on leaves of susceptible cultivars than on leaves of resistant cultivars.

DISTRIBUTION: Generally distributed wherever field beans are grown.

SYMPTOMS: Dark brown necrotic spots of various sizes, that are not water soaked and do not have a halo, occur on leaves. Marginal chlorosis surrounding lesions is also absent. Brown spots also occur on pods, frequently with a twisting of the pods at the point of infection.

CONTROL:
1. Plant healthy seed treated with a seed-protectant fungicide.
2. Rotate beans with nonsusceptible crops.
3. Do not plant field beans adjacent to lima beans (*Phaseolus limensis* Macfad.) fields since lima beans are quite susceptible to *P. syringae*. Inoculum may be disseminated easily from one field to the next.
4. Resistant germ plasm has been identified.

Common Blight

CAUSE: *Xanthomonas campestris* pv. *phaseoli* (Smith) Dye (syn. *X. phaseoli* (Smith) Dowson) is seedborne in both resistant and susceptible Phaseolus genotypes and overwinters in infested bean residue. However, crop debris has been reported to not constitute a source of primary inoculum in Michigan where pathogenic bacterium was never isolated over a 10-year period. Survival also occurs on kudzu *Pueraria thunbergiana* (Sieb. & Zucc.) Benth. in the southern United States. The bacterial population is greater on leaves than on stems or pods. Additionally bacteria survive in higher numbers in no-till tillage systems than in a conventional tillage system using a disk or plow. Internally infected seed is the main source of primary inoculum but externally infected seed is also a source. Infected seed gives rise to infected plants. Bacteria have been reported to survive up to six years on seed. Bacteria are blown or splashed onto leaves where they enter through natural openings or wounds caused by windblown soil or other means.

Bacteria exude from initial infections and are spread by rain, wind, and contact between plants and machinery. Disease development is favored by warm, humid conditions.

DISTRIBUTION: Generally distributed where field beans are grown under warm, humid conditions.

SYMPTOMS: Plants growing from internally infected seed are stunted and gradually die. Seedlings may not emerge and their cotyledons have reddish lesions on them. Small water-soaked spots appear on the underside of the leaf. As spots develop, dried bacterial ooze usually can be seen in the center. The spots or lesions may turn brown and coalesce, forming large dead areas, causing defoliation.

Small water-soaked spots may also occur on stems of seedlings. The spots enlarge and become reddish. On older plants the spots lengthen and extend lengthwise along the stem. Often a brownish lesion may encircle the stem at a lower node about the time pods are half mature, causing the plant to fall over.

Similar water-soaked spots appear on pods. Each spot consists of concentric red-brown or brick red rings. Spots eventually become dry and sunken with a crust of dried yellow bacterial ooze. Bacteria produced on pods may infect seed through pod hinges. Infected seed may have a discolored hilum and surrounding tissue that appears to have a varnished appearance. Infected seed may sometimes appear healthy.

CONTROL:
1. Plant certified seed.
2. Treat seed with a seed-protectant bactericide.
3. Do not cultivate when beans are wet since bacteria may be spread by the machinery.
4. Rotate beans every three to four years with cereals or other nonsusceptible crops.

5. Copper fungicides have limited value.
6. Plow under bean residue.
7. Disinfect seed equipment and storage facilities.
8. Plant the most tolerant cultivars; however, under tropical environment conditions tolerant cultivars may become completely susceptible.

Fuscous Blight

CAUSE: *Xanthomonas campestris* pv. *phaseoli* (Smith) Dye (syn. *X. phaseoli* (Smith) Dowson var. *fuscans* (Burk.) Starr & Burk.). According to Fahy and Persley the causal bacterium for common blight and fuscous blight are considered the same. Therefore, the same cause and identical symptoms indicate that common blight and fuscous blight are basically the same disease. However, because some researchers consider the causal organisms to be different, common blight and fuscous blight are discussed as two separate diseases. The bacterium is seedborne and overwinters in infected residue. Bacteria have been reported to survive up to six years on seed. Disease development is favored by warm, wet weather. The bacterium can grow epiphytically on leaves of nonhost plants, which may be significant in its epidemiology.

DISTRIBUTION: Generally distributed wherever beans are grown under warm, wet conditions.

SYMPTOMS: Symptoms of fuscous blight are identical to common blight. Positive identification of the disease is possible only when the causal organism is positively identified.

CONTROL: Control measures are the same as for common blight.

Halo Blight

This disease is also called bean halo blight.

CAUSE: *Pseudomonas syringae* pv. *phaseolicola* (Burk.) Young, Dye & Wilkie (syn. *P. phaseolicola* (Burk.) Dowson) is primarily seedborne and occasionally overwinters in infected residue. Susceptible cultivars are more likely to serve as inoculum sources than resistant cultivars. Infected seed gives rise to infected cotyledons. Bacteria exude from these initial infections and are spread by rain, wind, and contact between plants and machinery. Bacteria may also be splashed or blown from residue onto leaves where they enter through natural openings or wounds caused by windblown soil or other means. Disease development is favored by cool (optimum temperature 21°C), wet weather.

DISTRIBUTION: Generally distributed where field beans are grown under cool, humid conditions.

SYMPTOMS: Plants growing from internally infected seed are stunted and gradually die. Seedlings may not emerge and their cotyledons have reddish lesions on them. Initially small water-soaked spots appear on the underside of leaves. The spots turn red-brown, with surrounding tissue forming a large yellow-green halo. Spots rapidly enlarge and coalesce, forming large brown, dead areas and causing defoliation. Dried bacterial ooze that is a light cream color is present in the center of each spot. Upper trifoliate leaves of plants recently infected with halo blight bacteria have a yellow color.

Similar reddish spots develop on young stems and pods but extend lengthwise along the stem. Infections at nodes may girdle the stem when pods are half mature, causing the stem to break at the diseased node. Spots on pods are reddish with concentric rings. The pod hinge may discolor with bacterial exudate evident. Spots become dry and sunken. Seed may become infected through the hinge and become discolored. Seed from symptomless pods may also be infected.

CONTROL:
1. Apply copper fungicides to foliage. Copper fungicides are more helpful in controlling halo blight than common blight.
2. Plant certified seed.
3. Do not cultivate soil when beans are wet.
4. Rotate beans every three to four years with cereals or other nonsusceptible crops.
5. Plow under bean residue.
6. Disinfect seed equipment and storage facilities.
7. Some cultivars are less susceptible than others.
8. Resistant germ plasm has been identified.

Unnamed Bacterial Leafspot

CAUSE: *Erwinia nulandii* Schuster et al.

DISTRIBUTION: Nebraska.

SYMPTOMS: Initially leaves have small yellow spots that eventually enlarge and coalesce to form irregular-shaped necrotic lesions. Eventually leaf tissue collapses. White bean seed has a pink discoloration.

CONTROL: No control is reported.

Unnamed Bacterial Leafspot

CAUSE: *Pseudomonas blatchfordae* Nov. sp. The bacterium overwinters in bean straw on soil.

DISTRIBUTION: Nebraska.

SYMPTOMS: Symptom development is most rapid at 24°C. The oldest leaves appear to be the most susceptible to infection and subsequent symptom expression. Initially small yellow spots accompanied by water soaking appear first on the oldest leaves. The spots increase in size and become surrounded by a narrow yellow zone with interior areas turning orange to brown, giving the lesion a sunscalded appearance. Eventually the infected area becomes necrotic, resulting in defoliation.

CONTROL: No control is reported.

FUNGAL CAUSES

Alternaria Leaf Spot

CAUSE: *Alternaria alternata* (Fr.) Keissler (syn. *A. tenuis* Nees.). Disease development is more severe at high relative humidity and at lower temperatures of 16–24°C. Bean plants increase in susceptibility as they grow older and when grown in nitrogen- and potassium-deficient soils. See Alternaria Pod Flecking and Pod and Seed Coat Discoloration of White Beans.

DISTRIBUTION: United States.

SYMPTOMS: Small, brown, irregular-shaped lesions occur on leaves. The lesions expand and become gray-brown to dark brown in color and oval in shape with concentric zones. The older portion of the lesions sometimes falls out, leaving a shot-hole appearance. Lesions may coalesce to form large necrotic areas. Defoliation may occur.

CONTROL:
1. Ensure adequate soil fertility.
2. Application of foliar fungicides at the proper time will reduce disease severity.

Alternaria Pod Flecking and Pod and Seed Coat Discoloration of White Beans

This disease is also called black pod disease.

CAUSE: *Alternaria alternata* (Fr.) Keissler (syn. *A. tenuis* Nees.). The causal fungus overwinters in infected residue and has been reported to survive in infected seed stored at room temperatures. It can also be found growing on leaves of numerous weed species. The disease is most severe during cool, wet conditions similar to those favoring Alternaria leaf spot. Such conditions may occur in a closed foliage canopy.

DISTRIBUTION: New York and Ontario.

SYMPTOMS: Symptoms on pods appear initially as small, irregular, water-soaked flecks. Later these flecks coalesce to produce long streaks or irregular, large discolored areas that are reddish to dark brown or black. The infected areas are either sunken or raised and only a few cells deep. Usually a corky layer occurs below the lesions. Conidia are often found associated with flecks. Small, irregular brownish spots were also produced on stem and leaf tissues.

Symptoms of pod and seed coat discoloration of white beans varies. Pod discoloration varies from dark gray flecks to stipples to dark gray patches that later coalesce. Seeds within discolored pods show varying degrees of discoloration, which persists through processing.

CONTROL:
1. Cultivars differ in their reaction to *A. alternata*.
2. Certain foliar fungicides reduce disease severity.

Angular Leaf Spot

CAUSE: *Isariopsis griseola* Sacc. probably survives as mycelium in infected residue. The conidia are produced during wet weather and are disseminated by wind and rain to host tissue. The fungus may also be seedborne. Secondary infection is by conidia produced in lesions on the underside of the infected leaf and disseminating by wind or rain. Optimum temperature for infection is 24°C.

DISTRIBUTION: Eastern and southern United States. Angular leaf spot is considered of minor importance.

SYMPTOMS: Small brownish spots first occur on leaves. During moist weather *I. griseola* abundantly sporulates, giving the lesions a grayish color. Spots may be numerous enough to cause defoliation. Similar spots on pods are usually small but may enlarge and coalesce, thereby involving a large area of the pod. There is no correlation between disease severity on pods and percentage of seeds infected with pathogen. Seed infection normally occurs at the hilum; therefore, seeds become infected when located directly under lesions present at the pod suture.

CONTROL:
1. Rotate beans with resistant crops.
2. Plant disease-free seed.
3. Apply foliar fungicide.
4. Resistant germ plasm has been identified.

Anthracnose

CAUSE: *Colletotrichum lindemuthianum* (Sacc. & Magn.) Scribner is seedborne and overwinters in infected residue if plant material is kept dry. The fungus may survive as mycelium under the seed coat for years if seed is dry and stored under cool conditions (4°C). An alternating wet–dry cycle is detrimental to survival of the fungus, which will lose viability in several days. Infected seed gives rise to seedlings with lesions on the cotyledons. Conidia produced in lesions are wind and rain disseminated to healthy tissue. Long-distance spread of 3–5 meters during a rainstorm occurs when splashing raindrops are blown by gusting winds. Conidia are also carried on insects, machinery, and by other means. Conidia are covered by a sticky substance that enables them to adhere to whatever they touch. The disease is most severe during cool (14–18°C), wet weather. Temperatures above 25°C restrict disease development. Several races of anthracnose occur. These have been designated as alpha, beta, delta, epsilon, gamma, kappa, and lambda.

DISTRIBUTION: Eastern, midwestern, and southern United States. Anthracnose probably occurs where field beans are grown in cool, wet weather.

SYMPTOMS: Brown, oval, sunken cankers with purplish to brick red borders extend up and down the stem. The stems may be so weakened as to break over. Infection of leaves causes a few or many veins on the underside to become purple or red. Angular dead spots may appear on upper leaf surfaces, giving them a ragged appearance.

Lesions on pods start as numerous, small, elongated red-brown spots. Spots become somewhat circular and sunken with a rusty to brown border. Mature spots are 6 mm or larger in diameter. During moist weather, spots become pinkish due to formation of conidia.

Infected seed has dark, sunken lesions of various sizes that may extend through the seed coat into the cotyledons. When mature, the lesions have pinkish spore masses in their centers.

CONTROL:
1. Plant disease-free seed grown in areas where anthracnose does not occur.
2. Plant beans in the same field every three or four years.
3. Do not enter into fields when plants are wet.
4. Apply foliar fungicides before anthracnose becomes severe.
5. Plant a resistant cultivar.

Aphanomyces Root and Hypocotyl Rot

CAUSE: *Aphanomyces euteiches* Drechs. f. sp. *pisi* Pfend. & Hag. infects both peas and beans. *Aphanomyces euteiches* Drechs. f. sp. *phaseoli* Pfend. & Hag. infects only beans.

DISTRIBUTION: Not reported.

SYMPTOMS: *Aphanomyces euteiches* f. sp. *pisi* causes less injury than f. sp. *phaseoli*. Plants are stunted. Roots and lower stems are rotted. Necrosis of the hypocotyl ranges from streaks to complete destruction. Infrequently plant death occurs.

CONTROL:
1. Plant the most tolerant cultivars.
2. Resistant germ plasm has been identified.

Ascochyta Leaf Spot

CAUSE: *Ascochyta phaseolorum* Sacc. (syn. *A. boltshauseri* Sacc.) is seedborne. Disease is favored by cool temperatures and abundant moisture found at elevations greater than 1500 m.

DISTRIBUTION: Higher elevations in Central and South America.

SYMPTOMS: Leaf symptoms are dark gray to black zonate lesions that may contain small, black pycnidia. Defoliation may occur in severe cases. Lesions may also occur on peduncles, petioles, pods, and stems where girdling can cause plant death.

CONTROL:
1. Rotate beans with other crops.
2. Plant resistant cultivars.
3. Plant disease-free seed.
4. Apply foliar fungicides.

Ashy Stem Blight

CAUSE: *Macrophomina phaseoli* (Maubl.) Ashby is seedborne and survives as pycnidia in infected residue. The disease is most serious during periods of excess moisture and high temperatures.

DISTRIBUTION: Eastern, southern, and western United States.

SYMPTOMS: Sunken, red-brown lesions occur on the stem at or below soil level. Eventually the lesion extends down into the roots and up into the branches. As the lesion enlarges, it becomes gray at the center. Numerous pycnidia that appear as black objects about the size of a pinhead occur in the center of the lesion. The black color of the pycnidia contrasts with the gray center of the lesion.

Young plants that become infected usually die before seed is produced. Infection of an older plant produces superficial lesions that are usually not sunken. Frequently the disease is more pronounced on one side of a plant, causing the primary leaf on that side to droop and die. Any remaining leaves will turn yellow.

CONTROL:
1. Plant only healthy seed that has been treated with a seed-protectant fungicide.
2. Rotate beans every two to three years with a nonsusceptible crop.

Chaetoseptoria Leaf Spot

CAUSE: *Chaetoseptoria wellmanii* Stevenson may be seedborne. Disease occurs in areas of moderately cool temperatures and moist environments.

DISTRIBUTION: Central and South America.

SYMPTOMS: Symptoms occur on primary leaves soon after plant emergence. Circular lesions occur with light tan centers surrounded by a red-brown border. Small gray pycnidia may form in the lesions. Defoliation and yield reduction may occur.

CONTROL:
1. Rotate beans with other crops.
2. Plant disease-free seed.
3. Apply foliar fungicide.

Entyloma Leaf Spot

CAUSE: *Entyloma* sp.

DISTRIBUTION: Dominican Republic, Guatemala, and Honduras.

SYMPTOMS: Initially spots are water-soaked and later become gray-brown. Spots are more brown in color on upper leaf surfaces and more gray-blue on the lower leaf surface. Spots on mature leaves in advanced stage of disease show a deep brown necrosis that gives leaves a burned appearance. Spots are somewhat round or oval, and sometimes delimited by leaf veins, but tend to coalesce.

CONTROL: No control is reported.

Floury Leaf Spot

CAUSE: *Ramularia phaseoli* (Drummond) Deighton. Disease occurs in areas of moderate temperatures and moisture.

DISTRIBUTION: Central and South America.

SYMPTOMS: Initially symptoms occur on older leaves and progress to newer ones. Symptoms are white or floury, angular lesions that occur on lower leaf surfaces. The floury appearance is due to fungal growth of mycelium and spores. These lesions may coalesce and appear irregular. The upper leaf surface may have a light green discoloration but no floury appearance of mycelium and spores. Defoliation may occur.

CONTROL:
1. Rotate beans with other crops.
2. Apply foliar fungicides.

Fungi Reported on Seed

Colletotrichum dematium (Pers. ex. Fr.) Grove f. *truncata* (Schw.). Arx.
Fusarium semitectum Berk. & Rav.

Fusarium solani (Mart.) Appel. & Wr.
Phomopsis sp.
Trichothecium roseum Link

Fusarium Root Rot

This disease is also called dry root rot.

CAUSE: *Fusarium solani* (Mart.) Appel. & Wr. f. sp. *phaseoli* (Burk.)
Snyd. & Hans. can survive saprophytically in plant residue and as
chlamydospores in the soil. Chlamydospores are stimulated to germinate
by root excretions. The germ tube enters the roots and grows in the outer
layers of tissue, especially the cortex. Conidia and some cells of mycelium
growing on residue are eventually converted to chlamydospores. The fungus is disseminated in dust and plant debris mixed with seed, in soil on
machinery, and by any other means capable of transporting soil.

Beans planted in cold soil or subjected to cool weather after emergence tend to have severe root rot, whereas those planted and raised in
warm soil with adequate water tend to escape serious root rot. In cold
soil, root growth is restricted. Also herbicide and decomposing organic
matter may be toxicogenic in cold soil. Additionally, soil compaction
by tractor wheels and subsurface pans are among the root-restraining,
disease-preconditioning problems involved in bean root rot. The most
frequent preconditioning stress is intermittent drought, especially in
coarse-textured soils of low water-holding capacity and where soil compaction may confine roots to a small volume. Root rot and postemergence
damping off is greater in subsoiled or disked than in plowed treatments.
In one of three years, postemergence damping off is greater by applying
nitrogen broadcast preplant than by applying overhead irrigation.

Infection may be greater at lower inoculum levels when the lesion
nematode, *Pratylenchus penetrans* (Cobb) Sher. & Allen, is present.
However, there is no effect on final disease severity.

DISTRIBUTION: Generally distributed wherever field beans are
grown.

SYMPTOMS: Fusarium root rot normally causes little damage if
plants are stressed by other environmental factors that restrict root
growth. Initially slightly red to brown areas or streaks occur on the taproot and hypocotyl, often around the points of secondary root emergence.
This reddish discoloration increases in intensity and may eventually
cover the taproot and hypocotyl. Later in the season the red is replaced
by a brown discoloration and the roots become hollow and dry. If the
taproot is split open, the pith area is often a bright red. Lateral roots are
often destroyed and plants often develop a secondary root system near
the soil line. Most diseased roots do not extend beyond the plowed layer
of soil; many go no deeper than 7.6–15.2 cm.

Aboveground symptoms usually do not appear until roots are seriously decayed. Symptoms are similar to those caused by low soil moisture conditions or lack of nitrogen. Young plants will be stunted with yellow leaves. Older plants will have yellow leaves with some defoliation. Yield is reduced by reduction of seed weight; however, the number of pods per plant is normally not reduced. Root rot has been reported to be more severe on beans grown in soil infested with both *F. solani* (Mart.) Appel. & Wr. f. sp. *phaseoli* (Burk.) Snyd. & Hans. and *Pythium ultimum* Trow.

CONTROL:
1. Plant the most resistant cultivars.
2. Rotate beans every three to four years with crops such as alfalfa or small grains.
3. Plant cleaned seed free of any debris.
4. Use practices that condition soil for good moisture distribution and root penetration. Practices that increase soil compaction or cause drouth or poor drainage increase disease severity.
5. Plant rates high enough to insure a good stand.
6. Do not feed bean straw to animals since manure may spread the fungus.
7. Plow under bean refuse.
8. Close cultivation should be avoided so that newly formed roots are not cut off.
9. Subsoiling increases root volume and depth, thereby counter-acting effects of disease.
10. Plant into warm soil.
11. Irrigate. In combination with deep tillage and resistant culti-vars, yields can be maintained in infected soils.

Fusarium Yellows

This disease is also called Fusarium wilt.

CAUSE: *Fusarium oxysporum* Schlecht. f. sp. *phaseoli* Kendrick & Snyd. The fungus may be externally seedborne and infects through root and hypocotyl wounds.

DISTRIBUTION: Potentially wherever *Phaseolus vulgaris* is grown.

SYMPTOMS: The initial symptom is a yellowing of the lower leaves that progresses to upper leaves. Defoliation may occur. There is a reddish discoloration of the vascular system of the roots, stem, petioles, and peduncles. Early infection results in stunting.

CONTROL:
1. Plant tolerant or resistant cultivars.
2. Rotate beans with other crops.

3. Plant disease-free seed.
4. Apply fungicides as a seed treatment.

Gray Mold

CAUSE: *Botrytis cinerea* Pers. ex Fr., teleomorph *Botryotinia fucke-liana* (deBary) Whetz., overwinters as stroma and sclerotia on infected residue and other hosts, including weeds. Survival for extended periods may also occur as mycelium and possibly conidia. Conidia are produced starting early in the spring and throughout the summer when wet conditions occur. They are windborne to susceptible hosts. Apothecia of the teleomorph may also be produced from stroma or sclerotia. Ascospores are produced on the apothecia and are windborne to hosts.

Disease development is most severe under cool, humid conditions. Beans are infected at different stages of growth. Senescing cotyledons are colonized first. Young stem and leaf tissues can also become diseased before bloom and serve as inoculum sources within a field. Infected stems may continuously produce inoculum into the bloom period. Infection frequently occurs where the old blossom has fallen on the plant or has been retained at the tip of the pod.

DISTRIBUTION: Generally distributed wherever field beans are grown.

SYMPTOMS: All aboveground plant parts may show symptoms but pods are the most commonly infected part. Initially, water-soaked lesions occur. Under humid conditions, a white mold soon develops that is followed by gray fungal growth consisting of conidia and conidiophores. Tip wilt and decay of shoots may occur.

CONTROL:
1. Plant certified disease-free seed.
2. Rotate beans with nonsusceptible crops.
3. Apply a foliar fungicide if weather conditions favor disease development.
4. Do not plant beans too thickly since a thick canopy retards air circulation and favors disease development.

Mancha Gris (Gray Blotch)

This disease is also called gray spot.

CAUSE: *Cercospora vanderysti* P. Henn. Disease development is favored by cool, wet weather at elevations greater than 1500 m.

DISTRIBUTION: Colombia.

SYMPTOMS: Initially, numerous, slightly chlorotic, angular lesions occur on upper leaf surfaces. The spots are 2–5 mm in diameter and may be so numerous that the leaf has a mosaiclike appearance. Lesions may coalesce and cause defoliation. Eventually, a gray-white, powdery growth consisting of conidia and conidiophores occurs in the lesion on the upper leaf surface. The most distinguishing sign of the disease occurs in the lesions on the lower leaf surface where a gray mat or cushionlike fungal growth completely covers the lesions. The cushion consists of a dense growth of flexuous conidiophores and their conidia.

CONTROL:
1. Rotate beans with other crops.
2. Apply foliar fungicide.
3. Some breeding lines show resistance.

Pink Pod Rot

CAUSE: *Trichothecium roseum* Link is a common soil fungus that is saprophytic or weakly parasitic. In Canada, *T. roseum* is frequently isolated from seeds of soybean, pea, faba bean, kidney bean, and scarlet runner bean.

DISTRIBUTION: Red River Valley of North Dakota. The causal fungus has been occasionally isolated from bean roots showing root rot symptoms in southwestern Ontario.

SYMPTOMS: A powdery mold that is initially white but later turns pink is most frequently seen on senescent and mature pods. However, mold less frequently occurs on stems, petioles, and dead leaves. Seeds from diseased pods are shriveled and yellow or pink-yellow.

Inoculated pods were first water-soaked, developing lesions with chocolate-brown centers. Later a white, powdery mold developed in the lesions. The mold spread outward and eventually turned pink. Seeds from infected pods failed to germinate.

CONTROL: No control is reported.

Pod Rot, Seed Rot, and Root Rot

CAUSE: *Fusarium semitectum* Berk. & Rav. causes disease under humid conditions.

DISTRIBUTION: Brazil.

SYMPTOMS: Initially, pinpoint water-soaked lesions occur, that eventually become necrotic and rusty brown, varying in shape from circular to elongated, depending on the cultivar infected. Under humid conditions pods became soft and covered with fungal mycelium. Under less humid conditions, pods became brittle and dry.

White-seeded snap bean seed becomes red-brown, shriveled, and covered with mycelium. Light brown-seeded seed has blood-red, circular lesions.

Rotted roots are rusty brown and pulpy with a pruned appearance.

CONTROL: No control is reported.

Powdery Mildew

CAUSE: *Erysiphe polygoni* DC. ex Merat overwinters as cleistothecia on residue. Ascospores are formed in cleistothecia and are windborne to hosts. Conidia are formed on superficial mycelia and function as secondary inoculum. Cleistothecia are formed on older infection sites or on leaves as they reach maturity. Mycelium on infected residue may also function as a means of overwintering, especially in areas that have mild winters.

DISTRIBUTION: Worldwide.

SYMPTOMS: The most damage usually occurs to crops maturing late in the autumn and produced in the southern United States during the winter. Typically a white powdery material consisting of conidia and conidiophores is produced on all aboveground parts of the plant. Leaves turn yellow and may fall off. Pods and stems often turn a purplish color. Pods are often malformed, small, poorly filled, and fall off before any seed matures.

CONTROL:
1. Plant the most resistant variety adapted to an area.
2. Apply a foliar fungicide before disease becomes severe.

Pythium Root and Hypocotyl Rot

CAUSE: *Pythium aphanidermatum* (Edson) Fitzp., *P. myriotylum* Drechs., *P. irregulare* Buis., *P. splendens* Braun, and *P. ultimum* Trow., have all been reported to be incitants. Other *Pythium* spp. reported to be pathogenic to bean are *P. aristosporum* Vanterpool, *P. catenulatum* Matthews, and *P. dissotocum* Drechs. These fungi are most active as disease incitants in soils with high moisture. Optimum disease occurs at a soil temperature of 15°C and soil water potential of zero to −1 bar. Root rot and postemergence damping off is greater in soils that have been subsoiled or disked than in soils that have been plowed. In one of three years, postemergence damping off was greater by applying nitrogen broadcast preplant compared to applying nitrogen through overhead irrigation.

DISTRIBUTION: Not reported, but likely widespread.

SYMPTOMS: Damping off may occur. Initially hypocotyl symptoms occur 16–22 days after planting and appear as elongated, water-soaked areas on the hypocotyl that extend 3–5 cm above the soil line and one-fourth to three-fourths the distance around the hypocotyl. These symptoms are also present on the hypocotyl below the soil line. Most roots appear white and healthy. Within three weeks the water-soaked areas become drier and tan to red-brown with a slightly sunken surface. Later most of the belowground hypocotyl and fibrous roots are killed. After four weeks the belowground hypocotyl appears papery and red-brown. Adventitious roots may be produced above or at the soil line.

Plants may wilt and die, or if adventitious roots are produced, may be stunted or chlorotic, producing smaller and fewer pods. However, seed weight is not reduced. Root pruning that results in shortened roots may also occur. Secondary infection by *Fusarium* spp. may occur.

CONTROL:
1. Plant the most resistant cultivars.
2. Apply systemic fungicides to seed.
3. Plowing to a depth of 20–25 cm increased plant stand, vine weight, and yield in research plots.
4. Disease is suppressed by certain herbicides.

Pythium Wilt

CAUSE: *Pythium butleri* Subr. survives as oospores in soil and residue.

DISTRIBUTION: Generally distributed wherever field beans are grown.

SYMPTOMS: A water-soaked lesion occurs on the stem near the soil line and progresses upward into the lower branches. The lesion rarely extends below the soil line. The outer layer of tissue of the stem and branches becomes soft and watery, readily separating from the fibrous tissue of the stem. Leaves wilt and the plant quickly dies. Death of the plant is usually so rapid that wilted leaves first remain green and then rapidly become brown.

CONTROL: No control is usually necessary.

Rhizoctonia Root Rot

CAUSE: *Rhizoctonia solani* Kuhn anastomosis group (AG) 4 is the primary causal fungus. However, AG1 and AG2 have also been demonstrated to be pathogenic to *Phaseolus vulgaris* L. They survive as sclerotia in the soil and on infected residue. The fungus is also a vigorous saprophyte on dead plant tissue. Plant roots or hypocotyls are infected by mycelium growing from a precolonized substrate or by sclerotia germinating and forming infective hyphae. Infection likely only occurs early in the growing season. Infection occurs at a maximum temperature of 25°C and at low soil moisture.

Root rot, hypocotyl rot, and postemergence damping off are greater in soils that have been subsoiled or disked than in soils that have been plowed. Plowing reduces population of *R. solani* compared to disking. Additionally, in one of three years, postemergence damping off is greater by applying nitrogen broadcast preplant compared to applying nitrogen through overhead irrigation.

DISTRIBUTION: Generally distributed wherever field beans are grown.

SYMPTOMS: Damping off may occur. Seedlings are infected during emergence and become twisted and stunted. Reddish, sunken cankers occur on roots and the main stem. If a canker girdles the stem, the seedling will die. Lesions enlarge and the taproot and the stem belowground may be destroyed. The lesions then become more brownish than red. Infected plants are stimulated to produce numerous adventitious roots near the surface of the soil. Older plants are stunted and yellowed.

CONTROL:
1. Rotate field beans every three to four years.
2. Ridge soil around the base of bean plants and irrigate soon thereafter to promote adventitious roots.
3. Mechanical injury to seedlings during emergence enhances infection.
4. Black-seeded cultivars are more resistant than white-seeded cultivars. Seed coats of black-seeded cultivars adhere tightly to cotyledons, protecting the germinating seed from seed infection and preemergence damping off. Seed coats of white-seeded beans crack readily before emergence. Additionally black-seeded coats contain phenolic compounds that inhibit growth of *R. solani*. The resistance of three-week-old red kidney bean plants to *R. solani* is associated with the inability of the fungus to form infection cushions and penetrate the hypocotyl.
5. Plowing infested soil to a depth of 20–25 cm increases plant stand, vine weight, and yield in research plots.

Rust

CAUSE: *Uromyces appendiculatus* (Pers.) Unger var. *appendiculatus* (syn. *U. phaseoli* (Reben.) Wint. *typica* Arth.) is a macrocyclic autoecious rust. In the spring, teliospores germinate to produce basidiospores that are windborne to bean leaves. Pycnia are usually formed on the upper leaf surface and aecia are usually formed on the lower leaf surface. Urediospores are formed in uredia and are windborne to bean leaves, thereby providing secondary inoculum and functioning as a repeating cycle. Teliospores are eventually produced in the old uredia as the plant dies or matures. Disease development is favored by optimum temperatures between 17.5°C and 22.5°C with six to eight hours of leaf wetness. Several physiological races are known to exist.

DISTRIBUTION: Generally distributed wherever field beans are grown.

SYMPTOMS: Usually symptoms become most conspicuous later in the growing season but epiphytotics may occur earlier in the season. Pycnidia initially appear only on the upper leaf surface as chlorotic flecks that later become evident as a slightly raised white bump. Aecia develop first on the leaf underside and later occasionally on the upper leaf surface. Aecia are white rings or clusters of pustules. Uredia begin as small white spots that enlarge and rupture in a few days to expose orange-yellow urediospores.

Infected leaves turn yellow, then brown, and soon die, falling off the plant. Pods and stems may also become infected. Black telia replace urediospores in old pustules as the tissue dies.

CONTROL:
1. Rotate field beans with other crops. Grow beans every third year in the same soil.
2. Do not plant field beans adjacent to previously infected fields.
3. Plow under bean residue in the autumn.
4. Inspect fields during growing season to detect early symptoms of rust. Apply foliar fungicides if rust develops rapidly from early bloom to four weeks before harvest.
5. Plant resistant varieties adapted to an area. Resistance in some Jamaican varieties is correlated with mean hair density on both leaf surfaces.

Sclerotinia White Mold

This disease is also called white mold.

CAUSE: *Sclerotinia sclerotiorum* (Lib.) deBary (syn. *Whetzelina sclerotiorum* (Lib.) Korf & Dumont) survives five or more years as sclerotia in the soil. The fungus also overwinters as mycelium in seed but this is not as important a source of primary inoculum as sclerotia. Infection occurs in one of two ways: either as mycelium from sclerotia, which is considered secondary, or as ascospores, which are considered primary. During moist, humid weather, sclerotia germinate to produce mycelium that first colonizes dead plant tissue; then, using dead tissue as a substrate, the mycelium invades living plant tissue. However, this is not considered a very effective mode of infection.

Sclerotia may also germinate to produce one or many apothecia, but will not do so until they have been preconditioned. This preconditioning or physiological maturation occurs during the winter or noncropped season; however, freezing is not necessary. In semiarid areas carpogenic germination is usually initiated after the plant canopy has covered the soil surface. Apothecia are likely to be produced from sclerotia in an area that stays moist for a relatively long period of time, such as the side of an irrigation furrow or next to a stem shaded by the plant canopy. Optimum apothecia production occurs at a soil matrix potential of -0.25 bars and at a soil temperature of 15–18°C for 10–14 days. Ascospores are then produced in a layer of asci on the surface of the apothecium and are windborne to old blossoms, stems, or leaves, and cause infection. Insects and splashing water also disseminate the fungus. Temperatures of 25°C or greater, combined with relative humidity in excess of 35 percent, is detrimental to ascospore survival. Bean blossoms commonly serve as an energy source to support infection. Conformation of senescent blossoms may influence ascospore germination through availability of free moisture with nutrients leached from blossoms and held within blossom convolutions. The most common means of spread after initial infection is by contact of healthy plant parts with diseased tissues.

The disease is most severe later in the growing season during wet, humid weather or frequent irrigation and at an optimum temperature of 25°C. Epiphytotics commonly occur only after flowering. Disease severity is greater under a dense plant canopy. Injury from Sclerotinia wilt increases with increasing irrigation in Fusarium-free fields but is negligible in Fusarium-infected fields. *Sclerotinia sclerotiorum* is a pathogen of a large number of plant genera.

DISTRIBUTION: Generally distributed wherever field beans are grown.

SYMPTOMS: Symptoms appear on stems, leaves, and pods. Soft, watery, irregular areas occur first on stems just above the soil line, then on leaflet axils, leaves, and pods. These areas rapidly enlarge, and in a short time, dense, white, cottony masses of mycelium with small brown droplets in them cover the infected spots. Small, hard, black, sclerotial bodies appear in the mycelium. Main stems and branches are often encircled by the infection, then wilt and die. The affected plant parts dry out, become bleached, and punky. The white masses of mycelium turn light gray or brown and dry out, leaving the conspicuous sclerotia. Infected bean seed is orange and chalky textured.

CONTROL:
1. Rotate field beans with small grains.
2. Rows should be spaced 76 cm or wider to encourage air movement and dry out plants.
3. Apply foliar fungicide at early bloom stage of growth.
4. Clean out harvesting equipment between fields.
5. Some lines display a level of resistance.
6. Plant cultivars with open canopy structures.

Smut

This disease is also called leaf blister smut.

CAUSE: *Entyloma petuniae* Speg. survives in residue.

DISTRIBUTION: Caribbean and Central America.

SYMPTOMS: Gray-black blisters occur on upper leaf surface. The blisters contain subepidermal masses of black chlamydospores. Lesions are often delimited by leaf veins or veinlets. Usually the first or second trifoliate leaves are infected first.

CONTROL:
1. Rotate beans with other crops.
2. Destroy old plant residue.
3. Apply foliar fungicides.

Stem Rot

CAUSE: An unknown Basidiomycete that resembles *Athelia* spp.

DISTRIBUTION: Florida.

SYMPTOMS: Stem lesions are dry, firm, tan, and do not extend above the soil surface. Plants wilt and soon die. Dry rot is present on entire underground portions of stems that die and on plants that are not killed. Frequently only small lesions will develop on underground stems. Occasionally some plants will have only root rot but no stem lesions. Mycelial strands form on roots and underground stems of most infected plants.

CONTROL: No control is reported.

Southern Blight

CAUSE: *Sclerotium rolfsii* (Sacc.).

DISTRIBUTION: Generally distributed in warmer bean-growing areas.

SYMPTOMS: Brown, water-soaked lesions occur on the hypocotyl just beneath the soil line. Infection proceeds into the taproot, eventually destroying the cortex. The lower leaves become yellowed, followed by plant wilting and death. White mycelium develops around the hypocotyl and surrounding soil. Spherical, white sclerotia, 1–2 mm in diameter develop in mycelium. Eventually the sclerotia turn brown.

CONTROL:
1. Rotate beans with other crops.
2. Plant tolerant or resistant cultivars.

Web Blight

CAUSE: *Rhizoctonia solani* Kuhn (syn. *R. microsclerotia* Matz.), teleomorph *Thanatephorus cucumeris* (Frank) Donk., survives as sclerotia in soil and on infected residue. The fungus is disseminated by sclerotia being transported by wind, rain, machinery, and other means. Sclerotia and infected debris that are splashed onto plant parts by rain serve as the primary inoculum source. The sclerotia germinate under favorable conditions to form mycelium that infects the plant. Trifoliate leaves are infected by splashed inoculum but more frequently by advancing hyphae from infected tissues.

DISTRIBUTION: Costa Rica and southern United States; however, web blight is likely more widespread.

SYMPTOMS: Small, round, water-soaked spots are produced on leaves. Mycelial growth resembling spider webs occurs on bean stems, pods, and leaves. Numerous, small, brown sclerotia are imbedded or scattered throughout the mycelium.

   Spots on leaves are lighter in color than healthy tissue and appear as if they had been scalded. Spots become tan and surrounded by a dark border. Spots on young pods are light tan and irregular in shape but on mature pods they are dark brown and sunken, resembling those caused by anthracnose. Spots may coalesce to cover the entire pod. During moist, warm weather in the final stages of the disease, mycelium grows over all plant parts, binding leaves, petioles, flowers, and pods together.

CONTROL:
1. Rotate field beans with nonsusceptible crops.
2. Diseased plants should be destroyed as soon as possible after harvest.
3. Apply a foliar fungicide when disease occurs.
4. Mulch soil. This helps to prevent dissemination by splashing.
5. Plant the most tolerant cultivar adapted to your area.

White Leaf Spot

This disease is also called Mancha bianca.

CAUSE: *Pseudocercosporella albida* (Matta & Belliard) comb. nov. Disease is most severe under cool and wet conditions.

DISTRIBUTION: Central and South America.

SYMPTOMS: Initially whitish, angular spots that are restricted by veins occur first on the underside, then on the upper side of leaves. Only the lower leaves of resistant cultivars are normally infected; however, the upper leaves of susceptible cultivars are infected also. Spots enlarge, coalesce, and can cause defoliation in severe cases.

CONTROL: Plant resistant cultivars.

MYCOPLASMAL CAUSES

Machismo

CAUSE: A mycoplasmalike organism that is disseminated by the leafhopper *Scaphytopices fulginosus* Osborn. The causal organism is also graft transmitted but not mechanically or seed transmitted. The average period of incubation until symptom expression is 37 days.

DISTRIBUTION: Colombia.

SYMPTOMS: Symptoms occur at flowering and pod formation. Infected plants produce wrinkled, distorted pods with no seed formation. Plants may produce normal pods but seeds may germinate within the pod. In severe cases, buds proliferate, producing a witches' broom appearance. Plants infected later in the season normally produce normal yields.

CONTROL: No control is reported.

CAUSED BY NEMATODES

Root Knot Nematode

CAUSE: *Meloidogyne incognita* (Kofoid & White) Chitwood and *M. javanica* (Treub.) Chitwood. Infested soil contains eggs that hatch to produce larvae. The larvae move through the soil and penetrate the root tip. The larvae migrate through the root to the vascular tissues, become sedentary, and commence feeding. Excretions of the larvae stimulate cell proliferation, resulting in the development of syncythia from which the larvae derive their food. Females lay eggs that may be pushed out of the root into the soil.

DISTRIBUTION: Southern United States and California, primarily on light, sandy soils.

SYMPTOMS: Fleshy, irregular-shaped galls are produced on bean roots. Galls are usually larger than nodules and are more irregular in shape. Infected plants are stunted, yellowish, and generally appear unthrifty.

CONTROL:
1. Do not grow field beans in infested soil for at least three years.
2. Apply a nematicide to soil if economic conditions warrant it.

Root Lesion Nematode

CAUSE: *Pratylenchus scribneri* Steiner.

DISTRIBUTION: Generally distributed in warm bean-growing areas.

SYMPTOMS: Roots are greatly reduced and discolored.

CONTROL:
1. Rotate beans with other crops.
2. Apply nematicides.

_____VIRAL CAUSES_____

Bean Chlorotic Mottle

CAUSE: Bean chlorotic mottle virus (BCMV) is reservoired in several tropical weeds that serve as sources of inoculum. The virus is transmitted by the whitefly, *Bemisia tabaci* (Genn.).

DISTRIBUTION: Central and South America.

SYMPTOMS: Chlorotic mottled patches occur on leaves. Curling and deformation occur in some cultivars. Early infection may result in severe stunting and formation of a witches' broom.

CONTROL: Plant resistant cultivars.

Bean Common Mosaic

CAUSE: Bean common mosaic virus (BCMV) is seedborne and consti-tutes the major means of overwintering. BCMV is transmitted by several species of insects, including at least 11 species of aphids, from infected to healthy plants. BCMV is also mechanically transmitted and may be car-ried in the pollen of infected plants. Some isolates are seedborne. Several strains of BCMV are known to exist; each strain may cause different symptoms.

DISTRIBUTION: Generally distributed wherever field beans are grown.

SYMPTOMS: Differences in symptoms may vary greatly between plants and are influenced by cultivar, temperature, light, soil fertility, and moisture conditions. General symptoms include dwarfing, excessive branching, leaf cupping, and mottling.

The mosaic pattern may appear as dark bands along the major veins with a clearing of the leaf area along the margin and between the veins of the leaf. The pattern may appear only as dark green areas that vary in size from small granular flecks to large blisterlike areas in a generally chlorotic leaf. The darker areas are generally raised, giving the leaf a warty, puckered shape. Leaves showing only granular mottling have very little distortion and appear much like a normal leaf. Leaves having none of the more pronounced symptoms may be elongated, narrow, and cupped downward.

Plants growing from infected seed or that were infected early may be spindly, make very little growth, with only rare pod formation. Primary leaves of plants from infected seed rarely show a mosaic pattern.

An infected plant growing under low temperature and light intensity may not show mosaic symptoms. Severe symptoms consisting of local necrotic lesions on leaves occur under high light intensity and temperatures around 26.5°C.

Infected resistant varieties will display a systemic necrosis at high temperatures. At first young leaflets wilt slightly, leaves turn brown to black, then wilt, followed by the entire plant wilting and dying. The vascular system also becomes necrotic. The maturity of less severely infected plants is delayed.

CONTROL:
1. Plant only certified seed.
2. Plant resistant cultivars.
3. Rogue out infected plants if a small area is involved.

Bean Golden Mosaic

This disease is also called bean golden yellow mosaic.

CAUSE: Bean golden mosaic virus (BGMV) is transmitted by the whitefly *Bemisia tabaci* (Genn.). BGMV can also be transmitted mechanically but it is not seedborne. BGMV may be reservoired in numerous weeds.

DISTRIBUTION: Tropical countries.

SYMPTOMS: Infected plants have a yellow appearance. Newly emerged leaves have a bright yellow general mosaic pattern but older leaves have a less distinct mosaic pattern. Infected leaves curl downward. Plants of some varieties are stunted and have malformed pods.

CONTROL:
1. Plant tolerant or resistant varieties.
2. Control insects through application of insecticides.

Bean Rugose Mosaic

CAUSE: Bean rugose mosaic virus (BRMV) is transmitted mechanically and by the beetles *Cerotoma* spp. and *Diabrotica* spp.

DISTRIBUTION: Central and South America.

SYMPTOMS: Leaf symptoms include a light and dark green mosaic pattern often accompanied by severe leaf blistering, curling, and malformation, with a thickened or leathery appearance. Early infection results in severe stunting. Pods may also show a mosaic pattern and be malformed.

CONTROL:
1. Plant a resistant cultivar.
2. Plant virus-free seed.

Bean Yellow Mosaic

CAUSE: Bean yellow mosaic virus (BYMV) overwinters in perennial plants such as wild sweet clover. The virus is transmitted mechanically and by several species of aphids. It is not seedborne in bean fields.

DISTRIBUTION: Generally distributed wherever field beans are grown.

SYMPTOMS: Symptoms of bean yellow mosaic are usually more prevalent along the borders of fields. Initially leaflets droop at point of attachment to the stem, followed by development of small, angular pale spots ranging in diameter from 1.5–3.0 mm. Spots may coalesce, resulting in general chlorosis. Leaves become thickened, cupped, and brittle. Leaves may have a thickened granular appearance. Some strains may cause a purpling of leaf bases of lower leaves; plant death may follow.

In some bean cultivars, the pale, angular spots become more pronounced and are bright yellow. Leaves may be malformed and distorted. Infected plants may appear dwarfed, bunchy, and may die. Plant maturity is usually delayed and yield is reduced.

CONTROL:
1. Plant resistant cultivars.
2. Eliminate wild sweet clover from fence rows and ditch banks.

Cowpea Mild Mottle

CAUSE: Cowpea mild mottle virus (CMMV).

DISTRIBUTION: Tanzania.

SYMPTOMS: Leaves are mottled with chlorotic spots.

CONTROL: No control is reported.

Cowpea Severe Mosaic

CAUSE: Cowpea severe mosaic virus (CSMV).

DISTRIBUTION: Brazil.

SYMPTOMS: Leaves of infected plants have vein clearing, mottling, and chlorotic spots. Pods are mottled and distorted.

CONTROL: No control is reported.

Cucumber Mosaic

CAUSE: Cucumber mosaic virus (CMV) is seedborne.

DISTRIBUTION: Europe and the United States.

SYMPTOMS: Leaves have a mosaic appearance.

CONTROL: No control is reported.

Curly Dwarf Mosaic Disease

CAUSE: Bean curly dwarf mosaic virus (BCDMV) is transmitted by the Mexican bean beetle and the spotted cucumber beetle. It is also mechanically transmitted. Fifteen other species of legumes are also susceptible.

DISTRIBUTION: El Salvador.

SYMPTOMS: Plants are dwarfed with downward curling of leaves. Additionally, leaves have a mosaic pattern and are rugose. Inoculated plants have symptoms that range from mild mosaic to lethal top necrosis.

CONTROL: No control is reported.

Curly Top

CAUSE: Curly top virus (CTV) overwinters in a large number of perennial plants and is transmitted by the beet leafhopper *Circulifer tenellus* Baker.

DISTRIBUTION: Western United States.

SYMPTOMS: Bean plants are least tolerant to curly top when in the crook neck stage. Plants infected at this stage usually die. If a plant is in pod set when infected, the plant may live and seed will mature but no new pods will be set. Seed will be poorly developed and lightweight. If the plant has developed only primary leaves at the time of infection, the only symptoms of curly top will be the killing of the growing point and the subsequent yellowing and drying of the whole plant without production of any trifoliate leaves.

Plants infected in the early blossom stage may drop their blooms, become chlorotic, and die. The first symptom when older plants become infected is a downward cupping of the youngest trifoliate leaf. The growing point then dies or subsequent internodes are shortened, giving the plant a stunted, bushy appearance. The trifoliate leaves formed before infection become chlorotic, thickened, and cup downward. The whole plant becomes brittle and leaves and growing point are easily broken off.

CONTROL: Plant resistant cultivars.

Southern Bean Mosaic

CAUSE: Southern bean mosaic virus (SBMV) is transmitted by several beetle vectors and is seedborne. SBMV is also mechanically transmitted by growing plants in virus-infested soil. Incidence of infection in the latter case is higher with bean seeds with cracked coats than in those with intact coats.

DISTRIBUTION: South America.

SYMPTOMS: Either circular (1-3 mm in diameter), brown-red local lesions or systemic mottling and vein–banding symptoms occur on leaves. Additionally leaves may be blistered and malformed.

Pods have dark green, water-soaked, irregular-shaped blotches. The number and weight of seed may be reduced; however, the number of pods may be greater in infected than in noninfected plants.

CONTROL:
1. Plant virus-free seed.
2. Plant resistant cultivars.

Stunt

CAUSE: Peanut stunt virus (PSV). For further information, see Chapter 11 on diseases of peanuts.

DISTRIBUTION: North Carolina, Tennessee, and state of Washington.

SYMPTOMS: Diseased plants are stunted. Necrosis of the growing tip or of branches is common in some cultivars. Leaves are malformed, mottled, and crinkled. Pods are few, small, and malformed, with many having only a few small seeds.

CONTROL: No control is reported.

_____BIBLIOGRAPHY_____

Abawi, G. S.; Crosier, D. C.; and Cobb, A. C. 1977. Pod flecking of snap beans caused by *Alternaria alternata*. *Plant Dis. Rptr.* **61**:901–905.

Abawi, G. S.; Potlach, F. J.; and Molin, W. T. 1975. Infection of bean by ascospores of *Whetzelinia sclerotiorum*. *Phytopathology* **65**:673–678.

Adegbola, M. O. K., and Hagedorn, D. J. 1969. Symptomatology and epidemiology of Pythium bean blight. *Phytopathology* **59**:1113–1118.

Anderson, F. N.; Steadman, J. R.; Coyne, D. P.; and Schwartz, H. F. 1974. Tolerance to white mold in *Phaseolus vulgaris* dry edible bean types. *Plant Dis. Rptr.* **58**:782–784.

Bernier, C. 1975. Diseases of pulse crops and their control, in *Oilseed and Pulse Crops in Western Canada*. J. T. Harapick (ed.). Modern Press, Saskatoon, Saskatchewan.

Blad, B. L.; Steadman, J. R.; and Weiss, A. 1978. Canopy structure and irrigation influence white mold disease and microclimate of dry edible beans. *Phytopathology* **68**:1431–1437.

Burke, D. W., and Miller, D. E. 1983. Control of Fusarium root rot with resistant beans and cultural management. *Plant Disease* **67**:1312–1317.

Burke, D. W.; Miller, D. E.; Holmes, L. D.; and Barker, A. W. 1972. Counteracting bean root rot by loosening the soil. *Phytopathology* **62**:306–309.

Caesar, A. J., and Pearson, R. C. 1983. Environmental factors affecting survival of ascospores of *Sclerotinia sclerotiorum*. *Phytopathology* **73**:1024–1030.

Cafati, C. R., and Saettler, A. W. 1980. Transmission of *Xanthomonas phaseoli* in seed of resistant and susceptible Phaseolus genotypes. *Phytopathology* **70**:638–640.

Campbell, C. L., and Steadman, J. R. 1985. The relationship of *Sclerotinia sclerotiorum* ascospore germination on bean blossoms to disease development in a bean canopy microclimate. *Phytopathology* **75**:1369 (abstract).

Campbell, L. 1949. Gray mold of beans in western Washington. *Plant Dis. Rptr.* **33**:91–93.

Chupp, C., and Sherf, A. F. 1960. *Vegetable Diseases and Their Control.* Ronald Press, New York.

Claflin, L. E.; Stuteville, D. L.; and Armbrust, D. V. 1973. Wind-blown soil in the epidemiology of bacterial leaf spot of alfalfa and common blight of bean. *Phytopathology* **63**:1417–1419.

Coley-Smith, J. R. 1980. Sclerotia and other structure in survival. Pages 85–114 in *The Biology of Botrytis.* J. R. Coley-Smith, K. Verhoff, and W. R. Jarvis (eds.). Academic Press, New York, 317p.

Cook, G. F.; Steadman, J. R.; and Boosalis, M. G. 1975. Survival of *Whetzelinia sclerotiorum* and initial infection of dry edible beans in western Nebraska. *Phytopathology* **65**:250–255.

Coyne, D. P., and Schuster, M. L. 1970. "Jules," a great northern dry bean variety tolerant to common blight bacterium (*Xanthomonas phaseoli*). *Plant Dis. Rptr.* **54**:557–559.

Cupertino, F. P.; Costa, C. L.; Lin, M. T.; and Kitajimi, E. W. 1982. Infeccas natural do feyoeiro (*Phaseolus vulgaris* L.) Pelo virus do mosaico severs do feijao macassar. *Fitopathologia Brasileria* **7**:275–283 (in Portuguese).

Daub, M. E., and Hagedorn, D. J. 1981. Epiphytic populations of *Pseudomonas syringae* on susceptible and resistant bean lines. *Phytopathology* **71**:547–550.

Dhingra, O. D. 1978. Internally seedborne *Fusarium semitectum* and *Phomopsis* sp. affecting dry and snap bean seed quality. *Plant Dis. Rptr.* **62**:509–512.

Dhingra, O. D., and Kushalappa, A. C. 1980. No correlation between angular leaf spot intensity and seed infection in beans by *Isariopsis griseola*. *Fitopathologia Brasileria* **5**:149–152.

Dickson, M. H., and Abawi, G. S. 1974. Resistance to *Phythium ultimum* in white-seeded beans (*Phaseolus vulgaris*). *Plant Dis. Rptr.* **58**:774–776.

Echandi, E., and Hebert, T. T. 1970. An epiphytotic of stunt in beans incited by the peanut stunt virus in North Carolina. *Plant Dis. Rptr.* **54**:183–184.

Echandi, E., and Hebert, T. T. 1971. Stunt of beans incited by peanut stunt virus. *Phytopathology* **61**:328–330.

Fahy, P. C., and Persley, G. J. 1983. *Plant Bacterial Diseases: A Diagnostic Guide.* Academic Press, Inc. New York.

Galindo, J. J.; Abawi, G. S.; and Thurston, H. D. 1982. Variability among isolates of *Rhizoctonia solani* associated with snap bean hypocotyls and soils in New York. *Plant Disease* **66**:390–394.

Galindo, J. J.; Abawi, G. S.; Thurston, H. D.; and Galvez, G. E. 1982. Source of inoculum and development of web-blight of beans in Costa Rica. *Phytopathology* **72**:170 (abstract).

Galvez, G. E.; Mora, B.; and Alfaro, R. 1984. Integrated control of web-blight disease of beans (*Phaseolus vulgaris*). *Phytopathology* **74**:1015 (abstract).

Gilbertson, R. L.; Carlson, E.; Rand, R. E.; Hagedorn, D. J.; and Maxwell, D. P. 1986. Survival of *Xanthomonas campestris* pv. *phaseoli* in bean stubble and debris with three tillage systems. *Phytopathology* **76**:1078 (abstract).

Goode, M. J.; Hagedorn, D. J.; and Cross, J. E. 1985. Occurrence of snap bean bacterial blight pathogens on wild legumes. *Phytopathology* **75**:1287 (abstract).

Granada, G. A. 1979. Machismo, a new disease of beans in Colombia. *Phytopathology* **69**:1029 (abstract).

Groth, J. V., and Mogen, B. D. 1978. Completing the life cycle of *Uromyces phaseoli* var. *typica* on bean plants. *Phytopathology* **68**:1674–1677.

Hagedorn, D. J., and Binning, L. K. 1982. Herbicide suppression of bean root and hypocotyl rot in Wisconsin. *Plant Disease* **66**:1187–1188.

Hagedorn D. J., and Rand, R. E. 1979. Development of processing type beans (*Phaseolus vulgaris*) resistant to bacterial brown spot (*Pseudomonas syringae*). *Phytopathology* **69**:1030 (abstract).

Hagedorn, D. J., and Rand, R. E. 1986. Development and release of WIS(MDR) 147 bean breeding line. *Phytopathology* **76**:1067 (abstract).

Hagedorn, D. J.; Rand, R. E.; and Saad, S. M. 1972. *Phaseolus vulgaris* reaction to *Pseudomonas syringae*. *Plant Dis. Rptr.* **56**:325–328.

Hampton, R. O.; Silbernagel, M. J.; and Burke, D. W. 1983. Bean common mosaic virus strains associated with bean mosaic epidemics in the northwestern United States. *Plant Disease* **67**:658–661.

Hoch, H. C.; Hagedorn, D. J.; Pinnow, D. L.; and Mitchell, J. E. 1975. Role of Phythium spp. as incitants of bean root and hypocotyl rot in Wisconsin. *Plant Dis. Rptr.* **59**:443–447.

Howard, C. M.; Conway, K. E.; and Albregts, E. E. 1977. A stem rot of bean seedlings caused by a sterile fungus in Florida. *Phytopathology* **67**:430–433.

Hunter, J. E.; Dickson, M. H.; Boettger, M. A.; and Cigna, J. A. 1982. Evaluation of plant introductions of *Phaseolus* spp. for resistance to white mold. *Plant Disease* **66**:320–322.

Hutton, D. G.; Wilkinson, R. E.; and Mai, W. F. 1973. Effect of two plant-parasitic nematodes on Fusarium dry root rot of beans. *Phytopathology* **63**:749–751.

Johnson, K. B., and Powelson, M. L. 1983. Influence of prebloom disease establishment by *Botrytis cinerea* and environmental and host factors on gray mold pod rot of snap bean. *Plant Disease* **67**:1198–1202.

Katherman, M. J.; Wilkinson, R. E.; and Beer, S. V. 1980. The potential of four dry bean cultivars to serve as sources of *Pseudomonas phaseolicola* inoculum. *Plant Disease* **64**:72–74.

Kobriger, K., and Hagedorn, D. J. 1984. Additional Pythium species associated with bean rot complex in Wisconsin's Central Sands. *Plant Disease* **68**:595–596.

Lewis, J. A.; Lumsden, R. D.; Papavizas, G. C.; and Kantzes, J. G. 1983. Intergrated control of snap bean disease caused by *Pythium* spp. and *Rhizoctonia solani. Plant Disease* **67**:1241–1244.

Meiners, J. P.; Waterworth, H. E.; Lawson, R. H.; and Smith, F. F. 1977. Curly dwarf mosaic disease of beans from El Salvador. *Phytopathology* **67**:163–168.

Miller, D. E., and Burke, D. W. 1986. Reduction of Fusarium root rot and Sclerotinia wilt in beans with irrigation, tillage and bean genotype. *Plant Disease* **70**:163–166.

Morales, F. J., and Castano, M. 1985. Effect of a Colombian isolate of bean southern mosaic virus on selected yield components of *Phaseolus vulgaris. Plant Disease* **69**:803–804.

Muckel, R. D., and Steadman, J. R. 1981. Dissemination and survival of *Sclerotinia sclerotiorum* in bean fields in western Nebraska. *Phytopathology* **71**:244 (abstract).

Natti, J. J. 1971. Epidemiology and control of bean white mold. *Phytopathology* **61**:669–674.

Patel, P. N., et al. 1964. Bacterial brown spot of bean in central Wisconsin. *Plant Dis. Rptr.* **48**:335–337.

Pfender, W. F., and Hagedorn, D. J. 1981. Aphanomyces root and stem rot of snap beans. *Phytopathology* **71**:250 (abstract).

Pfender, W. F., and Hagedorn, D. J. 1982. *Aphanomyces euteiches* f. sp. *phaseoli*, a causal agent of bean root and hypocotyl rot. *Phytopathology* **72**:306–310.

Pieczarka, D. J., and Abawi, G. S. 1978. Effect of interaction between Fusarium, Pythium, and Rhizoctonia on severity of bean root rot. *Phytopathology* **68**:403–408.

Pieczarka, D. J., and Abawi, G. S. 1978. Influence of soil water potential and temperature on severity of Pythium root rot of snap beans. *Phytopathology* **68**:766–772.

Polach, F. J., and Abawi, G. S. 1975. The occurrence and biology of *Botryotinia fuckeliana* on beans in New York. *Phytopathology* **65**:657–660.

Prasad, K., and Weigle, J. L. 1976. Association of seed coat factors with resistance to *Rhizoctonia solani* in *Phaseolus vulgaris. Phytopathology* **66**:342–345.

Reeleder, R. D., and Hagedorn, D. J. 1981. Inheritance of resistance to *Pythium myriotylum* hypocotyl rot in *Phaseolus vulgaris* L. *Plant Disease* **65**:427–429.

Ribeiro, R. D L. D.; Hagedorn, D. J.; Durbin, R. D.; and Vchytil, T. F. 1979. Characterization of the bacterium inciting bean wildfire in Brazil. *Phytopathology* **69**:208–212.

Saad, S., and Hagedorn, D. J. 1969. Symptomatology and epidemiology of Alternaria leaf spot of bean *Phaseolus vulgaris*. *Phytopathology* **59**:1530–1533.

Saettler, A. W.; Cafati, C. R.; and Weller, D. M. 1986. Nonoverwintering of Xanthomonas bean blight bacteria in Michigan. *Plant Disease* **70**:285–287.

Saettler, A. W., and Stadt, S. J. 1981. Internal seed infection by *Pseudomonas phaseolicola* in susceptible and tolerant bean cultivars. *Phytopathology* **71**:252 (abstract).

Schieber, E., and Zentmyer, G. A. 1971. A new bean disease in the Caribbean area. *Plant Dis. Rptr.* **55**:207–208.

Schuster, M. L.; Blatchford, G. J.; and Schuster, A. M. 1980. A new bacterium, *Pseudomona blatchfordae*, nov. sp., pathogenic for bean. *Fitopathologia Brasileria* **5**:283–297.

Schuster, M. L.; Schuster, A. M.; and Nuland, D. S. 1981. A new bacterium pathogenic for beans (*Phaseolus vulgaris* L.). *Fitopatologia Brasileria* **6**:345–358.

Schwartz, H. F.; Galvez, G. E.; Schoonhoven, A. Van; Howeler, R. H.; Graham, P. H.; and Flor, C. 1978. *Field Problems of Beans in Latin America*. Centro Internacional de Agricultura Tropical, Cali, Colombia, 136p.

Schwartz, H. F.; Katherman, M. J.; and Thung, M. D. T. 1981. Yield response and resistance of dry beans to powdery mildew in Colombia. *Plant Disease* **65**:737–738.

Schwartz, H. G., and Steadman, J. R. 1978. Factors affecting sclerotium populations of, and apothecium production by, *Sclerotinia sclerotiorum*. *Phytopathology* **68**:383–388.

Shaik, M. 1985. Race-nonspecific resistance in bean cultivars to races of *Uromyces appendiculatus* var. *appendiculatus* and its correlation with leaf epidermal characteristic. *Phytopathology* **75**:478–481.

Sippell, D. W., and Hall, R. 1981. Effects of Fusarium and Phythium on yield components of white bean. *Phytopathology* **71**:564 (abstract).

Skiles, R. L., and Cardona-Alvarez, C. 1959. Mancha gris, a new leaf disease of bean in Colombia. *Phytopathology* **49**:133–135.

Steadman, J. R. 1983. White mold—a serious yield-limiting disease of bean. *Plant Disease* **67**:346–350.

Steadman, J. R.; Kerr, E. D.; and Mumm, R. F. 1975. Root rot of bean in Nebraska: Primary pathogen and yield loss appraisal. *Plant Dis. Rptr.* **59**:305–308.

Stockwell, V., and Hanchey, P. 1984. The role of the cuticle in resistance of beans to *Rhizoctonia solani*. *Phytopathology* **74**:1640–1642.

Sumner, D. R.; Smittle, D. A.; Threadgill, E. D.; Johnson, A. W.; and Chalfant, R. B. 1986. Interactions of tillage and soil fertility with root diseases in snap bean and lima bean in irrigated multiple-cropping systems. *Plant Disease* **70**:730–735.

Teakle, D. S., and Morris, T. J. 1981. Transmission of southern bean mosaic virus from soil to bean seeds. *Plant Disease* **65**:599–600.

Tu, J. C. 1981. Anthracnose (*Colletotrichum lindemuthianum*) on white bean (*Phaseolus vulgaris* L.) in southern Ontario: Spread of the disease from an infection focus. *Plant Disease* **65**:477–480.

Tu, J. C. 1981. Etiology of pod and seed coat discoloration of white beans. *Phytopathology* **71**:909 (abstract).

Tu, J. C. 1983. Epidemiology of anthracnose caused by *Colletotrichum lindemuthianum* on white bean (*Phaseolus vulgaris*) in southern Ontario: Survival of the pathogen. *Plant Disease* **67**:402–404.

Tu, J. C. 1984. Biology of *Alternaria alternata*, the causal fungus of black pod disease of white beans in southwestern Ontario. *Phytopathology* **74**:820 (abstract).

Tu, J. C. 1985. Pink pod rot of bean caused by *Trichothecium roseum*. *Can. J. Plant Pathol.* **7**:55–57.

Tu, J. C.; Sheppard, J. W.; and Laidlaw, D. M. 1984. Occurrence and characterization of the epsilon race of bean anthracnose in Ontario. *Plant Disease* **68**:69–70.

Van Bruggen, A. H. C.; Arneson, P. A.; and Whalen, C. H. 1983. Emergence of dry beans as affected by *Rhizoctonia solani*, soil moisture and temperature. *Phytopathology* **73**:1347 (abstract).

Van Bruggen, A. H. C.; Whalen, C. H.; and Arneson, P. A. 1986. Emergence, growth and development of dry bean seedlings in response to temperature, soil moisture and *Rhizoctonia solani*. *Phytopathology* **76**:568–572.

Walker, J. C. 1952. *Diseases of Vegetable Crops*. McGraw-Hill, New York.

Webster, D. M.; Atkin, J. D.; and Cross, J. E. 1983. Bacterial blights of snap beans and their control. *Plant Disease* **67**:935–940.

Webster, D. M.; Temple, S. R.; and Galvez, G. 1983. Expression of resistance to *Xanthomonas campestris* pv. *phaseoli* in *Phaseolus vulgaris* under tropical conditions. *Plant Disease* **67**:394–396.

Weller, D. M., and Saettler, A. W. 1980. Evaluation of seedborne *Xanthomonas phaseoli* and *X. phaseoli* var. *fuscans* as primary inocula in bean blights. *Phytopathology* **70**:148–152.

Yoshu, K., and Aamodt, D. 1978. Evaluation of bean varieties for resistance to *Pseudocercosporella albida* in highland Guatemala. *Phytopathology News* **12**:269 (abstract).

Zaumeyer, W. J., and Thomas, H. R. 1958. Bean diseases and their control. *USDA Farmers Bull. No. 1692*.

8

Diseases of Flax
(*Linum usitatissimum* L.)

FUNGAL CAUSES

Alternaria Flower and Stem Blight

CAUSE: *Alternaria lini* Dey is a saprophyte of plant residue. The fungus overseasons as mycelium and conidia in residue. During moist weather conidia are produced on residue and are windborne to hosts.

DISTRIBUTION: India and probably other flax-growing areas around the world.

SYMPTOMS: Apparently *A. lini* first infects moribund flowers or flowers that have died from other causes including maturity. From flowers, the infection spreads to stems and leaves, causing all infected parts to become a black-green color due to growth of mycelium and conidia on the plant surface.

CONTROL: No control is reported.

Alternaria Seedling Blight

CAUSE: *Alternaria linicola* Groves & Skolko survives primarily as a saprophyte on a wide range of plant residue. Conidia may also be seedborne on the outside of the seed coat. Only seedlings that have been weakened by another means are usually infected.

DISTRIBUTION: Canada and Europe.

SYMPTOMS: Seedlings are killed or roots will have a water-soaked, brown appearance similar to Pythium damping off.

CONTROL: No control is reported.

Anthracnose

CAUSE: *Colletotrichum lini* (Westerdyk) Tochinai survives most commonly as mycelium in seed and conidia on the seed coat. It also overwinters as mycelium, acervuli, and possibly conidia on infected residue. Seedlings are infected during cool, wet weather. Acervuli are eventually formed subepidermally in lesions produced from primary inoculum and rupture the epidermis to release conidia that are windborne to serve as secondary inoculum. Sporulation occurs throughout the growing season on plants that have died or on stem cankers. Seed becomes infected by conidia originating from stem lesions. Lodged plants are likely to have infected seed.

DISTRIBUTION: Generally distributed wherever flax is grown but is most common in cool, humid, flax-growing areas.

SYMPTOMS: Symptoms are most easily noticeable when plants are 5–7 cm tall. One or both cotyledons may be infected. Initially typical symptoms on cotyledons are small, circular, red-brown, zonate spots. The whole cotyledon quickly becomes brown and shriveled. Sometimes the seedling may be killed if the growing point is infected. Conidia are produced on infected cotyledons and are washed down to the soil line where the stem becomes infected. A reddish canker develops on the stem at or below the soil line and may girdle the stem, thereby killing the plant. Affected plants may be stunted.

Leaf spots, stem cankers, and acervuli can be observed on diseased plants throughout the growing season. Acervuli may be observed in older lesions during moist weather as a pink spore mass with dark hair-like objects projecting up from them. Sepals of flowers become infected and form a substrate on which *C. lini* grows to infect the bolls and eventually the seed.

CONTROL:
1. Plant only healthy seed that has been treated with a seed-protectant fungicide.
2. Plant resistant cultivars.
3. Rotate flax with other crops.

Browning and Stem Break

CAUSE: *Polyspora lini* Lafferty is primarily seedborne, overwintering as conidia and mycelium in or on the seed coat. It may also overwinter as conidia and mycelium in acervuli on infected residue. When infected seed germinates, conidia are produced on the seed coat on the cotyledon during warm, wet weather and are windborne to adjacent stems.

DISTRIBUTION: Generally distributed wherever flax is grown.

SYMPTOMS: Water-soaked spots are produced on the cotyledons and cankers are produced about 2.5 cm above the soil line on the stem. The plant may break over at the canker in the bud or flower stage. Plants may live after falling over but any seed that is set usually cannot be harvested.

The browning portion of the disease occurs as oval to elongated brown spots, about 6 mm long, on the upper stem. The spots are usually surrounded by a narrow purple border and do not coalesce unless they become very numerous. The fungus grows into the boll and infects the seed. Young seed is killed; however, *P. lini* grows into and survives in the seed coat of mature seed. Areas of plants that are heavily infected appear brown, hence the name browning.

CONTROL:
1. Plant only healthy seed that has been treated with a seed-protectant fungicide.
2. Rotate flax with other crops.
3. Plant flax early in the growing season so it may attain sufficient growth before disease becomes severe.
4. Plant resistant cultivars.

Foot Rot

CAUSE: *Phoma* sp. probably survives in infected residue as pycnidia, which produce conidia that are windborne and cause most primary infections. Survival is also by conidia on seed and mycelium in the seed coat. Much primary infection must evidently occur from conidia or possibly mycelium that is washed into soil. Foot rot becomes more severe under wet conditions when secondary infection occurs from conidia that were produced in pycnidia on recently killed plant parts.

DISTRIBUTION: Europe.

SYMPTOMS: Seedlings may yellow, wilt, and die. The stem at soil level and the roots become brown. Pycnidia that look like small black pinpoints are produced in dead aboveground plant parts. The most conspicuous symptoms are at flowering time when affected plants yellow and wilt. As in seedlings, the lower part of the stem and the roots are a light brown. Numerous small black pycnidia are formed within the brown area and the epidermis is easily detached from the stem. The lower part of the stem appears ragged. Eventually infected plants are killed and contrast sharply with the normal green color of healthy plants.

CONTROL:
1. Plant only healthy seed that has been treated with a seed-protectant fungicide.
2. Rotate flax with other crops.
3. Flax straw or residue should not be placed on soil that is to grow flax the following year.
4. Grow resistant or less susceptible cultivars.

Fusarium Secondary Rot

CAUSE: *Fusarium* spp. of which *F. roseum* (Lk.) emend. Snyd. & Hans. is probably the main incitant. A secondary infection by *Fusarium* spp. occurs in lesions originally caused by *Polyspora lini* Lafferty and pustules caused by *Melampsora lini* (Pers.) Lev. The *Fusarium* spp. overwinter as perithecia, chlamydospores, or saprophytic mycelium in infected residue.

DISTRIBUTION: Northern Ireland and Russia.

SYMPTOMS: Lesions caused by *P. lini* or the telia stage of *M. lini* are invaded secondarily by *Fusarium* spp. In both cases the symptoms are similar. The infected area becomes light brown with white mycelial strands growing on the outside of the stem under moist conditions. In the case of rust, this discolored area may extend 13 mm beyond the pustule. Sporodochia or the spore mass of conidia is produced in the center of the infected area, giving it a red to pink color.

CONTROL: No control is reported.

Fusarium Wilt

CAUSE: *Fusarium oxysporum* Schlecht. f. sp. *lini* (Bolley, Snyd. & Hans.) survives as chlamydospores in the soil or as a saprophyte in residue. Microconidia and macroconidia are also seedborne, being carried on the outside of seed, particularly weathered seed.

Infection hyphae enter plants primarily through the root hairs of young seedlings. Mycelium of the fungus grows into the xylem tissue, inhibiting uptake of water and causing the plant to wilt. During moist weather, both microconidia and macroconidia are produced on any part of the plant, particularly near the ground. Chlamydospores are formed from macroconidia, microconidia, and mycelium and can survive for several years in the soil. The fungus is disseminated by seed, by any means that will move soil, and by wind. Several physiologic races of the fungus are present in nature. Soil temperatures of about 25° C and higher and low soil moisture generally favor wilt development.

DISTRIBUTION: Generally distributed wherever flax is grown.

SYMPTOMS: Wilt symptoms are classified into four types: early or seedling wilt, late wilt, partial wilt, and one-sided or unilateral wilt. Seedling wilt is usually most common. Symptoms may appear at any stage of flax growth and may vary with plant age and cultivar, environmental conditions, and physiologic race involved.

Early wilt occurs when seedlings wilt and die from the cotyledonary stage until they are 15 cm tall. Plants wilt at the top and eventually become dry and brown. The stem becomes constricted at the ground level and the seedling eventually falls over. If the soil is moist, abundant mycelium and sporulation will cover the dead seedling.

Late wilt occurs from flowering through boll set. Stem tissues, both internal and external, become necrotic and brittle. Stem discoloration is a uniform light brown and is a darker color than premature ripening caused by root rot or other means.

Partial wilt occurs when a seedling wilts and dies except for roots and buds at the base of a stem. During cool weather, buds at the base of the stem develop into new shoots, following death of original shoots. With a return to hot weather, the lateral shoots wilt and the whole plant dies.

Unilateral wilt occurs when only one side of the stem is affected. All branches on that side become brown.

CONTROL: Plant resistant cultivars.

Gray Mold

CAUSE: *Botrytis cinerea* Pers. ex Fr. survives by sclerotia in the soil, growing saprophytically on plant residue, and as seedborne mycelium in the seed coat. Conidia are produced on mycelium and are windborne to healthy tissue. Gray mold is more severe under warm, wet conditions. Lodged flax is more subject to infection.

DISTRIBUTION: Europe and the United States.

SYMPTOMS: The first symptoms occur about the time the cotyledonary leaves unfold. Brown spots occur on the stem at the soil level. The seedling quickly wilts, falls over, and is overgrown by gray fungal growth, particularly during warm, wet weather.

Stem portions that vary in length from 1.5–7.5 cm are infected on older plants. These areas are light brown and soft, while the stem above these areas becomes yellow and eventually dies. Soon the entire plant dies and becomes covered with gray fungal growth during warm, wet weather. Eventually the hard, black sclerotia are formed that appear as relatively large black objects on the stem.

CONTROL:

1. Plant healthy seed that has been treated with a seed-protectant fungicide.
2. Plants too close together and excessive nitrogen may encourage lodging and more severe disease.

Pasmo

CAUSE: *Septoria linicola* (Speg.) Garassini overwinters as conidia in pycnidia and mycelium in infected residue and seed. The teleomorph is *Mycosphaerella linorum* (Wr.) Garcia-Rada; however, the perithecial stage is apparently unimportant in the life cycle of the fungus. When diseased seed is planted, some of the first leaves may be infected. However, most primary inoculum comes from pycnidia in residue or stubble. During warm, wet weather, conidia are exuded from pycnidia and are disseminated by wind or rain. Secondary inoculum comes from conidia produced in pycnidia on lesions of diseased plants. Lodged flax is more susceptible than standing flax. Early cultivars tend to be more susceptible than late cultivars.

DISTRIBUTION: Generally distributed wherever flax is grown.

SYMPTOMS: The longer the flowering, the less the expression of pasmo. This is possibly due to the fact that weather conducive to flowering (cool temperatures) is not conducive to disease expression. Initially circular green-yellow to dark brown lesions develop on cotyledons and then on lower leaves. Pycnidia develop on older lesions on cotyledons and leaves. Diseased leaves die and drop to the ground or stick tightly to stems.

Later in the season as flax ripens stem lesions develop first as small, brown, elongated lesions. The lesions enlarge and coalesce to form brown bands that encircle the stem and alternate with bands of healthy tissue. This gives a mottled appearance that is a good symptom of pasmo. Eventually stems turn brown and plants are defoliated if pasmo becomes severe. Pycnidia develop on stem lesions and appear as numerous small black specks scattered throughout the lesion. Flowers and young bolls may also be blighted. Lesions also form on the older bolls and seed may be shriveled. The slender stems or pedicels supporting the bolls may be weakened, so ripe bolls are broken off during wind or rain. If bolls do not fall off, seed may not develop properly.

CONTROL:

1. Plant resistant or tolerant cultivars.
2. Plant only healthy seed that has been treated with a seed-protectant fungicide.
3. Plow under crop refuse to bury inoculum.

Powdery Mildew

CAUSE: *Erysiphe polygoni* DC. ex Merat usually overwinters as cleistothecia on infected residue. However, frequently only the conidial stage occurs and cleistothecia are not present. In such cases, it is not known for certain how the fungus survives from growing season to season. It may overwinter on volunteer flax or other hosts. Primary infection may be by conidia or ascospores from perithecia. Secondary infection is by conidia being windborne from infected to healthy leaves.

DISTRIBUTION: Generally distributed wherever flax is grown; however, it is not considered an important disease.

SYMPTOMS: A white, powdery growth, which is the mycelium, conidiophores, and conidia of the fungus, develops on the stem, on both surfaces of the leaf, and on sepals. When the perithecial stage is present, small, round, black structures, which are the cleistothecia, are present within and on the white fungal growth.

CONTROL: Ordinarily no control is necessary. In the greenhouse, apply a sulphur fungicide as a foliar spray.

Rhizoctonia Seedling Blight and Root Rot

CAUSE: *Rhizoctonia solani* Kuhn is a fungus that survives in soil as sclerotia and mycelium or as a saprophyte in plant residue. The disease is most severe in warm, moist soil following summer fallow.

DISTRIBUTION: Generally distributed wherever flax is grown.

SYMPTOMS: Single plants or groups of seedlings in a row or circular area may be infected. Seedlings turn yellow, wilt, and die. The roots have red-brown lesions on them that appear to be somewhat superficial. Roots eventually turn dark brown or black, shrivel, and dry up.

Root rot symptoms occur upon flowering when plants appear to ripen prematurely. Few or no seeds are formed. Roots are brown to red-brown and appear to be stunted or rotted.

CONTROL:
1. Plant healthy seed that is free of cracks in the seed coat.
2. Plant as early as possible so plants are well developed when conditions conducive to root rot occur.
3. Plant flax on second crop rather than summer fallow land.
4. Treat seed with a seed-protectant fungicide.

Rust

CAUSE: *Melampsora lini* (Pers.) Lev. is an autoecious, long-cycled rust fungus that produces uredia, telia, pycnia, and aecia on the flax plant. The fungus overwinters as teliospores in telia on infected residue and as teliospores contaminating seed. In the spring, teliospores germinate to produce basidiospores that infect young flax tissues. Pycnia develop to produce pycniospores and receptive hyphae. Pycniospores are carried by insects or splashed by rain from one pycnium to the compatible receptive hyphae of other pycnia. Aecia are initiated on the opposite side of the leaf or stem and aeciospores are windborne to other healthy tissue. The uredia are initiated in aeciospore infections and urediospores are windborne to other flax plants. Secondary infections from urediospores account for most of the spread of the disease throughout the summer. Telia are formed around uredia as flax matures and provides the overwintering stage for the fungus. Several physiologic races of the fungus exist in nature.

DISTRIBUTION: Generally distributed wherever flax is grown.

SYMPTOMS: Telia occur mostly on stems but also on leaves and capsules. They are brown to black structures that are covered by the epidermis. Pycnia and aecia occur on leaves and stems early in the growing season as light to orange-yellow sori. Uredia are red-yellow and occur on leaves, stems, and bolls during the growing season.

CONTROL:
1. Plant resistant varieties.
2. Plant cleaned seed that has been treated with a seed-protectant fungicide.
3. Plant flax as early as possible in the growing season to escape early infection.
4. Rotate flax with other crops.

Sclerotinia Disease

CAUSE: *Sclerotinia sclerotiorum* (Lib.) deBary survives for several years in the soil as sclerotia. Additionally sclerotia produced on other hosts such as rapeseed may provide a source of primary inoculum. Sclerotia on or close to the soil surface germinate during wet conditions to form apothecia on which ascospores are produced in a layer of asci. Ascospores are windborne to host plants. Sclerotia eventually form from mycelium on and in stems and are returned to soil during harvest. *Sclerotinia sclerotiorum* has a wide host range.

DISTRIBUTION: Alberta and the warmer flax-growing areas of the United States and the USSR; however, it is not considered a serious disease.

SYMPTOMS: Plants are initially infected on the stem at or just above the soil line. Plants are quickly killed and lodged. White, fluffy, mycelial growth occurs on diseased stems. Plants are pale brown and dry up. Stems are easily shredded with the inside of stems containing elongated sclerotia. Eventually sclerotia of variable size and shape are formed on and in the stem but are more frequent in the pith cavity. These appear as large, white, spherical-shaped bodies that eventually turn black and are easily seen.

CONTROL:
1. Rotate flax with nonsusceptible crops such as small grains, corn, and so forth.
2. Control weeds in and around a flax field.
3. Cultural practices that promote rank, thick growth should be avoided.

Scorch

CAUSE: *Pythium megalacanthum* deBary survives in the soil as oospores. The life cycle is typical of most *Pythium* spp. The roots are infected during cool, wet soil conditions.

DISTRIBUTION: Northern Europe.

SYMPTOMS: Symptoms of scorch appear in circular areas in a field. Symptoms start at the bottom and progress up an infected plant. Leaves of plants are brown and shriveled about halfway up the stem. Leaves above this point are yellow with brown margins, while still higher, leaves are only partially yellow and limp. Leaves at the top of a plant may appear green and normal. If soils dry and warm up, plants may recover unless they are too severely infected.

CONTROL: Plant late when soils have warmed up.

Seed Rot, Seedling Blight, and Root Rot

CAUSE: Several fungi including *Pythium debaryanum* Hesse, *P. aphanidermatum* (Edson) Fitzp., *P. splendens* Braun, *P. vexans* deBary, *P. megalacanthum* deBary, *P. mamillatum* Meurs, *P. irregulare* Buis., *P.*

intermedium deBary, *Olpidium brassicae* (Woron.) Dang, *Thielaviopsis basicola* (Berk. & Br.) Ferr., *Polyspora lini* Lafferty, *Fusarium* spp., and *Rhizoctonia* spp. The most important fungi are the *Pythium* spp. Plant infection usually occurs under wet soil conditions and when mechanically injured seed is planted.

DISTRIBUTION: Generally distributed wherever flax is grown.

SYMPTOMS: Seeds may fail to germinate, become soft, mushy, and overgrown with white mycelium that will cause soil to adhere to the seed. Seeds that do not germinate are difficult to find in soil. Seedlings may be killed before they emerge and will be brown and water-soaked. Again, such plants would be difficult to find in the soil. Postemergence infection is characterized by stunted plants whose lower leaves become brown and necrotic. The roots are light brown and very soft to the touch. The entire root system is destroyed quickly if soils remain wet. Some fungi may cause more injury under cool soil temperatures whereas other fungi grow better at higher soil temperatures.

CONTROL:
1. Plant healthy, sound seed that is treated with a seed-protectant fungicide.
2. Rotate flax with other crops; however, this is a questionable practice since many fungi can infect a large number of host plants.

Thielaviopsis Root Rot

CAUSE: *Thielaviopsis basicola* (Berk. & Br.) Ferr. survives in infected residue and soil as chlamydospores. Chlamydospores germinate and infect only roots. Conidia are produced on diseased roots along with chlamydospores that are produced in the root. The fungus is disseminated mostly by any means that will transport soil.

DISTRIBUTION: Europe and the United States.

SYMPTOMS: Aboveground plant parts are stunted and yellowed. The roots are black and necrotic. Necrotic tissue contains chlamydospores that are dark brown and borne in short chains resembling an elongated spore with thick cells.

CONTROL:
1. Crop rotation. Do not grow flax in the same soil for several years.
2. Control weeds in a field since they may be a host for *T. basicola*.

MYCOPLASMAL CAUSES

Aster Yellows

CAUSE: The aster yellows mycoplasma survives in several dicotyledonous plants and in leafhoppers. The mycoplasma is transmitted primarily by the aster leafhopper, *Macrosteles fascifrons* Stal., and less commonly by *M. laevis* Rib. and *Endria inimica* Say.

DISTRIBUTION: Central and western North America.

SYMPTOMS: Symptoms are most conspicuous during and after flowering and appear only on lateral branches of plants infected late in the season. Diseased plants are stunted and chlorotic. Flower parts are distorted with petals remaining green and leaflike. Sometimes diseased and normal flowers are present on the same plant. No seed is ordinarily set.

CONTROL: No control is reported.

VIRAL CAUSES

Crinkle

CAUSE: Oat blue dwarf virus (OBDV) overwinters in several different species of plants. The virus is transmitted mainly by the adult aster leafhopper, *Mascrosteles facifrons* Stal. Immature leafhoppers may occasionally transmit virus.

DISTRIBUTION: North central United States and Canada.

SYMPTOMS: The only leaves affected are those formed after exposure of plants to viruliferous leafhoppers. A crinkle of the leaves occurs due to a swelling of lateral veins on margins of leaves, small indentations along upper surfaces of veins, and enations or pimples on lower leaf surface. Plants are also stunted with reduced boll development and seed set.

CONTROL: No control is reported.

Curly Top

CAUSE: Curly top virus (CTV) overwinters in a large number of perennial plants. CTV is transmitted by the beet leafhopper, *Circulifer tenellus* Baker.

DISTRIBUTION: Western North America.

SYMPTOMS: Infected seedlings turn bronze, then yellow, with leaves curled along the stem. Severely infected plants die prematurely while those that survive are stunted with a reduced number of tillers. Terminal leaves and flower parts are discolored, crinkled, and distorted with little or no seed set.

CONTROL:
1. Plant resistant cultivars.
2. Time planting to avoid high populations of the beet leafhoppers.

PHYSIOLOGIC OR ___WEATHER-ORIENTED PROBLEMS___

Boll Blight

CAUSE: Warm, dry weather following cool, moist weather.

DISTRIBUTION: Not known.

SYMPTOMS: Buds, flowers, and young bolls fail to develop.

CONTROL: No control is reported.

Heat Canker

CAUSE: High temperature at the soil line.

DISTRIBUTION: Generally distributed in semihumid plains and at high altitudes throughout the world.

SYMPTOMS: Cortical tissues collapse, resulting in seedling death or in sunken brown lesions on the stems. Plants that survive have an enlarged stem just above the canker. Cankers provide an entrance wound for secondary organisms to further rot stems.

CONTROL:
1. Plant early to avoid high soil temperature.
2. Drill rows north and south to provide maximum shading.

Top Dieback

CAUSE: Physiologic.

DISTRIBUTION: Not known. Affected plants may occur in a limited area or throughout a whole field.

SYMPTOMS: Top portion of flax plants turn brown after a hot spell during time that seed is ripening. Seeds are usually thin and lightweight.

CONTROL: No control is reported.

_____BIBLIOGRAPHY_____

Brooks, F. T. 1953. _Plant Diseases._ 2nd ed. Oxford University Press, London.

Christensen, J. J. 1954. The present status of flax diseases other than rust. _Adv. Agron._ **6:**161–168.

Ferguson, M. W.; Lay, C. L.; and Evenson, P. D. 1987. Effect of pasmo disease on flower production and yield components of flax. _Phytopathology_ **77:**805–808.

Frederiksen, R. A., and Goth, R. W. 1959. Crinkle, a new virus disease of flax. _Phytopathology_ **49:**538 (abstract).

Hoes, J. A. 1975. Diseases of flax in western Canada, *Oilseed and Pulse Crops in Western Canada*. Modern Press, Saskatoon, Saskatchewan.

Kenaschuk, E. O. 1976. Growing flax in Canada. *Canada Dept. Agric. Pub. 1577.*

Kommendahl, T.; Christensen, J. J.; and Fredericksen, R. A. 1970. A half century of research in Minnesota on flax wilt caused by *Fusarium oxysporum. Univ. Minn. Agric. Exp. Sta. Tech. Bull. 273.*

Mederick, F. M., and Piening, L. J. 1982. *Sclerotinia sclerotiorum* on oil and fibre flax in Alberta. *Can. Plant Dis. Surv. 62, 11p.*

Millikan, C. R. 1951. Diseases of flax and linseed. *Victoria (Australia) Dept. Agric. Tech. Bull. 9.*

Muskett, A. E. 1947. *The Diseases of the Flax Plant.* Northern Ireland Flax Development Committee, Belfast.

Diseases of Millet

The species included are foxtail millet or Italian millet, *Setaria italica* (L.) Beauv. (syn. *Chaeotchloa italica* (L.) Scrib.); browntop millet, *Panicum miliaceum* L.; Japanese millet or sawan millet, *Echinochloa crusgalli* var. *frumentacea* (Rozb.) Wright; jungle rice millet, *E. colonum* (L.) Link; pearl millet or bajra, *Pennisetum glaucum* (L.) R. Br.; and finger millet, ragi, or African millet, *Eleusine coracana* (L.) Gaertn. Certain diseases are reported to affect some millets. Any differences between millets in susceptibility to pathogens are noted in the text, if known.

BACTERIAL CAUSES

Bacterial Blight

CAUSE: *Xanthomonas coracanae* Desai, Thirumalachar & Patel.

DISTRIBUTION: Africa and India.

SYMPTOMS: Initially water-soaked, translucent, linear to elongate spots that develop parallel to the midrib occur on leaves. The spots are 5–10 mm long and pale yellow. Eventually they become brown and may coalesce to cover the entire leaf, causing it to wither.

CONTROL: No control is reported.

Bacterial Stripe

CAUSE: *Xanthomonas panici* (Elliott) Savul.

DISTRIBUTION: United States.

SYMPTOMS: Stripes on leaves are at first water-soaked and later turn brown. Dried scales of bacterial exudate may be present on the surface of the lesion.

CONTROL: No control is reported.

Yellow Leaf Blotch

CAUSE: *Pseudomonas* sp. is seed transmitted.

DISTRIBUTION: West Africa.

SYMPTOMS: Symptoms occur at the five-leaf stage as large, cream-yellow or light tan lesions up to 30 mm wide and over 150 mm long, usually only near the tips of leaves. Generally only the area near the tip of the leaf is affected. Young plants in the two- to three-leaf stage are sometimes severely stunted or killed. Older infected plants are not severely stunted. Maturing plants have large lesions on young leaves but outgrow the disease.

CONTROL: No control is reported.

Other Bacteria Reported on Millet

Bacterial spot—*Pseudomonas alboprecipitans* Rosen.
Bacterial leaf spot—*Pseudomonas syringae* van Hall.

FUNGAL CAUSES

Blast

CAUSE: *Pyricularia grisea* (Cke.) Sacc. Disease becomes severe during wet conditions.

DISTRIBUTION: Australia, India, and the USSR on proso millet, *Panicum miliaceum* L.

SYMPTOMS: Numerous small, brown flecks enlarge to form brown spindle-shaped spots on the leaves. Later spots develop ash-gray centers with brown margins, increase in size, and coalesce to give a blasted appearance. Spots are 3–8 mm long and 0.5–1.5 mm wide.

Culm nodes blacken, become fragile, and may lodge at these points. Brown to black spots appear at the panicle base, enlarge, and often girdle the neck below the panicles. The infected neck becomes shriveled and covered with mycelium, conidiophores, and conidia. Abundant conidia and conidiophores are formed on nodes and necks in wet weather.

When neck blast is severe, panicles fail to emerge. Spikelets become infected and blacken. When neck infection occurs early, the grains do not fill and panicles remain erect. Later infection results in lodging of partially filled panicles.

CONTROL: Apply foliar fungicide before disease becomes severe.

Blast

CAUSE: *Pyricularia setariae* Nishikado.

DISTRIBUTION: India on finger millet or ragi, *Eleusine coracana* L. Gaertn.

SYMPTOMS: Small, brown, circular to elongated spots occur on leaves and leaf sheaths that eventually develop into large elongated or spindle-shaped areas. The center of the spots are grayish with irregular red-brown margins. Spots may coalesce to involve a large part of the leaf blade. During humid conditions, an olive-gray fungal growth is seen in the middle of spots on the upper leaf surface.

Elongated, irregular patches occur just below the ear head, resulting in a condition known as rotten neck. The main stalk below the lowest spike becomes brown to black and sunken for 25–51 mm. An olive-gray fungal growth may also be seen in the center of these spots. The kernels may not develop and they may be lightweight and chaffy. A seedling blight phase also occurs, resulting in heavy losses.

CONTROL: Apply foliar fungicides before disease becomes severe.

Cercospora Leaf Spot

CAUSE: *Cercospora penniseti* Chupp. This fungus may be synonymous with *C. appii* Fres.

DISTRIBUTION: Georgia.

SYMPTOMS: Symptoms develop late in the growing season, usually in September or October. Spots on leaves are usually oval, 0.8–2.5 mm by 1.0–8.0 mm, but occasionally may be oblong to rectangular. Spots have a dark brown margin with a pale tan to gray or chalky white center dotted with rows of black conidiophore tufts. In moist weather, spots are covered with silvery layers of conidia.

Similar spots to those on leaves sometimes occur on stems and are 0.5–2.0 mm by 1.0–5.0 mm. Sometimes dark brown runners extend longitudinally from the ends of these spots for up to 3 mm.

CONTROL: Some cultivars appear to have more resistance than others; however, control is usually not necessary since the disease occurs too late in the season to cause much damage to foliage. Control may be desirable if millet is grown for seed.

Downy Mildew

This disease is sometimes called green ear.

CAUSE: *Sclerophthora macrospora* (Sacc.) Thirum., Shaw & Naras. survives as oospores in soil and infected residue. Oospores germinate to produce mycelium that may directly infect the host but more commonly sporangia are produced in which zoospores are formed. Zoospores are then liberated from sporangia and swim to meristematic tissue where they encyst and germinate to form a germ tube. Secondary infection occurs by production of sporangia that are borne on the ends of sporangiophores that have emerged through stomates. The sporangia are windborne to other hosts where zoospores are again released, germinate, and infect meristematic tissue. Oospores are eventually formed within diseased tissue. The disease is most severe under wet conditions.

DISTRIBUTION: Africa and India.

SYMPTOMS: Plants are stunted and have shortened internodes. There is a proliferation of the first glumes of the lower spikelets. Ultimately the whole inflorescence becomes involved, giving the plant a bushlike appearance. Ordinarily the entire plant is chlorotic or pale green.

CONTROL: No satisfactory control is known.

Downy Mildew

This disease is also called green ear disease. Symptoms are similar to those caused by *Sclerophthora macrospora* (Sacc.) Thirum, Shaw & Naras. and confusion regarding these two diseases appears in the literature.

CAUSE: *Sclerospora graminicola* (Sacc.) Schroet. can survive for several years in the soil as oospores. The fungus is disseminated by seed-borne oospores and by any means that will transport oospore-infested residue or soil. Oospores commonly germinate to produce one to many germ tubes that directly penetrate the meristematic tissue of young plants. Sporangia (conidia) are produced on the end of the conidiophores that have emerged through stomatal openings and are windborne to other hosts. Sporangia germinate to produce four zoospores that encyst and germinate to produce a germ tube that infects meristematic tissue. Oospores are eventually produced in infected tissue. The disease is most severe under moist conditions.

DISTRIBUTION: Generally distributed wherever millet is grown; however, downy mildew is a serious disease in Africa and India, but is found only on the weed *Setaria viridis* (L.) Beauv. in the United States.

SYMPTOMS: Plants are dwarfed with excessive tillering from the crown and development of axillary buds along the culm. Flower parts develop leaflike structures with no kernel development. A downy grayish growth, which is the conidia and conidiophores, develop on infected tissue during wet or humid weather. As plants approach maturity, leaves become brown, necrotic, and split or shred.

CONTROL:
1. Treat seed with a seed-protectant fungicide.
2. Control weeds that may serve as an alternate host.
3. Plant the most resistant cultivar.
4. Practice good fertilizer management. Phosphorus as superphosphate (16% P) decreased severity of downy mildew in field trials.

Ergot

CAUSE: *Claviceps fusiformis* Loveless. The causal fungus overseasons as sclerotia in the field and in grain storage. Sclerotia vary in shape (elongated to round), size (3.6–6.1 mm by 1.3–1.8 mm), color (light brown to dark brown), and compactness (hard to brittle). Sclerotia germinate to produce 1–16 stipes that terminate in a globular capitulum in which perithecia are produced. This coincides with flowering in millet. Ascospores infect inflorescences through stigmas and honeydew containing macroconidia is produced in four to six days. The macroconidia germinate to produce microconidia in two to three days. Several insect species function in the secondary spread of the fungus by becoming contaminated with conidia. Included are *Aphis indica, Dysdercus ungulatus, Monomorium salomonis, Musca domestica, Syrphus confractor, Tabanus rubidus, Vespa orientalis,* and *Vespa tropica,* with *M. domestica* and *A. indica* the most efficient disseminators.

DISTRIBUTION: India on pearl millet.

SYMPTOMS: Symptoms are typical of ergot found on other plants. Hard, black sclerotia or ergot bodies replace grain kernels.

CONTROL: Plant resistant cultivars.

Foot Rot

This disease is also called southern blight and wilt.

CAUSE: *Sclerotium rolfsii* Sacc. The disease is most severe under wet soil conditions. For further information on *S. rolfsii*, see soybeans.

DISTRIBUTION: Africa and India on finger millet or ragi. The aboveground symptoms generally occur wherever millet is grown.

SYMPTOMS: Symptoms develop when plants are six to eight weeks old. Single tillers and occasionally entire plants may be killed. Plants are stunted, chlorotic, and have a poorly developed root system. A brown to black discoloration is present on the crown and the root cortex is rotted. Later the basal portion of the stem turns brown and cortical tissue disintegrates and rots. Plants usually die at this stage of the disease. White mycelium is present on the crown and lower stem. Black sclerotia eventually develop within this mycelium.

Aboveground symptoms are the presence of white mycelial mats of the fungus that spread upward inside the leaf sheaths as much as 30 cm above the soil. Tissue beneath the mat becomes discolored and is eventually killed and shredded. Sclerotia form inside the leaf sheaths and around the base of the plant.

CONTROL: Plant the most resistant cultivar.

Head Mold

CAUSE: Several fungi including *Oidium tenellum* (Berk. & Curt.) Linder, *Helminthosporium stenospilum* Drechs., *H. rostratum* Drechs., *Curvularia lunata* (Wakker) Boed., and *Fusarium moniliforme* Sheldon. The disease is more severe following insect injury under humid, wet conditions. Plants become infected by *F. moniliforme* as they approach maturity, with disease severity increased by the presence of dew and other moisture.

DISTRIBUTION: Africa, India, and the United States.

SYMPTOMS: *Oidium tenellum* causes a mealy, white coating over the heads. Under magnification the coating appears as white hairs with white beads at their tips.

The *Helminthosporium* spp. and *C. lunata* cause a black wooly mat to grow over the head. *Fusarium moniliforme* initially causes an off-white mycelial growth on spikelets of immature heads that later turns pink or causes individual grains to be covered with an orange crust. Moldy spikelets are shaken off easily. Seed quality and subsequent germination are greatly reduced.

CONTROL: No acceptable control is known except timely harvest of millet grown for grain. It may be possible to avoid disease in wet millet-growing areas by timing planting so that heads mature after heavy rains have stopped.

Head Smuts

CAUSE: *Sphacelotheca destruens* (Schlect.) Stevenson & Johnson and *Ustilago crusgalli* Tracy & Earle are two smut fungi that survive in soil as chlamydospores.

DISTRIBUTION: Generally distributed wherever millet is grown.

SYMPTOMS: *Spacelotheca destruens* completely destroys the inflorescense except for the vascular strands. The sorus is covered with a grayish mycelial covering that does not persist long, exposing the red-brown chlamydospores that become windborne.
   Ustilago crusgalli forms sori in gall-like swellings on nodes and flowers that remain covered by host tissue. When exposed, chlamydospores are yellow to olive-brown in color.

CONTROL:
1. Treat seed with a seed-protectant fungicide.
2. Rotate millet with other crops.

Helminthosporiosis

CAUSE: *Helminthosporium nodulosum* Berk. & Curt. survives in infected residue in the soil. It is also seedborne. The disease is most severe at temperatures of 30–32°C and high moisture.

DISTRIBUTION: Africa and India on finger millet or ragi.

SYMPTOMS: Roots, culms, leaves, heads, and inflorescences are infected. Preemergence damping off may occur with seeds and seedlings. Numerous small, oval, brown spots commonly occur on upper leaf surfaces and gradually elongate to 1 mm by 10 mm. Other descriptions note the spots are pale yellow to black in color. Spots on leaf sheaths are larger than leaf spots, dark brown to nearly black, and usually occur at the junction of the leaf blade and sheath. Lesions on culms and internodes are similar to those on leaf sheaths. Infected heads dry up especially at the rachis.

CONTROL:
1. Plant the most resistant cultivar.
2. Treat seed with a seed-protectant fungicide.

Helminthosporium Leaf Spot

CAUSE: *Helminthosporium stenospilum* Drechs., *H. sacchari* (van Breda de Haan) Butler, and *H. setariae* Saw. Disease development caused by *H. setariae* is optimum at temperatures of 15–21°C. For further information see Sorghum, Chapter 17.

DISTRIBUTION: Georgia on pearl millet.

SYMPTOMS: Disease symptoms usually develop extensively during August through October after the plants head. Symptoms are extremely variable and lesions will vary from brown flecks, fine linear streaks, and small oval spots to large, irregular, oval, oblong, or almost rectangular spots measuring 0.5–3.0 mm by 1.0–10.0 mm. Sometimes large, elongated lesions are produced. The spots may expand and coalesce to form long streaks 2.5–5.0 mm by 20.0–90.0 mm. Lesions at first may be dark brown but later become tan or gray-brown with a less distinct dark brown border. During moist weather, a dense gray-brown growth consisting of conidiophores and conidia may be seen in the lesions. A black growth may also occur over individual grains in the head.

CONTROL: Some cultivars appear to have more resistance than others; however, control is usually not necessary since the disease occurs too late in the season to cause much damage to foliage production. Control may be desirable if millet is grown for seed.

Kernel Smuts

CAUSE: *Ustilago crameri* Koern. and *U. neglecta* Niessl. are two smut fungi that survive in the soil as chlamydospores. Dry soil favors the survival of chlamydospores. Chlamydospores germinate to produce a basidium on which either lateral branches or sporidia are formed that infect the germinating seedlings.

DISTRIBUTION: Africa, Asia, and the United States.

SYMPTOMS: Sori resemble enlarged kernels and are enclosed in floral bracts. The floral bracts rupture during the growing season or at harvest time to release yellow-brown to olive-green chlamydospores of *U. crameri* or purple-brown chlamydospores of *U. neglecta*.

CONTROL:
1. Treat seed with a seed-protectant fungicide.
2. Rotate millet with other crops.

Leaf Mold

CAUSE: *Curvularia* spp. including *C. affinis* Boed., *C. geniculata* (Tracy & Earle) Boed., *C. lunata* (Wakker) Boed., *C. maculans* (Bancroft) Boed., and *C. penniseti* (Mitra) Boed. It is not known for certain if these fungi are primary pathogens or only secondary organisms. While no pathogenicity tests have been made, the organisms are often the only fungi associated with various spots and blotches.

DISTRIBUTION: Not known for certain but leaf mold has been reported from India and the United States.

SYMPTOMS: Oval to oblong brown spots with yellow margins occur on leaves. Heads are also infected.

CONTROL: No control is necessary.

Leaf Spot

CAUSE: *Helminthosporium frumentacei* (Mitra) Ellis.

DISTRIBUTION: India on Japanese millet.

SYMPTOMS: Initially numerous, small, yellowish spots 1–2 mm in diameter occur on both sides of the leaf and on the leaf sheath. Gradually the spots increase in size and have a light brown center with a yellow margin. As the spots become older, they become dark brown. Spots elongate or coalesce to form long stripes.

CONTROL: No control is reported.

Long Smut

This disease is also called pearl millet smut and pearl millet kernel smut.

CAUSE: *Tolyposporium penicilliariae* Bref. (syn. *Ustilago penniseti* (Kunze) Rabh.) survives as chlamydospores in the soil, which are also the sole source of infection. Chlamydospores germinate during humid conditions to produce a basidium on which sporidia bud, germinate, and infect through the flowers. The flowers are most susceptible at an early stage of development before the stigma or anthers are visible. After pollination, flowers are virtually free from infection. Chlamydospores are formed instead of grain, released, and fall to the soil when the surrounding membrane ruptures.

DISTRIBUTION: Africa and Asia on pearl millet.

SYMPTOMS: The enclosing membrane and chlamydospores, called a sori, are brown to black, somewhat pear-shaped, and protrude from the floral bracts. Upon rupturing of the membrane, the exposed chlamydospores are green-brown.

CONTROL: Application of foliar fungicides shows promise of partial control in experimental plots.

Rhizoctonia Blight

CAUSE: *Rhizoctonia solani* Kuhn and *R. zeae* Voorhees survive in soil and infected residue as sclerotia. The sclerotia germinate to form infective hyphae. Sclerotia are disseminated by wind, splashing water, and so forth. The disease is more severe on millet types with low, dense growth habits, creating high-humidity conditions that are more conducive to disease development.

DISTRIBUTION: Georgia.

SYMPTOMS: Lesions occur mostly on leaf sheaths and the base of leaf blades. The lesions are irregular blotches 4–7 cm long. At first the lesions are water-soaked but become light tan with an indistinct brown border. Diffuse white to tan mycelium grows over the surface of the lesions and adjacent green tissue. Older lesions may be covered with tiny brown sclerotia that are usually associated with *R. solani*. *Rhizoctonia zeae* occurs in the same lesions and may be mixed with *R. solani*. It is distinguished by the presence of tiny, globular, red sclerotia on the lesion surface.

Apparently little damage is caused, although an occasional tiller may be killed. Rarely an entire plant may be so severely blighted that it has the appearance of being scalded.

CONTROL: No control is necessary.

Rust

CAUSE: *Puccinia substriata* Ell. & Barth. var. *indica* Ramachar & Cumm. (syn. *P. penniseti* Zimm.) is a macrocyclic heteroecious rust that has eggplant, *Solanum melongena* L., as an alternate host. Uredia and telia are produced on millet and the fungus probably overwinters as teliospores, at least in the United States. Pycnia are produced on the upper leaf surface of eggplant and aecia are produced on the lower surface.

DISTRIBUTION: India and the United States on pearl millet.

SYMPTOMS: Uredia occur as orange-brown pustules that rupture the leaf epidermis. Uredia occur on both leaf surfaces but are more abundant on the upper surface.

CONTROL:
1. Plant the least susceptible cultivar.
2. Remove eggplants from the vicinity of millet-growing areas.

Rust

CAUSE: *Puccinia substriata* Ell. & Barth var. *pennicillariae* (Speg.) stat. nov. (syn. *P. penniseti* Zimm.). The aecial stage is not presently known.

DISTRIBUTION: Africa.

SYMPTOMS: Typical brownish uredia occur on both leaf surfaces but are most abundant on the upper surface.

CONTROL: No control is reported.

Smut

CAUSE: *Melanopsichium eleusinis* (Kulkarni) Mundkur & Thirumalachar (syn. *Ustilago eleusinis* Kulk.). The fungus is not seedborne. Dissemination and infection likely occur by chlamydospores being windborne to flowers.

DISTRIBUTION: India.

SYMPTOMS: Symptoms are evident after flowering, with scattered kernels throughout the ear being affected. Diseased grains are transformed into galls that are five to six times the normal size of grain. At first the affected grains are slightly greenish, 2–3 mm in diameter, and project slightly beyond the glumes. Eventually the galls reach a diameter up to 16 mm and turn pink-green with cracks in the outer wall of the sorus. The sorus is eventually ruptured and chlamydospores are wind disseminated to developing flowers or fall to the soil surface.

CONTROL:
1. Some cultivars are apparently more resistant than others.
2. Rotate millet with other crops.

Other Smuts Reported on Millet

Neovossia barclayana Bref.
Sorosporium paspali McAlpine var. *verrucosum* Thirumalachar & Pavgi
Tilletia ajrekari Mundkur
Tolyposporium bullatum (Schrot.) Schrot.
T. senegalense Speg.
Ustilago paradoxa Syd. & Butl.

Zonate Eye Spot

CAUSE: *Helminthosporium rostratum* Drechs. survives as mycelium in infected residue. It may also be seedborne. Conidia are produced from mycelium in residue particularly during warm (30°C optimum), wet conditions. Conidia are windborne to healthy tissue. Conidia produced in spots during warm, wet conditions provide secondary inoculum.

DISTRIBUTION: Georgia on pearl millet. The disease is considered to be of minor importance.

SYMPTOMS: Lesions on leaves are generally small, 1–2 mm by 2–5 mm, and commonly limited laterally by leaf veins. Lesions may coalesce to form large, necrotic areas that extend across veins. At first lesions are dark brown but later become light brown, particularly on the older leaves. The necrotic center of all lesions may bleach to a straw color with a slight browning at the edges and some yellowing at the ends. *Helminthosporium rostratum* has also been associated with head mold.

CONTROL: No control is necessary as the disease occurs infrequently.

Zonate Leaf Spot

CAUSE: *Gloeocercospora sorghi* D. Bain & Edg. overwinters as sclerotia in residue on or above the soil surface. It is probable that most primary inoculum comes from this source. It is also seedborne as mycelium and possibly sclerotia that contaminate seed sources. In the spring sclerotia germinate to produce conidia that are splashed by rain or windborne to millet. During warm, wet weather sporulation occurs in lesions as pink sporodochia form above stomates. The conidia are borne in a pinkish, gelatinous mass and are splashed to healthy tissue. Sclerotia are eventually formed in lesions.

DISTRIBUTION: Georgia. The disease is of minor importance.

SYMPTOMS: Lesions on leaves appear as water-soaked spots that develop tan centers and dark brown borders and measure 2.0–3.5 mm by 3.5–5.0 mm. Spots enlarge to form roughly semicircular blotches that may cover half or more of the leaf width. The blotches are different shades of dark brown mottled with light tan spots that tend to be arranged in circles and give the appearance of incomplete tan rings alternating with dark brown rings. This zonation pattern is often absent on narrow-leafed strains of millet. During moist weather, tiny salmon-colored globules consisting of masses of spores are visible under magnification on both surfaces of the spots. As leaves die, lenticular black sclerotia appear in the dead tissue.

CONTROL: No control is necessary.

Other Fungi Recorded on Millet

Top rot (pokka boeng) — *Fusarium moniliforme* Sheldon.
Inflorescence blight — *Balansia claviceps* Speg.
Fusarium inflorescence blight and seedling blight — *Fusarium roseum* Lk. emend. Snyd. & Hans.

VIRAL CAUSES

Bajra (pearl millet) Streak

CAUSE: Bajra streak virus (BSV) is transmitted by the leafhopper *Cicadulina mbila* Naude. BSV is not mechanically transmitted. Corn, sorghum, finger millet, barley, oats, and wheat can also be infected.

DISTRIBUTION: India.

SYMPTOMS: Plants may be affected at all stages of growth. Long, parallel, chlorotic streaks develop the whole length of infected leaves.

Diseased plants develop poorly formed and filled ears. Plants infected early are stunted and produce almost empty heads.

CONTROL: No control is reported.

Panicum Mosaic Virus

CAUSE: Panicum mosaic virus (PMV) is mechanically transmitted. *Panicum* spp., *Digitaria sanguinalis* (L.) Scop., *Setaria italica* (L.) Beauv., and *Echinochloa crusgalli* (L.) Beauv. are susceptible. Different strains of PMV are present.

DISTRIBUTION: United States.

SYMPTOMS: Plants may be stunted and chlorotic. Other symptoms include a yellow-green mosaic, mottle, stunted deformed heads, stunted blasted heads, and plant death in severe cases. Some cultivars may be symptomless carriers.

CONTROL: Plant resistant cultivars. Generally vigorous-growing types are more susceptible whereas slow-growing plants with smaller leaves are more resistant.

DISEASES OF UNKNOWN ETIOLOGY

Brown Leaf Mottle

CAUSE: Unknown.

DISTRIBUTION: Unknown.

SYMPTOMS: Initially a faint brown irregular mottle of the leaf and sheath appears at about the time of rapid stem elongation. Symptoms become more pronounced as the plant matures. Affected tissue becomes necrotic before normal senescence and leaves on affected plants senesce sooner than healthy plants.

CONTROL: Unknown.

Red Leaf Spot

CAUSE: Unknown.

DISTRIBUTION: Unknown.

SYMPTOMS: Symptoms first appear about the time of flower initiation. Irregular or zonate water-soaked blotches appear on leaves that soon become mahogany in color. As plants mature, the affected areas coalesce and become necrotic in advance of surrounding tissue.

CONTROL: Unknown.

BIBLIOGRAPHY

Burton, G. W., and Wells, H. D. 1981. Use of near-isogenic host populations to estimate the effect of three foliage diseases on pearl millet forage yield. *Phytopathology* **71**:331–333.

Desai, S. G.; Thirumalachar, M. J.; and Patel, M. K. 1965. Bacterial blight disease of *Eleusine coracana* Gaertn. *Indian Phytopathol.* **18**:384–386.

Deshmukh, S. S.; Mayee, C. D.; and Kulkarni, B. S. 1978. Reduction of downy mildew of pearl millet with fertilizer management. *Phytopathology* **68**:1350–1353.

Elliott, C. 1923. A bacterial stripe disease of proso millet. *Jour. Agric. Res.* **26**:151–159.

Govindu, H. C.; Shivanandappa, N.; and Renfro, B. L. 1970. Observations on diseases of *Eleusine coracana* with special reference to resistance to the Helminthosporium disease, in *Plant Disease Problems*. International Symposium on Plant Pathology. Indian Phytopathology Society, New Delhi, pp. 415–424.

Keshi, K. C., and Mohanty, N. N. 1970. Efficacy of different fungicides and antibiotics on control of blast of ragi, in *Plant Disease Problems*. International Symposium on Plant Pathology. Indian Phytopathology Society, New Delhi, pp. 425–429.

Lee, T. A., Jr., and Toler, R. W. 1977. Resistance and susceptibility to panicum mosaic virus—St. Augustine decline strain in millets. *Plant Dis. Rptr.* **61**:60–62.

Luttrell, E. S. 1954. Diseases of pearl millet in Georgia. *Plant Dis. Rptr.* **38**:507–514.

Mehta, P. R.; Singh, B.; and Mathur, S. C. 1952. A new leaf spot of bajra (*Pennisetum typhoides* Stapf. & Hubbard) caused by a species of Pyricularia. *Indian Phytopathol.* **5**:140–143.

Misra, A. P. 1959. Diseases of millets and maize. *Indian Agriculturist* **3**:75–89.

Onesirosan, P. T. 1975. Head mold of pearl millet in southern Nigeria. *Plant Dis. Rptr.* **59**:336–337.

Pathak, V. N., and Gaur, S. C. 1975. Chemical control of pearl millet smut. *Plant Dis. Rptr.* **59**:537–538.

Ramakrishnan, R. S., and Soumini, C. K. 1948. Studies in cereal rust. I. *Puccinia penniseti* Zimm. and its alternate host. *Indian Phytopathol.* **1**:97–103.

Rangawami, G.; Prasad, N. N.; and Eswaran, K. S. S. 1961. Bacterial leafspot diseases of *Eleusine coracana* and *Setaria italica* in Madras State. *Indian Phytopathol.* **14**:105–107.

Safeeulla, K. M. 1970. Studies on the downy mildews of bajra, sorghum and ragi, in *Plant Disease Problems*. International Symposium on Plant Pathology. Indian Phytopathology Society, New Delhi. pp. 405–414.

Seth, M. L.; Raychaudhuri, S. P.; and Singh, D.V. 1972. Bajra (pearl millet) streak: A leafhopper-borne cereal virus in India. *Plant Dis. Rptr.* **56**:424–428.

Singh, S. D.; and Williams, R. J. 1980. The role of sporangia in the epidemiology of pearl millet downy mildew. *Phytopathology* **70**:1187–1190.

Sundaram, N. V.; Plamer, L. T.; Nagarajan, K.; and Prescott, J. M. 1972. Disease survey of sorghum and millets in India. *Plant Dis. Rptr.* **56**:740–743.

Thakur, R. P.; Ras, V. P.; and Williams, R. J. 1984. The morphology and disease cycle of ergot caused by *Claviceps fusiformis* in pearl millet. *Phytopathology* **74**:201–205.

Thakur, R. P., and Williams, R. J. 1980. Pollination effects on pearl millet ergot. *Phytopathology* **70**:80–84.

Thirumalachar, M. J., and Mundkur, B. B. 1947. Morphology and the mode of transmission of ragi smut. *Phytopathology* **37**:481–486.

Verma, O. P., and Pathak, V. N. 1984. Role of insects in secondary spread of pearl millet ergot. *Phytophylactia* **16**:257–258.

Weber, G. F. 1973. *Bacterial and Fungal Diseases of Plants in the Tropics*. University of Florida Press, Gainesville.

Wells, H. D. 1967. Effects of temperature of *Helminthosporium setariae* on seedlings of pearl millet, *Pennisetum typhoides*. *Phytopathology* **57**:1002.

Wells, H. D. 1978. Eggplant may provide primary inoculum for rust of pearl millet caused by *Puccinia substriata* var. *indica*. *Plant Dis. Rptr.* **62**:469–470.

Wells, H. D.; Burton, G. W.; and Hennen, J. F. 1973. *Puccinia substriata* var. *indica* on pearl millet in the southeast. *Plant Dis. Rptr.* **57**:262.

Young, G. V.; Lefebvre, C. L.; and Johnson, A. G. 1947. *Helminthosporium rostratum* on corn, sorghum and pearl millet. *Phytopathology* **47**:180–183.

Zummo, N. 1976. Yellow leaf blotch: A new bacterial disease of sorghum, maize, and millet in west Africa. *Plant Dis. Rptr.* **60**:798–799.

10

Diseases of Oats
(*Avena sativa* L.)

Bacterial Blight

CAUSE: *Pseudomonas avenae* Manns. Disease development is more severe at higher temperatures

DISTRIBUTION: Southeastern United States.

SYMPTOMS: Lesion shape varies from circular to linear. Initially lesions are water-soaked, then dry up, turning brown to black.

CONTROL: Plant later in autumn when temperatures are cooler.

Bacterial Stripe Blight

This disease is also called stripe blight.

CAUSE: *Pseudomonas syringae* pv. *striafaciens* (Elliott) Young, Dye & Wilkie (syn. *P. striafaciens* (Elliott) Starr. & Burk.) survives on seed and in infected residue for at least two years. During cool, wet weather bacteria are splashed or blown onto leaves. Warm, dry weather stops the spread of the bacteria and subsequent disease development.

DISTRIBUTION: Australia, Europe, North America, and South America.

SYMPTOMS: Bacterial stripe is widespread but is seldom a serious disease. It first appears on a leaf as sunken, water-soaked dots. If the dots are abundant the leaf may die. The dots enlarge into water-soaked stripes or blotches that may extend the length of the blade. These stripes often have narrow yellowish margins. As the stripes age, they become a translucent rusty brown. In moist weather, bacteria exude in droplets from the stripes and later dry, forming white scales on and around the stripes.

CONTROL:
1. Plant resistant cultivars if they are available.
2. Plow under infected residue.
3. Treat seed with a seed-protectant fungicide.

Black Chaff

CAUSE: *Xanthomonas campestris* pv. *translucens* (Jones, Johnson & Reddy) Dye (syn. *X. translucens* (Jones, Johnson & Reddy) Dows f. sp. *undulosa* (Smith, Jones & Reddy) Hagb.) overwinters in infected residue in the soil, on seed, and directly in the soil. Primary infection occurs by bacteria being disseminated from soil, host plants, and residue by splashing water and insects, particularly aphids. Primary infection may also occur by seedborne bacteria, but seed stored for six or more months is not considered an important source of inoculum. Bacteria enter the plant through natural openings such as stomata and through wounds. Free water is necessary for the bacteria to enter the plant; therefore, small spaces and grooves that contain such water may act as a reservoir for the bacteria. Secondary spread of bacteria occurs by plant to plant contact, rain, and sucking and chewing insects.

DISTRIBUTION: Generally distributed wherever oats are grown.

SYMPTOMS: Symptoms of black chaff on oats are more obvious on leaves than heads and usually occur after several days of damp or rainy weather. The inner portion of the glumes will have brown or black spots. In moist weather, tiny, yellow beads of bacteria ooze to the surface of these discolored areas.

Small water-soaked spots occur on tender green leaves and sheaths of older plants and sometimes on seedlings. The spots enlarge and coalesce, becoming glossy, olive green, translucent stripes or streaks of various lengths that later turn yellow-brown. Stripes may extend the length of a leaf and are usually narrow, limited by leaf veins. Occasionally a spot may become large and blotchlike, causing the leaf to shrivel, turn light brown, and die. Severely diseased leaves die back from the tips. Under humid conditions, early in the morning, droplets of bacterial exudate may be seen on the surface of diseased spots. The droplets dry into hard, yellowish granules that may be removed easily as dry flakes from the leaf surface. Grain is not destroyed but it may be brown, shrunken, and carry bacteria to infect next year's crop.

CONTROL:
1. Do not plant seed from diseased plants. Seed should be cleaned to remove light-weight infected kernels.
2. Treat seed with a seed-protectant fungicide.
3. Rotate oats with other crops, preferably noncereals.

Halo Blight

This disease is also called blade blight.

CAUSE: *Pseudomonas syringae* pv. *coronafaciens* (Elliott) Young, Dye & Wilkie (syn. *P. coronafaciens* (Elliott) Stevens) can survive for at least two years on infected seed and residue in the soil. During wet weather, in the spring, bacteria are splashed onto the first leaves where they enter the plant by wounds or natural openings such as stomates. The bacteria are spread in cool, wet weather by plants rubbing against each other during a stormy or windy period. The bacteria are also spread by insects, particularly aphids, that feed on diseased tissue and inadvertently carry the bacteria to a healthy plant. Warm, dry weather stops the spread of the bacteria and subsequent disease development.

DISTRIBUTION: Australia, Europe, North America, and South America.

SYMPTOMS: The symptoms of halo blight occur earlier in the season than any of the diseases caused by leaf-spotting fungi. Halo blight is first noticeable on leaves as small, light green, oval to oblong, water-soaked spots with a slightly sunken center. The spots change color from light green to yellow to light brown. The tissue around the spots forms a halo that is water-soaked and light yellow. Severely infected leaves turn brown and die back from the tip due to an increased number of lesions that eventually coalesce. Portions of leaves of highly susceptible cultivars may die when lesions are not present.

CONTROL:
1. Plant resistant cultivars.
2. Treat seed with a seed-protectant fungicide.
3. Plow under infected residue to reduce inoculum.

FUNGAL CAUSES

Anthracnose

CAUSE: *Colletotrichum graminicola* (Ces.) G. W. Wilson overwinters as mycelium or spores on infected residue. It is not thought to be seed-borne on oats. Conidia are produced from mycelium during wet weather at an optimum temperature of 25°C and are windborne to hosts.

DISTRIBUTION: Generally distributed wherever oats are grown but it is most severe when oats are grown on sandy soils that are low in fertility.

SYMPTOMS: Red-brown, elongated, lens-shaped spots are produced on leaves. Spots are often formed on a leaf blade at its junction with the sheath. After the death of leaf tissue acervuli that appear as dark brown objects are found scattered throughout the spot. In severe infections much of the leaf is covered with acervuli. Infected basal and crown tissue becomes bleached, then turns brown, and is covered with acervuli. Under magnification, the acervuli will look hairy or have a pin cushion appearance. The roots and stem may also be infected. Infection of panicles results in lightweight and shriveled grain. Later in the season, severely infected plants are very small and have few tillers.

CONTROL:
1. Make sure that soils have adequate fertility.
2. Rotate oats with legumes.

Cephalosporium Stripe

CAUSE: *Cephalosporium gramineum* Nis. & Ika. survives as conidia and mycelium in residue within the top 8 cm of soil for up to five years. Conidia serve as primary inoculum and infect roots through mechanical injuries caused by soil heaving and insects during the winter and early spring. Infection is most severe in wet, acid soils (pH 5.0) due to a combination of increased fungal sporulation and root growth. The subcrown internode can also be infected just as the seed is germinating. After infection, conidia enter xylem vessels and are carried upward in the plant where they lodge and multiply at nodes and leaves. No further damage is done to the roots. The fungus prevents water movement up the plant and also produces metabolites that are harmful to the plant. At harvest time, *C. gramineum* is returned to the soil in infected residue where it is a successful saprophytic competitor with other soilborne microorganisms.

DISTRIBUTION: Great Britain, Japan, and most oat-growing areas of North America.

SYMPTOMS: Oats are ordinarily not severely infected. Infected plants are scattered throughout a field and are usually more numerous in lower, wetter areas. Infected plants are dwarfed. During jointing and heading, distinct yellow stripes, one to four per leaf (usually one or two), develop on leaf blades, sheaths, and stems, normally occurring the length of the plant. Thin brown lines consisting of infected veins occur in the middle of a stripe. The stripes eventually become brown, are highly visible on green leaves, and are still noticeable on yellow straw. Toward harvest the culm at or below the node may become dark. Heads of infected plants are white and do not contain seed, or if seed is present, it is usually shriveled.

CONTROL:
1. Rotate oats for at least two years with a noncereal.
2. Infected residue should be plowed deeper than 8 cm if feasible.

Covered Smut

CAUSE: *Ustilago kolleri* Wille is seedborne and constitutes the only source of inoculum. After seed is planted, mycelium infects young oat shoots. Oat plants cease to be susceptible to infection when the first leaves have emerged more than 1 cm beyond the sheaths. Once inside the seedlings, mycelium grows systemically, keeping pace with the growing tip of the oat plant, finally entering the young, developing kernels. By heading time, the oat kernels and hulls are completely replaced with chlamydospores. Chlamydospores produced in diseased seeds are scattered by the wind or harvesting operations. Some of the spores lodge either outside or inside hulls where they remain dormant until the seed is planted. Other spores germinate immediately after being windborne to healthy kernels and grow into hulls or seed coats of the kernels. The mycelium remains inactive until seed is planted.

The soil temperature and moisture when oat seeds are sprouting, have an influence on infection. Infection occurs with between 5 percent and 60 percent moisture content, with 35–40 percent being optimum, and at a soil temperature between 5°C and 30°C, with 15–25°C being optimum.

DISTRIBUTION: Generally distributed wherever oats are grown.

SYMPTOMS: As a smutted oat panicle emerges from its enclosing sheath a dark brown to black powdery mass of chlamydospores contained within a white-gray membrane has replaced the grains and often the awns and glumes. Smutted panicles do not spread as much as healthy ones. All spikelets and panicles on an infected plant are usually smutted.

The white-gray membrane is more persistent than that of loose smut and remains intact until the chaff dries or until the grain is harvested. The persistence and the extent of the damage to the panicle varies with the oat cultivar. Smutted plants are generally shorter than healthy ones.

CONTROL:
1. Plant resistant cultivars.
2. Treat seed with a systemic seed-protectant fungicide.

Crown Rust

This disease is also called oat leaf rust.

CAUSE: *Puccinia coronata* Cda. is a heteroecious long-cycled rust fungus that has species of buckthorn (*Rhamnus carthartica* L. and *R. lanceolata* Pursh) as an alternate host to oats and some grasses. *Puccinia coronata* causes crown rust of oats by following one of two life cycles. The first occurs when urediospores are produced in uredia on oats, are windborne and infect other oat plants. Urediospores are produced every 8–10 days depending on weather conditions. The cycle will continue indefinitely as long as green oat plants are available. Urediospores produced in northern oat-producing areas will then be blown south in summer and autumn to infect oats and grasses in Mexico and the southern United States. The urediospores will continue to recycle on oats in southern oat-producing areas until they are wind disseminated to infect oats in the north during the late winter and spring.

The second and complete life cycle occurs in climates having cold winters where species of buckthorn are present. As oats ripen, telia that contain teliospores are formed, usually in or around the uredium. Teliospores may also form during periods of drouth, excessive moisture, or high temperatures. The teliospores are the means of survival through the winter. In the spring they germinate to form basidiospores that are windborne and only infect buckthorn. At the point of infection by the basidiospore, a pycnium is formed on the upper leaf surface.

The pycnia are the means by which new races of crown rust arise and function in exchanging genetic material between different strains of the fungus. Each pycnium produces pycniospores and special mycelium called receptive hyphae. The pycniospores are exuded out of the pycnium in a thick, sticky, sweet liquid that is attractive to insects. Pycniospores are splashed by water or carried by insects from one pycnium to another where they become attached to the receptive hyphae. The nucleus from the pycniospores enter into the receptive hyphae. This results in the formation of an aecium on the lower leaf surface. Aeciospores differ genetically from either pycniospores or receptive hyphae, are windborne, and infect only oats. Each infection gives rise to a uredium in which urediospores are formed. The urediospores then serve as the repeating stage, reinfecting only oats. Urediospores are produced at an optimum temperature of 21°C with dew, fog, or some form of moisture. The higher relative survival ability of isolates of some races compared to isolates of other races on *Avena sterilis* L. is due to a shorter period between inoculation and appearance of uredia, higher infection density, higher yield of urediospores per uredium, and extended longevity of urediospores. Telia are again produced as the oats mature, thus completing the life cycle. Numerous physiologic races exist.

DISTRIBUTION: Generally distributed wherever oats are grown.

SYMPTOMS: The uredia appear primarily on leaves as bright orange-yellow, round to oblong pustules. There are usually no bits of loose epidermis around the pustules, thus distinguishing crown rust from stem rust. Similar pustules may occur on sheaths, stems, and panicles. The pustules burst open to release the urediospores, which look like orange powder. Under favorable conditions for infection, the uredia become numerous and may coalesce. Stems may be so weakened that severe lodging may result.

As the oat plants begin to ripen, telia sometimes form in a ring around old uredia or they may develop independently of uredia. Telia remain covered indefinitely by the epidermis, giving them a gray-black, oblong, and slightly raised appearance.

Pycnia on buckthorn leaves appear as bright orange to yellow spots with sometimes a small drop in the middle consisting of pycniospores and liquid. Opposite the pycnial spots, usually on the underleaf surface, are the aecia that resemble raised, orange cluster cups.

CONTROL:
1. Plant resistant cultivars.
2. Plant oats as early as possible in the spring. This helps the crop to escape the rust.
3. Eradicate buckthorn shrubs within 1.5 km of oat fields.

Downy Mildew

CAUSE: *Sclerophthora macrospora* (Sacc.) Thirum., Shaw & Naras. (syn. *Sclerospora macrospora* Sacc.) survives as oospores in leaf and stem residue in soil. When residue decays, the oospores are liberated into the soil where they may persist for still longer periods of time. Oospores are disseminated by seed and in soil and residue by wind or water. Oospores germinate in saturated soil to form sporangia. Zoospores are released from sporangia and swim for a short distance to a host where they encyst, then germinate to form a germ tube that penetrates the plant. Disease is most severe during wet, cool environmental conditions.

DISTRIBUTION: Generally distributed wherever oats are grown but usually only occurs in wet areas of a field.

SYMPTOMS: Symptoms are most noticeable at heading time. Infected plants tend to be stiff, upright, and remain green for several days after healthy plants are ripe. Severely diseased plants are dwarfed, normally less than a third the height of healthy plants, and tiller excessively. The upper leaves may be curled about the panicles, which are often curled and twisted to the extent that they are reduced to a cluster of frayed and tangled spikelets. Leaves will also appear leathery.

Less severely infected plants may be slightly dwarfed. They appear stiff and produce distorted heads that may stand above healthy heads. The flag leaf may be curled and twisted about a poorly developed head. Heads may be deformed to the extent that no spikelets are recognizable and no viable seed is produced.

CONTROL: Drain areas in fields that are likely to be flooded.

Ergot

CAUSE: *Claviceps purpurea* (Fr.) Tul. survives as sclerotia on or in the soil. Sclerotia germinate before flowering to produce stromatic heads in which perithecia are produced at the end of a stipe or stalk. Ascospores are formed in perithecia and are windborne to young flowers. Ascospores germinate and the infective hyphae grow into the ovaries. Mycelium growing in the ovaries produce conidia in a sweet, sticky liquid called honeydew on the outside of the flowers. Insects are attracted to honeydew and inadvertently disseminate conidia from diseased to healthy flowers. Eventually sclerotia replace oat kernels.

DISTRIBUTION: Generally distributed wherever oats are grown; however, ergot is relatively uncommon on oats.

SYMPTOMS: Ergot is easily recognized by the purple-black, horn-shaped sclerotia that are produced in place of seed. One or several kernels on a head may be replaced by sclerotia. The sticky honeydew may be covered with dust and foreign matter, giving infected flowers a gray-black appearance.

CONTROL:
1. Deep plowing will place sclerotia at a depth in the soil where they will not be able to release ascospores into the air. Sclerotia are also decomposed in a year or so by soil microorganisms.
2. Control grassy weeds around fields.

3. Rotate crops with crops such as legumes that are not susceptible to *C. purpurea.*
4. Clean seed to remove ergot bodies.

Eyespot

This disease is also called Cercosporella foot rot, stem break, foot rot, strawbreaker, and culm rot.

CAUSE: *Pseudocercosporella herpotrichoides* (Fron.) Deighton survives in infected residue as mycelium. The fungus can persist in soil for several years in the absence of grain crops. Conidia are produced during cool, damp weather either in the autumn or spring and infect crown and basal culm tissue. Roots are not infected.

DISTRIBUTION: Ireland. Other reports from Europe and the United States indicate that oats are resistant.

SYMPTOMS: This disease is most conspicuous near the end of the growing season by lodging of diseased plants. The lodging caused by eyespot tends to cause straw to fall in all directions. Lodging caused by wind or rain is characterized by the straw falling in one direction. Whiteheads, similar to symptoms produced in take all, are also present at maturity.

Eye-shaped or ovate lesions with white to tan centers and brown margins develop first on the basal leaf sheath. Similar spots form on the stem directly beneath those on the sheath and cause the lodging. Roots are not infected but a necrosis occurs around the roots in the upper crown nodes. Under moist conditions, the lesions enlarge and a black stromalike mycelium develops over the surface of the crown and base of the culms, giving the tissues a charred appearance. The stems then shrivel and collapse. Plants that do not collapse are yellowish or pale green with heads reduced in size and numbers.

Characteristic eyespot lesions may be found on oats but they are not as common as on barley, rye, and wheat. Oats will often have an indented line present on one or both sides of the culm in the area of the plant where infection has occurred. The indentation resembles the culm being pressed in with a thumb nail. The indentation extends up from the plant base 2.5–5.0 cm and sometimes causes a split in the culm. The fungus is often visible when infected stems are cut longitudinally. Oats are usually not stunted but will break off above the soil line when plants are pulled up.

CONTROL: No satisfactory control exists except rotation with noncereals.

Fusarium Foot Rot

CAUSE: *Fusarium culmorum* (W. G. Sm) Sacc. (syn. *F. roseum* (Lk.) emend. Snyd. & Hans. f. sp. *cerealis* (Cke.) Snyd. & Hans. 'Culmorum') survives in the soil as chlamydospores or as mycelium in infected residue.

DISTRIBUTION: Generally distributed wherever oats are grown.

SYMPTOMS: Seedlings are infected below or at the soil surface. Light brown to red-brown lesions may be noticed at the base of the leaf at or just below the soil surface. Some seedlings may also have these spots on the mesocotyl near the seed. Seedlings may damp off either preemergence or postemergence.

CONTROL:
1. Treat seed with a seed-protectant fungicide.
2. Plow under infected residue where this is feasible and erosion is not a problem.

Helminthosporium Foot Rot

This disease is also called head blight and Helminthosporium culm rot.

CAUSE: *Bipolaris sorokiniana* (Sacc.) Shoemaker (syn. *Helminthosporium sativum* Pammel, King & Bakke) overwinters as mycelium and conidia in residue and seed.

DISTRIBUTION: Generally distributed wherever oats are grown.

SYMPTOMS: Roots and basal portion of young seedlings are rotted. Leaves may droop and look wilted or the whole leaf may turn yellow and die. In older plants the disease appears as a blackening and weakening of the lower stems. Frequently the infected stems bend over and lodge on the ground but the upper portion of the infected stems may stay alive for some time after falling over. Often the live tips of fallen stems turn and grow upward.

CONTROL:
1. Treat seed with a seed-protectant fungicide.
2. Rotate oats with other crops, particularly legumes.

Helminthosporium Leaf Blotch

This disease is also called blotch, crown rot, Helminthosporium leaf spot, leaf blotch, leaf streak, oat leaf blotch, and oat leaf spot.

CAUSE: *Drechslera avenacea* (Curt. ex Cke.) (syn. *Helminthosporium avenae* Eidam), teleomorph *Pyrenophora avenae* Ito & Kuribay, overwinters as mycelium under the seed coat, as spores on the seed, and as mycelium or perithecia on infected residue. Seedborne infection results in spots on seedling leaves. Conidia produced in these lesions during wet weather are windborne to other leaves. Infections of older leaves may originate from conidia or ascospores produced in perithecia during wet weather on infected residue. Disease development is favored by cool, wet weather.

DISTRIBUTION: Africa, Asia, Australia, Europe, and North America.

SYMPTOMS: In the seedborne phase of the disease, oblong to elongate, light red-brown spots appear on seedling leaves soon after they emerge. Seedling leaves may also be twisted or contorted.

On older leaves, lesions start as small, brown flecks that develop into longitudinal strips of dead tissue. The outer edges of the spots merge from brown to yellow or reddish shades that frequently spread over the greater part of an infected blade. Sometimes the disease does not produce well-defined spots but causes a leaf to assume a withered appearance as if injured by adverse weather. As diseased leaves die, the color changes from green to yellow or gray and the brown of the diseased spot fades. When kernels are infected, they will turn brown at the basal end.

CONTROL:
1. Treat seed with a seed-protectant fungicide.
2. Rotate oats with other crops.
3. Plow under residue to reduce inoculum originating from the field.
4. Some cultivars are more susceptible than others.

Helminthosporium Seedling Blight and Crown and Lower Stem Rot

CAUSE: *Drechslera avenacea* (Curt. ex Cke.) (syn. *Helminthosporium avenae* Eidam). For more information, see Helminthosporium leaf blotch.

DISTRIBUTION: Generally distributed wherever oats are grown.

SYMPTOMS: Oat seedlings are stunted; initially the leaves turn yellow, then reddish, and eventually brown as severely infected plants die. Dark brown to black lesions develop on the lower parts of the stem and leaf sheaths. Seedling infections can continue on older plants or later infections of roots, crowns, and lower stems can develop. Infections are characterized by dark brown areas of decay near the oat crown. Severely diseased plants are unproductive, more subject to lodging than healthy plants, or they may die.

CONTROL:
1. Plant high-quality seed that has been harvested on time; delayed harvest can result in increased *H. avenae* damage to kernels.
2. Treat seed with a seed-protectant fungicide.
3. Rotate oats with other crops.

Loose Smut

This disease is also called black loose smut, black head, false loose smut, and naked smut.

CAUSE: *Ustilago avenae* (Pers.) Rostr. is seedborne, which constitutes the only source of inoculum. After seed is planted, mycelium infects the young oat shoots. Oat plants cease to be susceptible to infection when the first leaves have emerged more than 1 cm beyond the sheath. Once inside the seedlings, mycelium grows systemically, keeping pace with the growing tip of the oat plant and finally entering the young, developing kernels. By heading time, the oat kernels and hulls are completely replaced with chlamydospores. Chlamydospores produced in diseased heads are windborne to healthy heads usually before harvest. Some of the spores lodge either outside or inside the hulls where they remain dormant until the seed is planted. Other spores germinate immediately after being windborne to healthy kernels and grow into the hulls or into the seed coats of the kernels, remaining inactive there until seed is planted.

The soil temperature and moisture have an influence on infection when oat seeds are sprouting. Infection occurs between 5 percent and 60 percent moisture (35–40% optimum) and temperatures between 5°C and 30°C (15–25°C optimum).

DISTRIBUTION: Generally distributed wherever oats are grown.

SYMPTOMS: As a smutted oat panicle emerges from its enclosing sheath, a dark brown to black powdery mass of smut spores (chlamydospores) contained within a delicate, white-gray membrane has replaced the grains and often the awns and glumes. The thin membrane usually breaks and disintegrates soon after the oat panicles emerge. Smutted panicles do not spread as much as healthy ones. All spikelets and panicles on an infected plant are usually smutted. Smutted plants are shorter than healthy ones and are often overlooked at harvest because the naked mass of chlamydospores is quickly scattered by wind and rain, leaving a bare panicle that is difficult to see.

CONTROL:
1. Plant resistant cultivars.
2. Treat seed with a systemic seed-protectant fungicide.

Microdochium Root Rot

CAUSE: *Microdochium bolleyi* (Sprague) de Hoog & Herm.-Nyhof (syn. *Aureobasidium bolleyi* (Sprague) von Arx and *Gloeosporium bolleyi* Sprague). The fungus is mainly saprophytic.

DISTRIBUTION: California.

SYMPTOMS: Disease symptoms are mild and consist of a discoloration and necrosis primarily on the subcrown internode.

CONTROL: None necessary.

Powdery Mildew

CAUSE: *Erysiphe graminis* DC. ex Merat f. sp. *avenae* overwinters as cleistothecia on wild grasses and on infected oat residue. Ascospores form within cleistothecia in the spring and are windborne to oats. Conidia are produced as soon as the mycelium becomes established on the leaf surface, especially during cool, humid weather but without the presence of free water. Conidia account for most of the secondary inoculum and are windborne from plant to plant. Cleistothecia are formed on the leaf surface as the plant approaches maturity.

DISTRIBUTION: Generally distributed wherever oats are grown.

SYMPTOMS: Powdery mildew first appears as patches of white, fluffy growth on the lower leaves and leaf sheaths. As the disease progresses, the patches become powdery, turn gray, then brown. The powder is a combination of conidia and mycelium growing on the leaf surface. Eventually a large area of the leaves may be infected. Infected leaves may turn yellow and shrivel. As the plant nears maturity, small brown to black round objects, which are the cleistothecia, may be seen interspersed in the mycelium and spores. Yield may be reduced.

CONTROL: Control measures are usually not necessary to control powdery mildew on oats; however, the following are aids to control:
1. Apply a sulfur fungicide to foliage.
2. Some cultivars have adult resistance.

Pythium Seed and Seedling Rot

CAUSE: *Pythium debaryanum* Hesse, *P. irregulare* Buis., and *P. ultimum* Trow. survive in infected residue or soil as oospores. In moist soil, oospores germinate to form sporangia in which zoospores are produced. The zoospores swim to root tips or the seed where they germinate and infect.

DISTRIBUTION: Generally distributed wherever oats are grown.

SYMPTOMS: The seed and seedlings may be killed either preemergence or postemergence. Infected plants that are not killed initially are yellowed and stunted but may recover the green of a normal plant. However, an infected plant is usually never as vigorous as a noninfected plant. Water-soaked translucent areas occur on roots that later turn a red-brown.

CONTROL: Treat seed with a seed-protectant fungicide.

Scab

This disease is also known as Fusarium blight.

CAUSE: *Gibberella zeae* (Schw.) Petch, anamorph *Fusarium* spp. Although they are not as well documented, the Fusaria that cause scab of oats are probably the same as the causal fungi that cause scab of wheat. See Scab of Wheat in Chapter 23 for further information. *Gibberella zeae* overwinters as mycelium in residue and as perithecia on oats, barley, corn, wheat, and other grass residue that was infected the previous season. The fungus is also seedborne. Ascospores are produced in perithecia during warm, moist weather and are windborne to hosts. Conidia of *Fusarium* spp. are produced from mycelium and windborne to hosts. Additional spores are produced on infected heads that are capable of causing new infections. Wet autumn weather favors disease development on lodged grain or grain in swaths.

DISTRIBUTION: Generally distributed wherever oats is grown. However, oats is the least susceptible of the cereals.

SYMPTOMS: Scab on oats is not as readily detected as on wheat or barley because a hull covers the oat kernel. Hulls of infected spikelets are ash gray and may be partially covered by a pink mold. In severe cases the hulls may be shriveled and rough in appearance but they do not turn brown. Small black perithecia may be produced on hulls as the grain nears maturity.

If seed is heavily infested, seedlings may be killed before pre-emergence. Frequently seedlings are only stunted and many will turn yellow and die after emergence. Roots of infected seedlings are red-brown and rotted. Later in the season roots may be covered by pink mycelium. If a joint on the stem becomes infected, the plant portion above it becomes white and perithecia will be produced on the affected joint.

CONTROL:
1. Treat seed with a seed-protectant fungicide.
2. When oats follows corn, scab is more severe when corn stalks are left on soil surface than when they are plowed under.
3. Rotate oats with legumes rather than cereals.

Scoleocotrichum Leaf Blotch

CAUSE: *Scoleocotrichum graminis* Fckl. var. *avenae* Ericks. probably survives as mycelium in infected residue. During wet weather in the spring, conidia are produced that are windborne to hosts. Sporulation ceases during dry weather.

DISTRIBUTION: Generally distributed in most oat-growing areas; however, Scoleocotrichum leaf blotch is uncommon.

SYMPTOMS: Oblong to linear, red-brown to brown-purple blotches with definite margins develop on leaves. The necrotic area is dry and sunken. A good diagnostic characteristic is the rows or tufts of conidio-phores that emerge through the stomata. These structures can be seen with the aid of a hand lens.

CONTROL: Control measures are ordinarily not practiced but sanitation will reduce inoculum.

Septoria Disease of Oats

This disease is also called dark stem of oats, leaf spot of oats, Septoria black stem, Septoria blight, Septoria leaf blotch, speckled blotch, and speckled leaf blotch.

CAUSE: *Septoria avenae* Frank (syn. *S. tritici* f. sp. *avenae* (Desm.) Sprague), teleomorph *Leptosphaeria avenaria* Weber, overwinters as mycelium and pycnidia (micropycnidia) in residue. It is not thought to be seedborne except with very susceptible cultivars. Conidia (microspores) are windborne during cool, wet weather in the spring to hosts. Perithecia are not thought to be an important means of overwintering nor are ascospores thought to be important either as a means of primary inoculum or in the spread of the disease.

Pycnidia are formed in mature lesions if weather continues cool and wet after the initial infection. Conidia are disseminated mostly by rain or water and are primarily responsible for stem infections. Perithecia and ascospores are mainly formed in the spring but some perithecia may be formed in the summer and overwinter.

DISTRIBUTION: Africa, Australia, Europe, and North America.

SYMPTOMS: Blotches first appear on lower leaves and then spread up the plant. Blotches are round to elongate or diamond-shaped, yellow to light or dark brown, and surrounded by a band of dull brown that changes to yellow as it blends into the green of healthy tissue. Pycnidia appear as black dots scattered throughout the center of older blotches, giving them a speckled appearance. As blotches enlarge, the infected leaf tissue dies. After heading, infection at a leaf base spreads into an adjoining leaf sheath and will appear chocolate or red-brown.

Gray-brown to shiny black lesions develop on stems or culm tissue beneath infected leaf sheaths mostly above the upper two joints. Plants frequently lodge near maturity. Dark gray fungus mycelium fills the hollow areas of infected culms.

During cool, wet weather, yellow to brown lesions occur on outer glumes. Black or dark brown lesions may extend to lemma and palea and eventually to the groat of the kernel.

CONTROL:
1. Early cultivars tend to be most susceptible.
2. Plant certified seed.
3. Treat seed with a seed-protectant fungicide.
4. Rotate oats every three or four years with other crops where this is practical.
5. Plow under infected residue where this is feasible and erosion is not a problem.

Sharp Eyespot

This disease is also called Rhizoctonia root rot.

CAUSE: *Rhizoctonia solani* Kuhn and *R. cerealis* van der Hoeven. Survival is as sclerotia in soil or mycelium in infected residue of a large number of plant genera. Oats may be infected any time during the growing season. Roots and culms are infected by mycelium from germinating sclerotia or growing from a precolonized substrate, particularly in dry (less than 20% moisture-holding capacity), cool soils.

DISTRIBUTION: Generally distributed wherever oats are grown.

SYMPTOMS: Oats are one of the most susceptible cereals to *R. solani*. Diamond-shaped lesions that resemble eyespot, but are more susceptible than those of eyespot, occur on lower leaf sheaths. They have light tan centers with dark brown margins, frequently with dark mycelium visible on them. Dark sclerotia may develop in lesions and between culms and leaf sheaths.

Seedlings may be killed but plants may produce new roots to compensate for those rotted off. When roots are infected, plants may lodge and produce white heads. Diseased plants may appear stiff, have a grayish cast, and be delayed in maturity.

CONTROL: No control is very effective. Vigorous plants growing in well-fertilized soil are not as likely to become severely diseased as unthrifty plants.

Snow Mold

This disease is also called pink snow mold.

CAUSE: *Fusarium nivale* (Fr.) Ces., teleomorph *Calonectria nivalis* Schaff., oversummers as perithecia or mycelium in infected residue. Plants are infected in autumn by ascospores being windborne to hosts or by mycelium growing from previously infected residue. Primary infections occur on leaf sheaths and blades near soil level. The infection is spread by mycelial growth during cool, wet weather and beneath a snow cover. Secondary infection in the spring occurs when perithecia develop in late spring during cool, humid weather and produce ascospores. Conidia are also produced on mycelium, and both are windborne to plants.

DISTRIBUTION: Canada, central and northern Europe, and the United States.

SYMPTOMS: Infected plants have chlorotic and dry necrotic leaves. A good diagnostic characteristic is the pink color of mycelium and sporodochia on leaf and crown tissues as snow melts in the spring.

CONTROL: Plant resistant cultivars.

Stem Rust

CAUSE: *Puccinia graminis* Pers. f. sp. *avenae* Ericks. & E. Henn. is a heteroecious long-cycle rust that has the common barberry, *Berberis vulgaris* L., as the alternate host to oats. Rust is caused by *P. graminis avenae* in one of two ways. The first way is urediospores (called summer or repeating spores) are produced in uredia on oats several weeks before ripening. The urediospores are windborne and infect oats, producing urediospores under moist conditions and moderate temperatures. This cycle will continue indefinitely as long as green oat plants are available. Urediospores produced in northern oat-producing areas will then be blown south in summer and autumn to infect oats in Mexico and the southern United States. The urediospores will then continue to recycle on oats in southern oat-producing areas until they are disseminated by wind northward during the late winter and spring to infect oats in the north.

The second way rust is caused is for *P. graminis avenae* to go through its complete life cycle. Telia containing teliospores are formed on maturing oats. The teliospores are the means of survival through the winter. In the spring, teliospores germinate to form basidiospores (sometimes called sporidia) that are windborne and infect barberry bushes. At the point of infection by the basidiospore, a pycnium (sometimes called spermogonium) is formed on the upper leaf surface. The pycnia are the means by which new races of stem rust arise and function in the exchange of genetic material, thereby creating new races of the fungus. Each pycnium produces pycniospores and special mycelium called receptive hyphae. The pycniospores are exuded out of the pycnium in a thick, sticky, sweet liquid that is attractive to insects. Pycniospores are splashed by water or are carried by insects from one pycnium to another where they become attached to the receptive hyphae. The pycniospore germinates with the nucleus from the spore entering into the receptive hyphae. This eventually results in the formation of an aecium on the lower leaf surface. Aeciospores that differ genetically from either pycniospore or receptive hyphae are produced in the aecia and are windborne to oats where infection occurs. Each infection gives rise to a uredium in which urediospores are formed. The urediospores then serve as the repeating stage and are windborne to oats where new generations of urediospores are formed. Telia are once again produced as the oats mature, thus completing the life cycle.

DISTRIBUTION: Generally distributed wherever oats are grown.

SYMPTOMS: Uredia and telia occur on the stems, leaf sheaths, blades, and panicles of the oat plant. Uredial pustules are large, oblong, and dark red-brown. The epidermis of the leaves and culms is ruptured and pushed back around the pustule, giving it a jagged appearance and exposing the urediospores. The dark color and jagged appearance of the pustule distinguish stem rust from the relatively smooth margin and light orange-yellow color of crown rust. As the host plant approaches maturity, telia begin to form in and around the uredia, particularly on the stems and sheaths. Telia are black, usually oblong, and the teliospores are exposed by the epidermis rupturing.

Pycnia on barberry leaves appear as bright orange to yellow spots with a small drop of moisture sometimes in the middle of the spots. Opposite the pycnial spots and normally on the under side of the leaf are the aecia that resemble raised, orange clusters.

CONTROL:
1. Plant resistant oat cultivars.
2. Eradicate common barberry shrubs from the vicinity of oat fields.

Take-All

This disease is also called white head.

CAUSE: *Gaeumannomyces graminis* (Sacc.) von Arx & Oliv. var. *avenae* (E. M. Turner) Dennis survives as mycelium in infected plants or as mycelium and perithecia in infected residue in the soil. However, ascospores produced in perithecia are not considered important in the dissemination of the disease. In the autumn or spring, seedlings probably become infected by roots growing into the vicinity of colonized residue and coming in contact with mycelium. Plant to plant spread occurs by hyphae growing through soil from an infected to a healthy plant or by a healthy root coming in contact with an infected one. Ascospores are produced in perithecia during wet weather but apparently are not disseminated a great distance either by splashing water or wind. Take-all is usually more severe on oats grown in alkaline soils.

DISTRIBUTION: Generally distributed wherever oats are grown.

SYMPTOMS: The first symptoms are light to dark brown necrotic lesions on the roots. By the time an infected plant reaches the jointing stage, most of its roots are brown and dead. At this point many plants die, or if plants are still alive, they are stunted and the leaves are yellow.

The disease becomes most obvious as plants approach heading. The stand is uneven in height and plants appear to be in several stages of maturity. At heading infected plants have few tillers, ripen prematurely, and their heads are bleached and sterile. Roots are sparse, blackened, and brittle. Plants can be broken free of their crown easily when pulled from the soil. A very dark discoloration of the stem is visible just above the soil line, along with a mat of dark brown fungus mycelium under the lower sheath between the stem and inner leaf sheaths.

CONTROL: Rotate oats with a nongrass crop since *G. graminis* var. *avenae* is not a good saprophyte and does not survive long in the soil in the absence of a host.

Victoria Blight

This disease is also called Helminthosporium blight.

CAUSE: *Bipolaris victoriae* (Meehan & Murphy) Shoemaker (syn. *Helminthosporium victoriae* Meehan & Murphy) overwinters as conidia on seeds and as mycelium and spores in infected residue.

DISTRIBUTION: Occurs wherever oat cultivars with the Victoria type of crown rust resistance are grown. The disease is now rare.

SYMPTOMS: *Bipolaris victoriae* produces a toxin that spreads throughout the plant. Infected seed may rot in the soil before they germinate. Usually the plants are infected in the seedling stage as they emerge from the soil. The seedling leaves will have a dull blue-gray color with red-brown stripes running the length of the blades. Such stripes may cover half a blade. Later the seedling leaves develop a reddish color and the entire plant may die. Still later in the season the basal parts of stems may turn a brownish color. The lower joints and leaf sheaths become covered with abundant conidia and conidiophores of *B. victoriae*, giving them a dark, velvety appearance. Severe infection causes a root rot and death of stems, causing severe lodging and premature ripening of plants.

CONTROL:
1. All oat cultivars having the Victoria type of crown rust resistance are susceptible. Most other cultivars are highly resistant.
2. Treat seed with a seed protectant fungicide.
3. Rotate oats with other crops.

_____MYCOPLASMAL CAUSES_____

Aster Yellows

CAUSE: Aster yellows mycoplasma survives in several dicotyledonous plants and leafhoppers. Aster yellows mycoplasma is transmitted primarily by the aster leafhopper, *Macrosteles fascifrons* Stal., and less commonly by *Endria inimica* Say and *M. laevis* Rib. The leafhoppers first acquire the mycoplasma by feeding on infected plants and then fly to healthy plants and feed on them, thereby transmitting the mycoplasma. Different strains exist. Canadian workers have designated a western strain of the aster yellows mycoplasma as CAYA and three eastern strains as DAYA, MAYA, and NAYA. Strains differ in ability to infect oats.

DISTRIBUTION: Generally distributed throughout eastern Europe, Japan, and North America. Aster yellows is rarely severe on oats.

SYMPTOMS: Symptoms become most obvious between 25°C and 30°C. Seedlings will either die two to three weeks after infection, or if they survive, they will be stunted, leaves will generally yellow or have yellow blotches, and the heads will be sterile and may be somewhat distorted. Floral parts that appear normal may still be nonfunctional.

Infection of older plants causes leaves to become stiff and discolored from the tip or margin inward in shades of yellow, red, or purple. Root systems may not be well developed.

CONTROL: No control is practical.

CAUSED BY NEMATODES

Cyst Nematodes

CAUSE: *Heterodera hordecalis* Andersson and H. *latipons* Franklin.

DISTRIBUTION: Not reported; however, *H. latipons* has been reported from Israel.

SYMPTOMS: Plants are stunted and chlorotic. Cysts not described on roots.

CONTROL: Rotate with a noncereal or grass.

Needle Nematode

CAUSE: *Longidorus cohni* Heyns. The nematode builds up to high populations during the winter and drops during an increase in soil temperatures in the spring.

DISTRIBUTION: Israel.

SYMPTOMS: Terminal swellings occur on roots. Plants are stunted and chlorotic, causing large bare patches in the field.

CONTROL: Rotate with a noncereal or grass.

Oat Cyst Nematode

This nematode is also called the cereal cyst nematode.

CAUSE: *Heterodera avenae* Woll. The entire life cycle of the oat cyst nematode from egg, to larva, to adult can be completed in 9–14 weeks. The female enters roots and begins feeding. Shortly the females swell up and break through the roots but remain attached by a thin neck. Females are inseminated by males and several hundred eggs develop within a body. Eventually the female body forms a hard body or cyst that detaches from the roots and survives in soil for several years. The following spring, second stage larvae hatch from an average of 75 percent of the eggs.

DISTRIBUTION: Africa, Australia, Europe, Japan, Russia, southeastern Canada, and Oregon.

SYMPTOMS: *Hederodera avenae* is also a pathogen of barley, rye, and wheat but is most prevalent on oats. The lemon-shaped cysts are first white, then gradually become dark brown as the cyst hardens. Eggs and larvae are white and microscopic. The cyst itself is barely visible to the naked eye.

The first symptom on oats is poor growth in one or more spots in a field. Leaf tips of young, heavily infested plants are red or purple. The discolored leaves die off and plants become yellow. Roots are thickened and more branched than normal. Heavy infestations cause wilting, particularly during times of water stress, stunted growth, poor root development, and early plant growth.

CONTROL:
1. Some oat cultivars are reported to be resistant. However, resistant cultivars will vary from area to area because the oat cyst nematode is known to have up to 20 races. Not all races are ordinarily present in one locality.
2. Rotate oats with a legume crop. The number of nematodes will decline the longer infested soil is cropped to an alternate crop, and damage is less likely to occur when oats are again grown in infested soil.
3. Apply a nematicide before planting.

Root Gall Nematode

CAUSE: *Subanguina radicicola* (Grf.) Param. survives in host roots. Larvae penetrate roots and develop in cortical tissue, forming a root gall in two weeks. The mature females begin egg production within a gall. Eventually the galls weaken and release larvae that establish secondary infection. Each generation is completed in about 60 days.

DISTRIBUTION: Canada and northern Europe.

SYMPTOMS: Seedlings frequently have reduced top growth and chlorosis. Galls on roots tend to be inconspicuous and vary in diameter from 0.5–6.0 mm. Roots may be bent at the gall site. At the center of larger galls is a cavity filled with nematode larvae.

CONTROL: Rotate oats with a noncereal crop.

Root Lesion Nematodes

CAUSE: *Pratylenchus* spp. are nematodes that overwinter as eggs, larvae, or adults in host tissue or soil. Both larvae and adults penetrate roots where they move through cortical cells, the females simultaneously depositing eggs as they migrate. Older roots are abandoned and new roots are sought as sites for penetration and feeding.

DISTRIBUTION: Generally distributed wherever oats are grown.

SYMPTOMS: Plants in areas of a field will appear yellow and under moisture stress. Roots and crowns will be rotted when *Rhizoctonia solani* Kuhn infects through nematode wounds. New roots may become dark and stunted with a resultant loss in grain.

CONTROL: There is no practical control.

Stubby Root Nematodes

CAUSE: *Paratrichodorus* spp. are nematodes that survive in soil or on roots as eggs, larvae, or adults. These nematodes feed only on the outside of oat roots and can move relatively rapidly through soil at 5 cm per hour, especially in fine sandy soils.

DISTRIBUTION: Widely distributed in most agricultural soils.

SYMPTOMS: Roots are thickened, short, and stubby. Root tips may have brown lesions. Tops of plants may appear to grow poorly and the entire plant may be pulled easily from the soil due to lack of a fibrous sytem.

CONTROL: No control is reported.

VIRAL CAUSES

African Cereal Streak

CAUSE: African cereal streak virus (ACSV) is limited to the phloem where it induces a necrosis. It is transmitted by the delphacid leafhopper, *Toya catilina* Fennah. The natural virus reservoir is probably native grasses. It is not mechanically or seed transmitted. Disease development is aided by high temperatures.

DISTRIBUTION: East Africa.

SYMPTOMS: Initially faint, broken, chlorotic streaks begin near the leaf base and extend upward. The broken nature of young streaks is clearly defined. Eventually definite alternate yellow and green streaks develop along the entire leaf blade. Later the leaves become almost completely yellow. New leaves tend to develop a shoestring habit and die.

Young infected plants become chlorotic, severely stunted, and die. Seed yield is almost completely suppressed. Plants become soft, flaccid, and almost velvety to the touch.

CONTROL: No control is reported.

American Wheat Striate Mosaic

CAUSE: American wheat striate mosaic virus (AWSMV) is transmitted mainly by the leafhopper, *Endria inimica* Say, and occasionally by *Elymana virescens* F.

DISTRIBUTION: Central United States and Canada.

SYMPTOMS: Leaves have obvious striations consisting of yellow to white parallel streaks. Older leaves are stunted, chlorotic, then necrotic.

CONTROL: No control is reported.

Blue Dwarf of Oats

CAUSE: Oat blue dwarf virus (OBDV) overwinters in infected wild plants. At least 18 different species of plants are infected by the virus. It is transmitted from diseased to healthy plants during the process of feeding by adult aster leafhoppers, *Macrosteles fascifrons* Stal. Immature leafhoppers may occasionally transmit OBDV. OBDV is probably transported into northern oat-growing areas each year from southern areas.

DISTRIBUTION: North central United States and Canada.

SYMPTOMS: Infected plants are dwarfed, uniformly dark blue-green, and remain alive longer than healthy plants. The dwarfed condition may make it easy to overlook infected plants. The leaves, especially the flag leaf, and stems are stiffer, shorter, and stand out at a greater angle from the stem than those of healthy plants. Tillers appear in larger numbers and may form above the crown. Severely blasted heads produce little or no seed. The virus is limited to the phloem.

CONTROL: No control is reported.

Cereal Chlorotic Mottle

CAUSE: Cereal chlorotic mottle virus (CCMV) is a rhabdovirus transmitted by cicadellids.

DISTRIBUTION: Australia and northern Africa.

SYMPTOMS: Severe necrotic and chlorotic streaks occur on leaves.

CONTROL: No control is reported.

Oat Soilborne Mosaic

CAUSE: Oat soilborne mosaic virus (OSBMV) is caused by a soilborne virus that survives in the soil, possibly in the soil fungus *Polymyxa graminis* Led. for up to five years. It is composed of at least two different strains. Circumstantial evidence suggests that OSBMV is transmitted to roots by soil fungi that have invaded the roots for at least two weeks before the fungus is transmitted to the host.

DISTRIBUTION: Europe and North America.

SYMPTOMS: The leaves of less severely infected plants have light green to yellow dashes and streaks and sometimes a necrotic mottling that parallels the axis of the leaf. Other strains of the virus cause eyespot lesions that are spindle-shaped with light green to ash-gray borders and green centers. One strain causes severely infected plants to grow in small rosettes.

CONTROL: Plant resistant cultivars.

Wheat Streak Mosaic

CAUSE: Wheat streak mosaic virus (WSMV) survives in infected plants and has a wide host range that includes, barley, corn, oats, rye, wheat, and several annual and perennial grasses. WSMV is transmitted from plant to plant only by the feeding of the wheat curl mite, *Aceria tulipae* Keifer. Once a mite has picked up the virus from an infected plant, it carries it internally for several weeks. As plants mature, mites migrate to nearby volunteer cereals, grasses, or corn and will infect them with WSMV. However, some grasses are hosts for the mites and not the virus and vice versa; some are susceptible to both, while some are resistant to both.

During late summer or early autumn, the virus may be disseminated to volunteer cereals. The virus may be disseminated from volunteer cereals to cereals planted in the autumn. Mites may be windborne at least 2.4 km.

Only the young mites acquire the virus by feeding one or more minutes on infected plants. Neither the mite nor WSMV can survive longer than one or two days in the absence of a living plant. Active virus has been detected in dead plants or in seed.

DISTRIBUTION: Eastern Europe, western and central North America, and the USSR.

SYMPTOMS: The greatest disease severity ordinarily occurs with autumn-planted oats. Oats planted in the spring are generally not severely infected but may serve as a reservoir for the virus to be carried to other autumn-planted cereals. The first symptoms consist of light green to light yellow blotches, dashes, or streaks in the leaves parallel to the veins. Plants become stunted, show a general yellow mottling, and develop an abnormally large number of tillers that may vary considerably in height. Stunted plants with sterile heads may remain standing after harvest at the same height or shorter than stubble. As infected plants mature, the yellow-striped leaves turn brown and die.

CONTROL:
1. Destroy all volunteer cereals and grasses in adjoining fields two weeks before planting and three to four weeks in the planting field before it is to be used.
2. Plant winter oats as late as practical after the Hessian fly-free date.

Yellow Dwarf

This disease is also called red leaf.

CAUSE: Barley yellow dwarf virus (BYDV) overwinters in perennial grasses and is transmitted by at least 11 species of aphids. Aphids acquire BYDV by feeding on a diseased plant; once BYDV is acquired, aphids are capable of transmitting it for the rest of their lives. One very active aphid feeding for short periods on different plants is a more important vector than several stationary aphids. In the autumn, aphids migrate to winter hosts and spread BYDV to autumn-planted small grains as well as to perennial grasses. A major source of BYDV in Indiana is exogenous aphid populations moving from distant plants in wind currents, especially in the spring. Transmission from local grasses is sporadic and less common.

DISTRIBUTION: Africa, Asia, Australia, Europe, New Zealand, North America, and South America.

SYMPTOMS: BYDV on oats is first seen in plants along the edge of a field. Fields may have a patchy and uneven appearance, varying in size from a few to several km in diameter. The first signs of disease in individual plants is the development of small yellow-green blotches or spots usually near the tips of the leaves. These blotches can be observed more easily if the leaves are held up to light. The blotches enlarge, coalesce, and turn various shades of yellow-red, orange, red, or red-brown. A reddish cast may come to the entire tips of leaves. The orange or red color eventually involves entire blades, progressively from the tip to the base. Leaf margins may show curling inward. The green of the plant may be a darker green than that of a healthy plant. Plants infected early may be severely dwarfed and premature death may result. Heads of diseased plants may have many blasted spikelets depending on the cultivar.

CONTROL:
1. Plant tolerant cultivars. No completely resistant cultivars are known.

2. Plant only spring cultivars.
3. Soil should be at proper fertility level. Vigorously growing plants are more tolerant of BYDV than weaker ones.

PHYSIOLOGICAL DISORDERS

Blast

This disorder is also called blight, blindness, or white star.

CAUSE: Any factor that interferes with the normal development of the oat plant when heads are forming. Late planting, lack of moisture, high temperatures, nutrient imbalance, crowding, disease, or insect attacks, or a combination of factors can cause blast.

DISTRIBUTION: Generally distributed wherever oats is grown.

SYMPTOMS: Spikes have a blighted appearance and fail to develop. These spikelets can be recognized by light color, delicate texture, and lack of grain as soon as a head emerges from its boot. Ordinarily a few spikelets on the lower half of the head are blasted, but occasionally half, and rarely the entire head, is affected.

CONTROL: No control reported.

BIBLIOGRAPHY

Boewe, G. H. 1960. Disease of wheat, oats, barley and rye. *Illinois Nat. Hist. Surv. Circ. 48.*

Books, F. T. 1953. *Plant Diseases.* 2nd ed. Oxford University Press, London.

Brodney, V.; Wahl, I.; and Rotem, J. 1983. Factors affecting the survival of physiologic races of *Puccinia coronata avenae* on *Avena sterilis* in Israel. *Phytopathology* **73:**363 (abstract).

Clement, D. L.; Lister, R. M.; and Foster, J. E. 1986. ELISA-based studies on the ecology and epidemiology of barley yellow dwarf virus in Indiana. *Phytopathology* **76**:86–92.

Cohn, E., and Ausher, R. 1973. *Longidorus cohni* and *Heterodera latipons*, economic nematode pests of oats in Israel. *Plant Dis. Rptr.* **57**:53–54.

Elliott, C. 1927. Bacterial stripe of oats. *Jour. Agric. Res.* **35**:811–824.

Gough, F. J., and McDaniel, M. E. 1974. Occurrence of oat leaf blotch in Texas in 1973. *Plant Dis. Rptr.* **58**:80–81.

Harder, D. E., and Bakker, W. 1973. African cereal streak, a new disease of cereals in east Africa. *Phytopathology* **63**:1407–1411.

Huffman, M. D. 1955. Disease cycle of Septoria disease of oats. *Phytopathology* **45**:278–280.

Jons, V. L. 1986. Downy mildew (*Sclerophthora macrospora*) of wheat, barley, and oats in North Dakota. *Phytopathology* **70**:892 (disease note).

Lockhart, B. E. L. 1986. Occurrence of cereal chlorotic mottle virus in Northern Africa. *Plant Disease* **70**:912–915.

Leukel, R. W., and Tapke, V. F. 1954. Cereal smuts and their control. *USDA Farmers Bull. No. 2069.*

McKay, R. 1957. *Cereal Diseases in Ireland.* Arthur Guiness, Son & Co. Ltd., Dublin.

Poole, D. D., and Murphy, H. C. 1953. Field reaction of oat varieties to Septoria black stem. *Agron. Jour.* **45**:369–370.

Richardson, M. J., and Noble, M. 1970. Septoria species on cereals—a note to aid their identification. *Plant Pathol.* **19**:159–163.

Roane, M. K., and Roane, C. W. 1983. New grass hosts of *Polymyxa graminis* in Virginia. *Phytopathology* **73**:968 (abstract).

Scardaci, S. C., and Webster, R. K. 1982. Common root rot of cereals in California. *Plant Disease* **66**:31–34.

Schaad, N. W.; Sumner, D. R.; and Ware, G. O. 1980. Influence of temperature and light on severity of bacterial blight of corn, oats and wheat. *Plant Disease* **64**:481–483.

Simmonds, P. M. 1955. Root diseases of cereals. *Canada Dept. Agric. Pub. 952.*

Simons, M. D., and Murphy, H. C. 1952. Kernel blight phase of Septoria black stem of oats. *Plant Dis. Rptr.* **36**:448–449.

Simons, M. D., and Murphy, H. C. 1968. Oat diseases and their control. *USDA Agric. Res. Serv. Agric. Hdbk. No. 343.*

Sinha, R. C., and Benki, R. M. 1972. *American Wheat Striate Mosaic Virus.* Commonwealth Mycological Institute. Descriptions of Plant Viruses Set 6. No. 99. Kew, England.

Smith, N. A. 1977. Smuts of oats and their control. *Michigan State Coop. Ext. Serv. Plant Pathol. Newsletter No. 43.*

Timian, R. G. 1985. Oat blue dwarf virus in its plant host and insect vectors. *Plant Disease* **69**:706–708.

Walker, J. 1975. Take-all disease of Graminae, A review of recent work. *Rev. Plant Pathol.* **54**:113–144.

Western, J. H. 1971. *Diseases of Crop Plants.* The Macmillan Press Ltd. London.

11

Diseases of Peanut (*Arachis hypogaea* L.)

BACTERIAL CAUSES

Bacterial Wilt

This disease is also called brown rot and slime disease.

CAUSE: *Pseudomonas solanacearum* (Smith) Smith survives in the soil and may be seedborne. Infection usually occurs through wounds or lenticels on the roots. Disease development is favored by moist soils, heavy clay soils, and soils in which peanuts have been grown for several successive years. The bacterium has a wide host range.

DISTRIBUTION: Generally distributed wherever peanuts are grown. Bacterial wilt is of minor importance in the United States.

SYMPTOMS: Different symptoms of the disease may occur in different peanut growing areas. A symptom from the East Indies is sudden wilting, with leaves remaining attached to dead plants.

Symptoms of the disease in the United States are not as severe. There are a large number of dead roots. Bacterial colonies form throughout the root, main stem, and lower branches and are evident as streaks of brown or black discoloration. Dark brown spots can be seen in the xylem and pith when stems and roots are cut in cross section. Eventually infected tissue is blackened with extensive plugging and necrosis. If young plants are infected, pods will be invaded. Such pods remain small or become wrinkled and develop a spongy or soft decay. When mature plants are infected, there is usually no invasion of the fruit.

CONTROL:
1. Plant resistant cultivars.
2. Treat seed with a seed-protectant fungicide.
3. Plant on light, well-drained soils.
4. Rotate peanuts with resistant crops such as sweet potatoes, small grains, and certain legumes.

FUNGAL CAUSES

Alternaria Spot and Veinal Necrosis

CAUSE: *Alternaria alternata* (Fr.) Keissler.

DISTRIBUTION: India.

SYMPTOMS: Orange-brown necrotic spots occur in the interveinal areas of leaves and extend into veins and veinlets.

CONTROL: No control is reported.

Other Alternaria Reported on Peanuts

Alternaria tenuissima (Fr.) Wiltsh.—Alternaria leaf blight.
A. arachidis Kulk.—Alternaria leaf spot.

Anthracnose

This disease is also called Colletotrichum leaf spot.

CAUSE: *Colletotrichum arachidis* Sawada, *C. dematium* (Pers. ex Fr.) Grove, and *C. mangenoti* Chevaugeon. These fungi likely survive as mycelium in infected residue. Conidia are produced in acervuli and are splashed or windborne to plants.

DISTRIBUTION: Africa, Argentina, India, Taiwan, and the United States.

SYMPTOMS: Different symptoms have been described. When plants are infected by *C. mangenoti*, lesions that are visible on both leaflet surfaces but rarely on petioles or stems are brown-gray and elongate to circular. Lesions may be large and involve up to half the leaflet.

Other descriptions are scattered, circular to irregular lesions with gray-white centers surrounded by dark brown borders. Some workers describe lesions as initially small, water-soaked yellow spots that enlarge to 1–3 mm in diameter and become dark brown. These spots grow rapidly, become irregular, and spread over the entire leaflet. Spots may extend into petioles and branches, causing death of the entire plant.

CONTROL: Control measures are usually not warranted.

Aspergillus Crown Rot

CAUSE: *Aspergillus niger* Van Tiegh and *A. pulverulentus* (McAlp.) Thom survive in plant residue as saprophytes and in soil as mycelium. Conidia are thought to survive in the soil for a period of time. The fungi are also carried on the seed surface and in or under the tissues of the testae. Abundant spores are produced that are airborne for a long distance. Agents capable of transporting soil also serve to disseminate the fungi.

Most infection occurs within 10 days after germination as the elongating hypocotyl comes in contact with soilborne inoculum. Hyphae penetrate directly into the hypocotyl or cotyledon. Disease is more severe when plants are delayed in emergence or predisposed by high soil and air temperatures. The causal fungi grow best in warm, moist conditions. Disease is thought to be more prevalent in soils low in organic matter.

DISTRIBUTION: Generally distributed wherever peanuts are grown.

CONTROL: Seed may be infected when they are placed in a moist environment whereupon they are covered with sooty black masses of spores. More commonly, seed germinates and the elongating hypocotyl is infected. The hypocotyl becomes water-soaked, light brown, and covered by black conidia and mycelium. Eventually the tissue becomes dark brown, then turns lighter in color, and shreds. Necrosis and shredding of tissue may extend up into the branches.

The aboveground symptom is a rapid wilting of the entire plant, especially during dry weather. Recovery of plants may occur during high soil moisture and is related to growth of adventitious roots above the infection site. However, most plants succumb in less than 30 days.

CONTROL:
1. Treat seed with a seed-protectant fungicide.
2. Plant bunch-type cultivars.

Blackhull

CAUSE: *Thielaviopsis basicola* (Berk. & Br.) Ferr. (syn. *Chalara elegans* Nag, Raj & Kendrick) overwinters as chlamydospores in the endocarp of unharvested peanuts. The following season, as the pods deteriorate, chlamydospores germinate and produce mycelium that infect the new pods. Survival is also saprophytically in residue and probably by endoconidia in soil. The fungus is disseminated by any means that will transport soil. Optimum growth of *T. basicola* in vitro is 22–28°C. Blackhull is more severe in wet soils that are neutral or slightly alkaline.

DISTRIBUTION: Occurs in most countries where Spanish and Valencia peanuts are grown.

SYMPTOMS: Infections occur on the external schlerenchymatous shell tissue during the development of fruit and are first seen as tiny black dots. Eventually the shell becomes blackened due to an aggregation of chlamydospores in the developing shell. Dark, crusty patches develop where great numbers of lesions have coalesced. The fungus grows throughout the shell tissue and produces masses of chlamydospores. The internal shell tissue and testae of kernels often show brown discoloration.

CONTROL:
1. Plant only healthy seed treated with a seed-protectant fungicide.
2. Rotate peanuts with grain sorghum, corn, or small grains.
3. Leave soil fallow.
4. Plant as late as feasible in the growing season.

Botrytis Blight

This disease is also called gray mold leaf and stem rot.

CAUSE: *Botrytis cinerea* Pers. ex Fr.; teleomorph *Botryotinia fuckeliana* (deBary) Whetz. survives as sclerotia in soil and residue. Sclerotia germinate to form mycelium or support conidiophores and conidia on sclerotial surfaces. Conidia are abundantly produced on infested tissue and are disseminated by wind or water. Sclerotia are disseminated by any means that will transport soil. Disease is most severe during damp, cool weather (temperature below 20°C) in late autumn. Although *B. cinerea* requires no wounds or necrotic tissue to facilitate infection, senescing, frost-injured, and mechanically injured tissue is prone to colonization. Organic debris on the soil surface serves as an energy source for *B. cinerea* to aid the infection process.

DISTRIBUTION: Japan, Tanganyika, Rhodesia, Nyassaland, the United States, and Venezuela. However, Botrytis blight is seldom a severe problem since climatic conditions that favor disease are rarely present.

SYMPTOMS: Leaves, stems, and subterranean organs are infected. Infected tissues are rapidly decayed and are sparsely covered with dark gray mycelium, conidiophores, and conidia. The infection progresses rapidly down into pegs and fruit. Flattened, black, irregular-shaped sclerotia develop on decayed stems and pods.

CONTROL:
1. Regulate planting dates to prevent plants from growing or maturing during cool, wet weather.
2. Apply foliar fungicides.

Charcoal Rot

CAUSE: *Macrophomina phaseolina* (Tassi) Goid.; anamorph, *Rhizoctonia bataticola* (Taub.) Butler, (syn. *Sclerotium bataticola* Taub.), survives as mycelium and probably pycnidia in residue, as sclerotia in residue and soil, and as mycelium in seeds. Sclerotia can remain viable in dry soil for several years but rapidly lose viability in wet soil. Dissemination is by any means that will transport soil, residue, and seed. Disease is more severe at high soil temperatures (35°C) and low soil moisture.

DISTRIBUTION: Worldwide. The disease is usually of minor importance.

SYMPTOMS: Water-soaked necrotic areas develop on stems at the soil line. The infected areas become dull brown and extend up the stem into the branches and down into the roots. Plants wilt if the stem becomes girdled by a lesion or if roots are decayed. Partial defoliation may also occur. When plants die, a blackening or sooty appearance, caused by numerous sclerotia, occurs over the entire stem. Sclerotia develop profusely in infected plant parts. A few pycnidia that resemble small black pimples may also develop.

Root infection can occur independently of stem rot. At first, roots are blackened and later the tap root is completely rotted.

When peanuts are physically damaged before or after harvest, *M. phaseoli* will grow through shells into kernels. Sclerotia, evident as a black or sooty growth, occurs on internal and external shell surfaces. Infection of leaflets is rare. It consists of large marginal zonate spots in which pycnidia, resembling small black bumps, are found.

CONTROL:
1. Treat seed with a seed-protectant fungicide.
2. Apply soil fungicides to reduce fruit infection.

Choanephora Leaf Spot

CAUSE: *Choanephora* spp.

DISTRIBUTION: Uganda.

SYMPTOMS: Brown lesions originate at the leaflet margin and eventually spread over the entire leaflet. Faint concentric circles are present in the lesion. Abundant sporulation occurs on both leaflet surfaces.

CONTROL: No control is reported.

Cylindrocladium Black Rot

CAUSE: *Cylindrocladium crotalariae* (Loos) Bell & Sobers; teleomorph, *Calonectria crotalariae* (Loos) Bell & Sobers. The fungus probably survives as microsclerotia in residue and soil. Microsclerotia survive better buried in soil than on the soil surface. Primary infection likely originates either from germination of microsclerotia or from mycelium growing from residue. Large microsclerotia (150 μm and larger) induce root rot more efficiently than small microsclerotia. The fungus may be disseminated a considerable distance by windborne plant fragments containing microsclerotia. Long-distance dissemination may also occur as propagules on the surface of seed. During moist conditions, perithecia occur in diseased tissue and ascospores are disseminated by rain or insects; however, their epidemiological importance is thought to be limited to short-range spread.

A soil temperature of 25°C and a moisture content near field capacity is most conducive for infection and root rot. Disease is most severe in areas of heavy clay soils where the moisture-holding capacity is great and waterlogging is common when rainfall occurs. Soil temperatures of 5°C and below generally reduce populations of the fungus. However, at these low temperatures, survival is greater in very moist soil than in dry soil.

DISTRIBUTION: Australia, India, Japan, and the United States.

SYMPTOMS: Symptom expression is usually optimum when moisture stress follows a period of high rainfall earlier in the growing season. The moisture stress is thought to enhance aboveground symptoms due to absence of functional roots after infection and root rot. Plants are stunted, yellowed, and may die. Plants will wilt on warm days and may eventually collapse. However, diseased plants usually do not die but are debilitated. The taproot is blackened and somewhat shredded with few or no lateral roots. Immature pegs and pods also show various stages of blackening. Microsclerotia are formed within infected roots and appear as small, black structures about the size and appearance of ground pepper.

Perithecia that look like small, red-orange, spherical objects appear in dense clusters on stems, pegs, and pods just above or beneath the soil surface. Perithecia are especially abundant in moist areas under dense foliage. After about two to three weeks, during wet weather, ascospores are exuded from each perithecium as a viscous, yellow ooze.

CONTROL:
1. Remove all affected plants prior to harvest in small localized areas.
2. Disc under all diseased plants prior to digging and combining operation. Avoid passing through these areas with equipment during harvest.
3. Dig and remove all diseased plants where feasible and either burn completely or bury in a sanitary landfill.
4. Practice sanitation on all equipment since microsclerotia can be moved in soil left on equipment. Wash equipment with a strong stream of water immediately before equipment is moved from infested areas.
5. Rotate peanuts every three to five years with corn, small grains, or perennial grasses. Do not rotate peanuts with soybeans.
6. Plant resistant cultivars.

Diplodia Collar Rot

CAUSE: *Diplodia gossypina* (Cke.) McGuire & Cooper survives as mycelium and pycnidia in residue. During wet weather conidia are produced in pycnidia and are splashed by water or windborne to plants. The mycelium of the fungus in residue is also disseminated by running water, cultivation, or any means capable of transporting soil. Infection rarely occurs unless plants have been predisposed by heat injuries. The fungus rapidly colonizes heat-injured tissue and grows mostly intercellularly through the cortical parenchyma. After infection is initiated in moribund tissue, adjacent unwounded tissues are readily invaded. Hot, dry weather is reported to favor infection.

DISTRIBUTION: Worldwide; however, it is not considered an important disease.

SYMPTOMS: Plants of all ages are infected. Initially there is a rapid wilting of branches or the entire plant. Plants die within a few days in warm weather. Stem lesions become gray-brown to black and extend toward the taproot. Necrotic stems become shredded.   Infection of a branch usually results in the death of that branch. Numerous pycnidia develop in necrotic tissue and appear as tiny, black, pimplelike dots.

CONTROL:
1. Heat canker may be more prevalent in Runner than in Spanish cultivars because leaf development is greater in the Spanish cultivars, thus providing shade.
2. Do not rotate peanuts with cotton and soybeans.
3. Plow under infected residue.
4. Heat canker can be reduced by planting peanut rows so that plants tend to shade each other.
5. A finely clodded soil surface is most favorable for reducing reflective sunlight energy.

Early Leaf Spot

This disease is also called tikka, viruela, peanut cercosporosis, Mycosphaerella leaf spot, brown leaf spot, leaf spot, and early Cercospora leaf spot.

CAUSE: *Cercospora arachidicola* Hori; teleomorph, *Mycosphaerella arachidis* Deighton, survives as conidia, asci, and mycelium in residue or in the soil. Conidia are produced from mycelium and ascospores produced in asci. The spores are splashed or blown to peanut leaves. Penetration by germ tubes is through open stomata or directly through lateral faces of epidermal cells. Secondary infection occurs from conidia produced in leaf spots that are carried by wind, rain, insects, and machinery. Infection is favored by temperatures between 26°C and 31°C and long periods of high relative humidity. Perithecia are produced in diseased tissue later in the growing season. Disease severity is greater where peanuts follow peanuts in a rotation. Different races of the fungus are present.

DISTRIBUTION: Generally distributed wherever peanuts are grown.

SYMPTOMS: Symptoms occur earlier in the season than with late Cercospora leaf spot. However, *C. arachidicola* is more frequent on common cultivars of *Arachis hypogaea* while *Phaeoisariopsis personata* (Berk. & Curt.) von Arx is more common on wild species; therefore, differences in occurrence of two species may be more closely related to host differences than to time of the growing season. The notation early and late leaf spot may not be necessarily pertinent in all peanut areas where both pathogens occur.

Initially small chlorotic spots occur, which enlarge, become brown to black, subcircular, and 1–10 mm or more in diameter. A chlorotic halo around each spot or lesion is common but not always present. Halos are more distinct on the upper leaflet surface. Lesions may coalesce during periods of wet weather. Sporulation occurs on the upper lesion surface, giving a sooty appearance. Petioles and stems may be infected and will have dark, elongate, somewhat superficial lesions with indistinct margins. Defoliation may occur. Both species of Cercospora may be present in late-season infections.

CONTROL:
1. Rotate peanuts with other crops.
2. Plow under or remove infected crop residue. Destroy volunteer peanut plants.
3. Apply a foliar fungicide during the growing season.

Fusarium Root, Stem, and Pod Diseases

CAUSE: Several species of Fusarium including *F. oxysporum* Schlecht. ex Fr., *F. solani* (Mart.) Appel. & Wr., *F. roseum* Lk. emend. Snyd. & Hans., *F. tricinctum* (Corda) Sacc., and *F. moniliforme* Sheldon. These fungi survive as chlamydospores in the soil and mycelium in residue. Primary inoculum is supplied either by chlamydospores that germinate to produce hyphae or conidia produced from mycelium. Conidia are disseminated by wind, water, or any means that will move soil or residue. Several fusaria are apparently only parasitic on belowground plant parts without being pathogenic. Injury from *Fusarium* spp. is ordinarily not great if plants are in a vigorous growing condition.

DISTRIBUTION: Generally distributed wherever peanuts are grown.

SYMPTOMS: Seedlings that have not emerged have gray, water-soaked tissues that are often overrun with mycelium. *Fusarium solani* apparently causes a dry root rot. The lower taproot is brown to red-brown, withers, and often curls. Secondary roots become brown and slough off. Eventually the disease progresses up to the hypocotyl. By this time infected plants wilt. Sometimes adventitious roots develop above the diseased area and allow the plant to survive.

Older plants may also be infected by *F. solani*. Initially small, elongate, slightly sunken, brown lesions occur on roots just below the crown. Eventually the lesions girdle the root, causing cortical tissue to become shredded, accompanied by chlorosis, wilting, and death of plants.

Wilt caused by *F. oxysporum* is characterized by gray-green leaves that quickly wilt. Tissues beneath the cortex have a brown discoloration.

Fusarium-infected pods usually do not have typical diagnostic symptoms. However, a violet-white discoloration of pods is reported to be somewhat characteristic of Fusarium infection.

CONTROL: Some cultivars apparently have moderate resistance.

Groundnut Leaf Blight

CAUSE: *Myrothecium roridum* Tode ex Fr.

DISTRIBUTION: India.

SYMPTOMS: Lesions on leaves are round to irregular, 5–10 mm in diameter, gray, and surrounded by a chlorotic halo. Lesions eventually coalesce, giving leaves a blighted appearance. Black fruiting bodies that are frequently arranged in concentric rings form on lower and upper leaf surfaces.

CONTROL: No control is reported.

Late Leaf Spot

This disease is also called tikka, viruela, peanut cercosporosis, Mycosphaerella leaf spot, brown leaf spot, leaf spot, and late Cercospora leaf spot.

CAUSE: *Phaeoisariopsis personata* (Berk. & Curt.) von Arx (syn. *Cercospora personata* (Berk. & Curt.) Ellis & Everhart and *Cercosporidium personatum* (Berk. & Curt.) Deighton); teleomorph, *Mycosphaerella berkeleyi* W. A. Jenkins. The fungus survives as conidia, asci, and mycelium in crop residue and mycelium in the soil. Conidia produced on mycelium and ascospores in asci are splashed or blown to peanut leaves. Penetration by the germ tubes is through open stomata or directly through the lateral faces of epidermal cells. Secondary infection occurs from conidia produced in leaf spots that are disseminated by wind, rain, insects, and machinery. Infection is favored by temperatures between 26°C and 31°C and long periods of high relative humidity. Perithecia are produced in diseased tissue later in the growing season.

Disease severity is greater where peanuts follow peanuts in a rotation. Different races of the fungus are present.

DISTRIBUTION: Generally distributed wherever peanuts are grown.

SYMPTOMS: Symptoms occur later in the season than those of early leaf spot. However, *Cercospora arachidicola* Hori is more frequent on common cultivars of *Arachis hypogaea* while *P. personata* is more common on wild species; therefore, differences in occurrence of two species may be more closely related to host differences than to period of the growing season. The notation early and late leaf spot may not be necessarily pertinent in all peanut areas where both pathogens occur.

Initially small chlorotic spots occur that enlarge, become dark brown to black, subcircular, 1–10 mm or more in diameter, and usually do not have a yellow halo. If a halo is present it is usually on the upper leaf surface. The color of the spot on the lower leaf surface tends to be black and is a good diagnostic characteristic to separate symptoms from those of early Cercospora leaf spot. Initially sporulation occurs on the lower leaf surface and later will sparsely occur on the upper leaf surface. Stroma, on which spores are produced, are often arranged in concentric circles and are visibly raised above the lesion surface.

Petioles and stems also may be infected and will have dark, elongate, and somewhat superficial lesions with indistinct margins. Defoliation may occur.

CONTROL:
1. Rotate peanuts with other crops.
2. Plow under or remove infected residue. Destroy volunteer peanut plants.
3. Apply a foliar fungicide during the growing season.

Macrophoma Leaf Spot

CAUSE: *Macrophoma* sp. is a fungus that persists as mycelium and pycnidia in residue.

DISTRIBUTION: Europe.

SYMPTOMS: Lesions are dark and composed of firm necrotic tissue. Marginal necrosis occurs along the apical portions of leaflets. Black pimplelike objects, which are the pycnidia, may be scattered throughout the necrotic tissue.

CONTROL: No control is necessary.

Melanosis

CAUSE: *Stemphylium botryosum* Wallr., teleomorph *Pleospora herbarum* (Pers. & Fr.) Rab. Some workers consider melanosis to be caused by *Macrosporium* sp. and *Alternaria* sp.

DISTRIBUTION: Argentina. Melanosis is considered a minor disease.

SYMPTOMS: Symptoms are most prominent on the lower leaf surface. Spots are irregularly circular to oval (0.5–1.0 mm in diameter), or elongate (1.5 mm long). The spots are dark brown and are often numerous enough to give the impression that the leaflet is covered with fly specks. At first, spots are slightly submerged but eventually become elevated and crustlike with age. Defoliation does not occur even in severe cases.

CONTROL: Some cultivars are more susceptible than others. Control is usually not necessary.

Neocosmospora Root Rot

CAUSE: *Neocosmospora vasinfecta* Smith.

DISTRIBUTION: South Africa.

SYMPTOMS: Taproots and lateral roots are discolored and split with perithecia produced on the infected tissue. External discoloration of the stem occurs about 1 cm above the soil surface and internal discoloration of vascular bundles and pith occurs about 3 cm above the soil surface. Small dark lesions and larger dark cracks occur in the surface tissues of pods; the underlying parenchymatous tissue is discolored brown.

CONTROL: No control is reported.

Olpidium Root Rot

CAUSE: *Olpidium brassicae* (Woron.) Dang.

DISTRIBUTION: India and Texas. Olpidium root rot is a minor disease of peanuts.

SYMPTOMS: Roots are discolored.

CONTROL: No control is needed.

Pepper Spot and Leaf Scorch

CAUSE: *Leptosphaerulina crassiasca* (Sechet) Jackson & Bell probably overwinters as pseudothecia and mycelium in residue. Ascospores are produced in pseudothecia and are forcibly ejected to be windborne a considerable distance. Ascospores become closely attached to the leaflet surface and germinate when free water is available. Germination is favored by a temperature of 28°C and 100 percent relative humidity. Eventually numerous pseudothecia are produced in detached dead leaves.

DISTRIBUTION: Africa, Argentina, Madagascar, India, Taiwan, and the United States.

SYMPTOMS: Symptoms are of two types and occur only on leaves. Pepper spots are small, usually less than 1 mm in diameter, dark brown to black, irregular to circular in outline, and occasionally depressed. Lesions do not rapidly enlarge with age. Spots are more common on the upper leaf surface but also occur on the lower leaf surface. A few spots do not have an obvious deleterious effect; however, when spots are numerous, they tend to coalesce, giving the leaflet surface a netted appearance. In such cases, leaflets normally die and defoliation occurs. Perithecia are abundantly produced in detached leaflets.

Leaflets with scorch symptoms become chlorotic and then necrotic at discrete points along the margins. The necrotic tissue becomes dark brown with a chlorotic zone along the edges. Commonly, lesions develop from the tips of leaflets along a wedge-shaped front toward the petiole. Pseudothecia are abundantly formed in necrotic tissue. Necrotic tissue tends to fragment along leaflet margins, presenting a tattered appearance.

CONTROL:
1. Apply foliar fungicides.
2. Plant resistant cultivars.

Pestalotia Leaf Spot

CAUSE: *Pestalotia arachidicola* Ponte, Silva & Santos.

DISTRIBUTION: Brazil.

SYMPTOMS: Circular, necrotic spots up to 12 mm in diameter occur on leaves.

CONTROL: No control is reported.

Pestalotiopsis Leaf Spot

CAUSE: *Pestalotiopsis arachidis* Satya persists as mycelium in residue. Conidia are produced in acervuli.

DISTRIBUTION: India.

SYMPTOMS: Infected leaves have dark brown, circular spots surrounded by yellow halos. Spots on the lower leaf surface are marked by concentric rings and have prominent black objects, which are acervuli, scattered throughout.

CONTROL: No control is necessary.

Phomopsis Foliar Blight

CAUSE: *Phomopsis sojae* Lehman probably persists as pycnidia or mycelium in infected residue.

DISTRIBUTION: India and the United States.

SYMPTOMS: Necrotic lesions occur on the margin of leaflets that are brown to black, often with a chlorotic zone between healthy and necrotic tissue. The lesion starts at the leaflet tip and grows to the petiole as a V-shaped area. Pycnidia appear as small, black, pimplelike structures in rows parallel to midribs or smaller veins.

Discrete lesions have occurred in the center of leaflets. Such lesions are small (1–10 mm in diameter), circular to irregular in shape, and surrounded by a red-brown margin. The center of the lesion is white to tan with a papery texture. Pycnidia are present in the center.

CONTROL: No control is necessary.

Phyllosticta Leaf Spot

CAUSE: *Phyllosticta arachidis-hypogaea* Vasant Rao and *P. sojicola* Massal. The fungi survive as pycnidia and mycelium in infected residue. The disease is usually more prevalent early in the growing season.

DISTRIBUTION: Generally distributed wherever peanuts are grown. The disease is usually not considered important. More than one disease may be involved.

SYMPTOMS: Different types of symptoms may occur. Lesions are circular to oval, 1.5–5.0 mm in diameter with definite borders. The margin is red-brown with gray to tan centers. Sometimes the center of the lesion falls out, giving the leaflet a shot hole appearance.

Other symptom descriptions note that spots are tan to red-brown with a dark brown margin surrounded by a chlorotic halo. Such lesions are found predominantly near the tips of leaflets and extending along the midrib.

CONTROL: No control is necessary.

Phymatotrichum Root Rot

This disease is also called cotton root rot, Ozonium root rot, and Texas root rot.

CAUSE: *Phymatotrichum omnivora* (Dug.) Henneb. (syn. *P. omnivorum* (Shear) Duggar) survives as sclerotia at soil depths between 30 cm and 75 cm for several years and on weeds. Disease is most severe in alkaline, poorly aerated, black clay soils. Dissemination is primarily by machinery moving infested soil. The fungus is very susceptible to freezing temperatures.

DISTRIBUTION: Southwestern United States. Phymatotrichum root rot is not an important disease of peanuts.

SYMPTOMS: Plants are stunted, chlorotic, and die. Roots are tan with strands of whitish to tan mycelium present.

CONTROL: Control is usually not necessary; however, the following are aids to control:
1. Plow under green manure crops.Crop residues can also be plowed under if nitrogen is applied to promote rapid decay.
2. Rotate peanuts with cereals or grasses.
3. Plow deeply during hot, dry weather, thus exposing and killing the fungus.
4. Provide adequate fertility to ensure vigorous peanut growth.

Powdery Mildew

CAUSE: *Oidium arachidis* Chorin. Disease development is optimum at 25°C.

DISTRIBUTION: Israel, Mauritius, Portugal, and Tanganyika.

SYMPTOMS: Symptoms first occur in midsummer. Portions of the upper leaflet surfaces are covered with a white, powdery growth consisting of the conidia and conidiophores of the fungus. As the disease progresses, the upper leaflet surface becomes covered with large spots that have brown, necrotic centers.

CONTROL: Some cultivars are more resistant than others.

Pythium Diseases

CAUSE: *Pythium myriotylum* Drechs. and other *Pythium* spp. survive as oospores and mycelium in soil and residue. Oospores germinate to produce mycelium that may infect plants directly or indirectly by forming sporangium. Zoospores form within sporangium, are released, and swim through soil water to a host where they encyst. The encysted zoospores germinate to produce hyphae that infect plants. Dissemination is by any means that will transport soil such as water and cultivation equipment.

Peanut diseases caused by *Pythium* spp. are more severe under high soil moisture and temperature conditions. The mite *Caloglyphus rodionovi* Zach. is associated with pod rot by enhancing pathogen spread to adjacent healthy pods and introducing propagules to the peanut pod surface or interior. It is thought that the primary role of mites is as a disseminating agent and not a wounding agent.

DISTRIBUTION: Pythium diseases have been reported from several countries and probably occur wherever peanuts are grown.

SYMPTOMS: Pod rot: Initially pods have a light browning with extensive water-soaking of the tissue. Eventually the entire pod appears watery and shows a brown to black necrosis. Immature pods are usually completely destroyed. Seeds in mature fruit show various degrees of water-soaking and brown to black necrosis. Pegs may become infected as they contact wet soil, causing rotted and blackened peg tips.

Damping off: Initially plants become wilted. Water-soaked necrotic tissue can be seen on the hypocotyl and lateral branches near the soil line. An elongate, sunken, tan lesion may partially or completely circle the stem and extend upward 2–4 cm above the soil line. Infected seedlings frequently topple over at the soil line.

Vascular wilt: At first one or more branches may wilt. Seldom does the entire plant wilt. Soon the foliage on the wilted branches becomes chlorotic and scorched, begining at the leaflet margin and extending inward until the entire leaflet and eventually the entire leaf is dry and crinkled. Petioles often remain green even if petiolules become dry. The vascular system in the taproot-hypocotyl region shows a brown to black discoloration.

CONTROL:
1. Reduce irrigation and allow top soil to dry.
2. Treat seed with a seed-protectant fungicide.
3. Treat soil with a combined fungicide/nematicide.
4. Gypsum has been applied to soil with varying success. Deep plow to bury organic matter along with gypsum application.

Rhizoctonia Diseases

CAUSE: *Rhizoctonia solani* Kuhn survives as sclerotia in soil or residue and as a saprophyte in residue. The teleomorph is *Thanatephorus cucumeris* (Frank) Donk. (syn. *Corticium vagum* Berk. & Curt., *C. solani* (Prill. & Delacr.) Bourd. & Galz. and *Pellicularia filamentosa* (Pat.) Rogers). Most *R. solani* isolates that infect peanut are in anastomosis group AG4 and rarely in AG2.

Rhizoctonia solani is seedborne and is disseminated by any means that will move soil containing sclerotia and mycelium. Basidiospores are either waterborne or windborne. Infection occurs either through wounds or directly through the epidermis. Disease development is favored by temperatures between 19°C and 36°C and moderate soil moisture.

DISTRIBUTION: Generally distributed wherever peanuts are grown.

SYMPTOMS: Plants in field are wilted with one to several branches dying. Pegs and small pods become brown or black at the tips and rot and wither. Pods during all stages of development are subject to infection. Discoloration ranges from slight superficial russeting to browning of the entire pod and decay of the contents. Lesions on pegs are brown to black and vary from slight to extensive sunken areas. Infected seed have discolored, faded, or stained seed coats. Cultivars with pink testae have light brown to gray discoloration.

Seedlings may damp off preemergence. On emerged plants, lesions on hypocotyls are sunken, elongate, dark brown areas, 2–3 cm long. A rapid browning of the hypocotyl sometimes occurs. Similar lesions develop on taproots along with lateral roots.

On older plants sunken dark brown cankers occur on primary roots with browning of the secondary roots. On the stem, dry, sunken, dark brown lesions, several centimeters long, occur near the soil line. The stem is eventually girdled, with the infected area thinner than the non-infected portion. With excess moisture, branches may become infected, and appear brown and shredded.

Lower leaves have temporary brown, speckled, or blotchy areas. These symptoms soon disappear as the leaves die.

CONTROL:
1. Plow under residue.
2. Direct soil away from plant during cultivation.
3. Do not rotate peanuts with beans, soybeans, cotton, or southern peas.
4. Treat seed with a seed-protectant fungicide.
5. Apply fungicide to soil where disease potential is great.

Rhizopus Seed and Preemergence Seedling Rot

CAUSE: *Rhizopus arrhizus* Fischer, *R. oryzae* Went & Prin.-Geerl., and *R. stolonifer* (Ehr. ex Fr.) Vuill survive in residue and soil. The fungi may go through one of two life cycles. As the fungi are heterothallic, one cycle entails sexual reproduction and requires compatible and physiologically distinct mycelium. When compatible mycelium fuse, a zygospore is formed that is capable of survival for lengthy periods of time. The zygospore germinates to form a sporangium in which sporangiospores (chlamydospores) are formed. The sporangiospores may also survive for a period of time or germinate to again form a sporangium in which more sporangiospores are formed. Sporangiospores may be airborne a great distance.

The second life cycle is the germination of sporangiospores to produce sporangium in which still more sporangiospores are formed. Peanut seed and seedlings are infected by mycelium from germinating sporangiospores.

Rhizopus spp. are commonly seedborne on peanut seed. The highest population of *Rhizopus* spp. is in the upper 15 cm of the soil profile. Apparently physiologic strains of the fungi exist that vary considerably in response to temperature.

DISTRIBUTION: Generally distributed wherever peanuts are grown.

SYMPTOMS: Seed and preemerged seedlings are reduced to a dark brown to black, rotted pulpy mass about 36-96 hours after planting. Frequently a mat of mycelium with adhering soil particles envelops each seed. In about five days the seed is indistinguishable from the surrounding soil. Decay is most rapid when infected seed are planted. Cotyledons are usually invaded first, followed by destruction of the primary root. Partial destruction of plumule and cotyledonary laterals results in a stunted seedling.

CONTROL:
1. Plant only healthy seed treated with a seed-protectant fungicide.
2. Deep plow to bury organic litter 7.5–15 cm deep.
3. Control soilborne insects to prevent pod damage.
4. Harvest before pods and fruits become overly mature.
5. Cure as rapidly as possible after harvest and store in cool, dry storage facilities.

Rust

This disease is also known as peanut leaf rust.

CAUSE: *Puccinia arachidis* Speg. is a rust fungus where only the uredial and telia stages are known. The telial stage is uncommon in most areas where peanut rust is found. Urediospores do not survive long in infected residue; therefore, overseasoning probably occurs on plants where there is continuous cultivation of peanuts and on volunteer peanut plants. In North America, the fungus probably overwinters in the West Indies and tropics and is disseminated by windborne urediospores to the southern United States. Other long-range dissemination includes movement of infected residue, pods, and seeds that are externally contaminated with urediospores. Short-range dissemination is by wind, splashing water, and insects. Infection occurs at temperatures in the 20–30°C range and in the presence of moisture. Dew is sufficient to promote infection.

DISTRIBUTION: Most peanut-growing areas of the world.

SYMPTOMS: Plants of all ages are susceptible to infection and pustules develop on all aerial plant parts except flowers. Initially whitish flecks appear on lower leaflet surfaces 8–10 days after inoculation. A few hours later, yellow-green flecks appear on the upper leaflet surface opposite the lower leaflet pustules; however, uredia are more common on the lower leaflet surface. Uredial pustules become visible within the whitish flecks on the lower leaf surface. At first, pustules are yellow, then change to orange to tan to brown. Then they enlarge and rupture within 2 days. Uredia are 0.5–1.4 mm in diameter, circular, and often surrounded by a light green to tan margin. Infection sites may coalesce, causing irregular-shaped patches of uredia. Eventually the tissue surrounding infection sites becomes necrotic and dries up in irregular patches. Infected leaflets may curl, but tend to remain attached to the plant. In highly susceptible genotypes, the original pustules may be surrounded by secondary pustules.

CONTROL:
1. Some cultivars are resistant.
2. Apply foliar fungicides.
3. Do not plant successive peanut crops in same field where viable urediospores are present.
4. Eradicate volunteer plants in a field.
5. Plant at a time that will avoid infection from outside sources.
6. Plant at a time that will avoid environmental conditions conducive to rust buildup.

Scab

CAUSE: *Sphaceloma arachidis* Bit. & Jenk. persists as mycelium in infected residue. Conidia are produced in acervuli. Disease may occur under both dry and humid conditions.

DISTRIBUTION: Argentina and Brazil.

SYMPTOMS: Small (1 mm in diameter), round to irregular spots with sunken centers and raised margins occur on both leaf surfaces beside the principal vein. On the upper leaf surface, spots are tan with narrow brown margins. During humid conditions lesions may be covered with gray velvety growth, which is the conidia and conidiophores of the fungus. After conidia fall away, acervuli become evident as small brown to black objects in the lesion. Spots on the lower leaf surface are pink-brown to red and may have a brown margin.

Petioles and branches have numerous lesions that are up to 3 mm in length. Lesions coalesce to cover extensive areas and appear like cankerous growths, causing distortion of branches and petioles, making them appear wavy or sinuous. Plants look burned and are covered with scabs.

CONTROL: Plant resistant cultivars.

Sclerotinia Blight

CAUSE: *Sclerotinia minor* Jagger and rarely *S. sclerotiorum* (Lib.) deBary. When *S. sclerotiorum* is found, it is in conjunction with *S. minor*. *Sclerotinia minor* overwinters in the soil as sclerotia. Occasionally the fungus may be seedborne. Movement over a short-distance is by infested residue or hay. Long-distance dissemination may be as sclerotia on the surface of contaminated seed. During periods of cool temperatures (18°C) and high relative humidity, sclerotia germinate to form fast-growing white mycelium that infects plant tissue near or in contact with the soil. Germination of sclerotia is favored by a soil pH of 6.0–6.5 and the presence of plant debris within the canopy. Mycelium that emerge from sclerotia at a pH of 6.0–6.5 in the presence of remoistened leaves may be capable of direct infection without an exogenous food base. Apothecia of *S. minor* are rarely observed during the growing season but are common in the spring and autumn on the soil surface. Asci are formed in a layer on the apothecial surface. The ascospores are liberated from the asci and are windborne to nearby plants and germinate under moist conditions. Senescing or mechanically injured leaflets are easily colonized by *S. minor*. However, tissue of this nature is not a prerequisite for infection.

DISTRIBUTION: Widely distributed in most peanut-growing countries of the world.

SYMPTOMS: All parts of the plant including branches, leaves, roots, pegs, and pods are subject to infection. The initial symptom is a rapid wilting of the tips of infected branches. Cankers that are about 2.5 cm wide occur at the soil line on stems and may eventually completely girdle the stem. The diseased area becomes sunken and dry and may have a white, cottony, mycelial growth. Large, black sclerotia will develop either in the mycelium on the stem surface or within the stem. Infection of aerial parts is uncommon.

Infected branch and peg tissue becomes shredded. Severe peg shredding results in pod loss.

CONTROL:
1. Some cultivars are less susceptible than others.
2. Application of foliar fungicides gives partial control.

Southern Stem Rot

This disease is also called foot rot, root rot, Sclerotium blight, Sclerotium rot, Sclerotium wilt, stem rot, southern blight, and white mold.

CAUSE: *Sclerotium rolfsii* Sacc., teleomorph: *Athelia rolfsii* (Curzi) Tu & Kimbrough (syn. *Corticium rolfsii* (Sacc.) Curzi and *Pellicularia rolfsii* West). *Sclerotium rolfsii* survives in soil as sclerotia and as a saprophyte on residue. It is most active near the soil surface possibly due to the presence of an adequate food base and an absence of competitive or antagonistic fungi. Antagonism may be increased by high soil moisture.

Infection occurs either by sclerotia germinating under low relative humidity to produce infection hyphae or mycelium growing through the soil from a precolonized substrate. It has been suggested that sclerotia may not have sufficient reserve energy to establish its mycelium in a living host in the absence of a substrate for saprophytic growth. Growth of mycelium from a food base may therefore seem a more important mode of infection. Eventually sclerotia are formed in mycelium on infected tissue. Temperatures between $-2°C$ and $-10°C$ kill mycelium and germinating sclerotia but not dormant mature sclerotia. Disease is favored by warm and moist environmental conditions that may be accentuated under a dense foliar canopy.

DISTRIBUTION: Generally distributed wherever peanuts are grown.

SYMPTOMS: Stem infection is the most frequent manifestation of the disease. Wilting is caused by infection of stems at the soil line. Initially a branch suddenly wilts, with leaves becoming chlorotic then brown as they dry up. Adjacent branches then become wilted. The infected stem near the soil line is overgrown with white mycelium, particularly if a canopy of leaves is present to maintain high humidity. Mycelium will also grow over the soil surface and organic debris. Eventually all branches on a plant will wilt but dead plants will tend to remain upright. The infected areas of the stem become shredded and numerous sclerotia are produced in the mycelium covering the stem. Sclerotia are spherical and initially are white, velvety appearing, and soft. Eventually they become light brown and brown.

Excessive moisture will prevent mycelium from growing on the outside of the stem. Stem bases are covered with elongate, eroded lesions that are tan to reddish. During dry weather, brown lenticular lesions may occur on stems just below the soil surface.

Peg infection causes light to dark brown lesions (0.5–2.0 cm long) to form, which eventually cause tissue shredding and pod loss. Lesions on young pods of Spanish peanuts are orange-yellow to light tan. Older pods have light brown to black, zonate lesions. Severely decayed kernels are shriveled and lightly covered with mycelium.

CONTROL:
1. Plant resistant cultivars.
2. Bury surface residue by deep plowing.

3. Plant in a flat or slightly raised bed area.
4. Move soil away from the row during cultivation to prevent accumulation of debris around the base of plants.
5. Control early leaf spot and late leaf spot to prevent leaf drop and subsequent accumulation of dead leaves at the base of plants.
6. Apply fungicides to soil before planting.
7. Rotate peanuts with small grains, corn, or cotton.
8. Apply transparent polyethylene sheets to soil in off-season to raise soil temperature where practical.

Verticillium Wilt

This disease is also called Verticillium pod rot.

CAUSE: *Verticillium albo-atrum* Reinke & Berth and *V. dahliae* Kleb. V*erticillium dahliae* is considered the primary causal agent.

Verticillium dahliae survives as microsclerotia that are formed on all infected plant parts. Weeds may also play a part in the fungal persistence in some areas where peanuts are grown. Upon decomposition, microsclerotia are released into the soil or remain embedded in bits of residue for up to several years. Fungal propagules are generally more numerous in the upper strata of soils. Peanut root exudates stimulate them to germinate and infect the root system.

Dissemination is primarily by the movement of soil and infected residue, by machinery, water, wind, and possibly seed.

DISTRIBUTION: Verticillium wilt is generally found in all peanut-growing countries but is not considered a serious problem.

SYMPTOMS: Symptoms usually occur at flowering time as a dull green or chlorotic discoloration of portions of lower leaflets. Eventually many leaflets over the entire plant become withered, brown, and fall from the plant. If adequate moisture is present, infected plants remain alive but are stunted with sparse foliage and are relatively unproductive. Brown to black vascular discoloration can be found in the root, stem, and petioles in advanced stages of the disease. Rot of peanut fruit is characterized by dark, blackened, rotted pods that are usually sprinkled with white powdery patches, which are masses of conidia.

CONTROL:
1. Plant bunch-type peanuts that are more resistant than Valencia and Spanish types.
2. Rotate infested soil to grass, grain sorghum, or alfalfa.
3. Practice clean fallow with occasional plowing during dry periods to deplete amount of soilborne inoculum.

4. Field sanitation, such as burning or removing infested plant debris, would reduce inoculum.
5. Irrigate infested fields frequently to reduce stress.
6. Practice good weed control.
7. Plant resistant cultivars where adapted.

Web Blotch

This disease is also called Ascochyta leaf spot, muddy spot, net blotch, Phoma leaf spot, and spatselviek.

CAUSE: *Phoma arachidicola Marasas, Pauer, & Boerema (syn. Ascochyta adzamethica Schoschiaschvili and A. arachidis Woron.).* The teleomorph has been variously reported as *Mycosphaerella arachidicola* Coch, *M. argentinensis* Frezzi, *Didymosphaeria arachidicola* (Khokh.) Alcorn, Punith. & McCarthy, and *Didymella arachidicola* (Khokh.) Tomilin. The causal fungus likely overwinters as pycnidia, pseudothecia, chlamydospores, and clusters of chlamydospores frequently called microsclerotia. Ascospores and pycnidiospores both serve as primary inoculum. Chlamydospores have also been reported to serve as primary inoculum under experimental conditions. Disease is most severe during cool temperatures and high relative humidity.

DISTRIBUTION: Africa, Asia, Australia, North America, and South America.

SYMPTOMS: Several different symptoms may be present. The most distinct symptom is a webbing or netting pattern caused by fungal strands that grow just beneath the waxy cuticle on the leaf surface. Brown fungal strands radiate from the point of infection in a random fashion, producing the web pattern. Webbing may occur without blotch symptoms but more commonly occurs with blotches.

Blotches are circular, dark brown, with a lighter colored, irregular-shaped margin. Under moist conditions, the blotch area is surrounded by a grayish margin. Varying degrees of webbing may surround the blotch. Another type of symptom that develops under conditions favorable for rapid disease development is a tan spot surrounded by a gray margin made up of fungal strands growing out from the dark center of the spot.

CONTROL:
1. Apply a foliar fungicide during the growing season.
2. Plow under crop residue.
3. Rotate peanuts with other crops.

Yellow Mold

CAUSE: *Aspergillus flavus* Lk. ex Fr. and *A. parasiticus* Speare persist as condia and hyphal fragments in soil or residue. The fungi are disseminated by wind, water, and any means that will transport soil. They are also seedborne. The fungi will grow on practically any plant material in, on, or aboveground. They also grow reasonably well in any type of soil. *Aspergillus flavus* and *A. parasiticus* will infect over a wide range of soil moisture and at an optimum temperature of about 32 C. Populations increase in the lower half of the plow layer following corn. Infection of mature peanuts usually takes place after digging when the moisture content of the peanut is between 12 percent and 35 percent. The fungus may enter the pods through wounds or other injuries made at harvest. The fungi produce the toxin called aflatoxin.

DISTRIBUTION: Generally distributed wherever peanuts are grown.

SYMPTOMS: Cotyledons of germinating seed are invaded first, then the emerging radicle and hypocotyl, which may decay rapidly. Seed and seedlings that have not emerged are shriveled, dried, and become a brown to black mass within four to eight days after planting. Infected plant parts may be covered with a yellow-green growth, which is the spores of the fungus.

Peanuts may be infected before, during, and after harvest, when mature plants are brought aboveground. The fruit becomes highly susceptible to infection, especially if the moisture content of the peanut is high and it has been injured. The peanut will be covered with masses of yellow-green spores.

CONTROL:
1. Use an inverter during the digging operation. Inverting allows peanuts to dry faster and minimizes contact of the soil by fruit.
2. After peanuts are combined, dry them below 12 percent moisture content as soon as possible. Do not allow a wagon full of peanuts to stand without proper drying.
3. Minimize injury to pods during harvest.
4. Store peanuts in cool, dry facilities.

Zonate Leaf Spot

Cause: *Cristulariella moricola* (Hino) Redhead (syn. *C. pyramidalis* Waterman & Marshall). The sclerotial stage is *Sclerotium cinnamomi* Sawada. The fungus has been associated with leaf spots on several deciduous trees.

DISTRIBUTION: Georgia. The disease is considered of minor importance.

SYMPTOMS: Symptoms occur late in the season. Necrotic spots that vary between 1–13 mm in diameter occur on leaves. Small lesions have a light brown center surrounded by a darker brown ring of necrotic tissue. Large lesions have a zonate appearance on both leaf surfaces.

CONTROL: No control is reported.

MYCOPLASMAL CAUSES

Witches' Broom

CAUSE: A mycoplasmalike organism.

DISTRIBUTION: China, India, Indonesia, Japan, Taiwan, and Thailand.

SYMPTOMS: Plants are bushy due to excessive proliferation of axillary shoots. Leaflets are chlorotic and smaller than normal. Pegs tend to grow upward. Yields may be greatly reduced.

CONTROL: No control is reported.

CAUSED BY NEMATODES

Kalahasti Malady or Pod Lesion Nematode

CAUSE: *Tylenchorhynchus brevilineatus* Williams (syn. *T. indicus* Siddigi).

DISTRIBUTION: India.

SYMPTOMS: Small, brown-yellow lesions occur on the pegs and pod stalks and on young developing pods. The margins of the lesions are slightly elevated. Pod stalks are greatly reduced in length, and in advanced stages of the disease, the pod surface is completely discolored.

Plants are stunted and root growth is reduced at maturity. Pods are reduced in size but kernels from such pods are apparently healthy.

CONTROL: Application of nematicide in research plots has been effective.

Ring Nematode

CAUSE: *Criconemella ornata* (Raski & Luc). The nematode is thought to be ectoparasitic.

DISTRIBUTION: United States.

SYMPTOMS: Plants are chlorotic. Roots, pods, and pegs are discolored with brown necrotic lesions that are superficial if small, but extend deep into tissue if large. Young roots and root primordia may be killed. Pod yields are reduced.

CONTROL:
1. Rotate peanuts with nonhost crops.
2. Apply nematicides.

Root Knot Nematodes

CAUSE: The peanut root knot nematode *Meloidogyne arenaria* (Neal) Chitwood and the northern root knot nematode *M. hapla* (Kofoid & White) Chitwood. Additionally the Javanese root knot nematode, *M. javanica* (Treub.) Chitwood has been reported to attack peanuts. Infested soil contains eggs that produce larvae. The larvae move through the soil and penetrate the root tip. The larvae migrate through the root to the vascular tissues, become sedentary, and commence feeding. Excretions of the larvae stimulate cell proliferation, resulting in the development of syncythia from which the larvae derive their food. Females lay eggs that may be pushed out of the root into the soil.

DISTRIBUTION: United States.

SYMPTOMS: Mild symptoms caused by *M. arenaria* may show a slight yellowing of the foliage and imperceptible stunting of the plant. Severe infection causes plants to be noticeably stunted, yellowed, wilt, and die during dry weather. Roots and pegs have various-size galls at many points along their lengths. Galls may be several times the diameter of a normal root or peg. Pods develop knobs, protuberances, or small warts. Galls on roots, pegs, and pods sometimes begin to deteriorate by maturity. The necrotic tissue due to the deteriorating gall may be colonized by a number of fungi.

The northern root knot nematode causes similar aboveground symptoms. Galls are usually smaller than those caused by *M. arenaria*. Additionally infected roots tend to branch near the point of invasion, resulting in a dense, bushy root system. Total crop loss may occur when infection by both nematodes is severe.

CONTROL:
1. Rotate peanuts every two to three years with a nonlegume.
2. Apply nematicides either banded at time of planting or as a fumigant, using plow-sole method during deep plowing soil prior to planting.

Root Lesion Nematodes

CAUSE: *Pratylenchus brachyurus* (Godfrey) Filipjev & Schwer.-Stek. and *P. coffeae* (Zimmerman) Schwer.-Stek. Both adults and larvae can infect roots, pegs, and pods, with the higher nematode populations usually present on pegs and pods. The nematodes directly penetrate the organ and feed on parenchyma tissue.

Nematode numbers may increase on the roots of other plants, especially rye, during winter months.

DISTRIBUTION: United States.

SYMPTOMS: Symptoms appear in circular areas in a field. Plants are stunted and yellowed. Roots are smaller and discolored. Pegs have brown lesions. Pods have brown to black angular lesions. Initially the spots are small, tan, with a dark center. Spots may be so numerous as to discolor the entire pod. Damage caused by nematodes allows secondary fungi to enter, causing further discoloration and decomposition of roots, pods, and pegs.

CONTROL:
1. Apply nematicides either banded at time of planting or as a fumigant, using plow-sole method during deep plowing soil prior to planting.
2. Fallowing reduces nematode population.

Sting Nematodes

CAUSE: *Belonolaimus longicaudatus* Rau and *B. gracilis* Steiner are nematodes with a wide host range. The nematodes feed on the outside of the plant without becoming attached to belowground parts. Nematodes feed at root tips and along the sides of succulent roots and other belowground parts. They are restricted to soils with 84–94 percent sand and occur mostly in the upper 30 cm of soil. They are most active at temperatures between 20°C and 34°C.

DISTRIBUTION: United States.

SYMPTOMS: Plants are stunted, chlorotic, and have stubby, sparse roots. Roots and pods may have small, dark, necrotic spots.

CONTROL:
1. Rotate peanuts every two to three years with a nonlegume.
2. Apply nematicides either banded at time of planting or a fumigant, using plow-sole method during deep plowing soil prior to planting.

Miscellaneous Nematodes Reported on Peanuts

Aphasmatylenchus straturatus Germani
Aphelenchoides arachidis Bos.
Helicotylenchus sp.
Paratrichodorus christei (Allen) Siddigi
Radopholus similis Thorne
Rotylenchus reniformis Lindford & Olivera
Tylenchorhynchus sp.
Xiphinema americanum Cobb and *X. diversicaudatum* (Micoletzky) Thorne

RICKETTSIAL CAUSES

Rugose Leaf Curl

CAUSE: A rickettsia-like organism. The causal organism is transmitted by leafhoppers.

DISTRIBUTION: Australia.

SYMPTOMS: Leaflets are distorted and puckered.

CONTROL: No control is reported.

VIRAL CAUSES

Bean Yellow Mosaic

CAUSE: An isolate of bean yellow mosaic virus (BYMV). BYMV is transmitted mechanically and by the cowpea aphid, *Aphis craccivora* Koch, in a nonpersistent manner. Arrowleaf clover, *Trifolium vesiculosum* Savi, is a reservoir for BYMV.

DISTRIBUTION: United States, particularly Georgia and Texas.

SYMPTOMS: Initial symptoms are chlorotic rings and spots of the leaves. After two to three weeks the symptoms disappear.

CONTROL: No control is reported.

Cowpea Chlorotic Mottle

CAUSE: Cowpea chlorotic mottle virus (CCMV).

DISTRIBUTION: Georgia.

SYMPTOMS: Symptomless; however, CCMV is reported to have significant disease potential when in mixed infections with other viruses.

CONTROL: No control is reported.

Cowpea Mild Mottle

CAUSE: Cowpea mild mottle virus (CMMV) is transmitted by sap and the whitefly, *Bemisia tabaci* (Genn.).

DISTRIBUTION: India. It is not considered a serious disease.

SYMPTOMS: The youngest leaves show a vein clearing followed by a downward rolling. Later leaves and petiole become necrotic and defoliation occurs. Plants are severely stunted and rarely produce pods. Plants in a field will show vein banding of youngest leaves and necrosis of older leaves.

CONTROL: Avoid planting peanuts near soybean or cowpea fields.

Cucumber Mosaic

CAUSE: A strain of cucumber mosaic virus (CMV-CA). The virus is transmitted by seed and the aphid *Macrosiphum euphorbiae* (Thomas). The virus infects 31 species of plants in 6 families.

DISTRIBUTION: China.

SYMPTOMS: Chlorotic spots occur on young emerging leaves. Other symptoms are chlorosis of young expanded leaves that are smaller and rolled and a mosaic or mottling of some leaves. Infected plants are stunted and are about one-half to two-thirds the size of a healthy plant.

CONTROL: No control is reported.

Groundnut Crinkle

CAUSE: Groundnut crinkle virus (GCV) is transmitted mechanically and possibly by whiteflies.

DISTRIBUTION: Ivory Coast.

SYMPTOMS: Plants are slightly stunted with a slight reduction in leaf size. Leaves show faint mottling and crinkling of lamia.

CONTROL: No control is reported.

Groundnut Eyespot

CAUSE: Groundnut eyespot virus (GEV).

DISTRIBUTION: West Africa. It is not considered a serious disease.

SYMPTOMS: Leaves show yellow eyespots. Vein banding along main veins may occur.

CONTROL: No control is reported.

Groundnut Streak

CAUSE: Groundnut streak virus (GSV).

DISTRIBUTION: Ivory Coast.

SYMPTOMS: Seedlings have necrotic streaks along leaflet veins. As plants mature, the streaks disappear.

CONTROL: No control is reported.

Marginal Necrosis

CAUSE: Possibly a virus. The disease agent is seed transmitted.

DISTRIBUTION: Papua and New Guinea.

SYMPTOMS: Leaflets have marginal necrosis and crinkling.

CONTROL: No control is reported.

Peanut Clump

CAUSE: Peanut clump virus (PCV). Two different isolates have been identified: the West African and Indian. *Polymyxa graminis* Led. transmits the West African PCV and is thought to transmit the Indian isolate. Both isolates are seed transmitted.

DISTRIBUTION: India and West Africa.

SYMPTOMS: In a field the disease occurs in spots that enlarge in succeeding years. Plants are stunted and new quadrifoliates have mosaic mottling and chlorotic rings. Subsequently, infected leaves are dark green with faint mottling. Plants eventually become bushy with small, dark, green leaves. Several flowers are produced on erect petioles. The number and size of pods are reduced, resulting in smaller-size seeds. Root systems are smaller. Infected roots are dark with an epidermal layer that easily peels off.

CONTROL:
1. Plant virus-free seed.
2. Plant resistant cultivars in India.

Peanut Green Mosaic

CAUSE: Peanut green mosaic virus (PGMV) can be mechanically transmitted.

DISTRIBUTION: India.

SYMPTOMS: Initially chlorotic spots and vein clearing occur on young quadrifoliates. Later a severe mosaic occurs.

CONTROL: No control is reported.

Peanut Mild Mottle

CAUSE: Peanut mild mottle virus (PMMV). The virus is transmitted mechanically and by the aphids *Aphis craccivora* Koch and *Myzus persicae* Sulzer in a nonpersistent manner.

DISTRIBUTION: China.

SYMPTOMS: Plants display a mild mottle of leaves. Top leaflets frequently have distinct chlorotic spots or ring spots. After several days spots disappear and leaves show the typical mottling with dark green islands on a light green background. Most leaf symptoms disappear during high temperatures. Plants are not stunted but yields are reduced.

CONTROL: No control is reported.

Peanut Mottle

CAUSE: Peanut mottle virus (PMV) is disseminated mechanically and by several aphid species. It is also seedborne at a low frequency; however, this is an important means of spread. The virus is located in the embryo and not in the seed coat or cotyledon. Wild peanut, *Arachis chacoense* Krap., serves as a reservoir for PMV. Different strains of PMV exist.

DISTRIBUTION: Worldwide.

SYMPTOMS: Affected plants have a leaf mottling, with the leaflets curling upward and a depression of interveinal tissue. A mild mottle occurs on the youngest leaves that is best observed by transmitted light. Symptoms tend to become obscure as plants mature, particularly during hot, dry weather. Kernels are discolored and pods may be smaller than normal with gray to brown patches. Yield is reduced. Nodulation is reduced.

Symptom expression varies with strains of Rhizobium producing nodules on the plant. Plants harboring an ineffective Rhizobium strain show more severe symptoms than plants harboring an effective strain.

CONTROL: Plant the most tolerant cultivar.

Peanut Stripe

CAUSE: Peanut stripe virus (PStV). The virus is transmitted mechanically, by aphids, particularly *Aphis craccivora* Koch, in a nonpersistent manner and through seed.

DISTRIBUTION: Primarily in research areas of Florida, Georgia, North Carolina, Texas, and Virginia, and in the Philippines and Thailand.

SYMPTOMS: Initially there is a discontinuous dark striping or banding along lateral veins of leaves that in many cases resembles sergeant stripes. Later the striping fades and a mild oak-leaf pattern occurs.

Infrequently dark green circular areas not associated with veins will develop. This is called peanut blotch and is considered a symptom variant of PStV.

CONTROL: The University of Georgia has issued the following guidelines for controlling PStV in Georgia:
1. With the exception of breeding and variety performance tests, only Georgia-certified seed shall be used to plant research plots.
2. Seed from research plots shall not be processed through commercial cleaning and shelling operations.
3. Peanuts from experimental plots shall be processed or sold for processing only and are not to be used for seed except in breeding tests.
4. All residual peanut seed and debris shall be removed from harvesting, handling, and transporting equipment before such items are removed from areas contaminated with PStV to clean areas.

5. All research plots containing plants inoculated with virus shall be grown under screen cages with proper precautions to prevent spread and shall be planted only on experiment station land.
6. Breeders shall release only seed that is free from PStV.
7. Virus-free seed to be retained for future seed production shall not be planted in areas where PStV-infected peanuts have been grown previously.
8. Virus-free seed shall not be planted in proximity to leguminous crops, or other hosts, and rigid weed control shall be practiced in and around virus-free plots.
9. Researchers shall not import peanuts unless they are either tested for PStV before planting or grown in screened isolation for one growing season to allow for visual inspection and serological assay if necessary.
10. Peanut-breeding plots shall be isolated from other research plots.
11. The Uniform Peanut Performance Trial shall be planted at one university farm exclusively.
12. Other legume crop breeding nurseries shall be separated from peanut research plots.
13. All seed shall be removed from PStV-infected research plots and plots shall be fumigated.
14. Infected seed except breeders' seed shall be destroyed.
   Other:
15. Some peanut accessions are resistant.

Peanut Stunt

CAUSE: Peanut stunt virus (PSV) overwinters primarily in white clover, *Trifolium repens* L., and in crown vetch, *Coronella varia* L., and other legumes. It can be transmitted by mechanical means, grafting, and dodder. Three species of aphids transmit PSV in a nonpersistent manner. The virus is also rarely seedborne but this is not thought to be a factor in transmission. PSV has a wide host range.

DISTRIBUTION: Europe, Japan, Morocco, Sudan, and the United States.

SYMPTOMS: Plants are severely dwarfed with malformation of foliar parts and suppression of fruit development. Stunting may involve the entire vine or only a portion of the plant. Mature normal leaves range from a few to one-half of the leaves on a stem. Apical to the normal leaves are stiff, erect, and pointed leaves with light green leaflets that are less than half normal size. Discoloration is variable; chlorosis of the entire leaf or plant part may occur or there may be a mild mottling, or chlorosis with a dark green vein banding, or leaves may be slightly chlorotic compared to nonaffected leaves. Additionally affected leaves may show upward curling. Fruits are small, poorly shaped, and the pericarps may be split.

CONTROL:
1. Do not plant peanuts in the vicinity of legumes.
2. Rogue out infected plants.

Peanut Yellow Mottle

CAUSE: Peanut yellow mottle virus (PYMV).

DISTRIBUTION: Nigeria.

SYMPTOMS: Young plants have a bright yellow mottle. Yields are reduced.

CONTROL: No control is reported.

Rosette

CAUSE: Groundnut rosette virus (GRV). Some workers attribute the disease to two viruses: the groundnut chlorotic rosette virus (GCRV) and the groundnut green rosette virus (GGRV). Both are transmitted by the aphid *Aphis craccivora* Koch in a persistent manner. Additionally, both are transmitted mechanically to other hosts. The rosette virus is not seed transmitted.

DISTRIBUTION: Africa, specifically Gambia, Senegal, Madagascar, Nigeria, Sierra Leone, Uganda, and Tanganyika. Additionally, rosette has been found in Java.

SYMPTOMS: There are two main types of rosette. One is called chlorotic rosette, which predominates in eastern and southern Africa. The plant is extremely stunted and rosetted and the leaves are uniformly chlorotic.

The other type of rosette is called green rosette and is most common in West Africa. The plants are stunted and rosetted but the leaves are a normal green except for an occasional chlorotic fleck.

Rosette is characterized by an overall stunting of the plant. Petioles and internodes are shortened, giving the plant a typical rosette or clumped appearance. Some leaves, especially the younger ones, are more or less chlorotic and faintly mottled. The first leaves formed after infection are pale yellow, with dark green veins. Successive leaves are smaller, curled, distorted, uniformly yellow and without green veins.

Eventually most of the leaves turn green and eventually appear normal. Early infection causes small sessil flowers that do not open.

CONTROL:
1. Plant a resistant cultivar.
2. Eradicate volunteer plants and plants missed during harvest.
3. Plant early in the growing season at a high plant density.

Spotted Wilt

CAUSE: Tomato spotted wilt virus (TSWV) overwinters in weeds and is transmitted by thrips. The virus is also graft transmissable but not seed transmitted.

DISTRIBUTION: Australia, Brazil, India, Nigeria, South Africa, and the United States.

SYMPTOMS: Plants are severely stunted to less than half the size of healthy ones if early infection occurs. Distinct ring spots appear on older leaves and new leaves are small and rounded. Leaves are closer together than normal, resulting in a stubby, bunchy growth. Growing points of stems are yellow and stunted and the number of flowers is reduced. Certain isolates of TSWV cause a bud necrosis that may spread to petioles and stems, sometimes causing death of the plant. A general chlorosis of plants also occurs. Pods are reduced in size and distorted in shape.

CONTROL:
1. Plant at time of year to avoid thrips.
2. Plant the most tolerant cultivars.
3. Plant at an increased plant density.

BIBLIOGRAPHY

Adams, D. B., and Kuhn, C. W. 1977. Seed transmission of peanut mottle virus in peanuts. *Phytopathology* **67**:1126–1129.

Baard, S. W., and Laubscher, C. 1985. Histopathology of black-hull incited by *Thielaviopsis basicola* in groundnuts. *Phytophylactica* **17**:85–88.

Baard, S. W., and Van Wyk, P. S. 1985. *Neocosmospora vasinfecta* pathogenic to groundnuts in South Africa. *Phytophylactica* **17**:49–50.

Balasubramanian, R. 1979. A new type of alternariosis in *Arachis hypogaea* L. *Curr. Sci.* **48**:76–77.

Bays, D. C., and Demski, J. W. 1986. Bean yellow mosaic virus isolate that infects peanut (*Arachis hypogaea*). *Plant Disease* **70**:667–669.

Behncken, G. M. 1970. The occurrence of peanut mottle virus in Queensland. *Aust. Jour Agric. Res.* **21**:465–492.

Bell, D. K.; Locke, B. J.; and Thompson, S. S. 1973. The status of Cylindrocladium black rot of peanut in Georgia since its discovery in 1965. *Plant. Dis. Rptr.* **57**:90–94.

Bell, D. K., and Sobers, E. K. 1966. A peg, pod and root necrosis of peanuts caused by a species of Calonectria. *Phytopathology* **56**: 1361–1364.

Black, M. C., and Beute, M. K. 1984. Relationships among inoculum density, microsclerotium size and inoculum efficiency of *Cylindrocladium crotalariae* causing root rot of peanuts. *Phytopathology* **74**:1128–1132.

Chappell, W. E. 1977. Diseases in peanuts, in *Pest Management Guide for Peanuts*. Virginia Polytechnic Institute and State University Pest Management Series 4.

Chohan, H. S., and Gupta, V. K. 1968. Aflaroot, a new disease of groundnut, caused by *Aspergillus flavus*. *Indian Jour Agric. Sci.* **38**:568–570.

Clinton, P. K. S. 1962. The control of soilborne pests and diseases of groundnuts in the Sudan central rainlands. *Emp. Jour Exp. Agric.* **30**:145–154.

Culver, J. N. 1987. Resistance to peanut stripe virus in Arachis germ plasm. *Phytopathology* **77**:640 (abstract).

Demski, J. W., and Lovell, G. R. 1985. Peanut stripe virus and the distribution of peanut seed. *Phytopathology* **69**:734–738.

Demski, J. W.; Reddy, D. V. R.; and Lowell, G., Jr. 1984. Peanut stripe, a new virus disease of peanut. *Phytopathology* **74**:627 (abstract).

Diener, U. L. et al. 1965. Invasion of peanut pods in the soil by *Aspergillus flavus*. *Plant Dis. Rptr.* **49**:931–935.

Dubern, J., and Dollett, M. 1979. Groundnut crinkle, a new virus disease observed in Ivory Coast. *Phytopathol. Zeitschr.* **95**:279–283.

Garren, K. H., and Jackson, C. R. 1973. Peanut diseases, in *Peanuts, Culture and Uses*, pp. 429–494. Peanut Research and Education Association, Stillwater, OK.

Griffin, G. J.; Garren, K. H.; and Taylor, J. D. 1981. Influence of crop rotation and minimum tillage on the population of *Aspergillus flavus* group in peanut field soil. *Plant Disease* **65**:898–900.

Grinstein, A.; Katan, J.; Abdul Razik, A.; Zeydan, O.; and Elad, Y. 1979. Control of *Sclerotium rolfsii* and weeds in peanuts by solar heating of soil. *Plant Dis. Rptr.* **63**:1056–1059.

Halliwell, R. S., and Philley, G. 1974. Spotted wilt of peanut in Texas. *Plant Dis. Rptr.* **58**:23–25.

Hau, F. C.; Beute, M. K.; and Smith, T. 1982. Effect of soil pH and volatile stimulants from remoistened peanut leaves on germination of sclerotia of *Sclerotinia minor*. *Plant Disease* **66**:223–224.

Hsi, D. C. H. 1965. Blackhull disease of Valencia peanuts. *New Mexico State Univ. Res. Rpt. 110.*

Hsi, D. C. H. 1976. Chemical control of peanut blackhull and its causal fungus, *Thielaviopsis basicola. New Mexico State Univ. Agric. Exp. Sta. Bull. 664.*

Jackson, C. R. 1962. Aspergillus crown rot in Georgia. *Plant Dis. Rptr.* **46**:888–892.

Jackson, C. R., and Bell, D. K. 1969. Diseases of peanut (groundnut) caused by fungi. *Univ. Georgia Coll. Agric. Exp. Sta. Res. Bull. 56.*

Johnson, G. I. 1985. Occurrence of *Cylindrocladium crotalariae* on peanut (*Arachis hypogaea*) seed. *Plant Disease* **69**:434–436.

Kucharek, T. A. 1975. Reduction of Cercospora leafspots of peanut with crop rotation. *Plant Dis. Rptr.* **59**:822–823.

Kuhn, C. W., and Demski, J. W. 1987. Latent virus in peanut in Georgia identified as cowpea chlorotic mottle virus. *Plant Disease* **71**:101 (disease notes).

Lana, A. F. 1980. Properties of a virus occurring in *Arachis hypogaea* in Nigeria. *Phytopathol. Zeitschr.* **97**:169–178.

Lutterell, E. S. 1981. Peanut web blotch: Symptoms and field production of inoculum. *Phytopathology* **71**:892 (abstract).

Pataky, J. K., and Beute, M. K. 1983. Effects of inoculum burial, temperature, and soil moisture on survival of *Cylindrocladium crotalariae* microsclerotia in North Carolina. *Plant Disease* **67**:1379–1302.

Phipps, P. M., and Beute, M. K. 1977. Influence of soil temperature and moisture on severity of Cylindrocladium black rot in peanut. *Phytopathology* **67**:1104–1107.

Porter, D. M. 1970. Peanut wilt caused by *Pythium myriotylum. Phytopathology* **60**:393–394.

Porter, D. M.; Garren, K. H.; Mozingo, R. W.; and Van Schaik, P. H. 1971. Susceptibility of peanuts to *Leptosphaerulina crassiasca* under field conditions. *Plant Dis. Rptr.* **55**:530–532.

Porter, D. M.; Smith, D. H.; and Rodriguez-Kubana, R. (ed.) 1984. *Compendium of Peanut Diseases.* American Phytopathological Society, St. Paul, MN, 73p.

Purss, G. S. 1962. Peanut diseases in Queensland. *Queensland Agric. Jour.* **88**:540–543.

Reddy, D. D. R.; Subrahmanyam, P.; Sankara Reddy, G. H.; Raja Reddy, C.; and Siva Rao, D. V. 1984. A nematode disease of peanut caused by *Tylenchorhynchus brevilineatus. Plant Disease* **68**:526–529.

Reddy, D. V. R.; Wongkaew, S.; and Santos, R. 1985. Peanut mottle and peanut stripe virus diseases in Thailand and the Philippines. *Plant Disease* **69**:1101 (disease note).

Rothwell, A. 1962. Diseases of groundnuts in southern Rhodesia. *Rhodesia Agric. Jour* **59**:199–201.

Rowe, R. C., and Beute, M. K. 1973. Susceptibility of peanut rotational crops (tobacco, cotton and corn) to *Cylindorcladium crotalariae. Plant Dis. Rptr.* **57**:1035–1039.

Rowe, R. C.; Johnston, S. A.; and Beute, M. K. 1974. Formation and dispersal of *Cylindrocladium crotalariae* microsclerotia in infected peanut roots. *Phytopathology* **64**:1294–1297.

Sandborn, M. R., and Melouk, H. A. 1983. Isolation and characterization of mottle virus from wild peanut. *Plant Disease* **67**:819–821.

Sharma, N. D. 1974. A previously unrecorded leafspot of peanut caused by *Phomopsis* sp. *Plant Dis. Rptr.* **58**:640.

Shew, H. D., and Beute, M. K. 1979. Evidence for the involvement of soilborne mites in Pythium pod rot of peanut. *Phytopathology* **69**:204–207.

Smith, D. H. 1972. *Arachis hypogaea*: A new host of *Cristulariella pyramidalis*. *Plant Dis. Rptr.* **56**:796–797.

Sobers, E. K., and Littrell, R. H. 1974. Pathogenicity of three species of Cylindrocladium to select hosts. *Plant Dis. Rptr.* **58**:1017–1019.

Subrahmanyam, P.; Reddy, L. J.; Gibbons, R. W.; and McDonald, D. 1985. Peanut rust: A major threat to peanut production in the semiarid tropics. *Phytopathology* **69**:813–819.

Wadsworth, D. F., and Melouk, H. A. 1985. Potential for transmission and spread of *Sclerotinia minor* by infected peanut seed and debris. *Plant Disease* **69**:379–381.

Weber, G. F. 1973. *Bacterial and Fungal Diseases of Plants in the Tropics*. University of Florida Press, Gainesville.

Wills, W. H., and Moore, L. D. 1973. Pathogenicity of *Rhizoctonia solani* and *Pythium myriotylum* from rotted pods to peanut seedlings. *Plant Dis. Rptr.* **57**:578–582.

Wongkaew, S., and Peterson, J. F. 1983. Peanut mottle virus symptoms in peanuts inoculated with different Rhizobium strains. *Plant Disease* **67**:601–603.

Woodard, K. E., and Jones, B. L. 1983. Soil populations and anastomosis groups of *Rhizoctonia solani* associated with peanut in Texas and New Mexico. *Plant Disease* **67**:385–387.

Xu, Z., and Barnett, O. W. 1984. Identification of a cucumber mosaic virus strain from naturally infected peanuts in China. *Plant Disease* **68**:386–389.

Xu, Z.; Yu, Z.; Liu, J.; and Barnett, O. W. 1983. A virus causing peanut mild mottle in Hubei Province, China. *Plant Disease* **67**:1029–1032.

12

Diseases of Rapeseed (Canola) and Mustard

Included are summer turnip rapeseed, *Brassica campestris* L.; Argentine-type rapeseed, *B. Napus* L.; Oriental and brown mustard, *B. juncea* L.; and yellow mustard, *B. hirta* Moench.

BACTERIAL CAUSES

Bacterial Wilt and Rot

CAUSE: *Erwinia carotovora* pv. *atroseptica* (van Hall) Dye (syn. *E. atroseptica* (van Hall) Jennison).

DISTRIBUTION: Mexico.

SYMPTOMS: The central part of the stem may be partially or completely rotted at flowering or before, causing an infected plant to wilt. Wilting is associated with a mottled leaf appearance. The rot starts where larvae of *Hylemyia* sp. damages the stem at soil level or where branching occurs.

CONTROL: No control is reported.

Black Rot

CAUSE: *Xanthomonas campestris* (Pammel) Dowson is seedborne and overwinters in infected residue in the soil. Bacteria are usually splashed by water and enter plants through stomates. Little is known of the life cycle of *X. campestris* as it relates to disease development on rapeseed and mustard.

DISTRIBUTION: Canada; however, black rot is not a common disease.

SYMPTOMS: Leaf tissue becomes chlorotic with veins in the chlorotic area dark in color.

CONTROL: The disease is not important enough to warrant control measures.

FUNGAL CAUSES

Alternaria Black Spot

This disease is also called gray leaf spot, pod drop, and dark leaf spot.

CAUSE: *Alternaria brassicae* (Berk.) Sacc. (syn. *A. brassiciola* (Schw.) Wiltsh.) and *A. raphani* Groves & Skolko survive in infected residue, probably as mycelium, and are seedborne as conidia and mycelium on the outside of the seed coat. Primary inoculum is from conidia that are produced on infected residue and are windborne to host tissue. Infected seed may cause preemergence or postemergence damping off. Hypocotyls and cotyledons of surviving seedlings may be infected with conidia produced on damped-off seedlings. Secondary inoculum also comes from conidia produced on leaf spots. Black spot becomes most severe under humid, warm conditions that occur during midseason.

DISTRIBUTION: Canada, Europe, Mexico, and the United States.

SYMPTOMS: Initially brown to black dots occur on leaves. Under humid conditions the dots enlarge to form gray, circular spots containing concentric rings with a purple or black border. Under dry conditions the spots may remain small and black or become gray with a black border. Tissue surrounding spots becomes chlorotic. Severe infection may result in defoliation.

On stems and pods, dots enlarge into lesions that are either entirely black or black with a gray center. Pods with infected pedicels fail to develop and drop off the plant. Severely spotted pods contain shrunken, infected seed. Such pods dry, shrink, and may split open prematurely, dropping the seed on the ground.

CONTROL:
1. Plant clean, disease-free seed.
2. Rotate rapeseed or mustard for a minimum of three years with noncrucifers.
3. Plow under infected residue.
4. Control volunteer mustard, rapeseed, and cruciferous weeds.

Blackleg

This disease is also called root and collar rot and stem canker.

CAUSE: *Leptosphaeria maculans* (Desm.) Ces. & de Not., anamorph: *Phoma lingam* (Tode ex. Fr.) Desm., overwinters as perithecia and pycnidia in infected residue and as mycelium in seed. During wet weather in the spring or during high relative humidity, ascospores are produced in perithecia after a few hours and are windborne. Pycnidiospores or conidia are produced in pycnidia and are splashed by rain to plants. Pycnidiospores are also possibly disseminated by adhering to seed after threshing and by the wind to neighboring fields during combining. Pycnidia are produced on infected plants throughout the growing season while perithecia develop as plants mature.

Two strains of *L. maculans* that vary considerably in virulence occur. The prevalent strain is weakly virulent, produces ascospores late in the growing season, and infects plants near maturity, thereby causing little injury. The second strain is highly virulent and sporulates early in the spring. A third strain has been found on stinkweed, *Thlaspi arvense* L., in Canada but does not seem to be important in causing blackleg on rapeseed or mustard.

Disease is more severe at 18°C than at 12°C.

DISTRIBUTION: Generally occurs throughout the world wherever crucifers are grown.

SYMPTOMS: Symptoms cannot be detected in the field until September in Canada. This is due to the ascospore discharge of the prevailing strain starting after plants are past the six-leaf stage. Younger tissue is more susceptible to infection than older tissue. Leaves, stem, and roots become infected. In the spring leaves have inconspicuous pallid areas. Eventually they become lesions that are gray-white, irregular in shape, and dotted with pycnidia that appear as small black specks.

Stems have poorly defined white to gray or black lesions that commonly develop near the soil line. Often these lesions have purplish borders. Pycnidia frequently form in stem lesions, commonly on lower stem portions and during wet weather, and exude a pinkish exudate containing conidia. Severely infected stems may have cankers over a large area and become girdled. Such plants ripen prematurely and produce light, shriveled seed. Lesions near the base of stems extend to the root system where black cankers are formed. Infected roots disintegrate, further weakening plants.

Stem canker phase can originate from a leaf lesion by systemic growth of *L. maculans* through lamina and petiole into the main stem. Direct infection of petiole and stems is unlikely to occur except after some form of injury has occurred.

CONTROL:

1. Rotate rapeseed and mustard with nonsusceptible crops such as cereals.
2. Do not plant rapeseed or mustard near fields that were infected the previous season.
3. Plow under infected residue as *L. maculans* is relatively short-lived, largely disappearing in two years.

Brown Girdling Root Rot

This disease is similar to Footrot in this chapter because of the number of similar pathogens involved. However, brown girdling root rot is hereby treated as a separate disease from footrot because of its separate treatment in the literature and apparent dissimilarity of symptoms. This disease is also called Rhizoctonia root rot; the postemergence seedling blight phase is called wirestem.

CAUSE: The most frequently isolated fungi are *Rhizoctonia solani* Kuhn, *Pythium ultimum* Trow., and *Fusarium roseum* Lk. emend. Snyd. & Hans.; however, *R. solani* is considered the primary cause of the root rot syndrome. Two anastomosis groups (AG) of *R. solani* are involved, AG2 and AG4. Isolates of AG2 are more virulent than AG4.

DISTRIBUTION: Alberta and British Columbia.

SYMPTOMS: Preemergence and postemergence damping off may occur. Later distinct brown to red-brown lesions occur on taproot. Several lesions may coalesce and girdle the taproot or a single lesion may grow to girdle the taproot.

CONTROL: It is assumed controls would be similar as those listed for footrot.

Damping Off

This disease is also called seedling blight.

CAUSE: Several fungi including *Pythium* spp., *Phytophthora* spp., *Fusarium* spp., and *Rhizoctonia* spp. Disease severity and incidence is greater in heavy, wet soils, low-lying areas, and during cool, wet weather. The causal fungi are weak pathogens that infect young, succulent tissue.

DISTRIBUTION: Generally distributed wherever rapeseed and mustard is grown.

SYMPTOMS: Seeds fail to germinate or seedlings are killed either preemergence or postemergence. Postemergence damping-off is characterized by constriction, tapering, and brown lesions on the hypocotyl. The seedling rots off at the soil line, causing it to fall over. Seedlings may emerge and not grow any further at the one- to four-leaf stage. Eventually they either die or resume growth.

CONTROL:
1. Plant seed when soil temperature has warmed up.
2. Seed should be planted at a depth of 1–25 mm in a firm seed bed.
3. Treat seed with a seed-protectant fungicide.

Downy Mildew

When downy mildew occurs together with white rust caused by *Albugo candida* (Pers. ex Chev.) Kuntze, it is called staghead.

CAUSE: *Peronospora parasitica* (Pers. ex Fr.) Fr. mainly overwinters systemically as mycelium in infected perennial or overwintering plants. It likely also survives as oospores in soil and residue. Systemically infected plants produce conidia (sporangium) on leaves that are windborne to other hosts. Typically zoospores are not formed in conidia but the mode of germination is by a conidium directly forming a germ tube. Oospores also germinate directly to form a germ tube or a conidiophore on which a large conidium is formed. Likewise, zoospores are also not thought to be commonly formed. Secondary inoculum is provided by conidia produced on leaves.

When downy mildew occurs together with white rust, the effect of infection by both causal fungi is synergistic as disease development is greater than infection by either alone. Cool, moist weather favors disease development.

DISTRIBUTION: Generally distributed wherever rapeseed is grown. However, downy mildew is usually more severe in northern rapeseed-growing areas.

SYMPTOMS: Spots appear on leaves, stems, and seed pods as small, purple, irregular areas. The leaf spots enlarge to form yellow areas on upper surfaces with a white, moldy growth and underneath are the conidia and conidiophores. The white, moldy growth occurs on blisters and stagheads.

CONTROL:
1. Grow resistant Argentine-type cultivars whenever possible. Resistant cultivars of turnip rape may be available.
2. Control volunteer turnip rape and wild mustard.
3. Plant only cleaned seed.
4. Rotate rapeseed and mustard with nonsusceptible crops.

Foot Rot

Foot Rot is similar to Brown Girdling Root Rot in this chapter because of the number of similar pathogens involved. However, foot rot is treated as a separate disease from brown girdling root rot because of its separate treatment in the literature and apparent dissimilarity of symptoms. This disease is also called late root rot.

CAUSE: *Rhizoctonia solani* Kuhn and *Fusarium roseum* Lk. with *R. solani* the major pathogen. *Fusarium roseum* survives as chlamydospores in the soil, mycelium in residue, and often as perithecia of its teleomorph, *Gibberella zeae* (Schw.) Petch on crop residue. *Rhizoctonia solani* survives as sclerotia and mycelium in residue. Both fungi have a wide host range.

Other fungi associated with this disease complex are *F. solani* (Mart.) Appel. & Wr., *F. tricinctum* (Corda) Sacc., *Leptosphaeria maculans* (Desm.) Ces. & de Not., *Pythium debaryanum* Hesse, *P. polymastum* Drechs., and *Alternaria alternata* (Fr.) Keissler. The *Fusarium* spp. are sometimes seedborne.

DISTRIBUTION: Generally distributed wherever rapeseed and mustard are grown; however, disease incidence is usually sporadic.

SYMPTOMS: Prematurely ripened plants may be found singly or in patches during the summer. A hard, brown lesion that has a rough surface and is clearly defined, sometimes with a black border, is present at the base of the stem. During wet conditions, salmon-pink spore masses may be present on lesions. A white mycelial growth may also be present on discolored root tissue. A severe infection will girdle the stem, killing the plant.

CONTROL:
1. Rotate rapeseed and mustard with cereals; however, it is not certain if cereals experience the same disease syndrome that occurs on rapeseed and mustard.
2. Plant cleaned, healthy, sound seed as early as practical.
3. *Rhizoctonia solani* is most destructive in a loose, well-worked soil; therefore, the seed bed should have firm soil.
4. Control cruciferous weeds and volunteer plants in and adjacent to fields.

Leaf Spot

CAUSE: *Alternaria alternata* (Fr.) Keissler.

DISTRIBUTION: Canada on Polish-type rapeseed.

SYMPTOMS: The symptoms described here were on artificially inoculated plants. Grayish to brownish necrotic spots develop on leaves. Spots enlarge, resulting in a drying and dropping out of affected areas. Flowers are blighted.

CONTROL: No control is reported.

Pod Drop

CAUSE: *Cladosporium* sp. and *Alternaria alternata* (Fr.) Keissler. Dead or dying petals are infected and provide a food base for the fungi to grow into the pedicel.

DISTRIBUTION: Canada.

SYMPTOMS: The pedicel is blackened just at or below its point of attachment to the pod. When pedicels are severely infected, pods fail to fill normally. The pedicel may be so weakened that the pod drops to the soil. The tip of the pedicel then shows the typical blackening.

CONTROL: Summer turnip rapeseed is less susceptible than the Argentine type. Oriental and brown mustards are also infected.

Sclerotinia Stem Rot

CAUSE: *Sclerotinia sclerotiorum* (Lib.) deBary overwinters as sclerotia either in soil or a contaminant of seed lots. Infrequently seed itself may be infected. Sclerotia germinate in one of two ways. In the spring sclerotia germinate to produce mycelium that grows a short distance across moist soil and infects plants at the soil line. Later in the growing season, usually about flowering time, sclerotia will germinate to produce one to several cup-shaped apothecia. Ascospores are borne in asci on the apothecium surface and are disseminated by wind, pollen, or insects to host plants. Infection by ascospores originates in leaf axils. Dead or dying blossoms become infected by ascospores and fall onto leaves and stems, thus initiating infection in another manner. Sclerotia are eventually formed from mycelium in diseased stems. Sclerotia have a dormancy mechanism that requires certain physical factors to occur before germination. Sclerotia from Saskatchewan required 2–6 weeks incubation at 10°C followed by 4–6 weeks at 20°C in diffuse light for stipe initiation and cap expansion.

Sclerotinia stem rot is most severe under warm, wet weather conditions. Lodged rapeseed and mustard provide conditions conducive to disease.

DISTRIBUTION: Generally distributed wherever rapeseed and mustard are grown.

SYMPTOMS: The first symptom is the presence of prematurely ripened plants scattered in or grouped among green plants. Such plants are easily pulled up. Pale gray stem lesions with faint concentric markings develop either from the soil line or axils of branches or leaves. At first such lesions are water-soaked; eventually lesions will expand to girdle the stem, killing the plant. Such stems appear bleached with a chalky white appearance and tend to shred longitudinally. Sclerotia that appear as hard, black, grain-sized bodies are found inside the stalk of killed plants, usually near the base. Under moist conditions sclerotia may occur in any part of the stem or pods.

CONTROL:
1. Rotate rapeseed and mustard at least every four years with cereals and grasses.
2. Plant only cleaned seed that has had sclerotia removed.
3. Plow under infected residue where possible and where erosion is not a problem.
4. Control volunteer rapeseed, mustard, and susceptible weeds between crops.

Storage Molds

CAUSE: *Penicillium* spp. mainly *P. verrucosum* var. *cyclopium* (West-ling) Sampson, Stalk & Hadlock; and *Aspergillus* spp., mainly *A. amstelodami* (Mang.) Thom & Church, *A. repens* (Corda) Sacc., and *A. sejunctus* Bain & Sartory. Other reported fungi are *A. candidus* Link ex Fr., *A. fumigatus* Fres., *A. versicolor* (Vuill.) Tiraboschi, *A. wentii* Wehmer., *Wallemia sebi* (Fres.) von Arx, *Cephalosporium acremonium* Cda., *Cladosporium cladosproioides* (Fres.) DeVries and *Alternaria* spp.

White Leaf Spot and Gray Stem

This disease is also called black blight, black stem ring spot, and ring spot.

CAUSE: *Mycosphaerella brassicicola* (Fr. ex Duby) Lindau overwinters as thick-walled mycelium and probably perithecia in residue. It is not seedborne to any extent. In the spring, conidia are produced on residue from mycelium and are windborne. Perithecia form on diseased tissue late in the season under conditions of low night temperatures and heavy dew. Ascospores are thought to be a means of spread of *M. brassicicola* in Europe and the United States but not in Canada. Most secondary spread is thought to be by conidia that are produced under cool, wet conditions.

DISTRIBUTION: Canada, Europe, and the United States.

SYMPTOMS: Circular (1–2 cm in diameter), whitish leaf spots are produced on leaves. Later in the season, large, elongate, purple to gray lesions with lighter centers speckled with numerous perithecia appear on stems and pods. Lesions may coalesce and completely blacken large portions of the plant.

CONTROL: The disease is not considered important because it devel-ops too late in the life of the plant; however, the following are aids to control:
1. Rotate rapeseed and mustard with noncruciferous crops.
2. Control weeds.
3. Plant healthy, cleaned seed.

White Rust

This disease is also called staghead either as a synonym only for white rust or to refer to the white rust-downy mildew complex. White rust is usually associated with downy mildew caused by *Peronospora parasitica* (Pers. ex Fr.) Fr.

CAUSE: *Albugo candida* (Pers. ex Chev.) Kuntze (syn. *A. cruciferarum* S. F. Gray) overwinters as oospores in residue, soil, and a contaminate of seed lots. In the spring, the oospore germinates to produce a sessile vesicle or a short exit tube that terminates in a vesicle. Zoospores are released from the vesicle and swim to host tissue where they encyst and germinate, infecting the tissue. The mycelium becomes systemic. Secondary infection is by sporangia being produced in leaf blisters and windborne to healthy tissue. Sporangia germinate by producing zoospores. Zoospores encyst and germinate to cause localized infections. Oospores eventually develop in swollen or hypertrophied tissue. White rust is most severe under wet weather conditions.

DISTRIBUTION: Generally distributed wherever rapeseed and mustard are grown.

SYMPTOMS: The most obvious symptom is a swelling and deformation of the terminal parts of flower stalks, resulting in the spiny staghead symptom. At first the stagheads are green but become brown and hard as they dry up. Portions of individual flowers may also become distorted. Raised green or brown blisters occur on the undersides of infected leaves. Initially the pustules are smooth but rupture to release chalky-appearing masses of sporangia. Similar blisters may appear on surfaces of green stagheads.

CONTROL:
1. Grow resistant Argentine-type cultivars wherever possible. Resistant cultivars of turnip rapeseed may be available.
2. Control volunteer turnip rapeseed and wild mustard.
3. Plant only cleaned seed.
4. Rotate rapeseed and mustard with nonsusceptible crops.

MYCOPLASMAL CAUSES

Aster Yellows

CAUSE: Aster yellows mycoplasma survives in several dicotyledonous plants and leafhoppers. Aster yellows mycoplasma is transmitted primarily by the aster leafhopper, *Macrosteles fascifrons* Stal. and less commonly by *Endria inimica* Say and *M. laevis* Rib.

DISTRIBUTION: Generally distributed where rapeseed and mustard is grown. Aster yellows has not been of major importance.

SYMPTOMS: Plants are infected systemically but symptoms usually occur only on the inflorescences. Flowers and pods become distorted and sterile, frequently forming small, blue-green, hollow, bladderlike structures instead of normal seed pods.

CONTROL: There is no practical means of control.

VIRAL CAUSES

Mosaic

CAUSE: Turnip mosaic virus (TMV) is transmitted mechanically and by the green peach aphid, *Myzus persicae* Sulzer. The virus can also be transmitted by several other species of aphids.

DISTRIBUTION: Alberta.

SYMPTOMS: Mosaic pattern on leaves.

CONTROL: No control is reported.

DISORDERS OF UNKNOWN CAUSE

Brown Girdling Root Rot

CAUSE: Unknown.

DISTRIBUTION: Western Canada.

SYMPTOMS: Symptoms are confined only to roots. Light brown lesions occur on the taproot and bases of the larger lateral roots about

5 cm or more below the soil surface. Lesions may expand, coalesce, and girdle the taproot. Such lesions become dark brown and sunken with vertical streaks. Lesions may expand up to the soil line but rarely beyond. If girdling occurs more than 5 cm below the soil line, some root regeneration may occur if there is sufficient moisture.

CONTROL: Argentine cultivars have some field tolerance.

Viruslike Disorder

CAUSE: Unknown, but probably of genetic or physiologic origin.

DISTRIBUTION: India.

SYMPTOMS: Affected plants are stunted with small, puckered, leathery, twisted leaves that are darker green than the color of normal leaves. Flowers are poorly developed and fruits contain few seeds. Longitudinal fissures that ooze a water-soluble yellow gum are present on aerial plant parts.

CONTROL: No control is reported.

BIBLIOGRAPHY

Anon. 1967. Diseases of field crops in the prairie Provinces. *Canada Dept. Agric. Res. Branch, Ottawa, Pub. 1008.*

Berkenkamp, B., and Vaartnou, H. 1972. Fungi associated with rape root rot in Alberta. *Can. Jour. Plant Sci.* **52**:973–976.

Calman, A. I.; Tewari, J. P.; and Mugala, M. 1986. *Fusarium avenaceum* as one of the causal agents of seedling blight of canola in Alberta. *Phytopathology* **70**:694 (disease notes).

Davidson, J. G. N. 1976. Plant Diseases in the Peace River Region in 1976. Agric. Canada Res. Sta. Beaverlodge, Alberta. Mimeo.

Davidson, J. G. N. (Undated). Disease Control in Rapeseed. Agric. Canada Res. Sta. Beaverlodge, Alberta. Mimeo.

Dueck, J. 1981. Dormancy in sclerotia of *Sclerotinia sclerotiorum*. *Phytopathology* **71**:214 (abstract).

Fucikovsky, L. 1979. Bacterial disease of rape and carrot in Mexico. *Phytopathology* **69**:915 (abstract).

Gladders, P., and Musa, T. M. 1980. Observations on the epidemiology of *Leptosphaeria maculans* stem canker in winter oilseed rape. *Plant Pathol.* **29**:28–37.

Hammond, K. E.; Lewis, B. G.; and Musa, T. M. 1985. A systemic pathway in the infection of oilseed rape plants by *Leptosphaeria maculans*. *Plant Pathol.* **34**:557–565.

Henry, A. W. 1974. Bacterial pod spot of rape in Alberta. *Can. Plant Disease Surv.* **54**:91–94.

Humpherson-Jones, F. M. 1985. The incidence of *Alternaria* spp. and *Leptosphaeria maculans* in commercial brassica seed in the United Kingdom. *Plant Pathol.* **34**:385–390.

Hwang, S. F.; Swanson, T. A.; and Evans, I. R. 1986. Characterization of *Rhizoctonia solani* isolates from canola in west central Alberta. *Plant Disease* **70**:681–683.

Kruger, W., and Wittern, I. 1985. Epidemiological investigations in the root and collar rot of soil seed rape caused by *Phoma lingam*. *Phytopathol. Zeitschr.* **113**:125–140.

McGee, D. C. 1977. Blackleg (*Leptosphaeria maculans* (Desm.) Ces et de Not.) of rapeseed in Victoria: Sources of infection and relationships between inoculum, environmental factors and disease severity. *Aust. Jour. Agric. Res.* **28**:53–62.

McGee, D. C., and Petrie, G. A. 1978. Variability of *Leptosphaeria maculans* in relation to blackleg of oilseed rape. *Phytopathology* **68**:625–630.

McGee, D. C., and Petrie, G. A. 1979. Seasonal patterns of ascospore discharge by *Leptosphaeria maculans* in relation to blackleg of oilseed rape. *Phytopathology* **69**:586–589.

Mathur, R. S., and Singh, B. R. 1975. A viruslike disorder of yellow mustard in India. *Plant Dis. Rptr.* **59**:174–175.

Mills, J. T., and Sinha, R. N. 1980. Safe storage periods for farm-stored rapeseed based on mycological and biochemical assessment. *Phytopathology* **70**:541–547.

Petrie, G. A. 1974. Diseases, in *Rapeseed: Canada's "Cinderella" Crop.* 3rd ed. R. K. Downey, A. J. Klassen, and J. McAnsh (ed.). Pub. No. 33. Rapeseed Association, Canada.

Petrie, G. A. 1975. Diseases of rapeseed and mustard, in *Oilseed and Pulse Crops in Western Canada.* J. T. Harapiak (ed.). Modern Press, Saskatoon, Saskatchewan.

Petrie, G. A., and Dueck, J. 1977. Diseases of rape and mustard, in *Insect Pests and Diseases of Rape and Mustard.* L. Burgess, J. Dueck, G. A. Petrie, and L. G. Putnam (eds.). Pub. No. 48, Rapeseed Association, Canada.

Ponce, F., and Mendoza, C. 1983. Fungal diseases and pests of the rapeseed *Brassica napus* L. and *B. campestris* L. in the high valleys of Mexico. *Phytopathology* **73**:124 (abstract).

Rao, D. V.; Hiruki, C.; and Chen, M. H. 1977. A mosaic disease of rape in Alberta caused by turnip mosaic virus. *Plant Dis. Rptr.* **61**:1074–1076.

Sippell, D. W.; Davidson, J. G. N.; and Sadasivaiah, R. S. 1985. Rhizoctonia root rot of rapeseed in the Peace River region of Alberta. *Can. Jour. Plant Pathol.* **7**:184–186.

Stelfox, D.; Williams, J. R.; Soehngen, V.; and Topping, R. C. 1978. Transport of *Sclerotinia sclerotiorum* ascospores by rapeseed pollen in Alberta. *Plant Dis. Rptr.* **62**:576–579.

Vaartnou, H., and Tewari, I. 1972. *Alternaria alternata*, parasitic on rape in Alberta. *Plant Dis. Rptr.* **56**:676–677.

Vanterpool, T. C. 1974. *Pythium polymastum* pathogenic on oilseed rape and other crucifers. *Can. Jour. Bot.* **52**:1205–1208.

Williams, J. R., and Stelfox, D. 1979. Dispersal of ascospores of *Sclerotinia sclerotiorum* in relations to Sclerotinia stem rot of rapeseed. *Plant Dis. Rptr.* **63**:395–399.

13

Diseases of Red Clover
(*Trifolium pratense* L.)

—————————BACTERIAL CAUSES—————————

Bacterial Leaf Spot

This disease is also called bacterial blight and bacterial leaf blight.

CAUSE: *Pseudomonas syringae* van Hall survives in infected residue. During cool, wet weather at any time of the growing season bacteria are disseminated by splashing rain and enter leaves either through wounds or natural openings.

DISTRIBUTION: Europe and North America.

SYMPTOMS: This disease affects stems, petioles, petiolules, stipules, and flower pedicles but it is most conspicuous on the leaflets. At first tiny, translucent dots appear on the lower leaf surface. The spots enlarge to fill the area between leaf veins and become black except for the margins, which remain water-soaked. Leaf tissue beyond the spot is yellow-green. During wet weather, a milky white bacterial exudate may develop in the spots and appear as a thin film or drop. When the exudate dries, it forms a thin, crusty film that glistens in the light. Leaves may become tattered and frayed by wind tearing out the dead portions. Lesions on petioles and stems are dark, elongated, and slightly sunken.

CONTROL: None is recommended since the disease is not considered important.

FUNGAL CAUSES

Black Patch Disease

CAUSE: *Rhizoctonia leguminicola* Gough & Elliott is seedborne and survives as mycelium in residue.

DISTRIBUTION: Southern United States.

SYMPTOMS: Black patch occurs in scattered patches in a field or only on scattered plants. The leaves, stems, flowers, and seeds are infected. Leaf symptoms are lesions that vary in color from brown to gray-black and usually have concentric rings. Large areas of a leaf may be affected. Often all the bottom leaves on a plant may be killed. Black, coarse, aerial mycelial growth occurs on the stems and petioles, frequently girdling the stems. Dark lesions and dark coarse mycelium are evident on the flowers and seeds. Seedlings are often blighted and overgrown with black mycelium.

CONTROL:
1. Treat seed with a seed-protectant fungicide.
2. Cut hay earlier than normal to reduce foliage loss.
3. Rotate red clover with alternate crops.

Brown Root Rot

CAUSE: *Phoma sclerotioides* (Preuss) Sacc. (syn. *Plenodomus meliloti* Dearn & Sanford) overwinters in residue in the soil.

DISTRIBUTION: Canada.

SYMPTOMS: Chlorotic, stunted plants occur in the spring and frequently die after beginning growth. Circular, brown, necrotic areas occur on roots and eventually spread into the crown. Numerous pycnidia occur on diseased areas.

CONTROL: Rotate red clover with nonlegumes.

Common Leaf Spot

This disease is also called Pseudopeziza leaf spot.

CAUSE: *Pseudopeziza trifolii* (Biv.-Bern.: Fries) Fckl. survives either as apothecia or as mycelium in residue. During cool, wet weather ascospores are produced that are windborne to healthy leaves.

DISTRIBUTION: Generally distributed wherever red clover is grown.

SYMPTOMS: Very small, dark spots that vary in color from olive to red-brown, purple, or black develop on either leaf surface. The shape of the spots may be angular to round and portions of the margin may jut out like a finger or branch, giving the spot a starlike appearance. An apothecium resembling a raised cushion may be observed, especially under magnification, in the center of spots on the underside of the leaf during wet weather. Later the apothecium may dry up, and become dark and difficult to observe. While common leaf spot is mostly confined to leaves, occasionally small, elongated, dark streaks may occur on petioles.

CONTROL: No practical control is available.

Crown Wart

CAUSE: *Physoderma trifolii* (Pass.) Karling (syn. *Urophlyctis trifolii* (Pass.) Magn.) survives in the soil and residue as resting sporangia. Sporangia germinate during wet soil conditions and liberate zoospores that swim to developing buds of the crown branches. The zoospore germinates to form an infective hypha that penetrates the plant.

DISTRIBUTION: Europe, North America, and South America, primarily in warmer red clover-growing areas.

SYMPTOMS: Crown wart occurs mostly in the south central United States and central Europe in excessively wet soils. Irregularly shaped galls form around the crowns of plants at or just below the soil level. Galls are first produced in late spring and continue to increase in size during the summer. Infected plants wilt in hot weather.

CONTROL: No control is practical since this is considered to be a minor disease.

Curvularia Leaf Blight

CAUSE: *Curvularia trifolii* (Kauf.) Boed. The disease is most severe during warm, moist weather.

DISTRIBUTION: Worldwide. Curvularia leaf blight is considered of minor importance.

SYMPTOMS: Water-soaked, angular, yellow to brown lesions develop on leaflets. Lesions often develop on petioles, resulting in necrotic leaves that remain attached to the plant.

CONTROL: No control is usually necessary.

Cylindrocladium Root and Crown Rot

CAUSE: *Cylindrocladium crotalariae* (Loos) Bell and Sobers.

DISTRIBUTION: Florida.

SYMPTOMS: Infected plants die. Black lesions occur in crowns and roots.

CONTROL: No control is reported.

Downy Mildew

CAUSE: *Peronospora trifoliorum* deBary survives in crown buds and possibly dead leaves. Disease occurs during wet or humid weather.

DISTRIBUTION: Downy mildew probably occurs wherever red clover is grown.

SYMPTOMS: Leaves at top of plants are light green with a gray fuzz consisting of mycelium on the underside. Leaves of severely diseased plants are twisted and stem growth is retarded.

CONTROL:
1. Harvest early.
2. Rotate with nonlegumes.

Fusarium Root Rot

This disease is also called common root rot.

CAUSE: *Fusarium* spp., including *F. oxysporum* Schlecht. ex Fr., *F. avenaceum* (Fr.) Sacc., *F. acuminatum* Ell. & Ev., and *F. solani* (Mart.) Appel. & Wr. are the fungi most commonly isolated from root and crown rots. However, several other fungi are also associated with root rot of red clover. The fusaria are usually present in soil and persist as chlamydospores and saprophytic mycelium in residue. The fusaria are generally considered to be weak pathogens and attack a plant after it has been weakened by some other cause such as drouth, low fertility, insects, other diseases, improper management, or winter injury. Wounding of roots results in localized increases in incidence and severity of necrosis caused by *F. acuminatum* and *F. avenaceum*. The main effect of wounding is not to provide an opening through which the fungus could enter but to accelerate fungal penetration and increased development. Additionally wounding alters host-pathogen interaction following penetration by favoring formation of distributive hyphae rather than chlamydospores.

DISTRIBUTION: Generally distributed wherever red clover is grown.

SYMPTOMS: Red clover plants may be killed in all stages of development but stand loss is most conspicuous during the second year. Plants appear unthrifty, stunted, yellowish, and wilt during hot, dry weather. Roots and crowns appear spongy or soft with a discoloration that varies from light brown to red-brown to dark brown. Lesions develop on the surface of larger roots and feeder roots may be pruned off. The inner core of the taproot may be rotted.

CONTROL: In general, practice good management to promote vigorous plant growth. This includes the following practices:
1. Plant certified disease-free seed of adapted cultivars. Some cultivars tend to be less affected by root rot than others.

2. Avoid overgrazing, particularly late in the growing season.
3. Get a soil test to determine the deficiencies or excesses of soil of nutrients.
4. Soil should have a pH between 6.2 and 7.0.

Gray Stem Canker

CAUSE: *Ascochyta caulicola* Laub.

DISTRIBUTION: Canada. This is not considered a serious disease.

SYMPTOMS: Silver-white cankers occur on stems, leaf stalks, and midribs. Lower stems may be girdled. Fruiting bodies of the fungus resembling black dots are present in the center of the cankers. Severely diseased stems may be stunted, swollen, and twisted with a few small leaves.

CONTROL: No control is reported.

Myrothecium Leaf Spot

CAUSE: *Myrothecium roridum* Tode ex Fr. and *M. verrucaria* (Alb. & Schw.) Ditm. ex Fr. These fungi are soil inhabitants and widespread facultative parasites. Disease development is favored by wounds and hot, humid weather.

DISTRIBUTION: Pennsylvania.

SYMPTOMS: Initially small, dark, water-soaked lesions occur on both leaf surfaces and eventually become necrotic, extending to both leaf surfaces. At first the lesions are circular but become irregular and coalesce as they enlarge. Lesions are usually dark brown to black but occasionally form concentric light and dark brown rings, giving a targetlike effect. A slight chlorosis may sometimes be seen around lesions and some lesions may be restricted by leaf veins. Symptoms appear most rapidly at wounds but may develop on nonwounded tissue.

Sporodochia develop on all parts of the lesion and on both leaf surfaces, often in several concentric rings. As lesions coalesce, the leaf becomes flaccid and necrosis may extend to the petioles. If humid conditions persist for five or more days, the leaves curl and die.

CONTROL: No control is reported.

Myrothecium Root Rot

CAUSE: *Myrothecium roridum* Tode ex Fr. and *M. verrucaria* (Alb. & Schw.) Ditm. ex Fr. See Myrothecium leaf spot. Both fungi are thought to be primary pathogens rather than secondary wound parasites.

DISTRIBUTION: Pennsylvania and Wisconsin.

SYMPTOMS: In slant-board evaluation, brown, water-soaked rots with poorly delineated margins occurred three days after inoculation. Foliage symptoms were chlorosis, purpling of leaflet margins, and death of leaves and petioles.

In greenhouse evaluations, rots were dark brown and occurred across entire roots with occasional streaks extending upward in the vascular system. Rotted root portions remained firm and were not water-soaked. Foliar symptoms included stunting in addition to leaf and petiole symptoms.

CONTROL: No control is reported.

Northern Anthracnose

This disease is also called clover scorch.

CAUSE: *Kabatiella caulivora* (Kirch.) Karak probably overwinters as mycelium and acervuli in residue. It may also be seedborne. During cool (20–25°C), wet weather, conidia are produced that are splashed or blown to healthy plants.

DISTRIBUTION: Cooler red clover-growing areas of Asia, Europe, and North America.

SYMPTOMS: From a distance a field of infected plants will appear to be scorched, as if by fire. The first symptom is elongated, dark brown to black lesions, which appear on leaf petioles and stems. The lesions become light colored with dark margins and may crack in the center; lesions may also girdle and kill the stems, causing a dying and browning of the foliage. The petiole and stem lesions cause leaves to wilt and the leaves or flower heads to droop like a shepherd's crook. Affected parts dry out rapidly and become so brittle that leaflets are easily broken off.

Small, irregular acervuli that are essentially colorless and lack setae can be observed in the deeper cracks and depressions along with the white mass of conidia and conidiophores. A hand lens is necessary for viewing.

CONTROL: Plant resistant cultivars.

Powdery Mildew

CAUSE: *Erysiphe polygoni* DC. ex Merat overwinters as cleistothecia scattered on residue surface. Ascospores are produced within asci in the cleistothecia and provide primary inoculum. The ascospores are blown to leaves and can infect at any time during the growing season but most commonly infect red clover during the cool nights and warm days of late summer and early autumn. Long periods of dry weather favor disease development. Conidia are produced on leaf surfaces and are the primary source of secondary inoculum. Cleistothecia are formed again in the autumn.

   Erysiphe polygoni is probably an aggregate species consisting of several physiologic races. It infects at least 359 species of plants in 154 different genera.

DISTRIBUTION: Generally distributed wherever red clover is grown in the temperate zones of the world.

SYMPTOMS: Powdery mildew is ordinarily not serious on the first hay crop but can cause reductions in yield and hay quality for red clover harvested later in the growing season. At first fine white mycelium that is barely visible develops on the upper surface of leaves. The patches enlarge and merge, giving the appearance of white flour or powder covering the leaf surface. The same symptoms can also occur on lower leaf surfaces. Infected leaves may turn yellow and wither prematurely but most remain green. An entire field may appear white if there is a severe attack.

CONTROL: Plant resistant cultivars.

Pseudoplea Leaf Spot

This disease is also called pepper spot and burn.

CAUSE: *Pseudoplea trifolii* (Rostr.) Petr. overwinters as perithecia or mycelium in residue that normally consists primarily of leaves. Ascospores are released during cool, wet weather and are windborne to healthy plants.

DISTRIBUTION: Generally distributed wherever red clover is grown but it is most common in temperate, humid areas.

SYMPTOMS: Pseudoplea leaf spot can occur throughout the growing season but is most likely to occur during cool, wet weather in the spring or autumn. Tiny, sunken, black spots develop on both surfaces of leaves and on petioles. The spots do not enlarge very much and they reach a diameter of only a few millimeters; however, they may become very numerous. Spots eventually turn gray with a red-brown border. Severely infected leaves and petioles become yellow, then turn brown and wither, forming a mass of dead leaves. Parts of the flower and flower stalk may also become infected and die.

CONTROL: There is no suitable control other than to cut hay before the disease becomes too severe.

Pythium Blight

CAUSE: *Pythium* spp., particularly *P. debaryanum* Hesse, survive in soil as oospores. Under moist soil conditions, the oospores germinate to produce a sporangium in which zoospores are produced. Zoospores are released and swim through free water to root tips or seeds where they germinate.

DISTRIBUTION: Generally distributed wherever red clover is grown.

SYMPTOMS: Only young plants are infected. Seedlings that are more than five days old are apparently almost immune and older plants are rarely infected. Symptoms are not obvious as the most common form of the disease occurs as a preemergence damping off. Seeds and seedlings are soft and mushy. When infected seeds and seedlings are placed in a moist chamber they will be overgrown with white mycelium. Small seedlings that have emerged will have a brown discoloration of the roots that will also be very soft.

CONTROL: Treat seed with a seed-protectant fungicide.

Rust

CAUSE: *Uromyces trifolii* (Hedw. f. ex. DC.) Lev. var. *fallens* (Desm.) Arth. is an autoecious long-cycled rust fungus. It overwinters mainly as teliospores on red clover in the southern United States. Urediospores are blown north each summer. In southern red clover-growing areas, infection can occur in the spring, but in the north, infection usually does not occur until late summer or autumn. Disease development is optimum during moist conditions with temperatures between 16°C and 22°C.

DISTRIBUTION: Generally distributed wherever red clover is grown in humid or semihumid areas.

SYMPTOMS: Rust is most important in new seedings and in stands left for seed. When rust is severe it is usually in those hayfields where cutting is delayed beyond the proper time.

The uredial stage is the symptom most likely to be seen by a grower. This will appear as small red-brown pustules on the petioles, stems, and leaves, especially on the leaf underside. Each pustule contains thousands of spores that appear as a red to brown dust or powder. The spores are blown by wind and infect healthy plants. Seriously rusted leaves may yellow and drop. The telia stage is similar in appearance to the uredia but is darker in color. Telia occur in old uredia or independently. The aecial stage appears as swollen, white to yellow pustules on stems, petioles, and leaves. When viewed under magnification, aecia appear as clusters of tiny cuplike structures. The aecia may cause distortion of affected leaves and petioles.

CONTROL: Some cultivars are more resistant than others.

Sclerotinia Root Rot and Crown Rot

This disease is also called Sclerotinia crown and stem rot.

CAUSE: *Sclerotinia trifoliorum* Eriks. survives for several years in the soil as sclerotia. The sclerotia germinate in the autumn to produce apothecia in which ascospores are produced. The ascospores are wind-borne to leaves of hosts.

DISTRIBUTION: Generally distributed in regions with mild winters or heavy snow cover. It is most important in the United States in the southern red clover areas.

SYMPTOMS: This disease occurs in patches that may merge to form large dead areas. Symptoms first appear in the autumn as small, brown spots on leaves and petioles. Disease development is optimum during wet weather at temperatures between 13°C and 18°C. Leaves turn light brown, wither, and become overgrown with white mycelium that spreads to the crowns and roots. By late winter or early spring, crowns and basal portions of young stems have a brown soft rot that extends into the roots. As stems and petioles are killed they are overgrown with white mycelium. Black sclerotia of various sizes form within the mycelium and are imbedded or attached to the surface of diseased tissue. When infected tissue decomposes, the sclerotia are released into the soil.

CONTROL: Rotate red clover with a nonsusceptible host. The sclerotia will eventually decompose or the causal fungus will die out in the absence of a susceptible host.

Sooty Blotch

This disease is also called black blotch.

CAUSE: *Cymadothea trifolii* (Pers. ex Fr.) Wolf overwinters as a stroma. Sometime during the growing season, perithecia form within the overwintered stroma. Ascospores are liberated from the perithecia and are windborne to healthy leaves. Infection occurs in the spring in the southern United States and in late summer and autumn in the northern United States.

DISTRIBUTION: Generally distributed in the temperate zones of the world.

SYMPTOMS: The first symptoms usually occur on the lower leaf surface as tiny olive-green dots. The dots enlarge and become thicker and darker, eventually looking like a velvety, black, elevated cushion. In the fall the cushions are replaced by other black areas that have a shiny surface. Chlorotic spots that turn necrotic appear on the upper leaf surface opposite a cushion. In a severe infection the entire leaf may turn brown, die, and fall off.

CONTROL: There is no practical control available.

Southern Anthracnose

CAUSE: *Colletotrichum destructivum* O'Gara overwinters as ascervuli on stems, crowns, and roots. During warm (28°C optimum), wet weather, conidia are produced that are splashed or blown to healthy plants.

DISTRIBUTION: Africa, southern Europe, and southern United States.

SYMPTOMS: Symptoms closely resemble those of northern anthracnose. However, southern anthracnose is frequently found on the upper part of the taproot and crown while northern anthracnose is not. Also, it is more likely to be present on the new growth of the second crop in the northern and midwestern part of the United States. It can be present in the spring on the young, succulent parts of stems and petioles of the first crop in the southern United States. It most commonly develops on young tissue but is not limited to it.

Dark brown spots of irregular shape develop on the leaves. These may vary in size from a barely visible dot to a general necrosis of the entire leaf. Petioles are frequently infected and become dark brown. The attached leaflets droop. The first symptoms on petioles and stems are water-soaked spots that become elongated, dark brown to black, with a gray or light brown center. Small black setae that resemble tiny black spines when viewed under magnification can be seen in the center of older sunken lesions. Stems may be killed when girdled by a lesion.

Diseased crown tissue is blue-black. Usually the entire plant will wilt and die when crowns are infected. Infected crowns and stems become brittle so plants break off at ground level.

CONTROL:
1. Grow adapted, resistant cultivars.
2. Use clean, certified seed from disease-free plants.

Spring Black Stem

This disease is also called Ascochyta leaf spot.

CAUSE: *Phoma trifolii* E. M. Johnson & Valleau overwinters as pycnidia in residue and may be seedborne. Spores are produced in pycnidia, become windborne, and infect plants during cool, wet weather in the autumn and spring.

DISTRIBUTION: Europe, North America, and South America.

SYMPTOMS: The most conspicuous symptom is on stems. At first, small, dark, brown to black lesions that are variable in size and shape appear on stems. Spots merge to form large black areas, especially on the lower part of the stem. Young shoots may be girdled and killed. Blackening may increase when clover is not cut at the proper time or is left for seed.

Leaves have dark brown to black spots that are irregular in size and shape. Some spots may have gray centers. Leaf infection together with stem infections may result in severe defoliation.

CONTROL:
1. Plant disease-free seed.
2. Where feasible, plow under infected residue.
3. Cut hay before defoliation becomes too severe in order to save leaves.

Stagnospora Leaf Spot

This disease is also called gray leaf spot.

CAUSE: *Stagnospora meliloti* (Lasch) Petr., teleomorph *Leptosphaeria pratensis* Sacc. & Briard, overwinters as mycelium, pycnidia, and perithecia in residue. The fungi are poor soil saprophytes and do not survive in the absence of infected residue. During wet weather in the spring, both conidia and ascospores are produced and are splashed or windblown to hosts. Secondary spread is mostly by conidia under wet conditions.

DISTRIBUTION: Generally distributed but most common in warm, humid areas or where red clover is grown under irrigation.

SYMPTOMS: Spots are 3–6 mm in diameter and circular to irregular in shape. The center of the spot is pale buff to almost white with a light to dark brown margin. In some instances the spots tend to show faint concentric zones the same color as the margin. Older spots have small dark spots, which are the pycnidia that are scattered throughout the affected area.

CONTROL: Rotate red clover for two to three years with a nonsusceptible host. Many legumes, including alfalfa and many other clovers, are susceptible.

Stemphylium Leaf Spot

This disease is also called target spot, zonate leaf spot, and ring spot.

CAUSE: *Stemphylium sarcinaeforme* (Cav.) Wiltshire and *S. botryosum* Wallr. are fungi that overwinter as mycelium in residue. At any time in the growing season conidia will be produced and are windborne to leaves. Infected leaves fall to the soil at harvest and provide inoculum for next year's infection. Disease development is optimum during moist weather between 20°C and 24°C. Sometimes the teleomorph of *S. botryosum, Pleospora herbarum* (Pers. & Fr.) Rab., will be present as perithecia that produce ascospores the following spring.

DISTRIBUTION: Generally distributed wherever red clover is grown, particularly in humid weather.

SYMPTOMS: Symptoms are usually more severe in late summer and autumn. Symptoms are confined to leaflets and are most severe in dense stands. At first, small, dark brown spots appear that later enlarge and develop into targetlike spots with alternate light and dark brown rings. The entire leaf becomes wrinkled, dark brown, with a sooty appearance, but usually remains attached to the plant. Sunken brown lesions infrequently appear on stems, petioles, and pods.

CONTROL: There is no satisfactory control, but selection of resistant cultivars may be possible. Harvest early to reduce losses.

Summer Black Stem

This disease is also called angular leaf spot and Cercospora leaf and stem spot.

CAUSE: *Cercospora zebrina* Pass. is seedborne and probably overwinters as mycelium in residue.

DISTRIBUTION: Europe, central and eastern North America.

SYMPTOMS: Summer black stem occurs later in the summer and early autumn. Leaf spots are dark brown, angular, and more or less delimited by veins of the leaf. Older spots may appear ash-gray due to sporulation of *C. zebrina* in the diseased tissue. Sunken dark brown lesions occur on stems and petioles. The lesions may merge to form extensive dark areas on lower parts of the stem. Seeds are also infected.

CONTROL:
1. Treat seeds with a seed-protectant fungicide.
2. Harvest hay crop before defoliation becomes severe. This also helps to reduce inoculum.

Winter Crown Rot

This disease is also called snow mold.

CAUSE: *Coprinus psychromorbidus* Redhead & Traquair. This fungus was previously called the low-temperature basidiomycete (LTB). Disease occurs at temperatures near 0°C at the soil surface under a snow cover.

DISTRIBUTION: Canada.

SYMPTOMS: Young stands have irregular areas of dead plants. Older stands have scattered plants with rotted crowns. White fungal threads are present at the edge of melting snow.

CONTROL: Rotate with nonlegume plants.

MYCOPLASMAL CAUSES

Phyllody

CAUSE: Mycoplasmalike organism that survives in biennial and pe-rennial plants. The organism is disseminated by leafhoppers.

DISTRIBUTION: Canada and Europe.

SYMPTOMS: Floral parts are turned into leaflike organs. Leaves are reduced in size with mild chlorosis. Newly emerged leaves are slightly deformed while old leaves may have a bronze color.

CONTROL: No control is reported.

Proliferation

CAUSE: Mycoplasmalike organism that probably overwinters in bien-nial and perennial plants. Vectors are not known for certain but leafhop-pers may be involved.

DISTRIBUTION: Alberta.

SYMPTOMS: Foliar growth is profuse, giving a witches' broom ap-pearance. Flowers may be transformed into leaflike appendages.

CONTROL: No control is reported.

Witches' Broom

CAUSE: Mycoplasmalike organisms.

DISTRIBUTION: Western Canada.

SYMPTOMS: Plants are severely dwarfed with numerous, short, spindly shoots originating from crown and axillary buds along the stem and giving an overall bunchy appearance. Leaves are small and chlorotic. Plants eventually die.

CONTROL: No control is reported.

Yellow Edge

CAUSE: Mycoplasmalike organism that overwinters in biennial and perennial plants. Dissemination is by leafhoppers.

DISTRIBUTION: Ontario and Quebec.

SYMPTOMS: Initially new growth has a mild chlorosis. Later a more pronounced chlorosis occurs along leaf margins. Chlorosis may change to a red-brown discoloration. Clusters of small leaves with short petioles and chlorotic margins develop at the nodes of creeping stems. Some plants do not have creeping stems but have numerous small leaves on upright stems, giving a witches' broom appearance. Flower size and production may be reduced. The entire plant is normally stunted.

CONTROL: No control is reported.

—————CAUSED BY NEMATODES————

Clover Cyst Nematode

CAUSE: Clover cyst nematode, *Heterodera trifolii* (Goff.) Oostenbrink.

DISTRIBUTION: Europe and probably other red clover-growing areas.

SYMPTOMS: Infected plants are stunted.

CONTROL: Follow general nematode control practices.

Root Knot Nematode

CAUSE: *Meloidogyne hapla* (Kofoid & White) Chitwood.

DISTRIBUTION: Probably widely distributed wherever red clover is grown.

SYMPTOMS: Plants are stunted and may wilt during moisture stress. Leaves frequently have a purple underside. Roots have galls, causing them to be excessively branched and brittle.

CONTROL:
1. Rotate red clover with nonhosts.
2. Apply nematicides if practical.

VIRAL CAUSES

Alfalfa Mosaic

CAUSE: Alfalfa mosaic virus (AMV) is disseminated by several species of aphids.

DISTRIBUTION: Canada and the United States. Probably distributed wherever alfalfa and red clover are grown.

SYMPTOMS: Leaves are somewhat distorted with green and yellow mosaic pattern.

CONTROL: See alfalfa mosaic of alfalfa.

Bean Yellow Mosaic

CAUSE: Bean yellow mosaic virus (BYMV) overwinters in infected legumes and is transmitted by aphids, especially the pea aphid, *Macrosiphum pisi* (Kaltenbach). BYMV probably interacts with *Fusarium* spp. to increase root rot and cause premature stand decline.

DISTRIBUTION: Africa, Asia, Europe, North America, and South America.

SYMPTOMS: A systemic mottling is the most common symptom but other symptoms may include vein yellowing, distortion, vein necrosis, and systemic necrosis.

CONTROL:
1. There is some evidence that resistant cultivars can be developed. Some cultivars are not as susceptible as others.
2. Avoid growing red clover near garden peas and beans.

Clover Yellow Mosaic

CAUSE: Clover yellow mosaic virus (CYMV) is seedborne.

DISTRIBUTION: Canada and probably other red clover areas.

SYMPTOMS: Plants are frequently stunted with a bushy appearance. Leaves may have a mosaic appearance with yellowing of leaf veins.

CONTROL: Use disease-free seed. Other general control measures may be useful.

Pea Common Mosaic

This disease is also called red clover mosaic and pea mosaic virus.

CAUSE: Pea common mosaic virus (PCMV) is reservoired in several legumes and is transmitted by aphids, particularly the pea aphid *Macrosiphum pisi* (Kaltenbach) and also *Myzus persicae* Sulzer and *Aphis rumicis* Linne. The aphids acquire the virus after feeding five minutes or more on diseased plants. The virus is not thought to be seedborne.

DISTRIBUTION: Canada, Europe, and the United States.

SYMPTOMS: A yellow mottling consisting of various kinds of chlorotic streaks and spots is located on and between the veins.

CONTROL:
1. There is some evidence that resistant cultivars can be developed. Some cultivars are not as susceptible.
2. Avoid growing red clover near garden peas and beans.

Red Clover Vein Mosaic

CAUSE: Red clover vein mosaic virus (RCVMV) overwinters in red clover and other legumes. It is transmitted by the pea aphid, *Macrosi-*

phum pisi (Kaltenbach), and other aphids. The pea aphid can acquire the virus in 1 hour of feeding but loses infectivity in 24 hours. Seed transmission is suspected but has not been demonstrated. Infected red clover plants serve as reservoirs for the virus that also causes pea stunt.

DISTRIBUTION: United States, particularly the eastern and north central states.

SYMPTOMS: Symptoms are most conspicuous in new growth leaves. The first symptom is a faint yellowing of the leaf veins. Gradually the yellowing becomes more chlorotic until veins and adjacent tissue become white. Yields are reduced and plants become more susceptible to root rot organisms, particularly *Fusarium* spp. Symptoms become masked during warm weather.

CONTROL:
1. There is some evidence that resistant cultivars can be developed. Some cultivars are not as susceptible.
2. Avoid growing red clover near garden peas and beans.

BIBLIOGRAPHY

Chilton, S. J. P. et al. 1943. Fungi reported on species of Medicago, Melilotus and Trifolium. *USDA Misc. Pub. 499.*

Cunfer, B. M.; Graham, J. H.; and Lukezic, F. L. 1969. Studies on the biology of *Myrothecium roridum* and *M. verrucaria* pathogenic on red clover. *Phytopathology* **59:**1306–1309.

Dickson, J. G. 1956. *Diseases of Field Crops.* 2nd ed. McGraw-Hill Book Co. Inc., New York.

Engelke, M. C.; Smith, R. R.; and Maxwell, D. P. 1975. Evaluating red clover germplasm for resistance to leaf rust. *Plant Dis. Rptr.* **59:**959–963.

Hanson, E. W. 1959. Relative susceptibility of seven varieties of red clover to diseases common in Wisconsin. *Plant Dis. Rptr.* **43:**782–786.

Hanson, E. W., and Hagedorn, D. J. 1961. Viruses of red clover in Wisconsin. *Agron. Jour.* **53:**63–67.

Hanson, E. W., and Kreitlow, K. W. 1953. The many ailments of clover. *USDA Yearbook* Washington, D.C., pp. 217–228.

Horsfall, J. G. 1930. A study of meadow-crop diseases in New York. *N. Y. (Cornell) Agric. Exp. Sta. Mem. 130.*

Kahn, M. A.; Maxwell, D. P.; and Smith, R. R. 1978. Inheritance of resistance to red clover vein mosaic virus in red clover. *Phytopathology* **68:**1084–1086.

Kreitlow, K. W.; Graham, J. H.; and Garber, R. J. 1953. Diseases of forage grasses and legumes in the northeastern states. *Pennsylvania Agric. Exp. Sta. and Reg. Pasture Lab., USDA Bull.*

Leath, K. T., and Barnett, O. W. 1981. Viruses infecting red clover in Pennsylvania. *Plant Disease* **65**:1016–1017.

Leath, K. T.; Bloom, J. R.; Hill, R. R.; Kaufman, T. D.; and Byers, R. A. 1983. Clover cyst nematode on red clover in Pennsylvania. *Phytopathology* **73**:369.

Leath, K. T., and Kendall, W. A. 1983. *Myrothecium roridum* and *M. verrucaria* pathogenic to roots of red clover and alfalfa. *Plant Disease* **67**:1154–1155.

McVey, D. V., and Gerdemann, J. W. 1960. Host-parasite relations of *Leptodiscus terrestris* on alfalfa, red clover and birdsfoot trefoil. *Phytopathology* **50**:416–421.

Martens, J. W.; Seaman, W. L.; and Atkinson, T. G. (ed.). 1984. *Diseases of Field Crops in Canada*. Canadian Phytopathological Society, Ottawa, Canada, 160p.

Murray, G. M.; Maxwell, D. P.; and Smith, R. R. 1976. Screening trifolium species for resistance to *Stemphylium sarcinaeforme*. Plant Dis. Rptr. **60**:35–37.

Rao, A. L. N., and Hiruki, C. 1985. Clover primary leaf necrosis virus, a strain of red clover necrotic mosaic virus. *Plant Disease* **69**:959–961.

Roberts, D. A., and Kucharek, T. A. 1983. *Cylindrocladium crotalariae* associated with crown and root rots of alfalfa and red clover in Florida. *Phytopathology* **73**:505 (abstract).

Smith, O. F. 1940. Stemphylium leaf spot of red clover and alfalfa. *Jour. Agric. Res.* **61**:831–846.

Stutz, J. C,; Leath, K. T., and Kendall, W. A. 1985. Wound-related modifications of penetration, development, and root rot by *Fusarium roseum* in forage legumes. *Phytopathology* **75**:920–924.

14

Diseases of Rice (*Oryzae sativa* L.)

Bacterial Blight

This disease is also called kresek.

CAUSE: *Xanthomonas campestris* pv. *oryzae* (Ishiyama) Dye (syn. *X. oryzae* (Uyeda & Ishiyama) Dowson) survives in the temperate regions of the world in the rhizosphere of weed hosts, infected straw piles, and base of stems and roots of rice stubble. In the tropics the bacterium survives on weeds, secondary growth from rice stubble, and in irrigation water in canals and rice fields. Glumes become readily infected but infected seed harbors viable bacteria for only two months, indicating seeds may not be an important source of inoculum.

Bacteria are locally disseminated by wind and water, particularly during a storm. Long-range dissemination is by irrigation water. The bacteria enter the plant through hydathodes, growth cracks caused by emergence of roots at the base of the leaf sheath, and other wounds that are made by wind whipping the leaves during a storm. Disease is also associated with root injury when seedlings are pulled from seedbeds. Bacteria do not enter through stomates. Young seedlings are more susceptible than older plants.

Once inside a plant, the bacteria affect the vascular tissue, cutting off water and causing wilting. Bacteria ooze from lesions whenever moisture is present. The ooze dries and falls into the paddy water. Several races of the bacterium exist.

DISTRIBUTION: Africa, Asia, Caribbean, Central and South America.

SYMPTOMS: Water-soaked stripes appear along one or both margins of the upper parts of leaf blades, soon enlarge, and turn yellow. Eventually lesions may cover the entire blade, turning it white then gray due to the growth of saprophytic fungi. Lesions may extend to the lower end of the leaf sheath. Lesion development increases as temperature increases and at a relative humidity of 70 percent. Discolored spots surrounded by water-soaked areas appear on glumes of green grain but appear gray or light yellow in mature grain.

The kresek or wilting symptoms appear two to three weeks after transplanting only in the tropics due to the practice of cutting off leaf tips before transplanting. Infected leaves become gray-green and roll along the midrib. Eventually the leaf withers. The bacteria reach the growing point through the vascular system and infect the base of other leaves, thereby killing the plant.

An additional symptom in the tropics is the production of pale yellow leaves in older plants. Older leaves remain green but the youngest leaf is uniformly pale yellow or whitish. Sometimes a broad green-yellow stripe may occur on the leaf with little further growth of the infected tiller.

Epiphytotics that occur after flowering have no measurable effect on grain yield or yield components. Epiphytotics that begin before flowering significantly reduce yield.

CONTROL: Plant resistant cultivars. Resistance varies according to age of the plant and race of the causal bacterium.

Bacterial Foot Rot

CAUSE: A bacterium similar to *Erwinia chrysanthemi* Burkholder, McFadden & Dimock. Iris plants are commonly found along canals, ponds, or irrigated rice fields, particularly in Japan. Such plants probably serve as overwintering hosts and sources of infection. Bacteria are liberated into water from diseased irises and are spread to rice plants by irrigation water. Other annual or perennial plants may also serve as sources of inoculum.

DISTRIBUTION: Asia.

SYMPTOMS: Diseased tillers display symptoms of either sheath rot or wilting of younger leaves. A dark brown rot occurs in the culms and bacterial ooze is present in the internodes. Sometimes decay is found on plants without detectable sheath rot symptoms. Diseased culms are soft and pull apart easily. A black discoloration occurs at the nodes. When decay is restricted below certain nodes, many new roots develop from the upper uninfected nodes, thereby allowing new leaves to develop from these nodes and allowing panicles to keep growing.

Systemic crown infections likely result from infection during the early tillering stage, whereas the sheath rot and restricted culm decay result from infections that take place later than the maximum tillering stage.

CONTROL: No control is reported.

Bacterial Glume Blotch

This disease is also called bacterial sheath rot.

CAUSE: *Pseudomonas syringae* pv. *syringae* van Hall (syn. *P. oryzicola* Klement). It has been reported in several Asian countries and Hungary as the cause of bacterial sheath brown rot.

DISTRIBUTION: Asia and Australia.

SYMPTOMS: Symptoms vary and range from scattered florets in a panicle exhibiting small (1–2 mm in size), dark brown areas that are surrounded by green to light brown tissue to the dark brown discoloration of complete florets. In severe cases 75 percent of the panicle may be affected, although florets at the base of the panicle are rarely completely discolored. Lesions also occur on florets just emerging from the boot.

Light green to dark brown lesions without definite margins occur on the flag leaf sheath and the veins are darker than the interveinal tissue. In severely affected plants, the stem may collapse at this point, although generally infection does not extend through to the stem tissue. Bacterial ooze occurs on infected tissue under warm, moist conditions.

CONTROL: The disease is not considered to be economically important; therefore, control measures are not recommended.

Bacterial Halo Blight

CAUSE: *Pseudomonas syringae* pv. *oryzae* pv nov. The causal bacterium is also pathogenic to barley, oats, and beans.

DISTRIBUTION: Japan.

SYMPTOMS: Disease symptoms develop in the tillering stage and disappear in the boot stage. Leaves have small, brown lesions surrounded by a large yellow halo.

CONTROL: No control is reported.

Bacterial Leaf Streak

CAUSE: *Xanthomonas campestris* pv. *translucens* (Jones, Johnson & Reddy) Dye, (syn. *X. translucens* (Jones, Johnson & Reddy) f. sp. *orizicola* (Fang et al.) Bradbury) is present year-round in the tropics either on rice or in water. The disease-caused by the bacterium is not present in the temperate areas.

Bacteria are disseminated by irrigation water and by storms. Bacteria enter plants through stomates but do not become systemic. After lesions develop, bacterial exudate develops on the surface of lesions during moist conditions at night. The bacteria are further disseminated by wind whipping leaves back and forth. Under dry conditions the exudates dry up and fall into the water.

DISTRIBUTION: Tropical Asia.

SYMPTOMS: Initially, watery, dark green, translucent stripes 3–5 mm long are confined between larger veins. The lesions enlarge and bacterial exudates appear on lesion surfaces under moist conditions. Susceptible cultivars may have a yellow halo around the lesions. During dry conditions, exudate dries to form numerous small yellow beads. Older lesions become light brown and entire leaves turn brown and die. Leaves then become bleached and gray-white and are colonized by saprophytic organisms.

CONTROL: Plant resistant cultivars.

Bacterial Sheath Brown Rot

The causal bacterium may be the principal causal agent of dirty panicle disease.

CAUSE: *Pseudomonas fuscovaginae* sp. nov. is seedborne. Several grasses are also hosts. Grain discoloration is more severe under high humidity and rainfall and/or in fields with soil nutrient imbalances.

DISTRIBUTION: Africa, Asia, Central and South America.

SYMPTOMS: Longitudinal, necrotic, red-brown stripes extending the full length of the sheath and often the entire length of the leaf lamina. Other sheath symptoms are small (1–5 mm), brown, elongated spots that expand and coalesce to form indistinct blotches on the sheath and sometimes the culm. Symptoms on seedlings are normally limited to brown, water-soaked necrosis of the sheaths with occasional necrotic brown stripes on leaves. Seedlings often die.

When older sheaths are affected, the flag leaf sheath enclosing the emerging panicle sometimes shows water-soaking, blotching, and brown necrosis. Developing florets are discolored brown and the panicle only partially emerges from boot. Those florets that do emerge frequently do not fill normally with grain, are spotted or totally discolored, and sterile. In severe cases the flag leaf sheath and/or collar are necrotic and dry.

Roots may also have a brown discoloration.

CONTROL: No control is reported.

Bacterial Sheath Stripe

This disease is sometimes called bacterial brown stripe and bacterial stripe. There appear to be some similarities between bacterial sheath stripe and bacterial stripe in the literature. The two diseases may be the same; however, the symptoms appear to be somewhat dissimilar and the causal organisms are identified in the literature as different. Therefore, bacterial sheath stripe and bacterial stripe are treated as two separate diseases here.

CAUSE: *Pseudomonas syringae* pv. *panici* (Elliott) Young, Dye & Wilkie (syn. *P. panici* (Elliott) Goto & Okabe). Disease is more severe under wet or humid conditions.

DISTRIBUTION: Japan, Philippines, and Taiwan.

SYMPTOMS: Only immature plants become infected. Initially water-soaked, longitudinal stripes form on lower portions of leaf sheaths. Lesions elongate to 10 cm long and 1 mm in width. Lesions often coalesce to form wide lesions that may involve the entire leaf sheath. Bacterial exudate may form on the surface of the lesion under humid conditions. Infection may progress to unfolding leaves and cause a bud rot. Plants may be stunted or killed.

CONTROL: No control is reported.

Bacterial Stripe

This disease is also called bacterial brown stripe.

CAUSE: *Pseudomonas avenae* Manns., (syn. *P. alboprecipitans* Rosen and *P. setariae* (Okabe) Savulescu). The bacterium are seedborne and survive up to eight years in seed.

DISTRIBUTION: Africa, Asia, Latin America, and Portugal.

SYMPTOMS: Symptoms occur only on seedlings that are primarily being grown in nursery boxes. Leaves and leaf sheaths develop water-soaked, brown, longitudinal stripes. Growth of seedlings with only the brown stripes is not affected in the box. Other seedlings are stunted or curved due to curving of leaf sheath and abnormal elongation of mesocotyl. Leaves normally fail to open on time on these plants.

Severely affected seedlings are dwarfed, often turn yellow, wilt, and die. A heavy bacterial ooze occurs at a cut across brown lesions. Infected seedlings die shortly after transplanting.

CONTROL: No control is reported.

FUNGAL CAUSES

Aggregate Sheath Spot

This disease is also called brown sclerotial disease. Aggregate sheath spot and sheath spot may be the same disease. However, the descriptions of the etiology and symptoms appear to be sufficiently different to consider these as two different diseases for now.

CAUSE: *Rhizoctonia oryzae* Ryker & Gooch (syn. *R. oryzae-sativae* (Saw.) Mordue and *Sclerotium oryzae-sativae* Sawada), teleomorph *Ceratobasidium* sp. The causal fungus probably survives as sclerotia in residue and soil. Initial infections occur by sclerotia floating on the paddy water to the rice plants. *Rhizoctonia oryzae* causes disease under warm weather conditions. Generally disease is more severe at lower nitrogen levels in the soil.

DISTRIBUTION: California, China, Japan, India, Thailand, and Vietnam.

SYMPTOMS: Lesions first appear on the lower leaf sheaths at the waterline during the tillering stage. Lesions are circular to elliptical with gray-green to straw-colored centers surrounded by distinct brown margins. A strip of necrotic cells runs down the middle of the lesion center. Frequently additional margins form around the initial lesion, producing a series of concentric bands. Lesions are 0.5–4 cm long.

The disease progresses to the upper leaf sheaths and sometimes to the lower portion of leaf blades. Lesions may coalesce and cover the entire leaf sheath, killing the leaves. Such leaves first turn bright yellow. Under favorable conditions, the flag leaf and panicle rachises can be infected, killing entire tillers. The culm may also be infected.

As leaf sheaths become rotted, abundant brown sclerotia are produced in and on diseased tissue. Sclerotia inside sheaths are cylindrical to rectangular in shape and are readily visible through the diseased tissue. Sclerotia produced on the surface are irregularly globose in shape.

CONTROL: Tall cultivars are more resistant than semidwarf cultivars.

Bakanae Disease

This disease is also called white stalk and palay lalake (man rice).

CAUSE: *Fusarium moniliforme* Sheldon, teleomorph *Gibberella fujikuroi* (Sawada) Woollenw. survives between growing seasons as mycelium and conidia on or in seed. It can also be soilborne but this is not considered to be an important source of inoculum. When seed is planted, the fungus infects the seedling and grows systemically within the plant as microconidia and mycelium within the xylem tissue. The fungus produces either fusaric acid or gibberellin, depending on growth conditions. Fusaric acid stunts plants while gibberellin causes plants to elongate abnormally. Elongated seedlings develop only under damp soil conditions and stunting occurs in dry soil. Thus, infected plants will either be stunted or elongated, depending on soil moisture conditions. Later in the growing season at flowering time, conidia are produced on diseased or dead culms and are windborne to flowers or developing grain. The spores either infect grain or contaminate it.

DISTRIBUTION: Generally distributed wherever rice is grown.

SYMPTOMS: The most common symptom is the abnormal elongation of infected plants in the seedbed or field. Infected seedlings are thin, chlorotic, and are several centimeters taller than normal plants. Severely infected seedlings may die before or after transplanting.

In the field, infected plants are much taller than normal and have chlorotic flag leaves. Infected plants have few tillers and leaves die in a relatively short time. Live plants have empty panicles. A white or pink fungus growth may appear on the lower part of the plant.

CONTROL:
1. Treat seed with a systemic seed-protectant fungicide.
2. Plant a resistant cultivar.

Basal Node Rot

CAUSE: *Fusarium oxysporum* Schlecht. ex Fr. The disease is prevalent in fields where rice is grown in rotation with pasture grass and in the second and third year of successive rice cultivation.

DISTRIBUTION: Brazil.

SYMPTOMS: Affected fields contain patches of dull green plants with delayed growth. Leaves of small plants are dull green. A black discoloration occurs at the underground basal node where adventitious roots are initiated. This discoloration often extends to the mesocotyl and adventitious roots. Severely diseased plants have poorly developed roots and few tillers. Plant death is rare.

CONTROL: No control is reported.

Blast

This disease is also called Pyricularia blight and rotten neck.

CAUSE: *Pyricularia oryzae* Cav. overwinters as mycelium and conidia on infected residue and seed in the temperate and probably the tropical regions of the world. However, survival is not as important in the disease cycle of the fungus in the tropics since airborne conidia are present throughout the year. The fungus may also survive on other winter cereals, diseased rice plants, and weed hosts.

Conidia are produced from mycelium and become windborne at night in the presence of dew or rain. High humidity then favors development of lesions after infection. Plants of all ages are susceptible. Generally rice cultivars that are adapted to temperate climates are more likely to be predisposed to infection at temperatures of 18–20°C than tropical cultivars. Rice plants are also more susceptible when grown in dry soil and resistant when grown under flooded conditions. Additionally close planting or heavy application of nitrogen increases susceptibility to infection. Secondary inoculum is provided by conidia that are produced on lesions about six days after inoculation in the presence of high relative humidity of 93 percent or more. The conidia are then primarily disseminated by wind but can also be disseminated by water, residue, or seed. Several physiologic races of *P. oryzae* exist.

DISTRIBUTION: Generally distributed wherever rice is grown.

SYMPTOMS: Symptoms of blast can be divided into two general phases depending on when the plant was infected. The leaf blast phase occurs between the seedling and later tillering stage; such plants are frequently killed. The panicle blast phase occurs after flowering. Spots are produced on leaves, nodes, panicles, and grains, but seldom on leaf sheaths.

Spots on leaves begin as small, water-soaked, whitish, grayish, or bluish dots. Spots enlarge quickly under moist conditions to 1.0–1.5 cm long and 0.3–0.5 mm wide. Mature leaf spots are elliptical, pointed at both ends with a gray to whitish center surrounded by a brown to red-brown margin. Spots may vary in shape and color depending on culti-var, environment, and age of spots. Resistant cultivars have only brown specks; moderately resistant ones have small, roundish lesions with necrotic centers and brown margins.

Some cultivars may be infected later in the growing season where the flag leaf attaches to the sheath. The lesion enlarges downward on the sheath and upward on the leaf, becoming gray at the point of attachment. The flag leaf may break off and become detached from the plant.

Nodes at or near flood level, generally the second or third node from the soil line, may be infected. The base of the sheath turns black due to production of conidia, with the node eventually breaking apart and remaining connected by a few vascular strands. The plant above the infected node dies.

Any portion of the panicle, panicle branches, and glumes may become infected, causing brown lesions. If the base of the panicle is infected, rotten neck or neck rot symptoms occur, causing the stem below the panicle to snap.

CONTROL:
1. Plant resistant cultivars; however, cultivars may not be resis-tant to all races of *P. oryzae*.
2. Plant only clean, healthy seed treated with a seed-protectant fungicide.
3. Apply small increments of nitrogen at any one time.
4. Manage flood water to maintain adequate but not excessive depth. Flooding is an effective way of controlling blast.
5. Avoid excessive plant populations.
6. Apply a foliar fungicide if disease conditions warrant its use.
7. Destroy previous crop residue as soon after harvest as feasible.
8. Rotate rice with other crops if feasible.
9. Control grasses and other weeds.

Brown Blotch

CAUSE: *Cercospora oryzae* Miyake. See Narrow Brown Leaf Spot.

DISTRIBUTION: Louisiana and Texas.

SYMPTOMS: Initially small, irregularly-shaped brown blotches occur on the leaf sheath usually 3–4 cm below the ligule. Eventually the blotch spreads to cover the entire leaf sheath. Leaves are prematurely desiccated and yields are reduced.

CONTROL: See Narrow Brown Leaf Spot.

Brown Leaf Spot

This disease is also called sesame leaf spot.

CAUSE: *Bipolaris oryzae* (Breda de Haan) Shoem. (syn. *Helminthosporium oryzae* Breda de Haan), teleomorph *Cochliobolus miyabeanus* (Ito & Kuribayashi) Drechs. ex Dastur, overwinters primarily as mycelium in seed. The following spring, such seeds give rise to infected seedlings with spots that develop on roots and coleoptiles. Leaves are usually not infected from seedborne inoculum due to their rapid growth at this time. Conidia are produced on spots that develop on the coleoptile and are windborne to leaves where secondary infection occurs. Leaf spots mostly occur from this secondary infection. Plants that grow in soils with nutritional deficiencies or in poorly drained soils where nutrient uptake is hindered are more susceptible to infection. Disease development is favored by high relative humidity (86–100%) with temperatures between 20°C and 25°C. Leaves must be wet for 8–24 hours.

DISTRIBUTION: Generally distributed in rice-growing areas of Africa, Asia, and the western hemisphere.

SYMPTOMS: The first symptoms on coleoptiles are pale, yellow-brown spots or streaks. Young roots will have blackish lesions. Young spots on leaves are small, circular, dark brown, or purple-brown. Typical spots are oval and about the size and shape of sesame seeds, and usually evenly distributed over the leaf surface. When fully developed, spots are brown with a gray or whitish center. On very susceptible cultivars, the spots may be larger, up to 1 cm long. Generally spots have a bright yellow halo surrounding them. Sometimes the spots may be so numerous that leaves are killed.

Black or dark brown spots appear on glumes and may cover the entire surface. During humid weather, these spots may have a velvet appearance due to production of conidia. Grain will have dark spots on the endosperm.

CONTROL:
1. Soil should have balanced fertility.
2. Rotate rice with other crops.
3. Plant healthy seed that has been treated with a seed-protectant fungicide.
4. Apply foliar fungicides to plants during the growing season.
5. Plant the most resistant and adapted cultivar.

Crown Sheath Rot

This disease is also called Arkansas foot rot, black sheath, and brown sheath rot.

CAUSE: *Ophiobolus oryzinus* Sacc. It is not known for certain how the fungus overwinters but it is probably as perithecia and mycelium in infected residue. *Ophiobolus oryzinus* is also reported to be seedborne to a slight extent. During the summer, in the presence of moisture, ascospores are windborne to host plants.

DISTRIBUTION: Africa, India, Japan, and the United States.

SYMPTOMS: Symptoms appear late in the growing season. A brown discoloration occurs on sheaths from the crown to the waterline or above. Perithecia are produced in the discolored area but are not obvious because only the peaks protrude above the surface of the epidermis. Red-brown mycelial mats may be found on the inner surface of the diseased sheaths. When sheaths become heavily infected, the leaf blades will die. At maturity, straw has a dull brownish cast. Culms may also be infected but perithecia are not produced. Infected plants produce few panicles and frequently only one panicle is produced.

CONTROL:
1. Plant resistant varieties.
2. Treat seed with a seed-protectant fungicide.

Downy Mildew

This disease is also called yellow wilt.

CAUSE: *Sclerophthora macrospora* (Sacc.) Thirum., Shaw & Naras. survives as oospores in infected leaves. Oospores germinate in the spring by forming a sporangium. The sporangium then germinates by a division of cell contents that form into zoospores. Zoospores are liberated into the water and swim to a germinating rice seed. Infection is optimum between 18°C and 20°C. Oospores and sporangia are produced on diseased plants and provide secondary inoculum.

DISTRIBUTION: Australia, China, India, Italy, Japan, and Arkansas. Downy mildew is not considered a serious disease.

SYMPTOMS: Symptoms are most obvious at flowering time. Initially the youngest leaves of infected plants have chlorotic or whitish spots. Leaf sheaths may be distorted and twisted. Panicles may become distorted due to failure to emerge from a twisted sheath. Additionally the panicle is smaller than normal and remains green, with the floral parts either bare or having a hairy appearance.

CONTROL: Treat seed with a seed-protectant fungicide.

Eyespot

This disease has also been called brown spot type disease.

CAUSE: *Drechslera gigantea* (Heald & Wolf) Ito. (syn. *Helminthosporium giganteum* Heald & Wolf). Grasses such as *Cynodon dacytlon* L. and *Eleusine indica* (L.) Gaertn. may be alternate hosts. Prolonged leaf wetness favors disease.

DISTRIBUTION: Colombia, Guatemala, Guyana, Honduras, Mexico, Panama, and Peru.

SYMPTOMS: Initially spots on leaves are small, water-soaked, olivaceous dots or rings. A yellow halo frequently occurs with young lesions but later disappears. Later these lesions develop into minute, longitudinally elongated, oval-shaped spots, 1–4 mm long and 0.5–1 mm wide, with white to straw-colored necrotic centers delimited by narrow, dark brown margins.

Under favorable conditions, successive development and coalescence of lesions produces a characteristic zonation which is "rough" compared to a more regular wavy pattern characteristic of leaf scald.

CONTROL: Plant resistant cultivars.

False Smut

This disease is also called green smut.

CAUSE: *Ustilaginoidea virens* (Cke.) Tak. overwinters in the temperate regions as sclerotia and chlamydospores. Most primary inoculum comes from sclerotia germinating to produce a stromata on the tip of the stalk. Ascospores are then produced in perithecia within the stromata and are windborne to flowers of rice. Rice is infected either at very early flowering or when grain is mature.

The individual grains are transformed into spore balls in which one to four sclerotia are produced. Secondary infection is by chlamydospores. Chlamydospores are formed on the spore ball and are windborne to other hosts where they may germinate to form germ tubes on which conidia are borne. Optimum germination occurs at 28°C. Chlamydo-spores do not free easily from smut balls because of a sticky material that holds the spores in place. High moisture favors development of false smut.

DISTRIBUTION: Generally distributed in most rice-growing areas; however, the disease causes little damage.

SYMPTOMS: A few to many individual grains of the panicle are transformed into ball-like objects that have a velvety appearance. At first the balls are small and visible between glumes but continue to grow and enclose the floral parts, reaching a diameter of 1.0 cm or more. The young balls are slightly flattened, smooth, yellow, and covered by a membrane. With continued growth, the membrane of the ball bursts and the color changes to orange and later yellow-green or green-black. The color is due to mycelium and powdery dark green spores. At this latter stage the surface of the ball cracks. When the ball is cut open, it is white and consists of mycelium together with floral and other parts of the host.

CONTROL: Apply a foliar fungicide a few days before heading.

Fusarium Diseases

CAUSE: *Fusarium equiseti* (Corda) Sacc., *F. semitectum* Berk. & Rav., *F. nivale* (Fr.) Ces., and *F. culmorum* (W. G. Sm.) Sacc.

DISTRIBUTION: Likely wherever rice is grown.

SYMPTOMS: *Fusarium culmorum* causes brown necrotic spots and tips in leaf blades of seedlings. These range from minute spots that are irregular in shape to rectangular spots, 1–2 mm wide by 4–6 mm long, and covering 3–5 percent of the leaf blade. Adult plants are usually not infected.

Fusarium equiseti causes discoloration of vascular tissues of culm. Discoloration may extend from base to third node. Infected culms are initially reddish but become brown to black.

Fusarium semitectum causes disease of panicles and culms. Panicles in the evolving sheath at the boot stage have irregular lesions that are initially yellowish but become dark brown. Necrosis develops that partially or totally covers the surface of kernels. A similar necrosis develops from flowering to ripening. Infected culms have a reddish to brownish vascular discoloration extending from the base to the third or fourth internode.

Fusarium nivale causes a necrosis of panicles and leaves.

CONTROL: No control is reported.

Helminthosporium Leaf Spot

CAUSE: *Helminthosporium rostratum* Drechs.

DISTRIBUTION: India.

SYMPTOMS: Symptoms are most prevalent in August. Small, light colored, oval to linear spots, 1.0–1.5 mm by 2.0–3.0 mm, occur on leaves.

CONTROL: No control is reported.

Kernel Smut

This disease is also called black smut and bunt.

CAUSE: *Tilletia barclayana* (Bref.) Sacc. & Syd. is the name now generally accepted for the causal fungus of kernel smut. Some confusion exists in the literature as to the proper identification of the causal fungus. Other fungi identified as the causal fungus are *T. horrida* Tak., *Neovossia barclayana* Bref., and *N. horrida* (Tak.) Padw. & Khun.

The fungus survives as chlamydospores in the endosperm, on the surface of seeds, and in the soil. The chlamydospores germinate in the presence of moisture to produce a promycelium on which several sporidia are borne in a whorl. The sporidia then produce secondary sporidia directly or from mycelium produced by the primary sporidia. The secondary sporidia are windborne and infect flowers that are opening, eventually producing smutted grain. However, the embryo is not destroyed, only the endosperm. Chlamydospores may be liberated from grain after a rain causes the spores to swell and break out of the hull, or at harvest, infesting soil and healthy seed. Diseased seed may also remain intact until planted the following spring. The causal fungus apparently does not become systemic through infecting seedlings. Disease occurs through local infection of florets. Abundant moisture during flowering increases disease incidence and severity.

DISTRIBUTION: Generally distributed wherever rice is grown.

SYMPTOMS: Symptoms are obvious only at maturity. Generally only a few grains in a panicle and often only a part of the grains, usually the endosperm, are diseased. After a rain or dew the mass of chlamydospores may swell and burst through the glume, appearing as tiny, black pustules or streaks. In severe infections, a short, beaklike outgrowth is produced by the rupturing glumes. The chlamydospores sometimes scatter onto leaves or grains of surrounding plants and appear as a black powder. Chlamydospores that do not break out can be seen through the wet hulls.

If not severely infected, seeds may germinate but seedlings are stunted. Smut spores give milled rice a grayish color.

CONTROL:
1. Plant the most resistant cultivars.
2. Plant earlier maturing cultivars as early as feasible in the growing season.
3. Avoid high rates of nitrogen fertilizer.

Leaf Scald

CAUSE: *Rhynchosporium oryzae* Hashioka & Yokogi, teleomorph *Metasphaeria albenscens* von Thuemen. The fungus probably overseasons in or on rice seeds. Disease development is favored by wet weather. It is a weak pathogen that usually infects a plant predisposed by other factors or through a wound. Seeds are probably the primary source of inoculum.

DISTRIBUTION: West Africa, Central America, southeast Asia, and the United States.

SYMPTOMS: Leaves, leaf sheaths, culms, and grains may display symptoms in all stages of development. Lesions usually occur near the tips of mature leaves. At first oblong or diamond-shaped, water-soaked blotches that are somewhat restricted by the veins develop into large olive-colored areas. These are encircled by alternate dark and light brown halos. Later the lesions become a gray-olive color and have a characteristic zonation of light and dark brown rings. A large portion of the leaf may be affected so that the leaf is killed and turns tan.

CONTROL:
1. Plant the most resistant cultivar.
2. Apply a foliar fungicide.
3. Apply a seed-treatment fungicide.

Leaf Smut

CAUSE: *Entyloma oryzae* H. & P. Syd. overwinters as chlamydospores in a sorus on leaves. The following summer, in the presence of moisture and at an optimum temperature of 28–30°C, the chlamydospores germinate to produce a promycelium on which sporidia are formed. The sporidia may bud to form secondary sporidia that are windborne to leaves and germinate in the presence of water.

DISTRIBUTION: Generally distributed wherever rice is grown; however, it is considered a minor disease.

SYMPTOMS: Leaf smut occurs late in the growing season. Small,
black, slightly raised spots, which are the sori, occur on both sides of the
leaf. The sori are generally rectangular in shape. While sori may become
numerous, they remain distinct from each other. Heavily infected leaves
may turn yellow, split, and sometimes die. The sorus is covered by the
epidermis and appears as a carbonaceous mass as the chlamydospores
are held tightly together. After soaking in water for a few minutes,
the epidermis ruptures, revealing the black mass of chlamydospores
beneath. The sorus has been mistaken for a sclerotium.

CONTROL: No control measures are necessary as leaf smut is not
considered a serious disease.

Narrow Brown Leaf Spot

This disease is also called Cercospora leaf spot.

CAUSE: *Cercospora oryzae* Miyake, teleomorph *Sphaerulina oryzina*
Hara. Little is known of how the fungus survives or is disseminated.

DISTRIBUTION: Generally distributed wherever rice is grown except
in Europe.

SYMPTOMS: Infection usually does not become severe until later in
the growing season as rice approaches maturity. Symptoms are most
common on leaves but may also occur on leaf sheaths, glumes, and pedicels. Leaf spots are long, narrow, and tan or brown. Generally spots occur
only in interveinal areas and are 1.5–3.0 mm wide and 3–13 mm long.
   Spots on glumes and pedicels are smaller and darker than on leaves.
Spots on leaf sheaths are light to dark brown with indefinite borders,
7–50 mm long, and may encircle the plant.

CONTROL:
1. Plant resistant cultivars.
2. Plant early-maturing cultivars early in the growing season to
   escape the disease.

Phoma Seedling Blight

CAUSE: *Phoma glomerata* (Cda.) Wr. & Hochapfel is seedborne.

DISTRIBUTION: Ghana.

SYMPTOMS: Seed is covered with a pale brown to brown mycelium. Pycnidia may be produced singly or in groups.

Seed may damp off within seven days after germination. Brown lesions occur on coleoptiles, leaves, and culms of emerged seedlings. Roots show different degrees of rotting, with pycnidia developing on them. Eventually seedlings are killed.

CONTROL: No control is reported.

Pithomyces Glume Blotch

CAUSE: *Pithomyces chartarum* (Berk. & Curt.) Ellis.

DISTRIBUTION: India.

SYMPTOMS: Diseased plants occur in patches. All glumes on a head may be diseased, with terminal seeds most often infected. Initially light brown lesions occur on glumes of individual grains. In most cases, lesions develop toward the tip rather than toward the base. Lesions enlarge and sometimes cover the whole surface of the grain. The enlarged spots are gray in the center surrounded by a dark brown ring. Infected glumes are grayer than healthy glumes. Short, dark conidiophores bearing conidia may be observed on infected plant parts.

Grains enclosed by infected glumes remain small, have an undeveloped endosperm, and germinate poorly. Grains are shriveled, discolored, and lightweight.

CONTROL: No control is reported.

Pythium Damping Off

CAUSE: *Pythium spinosum* Sawada apud Sawada & Chen, *P. irregulare* Buis., and *P. sylvaticum* Campbell & Hendrix are the main causal agents of damping off in the early growth stage of rice seedlings in nursery flats. Other *Phythium* spp. contributing to early damping off are P. *aphanidermatum* (Edson) Fitzp., *P. myriotylum* Drechs., *P. splendens* Braun, *P. ultimum* Trow., and *P. vexans* deBary. Disease is favored by low temperature and a soil pH of 6.0 at the early growth stage of rice seedlings. The main causal agents are pathogenic to rice seedlings at emergence but nonpathogenic to seedlings that are older than the first leaf stage.

At the two- to three-leaf stage, *P. graminicola* Subr. is the main agent causing damping off. Disease caused by *P. graminicola* occurs at the middle and latter growth stages of rice seedlings grown in nursery flats.

DISTRIBUTION: Damping off tends to be more of a problem in the temperate rice-growing areas of the world. It is generally not a problem in the tropics.

SYMPTOMS: Seedlings at the four- to five-leaf stage are chlorotic and stunted. Roots are necrotic and undeveloped. Severely diseased plants are killed.

CONTROL: Treat seed with a seed-treatment fungicide.

Rust

CAUSE: *Puccinia graminis* Pers. f. sp. *oryzae* Frag. The life cycle and symptoms are similar to those caused by *P. graminis* on other cereals. Rust is a minor disease on rice and causes no economic loss.

Sheath Blight

CAUSE: *Rhizoctonia solani* Kuhn, teleomorph *Thanatephorus cucumeris* (Frank) Donk. Other anastomosis groups (AG), Sasaki type fungi that have been attributed to be the causal agent are *Corticium sasakii* (Shirai) Matsumoto, *Hypochnus sasakii* Shirai, *Pellicularia filamentosa* (Pat.) Rogers, and *P. sasakii* (Shirai) Ito.

The causal fungus survives as sclerotia in the top 0–1 cm of the soil for up to two years. As soil depth increases, the number of sclerotia decreases. Sclerotial bodies and sometimes the basidial stage that are both highly virulent on rice have been observed on weeds. Thus, weeds are also hosts of the causal fungus and are an inoculum source to infect rice. Survival also occurs by mycelium in residue.

Sclerotia float on top of the water and germinate to form infection hyphae that penetrate the plant. Sclerotia tend to accumulate around the rice plant at the water-plant interface. Therefore, initial infection occurs near the waterline. Mycelium grows rapidly on or inside the tissues, initiating secondary spots. Infection spreads by means of runner hyphae to upper plant parts. Eventually sclerotia are formed in the center of spots that are easily detached to reinfest soil.

Infection and disease development is optimum at temperatures between 28°C and 32°C, and at high relative humidity (near 95%). Close planting and heavy fertilization increases the crop canopy, thereby raising the humidity and increasing incidence of the disease. In the United States, disease develops more extensively in semidwarf than in closely related standard height lines because of the shorter distance between the waterline and the panicles.

DISTRIBUTION: Generally distributed wherever rice is grown.

SYMPTOMS: Symptoms may be found on young plants but the initial symptoms are lesions on lower leaf sheaths after plants have reached the late tillering or early internode elongation stage. Spots or lesions appear mainly on the leaf sheath but may extend into the leaf blade if conditions are favorable for disease development. Initially spots appear near the waterline and are green-gray, elliptical, and about 1 cm long. Eventually the spots enlarge to 2–3 cm long and become gray with irregular purple-brown margins. Lesions on the upper plant coalesce to encompass entire leaf sheaths and stems. Brown sclerotia appear in the center of the spot and are easily detached. During humid conditions, mycelium may grow upward to infect other leaf sheaths. In severe cases, all leaves of a plant may be blighted. This allows increased sunlight penetration and decreased humidity. Lesions then dry, becoming white, tan, or gray, with brown borders.

Plants produce poorly filled grain. Additional losses result from increased lodging or reduced ratoon production.

CONTROL:
1. Fungicides and other chemicals are effective in controlling sheath blight.
2. Plant the recommended plant population.
3. Indica types are more tolerant than Japonica types. In the United States standard height lines become less diseased than semidwarf lines.

Sheath Rot

CAUSE: *Sarocladium oryzae* Gams & Hawksworth (syn. *Acrocylindrium oryzae* Sawada) survives as mycelium in infected residue. It is also seedborne. Conidia are windborne to rice where wounding by insects aids infection. High insect populations injure the leaf sheaths enclosing the panicles, thereby retarding emergence of the panicles and facilitating development and spread of the disease. The fungus gains entry through stomata and injuries and grows intercellularly in the vascular bundles and sheath mesophyll tissues. Maximum disease occurs when the minimum temperature is between 17°C and 20°C and the minimum relative humidity is between 40 percent and 56 percent at flowering time.

DISTRIBUTION: Southeastern Asia and the United States.

SYMPTOMS: Sheath rot occurs on the top leaf sheaths enclosing the young panicles, primarily the flag leaf sheath. Initially lesions start as oblong or somewhat irregular spots, 5–15 mm long, that have gray centers with brown margins or may be gray-brown throughout. Lesions enlarge, coalesce, and may cover most of the leaf sheath. Panicles may fail to emerge, or only partially emerge, then rot. A whitish, powdery growth may be found inside affected sheaths.

CONTROL: Some cultivars have at least partial resistance. It has been reported that excessive nitrogen increases disease incidence and increased potassium decreases disease incidence.

Sheath Spot

This disease is also called brown bordered leaf spot, Rhizoctonia sheath spot, and sheath blight. Aggregate sheath spot and sheath spot may be the same disease. However, the descriptions of the etiology and symptoms appear to be sufficiently different to consider these as two separate diseases for now.

CAUSE: *Rhizoctonia oryzae* Ryker & Gooch (syn. *R. oryzae-sativae* (Saw.) Mordue and *Sclerotium oryzae-sativae* Sawada). It is not known how the fungus survives. However, it is likely that *R. oryzae* overwinters as mycelium or sclerotia in residue and possibly the soil. Wild grasses have also been observed to act as overwintering hosts.

Dissemination is likely by wind or waterborne soil and residue. The disease is most severe along levees and other places where growth of rice is dense and where conditions of high humidity and temperatures of 29–32°C exist.

DISTRIBUTION: Japan, the United States, and Vietnam.

SYMPTOMS: Symptoms first appear during late tillering or early heading stage. Spots first appear on the sheath near the waterline and may occur on any of the leaf sheaths, although they are usually confined to the lower ones. Spots are confined almost entirely to leaf sheaths but do occasionally occur on leaves; none has been observed on the stems.

Spots initially appear as a red-brown discoloration that enlarges to an elliptical shape. As spots mature, they have a bleached, straw-colored center with a wide, red-brown margin. Spots average 1–3 cm in length but may reach 10 cm. Lesions may enlarge and coalesce to form large discolored areas that kill the leaves and cause lodging.

CONTROL:
1. Plant resistant cultivars.
2. Avoid excessive seeding rates.
3. Use proper timing of nitrogen-top dress applications.
4. Control grasses and other weeds.

Stackburn Disease

CAUSE: *Alternaria padwickii* (Ganguly) Ellis overwinters in soil and infested residue probably as sclerotia and mycelium. It is also seedborne.

Conidia are windborne to rice where infection is more likely to occur on a wounded leaf since *A. padwickii* is a weak pathogen.

DISTRIBUTION: Generally distributed wherever rice is grown.

SYMPTOMS: A few oval or circular spots usually occur on leaves and vary in size from 3–10 mm long. At first the center of the spot is tan, then becomes white with black dots (sclerotia) scattered throughout it. The spot is encircled by usually one, sometimes two, narrow, dark brown rings.

Glumes have tan to whitish spots with scattered sclerotia in them, surrounded by a relatively wide, dark brown margin. Infected kernels are discolored, shriveled, and brittle.

Roots and coleoptiles of germinating seedlings may be infected and have dark brown to black lesions with scattered sclerotia throughout. Infected seedlings may outgrow the infection or wither and die.

CONTROL: No satisfactory control is known.

Stem Rot

CAUSE: The causal fungus has several stages. The sclerotial stage is *Sclerotium oryzae* (Catt.) Krause & Webster; the anamorph has been referred to as *Helminthosporium sigmoideum* Cav., *Nakataea sigmoidea* (Cav.) Hara, and *Vakrabeeja sigmoidea* (Cav.) Sub., the teleomorph is *Leptosphaeria salvinii* Catt. Additionally a similar fungus, *H. sigmoideum* Cav. var. *irregulare* Cralley et Tullis, was found to be causing the same disease. However, its teleomorph is *Magnaporthe salvinii* Krause & Webster. The two perfect stages are rarely found. Some workers consider *L. salvinii* and *M. salvinii* to be synonymous.

The fungi overwinter as sclerotia in the top 5–10 cm of soil and in residue. During tillage operations, the sclerotia float on the surface of the water, germinate, and infect plants primarily through wounds. Sclerotia develop between the leaf sheath and culm. After harvest, the fungi continue to grow saprophytically even if little stem rot is present, producing large numbers of sclerotia. Rice tissue parasitically infected by *S. oryzae* remains the most important source of sclerotia. Inoculum levels are the greatest where large amounts of infested residues are left on or near the soil surface. Colonization of uninfected rice residue by mycelium from sclerotia is not an important source for production of new sclerotia. High nitrogen levels increase the incidence and severity of stem rot.

DISTRIBUTION: Burma, India, Japan, Philippines, Sri Lanka, Vietnam, Arkansas, California, Louisiana, and Texas.

SYMPTOMS: Symptoms appear late in the growing season, generally within a week after a field has been drained. Small, blackish, irregular lesions occur on the outer leaf sheath near the waterline. The lesions enlarge as the disease progresses. Eventually the leaf sheath becomes rotted and numerous sclerotia form as the fungus grows between the inner leaf sheath and culm. One or two internodes of the stem rot collapse, with only the epidermis remaining intact. Dark gray mycelium may be seen in the hollow stem, and numerous, small, black sclerotia occur on the inner surface. The next internodes may appear to be healthy. Infected stems lodge.

Occasionally in the tropics, leaf sheaths of transplanted seedlings rot at the waterline, causing the leaves above them to die.

CONTROL:
1. Plant early-maturing cultivars early in the growing season to escape disease injury.
2. Application of potassium fertilizer reduces stem rot severity.
3. Drain water at early stages of infection and keep soil saturated but not covered with water until rice has almost matured.
4. Crop rotation may have some value but the fungus can survive in soil for up to six years.
5. Foliar fungicides show promise of disease control.
6. Inoculum level and stem rot severity can be minimized by complete destruction or removal of rice straw infested with *S. oryzae*.

Water Mold Disease

CAUSE: Several aquatic phycomycetes including *Pythium dissotocum* Drechs., *P. spinosum* Sawada apud Swada & Chen, and *Achlya klebsiana* Pieters. The disease occurs in water-seeded rice.

DISTRIBUTION: Likely widely distributed.

SYMPTOMS: Germinating seed are rotted and seedlings are weakened or killed, resulting in losses through reduced density, irregular stands, and poor growth of infected seedlings.

CONTROL: Apply seed-treatment fungicide.

White Leaf Streak

CAUSE: *Romularia oryzae* Deighton & Shaw.

DISTRIBUTION: Africa and Asia.

SYMPTOMS: White to gray lesions, 0.5 mm by 1.0–2.5 mm, with narrow, brown borders develop on both leaf surfaces. However, lesions on upper leaf surface may be entirely white with no border and lesions on lower leaf surface may be entirely brown.

CONTROL: No control is reported.

MYCOPLASMAL CAUSES

Giallume Disease

CAUSE: A mycoplasmalike organism.

DISTRIBUTION: Italy.

SYMPTOMS: Plants are stunted with fewer than normal tillers. Foliage is yellowed.

CONTROL: No control is reported.

CAUSED BY NEMATODES

Rice Root Nematode

CAUSE: *Hirschmanniella oryzae* Ichinohe. The causal nematode is an endoparasite found in aquatic environments. The nematode is associated with lowland rice and is adapted to flooded paddies and marshes. It infects and reproduces on sedges and grasses. It remains dormant when paddies are dry. It occurs in parenchyma tissue but not in thin lateral roots where there are no air channels. It moves out of roots into soil soon after seed forms or growth ceases.

A total of seven species of *Hirschmanniella* spp. are reported to infect rice.

DISTRIBUTION: The nematodes are likely found worldwide wherever rice is grown.

SYMPTOMS: Roots become necrotic and discolored. Plants are stunted, with a reduction in number of tillers, panicles, top growth, and yield. There is also a delay in tillering and flowering.

CONTROL: No control is reported.

Stem Nematode

CAUSE: *Ditylenchus angustus* (Butler) Filipjev survives as adults in dry soil and infected residue. During periods of high humidity the nematodes climb up seedlings, enter young, growing tissue, and work their way between leaves and sheaths where they feed as ectoparasites. As plants grow, the nematodes move up to new tissue. All reproduction occurs only on the rice plant.

DISTRIBUTION: Bangladesh, Egypt, India, and southeast Asia.

SYMPTOMS: A chlorosis and malformation of the leaves may occur about two months after planting. A few brown spots appear on leaves and sheaths that later become darker brown along with upper internodes of the stem.

Panicles have few or no filled spikelets toward the base and are twisted or distorted. In some instances, two or three distorted ears may be surrounded by one sheath. There may also be more than the usual amount of branches on an infected stem.

CONTROL: Some cultivars are less susceptible than others but there is not a good means of control.

White Tip

CAUSE: *Aphelenchoides besseyi* Christie survives on seed for up to two years, thus providing the major source of inoculum the following spring. It does not overwinter in the soil but can be disseminated through soil from infected to healthy seedlings by water. The nematode is mostly an ectoparasite, occurring within folded leaves in the early stages of plant growth, then entering spikelets and finally glumes. The optimum temperature for nematode development is 28°C with a relative humidity of 70 percent or above.

DISTRIBUTION: Africa, Australia, Cuba, Japan, southeastern Asia, and the United States.

SYMPTOMS: The most obvious symptom is the chlorotic or white leaf tip, up to 5 cm long, that becomes evident at the beginning of elongation. The white areas later become brown and tattered. Plants are stunted and lack vigor. Additionally panicle length and number of spikelets are reduced. Affected panicles have a high percentage of sterility and small and distorted glumes and grains. The upper leaves, in particular the flag leaf, are twisted so that emergence of panicles from the boot is often incomplete. Panicles of severely diseased tillers mature late. Leaves are darker green than normal and plants give rise to tillers at higher nodes.

CONTROL:
1. Plant resistant cultivars.
2. Plant in water and keep the field flooded.

VIRAL CAUSES

African Cereal Streak

CAUSE: African cereal streak virus (ACSV) is limited to the phloem where it induces a necrosis. It is transmitted by the delphacid leafhopper *Toya catilina* Fennah. The natural virus reservoir is probably native grasses. ACSV is not mechanically or seed transmitted. Disease development is aided by high temperatures.

DISTRIBUTION: East Africa.

SYMPTOMS: Initially faint, broken, chlorotic streaks begin near the leaf base and extend upward. Eventually definite alternate yellow and green streaks develop along the entire leaf blade. Later the leaves become almost completely yellow. New leaves tend to develop a shoestring habit and die. Streaks are more broken than on other cereals, resulting in a somewhat mottled appearance.

Young infected plants become chlorotic, severely stunted, and die. Seed yield is almost completely suppressed. Plants become soft, flaccid, and almost velvety to the touch.

CONTROL: No control is reported.

Black-streaked Dwarf Disease

CAUSE: Black-streaked dwarf virus (BSDV) is transmitted by the planthoppers *Laodelphax striatellus* Fallen, *Unkanodes sapporonus* (Mats.), and *Ribautodelphax albifascia* (Mats.) in a persistent manner. Several grasses serve as hosts for BSDV.

DISTRIBUTION: Japan.

SYMPTOMS: Galls appear as waxy, elongated, irregularly-shaped protuberances extending along the major veins on the lower leaf surface, outer surface of leaf sheaths, and culms. The protuberances may form gray to dark brown streaks of various lengths on older leaves. The proximal portion of the leaf blade is often twisted.

Plants are severely stunted with more tillers than normal if infected early. Foliage is darker than normal and few or no panicles are formed. Grain often has dark brown blotches.

CONTROL: Plant resistant cultivars.

Dwarf Disease

CAUSE: Rice dwarf virus (RDV) is transmitted by the planthopper *Nephotettix nigropictus* Stal.

DISTRIBUTION: Japan and Nepal.

SYMPTOMS: Diseased plants are stunted to various degrees. Leaves are a dark green and have large numbers of discrete translucent spots that often coalesce to form longitudinal lines parallel to the midrib. Spots are also present on leaf sheaths.

Stunting depends on the growth stage at which plants are infected. Plants infected at early growth stages are stunted the most. Both root and shoot systems are reduced and panicles are either absent or reduced in size.

CONTROL: No control is reported.

Grassy Stunt

CAUSE: Rice grassy stunt virus (RGSV). However, there is some question as to whether the causal organism is a mycoplasmalike organism.

The causal organism persists in and is transmitted only by the rice brown planthopper, *Nilaparvata lugens* (Stal.). The organism also probably survives in rice, *Oryza sativa*, year-round and within several other species of rice. At least two different strains of RGSV are known to occur. These strains have been designated as RGSV-1 and RGSV-2.

DISTRIBUTION: Ceylon, India, Malaysia, Philippines, and Thailand.

SYMPTOMS: Symptoms caused by RGSV-1 are as follows: Plants are severely stunted, have excessive tillers, and have an erect growth habit. Leaves are short and narrow, chlorotic, and covered with numerous rusty-colored spots. Young leaves of some cultivars may be mottled or striped. Stripes are yellow, white, narrow, have diffuse margins, and are parallel to the midrib. They may be located at the basal portion of the leaf or extend the entire length of the leaf. The numerous small tillers give an infected plant a grassy appearance. Infected plants normally survive but produce few or no panicles.

Symptoms caused by RGSV-2 initially are stunting, yellow to orange discoloration and rusty spotting of lower leaves and narrowing of the leaf blades. Young emerging leaves are pale green to almost entirely chlorotic. These leaves also tend to have striping and mottling symptoms. Leaves remain yellow even when adequately fertilized; however, mature leaves of plants infected with RGSV-1 turn dark green when nitrogen fertilizer is applied. Severely infected seedlings produce few tillers and eventually die.

CONTROL: Some cultivars show a degree of resistance. Resistance is governed by a single dominant gene.

Grassy Stunt B

CAUSE: There is question if the causal organism (GSB) is a virus, perhaps a strain of rice grassy stunt virus, or a mycoplasmalike organism. GSB is transmitted by the rice brown planthopper *Nilaparvata lugens* (Stal.) in a persistent manner.

DISTRIBUTION: Taiwan.

SYMPTOMS: Symptoms vary somewhat depending on cultivar. Characteristic symptoms are excessive tillering with or without stunting. Initially young leaves are narrower than normal and are green with faint to conspicuous chlorotic stripes on both sides of the midrib. Later the leaves turn pale green with conspicuous mottling. Still later leaves may turn yellow and stripes may disappear. Mildly stunted plants live to maturity but produce only a few filled panicles.

CONTROL: No control is reported.

Grassy Stunt Y

CAUSE: There is question if the causal organism (GSY) is a virus, perhaps a strain of rice grassy stunt virus, or a mycoplasmalike organism. GSY is transmitted by the rice brown planthopper, *Nilaparvata lugens* (Stal.), in a persistent manner.

DISTRIBUTION: Taiwan.

SYMPTOMS: Symptoms vary somewhat depending on cultivar. Tillering varies from reduced in winter to increased in summer. Plants are slightly to markedly stunted. Leaves initially may be somewhat chlorotic; they may be narrow and stiff, with an erect growth habit. Small, dark brown spots may occur on leaves of some cultivars and conspicuous yellow-white stripes may occur along the leaf veins of other cultivars. Diseased plants are apparently not killed but produce empty panicles.

CONTROL: No control is reported.

Hoja Blanca

This disease is also called chlorosis, cinta blanca (white band), raya (stripe), raya blanca (white stripe), rayadilla (striped), and white leaf.

CAUSE: Hoja blanca virus is transmitted by the planthopper *Sogatodes oryzicola* (Muir). Once the virus is acquired by the insect, it is transmitted through eggs from infective females to their progenies for several generations. Little is known of how the virus overwinters; however, since hoja blanca is most prevalent in the tropics, there are probably infected rice plants growing year-round. Additionally jungle rice, *Echinochloa colonum* (L.) Link, is also a host.

DISTRIBUTION: Western hemisphere. In the United States, it has been reported from Florida, Louisiana, and Mississippi.

SYMPTOMS: Initially one or more white stripes or white mottling may occur on a leaf or the entire leaf may turn white. Diseased and normal tillers may be produced by the same plant. Often the second crop tillers of an infected plant show no symptoms.

   Diseased plants are stunted and panicles may fail to reach their normal size or emerge from the sheath. Hulls are often distorted, turn brown, and dry out. The flower parts are sterile or absent, thereby producing few seeds and causing the head to remain upright instead of bending over.

CONTROL: Plant resistant cultivars.

Necrosis Mosaic Disease

This disease is also called yaika-sho and eso mosaic.

CAUSE: Rice necrosis mosaic virus (RNMV). RNMV is sap transmissible and is soilborne. The weeds *Ludwigia perennis* Linn., and *Brachiaria ramosa* (L.) Haines are hosts and potential reservoirs for the virus. The mechanism of natural spread is unknown. Infected plants occur mostly in upland seedbeds.

DISTRIBUTION: India and Japan.

SYMPTOMS: Initially a mosaic mottling occurs on the lower leaves at about the maximum tillering stage after transplanting in the paddy field. Mottling consists of light green to yellow streaks that are oval to oblong, 1 mm wide and 1 mm to 10 cm long. Later streaks may coalesce to form irregular-shaped patches that eventually spread to upper leaves. When several patches are present, a leaf turns yellow. Mottling may also occur on the culm.

   Plants are slightly reduced in size; but this is not very noticeable. The number of tillers is also reduced; those present tend to lie flat at their bases, resulting in a spreading growth habit. Later a few elongated, necrotic lesions appear on the surface of leaf sheaths as well as on basal portion of culms. There is also a reduction in panicles and fewer and lighter grains.

CONTROL: No control is reported.

Rice Gall Dwarf

CAUSE: Rice gall dwarf virus (RGDV) is transmitted by leafhopper *Recilia dorsalis* Motsch., and green rice leafhoppers *Nephotettix nigropictus* Stal., *N. cincticeps* Uhler, *N. malayanus* Ishihara & Kawase, and *N. virenscens* Distant in a persistent manner. *Nephotettix* spp. have about a two-week incubation period. RGDV is not known to be seed transmitted.

DISTRIBUTION: Thailand.

SYMPTOMS: Infected plants are severely stunted. Leaves are dark green with the appearance of galls on the outer surface of leaf blades and leaf sheaths. Galls or vein swellings occur on the under surfaces of leaf blades as well as the outer sides of leaf sheaths. Galls are initially light green and translucent but eventually turn white. Galls are more abundant in early-infected plants. Galls vary from 0.4–8 mm in length, with most less than 2 mm long, and from 0.4–0.5 mm in width. Leaf tips may be twisted.

CONTROL: No control is reported.

Rice Orange Leaf Disease

CAUSE: Possibly a mycoplasma that is disseminated by the leafhopper *Recilia dorsalis* Motsch., and adults and nymphs of the leafhopper *Inazuma dorsalis* Motsch. The causal organism is persistent in the vectors.

DISTRIBUTION: Malaysia, Philippines, Sri Lanka, Thailand, and probably other countries in southeast Asia.

SYMPTOMS: Orange discoloration and rolling of leaves followed by a quick dying of plants.

CONTROL: No control is reported.

Rice Ragged Stunt

CAUSE: Rice ragged stunt virus (RRSV) is transmitted by the brown leafhopper *Nilaparvata lugens* (Stal.). Vectors must have an acquisition period of 24 hours or longer. The vectors are most efficient at transmitting RRSV five days after hatching and less efficient as the insect grows older. It is not mechanically or seed transmitted.

DISTRIBUTION: Philippines, India, Indonesia, and Sri Lanka.

SYMPTOMS: Diseased plants are stunted to various degrees at all growth stages. The disease is systemic. At early growth, leaves appear ragged with irregularly edged portions. The irregular edge consists of 1–17 breakings that are notches or indentations observable before leaf blades unroll. The ragged edges usually occur on one side of a leaf blade but occasionally appear on both sides and on the leaf sheath. The ragged portions vary in length and occupy only a part of the leaf blade. The ragged area is usually chlorotic, becoming yellow to brown-yellow, and eventually disintegrating. The ragged leaves become fewer as the plants become older.

The top portion of leaf blades is often twisted, resulting in a spiral shape of one or more turns. Vein swellings, caused by proliferation of phloem cells, often appear on the upper portion of the outer surface of the leaf sheaths, usually near the collar and less often on the lower surface of leaf blades and on culms. Vein swellings are 1 mm or less wide and 1 mm to more than 100 mm long. They are pale yellow or, infrequently, brown.

At booting stage, flag leaves are shortened, often twisted, malformed, or ragged. Diseased plants flower late. Tillers often generate nodal branches, which were secondary or tertiary tillers produced at upper nodes. Nodal branches often bear small panicles that are more numerous than on healthy plants.

CONTROL: No acceptable control is reported.

Rice Yellow Mottle

CAUSE: Rice yellow mottle virus (RYMV) is mechanically transmitted but it is not seedborne. Different strains exist. Disease development is more severe in younger than in older infected plants.

DISTRIBUTION: Africa.

SYMPTOMS: Initially scattered, orange-yellow spots occur either on the youngest leaves of recently planted rice or on mature rice. Eventually oldest leaves are yellow at the lower part of the plant and orange at the top. The young leaves show a very thin mottle ranging from yellow to dark green. Artificially and naturally infected plants may be dwarfed. Young seedlings may be killed. Additionally there is reduced tillering, malformation, partial emergence, and sterility of panicles. Yields are reduced depending on age of plants at infection.

CONTROL: Sources of resistance have been identified.

Stripe Disease

CAUSE: Stripe virus (SV) is transmitted by the planthoppers *Laodelphax striatellus* Fallen, *Unkanodes sapporonus* (Mats.), and *Ribautodelphax albifascia* (Mats.). SV multiplies and persists in the vectors. Several other plants are virus hosts.

DISTRIBUTION: Japan, Korea, and Taiwan.

SYMPTOMS: Young plants that become infected display more serious symptoms than older plants that become infected. Emerging leaves fail to unfold properly. Frequently leaves will emerge without unfolding, then elongate and become twisted and droop. Leaves are generally chlorotic and often have a wide chlorotic stripe with diffuse margins. Frequently a gray necrotic streak appears in the chlorotic area. The streak enlarges and eventually kills the leaf.

Leaves that later unfold properly have an irregular chlorotic mottling that appears as a stripe the length of the leaf blade. Mottling may also occur on the leaf sheath. Early infection may cause severe plant stunting or death. Plants infected later may be only slightly stunted. The number of tillers is somewhat reduced. The number of panicles is also reduced and they carry malformed spikelets that do not emerge properly from leaf sheaths. Grain may be malformed.

CONTROL: Plant resistant cultivars.

Transitory Yellowing Disease

This disease is also called brown wilt.

CAUSE: Transitory yellowing virus (TYV) is transmitted by *Nephotettix apicalis* Motschulsky, *N. cincticeps* Uhler, and *N. impicticeps* Ishihara. TYV persists and multiplies within the vector; however, it is not infective when the temperature is below 16°C or above 38°C.

DISTRIBUTION: Taiwan.

SYMPTOMS: Leaves become yellow, beginning at the distal portion of the lower leaves. Therefore, color is more intense in the lower leaves than the upper ones. Brown, rusty flecks or patches may appear on discolored leaves. Yellowing may be very indistinct in lower leaves, which soon roll and wither. Only a few uppermost leaves may live. Infected plants have a reduced root system and produce few or no panicles.

Diseased plants often recover under greenhouse conditions. Normal leaves may be produced at later growth stages, following approximately a month of yellow leaves. Consequently the appearance of diseased plants may become normal after yellow leaves have fallen off.

CONTROL: No control is reported.

Tungro

The following diseases are also believed to be tungro: accepta na pula (red disease), cadang-cadang (yellowing), leaf yellowing, mentek disease, panyakit merah (red disease), penyakit habang, rice dwarf, stunt disease, and yellow orange leaf.

CAUSE: Two types of tungro virus particles occur: spherical particles (RTSV) and bacilliform particles (RTBV). RTSV is known as a latent virus and acts as a helper for the transmission of RTBV by leafhopper vectors. RTBV causes the tungro symptoms and RTSV enhances them. Aside from tungro, a disease complex associated with RTBV and RTSV, RTSV also spreads and causes disease as an independent pathogen.

Tungro viruses survive in infected rice plants year-round. However, when one crop is grown a year, survival is in stubble of infected plants that ratoon and numerous grassy weeds. The viruses are transmitted primarily by the green rice leafhopper *Nephotettix virenscens* Distant. The insect feeds for at least 30 minutes on an infected plant to acquire the virus and can transmit it by feeding for 15 minutes on a healthy plant. The viruses are semipersistent and are retained for only two to three days in the insect; then they must be reacquired. The RTSV particles can be transmitted by *N. virenscens* but RTBV is transmitted only when RTSV is acquired before or at the same time. This indicates RTBV is dependent on RTSV for transmission. Other vectors are *N. malayanus* Ishihara & Kawase, *N. nigropictus* Stal., *N. parvus* Ishihara & Kawase, and *Recilia dorsalis* Motsch. Greater disease spread occurs during the wet season due to the greater numbers of vectors. Disease incidence generally increases with increase in plant spacing. Wider spacing usually results in larger plants than close spacing, and a large plant is more likely to become infected because of its large surface area. A larger plant has more tissue for virus multiplication and vector feeding, so the changes of virus spread from it are greater than from a smaller plant.

DISTRIBUTION: Bangladesh, India, Indonesia, Malaysia, Thailand, and Philippines.

SYMPTOMS: The virus causes severe damage only in areas where host plants and vectors multiply year-round. The most obvious symptoms are stunting and leaf discoloration that varies from several shades of yellow to orange. Discoloration starts at leaf tip and may sometimes extend to the lower part of the leaf. Young leaves may be mottled and older leaves have rusty-colored specks of various sizes. The number of tillers may be reduced.

Discoloration of leaves in moderately resistant cultivars may gradually disappear at later stages. Susceptible cultivars may be severely stunted, discolored, and die. Three strains of tungro virus are known to exist, S, M, and T; each produces distinct symptoms on certain rice cultivars. The S strain causes interveinal chlorosis, giving the appearance of yellow stripes. The M strain produces mottling. The T strain causes narrow leaf blades in some cultivars and causes striping in other cultivars.

CONTROL:
1. Plant resistant cultivars.
2. Treat seed with a systemic insecticide.
3. Adjust planting date to correspond to low numbers of vectors.

Unknown Virus-Caused Disease

Disease may be rice wilted stunt disease.

CAUSE: Unknown, but transmitted by a brown planthopper. Causal agent is possibly a virus and persists in the vector for a lengthy time, even after molting.

DISTRIBUTION: Philippines, particularly where brown planthopper populations are high.

SYMPTOMS: Initial symptoms occur 7–14 days after inoculation. Infected plants are stunted. Leaves are yellow and normally narrow and erect. This gives an infected plant an erect growth plant. Irregular brownish blotches and chlorotic streaks resembling a mosaic-type of mottling may be observed on the discolored leaves. Infected plants die prematurely.

CONTROL: No control is reported.

Waika Disease

CAUSE: Rice waika virus (RWV) is transmitted by the green rice leafhoppers *Nephotettix cincticeps* Uhler, *N. virenscens* Distant, *N. migropictus* Stal., and *N. malayanus* Ishihara & Kawase. Different strains designated as C and S are present in nature. RWV may be related to or the same as rice tungro spherical virus. RWV overwinters in ratoons.

DISTRIBUTION: Japan.

SYMPTOMS: Symptoms caused by RWV somewhat resemble those caused by tungro virus. Symptoms of strain C occur after the booting stage. Infected plants have slight stunting, slight discoloration and drooping of leaf blades, poor root systems, and delay of flowering time.
   Symptoms of strain S are leaf discoloration and appearance of irregularly-shaped brown blotches on discolored leaves. Plants are severely stunted with few tillers, delayed flowering, and a reduction in yield.

CONTROL: Plant resistant cultivars.

Wilted Stunt Disease

CAUSE: There is a question if the causal organism (GSW) is a strain of rice grassy stunt virus or a mycoplasmalike organism. GSW is transmitted by the rice brown planthopper *Nilaparvata lugens* (Stal.) in a persistent manner.

DISTRIBUTION: Taiwan.

SYMPTOMS: Symptoms vary somewhat depending on cultivar. Rusty spotting and yellowings occur on the lower basal leaves. The central, newly unfolded leaves are pale. The rusty spotting and yellowing of leaves later intensifies and plants wilt. Infected plants produce fewer tillers, the number reducing in winter but increasing slightly in summer. Leaf twisting and trapping of unfolded leaves also occurs in some cultivars. On other cultivars leaves are pale green, shortened, narrow, usually curled inward, and brittle. Tillering may increase during summertime. On other cultivars, young leaves are pale green to pale yellow and exhibit conspicuous mottling. Sometimes vague, chlorotic stripes form along veins; however, these eventually disappear. Plants may be severely stunted with few tillers. Eventually leaves dry from the basal portion.
   Severely infected plants die. Others that survive produce no or very few poorly developed panicles.

CONTROL: No control is reported.

Yellow Dwarf

CAUSE: Yellow dwarf virus. Recent evidence suggests the causal agent may be a mycoplasmalike organism. It overwinters in leafhoppers and in the wild grass *Alopecurus aequalis* Sobol. The causal agent is disseminated primarily by the leafhopper *Nephotettix cincticeps* Uhler and also by *N. impicticeps* Ishihara and *N. apicalis* Motschulsky. The leafhoppers acquire the causal agent by feeding on diseased plants for 1–3 hours. There is a 20–39-day incubation period, after which the inoculation feeding time is usually less than 1 hour. The latent period in rice is about 1 month in warm weather and 3 months in cool weather.

DISTRIBUTION: Tropical Asia.

SYMPTOMS: Infected plants are generally chlorotic, stunted, and produce numerous tillers. The uniformly pale green or pale yellow chlorosis first appears on the newest leaves and continues on the successive leaves. Sometimes only faint mottling occurs. Infrequently early-infected plants may die but usually they survive, producing abnormal or no heads.

CONTROL:
1. Plant a resistant cultivar.
2. Apply an insecticide to kill overwintering insects.

Yellow Mottle Disease

CAUSE: Yellow mottle virus (YMV) is transmitted mechanically and by adult beetles, *Sesselia pusilla* (Gerstaecker).

DISTRIBUTION: Kenya.

SYMPTOMS: Infected plants are stunted with fewer tillers than normal plants. Leaves are crinkled, mottled, with yellowish streaks. Infrequently leaves may be spirally twisted. Panicles may only partially emerge and may be malformed. Infected plants may die.

CONTROL: No control is reported.

DISORDERS OF ABIOTIC OR UNKNOWN CAUSE

Panicle Blight

CAUSE: Unknown.

DISTRIBUTION: Not reported.

SYMPTOMS: Aborted florets and poor yields.

CONTROL: No control is reported.

Straighthead

CAUSE: The exact cause is not known, but may be due to unfavorable soil conditions aggravated by prolonged flooding. It is more severe on sandy loam soils but seldom occurs on clay soils. It has often occurred when soil contained too much undecayed plant material that had been plowed under. In some limited areas it has been caused by an accumulation of arsenic as a result of repeated applications of cotton insecticides.

DISTRIBUTION: Colombia, Japan, Portugal, and the United States.

SYMPTOMS: Rice heads remain upright at maturity because the few grains formed are too light to bend them over. Diseased heads often contain no fertile seed and hulls are distorted into a crescent of parrot beak form, especially on long-grain cultivars. One or both hulls and other flower parts are missing. In severe cases, heads are smaller than normal and emerge slowly or incompletely from the boot. Affected plants remain green and may produce shoots from the lower nodes. Affected seeds may have a low abnormal germination.

CONTROL:
1. Plant resistant cultivars. No cultivar is immune or highly resistant.
2. Drain fields just prior to the stem elongation stage.

Sekiguchi Lesion

CAUSE: Sekiguchi lesion is conditioned by a single recessive gene, sl, and can be induced by a number of biotic and abiotic agents. Sekiguchi lesion is likely a manifestation of a flaw in the biological mechanisms that regulate the hypersensitive response of plants homozygous for the sl gene to pathogenic agents.

DISTRIBUTION: Japan and the United States.

SYMPTOMS: Initially lesions are 1–2 mm in diameter, water-soaked gray spots that enlarge rapidly into diurnally zonate orange-brown areas. Gradually lesions coalesce until the whole plant is affected.

CONTROL: No control is reported.

——————————BIBLIOGRAPHY——————————

Ahn, S.-W. 1980. Eyespot of rice in Colombia, Panama, and Peru. *Plant Disease* **64**:878–880.

Anjaneyulu, A.; Shukla, V. D.; Rao, G. M.; and Singh, S. K. 1982. Experimental host range of rice tungro virus and its vectors. *Plant Disease* **66**:54–56.

Atkins, J. G. 1972. Rice diseases. *USDA Farmers Bull. No. 2120.*

Attere, A. F., and Fatokum, C. A. 1983. Reaction of *Oryza glaberrima* accessions to rice yellow mottle virus. *Plant Disease* **67**:420–421.

Bajet, N. B.; Aguiero, V. M.; Daquioag, R. D.; Jonson, G. B.; Cabunagan, R. C.; Mesina, E. M.; and Hibino, H. 1986. Occurrence and spread of rice tungro spherical virus in the Philippines. *Plant Disease* **70**:971–973.

Bakr, M. A., and Miah, S. A. 1975. Leaf scald of rice, a new disease in Bangladesh. *Plant Dis. Rptr.* **59**:909.

Baravidan, M. R.; Mew, T. W.; and Aballa, T. 1982. Some studies on bacterial stripe of rice. *Philipp. Phytopathol.* **18**:9 (abstract).

Bockus, W. W.; Webster, R. K.; and Kosuge, T. 1978. The competitive saprophytic ability of *Sclerotium oryzae* derived from sclerotia. *Phytopathology* **68**:417–421.

Bockus, W. W.; Webster, R. K.; Wick, C. M.; and Jackson, L. F. 1979. Rice residue disposal influences overwintering inoculum level of *Sclerotium oryzae* and stem rot severity. *Phytopathology* **69**:862–865.

Cabauatan, P. Q. 1882. An unknown disease of rice transmitted by the brown planthopper *Nilaparvata lugens* in the Philippines. *Philipp. Phytopathol.* **18**:13 (abstract).

Cabauatan, P. Q.; Hibino, H.; Labis, D. B.; Omura, T.; and Tsuchizaki, T. 1984. Rice grassy stunt virus 2: A new strain of rice grassy stunt in the Philippines. *Philipp. Phytopathol.* **20**:9 (abstract).

Chattopadhyay, S. B., and Dasgyotam, C. 1959. *Helminthosporium rostratum* Drechs. on rice in India. *Plant Dis. Rptr.* **43**:1241–1244.

Chen, C. C., and Chiu, R. J. 1982. Three symptomatologic types of rice virus diseases related to grassy stunt in Taiwan. *Plant Disease* **66**:15–18.

Choong-Hoe, K.; Rush, M. C.; and Mackenzie, D. R. 1985. Effect of water management on the epidemic development of rice blast. *Phytopathology* **75**:501. (abstract).

Cother, E. J. 1974. Bacterial glume blotch of rice. *Plant Dis. Rptr.* **58**:1126–1129.

Danquah, O. A. 1975. Occurrence of *Phoma glomerata* on rice (*Oryza sativa*)—a first record in Ghana. *Plant Dis. Rptr.* **59**:844–845.

Deighton, F. C., and Shaw, D. 1960. White leaf streak of rice caused by *Ramularia oryzae* sp. nov. *Brit. Mycol. Soc. Trans.* **43**:516–518.

Fauquet, C., and Thouvenel, J. C. 1977. Isolation of the rice yellow mottle virus in Ivory Coast. *Plant Dis. Rptr.* **61**:443–446.

Ghosh, S. K. 1981. Weed hosts of rice necrosis mosaic virus. *Plant Disease* 65:602–603.

Goto, M. 1979. Dissemination of *Erwinia chrysanthemi*, the causal organism of bacterial foot rot of rice. *Plant Dis. Rptr.* 63:100–103.

Goto, M. 1979. Bacterial foot rot of rice caused by a strain of *Erwinia chrysanthemi*. *Phytopathology* 69:213–216.

Gunnell, P. S., and Webster, R. K. 1983. Sheath blight of rice in California. *Phytopathology* 73:796 (abstract).

Gunnell, P. S., and Webster, R. K. 1984. Aggregate sheath spot of rice in California. *Plant Disease* 68:529–531.

Gunnell, P. S., and Webster, R. K. 1985. The effect of cultural practices on aggregate sheath spot of rice in California. *Phytopathology* 75:1340 (abstract).

Gunnell, P. S., and Webster, R. K. 1985. The perfect state of *Rhizoctonia oryzae-sativae:*Causal organism of aggregate sheath spot of rice. *Phytopathology* 75:1383 (abstract).

Harder, D. E., and Bakker, W. 1973. African cereal streak, a new disease of cereals in East Africa. *Phytopathology* 63:1407–1411.

Hibino, H. 1983. Transmission of two rice tungro-associated viruses and rice waika virus from doubly or singly infected source plants by leafhopper vectors. *Plant Disease* 67:774–777.

Hibino, H.; Cabauatan, P. Q.; Omura, T.; and Tsuchizaki, T. 1985. Rice grassy stunt virus strain causing tungrolike symptoms in the Philippines. *Plant Disease* 69:538–541.

Hibins, H.; Roechan, M.; and Sudarisman, S. 1978. Association of two types of virus particles with penyakit habang (tungro disease) of rice in Indonesia. *Phytopathology* 68:1412–1416.

Imolehin, E. D. 1983. Rice seedborne fungi and their effect on seed germination. *Plant Disease* 67:1334–1336.

Inoue, H. 1978. Strain S, a new strain of leafhopper-borne rice waika virus. *Plant Dis. Rptr.* 62:867–871.

Inoue, H., and Omura, T. 1982. Transmission of rice gall dwarf virus by the green rice leafhopper. *Plant Disease* 66: 57–59.

John, V. T.; Heu, M. H.; Freeman, W. H.; and Manandhar, D. N. 1979. A note on dwarf disease of rice in Nepal. *Plant Dis. Rptr.* 63:784–785.

Kato, S.; Nakanishi, T.; Takahi, Y.; and Nakagami, K. 1985. Studies on Pythium damping-off of rice seedlings: (1) Pythium species associated with damping-off in the early growth stage of rice seedlings in nursery flats. *Phytopathol. Soc. Japan Ann.* 51:159–167.

Kato, S.; Nakanishi, T.; Takahi, Y.; Nakagami, K.; and Ogawa, M. 1985. Studies on Pythium damping-off of rice seedlings: (2) Pythium species associated with damping-off at the middle and latter growth stages of rice seedlings in nursery flats. *Phytopathol. Soc. Japan Ann.* 51:168–175.

Kauffman, H. E., and Reddy, A. P. K. 1975. Seed transmission studies of *Xanthomonas oryzae* in rice. *Phytopathology* 65:663–666.

Kondaiah, A.; Rao, A. V.; and Srinivasan, T. E. 1976. Factors favoring spread of rice "tungro" disease under field conditions. *Plant Dis. Rptr.* 60:803–806.

Krishna, P. G., and Rush, M. C. 1983. Role of *Pythium spinosum* in the fungal complex causing the water-mold disease of water-seeded rice. *Phytopathology* **73**:502 (abstract).

Kuwata, H. 1985. *Pseudomonas syringae* pv. *oryzae* pv, nov., causal agent of bacterial halo blight of rice. *Phytopathol. Soc. Japan Ann.* **51**:212–218.

Lee, F. N., and Rush, M. C. 1983. Rice sheath blight: A major rice disease. *Plant Disease* **67**:829–832.

Leu, L. S., and Yang, H. C. 1985. Distribution and survival of sclerotia of rice sheath blight fungus, *Thanatephorous cucumeris*, in Taiwan. *Phytopathol. Soc. Japan Ann.* **51**:1–7.

Lindberg, G. D. 1970. Loss of rice yield caused by stem rot. *Phytopathology* **60**:1300 (abstract).

Ling, K. C. 1975. *Rice Virus Diseases*. International Rice Research Institute, Los Banos, Philippines.

Ling, K. C.; Tiongco, E. R.; and Aguiero, V. M. 1978. Rice ragged stunt, a new virus disease. *Plant Dis. Rptr.* **62**:701–705.

Lozano, J. C. 1977. Identification of bacterial leaf blight in rice, caused by *Xanthomonas oryzae*, in America. *Plant Dis. Rptr.* **61**:644–648.

Marchetti, M. A. 1983. Potential impact of sheath blight on yield and milling quality of short-statured rice lines in the southern United States. *Plant Disease* **67**:162–165.

Marchetti, M. A.; Bollich, C. N.; and Vecker, F. A. 1983. Spontaneous occurrence of the Sekiguchi lesion in two American rice lines: Its induction, inheritance and utilization. *Phytopathology* **73**:603–606.

Marin-Sanchez, J. P., and Jimenez-Diaz, R. M. 1982. Two new Fusarium species infecting rice in southern Spain. *Plant Disease* **66**:332–334.

Mew, T. W., and Rosario, M. B. 1980. Bacterial foot rot of rice. *Philipp. Phytopathol.* **16**:5 (abstract).

Morinaka, T.; Putta, M.; Chettanachit, D.; Parejarearn, A.; Disthaporn, S.; Omura, T.; and Inoue, H. 1982. Transmission of rice gall dwarf virus by cicadellid leafhoppers *Recilia dorsalis* and *Nephotettix nigropictus* in Thailand. *Plant Disease* **66**:703–704.

Mueller, K. E. 1974. *Field Problems of Tropical Rice*. International Rice Research Institute, Los Banos, Philippines.

Nuque, F. L.; Aguiero, V. M.; and Ou, S. H. 1982. Inheritance of resistance to grassy stunt virus in rice. *Plant Disease* **66**:63–64.

Omura, T.; Inoue, H.; Morinaka, T.; Saito, Y.; Chettanachit, D.; Putta, M.; Parejarearn, A.; and Disthaporn, S. 1980. Rice gall dwarf, a new virus disease. *Plant Disease* **64**:795–797.

Ou, S. H. 1973. *A Handbook of Rice Diseases in the Tropics*. International Rice Research Institute, Los Banos, Philippines.

Ou, S. H. 1986. *Rice Diseases*. 2nd ed. Commonwealth Mycological Institute, Kew, Surrey, England. Eastern Press Ltd., London, 391p.

Ou, S. H.; Nuque, F. L.; and Vergel de Dios, T. I. 1978. Perfect stage of *Rhynchosporium oryzae* and the symptoms of rice leaf scald disease. *Plant Dis. Rptr.* **62**:524–528.

Padwick, W. 1950. *Manual of Rice Diseases*. Commonwealth Mycological Institute, Kew, Surrey, England. Oxford University Press, England.

Pillay, M.; Schneider, R. W.; and Rush, M. C. 1986. Association of *Pythium* spp. with seedling disease and feeder root decline in rice. *Abstracts of Presentations 295*. American Phytopathological Society.

Prabhu, A. S., and Bedendo, I. P. 1983. Basal node of rice caused by *Fusarium oxysporum* in Brazil. *Plant Disease* **67**:228–229.

Reddy, A. P. K.; Mackenzie, D. R.; Rouse, D. I.; and Rao, A. V. 1979. Relationship of bacterial leaf blight severity to grain yield of rice. *Phytopathology* **79**:967–969.

Reddy, P. R., and Mohanty, S. K. 1981. Epidemiology of the kresek phase of bacterial blight of rice. *Plant Disease* **65**:578–580.

Rivera, C. T.; Ou, S. H.; and Pathak, M. D. 1963. Transmission studies on the orange-leaf disease of rice. *Plant Dis. Rptr.* **47**:1045–1048.

Saito, Y.; Chaimongkol, U.; Singh, K. G.; and Hino, T. 1976. Mycoplasmalike bodies associated with rice orange leaf disease. *Plant Dis. Rptr.* **60**:649–651.

Shahjahan, A. K. M.; Harahap, Z.; and Rush, M. C. 1977. Sheath rot of rice caused by *Acrocylindrium oryzae* in Louisiana. *Plant Dis. Rptr.* **61**:307–310.

Shakya, D. D.; Vinther, F.; and Mathur, S. B. 1985. Worldwide distribution of a bacterial stripe pathogen of rice identified as *Pseudomonas avenae*. *Phytopathol. Zeitschr.* **114**:256–259.

Shukla, V. D., and Anjaneyulu, A. 1981. Adjustment of planting date to reduce rice tungro disease. *Plant Disease* **65**:409–411.

Shukla, V. D., and Anjaneyulu, A. 1981. Plant spacing to reduce rice tungro incidence. *Plant Disease* **65**:584–586.

Singh, R. A., and Raju, C. A. 1981. Some observations on sheath rot of rice. *Philipp. Phytopathol.* **17**:5 (abstract).

Thomas, M. D.; Mayango, D.; and Oberly, W. 1985. Suppressions and elimination of *Rhynchosporium oryzae* by benomyl in rice foliage and seed in Liberia. *Plant Disease* **69**:884–886.

Walawala, J. J., and Davide, R. G. 1984. Pathogenicity, damage assessment, and field population pattern of *Hirschmanniella oryzae* in rice. *Philipp. Phytopathol.* **20**:39–44.

Whitney, N. G. 1980. Brown blotch, a new disease of rice. *Phytopathology* **70**:572 (abstract).

Zaragoza, B. A., and Mew, T. W. 1979. Relationship of root injury to the "Kresek" phase of bacterial blight of rice. *Plant Dis. Rptr.* **63**:1007–1011.

Ziegler, R. S., and Alvarez, E. 1987. Bacterial sheath brown rot of rice caused by *Pseudomonas fuscovaginae* in Latin America. *Plant Disease* **71**:592–597.

15

Diseases of Rye
(*Secale cereale* L.)

BACTERIAL CAUSES

Bacterial Blight

This disease is called black chaff on other crops and bacterial leaf blight.

CAUSE: *Xanthonomonas campestris* pv. *translucens* (Jones, Johnson & Reddy) Dye (syn. *Xanthomonas translucens* (Jones, Johnson & Reddy) Dows f. sp. *secalis* (Reddy, Godkin & Johnson) Hagborn) is seedborne and overwinters in residue. Initial infection comes from seedborne bacteria that are disseminated by leaf to leaf and plant to plant contact, splashing rain, and possibly insects. Bacteria enter into plants through natural openings such as stomates and wounds.

DISTRIBUTION: Australia and North America.

SYMPTOMS: Symptoms occur after several days of damp or rainy weather. Infected plants may be stunted but this is not usually noticed until plants mature. Small water-soaked spots occur on the youngest leaves and sheaths of older plants and sometimes on seedlings. The spots enlarge and may coalesce, becoming glossy and translucent, eventually turning yellow or brown with an irregular margin. The spots when coalesced do not become as stripelike as bacterial blight on barley.

Severely diseased leaves die back from the tip. Under humid conditions, early in the morning, droplets of milky bacterial exudate may be seen on the surface of diseased spots. The droplets dry into hard, yellowish granules that may be removed easily as dry flakes from the leaf surface.

Water-soaked areas may occur on chaff of rye heads but bacterial exudate is not plentiful. Grain is not destroyed but it may be brown, shrunken, and carry bacteria to infect next year's crop. If a flag leaf is infected, the head may not emerge from the boot but break through the side of the sheath and be destroyed and blighted.

Bacterial blight on rye can be diagnosed in the field in the same manner as bacterial blight on barley. Cut a newly developed spot crosswise. Either piece will exude a bead of milky ooze at the cut edge.

CONTROL:

1. Do not plant seed from diseased plants.
2. Treat seed with a fungicide as a precaution even though bacterial blight is not known to be present.
3. Rotate rye with other crops. The causal bacterium that infects rye does not infect barley.

Halo Blight

CAUSE: *Pseudomonas syringae* pv. *coronafaciens* (Elliott) Young, Dye & Wilkie (syn. *P. coronafaciens* (Elliott) Stevens) may be seedborne. Disease development is most severe under cool, moist conditions.

DISTRIBUTION: Generally distributed wherever rye is grown; however, the disease is rarely severe enough to be economically important.

SYMPTOMS: Initially minute, brown spots occur on leaf sheaths, blades, or along leaf margins often in association with frost injury. Yellow halos up to 15 mm in diameter surround the spots in a few days. Eventually the center of a spot becomes necrotic with a narrow chlorotic halo. Lesions may coalesce, causing the entire leaf to turn brown and giving a scalded appearance that is most common from flowering to maturity.

Small halos may occur on glumes but usually the entire floret is blighted. Individual florets on heads are sterile and turn white on early tillers. On later tillers, all florets of some heads are infected and produce no seed or greatly shriveled grain. Later in the season, as temperatures increase, few or no new lesions are formed.

Atypical symptoms reported from Virginia are as follows: Plants were slightly to severely stunted. Blades on some plants had chlorotic and necrotic lesions but no halo blight was observed. On other plants, entire blades were yellowed. Leaves of stunted plants often had chlorotic veins.

CONTROL: No control is reported.

FUNGAL CAUSES

Anthracnose

CAUSE: *Colletotrichum graminicola* (Ces.) G. W. Wilson is an excellent saprophyte and overwinters as mycelium and conidia on winter cereals or residue. Conidia are produced during wet weather at an optimum temperature of 25°C and are windborne to healthy plants.

DISTRIBUTION: Generally distributed wherever rye is grown.

SYMPTOMS: Rye is one of the most susceptible cereals to infection by *C. graminicola*. Symptoms are most severe during wet, warm (25°C optimum) weather. Anthracnose is most likely to occur when rye is grown in a rotation with another cereal on coarse soils that are low in fertility.

Symptoms become apparent toward plant maturity. Premature ripening or whitening, a general reduction in plant vigor, or dying are gross symptoms. Specific symptoms become manifest most often on the lower part of the plant because primary infection starts there. The crown and bases of some stems become bleached and later turn brown. Later small, black, raised spots, which are acervuli, develop on the surface of the lower leaf sheaths and culms. Acervuli may develop on leaves of dead plants when moisture is plentiful. Round to oblong lesions bearing acervuli may occur on green leaves. Under magnification, acervuli will appear as a clump of spines. Kernels may be shriveled if infection occurs early. A seedling and crown infection may occur under severe disease conditions.

CONTROL:
1. Rotate rye with a noncereal or grass because *C. graminicola* can infect a wide range of grasses and cereals. Legumes are not susceptible and would be a suitable crop in a rotation.
2. Improve soil fertility.

Cephalosporium Stripe

CAUSE: *Cephalosporium gramineum* Nis. & Ika. (syn. *Hymenula cerealis* Ell. & Ev.), survives as conidia and mycelium in residue and in the top 8 cm of soil for up to five years. Conidia serve as a primary inoculum and infect roots during winter and early spring through mechanical injuries caused by soil heaving and insects. Infection is most severe in wet, acid soils (pH 5.0) due to increased fungal sporulation and root growth. The subcrown internode can also be infected just as the seed is germinating. After infection, conidia enter xylem vessels and are carried upward where they lodge and multiply at nodes and leaves, not doing any further damage to roots. The fungus prevents water movement up the plant and also produces metabolites that are harmful to the plant. At harvest time, *C. gramineum* is returned to the soil in residue where it is a successful saprophytic competitor with other soil organisms.

DISTRIBUTION: Great Britain, Japan, and in most winter rye-growing areas of North America.

SYMPTOMS: Winter rye is most severely infected. Spring rye is susceptible but apparently escapes the disease and usually does not show symptoms. Infected plants are scattered throughout a field, and are more numerous in lower, wetter areas. Infected plants are dwarfed. During jointing and heading, distinct yellow stripes that number one to four per leaf but usually one or two, develop on leaves, sheaths, and stems and usually occur the length of the plant. Thin, brown lines consisting of infected veins usually occur in the middle of a stripe. The stripes eventually become brown, are highly visible on green leaves, and are still noticeable on yellow straw. Toward harvest, the culm at or below the node may become dark. Heads of infected plants are white and do not contain seed, or if seed is present, it is usually shriveled.

CONTROL:
1. Plant cultivars that are tolerant; no cultivars are resistant.
2. Rotate rye for at least two years with a noncereal.
3. Plant later in the autumn or when the soil temperature at 10 cm soil depth is below 13°C. Such plants apparently have limited root growth, thus reducing the number of infection sites.
4. Infected residue should be plowed deeper than 8 cm where feasible.

Common Bunt

Cause, distribution, symptoms, and control are similar to those of wheat. For further information see Common Bunt on Wheat, Chapter 23.

Common Root Rot

CAUSE: *Bipolaris sorokiniana* (Sacc.) Shoemaker (syn. *Helminthosporium sativum* Pammel, King & Bakke) and *Fusarium* spp. survive in soil as saprophytes on residue. Additionally conidia of *B. sorokiniana* and chlamydospores of *Fusarium* spp. survive for several months in soil and are seedborne. Primary infections occur on coleoptiles, primary roots, and subcrown internodes. The infected plant normally does not die as long as it can produce new roots. Conidia are produced when infection progresses aboveground.

Plants under stress due to drouth, warm temperatures, lack of nutrition, and insect injury are most subject to infection. Moisture is required for infection, but once initiated, disease development requires warm temperatures and moisture stress.

DISTRIBUTION: Generally distributed wherever rye is grown.

SYMPTOMS: Seedlings may be killed either preemergence or postemergence, particularly in dry soil when inoculum is seedborne. Surviving seedlings have brown lesions on coleoptiles, roots, and culms. Darkening of the subcrown internode is usually caused by *B. sorokiniana*. *Fusarium* spp. infect secondary roots as they emerge from the crown. Infections of the crown and foot usually kill plants. Diseased plants occur in random patches, are stunted, lighter green, and mature early. Additionally, diseased plants have few tillers, and the heads are bronzed, bleached, or white-headed and contain shriveled seed. There may be a browning of root systems that can be observed only when roots are washed. *Fusarium* spp., in particular, cause roots, culm bases, and lower nodes to dry and darken. Diseased plants then become brittle and break off easily near the soil level.

CONTROL:
1. Treat seed with a seed-protectant fungicide.
2. Soil should have proper fertility to ensure growth of vigorous plants that are able to produce new roots and overcome root rot.
3. Plant rye in late autumn.

Downy Mildew

CAUSE: *Sclerophthora macrospora* (Sacc.) Thirum, Shaw & Naras. survives for years as oospores in residue or soil when infected tissues eventually decay. It is an obligate parasite and cannot grow saprophytically on dead plant tissue. Oospores are disseminated on seed and in soil residue carried by wind or water. Oospores germinate in wet soil to produce sporangia. Zoospores are then produced within sporangia, released, and swim through soil water to seedlings. Infection occurs by a zoospore germinating to form a germ tube that penetrates the plant. Oospores may survive for months in dry soil and germinate to form only a germ tube that directly penetrates a host. Infection occurs over a wide temperature range of 7–31°C. Following infection, *S. macrospora* develops systemically within the plant, particularly in systemic tissue.

DISTRIBUTION: Generally distributed wherever rye is grown. The disease is most severe in wetter areas in a field.

SYMPTOMS: Downy mildew occurs only in localized areas of fields where seedlings have been growing in flooded or water-logged soil for 24 hours or longer.   Plants are dwarfed, deformed, and may tiller excessively. Infected plants have leathery, stiff, thickened leaves, stems, and heads. Affected parts are twisted and distorted and no seeds are formed in severely diseased plants. In less severely diseased plants, dwarfing may be slight, with one or more of the upper leaves stiff, upright, or variously curled and twisted. The heads and stems may not be deformed. Numerous round, yellow-brown oospores may be found in diseased tissue examined under magnification.

CONTROL:
1. Provide proper soil drainage where possible.
2. Control grassy weeds that may serve as hosts.
3. Plant cleaned seed from disease-free plants to ensure no infested residue is disseminated with seed.

Ergot

CAUSE: *Claviceps purpurea* (Fr.) Tul. survives as sclerotia on or in the soil. Just before blossoming, a sclerotium germinates to produce one or many stipes (stalks), each bearing a stromatic head in which perithecia are formed. Eventually ascospores are produced in the perithecia and are windborne to flowers. Ascospores germinate to produce a germ tube that grows into the young ovary. Conidia are then produced from mycelium growing in the ovary in a sweet, sticky liquid called honeydew. Insects are attracted to honeydew and inadvertently disseminate conidia to healthy flowers as a means of secondary spread. The sclerotial body replaces the kernel in the infected flower. While the sclerotia usually occur after the conidial stage, under some conditions, particularly in tropical areas, no sclerotia are produced. As the sclerotia enlarge, the floral bracts spread apart and the dark sclerotial bodies are evident as they protrude beyond the floral bracts.

   Several grasses and cereals are susceptible to *C. purpurea*; however, ergot is usually more severe on rye. Rye is predisposed to ergot by barley yellow dwarf virus infection.

DISTRIBUTION: Generally distributed wherever rye is grown.

SYMPTOMS: Few to many sclerotia occur on a spike or panicle. The sclerotia or ergot bodies are recognized as purple-black, horn-shaped bodies that are produced in place of seed. Prior to sclerotial formation, honeydew and conidia accumulate in liquid droplets or adhere to the surface of the floral structures. Insects feeding on honeydew are conspicuous around the infected spikes. Saprophytic fungi grow on the honeydew and give the infected spike a black or sooty appearance.

CONTROL:
1. Deep plowing will place sclerotia in soil where they will not effectively germinate and eventually decompose.
2. Control grassy weeds since they may be infected by *C. purpurea* and serve as a source of inoculum.
3. Rotate rye with crops such as legumes or corn that are not susceptible to *C. purpurea*.
4. Clean seed to remove any ergot bodies present.

Eyespot

This disease is also called Cercosporella footrot, stem break, foot rot, strawbreaker, and culm rot.

CAUSE: *Pseudocercosporella herpotrichoides* (Fron) Deighton (syn. *Cercosporella herpotrichoides* Fron.) survives for several years as mycelium in residue. Conidia are produced during cool, damp weather in the autumn or spring and infect crown and basal culm tissue but not roots.

DISTRIBUTION: Generally distributed where rye is grown in cool, moist climates.

SYMPTOMS: Winter rye is more likely to be infected than spring rye. This disease is most conspicuous near the end of the growing season by lodging of diseased plants and presence of whiteheads. Initially eye-shaped or ovate lesions with white to tan centers and brown margins develop on the basal leaf sheath. Similar spots form on the stem directly beneath those on the sheath and cause the lodging. Roots are not infected but a necrosis occurs around the upper crown nodes. Under moist conditions, lesions enlarge and a black stromalike mycelium develops over the surface of the crown and base of culms, giving the tissues a charred appearance. Stems either shrivel and collapse, or plants are yellowish or pale green with reduced size and number of heads. Early infection causes individual culms and plants to be killed before maturity.

CONTROL: No satisfactory control exists except to rotate rye with noncereals or grass crops such as legumes.

Glume Blotch

CAUSE: *Septoria nodorum* Berk., teleomorph *Leptosphaeria nodorum* Muller, survives as mycelium in live plants and seed, and as pycnidia on residue for two to three years. During warm (20–27°C), wet weather in the autumn or spring, pycnidiospores are exuded from pycnidia within a gelatinous drop (cirrhi) that protects them from drying out and radiation. Pycnidiospores are disseminated by splashing and blowing rain to lower leaves. Infection requires six or more hours of wetness. New spores are produced in 10–20 days. Ascospores are produced in perithecia as rye matures in late summer or early autumn and are windborne.

DISTRIBUTION: Generally distributed wherever rye is grown.

SYMPTOMS: Leaves and stems are infected but not heads. Nodes turn brown, shrivel, and are speckled with black pycnidia. Such infections often cause straw to bend over and lodge just above the nodes.

Light brown spots occur on leaves that are similar in appearance to Septoria leaf blotch. A brown margin may surround leaf spots and pycnidia will be present on both surfaces of a diseased leaf. If a flag leaf is infected, the head may be deformed. Infection of the leaf sheath causes a dark brown lesion that may include most of the sheath. Severely infected plants are stunted.

CONTROL:
1. Plant certified, disease-free seed that has been cleaned and treated with a seed-protectant fungicide.
2. Plant resistant rye cultivars if they are adapted to the area.
3. Plow under infected residue where this is feasible.
4. Apply a foliar fungicide if conditions warrant it.
5. Rotate rye every third or fourth year with a nonsusceptible crop such as a legume.

Leaf Rust

This disease is also called brown rust.

CAUSE: *Puccinia rubigo-vera* (DC.) Wint. f. sp. *secalis* (Eriks.) Carl. is a heteroecious, long-cycle rust that has species of alkanet, *Anchusa* spp. as alternate hosts to rye and other *Secale* spp. However, the pycnial and aecial stages are rarely found in the United States.

The rust commonly overwinters as urediospores and mycelium in leaf tissue. Because of this overwintering, losses are greater in southern rye-growing areas where urediospores overwinter in greater numbers. During warm, wet weather in spring, urediospores are produced within uredia and are windborne. If warm, moist weather conditions continue, new uredia will be produced in 7–10 days. As the rye plants mature, telia are formed in which teliospores are produced. The teliospores will either germinate upon maturity in the autumn and produce basidiospores or overwinter and germinate in the spring. In the United States, *Anchusa* spp. have rarely been observed to be infected in nature. Therefore, the teliospores stage is not considered to be important in the etiology of the fungus, and mycelium and urediospores perform the overwintering function.

In Europe, teliospores will germinate in the autumn and produce basidiospores or overwinter and germinate the following spring. Aecia are observed on *Anchusa* spp. from spring until autumn. Therefore, pycnia and pycniospores are probably produced in autumn or more likely in the spring. The resulting aeciospores are windborne to rye and uredia are produced. However, mycelium and urediospores also overwinter and the aecial stage is also not thought to be essential in Europe. The disease is most severe under rainy conditions or cool, dewy nights and warm, wet days.

DISTRIBUTION: Generally distributed wherever rye is grown.

SYMPTOMS: Leaf rust of rye is similar in appearance to leaf rust of wheat and barley, and crown rust of oats. Uredia will appear as small, oval, orange-brown pustules on both leaf surfaces. As uredia rupture, powdery, red-brown urediospores are exposed. Similar pustules may also develop on leaf sheaths and stems. These are more elongated, resembling a short pustule of stem rust. Severely rusted plants may have pustules on necks and glumes also.

Telia appear toward plant maturity as small, elongated, dark gray pustules that do not immediately break through the epidermis. Several telia may form a circle. The pycnial and aecial stage on *Anchusa* spp. resemble those found on the alternate hosts of other leaf rusts.

CONTROL: There is no satisfactory control. Some rye cultivars apparently have more resistance than others. However, since rye is cross-pollinated, cultivars are not uniformly pure and do not display a high amount of resistance.

Leptosphaeria Leaf Spot

CAUSE: *Leptosphaeria herpotrichoides* de Notaris overwinters as mycelium in residue and ascospores in asci from pseudothecia on straw. Free water must occur on leaves for more than 48 hours for infection to proceed.

DISTRIBUTION: Canada, Europe, and the United States. It is considered a minor disease.

SYMPTOMS: Irregular, diffuse, yellow to tan spots occur on leaves.

CONTROL: Some cultivars are more resistant than others.

Loose Smut

CAUSE: *Ustilago tritici* (Pers.) Rostr. is seedborne, surviving as dormant mycelium within the seed embryo. A smutted head emerges from the boot a day or two earlier than heads of healthy plants. Chlamydospores are windborne to flowers of healthy heads. During cool temperatures of 16–22°C and some moisture provided by dews or light showers, chlamydospores germinate. The resulting germ tube penetrates the ovary of the flower and probably the stigma, with subsequent mycelial growth in the germ or embryo of the developing seed. As the grain matures, the loose smut fungus becomes dormant with the embryo of the seed until the following growing season. When infected seed germinates, mycelium grows systemically within the plant and replaces healthy kernels with smutted heads filled with chlamydospores.

DISTRIBUTION: Generally distributed wherever rye is grown; however, loose smut on rye is relatively uncommon.

SYMPTOMS: Infected seed does not show any outward symptoms and germination is not ordinarily affected. Infected seed gives rise to plants with smutted heads; however, the entire head of rye is seldom affected, and usually only a few kernels are affected. Smutted heads emerge a couple of days earlier than healthy heads. The brown to dark brown spore mass is enclosed within a fragile, gray membrane that soon ruptures, releasing the spores to be windborne to flowers. Soon an erect naked rachis that protrudes above the healthy plants is all that remains of a smutted head. At maturity, the barren rachis remains erect among the reclining healthy ones.

CONTROL: Since loose smut is uncommon on rye, control measures are usually not necessary. However, treating seed with a systemic fungicide will destroy the dormant mycelium in the ovary of the seed.

Pink Snow Mold

CAUSE: *Fusarium nivale* (Fr.) Ces., teleomorph *Calonectria nivalis* Schaff., oversummers as perithecia on lower leaf sheaths. In the autumn, leaf sheaths and blades near the soil level are infected by ascospores from perithecia and mycelium from residue. Roots are also infected. During cool, wet periods, with or without snow cover, mycelium grows from infected to healthy plants. In the spring, secondary infection may occur from conidia or ascospores. Disease is most severe under a heavy snow cover.

DISTRIBUTION: Central and northern Europe and North America.

SYMPTOMS: Damage usually corresponds to the pattern of snow cover. Pink mycelium and sporodochia are visible on living and dead tissue and on the soil surface at snow melt. Leaves of infected plants become chlorotic, then necrotic. Lesions that are somewhat rectangular and brown, surrounded by a darker brown band, occur on first and second leaves. Necrotic leaves remain intact and do not disintegrate. Unless crowns are infected, plants may recover during warm, dry weather despite extensive leaf infection.

CONTROL:
1. Plant winter rye later in the autumn.
2. Rotate rye with a noncereal crop such as a legume.
3. Plant rye in the spring. Most of the plant growth will be under conditions unfavorable to infection.

Powdery Mildew

CAUSE: *Erysiphe graminis* DC. ex Merat f. sp. *secalis* Em. Marchal overwinters as cleistothecia on residue. In areas where leaf tissue survives, the fungus may overwinter as mycelium. Ascospores serve as the primary inoculum during the spring in northern rye-growing areas. Most ascospores form within cleistothecia in the spring and are windborne. Conidia are produced almost as soon as mycelium becomes established on the leaf surface, especially during humid, cool weather but without the presence of free water. Conidia account for most of the secondary inoculum and spread of the disease during the growing season. Cleistothecia are formed on leaf surfaces as plants approach maturity. Powdery mildew ceases to be a problem when the weather becomes dry and warm later in the growing season.

DISTRIBUTION: Generally distributed wherever rye is grown.

SYMPTOMS: Rye is the most resistant of the cereals to powdery mildew. Disease is most severe on tender, rank-growing plants that have been heavily seeded or have had heavy applications of nitrogen fertilizer. Superficial mycelium and conidia appear as light gray or white spots on the upper surface of leaves, sheaths, and floral bracts. Affected plant parts appear as if they have been dusted with a gray powder. As the plant matures, spots enlarge, yellow, and darken. Most fungal growth occurs on the upper leaf surface and infrequently on the lower surface. Numerous small, round, dark cleistothecia develop on the affected areas and can be readily observed with magnification.

CONTROL: Most cultivars of rye are resistant to powdery mildew but foliar fungicides can be applied in unusual cases where the disease becomes severe.

Platyspora Leaf Spot

CAUSE: *Platyspora pentamera* (Karst.) Wehm. survives as perithecia in residue. During wet weather in the spring, ascospores are produced and are windborne to hosts. There must be 24–72 hours of continual moisture in order for infection to occur. Perithecia are produced in straw as rye matures.

DISTRIBUTION: North central United States and Canada.

SYMPTOMS: Nondescript yellow-brown spots are randomly produced on leaves. Dark perithecia are eventually produced as plants mature.

CONTROL: No specific controls are necessary.

Scab

Cause, distribution, symptoms, and control are similar to those on wheat. For further information see Scab on Wheat in Chapter 23.

Scald

This disease is also called Rhynchosporium leaf scald.

CAUSE: *Rhynchosporium secalis* (Oud.) Davis overwinters as stroma in lesions of perennial grasses or winter rye infected in the autumn and in residue. During cool, humid weather in the spring or autumn, conidia are produced on the stroma and are windborne to hosts. Secondary inoculum consists of conidia produced during cool, humid weather.

DISTRIBUTION: Scald occurs to some extent wherever rye is grown.

SYMPTOMS: Young spots are dark blue-gray, water-soaked blotches that occur mainly on leaves and occasionally on sheaths. Spots become oval or lens-shaped, with a light tan or white center surrounded by a straw-colored border that, in turn, is surrounded by a yellow-green halo whose margin gradually fades into the normal green of the leaf. Leaf spots enlarge in cool weather and may have a zonate appearance due to successive enlargements.

Spots at the base of a leaf may extend across the blade and down into the sheath, causing the death of the entire leaf. Elongated spots may cause the entire leaf to lose its color or a wide spot will cause leaf tissue beyond it to die.

CONTROL:
1. Rotate rye with resistant crops.
2. Plow under infected residue where feasible.
3. Destroy perennial grasses along the edge of a field.
4. Plant cultivars that are resistant or the most tolerant.

Septoria Leaf Blotch

CAUSE: *Septoria secalis* Prill & Del. survives as conidia within pycnidia in residue; and as mycelium in residue, lesions on live plants, and seed. There is some question whether *S. secalis* is the same as *S. tritici* Rob. ex Desm. Symptoms are similar but some reports state no secondary infection occurs with *S. tritici*. In the autumn, conidia are windborne and infect winter rye under moist conditions. During moist, cool conditions in the spring, rye is infected by conidia formed the previous season or the current spring.

DISTRIBUTION: Generally distributed wherever rye is grown.

SYMPTOMS: The first symptoms are light green to yellow spots that occur mostly between leaf veins. Eventually spots become light brown, irregular blotches with a speckled appearance due to presence of pycnidia within the blotch. Pycnidia are very small but are visible to the naked eye as tiny black or dark brown specks. After the leaf has died, blotches will be lighter than the surrounding tissue. Under moist conditions, leaves may die and the plant crowns may become infected, resulting in weakened and even dead plants.

CONTROL:
1. Apply a fungicide to foliage before disease becomes severe.
2. Plow under infected residue where feasible to aid in decomposition of infected tissue and to prevent conidia from being placed on soil surface where they can become more easily windborne.
3. Rotate rye with a nonsusceptible crop, preferably a legume.
4. Treat seed with a systemic seed-protectant fungicide to kill fungi on seed or any mycelium that may have grown into the seed coat.

Sharp Eyespot

This disease is also called Rhizoctonia root rot.

CAUSE: *Rhizoctonia solani* Kuhn survives as sclerotia in soil or mycelium in residue. Rye may be infected any time during the growing season. Sclerotia germinate to form mycelium or mycelium grows from a precolonized substrate to infect roots and culms, particularly in dry (less than 20% moisture holding capacity) and cool soils. Only anastomosis group (AG) 4 has been reported to be pathogenic to rye seedlings.

DISTRIBUTION: Generally distributed wherever rye is grown.

SYMPTOMS: Diamond-shaped lesions that resemble those of eyespot occur on lower leaf sheaths. Lesions are more superficial than eyespot lesions and have light tan centers with dark brown margins. Frequently dark mycelium is visible on them. Dark sclerotia may develop in lesions and between culms and leaf sheaths.

Seedlings may be killed but plants produce new roots to compensate for those rotted off. When roots are infected, plants may lodge and produce white heads. Diseased plants may appear stiff, have a grayish cast, and are delayed in maturity.

CONTROL: No control is truly effective. Vigorous plants growing in well-fertilized soil are not as likely to become severely diseased.

Speckled Snow Mold

This disease is also called Typhula blight.

CAUSE: *Typhula incarnata* Lasch ex. Fr. and *T. ishikariensis* Imai var. *idahoensis* (Remsb.) Arsvold & Smith oversummer as sclerotia in residue or soil and as mycelium on live plants. Neither fungus survives well as a saprophyte on infected residue. Infection occurs optimally at a temperature of 1–5°C. Sclerotia germinate during wet weather in the autumn and form either a basidiocarp or mycelium. Infection occurs by mycelium or basidiospores formed on the basidiocarp. Sclerotia are formed either within necrotic tissue or in mycelium. Further infection occurs by mycelial growth under the snow cover. *Typhula* spp. growth is more closely associated with snow cover than *Fusarium nivale* (Fr.) Ces., the cause of pink snow mold.

DISTRIBUTION: Canada, central and northern Europe, Japan, and the northwestern United States.

SYMPTOMS: At snow melt, a gray-white mycelium is present on plants and soil. Numerous sclerotia appear within living tissues and scattered in mycelium growing over plant surfaces, giving the infected plants a speckled appearance. *Typhula incarnata* can infect plants either aboveground or belowground. Dead leaves are common, but unless the crown is infected, plants recover during warm, dry weather yet may never be as vigorous as healthy plants. Leaves killed by these two fungi will crumple easily in contrast to leaves killed by *F. nivale* and will be covered with gray to white mycelium.

CONTROL: Rotation with a legume will help to reduce inoculum in the soil.

Spot Blotch and Associated
Seedling and Crown Rots

CAUSE: *Bipolaris sorokiniana* (Sacc.) Shoemaker (syn. *Helminthosporium sativum* Pammel, King & Bakke and *H. sorokinianum* Sacc. apud Sorokin.) overwinters as mycelium in residue, on seedling leaves of winter rye, or it may be seedborne. During moist weather in the spring, conidia are produced on residue and are windborne to cause infection of seedling leaves.

DISTRIBUTION: Generally distributed wherever rye is grown; however, damage is usually not severe.

SYMPTOMS: Seedborne inoculum results in seedling blight and root and crown rot in dry, warm soils. The first symptoms are dark brown, to black spots near the soil line or at the base of the sheaths that cover the seedling leaves. Infection may progress, turning seedlings yellow and killing seedlings either preemergence or postemergence. Seedling leaves of infected plants are dark green with dark brown lesions near the soil line that eventually extend into the leaf blade. Infected seedlings are dwarfed with excessive tillering. Crowns and roots may appear dark brown and rotted. With severe infection, heads may not emerge completely and kernels are poorly filled.

Lesions of various sizes and shapes appear on lower leaves after warm, moist weather. Individual lesions are round to oblong, dark brown, with a definite margin. Spots coalesce to form blotches that cover large areas of the leaf. Older lesions are olive-colored due to sporulation of the fungus. Heavily infected leaves will dry up completely. Lesions on floral bracts and kernels vary from small black spots to dark brown discoloration of the entire surface that results in a blackened end of the kernel. Head blight occurring early causes sterility or death of individual kernels soon after pollination.

CONTROL:
1. Treat seed with a seed-protectant fungicide.
2. Spray with a foliar fungicide to control leaf blotch phase before disease becomes too severe.
3. Rotate rye with a nonsusceptible crop.

Stalk Smut

This disease is also called flag smut, leaf smut, stem smut, and stripe smut.

CAUSE: *Urocystis occulata* (Wallr.) Rabh. (syn. *Erysibe occulata* Wallr.) overwinters mainly as chlamydospores on seed. Infection infrequently occurs from soilborne chlamydospores, particularly with winter rye planted in dry soil. Chlamydospores on seed germinate to produce basidiospores that infect seedlings. The fungus develops with the culm. Sori form in the parenchyma tissue between the veins of the culm and infrequently in the leaf blade. At first, sori composed of large numbers of chlamydospores are covered by the epidermis. At harvest time the sori is ruptured and chlamydospores are windborne to contaminate seed or soil. Chlamydospores in the soil will germinate if moisture is present and infect rye seedlings. If no host seedlings are present, the population of chlamydospores in the soil decreases.

DISTRIBUTION: Generally distributed wherever rye is grown.

SYMPTOMS: Stalk smut becomes evident just before heading. Diseased plants are stunted and may be overlooked or hidden among normal plants. Diseased plants are darker green than normal and lighter green streaks may be observed in the upper leaves. Soon the light green streaks become long, lead-colored, stripes on leaves, sheaths, and stems. These stripes are the sori or spore mass that contains chlamydospores. Stripes then become black and the epidermis splits, exposing the mass of chlamydospores that resembles a dark brown or black dust. Rarely does an infected plant head out. If heads are produced, sori may be found on chaff. Affected parts may be twisted and distorted with the leaves split along the stripes. Usually every stem of an affected plant is smutted.

CONTROL:
1. Treat seed with a seed-protectant fungicide to kill chlamydospores that have contaminated the seed surface.
2. Rotate rye with another crop. If stalk smut is a problem, do not grow rye continuously in the same field.

Stem Rust

Cause, distribution, symptoms, and control are similar to those of wheat. For further information see Stem Rust of Wheat in Chapter 23.

Stripe Rust

CAUSE: *Puccinia striiformis* West. is a rust fungus not known to have an alternate host. It oversummers between harvest and emergence of autumn-planted rye as urediospores on residual green cereals and grasses. Mycelium and infrequently urediospores overwinter on cereals such as wheat, barley, and rye, and on grasses. Urediospores are formed during cool, wet weather and are windborne to healthy plants to cause infection. Little infection occurs above 15°C in summer. Teliospores are formed but are not known to function as overwintering spores.

DISTRIBUTION: In North America, stripe rust occurs at the higher elevations and cooler climates along the Pacific Coast and intermountain areas from Canada to Mexico. It also occurs under the same environmental circumstances in South America and in the mountainous areas of central Europe and Asia. Stripe rust is not as common on rye as it is on wheat and barley.

SYMPTOMS: The most severe symptoms occur during cool, wet weather in the early growth period of rye. Symptoms occur early in the spring before other rusts, especially in areas with mild winters. Yellow uredia appear on autumn foliage and early in the spring on new foliage. Uredia coalesce to produce long stripes between veins of the leaf and sheath. Small linear lesions occur on floral bracts. Telia develop as narrow, linear, dark brown pustules covered by the epidermis.

CONTROL: Many rye cultivars have at least partial resistance.

Take-All

CAUSE: *Gaeumannomyces graminis* (Sacc.) von Arx & Oliver Var. *tritici* Walker (syn. *Ophiobolus graminis* Sacc.) survives as mycelium in live plants or as mycelium and perithecia in residue. However, ascospores are not considered important in the dissemination of the disease. In the autumn or spring, seedlings commonly become infected by roots growing into the vicinity of residue precolonized by *G. graminis* var. *tritici*. Plant to plant spread of take-all occurs by hyphae growing through the soil from an infected to a healthy plant or by a healthy root coming in contact with an infected one. Ascospores are produced in perithecia during wet weather but apparently are not disseminated a great distance either by splashing water or wind.

DISTRIBUTION: Generally distributed wherever rye is grown. Take-all is less severe on rye than on wheat and barley.

SYMPTOMS: Take-all is usually more severe on rye grown in alkaline soils. The first symptoms are light to dark brown necrotic lesions on the roots. By the time an infected plant reaches the jointing stage, most of its roots are brown and dead. At this point many plants die, or if plants are still alive, they are stunted and the leaves are yellow.

The disease becomes most obvious as plants approach heading. The stand is uneven in height and plants appear to be in several stages of maturity. Infected plants at heading have few tillers, ripen prematurely, and their heads are bleached and sterile. Roots are sparse, blackened, and brittle. Plants can be easily broken free of their crown when pulled from the soil. A very dark discoloration of the stem is visible just above the soil line, along with a mat of dark brown fungus mycelium under the lower sheath between the stem and the inner leaf sheaths.

CONTROL: Rotate rye with a nongrass crop since *G. graminis* var. *tritici* is not a good saprophyte and does not survive long in the soil in the absence of a host. Corn, *Zea mays* L., roots can become infected and carry over inoculum for subsequent crops.

Yellow Leaf Spot

This disease is also called tan spot, leaf spot, and blight.

CAUSE: *Pyrenophora tritici-repentis* (Died.) Drechs. (syn. *P. trichostoma* (Fr.) Fckl.), anamorph *Drechslera tritici-repentis* (Died.) Shoemaker, overwinters as pseudothecia on residue. During wet weather, in the spring, ascospores are released to infect healthy plants. Conidia and hyphal fragments may also serve as primary inoculum. During wet weather, in the growing season, conidia are produced in older lesions and are windborne to hosts, thus serving as secondary inoculum. In the autumn, pseudothecia are produced on infected culms and leaf sheaths.

DISTRIBUTION: Generally distributed wherever rye is grown; however, rye is usually not severely affected.

SYMPTOMS: Symptoms first appear in spring on lower leaves and continue on upper leaves into early summer. Symptom development is favored by frequent rains and cool, cloudy, humid weather in the early growing season. At first, tan flecks appear on both sides of the leaf. The flecks eventually become tan, diamond-shaped lesions up to 12 mm long, with a yellow border and a dark brown spot in the center that is due to *P. tritici-repentis* sporulating. Lesions may coalesce to cause large areas of the leaf to die, usually from the tip. Pseudothecia will eventually develop on straw as dark raised bumps.

CONTROL: Ordinarily yellow leaf spot is not a problem on rye; however, the following will help to control the disease.
1. Apply a foliar fungicide before the disease becomes severe and if weather conditions favor development of the disease.
2. Plow under residue if feasible.

MYCOPLASMAL CAUSES

Aster Yellows

CAUSE: Aster yellow mycoplasma survives in several dicotyledonous plants and leafhoppers. Aster yellows mycoplasma is transmitted primarily by the aster leafhopper, *Macrosteles fascifrons* Stal., and less commonly by *Endria inimica* Say and *M. laevis* Rib. The leafhoppers first acquire the mycoplasma by feeding on infected plants, then fly to healthy plants and transmit the mycoplasma by again feeding on healthy plants.

DISTRIBUTION: Generally distributed throughout eastern Europe, Japan, and North America.

SYMPTOMS: Symptoms become most obvious between 25° and 30°C. Seedlings will either die two or three weeks after infection, or if they survive, they will be stunted, leaves will generally yellow or have yellow blotches, and the heads will be sterile and distorted.

Infection of older plants causes leaves to become somewhat stiff and discolored in shades of yellow, red, or purple, from the tip or margin inward. Root systems may not be well developed.

CONTROL: No control is reported.

_____CAUSED BY NEMATODES_____

Root Gall Nematode

CAUSE: *Subanguina radicicola* (Grf.) Param. survives in host roots. Larvae penetrate roots and develop in cortical tissue, forming a root gall in two weeks. The mature females begin egg production within a gall; eventually the galls weaken and release larvae that establish secondary infections. Each generation is completed in about 60 days.

DISTRIBUTION: Canada and northern Europe.

SYMPTOMS: Seedlings frequently have chlorosis and reduced top growth. Galls on roots tend to be inconspicuous and vary in diameter from 0.5–6.0 mm. Roots may be bent at the gall site. At the center of larger galls is a cavity filled with nematode larvae.

CONTROL: Rotate rye with noncereal crops.

Root Lesion Nematode

CAUSE: *Pratylenchus* spp. overwinter as eggs, larvae, or adults in the host tissue or soil. Both larvae and adults penetrate roots where they move through cortical cells, the females simultaneously depositing eggs as they migrate. Older roots are abandoned and new roots are sought as sites for penetration and feeding.

DISTRIBUTION: Generally distributed wherever rye is grown.

SYMPTOMS: Plants in areas of a field will appear yellow and under moisture stress. Roots and crowns will be rotted when *Rhizoctonia solani* infects through nematode wounds. New roots may become dark and stunted with resultant loss in set of grain.

CONTROL:
1. Plant in autumn when soil temperatures are below 13°C.
2. Soil fumigants could be used where the high costs warrant them.

Seed Gall Nematode

CAUSE: *Anguina tritici* (Steinbuch) Chitwood survives as larvae within seed galls for several years. When galls are planted along with rye seed, larvae are released into moist soil. The larvae move upward to the flower primordia in a water film. The nematodes then mature, copulate, and produce eggs. The seed galls develop from undifferentiated flower tissue interacting with nematodes. If fall development is retarded, larvae may be present in healthy-appearing seed. Galls are mixed with normal seed or fall to the soil where nematodes become dormant under dry conditions.

DISTRIBUTION: Eastern Asia, parts of Europe, India, and southeastern United States. This is not a common disease of rye.

SYMPTOMS: Rye plants prior to heading are swollen near the soil line and leaves will be twisted, wrinkled, or rolled. After heading, the distortions are not as obvious but plants are stunted and mature slowly. Heads are small and the dark seedlike gall forces the glumes to spread apart. Galls are dark brown and do not have the brush or embryo markings of normal seed.

CONTROL:
1. Plant clean seed.
2. Seed may be soaked for 10 minutes at 54°C.
3. Rotate rye for two years with a nonhost crop. Wheat is also susceptible.

Stubby Root Nematode

CAUSE: *Paratrichodorus* spp. survive in soil or on roots as eggs, larvae, or adults. These nematodes feed only on the outside of rye roots and can move relatively rapidly through soil at 5 cm per hour, especially through fine, sandy soils.

DISTRIBUTION: Widely distributed in most agricultural soils. Rye planted early in autumn in sandy soils is most severely diseased.

SYMPTOMS: Roots are thickened, short, and stubby. Root tips may have brown lesions. Tops of plants may appear to grow poorly and the entire plant may be pulled easily from the soil due to lack of a fibrous system.

CONTROL: No practical control is presently available.

VIRAL CAUSES

African Cereal Streak

CAUSE: African cereal streak virus (ACSV) is limited to the phloem where it induces a necrosis. ACSV is transmitted by the delphacid leafhopper *Toya catilina* Fennah. The natural virus reservoir is probably native grasses. It is not mechanically or seed transmitted. Disease development is aided by high temperatures.

DISTRIBUTION: East Africa.

SYMPTOMS: Initially faint, broken, chlorotic streaks begin near the leaf base and extend upward. Eventually definite alternate yellow and green streaks develop along the entire leaf blade. Later the leaves become almost completely yellow. New leaves tend to develop a shoe-string habit and die.
   Young infected plants become chlorotic, severely stunted, and die. Seed yield is almost completely suppressed. Plants become soft, flaccid, and almost velvety to the touch.

CONTROL: No control is reported.

Wheat Soilborne Mosaic

This disease is also called soilborne mosaic.

CAUSE: Wheat soilborne mosaic virus (WSBMV) survives in soil in *Polymyxa graminis* Led., a soilborne plasmodiophoraceous fungus that is an obligate parasite in the roots of many higher plants. The fungus enters root hairs and epidermal cells of roots as motile zoospores when soil is wet and at a low soil temperature (10–20°C). Once inside the plant, *P. graminis* replaces plant cell contents with plasmodial bodies that will either segment into additional zoospores or develop into thick-walled resting spores two to four weeks after infection. WSBMV is spread by any method such as tillage, water, or wind that will disseminate infested soil.

DISTRIBUTION: Argentina, Brazil, Egypt, Italy, Japan, and eastern and central United States. Symptoms are not as common on rye as on wheat.

SYMPTOMS: WSBMV is most common in low-lying areas of fields that tend to be wet. The disease is most severe on autumn-planted rye.

Symptoms are most prominent in the spring on lower leaves. They range from a light green to yellow leaf mosaic. Plants may be severely or slightly stunted. The youngest leaves and sheaths are mottled and develop parallel spots or streaks. Warm weather prevents development of disease symptoms.

CONTROL:
1. Rotate rye with noncereal crops.
2. Plant rye later in the autumn.

Wheat Streak Mosaic

CAUSE: The wheat streak mosaic virus (WSMV) survives in infected plants. It has a wide host range that includes barley, corn, oats, rye, and several annual and perennial grasses. WSMV is transmitted only by the feeding of the wheat curl mite *Aceria tulipae* Keifer. Once a mite has picked up the virus, it carries it internally for several weeks. As winter rye plants mature, mites migrate to nearby volunteer cereals, grasses, or corn and will infect them. However, some grasses are hosts for the mites and not the virus and vice versa; some are susceptible to both, whereas some are resistant to both. During late summer or early autumn, the virus is disseminated by mites from volunteer cereals and grasses to early-planted rye.

Only young mites acquire the virus by feeding 15 minutes or more on infected plants. Neither the mite nor the virus can survive longer than one or two days in the absence of a living plant.

DISTRIBUTION: Eastern Europe, western and central North America, and the USSR.

SYMPTOMS: Initially light green to light yellow blotches, dashes, or streaks occur on the leaves parallel to the veins. Plants become stunted, show a general yellow mottling, and develop an abnormally large number of tillers that vary considerably in height. Stunted plants with sterile heads may remain after harvest, standing at the same height or shorter than stubble. As infected plants mature, the yellow-striped leaves turn brown and die.

Heads may be sterile or partially sterile with shriveled kernels. In severe cases, plants may die before maturity. Feeding mites often cause leaf edges to curl tightly in toward the upper midvein.

CONTROL:
1. Destroy all volunteer cereals and grasses in adjacent fields two weeks before planting and three to four weeks before planting in the field to be used.
2. Plant rye as late as practical after the Hessian fly-free date. If rye emerges in October or later, it often escapes infection.

_____BIBLIOGRAPHY_____

Boewe, G. H. 1960. Diseases of wheat, oats, barley, and rye. *Illinois Nat. Hist. Surv. Circ. 48.*

Brooks, F. T. 1953. *Plant Diseases.* 2nd ed. Oxford University Press, London.

Cunfer, B. M., and Schaad, N. W. 1976. Halo blight of rye. *Plant Dis. Rptr.* **60**:61–64.

Cunfer, B. M.; Schaad, N. W.; and Morey D. D. 1978. Halo blight of rye: Multiplicity of symptoms under field conditions. *Phytopathology* **68**:1545–1548.

Degenhardt, K. J.; Harper, F. R.; and Atkinson, T. G. 1982. Stem smut of fall rye in Alberta: Incidence, losses and control. *Can. Jour. Plant Pathol.* **4**:375–380.

Harder, D. E., and Bakker, W. 1973. African cereal streak, a new disease of cereals in East Africa. *Phytopathology* **63**:1407–1411.

Hosford, R. M. 1978. Effects of wetting period on resistance to leaf spotting of wheat, barley, and rye by *Leptosphaeria herpotrichoides*. *Phytopathology* **68**:591–594.

Jedlinski, H. 1983. Predisposition of winter rye (*Secale cereale* L.) to ergot (*Claviceps purpurea* (Fr.) Tul.) by barley yellow dwarf virus infection. *Phytopathology* **73**:843 (abstract).

Leukel, R. W., and Tapke, V. F. 1954. Cereal smuts and their control. *USDA Farmers Bull. No. 2069.*

McBeath, J. H. 1985. Pink snow mold on winter cereals and lawn grasses in Alaska. *Plant Disease* **69**:722–723.

McKay, R. 1957. *Cereal Diseases in Ireland.* Arthur Guiness Son & Co. Ltd. Dublin.

Mains, E. B., and Jackson, H. S. 1924. Aecial stages of the leaf rusts of rye, *Puccinia dispersa* Eriks. and Henn. and of barley *P. anomala* Rostr. in the United States. *Jour. Agric. Res.* **28**:1119–1126.

Ploetz, R. C.; Mitchell, D. J.; and Gallaher, R. N. 1985. Characterization and pathogenicity of Rhizoctonia species from a reduced-tillage experiment multicropped to rye and soybeans in Florida. *Phytopathology* **75**:833–839.

Richardson, M. J., and Noble, M. 1970. Septoria species on cereals—a note to aid their identification. *Plant Pathol.* **19**:159–163.

Roane, C. W. 1984. Atypical symptoms in rye caused by *Pseudomonas syringae* pv. *coronafaciens*. *Phytopathology* **74**:792 (abstract)

Simmonds, P. M. 1955. Root diseases of cereals. *Canada Dept. Agric. Pub. 952.*

Walker, J. 1975. Take all disease of Graminae: A review of recent work. *Rev. Plant Pathol.* **54**:113–144.

Western, J. H. 1971. *Diseases of Crop Plants.* The Macmillan Press Ltd., London.

16

Diseases of Safflower
(*Carthamus tinctorius* L.)

BACTERIAL CAUSES

Bacterial Blight

This disease is also called bacterial leaf spot and stem blight.

CAUSE: *Pseudomonas syringae* van Hall persists in soil, residue, and as a parasite and pathogen on several other plant hosts. The bacterium is likely disseminated by seed, flowing water, or any means that will transport soil. The bacteria are splashed or blown onto plant surfaces. The disease is most severe during periods of warm rain or under sprinkler irrigation. Seeds, weeds, plant residue, and soil are possible sources of primary inoculum.

DISTRIBUTION: Western United States.

SYMPTOMS: Dark, water-soaked lesions occur on stems, leaf petioles, and leaves. Spots on leaves become red-brown necrotic spots with pale margins. Other descriptions state that tissue in the center of the leaf spot becomes translucent and surrounded by a dark brown to black margin. The terminal bud is often necrotic. Rotting of the interior tissues of petioles often extends below the soil line into roots. Severely infected plants die.

   Plants not severely infected will recover during drier conditions. Loss of the initial stem and branches can be compensated for by increased lateral branching.

CONTROL: No control is reported.

Stem Soft Rot

CAUSE: *Erwinia carotovora* pv. *carotovora* (Jones) Bergey et al. Disease occurs during wet weather.

DISTRIBUTION: Mexico.

SYMPTOMS: Plants wilt. Stems have a soft internal rot.

CONTROL: No control is reported.

FUNGAL CAUSES

Alternaria Leaf Spot

CAUSE: *Alternaria carthami* Chowdbury and *A. alternata* (Fr.) Keissler. Both fungi are seedborne and survive in residue for at least two years. Leaf spots in early stages of development have been reported to yield only *A. carthami*, whereas toward plant maturity *A. alternata* was predominant. Under greenhouse conditions, *A. carthami* is pathogenic on safflower at all growth stages. *Alternaria alternata* also infects healthy safflower plants but infections remain dormant until leaf senescence. *Alternaria solani* Sorauer is also reported to penetrate healthy leaves; however, it does not cause leaf spots and mycelium remains dormant until leaf senescence. Secondary spread is by windborne conidia. Disease development is favored by abundant moisture from dew or rain and optimum temperatures between 25°C and 30°C.

DISTRIBUTION: Generally distributed wherever safflower is grown.

SYMPTOMS: Seeds may rot or seedlings damp off. Aboveground symptoms first appear on the cotyledons of seedlings as brown spots up to 5 mm in diameter. When stems of emerging seedlings are infected, a brown discoloration appears on the stem just above the soil level. Stem infection may cause the complete collapse of a plant.

Conidia are produced in spots, giving them a dusty appearance. In more mature plants, leaves have small brown to dark spots with concentric rings 1–2 mm in diameter. Spots enlarge to about 1 cm and may coalesce to form large lesions. The center of the spots are lighter in color and have alternate light and dark brown rings. Fully mature spots tend to develop holes and may cause a cracking of the leaf. Flower heads are first infected at the base of the calyx. Infected buds do not open, but shrink and dry up instead. Seed may have brown, sunken lesions on the coat.

CONTROL:
1. Plant seed from disease-free plants.
2. Treat seed with a fungicide.
3. Plant the most resistant cultivar.
4. Place seeds in hot water at 50°C for 30 minutes.

Botrytis Head Rot

This disease is also called Botrytis flower blight.

CAUSE: *Botrytis cinerea* Pers. ex Fr. survives as sclerotia in soil or residue. Sclerotia either germinate to form mycelium or support conidiophores and conidia on their surfaces. Conidia are disseminated by wind or water and sclerotia are disseminated by any means that will transport soil. The disease is most severe during cool, wet weather during or after flowering time. Conidia are abundantly produced on infected tissue during moist weather and are windborne considerable distances to provide secondary inoculum and infect other flowers.

DISTRIBUTION: Generally distributed where safflower is grown in moist climates.

SYMPTOMS: At first infected seed heads change color from a dark to light green, followed by complete browning. The stage of bloom at time of infection determines whether seed will be normal or lightweight. During cool, wet weather, the infected floral parts will be covered with a gray, fuzzy mold consisting of the conidiophores and conidia of the fungus. The fungus progresses through the seed head and into the stem, rendering infected heads readily detachable from the plant.

CONTROL: No control is presently available; therefore, it is not advisable to grow safflowers in areas where environmental conditions are conducive to Botrytis head rot.

Cercospora Leaf Spot

CAUSES: *Cercospora carthami* Sundararaman persists in residue. Fungal growth and subsequent disease development is greatest under moist, humid conditions with little wind. Dew condensation early in the morning appears to be critical for spore germination and germ tube elongation. Conidia require three to four days to penetrate a leaf and are able to withstand three to four intervening 20-hour dry periods.

DISTRIBUTION: Australia, Africa, India, Middle East, southern USSR, and the southern United States. The disease is considered of minor importance and generally occurs in warmer safflower-growing areas.

SYMPTOMS: Symptoms can occur at almost any stage of growth. Round to irregular brown spots, 3–10 mm in diameter, occur mostly on lower leaves. Spots are slightly sunken, frequently with concentric rings and a yellowish margin. Leaves become distorted with interveinal necrosis. During moist conditions, spots have a grayish appearance due to production of conidia on both sides of spots. In Australia, leaf spots are reported to have a whitish center, a gray-brown margin, and remain confined to older leaves.

Stems and nodes may become infected along with bracts. Flower buds turn brown, wither, and die with no seed being set.

CONTROL: No control is necessary.

Colletotrichum Stem Rot

CAUSE: *Colletotrichum orbiculare* (Berk. & Mont.) Arx.

DISTRIBUTION: Australia.

SYMPTOMS: Brown lesions that eventually become necrotic occur on the base of stems. Plants may die.

CONTROL: No control is necessary.

Fusarium Wilt

CAUSE: *Fusarium oxysporum* Schlecht. f. sp. *carthami* Klisiewicz & Houston survives as chlamydospores in soil and residue. Mycelium and conidia may also persist in residue for a time. The fungus is also seed-borne internally as mycelium or as conidia contaminating the outside of the seed. Mycelium penetrates the roots and enters into the vascular tissue. Abundant conidia are produced on dead tissue and are windborne to other areas. When conidia fall to the soil, some may be converted to chlamydospores.

Disease severity is greater during high temperatures, moist weather, and in plants grown in acidic, light-textured soils that are high in nitrogen. Different physiologic races of the fungus exist.

DISTRIBUTION: India and the United States.

SYMPTOMS: Typical symptoms are a yellowing of leaves on one side of the plant. Unilateral yellowing also occurs on leaves and some seed heads. The yellowing begins on the lower leaves, followed by a wilting that progresses up the plant. Young plants are usually killed. Older plants may also die, but usually only branches on one side of the plant are killed. Seed heads are frequently blighted and distorted due to the fungus growing up the vascular system into the seed head. Infected heads have aborted seed. The vascular system of the roots and stems has a brown discoloration.

CONTROL:
1. Seed from infected plants should not be planted.
2. Treat seed with a seed-protectant fungicide.
3. Rotate safflowers with other crops.
4. Some safflower selections are resistant to Race 4.

Phytopthora Root Rot

CAUSE: *Phytophthora cryptogea* Pethyb. and Laff. (syn. *P. drechsleri* Tucker), *P. cactorum* (Lebert & Cohn), and *P. parasitica* Dast. have been reported on safflower. All species survive in residue and soil. Sporangia are formed that germinate directly by mycelium or by producing zoospores that swim to the host root, encyst, and germinate. Phytophthora root rot is most severe in wet soils, particularly in surface-irrigated fields or poorly drained areas. Serious losses do not usually occur until close to plant maturity and when soil temperatures are between 25°C and 30°C. Water stress before infection can predispose plants to root rot caused by *P. cryptogea*. Physiologic races of the fungi exist.

DISTRIBUTION: Australia and the western United States.

SYMPTOMS: Plants may be infected at all stages of growth. The lower stems of seedlings collapse, causing them to die. The first aboveground symptom on older plants is the leaves turn light green or yellow, then wilt, and die, eventually turning brown. In the early stage of wilting the roots become reddish; then roots and lower stems become dark brown to black, with the discoloration extending up the stem above the soil line.

CONTROL:
1. Plant the most resistant cultivar.
2. Provide good drainage. Where flood irrigation is used, avoid ponding of water.
3. It is more desirable to grow surface-irrigated safflowers on beds under furrow irrigation instead of flood irrigation.

Powdery Mildew

CAUSE: *Erysiphe cichoracearum* DC. ex Merat f. sp. *carthami.*

DISTRIBUTION: Afghanistan, India, Israel, and USSR.

SYMPTOMS: Typical gray, powdery mass of condia and conidiophores occur on leaf surfaces.

CONTROL: No control is reported.

Pythium Root Rot

CAUSE: *Pythium* spp., particularly *P. ultimum* Trow. and *P. splendens* Braun. Disease severity is greatest on safflowers grown under irrigation. Only young plants appear to be susceptible and susceptibility is related to elongation of the hypocotyl and first internode tissue.

DISTRIBUTION: Western United States.

SYMPTOMS: The hypocotyl and the first internode become water-soaked and soft with a light brown discoloration. Plants topple over with the continued rotting and collapse of these tissues.

CONTROL: Treat seed with a seed-protectant fungicide. Plants not grown under irrigation are infrequently infected.

Ramularia Leaf Spot

CAUSE: *Ramularia carthami* Zaprometov.

DISTRIBUTION: France, India, Israel, and USSR.

SYMPTOMS: Round and regular spots, 10 mm or more in diameter, occur on both sides of leaves. Spots are characterized by a whitish, dense mass of conidiophores, and conidia in their centers. Dry spots are brown. Yields are reduced and seed quality is affected.

CONTROL: Rotate safflower with other crops.

Rhizoctonia Blight

CAUSE: *Rhizoctonia solani* Kuhn. The optimum soil temperature for disease development is 20–25°C.

DISTRIBUTION: New Mexico and Texas.

SYMPTOMS: Dark cortical lesions occur slightly below or at the soil line on the seedling stem. In advanced stages of disease the lesions extend up the stem beyond the attachment point of the lowest leaves. Lesions frequently girdle the stem. Root development is reduced.

CONTROL: Plant resistant cultivars.

Rust

CAUSE: *Puccinia carthami* Cda. is an autoecious rust fungus whose five spore stages occur only on safflower. Teliospores are seedborne and also overwinter on residue. In the spring, teliospores germinate to produce a promycelium on which four basidiospores are formed. The basidiospores are violently discharged to reach host tissue. Pycnia are subepidermal, with the opening above the epidermis surface of cotyledons, hypocotyls, tender stems, and true leaves. Aecia are formed in four to seven days and in close association or closely grouped with pycnia. Aecial pustules in gross morphology are similar to uredia and have been referred to as a primary uredium and a uraecia. Uredia are formed scattered over the host tissue and are not closely grouped like aecia. As the pustules mature, telia may be formed in uredium or in a separate telia. Several physiologic races of the fungus exist.

DISTRIBUTION: Generally distributed wherever safflower is grown.

SYMPTOMS: The fungus, unlike most rust fungi, has two distinct pathological stages: a seedling phase and a foliage phase. The seedling phase results from infection by basidiospores originating from soil or seedborne teliospores. The foliage phase, which can occur at any growth stage, results from infection by windborne urediospores.

Infection of the seedling hypocotyl causes swelling and often bending and twisting toward one side. Chestnut-brown pustules, which are most commonly aecia, and rarely uredia, form on infected hypocotyls and less commonly on cotyledons and young leaves. Aecia will be grouped in clusters but uredia will be scattered. Affected seedlings may collapse at the soil surface or slightly below it and die. On older plants, a pronounced girdling and hypertrophy of the stem base may occur. During wind and rain storms these plants often break at the girdled areas.

Aeciospores or urediospores from seedling infections initiate foliar infections. Small, powdery, chestnut-brown pustules develop most abundantly on lower leaf surfaces and also on upper leaf surfaces. Pustules also develop on cotyledons and flower bracts. On highly susceptible varieties the pustules may reach a diameter of 1–2 mm and be surrounded by a secondary and frequently a tertiary ring of uredia. Small chlorotic or necrotic flecks may develop without sporulation on resistant varieties. The chestnut-brown pustules become black, due to formation of teliospores in them, toward the end of the growing season.

CONTROL:
1. Treat seed with a seed-protectant fungicide.
2. Rotate safflowers with other crops.
3. Plow under residue where erosion is not a problem.
4. Flooding soils for seven days reduces the amount of inoculum in the soils.
5. Plant resistant cultivars; however, cultivars resistant to one race of *P. carthami* may be susceptible to other races.

Sclerotinia Stem Rot

CAUSE: *Sclerotinia sclerotiorum* (Lib.) deBary survives in soil and residue as sclerotia. During wet weather, sclerotia germinate to form either several apothecia or mycelium. Ascospores are formed in a layer of asci on the apothecium and are windborne to hosts. The disease is most severe under humid conditions around the base of the plant.

DISTRIBUTION: Generally distributed wherever safflower is grown.

SYMPTOMS: Leaves turn yellow, wilt, turn brown, and shrivel. The entire plant may die, causing the plant top to curve downward. A white, cottony growth, which is the fungus mycelium, occurs on the base of the stem. Sclerotia, which appear as black, elongated objects, 2–12 mm long, form on the stems and in them and on adjoining roots. Roots are generally not infected or discolored. The cortical tissue of the lower stem often becomes shredded.

A head rot may also occur with the infected flower heads sometimes falling from plants, leaving the outer bracts in place. Sclerotia also form in the flower head. At maturity, infected heads contain few or no seeds, which are often replaced by a dusty-appearing residue.

CONTROL:
1. Do not include safflowers in a rotation with other plants susceptible to *S. sclerotiorum*.
2. Use soil fumigation on small areas.
3. Fields may be flooded for four weeks, then dried up prior to planting.

Verticillium Wilt

CAUSE: *Verticillium dahliae* Kleb. survives as microsclerotia in soil and residue. Verticillium wilt is more severe during cool, wet weather.

DISTRIBUTION: Generally distributed wherever safflower is grown.

SYMPTOMS: Plants may be infected at any stage of growth. Young plants may be stunted. Leaves are a darker green than those of healthy plants and may be crinkled between the veins. Cotyledons are yellowish, flaccid, and dry up soon.

In older plants, symptoms will occur on lower leaves first. Infected leaves may show unilateral growth, that is, one side of the leaf will grow faster than the other, causing the leaf to grow in the shape of a "C" or comma. Discoloration may also accompany unilateral growth, with the normal-appearing or faster-growing side a normal green while the other side becomes tan or light brown. A characteristic symptom is chlorotic areas appearing on the leaf margins and the principal veins, giving the leaf a mottled appearance. As the plants grow older, the chlorotic areas enlarge and become a light tan or whitish. Plants may die prematurely and usually do not produce viable seed. A dark discoloration occurs in the vascular strands, particularly at the nodes and in the petioles and leaf traces.

CONTROL: Verticillium wilt is ordinarily considered a minor disease. Do not rotate safflower with cotton, peanuts, or other susceptible crops.

MYCOPLASMAL CAUSES

Safflower Phyllody

CAUSE: Safflower phyllody mycoplasma is transmitted by the leaf hopper *Neoaliturus fenestratus* Distant. Other species in the family Compositae are also susceptible to the mycoplasma including tricolor chrysanthemum, *Chrysanthemum carinatum* Bergsmans ex. J. Ingram; strawflower, *Helichrysum bracteatum* (Venten.) Andr.; common sunflower, *Helianthus annuus* L.; and cape marigold, *Dimorphotheca sinuata* DC. Periwinkle, *Vinca rosea* L., is also susceptible.

DISTRIBUTION: Israel.

SYMPTOMS: Initially abnormal axilliary budding occurs, starting when the infected plant reaches a height of 20–30 cm. The main stem of an infected plant produces numerous secondary shoots, starting at the top and spreading rapidly throughout the entire main stem. These secondary shoots are very thin and may produce new branchings. Very small, yellowish leaves appear, and after several weeks, witches' broom symptoms develop. Many short stems appear on the infected flower heads instead of the typical tubular flowers of healthy plants. Flowers on diseased plants either have no seeds or seeds are sterile. In contrast, healthy plants produce one main stem and few lateral ones on the upper part of the central stem and from one to five colored flower heads.

CONTROL: Keep safflower fields free of the weed *Carthamus tenuis* L. because the leafhopper breeds in this weed.

VIRAL CAUSES

Chilli Mosaic

CAUSE: Chilli mosaic virus (CMV) is transmitted from infected to healthy plants by several species of aphids.

DISTRIBUTION: India.

SYMPTOMS: Light and dark green patches are scattered over leaves.

CONTROL: No control is reported.

Cucumber Mosaic

CAUSE: Cucumber mosaic virus (CMV) is transmitted from infected to healthy plants early in the growing season primarily by the green peach aphid, *Myzus persicae* Sulzer. CMV has a wide host range.

DISTRIBUTION: Generally distributed wherever safflower is grown.

SYMPTOMS: A light and dark green mosaic pattern occurs primarily on upper leaves and bracts. Some leaves may become blistered and distorted. The mosaic pattern is less distinct on lower leaves and becomes a scattering of light and dark green flecks. Infected plants may be stunted but usually develop to maturity.

CONTROL: No control is reported.

Severe Mosaic

This disease is also called turnip mosaic virus.

CAUSE: Turnip mosaic virus (TuMV). The virus is mechanically transmitted by the aphid *Myzus persicae* Sulzer. Weed species of *Brassica* spp. serve as sources for TuMV.

DISTRIBUTION: Sacramento Valley in California.

SYMPTOMS: Affected plants are generally stunted with reduced leaf and seed head size. Leaves and bracts of seed heads have small dark green and pale green areas, pale green veinbanding, distortion, and bronzing. Seed ovules rot, resulting in reduction of seed yield.
Some cultivars develop necrotic lesions. Plants with necrotic lesions have no mosaic.

CONTROL: No control is reported.

Tobacco Mosaic

CAUSE: Tobacco mosaic virus (TMV).

DISTRIBUTION: Morocco.

SYMPTOMS: Small, reddish local lesions occur on cotyledons when mechanically inoculated. Under conditions of high temperatures (28–32°C) and light intensity, necrotic ring and line patterns later appear on cotyledons. Later typical blotchy light and dark green mosaic patterns occur on all subsequent leaves up until flowering.

CONTROL: No control is reported.

_____BIBLIOGRAPHY_____

Ashri, A. 1961. The susceptibility to safflower varieties and species to several foliage diseases in Israel. *Plant Dis. Rptr.* **45:**146–150.
Beech, D. F. 1960. Safflower—an oil crop for the Kimberleys. *Western Australia Dept. Agric., Perth, Bull.* 2720.

Chakrabarti, D. K., and Basuchaudhary, K. C. 1978. Incidence of wilt of safflower caused by *Fusarium oxysporum* f. sp. *carthami* and its relationship with the age of the host, soil and environmental factors. *Plant Dis. Rptr.* **62**:776–778.

Claassen, C. E.; Schuster, M. L.; and Ray, W. W. 1949. New diseases observed in Nebraska on safflower, *Plant Dis. Rptr.* **33**:73–75.

Duniway, J. M. 1977. Predisposing effect of water stress on the severity of Phytophthora root rot in safflower. *Phytopathology* **67**:884–889.

Erwin, D. C.; Starr, M. P.; and Desjardins, P. R. 1964. Bacterial leaf spot and stem blight of safflower caused by *Pseudomonas syringae*. *Phytopathology* **54**:1247–1250.

Heritage, A. D., and Harrigan, E. K. S. 1984. Environmental factors influencing safflower screening for resistance to *Phytopthora cryptogea*. *Plant Disease* **68**:767–769.

Irwin, J. A. G., and Jackson, K. J. 1977. Safflower diseases in Queensland. *Queensland Agric. Jour.* **103**:516–520.

Jacobs, D. L.; Bergman, J. W.; and Sands, D. C. 1982. Etiology and epidemiology of bacterial leaf spot and stem blight of safflower in Montana. *Phytopathology* **72**:961 (abstract).

Klein, M. 1970. Safflower phyllody–a mycoplasma disease of *Carthamus tinctorius* in Israel. *Plant Dis. Rptr.* **54**:735–738.

Klisiewicz, J. M. 1974. Assay of Verticillium in safflower seed. *Plant Dis. Rptr.* **58**:926–927.

Klisiewicz, J. M. 1975. Survival and dissemination of Verticillium in infected safflower seed. *Phytopathology* **65**:696–698.

Klisiewicz, J. M. 1977. Effect of flooding and temperature on incidence and severity of safflower seedling rust and viability of *Puccinia carthami* teliospores. *Phytopathology* **67**:787–790.

Klisiewicz, J. M. 1977. Identity and relative virulence of some heterothallic Phytophthora species associated with root and stem rot of safflower. *Phytopathology* **67**:1174–1177.

Klisiewicz, J. M. 1980. Safflower germ plasm resistant to Fusarium wilt. *Plant Dis.* **64**:876–877.

Klisiewicz, J. M. 1981. Isolation of turnip mosaic virus from safflower. *Phytopathology* **71**:886 (abstract).

Klisiewicz, J. M. 1983. Etiology of severe mosaic and its effect on safflower. *Plant Disease* **67**:112–114.

Klisiewicz, J. M.; Houston, B. R.; and Peterson, L. J. 1963. Bacterial blight of safflower. *Plant Dis. Rptr.* **47**:964–966.

Knowles, P. F., and Miller, M. D. 1965. Safflower. *Univ. California Agric. Expt. Sta. Ext. Circ. 532.*

Lockhart, B. E. L., and Goethals, M. 1977. Natural infection of safflower by a tobamovirus. *Plant Dis. Rptr.* **61**:1010–1012.

Lopez, C. J. A., and Fucikovsky, Z. L. 1986. Safflower *Carthamus tinctorius* L. stem soft rot at Chapingo, Mexico. *Phytopathology* **76**:374 (abstract).

Mortensen, K., and Bergman, J. W. 1983. Cultural variance of *Alternaria carthami* isolates and their virulence on safflower. *Plant Disease* **67**:1191–1194.

Mortensen, K.; Bergman, J. W.; and Burns, E. E. 1983. Importance of *Alternaria carthami* and *A. alternata* in causing leaf spot diseases of safflower. *Plant Disease* **67**:1187–1190.

Peterson, W. F. 1965. Safflower culture in the west-central plains. *USDA Agric. Res. Serv. Agric. Info. Bull. No. 300.*

Raccah, B., and Klein, M. 1982. Transmission of the safflower phyllody mollicute by *Neolaiturus fenestratus*. *Phytopathology* **72**:230–232.

Rathaiah, Y., and Pavgi, M. S. 1971. Ability of germinated conidia of *Cercospora carthami* and *Ramularia carthami* to survive dessication. *Plant Dis. Rptr.* **55**:846–847.

Sackston, W. E. 1953. Foot and root infection by safflower rust in Manitoba. *Plant Dis. Rptr.* **37**:522–523.

Sackston, W. E. 1960. *Botrytis cinerea* and *Sclerotinia sclerotiorum* in seed of safflower and sunflower. *Plant Dis. Rptr.* **44**:664–668.

Stovold, G. 1973. *Phytophthora drechsleri* Tucker and *Pythium* spp. as pathogens of safflower in New South Wales. *Australian Jour. Exp. Agric. and Animal Husbandry* **13**:455–459.

Thomas, C. A. 1970. Effect of seedling age on Pythium root rot of safflower. *Plant Dis. Rptr.* **54**:1010–1011.

Thomas, C. A.; Klisiewicz, J.; and Zimmer D. 1963. Safflower diseases. *USDA Agric. Res. Serv. ARS 34-52.*

Thomas, C. A.; Rubis, D. D.; and Black, D. S. 1960. Development of safflower varieties resistant to Phytophthora root rot. *Phytopathology* **50**:129–130.

Weiss, E. A. 1971. Diseases of safflower, in *Castor, Sesame and Safflower*. Barnes and Noble, Inc., New York.

Zazzerini, A.; Cappelli, C.; and Panattoni, L. 1985. Use of hot-water treatment as a means of controlling *Alternaria* spp. on safflower seeds. *Plant Disease* **69**:350–351.

Zimmer, D. E. 1963. Spore stages and life cycle of *Puccinia carthami*. *Phytopathology* **53**:316–319.

Zimmer, D. E.; Klisiewicz, J. M.; and Thomas C. A. 1963. Alternaria leaf spot and other diseases of safflower in 1962. *Plant Dis. Rptr.* **47**:643.

Zimmer, D. E., and Thomas, C. A. 1967. Rhizoctonia blight of safflower and varietal resistance. *Phytopathology* **57**:946–949.

17

Diseases of Sorghum (*Sorghum bicolor* (L.) Moench)

BACTERIAL CAUSES

Bacterial Leaf Spot

This disease is also called bacterial spot.

CAUSE: *Pseudomonas syringae* pv. *syringae* van Hall (syn. *P. syringae* van Hall) overwinters in soil and on seed. It probably also survives on infected residue, Johnsongrass, and on sorghum plants remaining after harvest. Bacteria are disseminated by wind, rain, and insects. Infection occurs through stomates and wounds. Disease development is favored by wet, cool weather (12°C optimum).

DISTRIBUTION: Generally distributed wherever sorghum is grown.

SYMPTOMS: Spots first appear on lower leaves with infection moving up the plants as they approach maturity. Spots are circular to elliptical and 1–10 mm in diameter. Initially spots are dark green and water-soaked but soon become tan with a red border. Small lesions may be entirely red with somewhat sunken centers. Sometimes spots are so numerous that they coalesce to form large diseased areas, resulting in the death of the whole leaf. Lesions may also occur on leaf sheaths and seeds.

CONTROL:
1. Plant healthy seed that has been treated with a seed-protectant fungicide.

2. Rotate sorghum with nonsusceptible crops.
3. Plow under infected residue where soil erosion is not a problem.
4. Destroy Johnsongrass and any sorghum plants left after harvest.

Bacterial Leaf Streak

This disease is also called bacterial streak.

CAUSE: *Xanthomonas campestris* pv. *holcicola* (Elliott) Dye (syn. *X. holcicola* (Elliott)) overwinters in residue and plants left standing after harvest. Bacteria are locally disseminated by water and wind. Dissemination over a long distance is by infected residue and seed. Disease development is favored by warm, wet weather.

DISTRIBUTION: Argentina, Australia, India, Mexico, Nigeria, New Zealand, Philippines, South Africa, the United States, and the USSR.

SYMPTOMS: Streaks occur on leaves of plants of all ages as water-soaked and translucent, about 3 mm wide and 25–150 mm long. Initially only light yellow drops of bacterial exudate are present on the translucent streaks. This will eventually dry to thin white or cream-colored scales. Later, red-brown blotches appear that eventually enlarge and become red throughout the streak, causing the water-soaked and translucent areas to disappear. Portions of the streaks may broaden into elongated oval spots with tan centers and narrow red margins. Numerous streaks may join to form long, irregular areas that cover a large area of the leaf with necrotic tissue bordered by dark narrow margins between the red-brown streaks.

CONTROL:
1. Rotate sorghum with other crops.
2. Plow under infected residue where erosion is not a problem.
3. Plant healthy seed that has been treated with a fungicide.

Bacterial Leaf Stripe

This disease is also called bacterial stripe.

CAUSE: *Pseudomonas andropogonis* (Smith) Stapp is seedborne and survives in residue, soil, and on sorghum plants remaining in the field after harvest. Local dissemination is by wind, water, and insects. Long-range dissemination is by infested seed or residue. Infection occurs through stomates and injuries caused by wind and insects. Disease development is favored by warm (25–29°C), wet weather such as the cloudy, humid days following rain.

DISTRIBUTION: Argentina, Australia, China, Formosa, Nigeria, and the United States.

SYMPTOMS: Initially, small (1 cm), linear, interveinal lesions develop on leaves. Stripes develop later on leaves and sheaths that vary in color from tan-red to brick red to dark purple-red, depending on the infected cultivar. Color is continuous throughout the stripe. The stripes are usually restricted to the interveinal areas of apical portions of the lower leaves. The stripes range in length from 25 mm to the length of the leaf blade and sometimes coalesce to cover large areas of the leaf. Water-soaking of tissue adjacent to lesions is normally not observed.

Bacterial exudate may be found on the underside of affected portions of leaves and along leaf margins, drying and forming red crusts or thin scales that may be washed off by rain. Lesions may also occur on kernels, peduncles, and rachis branches and in the interior of the stalk.

CONTROL:
1. Rotate sorghum with nonsusceptible crops.
2. Plant healthy seed treated with a seed-protectant fungicide.
3. Where practical, delay planting to avoid wet climatic conditions.
4. Control weeds such as Johnsongrass.
5. Plow under residue where erosion is not a problem.
6. Some sorghum cultivars are more susceptible than others.

Erwinia Soft Rot

CAUSE: *Erwinia chrysanthemi* Burkholder, McFadden & Dimock. The mode of entry into a plant is probably by contaminated water entering into the whorl of young, growing plants, a situation that would occur during mild flooding. Disease is favored by temperatures of 30°C and higher.

DISTRIBUTION: Nebraska.

SYMPTOMS: Symptoms appear in about the fourth growth stage. Severe symptoms include death of the top four to five leaves; however, lower leaves appear normal. Dead tissue can be pulled easily from the whorl; inside the plant whorl is wet, necrotic tissue that has a putrid odor. Rot is limited to a node with only tissue above the node affected.

Less severe symptoms include necrotic stripes, pigmented stripes, or blotches on the upper leaves with no stalk rot or top necrosis.

CONTROL: No control is reported. The disease is not thought to pose a significant threat except in poorly drained fields where prolonged periods of high temperatures occur during the early stages of seedling development.

Yellow Leaf Blotch

CAUSE: *Pseudomonas* sp. is seedborne.

DISTRIBUTION: West Africa.

SYMPTOMS: Scattered cream to yellow to beige lesions, 25–35 mm long by 8–10 mm wide, occur on leaves. Occasionally lesions will be 200 mm long by 30 mm wide, slightly water-soaked, and have a tendency to follow the veins. Young plants infected at the two- to three-leaf stage are sometimes severely stunted and killed.

CONTROL: No control is reported.

_____FUNGAL CAUSES_____

Acremonium Wilt

CAUSE: *Acremonium strictum* Gams. (Syn. *Cephalosporium acremonium* Cda.). Infection occurs in the leaf blade or sheath where systemic colonization occurs. Soilborne infection has been reported from Egypt.

DISTRIBUTION: Argentina, Egypt, Honduras, India, and the United States.

SYMPTOMS: Plants are stunted and chlorotic. The vascular system has a brownish discoloration extending from leaf sheath into the vascular bundles of the stalk. Initially the lower leaves turn chlorotic and develop red-brown streaks. Such leaves eventually die. Other symptoms may include large areas of wilted tissue that develop along one axis of the leaf on either side of the midrib. In severely affected plants, the upper leaves and shoot die and foliar wilting occurs.

CONTROL: Most cultivars are resistant.

Anthracnose Leaf Blight

This disease is also called red leaf spot.

CAUSE: *Colletotrichum graminicola* (Ces.) G. W. Wilson (syn. *Dicladium graminicola* Ces., *C. lineola* Cda., and *C. cereale* Manns). The teleomorph is *Glomerella graminicola* Politis. The causal fungus is seedborne and survives in residue, soil, Johnsongrass, and other weeds. Spores are most likely to be produced during warm, wet weather in midsummer. During moist weather, later in the growing season, conidia are produced in lesions and are windborne to other hosts. Different physiologic races exist.

DISTRIBUTION: Generally distributed wherever sorghum is grown but is most severe in warm, humid climates.

SYMPTOMS: Spots on leaves are most likely to occur at midsummer. Spots are well defined, circular to oval, and up to 25 mm in diameter; however, they are commonly 3–6 mm in diameter. Spots develop small, circular, light tan centers with wide margins that are tan, orange, red, or black-purple, depending on the cultivar infected. Under high humidity conditions, spots increase in number to cover much of the leaf area. Elliptical to elongate lesions may cover the entire midrib length.

Setae may be present in spots as a blackish growth that appears under magnification as short stiff hairs. Pink spore masses may develop in these blackish areas during humid weather.

Plants may be defoliated and, in severe cases, they may die before maturity. On broom corn, production of heads may be poor and sugar content of sweet sorghum may be reduced.

The stalk rot phase is discussed under Red Rot.

CONTROL:
1. Plant resistant cultivars.
2. Plant healthy seed treated with a seed-protectant fungicide.
3. Control weeds in a field.
4. Plow under residue where erosion is not a problem.

Banded Leaf and Sheath Blight

This disease is also called Rhizoctonia blight.

CAUSE: *Rhizoctonia solani* Kuhn anastomosis group (AG) 1. Anamorph is *Thanatephorus cucumeris* (Frank) Donk. Initial inoculum probably originates from mycelium and sclerotia in residue or soil. Secondary spread is by mycelium contacting leaves, or windborne and rainborne mycelium and sclerotia. Basidiospores may also be important. Disease development occurs during moist conditions with high relative humidity and temperature.

Aerial blight of soybeans and sheath blight of rice are also caused by the same fungus.

DISTRIBUTION: Tropical areas of the world; however, it is normally not considered an important disease.

SYMPTOMS: Initially leaves and sheaths develop water-soaked, gray-green lesions. Later these become irregularly shaped (2–8 mm wide), tan to red-brown, with white centers. One or several lesions form horizontal bands across the leaf blade and sheath. Lesions may have a purple border. Lesions may be covered with fluffy white mycelium or brown mycelial mats of sclerotia. Individual sclerotia that appear as white to dark brown objects, 1–5 mm in diameter, can be found on older lesions.

CONTROL: Do not rotate sorghum with rice or soybeans.

Charcoal Rot

CAUSE: *Macrophomina phaseolina* (Tassi) Goid. (syn. *M. phaseoli* (Maubl.) Ashby and *Botryodiplodia phaseoli* (Maub.) Thirum.). The anamorph is *Rhizoctonia bataticola* (Taub.) Butler (syn. *Sclerotium bataticola* Taub.).

Macrophomina phaseolina survives primarily as sclerotia that are closely associated with residue. Pycnidia are not known to commonly occur but would afford another means of overwintering. The fungus is not a good soil saprophyte and is inhibited in growth by other soil microflora. During high soil temperatures (30°C and above), and low soil moisture (80% available soil moisture and below), the sclerotia germinate to form infection hyphae that penetrate belowground plant parts. Charcoal rot generally occurs in localized areas; therefore, it is probable that only sclerotia are disseminated by any means that carry soil or residue.

Stress created by drouth is necessary for infection by *M. phaseolina* to occur.

DISTRIBUTION: Generally distributed wherever sorghum is grown under hot, dry conditions.

SYMPTOMS: Charcoal rot will first occur on plants that grow in the driest soil in a field. The symptoms move up the stalk from the crown. At first there is a general water-soaking of the pith followed by a bright red to black color of the affected tissues. Last, tissues die; the stalk color fades; the lower stem dries and shreds; and numerous, small, black sclerotia form on the remaining vascular bundles. Affected plants then lodge. Plants may be destroyed in two or three days.

CONTROL:
1. Sweet sorghums and many forage sorghums are resistant but grain sorghums tend to be susceptible.
2. Rotate sorghum with other crops. Growing successive crops of sorghum builds up inoculum.
3. Excessively high stands tend to predispose plants.
4. Follow cultural practices that conserve soil moisture.

Cochliobolus Leaf Spot

CAUSE: *Cochliobolus bicolor* Paul & Parberry.

DISTRIBUTION: India.

SYMPTOMS: Lesions on leaves are circular with alternating concentric straw-colored bands. In severe cases, spots coalesce to cover a large area of the leaf surface.

CONTROL: No control is reported.

Covered Kernel Smut

CAUSE: *Sporisorium sorghi* (Ehrenberg) Link (syn. *Sphacelotheca sorghi* (Link) Clinton). The causal fungus overwinters primarily as teliospores on the outside of seed. When contaminated seed is planted, teliospores germinate with the seed, forming a four-celled promycelium bearing lateral sporidia. The sporidia germinate and infect the developing seedling. Mycelium grows systemically within the plant. Much variation in teliospore germination occurs. Sometimes teliospores germinate directly by only producing germ tubes. Germination may occur by both means. At heading, teliospores replace kernels in a membrane that ruptures at harvest, releasing the teliospores to contaminate seed or soil. Teliospores in the soil are apparently not important in infecting seedlings. Smut incidence decreases when seed is planted in wet, warm soils (15.5–32°C). Several physiologic races exist.

DISTRIBUTION: Generally distributed wherever sorghum is grown.

SYMPTOMS: Growth of the plant is generally not affected and infected plants appear normal until heading time. Usually all kernels are replaced by dark brown, powdery masses of chlamydospores, covered with a gray or brown membrane; however, an occasional normal kernel may occur. The membrane ordinarily remains intact until harvest time when it is ruptured by harvest equipment. Sori vary in size from those small enough to be covered by the glumes to those over 1 cm long.

CONTROL:
1. Treat seed with a protectant fungicide.
2. Plant seed from smut-free plants.
3. All commercial cultivars of sorgo, kaffir, durra, sudangrass, and broom corn are susceptible. Hegari, milo, gurno, and feterita types of sorghum are resistant to one or more races.

Crazy Top

CAUSE: *Sclerophthora macrospora* (Sacc.) Thirum., Shaw & Naras. (syn. *Sclerospora macrospora* Sacc.) survives as oospores in sorghum residue and other grasses. Oospores germinate in saturated soil in 24–48 hours to form sporangia in which zoospores are produced. Zoospores are then liberated from sporangia in saturated soil and swim to preemerged or just emerging sorghum seedlings where they germinate and produce systemic mycelium. Oospores are produced within infected leaves, usually within and adjacent to veins. Disease occurs over a wide range of temperatures.

DISTRIBUTION: Generally distributed where sorghum is grown. Crazy top occurs only when soil becomes flooded or water-logged before seedlings emerge and become established.

SYMPTOMS: Leaves are mottled, thickened, stiffened, twisted, or curled and the surface is covered with bumps. Heads are usually not produced. Heads that are produced have numerous leafy shoots.

CONTROL: Crazy top is rarely serious on sorghum and control measures are not necessary.

Curvularia Kernel Rot

CAUSE: *Curvularia lunata* (Wakker) Boed. Infection is favored by rain, high humidity, and temperatures between 24–30°C.

DISTRIBUTION: Mexico.

SYMPTOMS: Symptoms are observed on spikelets at maturity. Kernels are covered with a dense velvet-appearing fungal growth consisting of conidia and conidiophores.

CONTROL: No control is reported.

Ergot

This disease is also called honeydew and sugary disease.

CAUSE: *Claviceps sorghi* Kulk, et al. The anamorph is *Sphacelia sorghi* McRae. Survival is by sclerotia. These germinate to produce stromatic heads at the end of a stalk in which perithecia are formed. Ascospores are produced within the perithecia and are windborne to florets of the host. Both ascospores and conidia from collateral hosts may serve as primary inoculum. Florets are susceptible to infection from panicle emergence through fertilization of the ovary.

A sweet, sticky substance called honeydew, which consists of liquid and conidia, exudes from infected florets. This liquid is disseminated from flower to flower by insects, rain, and wind. Dissemination is favored by rainy, cloudy weather. Long-range spread is by sclerotia being mixed with seed.

Disease is favored by temperatures of 20–25°C and high humidity during anthesis.

DISTRIBUTION: Africa and Asia.

SYMPTOMS: Initially sticky, pinkish to brownish liquid drops exude from ovaries. The drops are sweet to the taste and attract numerous insects. A saprophytic fungus, *Cerebella volkensii* Ces., may overgrow the honeydew and convert it into a sticky, matted, black mass. If the mycoparasite is not present, fungal tissue in the ovary grows into a hard, grayish, elongated, and slightly curved, hornlike sclerotium.

CONTROL:
1. Plant sclerotia-free seed lots.
2. Locate seed fields in areas that are not conducive to disease.
3. Apply a foliar fungicide from the flag leaf stage to the end of the anthesis.

Fusarium Head Blight

CAUSE: *Fusarium moniliforme* Sheldon. Little is known for certain of the survival or dissemination of *F. moniliforme* as it relates to head blight of sorghum; however, it is probably similar to Fusarium stalk rot of corn. The fungus probably overwinters as mycelium in infected sorghum or corn residue. Conidia are produced and windborne to peduncles where they enter through cracks or wounds in the peduncle, rachis, or panicle branches. Prolonged wet weather preceding anthesis favors disease.

DISTRIBUTION: Generally distributed where sorghum is grown under warm, humid conditions.

SYMPTOMS: Several to all florets in a seed head may be killed. The entire seed head may be covered with a white to pinkish mycelial growth. When the panicle is split, a red, brown, or black discoloration is evident in the upper peduncle extending into the branches of the head. Sometimes the discoloration extends into the stalk, also discoloring the rind. The peduncle may be necrotic and break over. The rachis and rachis branches may become necrotic or have external reddening.

CONTROL: Plant resistant cultivars.

Fusarium Root and Stalk Rot

CAUSE: *Fusarium moniliforme* Sheldon, teleomorph *Gibberella fujikuroi* (Sawada) Wollenw. Other *Fusarium* spp. have been reported as causal fungi; however, *F. graminearum* Schwabe, teleomorph *G. zeae* (Schw.) Petch, is considered to be the predominant species. The fungi survive as chlamydospores and chlamydospore-like structures and mycelium in corn or sorghum residue. Conidia are produced on residue and are windborne to plants. *Fusarium moniliforme* is also seedborne. Mycelium may grow through soil to roots, where it may enter natural openings or wounds made by insects, machinery, and other means. Plants must be predisposed near maturity in order for Fusarium root and stalk rot to occur. Such predisposition may take the form of insect wounds, disease injury, poor fertility, or hot, dry weather. Usually disease is most severe during cool, wet weather following hot dry weather as plants near maturity. Disease is more severe in conventional tillage compared to ecofallow systems. A high nitrogen to potassium ratio may also increase disease incidence.

DISTRIBUTION: Generally distributed where sorghum is grown under hot, dry conditions. In the United States, the area from Kansas to Texas is most severely affected.

SYMPTOMS: Although stalk rot is ordinarily accompanied by root rot, symptoms may not be noticed under irrigation and adequate soil fertility. The cortical tissue of roots is first decomposed, then the vascular tissue. Older roots are often totally destroyed, leaving the plant with little anchorage. Such plants can be pulled out of the soil easily or toppled over by wind. Newly formed roots have lesions of various sizes and shapes that vary in colors from a light tan, brick red, pink, or black.

Stalk rot symptoms resemble those on corn. The fungus grows from the roots and crown into the stalk. At first leaves turn brown. The outside of the nodes may have dark lesions. Inside the stalk, the pith at first is water-soaked and eventually becomes disintegrated with only the vascular bundles intact that are pink, red, or red-purple. Later infected interior stalk tissue becomes a deep red color. A white to pink mycelium may grow on the outside of the nodes during damp weather. The stalk may break over near the soil.

CONTROL:
1. Provide a full moisture profile at planting and ensure adequate irrigation during the growing year.
2. Practice good weed and insect control.
3. Maintain a balanced soil fertility.
4. Plant the recommended plant population of hybrids with good stalk strength.
5. Practice an ecofallow system, especially in a winter wheat-grain sorghum-fallow rotation.

Gray Leaf Spot

This disease is also called angular leaf spot and cercosporiosis.

CAUSE: *Cercospora sorghi* Ell. & Ev. survives mostly as mycelium in residue, plants standing after harvest, grasses, and seed. Conidia are produced during warm, wet weather and are windborne or splashed by rain to hosts. Conidia are produced late in season in lesions to cause secondary infections.

DISTRIBUTION: Generally distributed wherever sorghum is grown in warm, humid, or wet climates.

SYMPTOMS: Ordinarily gray spot occurs too late in the growing season to cause much injury. However, if humid weather persists in midseason, considerable damage may occur. Initially small circular to elliptical, dark purple or red spots occur on leaves late in the growing season, usually after plants are mature. Later the spots elongate up to 30 mm until they are somewhat limited by the leaf veins and become rectangular. Lesions may be light to dark red, purple, or tan. The center of the spots becomes tan or brown but thick grayish mycelium, together with profuse production of conidia, eventually covers the spot; hence the name gray leaf spot. Sporulation is more prevalent on the lower surface of spots. During periods of high humidity, large areas of leaves, together with spots, may be covered with a thick mycelial growth. In severe infections leaf sheaths and upper stems may also become infected.

CONTROL:
1. Plant tolerant or resistant cultivars.
2. Rotate sorghum with small grains.
3. Plow under infected residue.
4. Eliminate wild sorghum and grasses that act as a pathogen host.

Green Ear

CAUSES: *Sclerospora graminicola* (Sacc.) Schroet. overwinters as oospores in soil or on seed. Oospores germinate to form germ tubes that infect sorghum leaves. Temperatures of 24–32°C for two days after planting are conducive to heavy infection. Secondary spread is by formation of sporangia on infected leaves that are windborne to other leaves. Sporangia infrequently germinate directly to form only a germ tube. More commonly, sporangia germinate indirectly to release three or more zoospores that eventually germinate to form a germ tube. Disease development is favored by high humidity and an optimum temperature of 17°C. Oospores are formed in diseased tissue.

DISTRIBUTION: Africa, Asia, Europe, and North America.

SYMPTOMS: The head is transformed completely or partially into a loose green head composed of a mass of small twisted leaves; hence the name green ear. The malformation is probably due to conversion of upper segments of the floral axis into small leafy shoots. These normally develop into grain. Sporangia that resembles a fuzz are produced on infected leaves during periods of high humidity.

CONTROL: No control is necessary.

Head Smut

NOTE: In some literature Sporosorium is spelled Sporisorium.

CAUSE: *Sporosorium reilianum* (Kuhn) Langdon & Fullerton (syn. *Sphacelotheca reiliana* (Kuhn) Clint.) survives as chlamydospores in soil. When seed is planted, chlamydospores in soil germinate to form a four-celled or branched promycelium that bear sporidia terminally and near the septa. Sporidia may sprout to form secondary sporidia or may germinate to form a germ tube that penetrates meristematic tissue of seedlings. Germination occurs best at temperatures between 27°C and 31°C during periods of moist soil. Seedborne chlamydospores apparently are not important in infection. The fungus develops only in actively growing meristematic tissue. Physiologic races are present.

DISTRIBUTION: Generally distributed wherever sorghum is grown.

SYMPTOMS: All or part of the panicle becomes incorporated into a single sorus. Parts of an infected panicle not included in a sorus usually show blasting or proliferation of individual florets. Sori may develop on foliage and culms in some sweet sorghums and sudangrass cultivars.

Infection first appears when the young head enclosed in the boot is completely replaced by a large smut gall surrounded by a whitish membrane. The membrane ruptures even before the head emerges, exposing a mass of dark brown to black, powdery chlamydospores intermingled with a network of long, thin, dark broomlike structures. Sometimes chlamydospores are not present, but witches' brooms, consisting of many small, rolled leaves, protrude from the heads of suckers at the nodes or joints.

CONTROL:
1. Plant resistant cultivars and hybrids.
2. Rotate susceptible sorghum cultivars and hybrids with nonsusceptible crops. Grow susceptible sorghum in the same field only once every four years.

Ladder Leaf Spot

CAUSE: *Cercospora fusimaculans* Atk.

DISTRIBUTION: Brazil, Colombia, Cuba, El Salvador, Honduras, Malawi, Mexico, Rwanda, Georgia, Texas, Venezuela, and Zambia.

SYMPTOMS: Scalariform or ladderlike lesions occur on leaves. Lesions are elliptical, almost rectangular, pale brown, segmented by lad-

derlike markings, dark bordered, and delimited by secondary and tertiary veins.

CONTROL: Levels of resistance exist in *Sorghum bicolor.*

Leaf Blight

This disease is also called Helminthosporium leaf blight, Helminthosporium blight, and northern leaf blight.

CAUSE: *Exserohilum turcicum* (Pass.) Leonard & Suggs. (syn. *Helminthosporium turcicum* Pass., *Bipolaris turcica* (Pass.) Shoemaker, and
Drechslera turcica (Pass.) Subram. & Jain). The teleomorph is *Trichometasphaeria turcica* Lutt. (syn. *Setosphaeria turcica* (Lutt.) Leonard &
Suggs). The causal fungus is seedborne and overwinters in residue and
soil as mycelium and conidia. Conidia can be converted into chlamydospores. Conidia are produced on residue at the soil surface during
moist weather at optimum temperatures of 18–27°C and are windborne
to hosts. Heavy dews particularly favor disease development. During
humid weather, conidia are abundantly produced in lesions and are
windborne to other hosts to serve as secondary inoculum.

DISTRIBUTION: Generally distributed wherever sorghum is grown;
however, it is not common in drier climates.

SYMPTOMS: Seed rot and seedling blight can occur in cold, wet soil.
Infected seedlings may die or develop into stunted plants. Small red-
purple or yellow-tan leaf spots develop in infected seedlings and may
enlarge, coalesce, and turn leaves purple-gray.
   Spots on older leaves occur as elliptical yellowish, grayish, or tan
lesions several centimeters long. Lower leaves are infected first and disease progresses up the plant. During humid conditions, lesions enlarge to
kill and wither large parts of the leaves, giving plants the appearance of
being injured by frost. The center of the lesions is usually grayish to tan,
depending on cultivar, with a red-purple or tan margin. In warm, humid
weather, profuse sporulation occurs, giving the lesion a dark greenish
or black appearance. Conidia are windborne to other hosts. Grain is not
infected but yield may be reduced.
   Trichometasphaeria turcica has been reported to cause a root rot when
seedlings were inoculated. Roots, crowns, and stalks rot, causing wilted
leaves and plants to die.

CONTROL:
1. Most grain and sweet sorghum cultivars have some resistance.
   In humid areas of the southern United States, most cultivars
   will sustain some leaf blight. Two types of resistance are known:

polygenic resistance characterized by a few small lesions, and monogenic resistance characterized by a hypersensitive fleck and little or no lesion development.
2. Leaf blight is usually more severe under minimum tillage conditions. Crop residue should be plowed under in the autumn soon after harvesting the crop if erosion is not a problem.

Leaf Spot

Both leaf spot and red spot are reported to be caused by *Helminthosporium rostratum*; however, the description of the symptoms are sufficiently different to warrant describing them as two different diseases.

CAUSE: *Helminthosporium rostratum* Drechs. survives as mycelium in infected residue and may be seedborne. Conidia are produced on residue during warm (30°C optimum), wet conditions and are windborne to hosts. Conidia are produced in spots during warm, wet conditions to provide secondary inoculum.

DISTRIBUTION: Africa and the United States, particularly in warm areas.

SYMPTOMS: Older leaves are more susceptible than young leaves. Small (1–2 mm by 2–5 mm) spots develop that have a tan center and purple margin. The spots are limited laterally by leaf veins but they may coalesce to form large necrotic areas.

CONTROL: The disease is of minor importance on sorghum and no control is necessary.

Long Smut

CAUSE: *Tolyposporium ehrenbergii* (Kuhn) Patouillard is disseminated by seeds, soil, and wind. Spores on the soil surface are the initial source of infection. They are picked up by wind during dry periods and deposited on flag leaves where they are washed down behind leaf sheaths. Floral parts from boot stage to anthesis are infected by sporidia or germinating teliospores. Seed or seedling infection does not occur. Systemic infection does not occur from seed or soilborne spores or mycelium but only during flowering by airborne spores.

DISTRIBUTION: Africa and Asia.

SYMPTOMS: Sori in ovaries are irregularly distributed, more or less cylindrical with tapered ends, up to 4 cm long and 5–10 mm wide, and surrounded by a firm, false membrane of fungal tissue.

CONTROL: No present control but resistant germ plasm does exist.

Loose Kernel Smut

CAUSE: *Sphacelotheca cruenta* (Kuhn) Potter (syn. *Sporisorium cruenta* (Kuhn) Langdon & Fullerton) overwinters as teliospores on the outside of seed. When seed is planted, teliospores germinate along with the seed by first forming a four-celled promycelium bearing lateral sporidia. The sporidia germinate and infect the developing seedling. Infection occurs over a wide range of soil moisture, pH, and at a temperature between 20°C and 25°C. Mycelium grows systemically within the plant. At heading, teliospores replace kernels and are enclosed within a membrane that soon disintegrates, releasing the spores. Secondary infection may occur when teliospores infect late developing heads, causing them to become smutted. Teliospores in the soil are apparently not important in infecting seedlings. Physiologic races exist.

DISTRIBUTION: Generally distributed wherever sorghum is grown but it is not a common disease.

SYMPTOMS: Infected plants are stunted and head out early. Abundant side branches (tillers) may develop. Normally all kernels in an infected panicle are smutted, but some may be transformed into leafy structures or escape infection completely. Individual kernels are replaced by small smut galls that are long (2.5 cm or longer), pointed, and covered with a thin membrane that usually breaks when the galls are full size. The powdery, dark brown to black teliospores are soon blown away, leaving a long, dark, pointed, and curved structure called a columella in the center of what was the gall. Sometimes the primary head may not be smutted but the tillers or the ratoon crop will be.

CONTROL: Control is the same as for covered kernel smut.

Oval Leaf Spot

CAUSE: *Ramulispora sorghicola* Harris probably survives as sclerotia on leaf residue on the soil surface. After periods of high humidity or rain, conidia are produced that are windborne or splashed by rain to leaves.

DISTRIBUTION: Africa, Asia, Caribbean, and Central America.

SYMPTOMS: Plants are infected at all stages of growth. Initially small, water-soaked spots appear with tan, brick red, or purple borders. These develop into roughly circular lesions (3–8 mm by 1.5–4 mm), with pink-gray to yellowish centers with dark red or tan borders up to 1 mm wide. Sometimes a few scattered, small, black sclerotia appear in spots on the underside of leaves.

Oval leaf spot can be confused with anthracnose. The diseases may be distinguished by the setae. In oval leaf spot, the setae are clear brown and arise from the sclerotia. In anthracnose, setae are black and do not arise from sclerotia.

CONTROL: Rotate sorghum with nonsusceptible crops.

Panicle and Grain Anthracnose

CAUSE: *Colletotrichum graminicola* (Ces.) G. W. Wilson. See description in Anthracnose Leaf Blight of Sorghum. Disease occurs on mature plants and is favored by cloudy, warm, humid weather. Conidia are produced on leaves and are washed behind leaf sheaths by rain. The germinating conidia enter the panicle.

DISTRIBUTION: Same as for anthracnose leaf blight.

SYMPTOMS: Normally only mature plants are affected. Initially the infection appears as water-soaked, discolored lesions that turn tan to black-purple. Young lesions appear as elliptic pockets or bars immediately beneath the epidermis. The interior of the panicle has a mottled appearance of red-brown interspersed with white tissue. Infected panicles are small, lightweight, and mature early. Small black streaks, which are the acervuli, may appear and extend to the seed.

Infected seeds have encircling dark brown to black streaks. Severely infected seeds may be totally discolored and either germinate poorly or cause seedling blight from the seedborne inoculum.

CONTROL:
1. Plant disease-free seed.
2. Apply seed-treatment fungicides.
3. Plant resistant cultivars.

Periconia Root Rot

This disease is also called milo disease and milo root rot.

CAUSE: *Periconia circinata* (Mangin) Sacc. probably survives as chlamydospores or thick-walled mycelial cells in roots and soil. Conidia also persist in soil. Chlamydospores germinate to produce either conidiophores on which conidia are borne or hyphae. Both chlamydospores and conidia germinate to produce infection hyphae that infects roots.

The fungus produces a toxin that kills host tissue in advance of mycelial growth. Dissemination is by chlamydospores that are moved with soil. Chlamydospores are eventually produced within infected tissue and returned to soil during harvest.

A pathogenic population of *P. circinata* depends on continuous growth of susceptible cultivars. The population of toxin-producing isolates is higher in the roots of susceptible genotypes than in the soil. However, pathogenic toxin-producing isolates are not isolated from the roots of resistant genotypes.

DISTRIBUTION: Primarily the sorghum-growing areas of the southern United States.

SYMPTOMS: The first symptoms may occur three to four weeks after planting in heavily infested soil. Initially plants are stunted; there is a slight rolling of leaves, with the older leaves turning light yellow at the tips and margins. Leaves continue to yellow with drying occurring until all leaves are affected and the plants eventually die, usually at heading time. Sometimes symptoms may not occur until heading time with a rapid development of symptoms. A dark red color occurs in the center of infected stalks that extends into the roots. Roots may be so severely rotted as to slough off from the plant. Diseased plants usually produce smaller heads.

CONTROL: Plant resistant cultivars and hybrids.

Phoma Leaf Spot

CAUSE: *Phoma insidiosa* Tassi overwinters as pycnidia on residue and seed. Conidia are produced in pycnidia during warm, wet weather in the spring and are blown or splashed to healthy leaves. Secondary spread is by formation of pycnidia on infected tissue and subsequent production of conidia.

DISTRIBUTION: Generally distributed wherever sorghum is grown under warm temperatures. Phoma leaf spot is considered a minor disease.

SYMPTOMS: Irregular or subcircular spots develop on leaves and are 1 cm or more in diameter. Spots are yellow-brown or tan to gray and may be indefinite in outline or have a narrow red-purple margin. Small, black dots, which are the pycnidia, develop scattered over dead leaf tissue or sometimes in groups or lines between veins. Pycnidia also develop on glumes and seed.

CONTROL: The disease is generally considered to be very minor; however, the following aid in control of the disease:

1. Plant clean, healthy seed that has been treated with a seed-protectant fungicide.
2. Plow under infected residue where erosion is not a problem.

Pokkah Boeng

This disease is also called top rot and twisted top.

CAUSE: *Fusarium moniliforme* var. *subglutinans* Wr. & Reink. The teleomorph is *Gibberella fujikuroi* var. *subglutinans* Edwards. It overwinters as mycelium and perithecia in residue. It may also be seedborne as mycelium and conidia. Conidia are produced from mycelium growing on residue during wet weather and are windborne or splashed to hosts. Conidia lodge in leaf sheaths or whorls where they germinate. During wet weather, the fungus also grows upward on the outside of the stalk. Metabolites are produced by the fungus that incite distortions in the plant.

DISTRIBUTION: Occurs in tropical or semitropical areas where humidity is high.

SYMPTOMS: Leaves near the top of the plant are deformed or discolored. Often leaves are so wrinkled that they do not unfold properly, resulting in a plant with a ladderlike appearance. Leaves also have wrinkled bases and numerous transverse cuts in the margins. In severe cases, the fungus may grow from the whorls and sheaths into the stalks, causing the top part of the plant to die. In mild cases, symptoms may have a mosaic appearance that resembles a virus. Stalks may bend over. There may be a knife-cut symptom that resembles narrow, uniform, transverse cuts in the rind as if a knife had cut out portions of rind. The latter symptom may be covered by leaf sheaths and may not be apparent until stalks are broken over at the cut.

CONTROL: Pokkah boeng is not considered a serious disease of sorghum and no control measures are necessary.

Pythium Root Rot

CAUSE: *Pythium arrhenomanes* Drechs. and *P. graminicola* Subr. are considered the primary causal organisms. Other *Pythium* spp. associated with root rot are *P. periplocum* Drechs. and *P. myriotylum* Drechs. Disease is favored by high soil temperatures and moisture throughout the growing season; however, infections increase at the boot stage and later. Plant death normally occurs during stress conditions at plant maturity or later. Seedling blight by *P. arrhenomanes* has been reported to occur at low (15°C) soil temperatures.

DISTRIBUTION: Not known for certain but likely widespread.

SYMPTOMS: Symptoms usually occur on larger adventitious roots. Red-brown to blackish sunken lesions occur on roots. Frequently at root death the entire root will become discolored.

Pythium arrhenomanes has been reported to cause a preemergence and postemergence seedling death.

CONTROL: Highly susceptible genotypes have been identified.

Pythium Seed and Seedling Blight

Cause, distribution, symptoms, and control are similar to those on corn. Resistance to *Pythium arrhenomanes* Drechs. has been identified in a grain sorghum cultivar.

Red Rot

This disease is also called anthracnose stalk rot, Colletotrichum rot, and red stalk rot.

CAUSE: *Colletotrichum graminicola* (Ces.) G. W. Wilson is discussed under Anthracnose Leaf Blight. Conidia are washed down behind leaf sheaths, germinate, and infect stalks anytime after jointing stage.

DISTRIBUTION: Generally distributed wherever sorghum is grown.

SYMPTOMS: Red rot occurs primarily in stalks of mature sorghum plants. The basal portion of the stalk becomes reddened or purple. When stalks are split lengthwise, the pith has alternate discolored and white areas, giving a marbled appearance throughout the infected area. Depending on the cultivar, the discolored areas range from tan to purplish red. A similar symptom occurs when the peduncle or upper stem below the head becomes infected. Cankers on stalk are bleached with surrounding areas that are reddish, tan, orange, or purple.

Stalks frequently break near the middle of the stalk or just below the seed head. Diseased plants may not break but produce small heads often with small seeds.

CONTROL:
1. Plant resistant cultivars and hybrids.
2. Rotate sorghum with nonsusceptible crops.

Red Spot

Both leaf spot and red spot are reported to be caused by *Helminthosporium rostratum*; however, the description of the symptoms is sufficiently different to warrant describing them as two different diseases.

CAUSE: *Helminthosporium rostratum* Drechs.

DISTRIBUTION: Mississippi.

SYMPTOMS: Symptoms resemble those of Anthracnose Leaf Blight; however, the lesions of red spot are not sunken. On sweet sorghum, red spot lesions of the leaf midrib are red-brown to black with tan or brown centers, irregular, and longitudinal (8–12 mm by 50 mm).

CONTROL: The disease is of little economic importance but symptoms could be mistaken for those of red rot in breeding programs that screen for anthracnose leaf blight.

Rhizoctonia Stalk Rot

CAUSE: *Rhizoctonia solani* Kuhn survives in soil and residue as sclerotia. Sclerotia are disseminated by any means that will move soil, such as by wind and running water. The sclerotia germinate to form infective hyphae that penetrate the host. Plants that are generally lacking in vigor are probably more susceptible to infection.

DISTRIBUTION: Widely distributed but of minor importance.

SYMPTOMS: Initially the pith becomes reddish with vascular bundles remaining as light streaks. Later brown sclerotia between 1 mm and 5 mm in diameter form on the outside of the stalk under the leaf sheath.

CONTROL: No control is necessary other than maintaining plants in a vigorous growing condition.

Rough Leaf Spot

This disease is also called Ascochyta spot and rough spot.

CAUSE: *Ascochyta sorghina* Sacc. is seedborne and overwinters as pycnidia in soil, and pycnidia and mycelium in residue. During wet weather, conidia are produced in pycnidia and ooze out where they are splashed or windborne to hosts. Later pycnidia are produced in spots and conidia are again formed during wet weather to be disseminated as secondary inoculum.

DISTRIBUTION: Africa, Asia, southern Europe, and the United States, particularly the humid sorghum growing areas of the southeastern states. The disease is generally minor.

SYMPTOMS: Initially small circular to oblong, light-colored spots with well-defined margins develop near the ends of leaves. The spots enlarge and coalesce to form large blotches covered by hard, black, raised pycnidia that give the spots a rough surface and appearance. These lesions may be grayish, yellow-brown, or purple-red and elongated parallel to the veins. As leaves mature the pycnidia may fall off or be washed off by rain so that only large areas of tan necrotic tissues remain that may have a reddish, tan, or no distinct border. Sometimes pycnidia appear on healthy green parts of the leaf surface. Lesions may also occur on leaf sheaths and stalks. Pycnidia may develop on glumes.

CONTROL:
1. Plant resistant grain sorghum and sweet sorghum cultivars.
2. Rotate sorghum with other crops.
3. Plow under infected residue where erosion is not a problem.

Rust

CAUSE: *Puccinia purpurea* Cke. (syn. *Uredo sorghi* Pass., *P. sanguinea* Diet., *Dicaeoma purpureum* (Cke.) Kuntze, and *P. prunicolor* H. Syd. et al.) is a rust fungus that probably overwinters as urediospores and mycelium in residue in warmer areas or on Johnsongrass. However, spores are relatively short-lived in the absence of a living host and thrive on ratooned and successively planted sorghum. *Oxalis corniculata* L. may be an alternate host, as it has been reported to be infected, producing abundant aecia. Infected perennial and collateral hosts and scattered sorghum plants serve as infection foci. Urediospores are windborne to hosts, and during periods of high humidity, urediospores are produced within uredia, producing secondary inoculum. Light drizzles and heavy dew especially favor infection. Two-celled teliospores develop as leaves mature and may germinate immediately or overwinter before germinating to produce a single basidiospore per cell. However, the function of the teliospore is generally not considered important in the life cycle of the fungus.

DISTRIBUTION: Generally distributed wherever sorghum is grown. Rust is most severe in areas of high humidity or rainfall.

SYMPTOMS: Rust normally appears on plants nearing maturity. The first symptoms are small, purple, red or tan flecks or spots on both leaf surfaces. In susceptible cultivars the spots enlarge and may cover much of the leaf surface. Small, raised, brownish pustules (uredia) develop in the spots. Uredia are elliptical, 2.0 mm in length, and lie parallel to and between veins. The epidermis of the uredia ruptures to expose reddish to dark brown urediospores. There is often a purple, reddish, or tan area surrounding each uredium. Telia develop later in the season, frequently within old uredia and commonly on the underside of the leaf. Telia are elliptical to linear, often seen as streaks or stripes, dark brown, and 1–3 mm in length. The epidermis may stay intact for a long time.

CONTROL: Forage sorghum types are usually more severely infected than grain sorghums. Rust usually appears too late in the growing season to cause much injury; therefore, control measures are usually not warranted. Some accessions have resistance.

Sooty Stripe

CAUSE: *Ramulispora sorghi* (Ell. & Ev.) Olive & Lefebvre (syn. *R. andropogonis* Miura) survives primarily as sclerotia in soil or residue. Sporodochia also are important for fungal survival when infected leaf residue is left on or above the soil surface throughout the winter. During warm, moist conditions in the growing season, conidia are produced by sclerotia or sporodochia that are windborne or splashed onto healthy tissues. Conidia produced in lesions are disseminated to function as secondary inoculum. Eventually sclerotia are produced in lesions.

DISTRIBUTION: Africa, Asia, South America, and the United States, particularly in warm, humid, sorghum-growing areas.

SYMPTOMS: Disease occurs on plants at all growth stages, but oldest leaves and sheaths are infected first. Initially small, water-soaked, oblong, red-purple or tan spots with a yellow halo appear on leaf laminae or sheaths. These develop into elongate to elliptical lesions with tan necrotic centers and red-purple to tan borders. Affected leaves often turn bright yellow due to broad yellow margins that surround lesions. Mature lesions average about 5 cm or longer in length. During warm, wet weather, the lesion centers become grayish due to production of conidia. Eventually the lesion centers become sooty due to formation of numerous small, black, superficial sclerotia. The sclerotia are wiped off easily and cling to fingers like soot.

CONTROL: Generally sooty stripe occurs too late to do much damage.
1. Rotate sorghum with other crops.
2. Treat seed with a seed-protectant fungicide.
3. Plow under infected residue.

Sorghum Downy Mildew

CAUSE: *Peronosclerospora sorghi* (Weston & Uppal) C. G. Shaw (syn. *Sclerospora sorghi* Weston & Uppal and *S. graminicola* var. *andropogonis-sorghi* Kulk.) survives for several years as oospores either in residue or soil after residue has decomposed. The fungus is also seed-borne as oospores on the surface of seeds and mycelium within seeds. Oospores germinate and infect young plant roots under low soil moisture conditions and a minimum temperature of 10°C. Mycelium then grows systemically inside the plant. Penetration of leaves occurs by germ tubes penetrating through stomatal openings. Conidia are produced only at night on leaf surfaces when long dew periods of several hours are associated with cool nights (18–21°C). Viable conidia may be carried several thousand meters but most remain within a few centimeters or a few meters from where they were produced and play no role in long-distance dissemination. Conidia generally cause local lesions but can also cause systemic infections if leaves are infected before they are fully developed. Eventually oospores are formed within systemically infected tissue.

Disease severity will increase more rapidly from year to year in sandy soils than in clay soils. Highest disease incidence apparently occurs at a soil temperature–soil moisture combination of 25°C and −0.2 bar, and a soil texture of 80 percent sand.

DISTRIBUTION: Africa, India, southeastern Asia, and the United States, particularly the Gulf and midwestern states.

SYMPTOMS: Seedlings are infected at germination and show a chlorotic mottle or are generally chlorotic and stunted. Conidia that look like white fuzz are produced on the underside of chlorotic leaves during humid weather.

As plants grow, vivid green and white stripes appear on leaves in late spring or early summer. Sometimes whole leaves may become chlorotic or may show one or two narrow green stripes. Conidia are usually not produced on these leaves.

Local infections appear as small, rectangular, chlorotic spots that soon become necrotic, giving a stippled or speckled appearance. Conidia are also produced in these local lesions.

Systemically diseased plants that develop late in the season have striped or mottled leaves. Plants may fail to head, produce sterile heads, or form partially affected heads. Leaf shredding occurs as the leaf cells between veins are destroyed, releasing oospores.

CONTROL:
1. Apply a systemic fungicide seed treatment.
2. Plant resistant sweet and grain sorghum hybrids.
3. Do not plant grain sorghum after sudangrass.
4. Plant only high-quality seed.

5. Do not plant sudangrass and sorghum sudan hybrids in areas where downy mildew is present since these very susceptible sorghums tend to increase the population of oospores in the soil.
6. Rotate sorghum with other crops not susceptible to P. sorghi.
7. Destroy sorghum stubble after harvest.
8. Deep plow to place oospores below seed depth.
9. Plant as early as possible.

Southern Leaf Blight

CAUSE: *Bipolaris maydis* (Nisik.) Shoemaker (syn. *Helminthosporium maydis* Nisik.) is discussed in the chapter on corn. The fungus does not sporulate readily on sorghum.

DISTRIBUTION: Generally distributed wherever sorghum is grown.

SYMPTOMS: Similar to corn. Symptoms are mostly spots on leaves. Occasionally lesions merge to cover large areas of the leaves.

CONTROL: Southern leaf blight is generally not considered an important disease on sorghum. Control measures are not considered necessary.

Southern Sclerotial Rot

CAUSE: *Sclerotium rolfsii* Sacc. The anamorph is *Corticium rolfsii* (Sacc.) Curzi. The causal fungus survives as sclerotia in soil and residue, and mycelium in residue. However, the fungus only survives in warmer climates as cold winter temperatures kill both sclerotia and mycelium. Sclerotia germinate to form mycelium that infect seeds and roots; however, the fungus is also thought to grow saprophytically from the soil to the lowest leaf sheath under continuous wet soil conditions.

DISTRIBUTION: Warmer sorghum-growing areas of the world. The disease is not important on sorghum.

SYMPTOMS: Initially a water-soaked lesion that becomes red, purple, or brown forms at base of the lowest leaf sheath in contact with the soil. Leaf blades eventually wilt and become necrotic. Typically one to three sheaths are affected on plants nearing bloom or older. Occasionally all sheaths may be diseased; in such cases the plant may be killed. Mats of white mycelium grow inside leaf sheaths to about 30 cm above the soil. Fans of mycelium merge from sheaths and spread over the base of the leaf for about 13 mm. Diseased sheaths are reddish and may be shredded. Young axillary shoots become blighted and covered with mycelium.

Numerous sclerotia, about 1–2 mm in diameter, white at first then brown, are produced inside sheaths. Dead shredded leaves at the base of plants have abundant sclerotia produced in and on them.

CONTROL: No control is necessary since the disease is of minor importance on sorghum.

Tar Spot

CAUSE: *Phyllachora sacchari* P. Henn. (syn. *P. sorghi* Hohn.) survives on weed hosts such as Johnsongrass and wild sorghums. Ascospores are formed within stromata on leaf surfaces and are windborne to other leaf surfaces. Disease is most severe during warm, wet weather.

DISTRIBUTION: India, Madagascar, Malaysia, Philippines, and Thailand.

SYMPTOMS: Small (1 mm or less in diameter), black, round to elongate, raised spots occur on leaves. The spots, composed of hard stromatic masses of causal fungus, extend through the leaf to both surfaces. The black stroma may be surrounded by a chlorotic or necrotic ring that may be reddish or yellowish. Lesions may coalesce and become large enough to harm the leaf.

CONTROL: Plant resistant cultivars.

Target Leaf Spot

CAUSE: *Bipolaris cookei* (Sacc.) Shoemaker (syn. *B. sorghicola* (Lefebvre & Sherwin) Shoem., *Drechslera sorghicola* (Lefebvre & Sherwin) Richardson & Fraser, and *Helminthosporium sorghicola* Lefebvre & Sherwin) survives primarily as mycelium in residue, particularly in leaf veins in delimiting lesions. The fungus also survives on leaves as chlamydospores formed from conidia. Sporulation occurs under humid conditions with conidia being airborne to host tissue. Disease development is favored by moist humid weather and temperatures of 19–33°C.

DISTRIBUTION: Cyprus, India, Israel, South America, Sudan, and the United States. The disease is of minor importance.

SYMPTOMS: Symptoms vary according to sorghum genotype. Initially disease occurs as red or gray dots that later develop into definite spots. Spots on leaves are well defined, delimited by leaf veins, and elongated. Spots range in size from small spots (2–3 mm) to large lesions (10–15 mm). Lesions coalesce to form large areas of necrotic tissue. Shape of spots will vary from rectangular to narrow and color will vary from brown, reddish with light-colored centers, tan, purple, or light brown.

CONTROL: No control is necessary.

Zonate Leaf Spot

This disease is also called Gloeocercospora leaf spot.

CAUSE: *Gloeocercospora sorghi* D. Baine & Edg. overwinters as sclerotia in residue on or above soil surfaces. Primary inoculum comes from this source as conidia that are splashed to infection sites. It is also seedborne as mycelium and possibly sclerotia. Sclerotia germinate to produce conidia that are splashed or windborne to sorghum. During warm, wet weather, sporulation occurs in lesions as pink sporodochia form above stomates. The conidia are borne in a pinkish, gelatinous mass and are splashed to healthy tissue. Sclerotia are eventually formed in leaf lesions in lines parallel to the veins.

DISTRIBUTION: Africa, Asia, Central and South America, United States, and West Indies. It is most severe in warm, wet climates.

SYMPTOMS: Both leaves and sheaths can be infected. Initially small, red-brown, water-soaked spots, sometimes with a narrow chlorotic halo, occur on leaves near the ground. The spots enlarge and elongate in a direction parallel with the veins and become dark red or tan. Eventually the spots occur as circular areas on leaves or semicircular patterns along the margin. Spots are typified by circular, red-purple bands, alternating with tan areas that give a concentric or zonate pattern with irregular borders. Spots vary in diameter from a few millimeters to several centimeters. Spots may cover the entire width of the leaf.

During warm, wet weather, conidia will be produced in lesions as a pinkish gelatinous spore mass that forms above a stomate. Sclerotia eventually appear as small, black, raised bodies in lines parallel to the veins.

Plants severely infected in the seedling stage may become defoliated or die. Older plants may have foliage destroyed before maturity with poorly filled seed.

CONTROL:
1. Plant resistant or tolerant cultivars or hybrids.
2. Plant only clean seed treated with a seed-protectant fungicide.

3. Rotate sorghum with other crops.
4. Plow under infected residue where erosion is not a problem.

_____MYCOPLASMAL CAUSES_____

Yellow Sorghum Stunt

CAUSE: A mycoplasma. It is not known how the causal agent over-seasons or is transmitted.

DISTRIBUTION: Southern and midwestern United States, including Alabama, Georgia, Kentucky, Louisiana, Mississippi, Ohio, and Texas.

SYMPTOMS: Infected plants are generally more prevalent at or near edges of fields. Plants are stunted, about half of normal height, with leaves bunched together at the tops. Leaves are rigid and curled adaxially about the blade axis. Leaves have pronounced puckering, resulting in undulating margins and a yellow-tinged cream color that makes plants conspicuous from a distance.

Affected plants seldom produce panicles and if panicles are produced they are normally barren. Plants infected when they are less than 0.3 mm high remain at that height but stalks continue to increase in diameter. These plants usually display other disease symptoms also.

CONTROL: Plant resistant cultivars.

CAUSED BY
_____NEMATODES_____

Cereal Root Knot Nematode

CAUSE: _Meloidogyne naasi_ Franklin overwinters as eggs. Second-stage larvae emerge from eggs and penetrate the root epidermis just above the root cap. Nematode secretions transform pericycle cells into giant cells. Two to three generations of nematodes are produced each growing season.

DISTRIBUTION: Kansas and Thailand.

SYMPTOMS: Plants in irregular-shaped areas in a field are stunted and chlorotic. Root galls may be elongated swellings or discrete knots. These galls are normally smaller than those caused by other root-knot nematodes. Roots are often curved in the shape of a hook, horseshoe, or a complete spiral.

CONTROL: None normally practiced.

Root Knot Nematodes

CAUSE: *Meloidogyne* spp. Certain grain sorghum cultivars have been reported to be highly susceptible to *M. incognita* (Kofoid & White) Chitwood.

DISTRIBUTION: Not known for certain but likely widely distributed.

SYMPTOMS: Plants in irregularly shaped areas are stunted and chlorotic with delayed blooming and reduced yields. Roots have galls, elongated swellings, or discrete knots or swellings. Root proliferation is common.

CONTROL: None is normally practiced.

Root Lesion Nematodes

CAUSE: *Pratylenchus hexincisus* Taylor & Jenkins and *P. zeae* Graham are endoparasitic.

DISTRIBUTION: Worldwide.

SYMPTOMS: Plants are stunted and chlorotic. Roots are poorly developed and have necrotic lesions. *Pratylenchus hexincisus* increases the severity of charcoal rot.

CONTROL: None is normally practiced.

Stunt Nematodes

CAUSE: *Merlinius* spp., *Quinisulcius* spp. and *Tylenchorhynchus* spp. are ectoparasitic.

DISTRIBUTION: Not reported.

SYMPTOMS: Plants may be stunted and chlorotic. The root system is poorly developed with fewer than normal feeder roots. Numerous root tips may be short and thickened.

CONTROL: No control is reported.

Other Nematodes Reported Associated with Sorghum

Acrobeles sp.
Aphelenchoides sp.
Aphelenchus avenae Bastian
Criconemella xenoplax Raski
Diphtherophora sp.
Dorylaimus sp.
Helicotylenchus dihystera (Cobb) Sher.
Hoplolaimus galeatus (Cobb) Thorne
Malenchus sp.
Mononchus sp.
Nothotylenchus sp.
Pratylenchus zeae Graham
Pseudohalenchus sp.
Psilenchus sp.
Rhabditis sp.
Rotylenchulus reniformis Lindford & Olivera
Tylencholaimus sp.
Tylenchus sp.
Xiphinema americanum Cobb

 VIRAL CAUSES

Brome Mosaic

CAUSE: Brome mosaic virus (BMV) persists in cereal crops and perennial grasses. BMV can survive for several months in dried tissue. It is transmitted primarily through sap transmission. Local spread likely occurs through plant contact and transfer of sap by machinery and wind.

Soil transmission by the dagger nematodes *Xiphinema coxi* Tarjan and *X. paraelongatum* Alther has been reported from Europe.

DISTRIBUTION: Europe and the United States.

SYMPTOMS: Leaves display narrow, interveinal chlorotic streaks that appear mottled from a distance. Mechanically inoculated plants may be stunted.

CONTROL: Plant resistant cultivars.

Maize Chlorotic Dwarf

CAUSE: Maize chlorotic dwarf virus (MCDV) overwinters in Johnsongrass and is transmitted in a semipersistent manner by the leafhoppers *Graminella nigrifrons* (Forbes) and *G. sonora* (Ball).

DISTRIBUTION: United States, particularly Mississippi, Louisiana, and Texas.

SYMPTOMS: Slight stunting and tertiary veinal chlorosis.

CONTROL: The disease is not of economic importance at present; therefore, control is not warranted.

Maize Dwarf Mosaic

CAUSE: Maize dwarf mosaic virus (MDMV) persists in several annual and perennial grasses. In the spring, aphids, particularly the corn leaf aphid *Rhopalosiphus maidis* (Fitch), and the greenbug *Schizaphis graminum* (Rondani), feed on overwintering hosts and acquire the virus. Twenty-three species of aphids are known to be a vector of strain A. Aphids can retain infective virus for 20 minutes after feeding and transmit in a styletborne nonpersistent manner during their movements between infected and healthy plants. High aphid populations often induce a high disease incidence through secondary infection. Different strains exist but only strains A and B have been observed naturally in sorghum. Strain A infects Johnsongrass and strain B is referred to as the non-Johnsongrass strain.

DISTRIBUTION: Generally distributed throughout the areas of the world where sorghum is grown.

SYMPTOMS: Symptoms of MDMV infection can vary according to sorghum genotype, virus strain, and temperature. Plants infected as seedlings are more severely affected than plants infected later in the growing season. Symptoms are most evident on the upper two or three leaves as an irregular mottling of dark and light green areas interspersed with longitudinal white or light yellow streaks. Some sorghums that have the gene for red pigmentation may have the mottling replaced by a red leaf symptom especially on abnormally cool nights. This causes a red leaf or a red stripe symptom that appears as elongated stripes with necrotic centers and reddish margins or infrequently as round spots.

Severely infected plants may die. Living, infected plants may be stunted with delayed flowering and subsequent failure to head or set seed.

Maize dwarf mosaic, strain A, causes pigmented necrotic lesions on panicle branches followed by excessive shrinkage of seed. These symptoms occur when temperatures of 16°C or less prevail four or more consecutive days while seed is in milk to soft dough stage.

CONTROL:
1. No sorghum cultivars can resist infection by all strains of the virus. However, some cultivars are resistant to some strains of the virus.
2. Control grassy weeds, particularly Johnsongrass, in a field.

Maize Dwarf Mosaic Head Blight

CAUSE: Maize dwarf mosaic virus is discussed under Maize Dwarf Mosaic.

DISTRIBUTION: Same as maize dwarf mosaic.

SYMPTOMS: Plants infected by virus at or near heading may fail to fill properly. Symptoms may be absent on vegetative parts. Normal and underdeveloped seed occur together in the same spikelet or panicle. The small seed may be restricted to one side of a head or follow patterns of anthesis. Affected seeds are lightweight, have chalky endosperms, shatter easily, and may be black due to growth of such fungi as *Alternaria* spp. Pigmented, necrotic lesions appear on branches of panicles and are most likely to develop on cool nights on sorghums that carry the gene for red pigmentation. Similar symptoms may occur from root or stalk rots except the lesions on branches of panicles are lacking.

CONTROL: Same as for maize dwarf mosaic.

Maize Stripe

CAUSE: Maize stripe virus (MStpV) is transmitted in a persistent manner by the planthopper *Peregrinus maidis* Ashm.

DISTRIBUTION: Australia and possibly India and Africa.

SYMPTOMS: Infected plants are chlorotic, flower early, and tiller excessively.

CONTROL: No control is reported.

Peanut Clump

CAUSE: Peanut clump virus (PCV) is soilborne with *Polymyxa graminis* Led. the suspected vector. PCV is also mechanically transmitted.

DISTRIBUTION: Upper Volta.

SYMPTOMS: *Sorghum arundinaceum* Moench. (great millet) is a symptomless natural host.

CONTROL: No control is reported.

Red Stripe Disease

CAUSE: Johnsongrass-infecting strain of sugarcane mosaic virus.

DISTRIBUTION: New South Wales, Australia.

SYMPTOMS: Symptoms vary with infected cultivar. Initially a mosaic pattern develops on leaves that eventually becomes interspersed with necrotic areas. Some cultivars develop a severe necrosis. Systemically infected leaves may also become necrotic. Plants may be stunted with a severe yellow mosaic.

CONTROL: Control Johnsongrass.

Sorghum Stunt Mosaic

CAUSE: Sorghum stunt mosaic virus (SSMV) is transmitted by *Graminella sonora* (Ball). It is not mechanically transmitted. Other hosts of SSMV include maize and wheat.

DISTRIBUTION: California.

SYMPTOMS: Plants are severely stunted. Leaves have chlorotic and necrotic mottling and streaking.

CONTROL: No control is reported.

Sugarcane Fiji Disease

CAUSE: A virus that is transmitted by the leafhopper *Perkinsiella saccharicida* Kirkaldy.

DISTRIBUTION: Philippines and possibly Nigeria.

SYMPTOMS: Leaves have galls and are stiff and malformed. Infected plants have shortened stalks with no formed heads or poorly formed ones and turn prematurely brown.

CONTROL: No control is reported.

Sugarcane Mosaic

CAUSE: Sugarcane mosaic virus (SCMV) is soilborne but the actual mechanism of soil transmission is unknown. The virus is transmitted by at least seven aphid species in a nonpersistent manner. Aphids normally acquire SCMV from sugarcane or perennial grasses and may become widely dispersed by wind. SCMV is also mechanically transmitted.

Several strains exist. SCMV has a wide host range of annual and perennial grasses.

DISTRIBUTION: Anywhere in the world where sorghum and sugarcane are grown in the same area.

SYMPTOMS: Virus-caused symptoms are related to distribution of aphid vectors. Edges of fields are frequently the first and only areas affected. Symptoms vary according to sorghum cultivar, SCMV strain, and environment.

Plants have light green mottling or mosaic, often with red lesions. However, the red lesions occur at 16°C and below. Infected plants are stunted with reduced number of heads, seeds, and head length. Flowering is delayed.

CONTROL: Plant a tolerant or resistant cultivar or hybrid.

DISORDERS OF
_____UNKNOWN CAUSE_____

Weak Neck

CAUSE: Originally the malady was associated with physiologic disorders resulting from inherent plant characteristics and environmental stresses in combine types of grain sorghum. The peduncle and rachis do not develop sufficient thick-walled tissue that ripen to support the developing head. Secondary organisms invade such tissue and further weaken it.

DISTRIBUTION: United States.

SYMPTOMS: The culm breaks below the head, resulting in poorly developed heads with lightweight seed. The upper culm tissues dry out, become spongy, and bleach to a tan color. Frequently the peduncle may be water-soaked and have an accumulation of a sticky exudate during wet weather.

CONTROL: Plant resistant cultivars in which grain ripens before culm weakens.

_____BIBLIOGRAPHY_____

Ali, M. E. K., and Warren, H. L. 1987. Physiological races of *Colletotrichum graminicola* on sorghum. *Plant Disease* **71**: 402–404.

Anon. 1975. Prevent and identify grain sorghum diseases. *World Farming* **17**:11–14.

Bandyopadhyay, R.; Mughogho, L. K.; and Satyanarayana, M. V. 1987. Systemic infection of sorghum by *Acremonium strictum* and its transmission through seed. *Plant Disease* **71**: 647–650.

Bell, D. K.; Harris, H.; and Wells, H. D. 1973. Rhizoctonia blight of grain sorghum foliage. *Plant Dis. Rptr.* **57**: 549–550.

Birchfield, W., and Anzalone, L., Jr. 1982. Grain sorghum root-knot and reniform nematode host reactions. *Phytopathology* **72**:355 (abstract).

Borges, O. L. 1983. Pathogenicity of *Drechslera sorghicola* isolates on sorghum in Venezuela. *Plant Disease* **67**: 996–997.

Castor, L. L., and Frederiksen, R. A. 1980. Fusarium head blight occurrence and effects on sorghum yield and grain characteristics in Texas. *Plant Direase* **64**:1017–1019.

Cook, G. E.; Boosalis, M. G.; Dunkle, L. D.; and Odvody, G. N. 1973. Survival of *Macrophomina phaseoli* in corn and sorghum stalk residue. *Plant Dis. Rptr.* **57**: 873–875.

Cuarezma-Teran, J. A.; Trevathan, L. E.; and Bost, S. C. 1984. Nematodes associated with sorghum in Mississippi. *Plant Disease* **68**:1083–1085.

Dean, J. L. 1968. Germination and overwintering of sclerotia of *Gloeocercospora sorghi*. *Phytopathology* **58**:113–114.

Doupnik, B., Jr.; Boosalis, M. G.; Wicks, G.; and Smika, D. 1975. Ecofallow reduces stalk rot in grain sorghum. *Phytopathology* **65**:1021–1022.

Edmunds, L. K.; Futrell, M. C.; and Frederiksen, R. A. 1969. Sorghum diseases, in *Sorgum Production and Utilization*. J. S. Wall and W. M. Ross (eds.). AVI Publishing Co., Westport, CT.

Edmunds, L. K., and Niblett, C. L. 1973. Occurrence of panicle necrosis and small seed as manifestations of maize dwarf mosaic virus infection in otherwise symptomless grain sorghum plants. *Phytopathology* **63**: 388–392.

Edmunds, L. K., and Zummo, N. 1975. Sorghum diseases in the United States and their control. *USDA Agric. Res. Serv. Agric. Hdbk. No. 468.*

El-Shafey, H. A.; Abd-El-Rahim, M. F.; and Refaat, M. M. 1979. A new Cephalosporium-wilt disease of grain sorghum in Egypt. *Egypt. Phytopathol. Congr., 3rd. Proc.*, pp. 514–532.

Forbes, G. A., and Collins, D. C. 1985. A seedling epiphytotic of sorghum in South Texas caused by *Pythium arrhenomanes*. *Plant Disease* **69**: 726. (disease note).

Forbes, G. A.; Ziv, O.; and Frederiksen, R. A. 1987. Resistance in sorghum to seedling disease caused by *Pythium arrhenomanes*. *Plant Disease* **71**:145–148.

Frederiksen, R. A. (ed.) 1986. *Compendium of Sorghum Diseases.* American Phytopathological Society, St. Paul, MN, 82p.

Frederiksen, R. A., and Rosenow, D. T. 1971. Disease resistance in sorghum. *26th Ann. Corn and Sorghum Res. Conf. Proc.* Chicago, pp. 71–82.

Frederiksen, R. A. 1980. Sorghum downy mildew in the United States: Overview and outlook. *Plant Disease* **64**: 903–908.

Janke, G. D.; Pratt, R. G.; Arnold, J. D.; and Odvody, G. N. 1983. Effects of deep tillage and roguing of diseased plants on oospore populations of *Peronosclerospora sorghi* in soil and on incidence of downy mildew in grain sorghum. *Phytopathology* **73**:1674–1678.

Jensen, S. G.; Mayberry, W. R.; and Obrigawitch, J. A. 1982. *Erwinia chrysanthemi* as a pathogen of grain sorghum. *Phytopathology* **72**: 990. (abstract).

Jensen, S. G.; Mayberry, W. R.; and Obrigawitch, J. A. 1986. Identification of *Erwinia chrysanthemi* as a soft-rot-inducing pathogen of grain sorghum. *Plant Disease* **70**: 593–596.

Jones, B. L. 1971. The mode of *Sclerospora sorghi* infection of *Sorghum bicolor* leaves. *Phytopathology* **61**: 406–408.

Jones, B. L. 1978. The mode of systemic infection of sorghum and sudangrass by conidia of *Sclerospora sorghi*. *Phytopathology* **68**: 732–735.

Lakshmanan, P.; Mohan, L.; Mohan, S.; and Jeyarajan, R. 1987. A new leaf spot of sorghum caused by *Cochliobolus bicolor* in Tamil Nadu, India. *Plant Disease* **71**: 651 (disease notes).

Langham, M. A.; Toler, R. W.; Alexander, J. D.; and Miller, F. R. 1984. Evaluation of *Sorghum bicolor* (L.) Moench accessions under natural infection with yellow sorghum stunt mycoplasma. *Phytopathology* **74**: 868 (abstract).

Leon-Gallegos, H. M., and Sanchez Castro, M. A. 1977. The occurrence in Mexico of *curvularia lunata* on sorghum kernels. *Plant Dis. Rptr.* **61**:1082–1083.

Mabry, J. E., and Lightfield, J. W. 1974. Long smut detected on imported sorghum seed. *Plant Dis. Rptr.* **58**: 810–811.

Manzo, S. K. 1976. Studies on the mode of infection of sorghum by *Tolyposporium ehrenbergii*, the causal organism of long smut. *Plant Dis. Rptr.* **60**: 948-952.

Mayhew, D. E., and Flock, R. A. 1981. Sorghum stunt mosaic. *Plant Disease* **65**: 84–86.

Natural, M. P.; Frederiksen, R. A.; and Rosenow, D. T. 1982. Acremonium wilt of sorghum. *Plant Disease* **66**: 863–865.

Odvody, G. N., and Dunkle, L. D. 1973. Overwintering capacity of *Ramulispora sorghi*. *Phytopathology* **63**:1530–1532.

Odvody, G. N., and Dunkle, L. D. 1975. Occurrence of *Helminthosporium sorghicola* and other minor pathogens of sorghum in Nebraska. *Plant Dis. Rptr.* **59**:120–122.

Odvody, G. N., and Dunkle, L. D. 1979. Charcoal stalk rot of sorghum: Effect of environment on host-parasite relations. *Phytopathology* **69**: 250–254.

Odvody, G. N., and Forbes, G. 1984. Pythium root and seedling rots. In *Sorghum Root and Stalk Rots: A Critical Review*. L. K. Mughogho, (ed.). International Crops Research Institute for the Semi-Arid Tropics, Pantancheru, India. pp. 31–35.

Odvody, G. N., and Madden, D. B. 1984. Leaf sheath blights of *Sorghum bicolor* caused by *Sclerotium rolfsii* and *Gloeocercospora sorghi* in South Texas. *Phytopathology* **74**: 264–268.

O'Neill, N. R., and Rush, M. C. 1982. Etiology of sorghum sheath blight and pathogen virulence on rice. *Plant Disease* **66**:1115–1118.

Penrose, L. J. 1974. Identification of the cause of red stripe disease sorghum in New South Wales (Australia) and its relationship to mosaic virus in maize and sugarcane. *Plant Dis. Rptr.* **58**: 832–836.

Pratt, R. G., and Janke, G. D. 1978. Oospores of *Sclerospora sorghi* in soils of south Texas and their relationships to the incidence of downy mildew in grain sorghum. *Phytopathology* **68**:1600–1605.

Pratt, R. G., and Janke, G. D. 1980. Pathogenicity of three species of Pythium to seedlings and mature plants of grain sorghum. *Phytopathology* **70**: 766–771.

Schuh, W.; Jeger, M. J.; and Frederiksen, R. A. 1987. The influence of soil temperature, soil moisture, soil texture, and inoculum density on the incidence of sorghum downy mildew. *Phytopathology* **77**:125–128.

Shafie, A. E., and Webster, J. 1979. *Trichometasphaeria turcica* as a root pathogen of *Sorghum bicolor* var. *feterita*. *Plant Dis. Rptr.* **63**: 464–466.

Singh, D. S., and Pavgi, M. S. 1982. Perpetuation of two foliicolous fungi parasitic on sorghum in India. *Phytopathol. Medit.* **21**: 41–42.

Tarr, S. A. 1962. *Diseases of Sorghum, Sudan Grass and Broom Corn.* Commonwealth Mycological Institute, Kew, England. University Press, Oxford.

Toler, R. W. 1985. Maize dwarf mosaic, the most important virus disease of sorghum. *Plant Disease* **69**:1011–1015.

Trimboli, D. S., and Burgess, L. W. 1983. Reproduction of *Fusarium moniliforme* basal stalk rot and root rot of grain sorghum in the greenhouse. *Plant Disease* **67**: 891–894.

Tuleen, D. M.; Frederiksen, R. A.; and Vudhivanich, P. 1980. Cultural practices and the incidence of sorghum downy mildew in grain sorghum. *Phytopathology* **70**: 905–908.

Wall, G. C.; Mughogho, L. K.; Frederiksen, R. A.; and Odvody, G. N. 1987. Foliar disease of Sorghum species caused by *Cercospora fusimaculans*. *Plant Disease* **71**: 759–760.

Wilson, J. M., and Frederiksen, R. A. 1970. Histopathology of the interaction of *Sorghum bicolor* and *Sphacelotheca reiliana*. *Phytopathology* **60**: 828–832.

Young, G. V.; Lefebvre, C. L.; and Johnson, A. G. 1947. *Helminthosporium rostratum* on corn, sorghum and pearl millet. *Phytopathology* **47**: 180–183.

Zummo, N. 1986. Red spot (*Helminthosporium rostratum*) of sweet sorghum and sugarcane, a new disease resembling anthracnose and red rot. *Plant Disease* **70**: 800 (disease note).

Zummo, N.; Bradfute, O. E.; Robertson, D. C.; and Freeman, K. C. 1975. Yellow sorghum stunt: A disease symptom of sweet sorghum associated with a mycoplasmalike body in the United States. *Plant Dis. Rptr.* **59**: 714–716.

Zummo, N., and Broadhead, D. M. 1984. Sources of resistance to rough leaf spot disease in sweet sorghum. *Plant Disease* **68**:1048–1049.

18

Diseases of Soybeans (*Glycine max* (L.) Merr.)

BACTERIAL CAUSES

Bacterial Blight

CAUSE: *Pseudomonas syringae* pv. *glycinea* (Coerper) Young, Dye & Wilkie (syn. *P. glycinea* (Coerper) Stapp) is seedborne, overwinters in residue, and possibly directly in soil. Bacteria associated with soybean leaf tissue overwinter on the soil surface if weather is cold and dry. Bacteria are spread by wind and water from soil, and cotyledons and leaves of diseased plants. Infected cotyledons from seedborne infection may be a major source of secondary inoculum. Bacteria are also spread by leaves rubbing together and during cultivation when foliage is wet. The major spread of bacteria occurs during windy, cool, wet weather; bacteria become dormant during warm, dry weather. Bacteria enter plants through wounds and stomates. Free water must be present for the bacteria to infect. Optimum temperature for infection is between 24°C and 26°C.

Cool, rainy weather favors development of bacterial blight. Several races exist.

DISTRIBUTION: Generally distributed wherever soybeans are grown.

SYMPTOMS: Symptoms may appear about five to seven days after a storm has occurred and are most common from early to midseason. Lesions first appear on cotyledon margins. As lesions enlarge the entire cotyledon collapses and becomes dark brown. Seedlings may be stunted, and if the growing point is infected, the seedling may be killed.

Leaf symptoms start as small, angular, water-soaked spots that turn yellow. The centers dry out and turn red-brown to black as the tissue dies. The spot is usually surrounded by a water-soaked margin bordered by a yellow halo. As spots increase in size, large areas of the leaf may fall out due to wind and rain, creating a ragged appearance. Severely infected leaves may drop off plants. Initially pod lesions are small and water-soaked but enlarge to cover much of the pod and become brown to black. Infected seeds may be shriveled, sunken, discolored, or may show no symptoms at all. Severely infected seeds in the field may be covered with a slimy bacterial growth. Seed may be infected during harvest or storage.

Systemic symptoms have been reported to consist of chlorotic mottling, yellowing, formation of narrow, elongated leaves and overall plant stunting.

CONTROL:
1. Plant disease-free seed.
2. Rotate soybeans with other crops.
3. Plow under infected residue where erosion is not a problem.
4. Do not cultivate when foliage is wet.
5. Apply a foliar fungicide during the growing season if conditions warrant it.
6. Treat seed with a seed-protectant fungicide.
7. Avoid planting highly susceptible cultivars.

Bacillus Seed Decay

CAUSE: *Bacillus subtilis* (Ehrenberg) Cohn. Losses are caused under warm (25–35°C), moist conditions in storage, in the field, and under experimental conditions.

DISTRIBUTION: Not known for certain, but probably occurs wherever soybeans are grown under warm and wet conditions.

SYMPTOMS: Seeds have a soft, mushy decay and are often covered with a slimy, rough to smooth, whitish to grayish bacterial growth.

CONTROL: No control is normally practiced.

Bacterial Crinkle Leaf Spot

CAUSE: Tentatively identified as *Pseudomonas syringae* van Hall.

DISTRIBUTION: United States.

SYMPTOMS: Young leaflets are severely distorted or crinkled and have necrosis of the veins and interveinal tissues. Lesions are not water-soaked and do not develop chlorosis or a halo.

CONTROL: Bacterial crinkle leaf spot is not economically important.

Bacterial Pustule

CAUSE: *Xanthomonas campestris* pv. *phaseoli* (Smith) Dye (syn. *X. campestris* pv. *glycines* (Nakano) Dye and *X. phaseoli* (Smith) Dowson var. *sojensis* (Hedges) Starr & Burkholder) overwinters in residue, seeds and the rhizosphere of wheat roots. Some weeds are also hosts. Disease develops during wet, warm weather at optimum temperatures of 30–33°C. Bacteria are spread by splashing water, leaf contact, and during cultivation when foliage is wet. Bacteria enter plants through leaf wounds or natural openings such as stomates.

DISTRIBUTION: Generally distributed wherever soybeans are grown in a warm, wet climate.

SYMPTOMS: New infections will develop throughout the growing season, regardless of temperature. The first symptoms on leaves are small, yellow-green spots with brown centers. Lesions lack the water-soaked appearance typical of most bacterial infections. Spots are most conspicuous on the upper leaf surface. The best diagnostic symptom is the presence of small raised pustules that develop in the center of the lesions on the lower leaf surface. Spots may merge, forming larger dead areas in which the tissue falls out, giving the leaf a ragged appearance. In severe cases, defoliation may occur. Pustules eventually rupture and dry. The pustule and the absence of a water-soaked appearance before the spot turns yellow distinguishes bacterial pustule from bacterial blight.

CONTROL:
1. Plant resistant cultivars.
2. Rotate soybeans with other crops.
3. Plow under infected residue where erosion is not a problem.
4. Plant disease-free seed.
5. Do not cultivate when foliage is wet.

Bacterial Tan Spot

CAUSE: *Corynebacterium flaccumfaciens* (Hedges) Dows. is seedborne and is spread throughout the growing season by wind-blown rain.

DISTRIBUTION: Iowa.

SYMPTOMS: Initially an oval to elongate chlorotic pattern starts at the leaf margin and progresses to the midrib. Infection continues to the petiole, leaving large necrotic areas that fall out during high winds. Leaflets with large or multiple lesions usually fall from the plants. Yield losses may occur with susceptible cultivars.

CONTROL: Plant resistant cultivars.

Bacterial Wilt

CAUSE: *Corynebacterium flaccumfaciens* pv. *flaccumfaciens* (Hedges) Dowson (syn. *C. flaccumfaciens* (Hedges) Dows.). The bacterium is seedborne and may overwinter in crop residue.

DISTRIBUTION: Not known for certain.

SYMPTOMS: Leaves are yellowed and eventually wilt. The vascular system is discolored. Infected seeds may appear normal or a bright yellow due to bacteria under the seed coat. Infrequently a small amount of yellow exudate may form on the hilum of an infected seed.

CONTROL: Follow control practices suggested for bacterial blight.

Bacterial Wilt

CAUSE: *Corynebacterium* sp. This organism resembles *C. flaccumfaciens* pv. *flaccumfaciens*.

DISTRIBUTION: Iowa.

SYMPTOMS: Infected plants are wilted and severely stunted. Pods are empty and abnormally formed. Lower leaves have marginal necrosis. Wilt does not occur after flower set.

CONTROL: No control measures are normally necessary; however, control measures could be followed that are suggested for bacterial blight.

Bacterial Wilt

CAUSE: *Pseudomonas solanacearum* (Smith) Smith is occasionally seedborne and overwinters in crop residue.

DISTRIBUTION: North Carolina and the USSR.

SYMPTOMS: Young plants may develop a rapid, severe wilt. Other plants wilt only slightly and are stunted.

CONTROL: Follow control practices recommended for bacterial wilt.

Pseudomonas Seed Decay

CAUSE: *Pseudomonas syringae* pv. *glycinea* (Coerper) Young, Dye & Wilkie. See Bacterial Wilt.

DISTRIBUTION: Generally distributed wherever soybeans are grown.

SYMPTOMS: Seeds may be symptomless. If symptoms occur, seeds may be shriveled, develop raised or sunken lesions, and become slightly discolored. Under high moisture conditions, seeds may be covered with a slimy growth.

CONTROL: See Bacterial Wilt.

Wildfire

CAUSE: *Pseudomonas syringae* pv. *tabaci* (Wolf & Foster) Young, Dye & Wilkie (syn. *P. tabaci* (Wolf & Foster) Stevens and *Bacterium tabacum* Wolf & Foster). The bacterium overseasons in residue and seed; however, alternate freezing and thawing kills it. It is spread by splashing water and wind. Water congestion of plant tissues caused by beating rains is often required for invasion and infection. Bacterial pustule lesions provide the entrance for *P. syringae* pv. *tabaci*; consequently wildfire is closely associated with bacterial pustule and a pustule can usually be found in the center of a wildfire lesion.

DISTRIBUTION: Brazil and the southern United States.

SYMPTOMS: Light brown, necrotic spots that vary in size and shape form on leaves and are usually surrounded by a large yellow halo with a definite margin. Sometimes smaller dark brown to black lesions may form without the halo. Lesions may enlarge during wet weather to form large dead areas that eventually tear away. Under severe disease conditions of wet weather, almost complete defoliation may occur.

CONTROL:
1. Plant resistant cultivars.
2. Rotate soybeans with other crops.
3. Plow under infected residue where erosion is not a problem.

4. Plant disease-free seed.
5. To prevent spread of the bacteria, do not cultivate soybeans when the foliage is wet.

FUNGAL CAUSES

Acremonium Wilt

CAUSE: *Acremonium* sp.

DISTRIBUTION: Not known for certain, but likely the same as *Phialophora gregatum* (Allington & Chamberlain) Gams.

SYMPTOMS: Similar to those caused by *P. gregatum*. A brown discoloration of vascular and pith tissues.

CONTROL: None reported.

Aerial Blight

This disease is also called Rhizoctonia foliage blight, foliage blight, and web blight.

CAUSE: *Rhizoctonia solani* Kuhn. During prolonged warm, humid weather, soil or plant parts containing propagules of the fungus are splashed onto lower plant leaves. For further information see Rhizoctonia root rot. Isolates that cause foliage blight on soybeans may not necessarily cause root and stem rot.

DISTRIBUTION: China, India, Malaysia, Mexico, Philippines, Puerto Rico, and the United States.

SYMPTOMS: The disease first appears on the lower leaves, pods, and stem. Symptoms gradually progress up the plant. Lesions may first appear as oval to linear water-soaked areas that become light brown, green-brown, or red-brown on stems and petioles. Later they turn tan, brown, or black, sometimes with red-brown margins.

When leaflets become diseased, infection usually starts at the base of the petiole with subsequent spread in a fan-shaped manner to the rest of the leaflet. Lesions are initially water-soaked and gray-green, with mycelium advancing on the leaf surface ahead of the lesion margin. Lesions become necrotic and dark brown to tan with a red-brown margin. Infected leaves droop. Mycelium grows from infected to healthy leaves, pods, and stems. Lesions on leaves vary from small spots to the entire leaf being affected. Diseased areas generally fall out during dry weather, giving the leaf a ragged appearance. Severely infected plants may be defoliated.

Stems and petioles have brown lesions. Pods have either small, brown spots or the entire pod may be affected. High humidity and warm temperatures encourage mycelial growth and sclerotial formation in lesions and favor disease development in general.

CONTROL:
1. Plant to avoid times of high moisture.
2. Do not plant too thick.
3. Apply foliar fungicide during reproductive phases.
4. Plant resistant cultivars.

Alternaria Leaf Spot

CAUSE: *Alternaria* spp. *Alternaria atrans* Gibson is known to be a weak parasite that invades leaves through wounds made by a mechanical injury such as aphid punctures and sunburn injury. *Alternaria tenuissima* (Fr.) Wiltsh. and *A. alternata* (Fr.) Keissler are frequently associated with seeds and with injury caused by the bean leaf beetle (*Ceratoma trifurcata* (Forster)). Also, *A. tenuissima* has been reported to cause a leaf spot in India and wilt in Kenya. *Alternaria* spp. are seedborne. Conidia produced by saprophytic mycelium are windborne to soybeans. Occasionally infection occurs on young plants; however, Alternaria leaf spot is generally a disease of plants nearing maturity. High levels of seed infection have been associated with wet years, frost injury, and stinkbug and bean leaf beetle damage.

DISTRIBUTION: Generally distributed wherever soybeans are grown.

SYMPTOMS: Spots on leaves and pods are brown with concentric rings that may be 6 mm in diameter or larger. Spots may merge, forming large dead areas that cause leaves to dry up and drop prematurely. Infected seeds are shrunken and green to brown and may not germinate; seed infected by *A. alternata* is small, shrunken with light to dark brown lesions.

CONTROL: The disease usually goes unnoticed, occurs too late in the growing season to cause yield reductions, and is not considered severe enough to warrant control measures.

Anthracnose

CAUSE: *Colletotrichum dematium* (Pers. ex Fr.) Grove f. *truncatum* (Schw.) v. Arx. (syn. *C. dematium* (Pers. ex Fr.) Grove f. *truncata* (Schw.) v. Arx., *C. truncatum* (Schw.), Andrus & W. D. Moore, *C. glycines* Hori, *C. caulivorum* Heald & Wolf, *C. viciae* Dearn & Overh., *Vermicularia truncata* Schw. and *V. polytricha* Cke.) and *Glomerella glycines* (Hori) Lehman & Wolf, anamorph *C. destructivum* O'Gara. The causal fungi overwinter as mycelium in seed, residue, and on weeds. *Colletotrichum gloesopriodes* (Penz.) Penz. teleomorph *G. cingulata* (Ston.) Spauld. & Schrenk., has been reported from Spain as a cause of anthracnose; however, it is not associated with most phases of the disease. Conidia are produced during warm, wet weather from mycelium growing either on infected cotyledons or residue in soil and are windborne to infect healthy plants. Disease tends to be more severe under narrow row conditions when plants are irrigated.

DISTRIBUTION: Generally distributed wherever soybeans are grown. Anthracnose is widespread and severe in the Midwest and southeastern United States.

SYMPTOMS: *Colletotrichum dematium* f. *truncatum* infects soybeans in all stages of development but *G. glycines* infects only older plants. *Colletotrichum dematium* f. *truncatum* grows from infected cotyledons onto stems and either causes numerous, small, shallow, elongated, red-brown lesions or large dark lesions that kill the young plant. Both fungi may cause preemergence or postemergence damping off especially during wet springs. In humid weather, cotyledons become water-soaked, wither, and fall off. Plants may be infected early without showing symptoms.

Stems, pods, and leaves of young plants may be infected without showing external symptoms until environmental conditions become more favorable for disease development. Symptoms on stems, pods, and petioles are irregularly shaped, brown to black areas that somewhat resemble symptoms of pod and stem blight. In advanced stages of disease, infected tissue is covered with acervuli. Acervuli under magnification resemble tiny pincushions that are covered with black spines. Foliar symptoms may develop after prolonged periods of high humidity and include leaf rolling, necrosis of laminar veins, petiole cankering, and premature defoliation. Infected plants may be shorter than healthy ones. Losses may occur on mature plants during wet weather when lower branches and leaves are killed. Young pods may also become infected and die. When pods or pedicels are infected early, a few small seeds or no seeds develop. Mycelium may completely fill the pod cavity.

Infected seeds may appear dark brown and shriveled or may show no visible signs of infection. Less severely infected seeds may be symptomless. Infected seed will result in reduced emergence.

CONTROL:
1. Plant seed from disease-free plants.
2. Treat seed with a seed-protectant fungicide.
3. Apply a foliar fungicide to plants during growing season.
4. Rotate soybeans with other crops.
5. Plow under infected residue.

Aspergillus Seedling Blight

CAUSE: *Aspergillus melleus* Yukawa is seedborne.

DISTRIBUTION: United States.

SYMPTOMS: Seedlings are stunted and chlorotic. Cotyledons are curled with deep necrotic lesions on the lower side. Russetting occurs on hypocotyls of a few plants.

CONTROL: No control is reported.

Black Root Rot

This disease is also called Calonectria root rot, Cylindrocladium root rot, and red crown rot.

CAUSE: *Calonectria crotalariae* (Loos) Bell & Sobers, anamorph *Cylindrocladium crotalariae* (Loos) Bell & Sobers, is thought to overwinter as microsclerotia. The fungus infects roots and grows inside the stem. Nematodes have been reported to increase injury by the causal fungus.

In Brazil, *Calonectria clavatum* Hodges & May causes identical symptoms to those caused by *Calonectria crotalariae*.

DISTRIBUTION: Cameroon, Japan, Korea, and southern United States.

SYMPTOMS: Individual or groups of plants may show symptoms. The first symptom at early pod set (R3–R4) is a yellowing of top leaves. Later interveinal tissue turns brown and defoliation occurs without wilting of the leaves. The inside of the stem is gray-brown and the entire root system becomes rotted. After stem tissue is killed, red-orange perithecia develop 2.5–7.5 cm above the soil line.

CONTROL: None reported.

Brown Spot

This disease is also called Septoria brown spot.

CAUSE: *Septoria glycines* Hemmi overwinters in residue and in diseased seeds as pycnidia and mycelium. During warm, wet weather conidia are produced on residue or cotyledons and on unifoliates infected by seedborne inoculum, and are splashed or blown to healthy leaves. Dry weather halts disease spread. Optimum disease development occurs at 28°C and high relative humidity.

DISTRIBUTION: Asia, Brazil, Canada, Europe, and the United States, particularly the Midwest.

SYMPTOMS: Angular red to brown spots that vary from the size of a pinpoint to about 5 mm in diameter appear on both surfaces of unifoliate leaves in the spring; however, spots are more pronounced on the under-leaf surface. Leaves become yellow and are prematurely shed. Spores produced on the primary leaves may spread and infect the trifoliate leaves, stems, and pods. Trifoliate leaves have numerous, irregular tan lesions that gradually turn dark brown. Later in the growing season, entire leaves may turn rusty brown and drop prematurely. Defoliation of the plant occurs from the bottom upward, and the lower portion of diseased plants may be bare of leaves before maturing. Brown lesions without characteristic size or shape form on pods, stems, and petioles. Petioles that have fallen to the soil may be covered with pycnidia that appear as scattered, tiny, black spots. This is a good indication of the presence of the causal fungus. Spores produced within a pycnidium are discharged during moist weather and start new infections on other plants. Areas with poor drainage favor spread of the disease. Seed weight is reduced but not the number of pods per plant or number of seeds per pod.

CONTROL:
1. Apply a foliar fungicide to plants during the growing season.
2. Plant a less susceptible cultivar.
3. Rotate soybeans with other crops.
4. Plant seed from disease-free plants.
5. Plow under infected residue.

Brown Stem Rot

This disease is also called Phialophora stem rot.

CAUSE: *Phialophora gregatum* (Allington & Chamberlain) Gams. (syn. *Cephalosporium gregatum* Allington & Chamberlain) survives as mycelium in residue, particularly woody stem tissue. Conidia are produced from mycelium that precolonized residue to a depth of 30 cm in the soil. Conidia are the main source of inoculum. The fungus has been reported to be seedborne as mycelium within the seed coat but not the embryo or cotyledon. Infections occur through the roots and lower stem early in the growing season with the mycelium growing upward in the xylem vessels. Conidia are thought to also be carried upward in the xylem vessels. Water and nutrient flow in the plant is thus interrupted. Disease development is greatest between 15°C and 27°C. Adequate moisture early in the growing season and moisture stress later increase disease severity.

DISTRIBUTION: Canada, Egypt, Mexico, and the United States.

SYMPTOMS: Two forms of the pathogen have been reported. All isolates cause browning of internal stem tissues but differ in their ability to cause foliar symptoms. Type I isolates cause chlorosis, necrosis, and wilt of foliage; type II isolates cause no foliar symptoms. Internal pith and vascular tissue is brown. Often there may be an internal browning only at the nodes while internodal tissue may be white. The browning (indicative of fungal growth) progresses upward in the stem during the growing season, moving most rapidly during cool weather. Maximum symptom development occurs between 15°C and 27°C. Usually the more extensive the discoloration, the greater the yield reduction. Warm weather with temperatures above 27°C suppresses the disease. Severely diseased plants may lodge. Later in the growing season, the lower part of the stem may show external browning with wilting and premature leaf drop.

Leaf symptoms are not ordinarily a reliable diagnostic tool since they are relatively uncommon and may be confused with other leaf disorders. Leaf symptoms develop when infected plants are subjected to high temperatures or drouth stress following a period of cool weather. About three weeks before maturity, leaves may wilt and tissue between the veins turns brown and dries rapidly, whereas tissue adjacent to veins remains green. Eventually the whole leaf dies. An infected field of soybeans turns brown, suggesting early frost in contrast to the yellow-green color of a normally maturing field. Since the disease is usually difficult to identify by visible outward symptoms, most growers blame a low yield on other causes.

CONTROL:
1. Do not plant soybeans, alfalfa, or red clover in infected soil for three years. Infected residue decomposes within this time and *P. gregatum* cannot survive in soil outside of precolonized residue.
2. Plant resistant cultivars where feasible.

Cercospora Leaf Spot and Blight

CAUSE: *Cercospora kikuchii* (Matsumota & Tomoyasu). See Purple Seed Stain. Disease is favored by extended periods of high humidity and temperatures of 28°C and 30°C.

DISTRIBUTION: Japan, Taiwan, Uganda, and the United States.

SYMPTOMS: Disease is more severe on early-maturing than late-maturing cultivars. In Arkansas, the initial symptoms are observed at the beginning of seed set through full seed set stages.

Upper leaves exposed to the sun have a light purple, leathery appearance. Later, red-purple, angular to irregular lesions occur on upper and lower leaf surfaces. These lesions vary in size from a pinpoint to irregular patches up to 1 cm in diameter and may coalesce to form large necrotic areas. Veinal necrosis may also occur. Severe disease may cause rapid chlorosis and necrosis, resulting in defoliation starting with the upper leaves. Green leaves may occur below the defoliated areas. This obvious blighting occurs over large areas, including entire fields.

Red-purple, slightly sunken lesions that are several millimeters long occur on petioles and stems. On susceptible cultivars, red-purple lesions that later become purplish black occur on pods.

CONTROL: Plant the least susceptible cultivars.

Cercospora Seed Decay

CAUSE: *Cercospora sojina* Hara. See Frogeye Leaf Spot.

DISTRIBUTION: See Frogeye Leaf Spot.

SYMPTOMS: Seeds have conspicuous light to dark gray or brown areas that vary from specks to large blotches covering the entire seed coat. Some lesions show alternating bands of light and dark brown. Sometimes brown and gray lesions diffuse into each other. Normally the seed coat cracks or flakes.

CONTROL: See Frogeye Leaf Spot.

Charcoal Rot

This disease is also called dry weather wilt and summer wilt.

CAUSE: *Macrophomina phaseolina* (Tassi) Goid. (syn. *M. phaseoli* (Maubl.) Ashby, *Rhizoctonia bataticola* (Taub.) Butler, *Sclerotium bataticola* Taub., and *Botryodiplodia phaseoli* (Maubl.) Thirum). Survival is by sclerotia and mycelium in dry soil and residue. Depth of burial has little effect on survival. It is also seedborne as ectophytic and endophytic hyphae and sclerotia; hyphae are in seed coat, endosperm, and embryo. Pycnidia are rarely produced on soybeans. The fungus does not survive longer than a few weeks in wet soil. It is a good colonizer of plant debris in soil but cannot persist long in the presence of other soil microflora. Sclerotia germinate on root surfaces and form several germ tubes that penetrate the root and grow in the xylem tissues. Plants wilt due to the xylem physically being plugged by mycelium or by the production of toxins or enzymes.

Seedling disease is greatest at high temperatures of 28–35°C with either sclerotia or mycelium equally effective as infective propagules. Disease development of older plants is also associated with hot, dry weather or when unfavorable environmental condition stops plant growth. The disease occurs on irrigated soybeans when water is withheld to promote maturity.

DISTRIBUTION: Generally distributed wherever soybeans are grown.

SYMPTOMS: Infected seedlings may show a red-brown discoloration at the soil line that eventually becomes ash-gray to black. Seedlings are more likely to die in warm and dry than in cool and moist soils. Leaves of severely infected plants turn yellow, wilt, and remain attached. Stem lesions may extend up from the soil line. The best diagnostic symptom occurs after midseason. When the epidermis of the stem is removed, small black bodies, which are the sclerotia, may be so numerous as to give a gray-black color to the tissues, resembling a sprinkling of powdered charcoal. When the lower portion of the plant is split open there will be black streaks in the woody part. The disease becomes more severe when soybeans are grown in the same field in successive years. Infected seeds have indefinite black spots and blemishes on the seed coat.

CONTROL:
1. Rotate soybeans with other crops. Two years out of soybeans is required to significantly reduce propagules.
2. Do not overplant. Crowded seedlings are more subject to infection.
3. Follow a good fertility program.
4. Plow under infected residue.

Choanephora Leaf Blight

CAUSE: *Choanephora infundibulifera* (Currey) Sacc. (syn. *C. trispora* (Thaxter) Sinha). Disease is severe under humid conditions.

DISTRIBUTION: Thailand.

SYMPTOMS: Distal halves of infected leaves initially develop a grayish color. Later the infected portion dries up and curls. Defoliation may occur during periods of high humidity. If humidity is low, only the affected area of the leaf drops off, leaving the unaffected portion intact.

Under severe defoliation plants may be stunted and seeds may be smaller than normal.

CONTROL: No control is reported.

Dactuliophora Leaf Spot

CAUSE: *Dactuliophora glycines* Leakey. Overseasoning is likely by sclerotia.

DISTRIBUTION: Zimbabwe.

SYMPTOMS: Spots appear on both sides of leaflets and are somewhat circular, broadly but distinctly zonate, and up to 4 mm in diameter. Spots may spread to leaf margins and become confluent.

CONTROL: No control is reported.

Downy Mildew

CAUSE: *Peronospora manshurica* (Naum.) Syd. ex Gaum. (syn. *P. sojae* Lehman & Wolf) overwinters either as oospores in residue or encrusted over seed. Infected seeds may give rise to systemically infected seedlings under cool conditions. Seedling hypocotyls are also infected by soilborne oospores. In either case mycelium grows systemically. Conidia are produced on the underside of leaflets and serve as secondary inoculum to infect other plants. Conidia are windborne to hosts. Disease development is favored by temperatures of 20–22°C optimum and high humidity. Several physiologic races exist.

DISTRIBUTION: Generally distributed wherever soybeans are grown.

SYMPTOMS: Early symptoms are indefinite yellow-green areas on the upper leaf surface. In time, the diseased areas enlarge, become grayish to dark brown and are surrounded by yellow-green margins. A gray fuzz, which is the mycelium and spores of the fungus, develops on the underside of the leaf in diseased areas, usually during periods of frequent dews or abundant rain. Severely infected leaves may turn yellow, then brown, and drop prematurely. Pod infections may occur and not be evident externally; however, the inside of the pod and infected seeds are encrusted with a white coating that consists mainly of thick-walled oospores. Infected seeds often appear dull white, have cracks in the seed coat, and may be smaller or lighter in weight than normal seeds.

Systemically infected plants may occur by planting oospore-encrusted seed. Such plants remain small with mottled gray-green leaves that curl downward at the edges. The underside of the leaf is covered with the typical fungal growth. Infected seeds usually are not killed, nor are the seedlings that develop from them, but they may serve as infection foci from which the disease spreads to healthy plants. Older leaves become resistant to infection.

CONTROL:
1. Grow resistant cultivars.
2. Rotate soybeans with other crops.
3. Plow under infected refuse.
4. Treat seed with a fungicide.

Drechslera Blight

CAUSE: *Drechslera glycini* Narayanasamy & Durairaj.

DISTRIBUTION: India.

SYMPTOMS: Circular to angular brown spots form near leaf margins. As the lesions enlarge, the center turns gray with a dark brown margin that is sometimes surrounded by a chlorotic halo. Lesions may develop along veins, causing the leaf to curl and dry.

CONTROL: No control is reported.

Frogeye Leaf Spot

This disease is also called Cercospora leaf spot.

CAUSE: *Cercospora sojina* Hara (syn. *C. daizu* Miura) overwinters as mycelium in seed and residue. When infected seed germinates, spores are produced in cotyledonary lesions under warm, moist conditions and are windborne to young leaves. Conidia are also produced on residue during warm, moist weather. All leaves may become infected as well as pods, stems, and seeds. Different races of the fungus exist.

DISTRIBUTION: Mostly in the southern United States and areas in the world where soybeans are grown in a similar climate.

SYMPTOMS: Frogeye is primarily a disease of the foliage; however, stems, pods, and seeds may also be infected. Small, circular to angular, red-brown spots are formed on the upper leaf surface. Spots enlarge to 1–

5 mm in diameter and have grayish centers surrounded by a narrow red-brown border. No yellow area surrounds lesions. Several spots may coalesce to form large irregular areas. On the lower leaf surface, spots are a darker brown or gray. Clusters of dark gray conidiophores develop in the center of each lesion, mostly on the underside of the leaf. Older spots become thin, paperlike, and transparent. Lesions are 1–5 mm in diameter and generally discrete. Spots may coalesce to form large necrotic areas. Heavily infected leaves may fall prematurely.

Infection of stems and pods occurs later in the growing season. Initially lesions on stems are elongated and reddish with a black border. Lesions then turn brown, then light gray, and eventually black when the causal fungus starts to sporulate. Pod lesions are circular, brown to gray, with a narrow dark brown ring. The fungus may grow through the pod wall to infect the seed. See Cercospora Seed Decay.

CONTROL:
1. Plant resistant cultivars.
2. Plant disease-free seed.
3. Rotate soybeans with other crops. Leave field out of soybeans for at least two years.
4. Apply foliar fungicides at the R3–R5 growth stages.
5. Plow under crop residue.

Fusarium Root Rot and Seedling Blight

CAUSE: *Fusarium* spp., including *F. solani* (Mart.) Appel. & Wr. and *F. oxysporum* Schlecht. ex Fr., are common soil fungi that survive in soil as chlamydospores and mycelium in residue. Survival of *Fusarium* spp. is better at low soil moisture. Seeds and roots of seedlings are infected during dry soil conditions.

DISTRIBUTION: Generally distributed wherever soybeans are grown.

SYMPTOMS: Infected seed may have poor germination, resulting in either preemergence or postemergence damping off, or late emergence and stunted plants. Dark brown lesions are confined to roots and lower portions of stems and may be confused with other root rots. Root systems of severely infected plants may be completely destroyed. Wilting is most frequently observed on seedlings or young plants when roots are rotted and soil moisture is low. Older plants are seldom killed but wilt when soil moisture is low and recover turgidity at night or when moisture again becomes adequate. Poor stands caused by Fusarium root rot are usually associated with poor seed quality, heavy rains, soil compaction, or wet soil after planting.

CONTROL:
1. Seedlings infected with *Fusarium* spp. and showing signs of wilting or death of lower leaves should not be cultivated until adequate soil moisture is available.
2. When plants are cultivated, soil should be ridged around the base of the plants. This will promote development of roots from the stem base above the diseased area. There roots are not as easily infected by the fungus and help the plants to recover rapidly.

Fusarium Wilt

This disease is also called Fusarium blight.

CAUSE: *Fusarium oxysporum* Schlecht. f. sp. *tracheiphilum* Armst. & Armst. Race 1 (syn. *F. tracheiphilum* E. F. Smith), *F. oxysporum* Schlecht. f. sp. *vasinfectum* (A & K.) Snyd. & Hans. Race 2, and *F. oxysporum* Schlecht. f. sp. *glycines* Armst. & Armst. survive as chlamydospores and mycelium in residue, soil, and weed hosts. Mycelium from either source penetrates roots and grows up the xylem tissue, thereby inhibiting flow of water to the top of the plant. Seedlings and young plants are predisposed to infection by soybean cyst nematode, root-knot nematode, and sting nematode.

DISTRIBUTION: Generally distributed wherever soybeans are grown.

SYMPTOMS: Symptoms generally appear about midseason when the weather becomes hot, 28°C optimum, and there is a moisture stress. Areas of infected plants consist of randomly scattered round to elongate patches in a field. Leaves on infected plants may wither and drop, or if they persist on the plant, they will become chlorotic. Wilting of stem tips and the presence of flaccid leaves are most common on young plants. The best diagnostic characteristic is a browning or blackening of the vascular system that can be seen when stems are split open.

CONTROL: Some cultivars have a level of resistance.

Leptosphaerulina Leaf Spot

CAUSE: *Leptosphaerulina trifolii* (Rost) Petr. is reported as a pathogen from India and *L. briosiana* (Poll.) Graham & Luttrell from Maryland.

DISTRIBUTION: India and Maryland.

SYMPTOMS: Gray, circular, necrotic lesions appear on both young and old leaves. Later the spots become irregular and dull white to gray. In Maryland small necrotic spots appear on leaves.

CONTROL: No control is reported.

Mycoleptodiscus Root Rot

CAUSE: *Mycoleptodiscus terrestris* (Gerd.) Ostazeski (syn. *Leptodiscus terrestris* Gerd.) is thought to survive as sclerotia in residue and soil. It is also a pathogen of alfalfa, red clover, and birdsfoot trefoil. Infection occurs in warm, wet soils. It is not known for certain but sclerotia probably germinate directly by forming mycelium that penetrate roots. Acervuli and conidia are produced in foliar infections but it is not known what function they perform in the disease cycle.

DISTRIBUTION: United States.

SYMPTOMS: Postemergence damping off of seedlings is characterized by a brown to black rotting of the roots. Older plants may also develop a root rot characterized by a red-brown decay of the cortical tissue of the taproot and a destruction of the secondary root system. This is a similar symptom to that caused by *Rhizoctonia solani* Kuhn. It is thought that *M. terrestris* occurs in association with other root rot fungi; therefore, distinct symptoms are difficult to diagnose and may be confused with other rots.

Leaf infections have occurred in growth chambers. Leaf spots are red to light brown and up to 3 mm in diameter. Acervuli form on leaves.

CONTROL: No control is reported.

Myrothecium Disease

CAUSE: *Myrothecium roridum* Tode. ex Fr. Disease development is favored by high temperature and humidity.

DISTRIBUTION: Southeast Asia, Brazil, India, and the United States.

SYMPTOMS: Spots on leaves at first are small, round, 1 mm in diameter, and brownish in color. Spots enlarge to 8–10 mm in diameter, become dark brown, and surrounded by translucent concentric zones. Sporodochia appear in translucent areas as small, white, erumpent structures that gradually turn dark green and ultimately appear as black dots surrounded by slight chlorosis. The necrotic tissue in the center of the spots falls out to give a shot-hole effect.

Small, white dots that later turn black due to sporodochia appear on flowers. Infected flowers turn brown and drop. Sporochia on pods first appear as white dots but later turn black and coalesce to form a black mass. Mature pods are less severely infected than young pods.

CONTROL:
1. Do not plant infected seed.
2. Treat seed with a seed-protectant fungicide.

Neocosmospora Stem Rot

This disease is also called Neocosmospora wilt.

CAUSE: *Neocosmospora vasinfecta* Smith. Disease severity is greater at higher temperatures. The fungus survives as ascospores in soil or embedded inorganic matter at low soil moisture. Disease could be related to soil nematode infections.

DISTRIBUTION: Japan, Nigeria, and the southeastern United States. The disease is not considered of economic importance.

SYMPTOMS: The primary symptom is a red-brown to dark brown discoloration that extends 20–25 cm upward from soil surface in the pith and xylem of the stem. Interveinal areas in leaves are brown with chlorotic margins, resulting in infrequent yellowing and shedding of lower leaves. Internal symptoms may resemble those of brown stem rot. External red-brown lesions (20 mm or less in length) may occur on stems but this is uncommon. Orange to red perithecia may occur on lower stems of dead plants.

CONTROL: No control is reported.

Phyllosticta Leaf Spot

CAUSE: *Phyllosticta sojaecola* Massal. (syn. *Phyllosticta glycineum* Tehon & Daniels), teleomorph *Pleosphaerulina sojaecola* (Massal.) Miura. Note the specific name sojaecola is sometimes spelled sojicola in the literature. The causal fungus may be seedborne and also overwinters as pycnidia and mycelium in residue. During wet weather, conidia are produced in pycnidia on residue and windborne to hosts. Disease is favored by cool, wet weather.

DISTRIBUTION: Borneo, Bulgaria, Canada, China, Europe, Japan, the USSR, and the United States.

SYMPTOMS: This disease can occur on any age plant but is most likely to be found on younger plants where it is often confused with herbicide injury or drouth stress. Leaf spots are round to oval, irregular, or frequently V-shaped and up to 2 cm in diameter. Lesions form on leaf margins of young plants and grow inward. At first, lesions are dull green and later become gray to tan with a dark brown to purple border. Numerous pycnidia appear as small black specks in older lesions. Small, narrow, gray lesions also form on petioles, stems, and pods. There may also be circular lesions with a purple to red-brown border on pods.

CONTROL:
1. Plant seed from disease-free plants.
2. Rotate soybeans with other crops.
3. Plow under infected crop residue.

Phyllosticta Pod Rot

CAUSE: *Phyllosticta sojaecola* Massal. See Phyllosticta Leaf Spot.

DISTRIBUTION: Germany and the United States.

SYMPTOMS: Spots on pods grow to 8 mm in diameter with dark purple-red borders surrounding lighter, brownish centers on which numerous dark pycnidia are found. Infection is most severe on pods developing in upper half of plants.

CONTROL: See Phyllosticta Leaf Spot.

Phymatotrichum Root Rot

CAUSE: *Phymatotrichum omnivorum* (Shear) Duggar survives for several years in the soil as sclerotia. Phymatotrichum root rot is most prevalent in alkaline soils.

DISTRIBUTION: Southwestern United States and northern Mexico.

SYMPTOMS: Plants in areas wilt rapidly and die. The root cortex is rotted and brown stands of fungus mycelium can be seen growing on the roots.

CONTROL:
1. Plowing under a green manure crop such as sweet clover lessens the disease severity in succeeding years.
2. Rotate soybeans with plants in the grass family.

Phytophthora Root Rot

CAUSE: *Phytophthora megasperma* (Drechs.) var. *sojae* A. A. Hildebrand (syn. *P. sojae* Kaufmann & Gerdemann and *P. megasperma* var. *glycinea* Kuan & Erwin) survives primarily as oospores and sometimes as mycelium in soil or residue. It is not known for certain how oospores germinate in wet soils to form a sporangium in which numerous zoospores are formed. The zoospores swim through soil water to roots where they encyst and germinate. Leaf infection also occurs when infected soil particles are splashed or blown onto leaves during a storm. Oospores eventually form in infected roots. Disease development is most favorable at soil temperatures of 15°C. Nitrogen may favor disease development. Plants may become infected later in the growing season even under dry conditions. However, the disease is most common in low areas of a field, poorly drained soils, compacted soils, and soils with a high clay content. It may also occur on well-drained soil during a wet growing season. Other root rot fungi and northern root knot nematodes may increase severity of root rot. Trifluralin herbicide increases disease severity. Several physiologic races exist in nature.

DISTRIBUTION: Asia, Canada, and the United States.

SYMPTOMS: Soybeans may be infected at any stage of development. Seed rot and preemergence damping off may occur. There is a yellowing followed by a wilting of the leaves that remain attached to the plant. One of the most characteristic symptoms is the stem rot phase, which is a brown discoloration of the stem and lower branches that extends from below the soil line on the stem up to 20 cm or more above the soil line. The taproot is dark brown and the entire root system may be rotted. Sometimes there may be no obvious symptoms but plants will be reduced in vigor and stunted. This latter symptom is often difficult to recognize unless there is a side-by-side comparison with a resistant cultivar. Infection of lateral branches near the soil may also occur. In such cases the fungus eventually grows into the stem, and since the roots may not be rotted, the disease may be confused with stem canker.

The disease pattern varies within a field. It may be roughly circular, corresponding to the poorly drained areas, or it may occur as dead or dying individual or groups of plants in a row. Infected plants may often be found more easily in the end rows than within a field.

CONTROL:
1. Plant resistant cultivars; however several physiologic races of the fungus exist and cultivars resistant to one race may be susceptible to others.
2. Apply systemic fungicide as a seed treatment and/or to the soil either as a band or in furrow treatment.
3. Plant in warm (18°C or more), well-drained soil.
4. Avoid planting susceptible cultivars in soils where disease has occurred.

Pod and Stem Blight

CAUSE: *Diaporthe phaseolorum* (Cke. & Ell.) Sacc. var. *sojae* (Lehman) Wehm. & Sacc. (syn. *D. sojae* Leh.), anamorph *Phomopsis sojae* Lehman and *P. longicolla* Hobbs sp. nov. are seedborne and overwinter as mycelium, perithecia, or pycnidia in residue. However, most primary inoculum originates from infested residue. Ascospores are produced in perithecia or conidia in pycnidia during wet weather and are windborne or splashed to hosts. Two types of conidia (alpha and beta) are produced. Infected seeds often fail to germinate in cool (15–20°C), wet soils. However, disease may also reduce emergence and establishment when seeds are incubated in dry soil. Low water potential evidently inhibits seedling growth more than growth of the causal fungi, particularly Phomopsis. If infected seeds germinate, they often give rise to infected seedlings that may serve as a source of inoculum. However, soil is probably a more important source of inoculum than seed. Disease development is more severe if warm (above 20°C), wet or humid weather occurs in late summer and autumn, or between physiological maturity (R6–R8) and harvest, and if soybeans remain in the field for a long time after maturity. High relative humidity of close to 100 percent for prolonged periods is essential for seed infection to take place. The process of seed infection can be resumed after periods of high humidity are interrupted by low humidity. Therefore, there could be cumulative effects of wet periods. Pods normally become infected at the yellow-pod (R7) stage or afterward. Seed infections are usually concentrated at lower plant nodes but may occur throughout the plant under wet conditions or when harvest is delayed. Disease tends to be more severe in early-planted or early-maturing cultivars.

DISTRIBUTION: Generally distributed wherever soybeans are grown.

SYMPTOMS: Soybeans are infected early in the growing season but symptoms of the disease appear on pods and stems of plants nearing maturity, especially if wet weather occurs at this time. The first symptom is the presence of pycnidia on the petioles that have fallen from the lower portion of a plant to the soil surface. This is a good indication that pod and stem blight may be a potential problem in the current growing season. The best symptom is the presence of numerous small, black pycnidia that are arranged in linear rows on the stem but scattered over the pod. During dry weather, pycnidia may be present only on the lower stem or on nodes.

Infected seed may sometimes have a white mold growing over it but often not show any external signs of the disease. Infected seed may damp off either preemergence or postemergence, resulting in a poor stand.

The disease becomes most severe if wet weather is present later in the growing season. Secondary infection occurs from spores produced in pycnidia. Plants infected with bean pod mottle virus (BPMV) are more likely to have high levels of *Phomopsis* sp. seed infection than plants uninfected with BPMV.

CONTROL:
1. Harvest seed as promptly as possible after maturity.
2. Apply a foliar fungicide from midflowering (R3–R4) to late pod stage (R6–R7).
3. Treat seed with a seed-protectant fungicide.
4. Plant seed from disease-free plants.
5. Plow under infected residue.
6. Rotate soybeans with other crops.

Pod Rot and Collar Rot

CAUSE: *Fusarium semitectum* Berk. & Rav. is seedborne.

DISTRIBUTION: India in seed received from the United States.

SYMPTOMS: Plants having rotted pods do not show any symptoms other than less vigorous growth compared to healthy plants. All pods on diseased plants show drying symptoms characterized by blackening that starts from the distal end. Severely infected pods produce no seeds, or if seeds are present, they are black, rotted, and shriveled. Infected pods are blackish on both inner sides especially where the seed rests. Seed coats and cotyledons of infected seeds are also black.

Seedlings initially have water-soaked, depressed, cream-colored lesions on cotyledons and hypocotyls. Lesions soon become dark brown to black and sometimes coalesce to form enlarged spots on cotyledons. Spots enlarge longitudinally on the hypocotyl to the radicle. Seedlings eventually die.

CONTROL: No control is reported.

Powdery Mildew

CAUSE: *Microsphaera diffusa* Cke. and Pk. (syn. *Erysiphe polygoni* DC. ex Merat, *E. glycines* Tai, and *Microsphaera* sp.) overwinters as cleistothecia on infected soybean residue. Primary infection occurs from ascospores that are produced in cleistothecia and are windborne to healthy plants. Secondary infection occurs from conidia produced on infected leaves and windborne to hosts. Disease development occurs later in the growing season during humid weather but without presence of free water. Cooler than normal temperatures have been observed to favor disease.

DISTRIBUTION: Brazil, Canada, China, Peru, South Africa, and the United States.

SYMPTOMS: Powdery mildew becomes evident about the middle of the growing season. Cotyledons, stems, pods, and leaf surfaces are affected. On leaves small colonies of thin, light gray mycelium are seen on the upper surface. In time, the whitened areas enlarge to cover much of the leaf surface and have a white, powdery appearance as though covered with flour. This is due to the production of large numbers of conidia and conidiophores. Infected plants may become chlorotic, display green islands and rusty patches, or luxuriant fungal growth may occur without other symptoms.

Yield losses of up to 26 percent have been reported.

CONTROL:
1. Plant resistant cultivars. Some cultivars are susceptible as seedlings but not as adult plants.
2. Apply a foliar fungicide.

Purple Seed Stain

This disease is also called purple spot, purple blotch, purple speck, and lavender spot.

CAUSE: *Cercospora kikuchii* (Matsumoto & Tomoyasu) overwinters as mycelium in the seed coat and residue. Infected seed gives rise to seedlings with infected cotyledons. Often the seed coat may slip off the germinating seed before cotyledons become infected. During warm, humid weather, conidia are produced on infected cotyledons and are windborne or splashed onto leaves, stems, and pods. The mycelium grows through pods and into the seed coat. Optimum temperature for infection is 20–24°C. A minimum of 8 hours of dew is necessary with 24 hours of dew optimum for infection to occur.

DISTRIBUTION: Generally distributed wherever soybeans are grown.

SYMPTOMS: The best diagnostic symptom is a pale to dark purple discoloration of the seed coat only. Cotyledons are generally not infected or discolored. Infected seeds can produce diseased seedlings with cotyledons that are shriveled, dark purple, and fall prematurely. However, there is no difference in incidence of purple stain between plants arising from purple stain and purple stain-free seed. The fungus can grow from the cotyledon into the stem, causing a necrotic area that may kill the plant. Late in the season, thickened and crusty-appearing red-brown spots may develop on infected leaves. There is no effect on maturity, lodging, and plant height. Yields are not reduced but the value of the seed is lowered from a grading standpoint. The purple discoloration does not affect seed used for processing since it disappears during heating.

Seedling emergence from purple stained seeds is lower than from healthy seeds. Prior infection with *C. kikuchii* reduces infection by other seedborne fungi, particularly *Diaporthe phaseolorum* var. *sojae*.

CONTROL:
1. Apply a foliar fungicide to plants during the growing season.
2. Treat seed with a seed-protectant fungicide.
3. In general, earlier-maturing cultivars are more susceptible than late-maturing cultivars.
4. Some cultivars are moderately resistant.

Pythium Seed and Seedling Rot

CAUSE: *Pythium aphanidermatum* (Edson) Fitz., *P. debaryanum* Hesse, *P. irregulare* Buis., *P. myriotylum* Drechs., and *P. ultimum* Trow. *Pythium aphanidermatum*, *P. debaryanum*, and *P. ultimum* survive in soil or residue as oospores and saprophytically as mycelium on residue. In cool (10–15°C), wet soil, oospores of *P. debaryanum* and *P. ultimum* germinate to form sporangium. Zoospores are formed in sporangia and swim through soil water to seed or roots of seedlings. At higher temperatures, oospores may germinate directly and form a germ tube. *Pythium aphanidermatum* causes disease at higher temperatures of 25–36°C. Seedlings are most susceptible; soybeans become progressively more resistant as they grow older.

DISTRIBUTION: The disease is most prevalent under wet soil conditions, but generally distributed wherever soybeans are grown.

SYMPTOMS: The disease may occur from the time seed is planted to the end of flowering. However, it is usually associated either with seed rotting in the soil or young seedlings being killed. Soybean seed planted in cool and wet soil is most subject to infection by these fungi.

Infected seeds may not germinate and may be soft and overgrown with other fungi and bacteria, giving the seed a fuzzy appearance. Seedling rot or blight results from infection by these fungi after the seed has germinated but before or just after the seedling has emerged through the soil. Roots infected by *P. ultimum* will have a brown color and a wet appearance. Other symptoms attributed to *P. ultimum* include swollen hypocotyls and lesions at the junction of the hypocotyl and primary root. Infected seedlings may also have a curling growth habit and reddish to brown lesions on the hypocotyls and cotyledons. If plants are infected after emergence, the plant will wilt, and if leaves are present, they will first have a gray-green color and then turn brown after a day or so. Such infected plants are easily pulled from the soil. As the soil dries, an infected root resembles a shoestring attached to the cotyledon. Diseased plants may stand singly, in small circular groups particularly in low spots in the field, or may occur uniformly over an entire field if there has been a period of rain.

Pythium debaryanum causes the growing point to be retarded, resulting in a baldhead symptom.

CONTROL:
1. Treat seed with a seed-protectant fungicide.
2. Do not plant carry-over seed or seed that has a high percentage of broken seed coats.
3. Plant high-quality, high-germinating seed.
4. Plant cultivars that are more resistant than others.

Red Leaf Blotch

This disease is also called Pyrenochaeta leaf blotch and Pyrenochaeta leaf spot.

CAUSE: *Pyrenochaeta glycines* Steward (syn. *Dactuliophora glycines* Leakey) likely overseasons as pycnidia and sclerotia. Dissemination may be by sclerotia carried in contaminated soil during farming operations. Sclerotia are splashed onto leaf surfaces where they germinate. Disease occurs during abundant rainfall and high humidity.

DISTRIBUTION: Cameroon, Malawi, Rwanda, Uganda, Zaire, Zambia, and Zimbabwe.

SYMPTOMS: Lesions are caused on leaves, petioles, pods, and stems throughout the growing season and are often associated with primary leaf veins. Initially dark red to brown, circular to angular lesions, 1–3 mm in diameter, appear on the unifoliate leaves. Later dark red spots develop on the upper surface and red-brown spots with dark borders develop on the lower surface of trifoliate leaves. During high humidity, a diffuse mycelial growth may surround the lesions. Lesions may enlarge and coalesce to form irregular blotches 3–10 mm in diameter with buff-colored centers and dark margins. Older blotches may be surrounded by chlorotic halos and merge to form large necrotic blotches up to 2 cm in diameter. Necrotic tissue frequently falls out, giving a shot-hole appearance. Diseased plants defoliate prematurely. Lesions on stem, petioles, and pods are ovoid, 1–5 mm long, and mauve to red-purple. These lesions appear below the uppermost leaf with symptoms. Sclerotia develop primarily on the lower leaf surface and pycnidia form mainly in blotches on the upper leaf surface. Yield loss is due to reduced seed size.

CONTROL: No control is reported.

Rhizoctonia Root Rot

CAUSE: *Rhizoctonia solani* Kuhn, teleomorph, *Thanatephorous cucumeris* (Frank) Donk. *Rhizoctonia solani* is a common soil fungus that survives in soil and residue as sclerotia or resting mycelium. It is an excellent saprophyte of different plant residues and survives for several years as saprophytic mycelium or sclerotia in absence of soybeans. During wet soil conditions, sclerotia germinate to form mycelium or mycelium grows from precolonized residue and infects seeds, roots, and hypocotyls. Disease development is most severe on young plants. Rainfall followed by cool and then warm weather is favorable for disease development. Optimal temperature for disease development is between 25°C and 29°C.

DISTRIBUTION: Generally distributed wherever soybeans are grown.

SYMPTOMS: The disease is first noticed by the presence of wilted and dead plants as the weather becomes warm in the early growing season. Typical symptoms are decay of lateral roots and localized brown to red-brown lesions on the hypocotyl and lower stem that do not extend above the soil line. The red-brown color is a good symptom to diagnose the disease; however, it is best observed immediately after removing the plant from the soil as the color fades upon exposure to air. The discoloration is usually limited to the cortical layer of the main root and hypocotyl and does not extend into the root or stem. Infected stems remain firm and dry. The disease pattern may occur as a single plant or a group of dead plants in a row or in circular areas where soil moisture may be higher than in other areas of the field. Although the disease is associated with young plants, older plants may die if there is a moisture stress and the hypocotyl and roots are sufficiently decomposed to limit uptake of water.

CONTROL:
1. Ridge soil around base of plants during cultivation to promote root growth above the diseased portion of the plant.
2. Apply seed-protectant fungicide.
3. Use good soil drainage.

Rust

CAUSE: *Phakopsora pachyrhizi* Sydow (syn. *P. meibomiae* (Arth.) Trotter, *P. sojae* Fujikuro, *P. sojae* Sawada, *P. sojae* (P. Henn.) Sawada, *P. vignae* Arthur, *P. vignae* (Bres.) Arth., *Physopella concors* Arthur, *Physopella meibomiae* Arthur, *Physopella meibomiae* (Arth.) Trotter, *Uredo concors* Arthur, *U. sojae* P. Henn., *U. sojae* (P. Henn.) Sydow, *U. vignae* Bresadola, *Uromyces mucunae* Rabenh., *Uromyces sojae* P. Hennings, *Uromyces sojae* (P. Henn.) Sydow, and *Uromyces sojae* Miura non Sydow). The pycnidiospore and aeciospore stages are unknown. Teliospores are formed but germination has not been observed. Urediospores are the probable overseasoning structures and may be found on other host plants in warm climates all year. Urediospores are disseminated by splashing rain and wind. Upon germination, the resulting germ tube either penetrates directly through the cuticle or through stomates. Free moisture for at least six hours and a temperature of 20°C is optimum for infection. *Phakopsora pachyrhizi* has a wide host range that includes several leguminous plants.

DISTRIBUTION: Central America, China, Japan, Korea, and Taiwan.

SYMPTOMS: Three infection types are recognized: TAN, RB, and O. The uredial stage of soybean rust is the commonly observed stage. At onset of disease, chlorotic or gray-brown spots appear on leaves and less frequently on petioles and young stem tissue. On leaves, the spots enlarge to about 0.4 mm square and become either red-brown (RB) or tan lesions (TAN) that are delineated by secondary or tertiary leaf veins. Type RB has 0–2 uredia on the abaxial surface while type TAN has 2–5 uredia on the abaxial surface. When young, the lesions of soybean rust are identical to those of bacterial pustule except for a chlorotic ring around the bacterial pustule lesion. As the bacterial pustule matures, a crack or split can be observed down its middle while the rust uredium has a distinct pore with whitish urediospores clumped around it that have been forced through the pore. Severely infected plants are defoliated and have reduced pod formation, seed number, and seed weight.

CONTROL:
1. Apply a foliar fungicide during the growing season.
2. Some accessions of soybean have resistance.

Scab

CAUSE: *Sphaceloma glycines* Kurata & Kuribayashi.

DISTRIBUTION: Japan.

SYMPTOMS: Leaves, stems, and pods are infected. Lesions normally occur on both surfaces and are centered on the veins. Lesions are somewhat circular, up to 4 mm in diameter, slightly raised and buff-colored, often fading to gray.

Stems have lesions that are tiny elliptical spots that coalesce to form large areas up to 2 cm long. They are buff-colored sometimes with redbrown margins.

Pods have lesions that are initially red to brown. At maturity, lesions become black with pale centers and red-brown margins. Seeds do not develop in infected pods.

CONTROL: Soybean accessions with resistance are known.

Sclerotinia Stem Rot

This disease is also called Whetzelinia stem rot.

CAUSE: *Sclerotinia sclerotiorum* (Lib.) deBary (syn. *Whetzelinia sclerotiorum* (Lib.) Korf & Dumont) overwinters in soil as sclerotia. The fungus may occasionally be seedborne particularly during a wet autumn. During periods of high moisture, sclerotia germinate to form either apothecia or mycelium. Several apothecia may be formed per sclerotium. The apothecia are formed at the end of a stalk (stipe) and asci containing eight ascospores are formed in a layer on the apothecal surface. The ascospores are windborne to nearby plants and germinate under moist conditions. Plants of all ages are infected. Sclerotia are formed in pith, on stems, and rarely in pods. During harvest, sclerotia are not separated from seed. Thus sclerotia may also be planted along with the seed the following spring. Disease is most severe in wet weather.

DISTRIBUTION: Brazil, Canada, Hungary, India, Nepal, South Africa, and the United States.

SYMPTOMS: During wet weather, seedlings may damp off and older plants (R2 and R3 growth stage) may wilt with the withered leaves remaining attached to the stem for a long time. Brown lesions may occur on seedlings of some cultivars. Plants are killed by the fungus girdling the stem. Sites of infection originate at stem nodes 10–50 cm above the soil line. Water-soaked lesions progress acropetally and basipetally from the node. Side branches may also be infected. A white cottonlike growth, which is the fungus mycelium, occurs over the stem surface. Sometimes mycelial growth occurs halfway up the plant on branches, pods, and stems. Numerous round to oblong, dark brown to black sclerotia are produced along with the mycelium in the stem and on the plant surface. On older plants, leaves may wilt and remain attached to the plant. Infected seed are discolored, flattened, and smaller than noninfested seed. Pods are rarely formed on the lower portion of infected stems.

CONTROL:
1. Practice any cultural method that promotes drying out of the canopy such as planting soybean cultivars that do not readily lodge and planting in rows wider than 50 cm.
2. Do not rotate soybeans with beans (*Phaseolus* spp.).
3. Reduce irrigation before R1 growth stage.
4. Plant the least susceptible cultivars.

Southern Stem Blight

This disease is also called sclerotial blight and southern blight.

CAUSE: *Sclerotium rolfsii* Sacc. overwinters as sclerotia in soil and mycelium in residue and seeds. Mycelium from sclerotia or residue and seed serves as primary inoculum. High soil temperatures, relatively dry soil, and low soil pH (3–6) favor disease development. High relative humidity stimulates aerial growth of mycelium on stems, branches, and leaves. Cool temperatures in winter kill sclerotia and mycelium.

DISTRIBUTION: Generally distributed wherever soybeans are grown but it is mostly likely to occur in tropics and subtropics. Southern stem blight is not considered an important disease on soybeans.

SYMPTOMS: Southern stem blight usually does not cause economic losses and occurs as localized areas of dead plants scattered across a field. The disease is most severe in sandy soils when soil and air temperatures are high. Preemergence damping off may occur. The best symptom is a white, cottonlike, mycelial growth on the stem near the soil surface. Abundant tan to dark brown sclerotia are formed that are about the size of a mustard seed on the surface of the mycelium. Leaf infection is rare and occurs when infested soil particles are carried to the leaves. Leaf spots are circular, 6–12 mm in diameter, brown, have concentric markings, and may have a clump of white mycelium with a sclerotium in the middle.

CONTROL:
1. Most soybean cultivars have some resistance or tolerance.
2. Rotate soybeans with other crops.
3. Plow infected soybean residue 12 cm or deeper.
4. Apply lime to acid soils to obtain a pH of 6.5–7.0.

Stem Canker

This disease is also called southern stem canker when caused by southern isolates of *Diaporthe phaseolorum* var. *caulivora*.

CAUSE: *Diaporthe phaseolorum* (Cke. & Ell.) Sacc. var. *caulivora* Athow & Caldwell, anamorph *Phomopsis sojae* Lehman, is seedborne and overwinters as perithecia within infected residue and in infected seeds. Infected seed will give rise to seedlings with infected cotyledons that serve as a source of further infection. Diaporthe and Phomopsis isolates from cotton leaves and seedlings may also incite stem cankers on soybeans. Thus cotton may serve as an alternate source of inoculum. In the spring, ascospores are produced within perithecia during wet weather and carried by wind to soybeans. Most infections occur on the lower leaf blades and petioles of a plant and do not take place directly through leaf scars or stem wounds. Infection does occur only in the lower plant part. Disease development is optimum during wet weather with a daily mean temperature of 21°C. Disease is most severe under no-till conditions. Populations from the northern and southern United States differ in pathogenicity and morphological characteristics. Seed infection occurs at warm (20°C and above) temperatures and high relative humidity (close to 100%) for prolonged periods at growth stages R6–R8. Isolates of *D. phaseolorum* var. *caulivora* from Iowa differ in morphology and pathogenicity compared to isolates from the southern United States.

DISTRIBUTION: Canada and the United States.

SYMPTOMS: Stem canker appears about midseason and continues until plant maturity. At first, small brown lesions occur on cotyledons that may extend to stems and kill seedlings. Later dead plants are observed with leaves still attached. On the lower part of the stem a small red-brown lesion occurs on scars after the petiole has fallen, and a brown, slightly sunken lesion will develop at the base of a branch or leaf petiole on one side of a stem. Eventually the stem is girdled and the upper part of the plant is killed, causing patches of scattered dead plants within a field with the leaves still attached. The lesions near the soil line can be confused with symptoms of Phytophthora root rot. However, lesions can occur higher on the stem. After plant death, perithecia may be seen as tiny black dots on the stem.

CONTROL:
1. Plant seed from disease-free plants.
2. Plow infected residue in the fall, if possible, to hasten decay of residue and reduce level of inoculum.
3. Rotate soybeans with other crops.
4. Treat seed with a seed-protectant fungicide.
5. Plant resistant cultivars.

Stemphylium Leaf Blight

CAUSE: *Stemphylium botryosum* Wallr.

DISTRIBUTION: India.

SYMPTOMS: Initially small, circular, necrotic spots with dark brown margins and gray centers appear on leaves of all ages. Later spots will enlarge and may coalesce, particularly along the leaf margins. A dark green fungal growth may occur on the surface of spots. Eventually leaves dry up and defoliation may occur.

CONTROL: No control is reported.

Target Spot

CAUSE: *Corynespora cassiicola* (Berk. & Curt.) Wei (syn. *Cercospora melonis* (Cooke) Lindau, *C. virginicola* Kawamura, *Helminthosporium vignae* L. Oliver and *H. vignicola* (Kawamura) L. Olive). Survival is for two or more years as mycelium and chlamydospores in residue and soil. It can also grow saprophytically on many different kinds of residue in soil. Infection of roots and stems occurs in moist soils at a temperature of 15–18°C. Conidia are splashed or blown to leaves and cause infection when free moisture is present or at a relative humidity of 80 percent and above. The fungus infecting hypocotyls, roots, and stems is thought to be caused by a different race of the fungus than the one infecting leaves, pods, and seeds.

DISTRIBUTION: Eastern Asia, Canada, Central America, and the United States.

SYMPTOMS: Roots and stems can be infected during cool, moist weather early in the growing season. No symptoms develop at temperatures above 20°C. Target spot is most noticeable on leaves. However, leaf infection generally occurs later in the growing season and only when the relative humidity is 80 percent or above or when free moisture is present.

Leaf lesions are red-brown and may vary from small spots to spots 12 mm or larger in diameter. The larger spots are generally zonate, suggesting a target. Spots are generally surrounded by a light green to yellow-green halo. Narrow, elongated spots may develop along veins of the upper leaf surface.

Spots on petioles and stems are dark brown and vary in shape from a small speck to an elongated spindle-shaped lesion. Spots on pods are generally round, 1 mm in diameter, with a slightly depressed, purple-black center and brown margins. Sometimes small, dark brown spots are formed on seeds.

Taproots, hypocotyls, and larger lateral roots have oval red-brown lesions that later become purple due to the fungus sporulating. Lesions will grow larger and coalesce as the plant matures.

CONTROL:
1. Grow resistant cultivars.
2. Treat seed with a seed-protectant fungicide.
3. Rotate soybeans with other crops.

Thielaviopsis Root Rot

This disease is also called black root rot.

CAUSE: *Thielaviopsis basicola* (Berk. & Br.) Ferr. (syn. *Thielavia basicola* Zopf.) survives as chlamydospores in residue. Chlamydospores germinate during moist, cool soil conditions and infect hypocotyls and roots. Chlamydospores are formed within infected tissues and returned to soil at harvest time. Seedborne infection is uncommon. The disease is most severe under cool (16–20°C) soil temperatures early in the season.

DISTRIBUTION: Canada, Germany, and the midwestern United States, particularly Michigan.

SYMPTOMS: The cortex of the hypocotyl below the soil line and the taproots and fibrous roots become dark brown and necrotic. Frequently a severe root rot will develop with subsequent wilting of foliage or plant death.

CONTROL:
1. Plant resistant cultivars.
2. The herbicide chloramben has been reported to increase injury.

Top Dieback

CAUSE: *Diaporthe phaseolorum* (Cke. & Ell.) Sacc. var. *caulivora* Athow & Caldwell. See Stem Canker.

DISTRIBUTION: See Stem Canker; however, distribution of this disease may differ from stem canker.

SYMPTOMS: Disease develops late in season and is distinct from stem canker. The five or six uppermost internodes die prematurely.

CONTROL: See Stem Canker.

Twin Stem Abnormality Disease

CAUSE: *Sclerotium* sp. and *Macrophomina phaseolina* (Tassi) Goid.

DISTRIBUTION: Brazil and the United States.

SYMPTOMS: Symptoms appear after cotyledons have opened and have three distinct severity forms. The most severe form is Type I.

Elongation of the first internode that bears the primary leaves, is completely inhibited. The apical meristem resembles a convex knob between the cotyledons. The cotyledons are large and spongy with a thickened hypocotyl. In 7–10 days two pairs of primary leaves with rudimentary petioles and internodes form. At maturity the plant is stunted and bushy and may have two main stems originating from the cotyledonary node.

In the moderately severe form (Type II), the first internode with underdeveloped primary leaves elongates to a limited extent. The original primary leaf primordials are necrotic. Two secondary pairs of primary leaves develop from the cotyledonary node and continue to grow. Adult plants are stunted and bushy and may have two main stems originating from the cotyledonary node.

In the less severe form (Type III), the first internode with underdeveloped primary leaves elongates normally or excessively. Leaves remain wrapped around each other and are bleached at the apex. When leaves open, the bleaching extends to the margin and inward, the apex is curved inward, and the entire leaf is wrinkled. The first trifoliate leaf is formed with little or no elongation of the second internode. One or two pairs of leaves may form at the cotyledonary node but generally remain rudimentary.

CONTROL: No control is reported.

Verticillium Pod Loss

CAUSE: *Verticillium nigrescens* Psthybr. High soil moisture levels are related to increased infection. Increasing incidence is related to a long history of cotton production.

DISTRIBUTION: Georgia.

SYMPTOMS: *Verticillium nigrescens* normally grows from one end of pedicle and rarely from other parts of the pod. The fungus can be isolated from flower and pods at all stages of development. Numbers of pods per plant are reduced but there is no vascular discoloration or wilting. Seed weight is also reduced.

CONTROL: No control is reported.

Verticillium Wilt

CAUSE: *Verticillium dahliae* Kleb. is possibly seedborne.

DISTRIBUTION: Not known for certain but *V. dahliae* has been isolated from soybeans growing in fields in Iowa.

SYMPTOMS: Vascular tissue in the lower portion of the stem is discolored. Leaves die but remain attached to plants. Such plants are scattered throughout a field.

Inoculated plants display sudden wilt symptoms 7–10 days after inoculation. Early wilt symptoms occur as apical, marginal, or unilateral wilting of primary leaves. Young tissues remain green and dry without first becoming chlorotic. Vascular discoloration is red-brown to black. No pith discoloration occurs and discoloration is restricted to vascular tissue.

CONTROL: Some cultivars are evidently resistant. Verticillium wilt is not considered important; possibly other diseases have been mistaken for it.

Yeast Spot

This disease is also called Nematospora spot.

CAUSE: *Nematospora coryli* Pegl. is a yeast fungus that is transmitted by and survives mainly in the adult green stinkbug, *Acrosternum hilare* (Say). There are other species of stinkbugs that are also known to transmit *N. coryli*. The stinkbug pierces the pod wall and feeds on the developing bean. During feeding, only yeast cells located in the region of the mouth parts are forced into the developing seed.

DISTRIBUTION: Africa, Brazil, and the United States.

SYMPTOMS: Yeast spot is primarily a disease of seeds. Seeds infected early may not mature or they may remain small and shriveled. Where infection occurs during pod formation, seeds may fail to develop and the pods may drop prematurely. On developing green seeds, infected tissue is slightly depressed and varies from a light cream or yellow to brown. There may be varying amounts of dead tissue extending into the infected embryo. On the surface of the dry seed coat, lesions appear as very small, discolored punctures. Seed shriveling caused by other diseases may be differentiated from that caused by yeast spot by observing these punctures. The outside of the pods display small, pinpoint, discolored areas.

CONTROL:
1. Plant disease-free seed.
2. Control stinkbugs by insecticides when nymphs average one per 30 cm of row.

_____MYCOPLASMAL CAUSES_____

Machismo

This disease is also called amachamiento and proliferación de yemas.

CAUSE: A mycoplasm. The causal agent is transmitted by the leafhopper *Scaphytopius fuliginosus* Osborn. It is not known to be mechanically transmitted or seedborne. Several weeds also serve as hosts.

DISTRIBUTION: Colombia and Mexico.

SYMPTOMS: The younger the plant when infected, the more severe the symptoms. Frequently a branch that is close to the ground will show symptoms but the rest of the plant will appear normal. Symptoms first occur at time of flowering or when pod formation begins. Pods may normally be in an upright position and rigid, curved, flat, and thin with no beans produced, or they may be transported into corrugated, leaflike structures resembling phyllody. Once floral parts have been transformed into leaflike structures, buds proliferate from leaf axils anywhere on the plant, causing a witches' broom.
   Sepals may be larger and duller in color than normal sepals, giving the appearance of small leaves. Flowers are smaller and remain closed. In these cases the pods or leaflike structures grow through the tips of the unopened flowers.
   Pods formed on diseased plants are slightly corrugated and are usually thicker and softer than those on healthy plants. Immature green seeds can germinate in the pod, causing roots to rupture through the pod wall.
   Infected plants have a darker green color and remain green longer than healthy plants, thereby causing difficulty in harvesting.

CONTROL:
1. Control weed hosts.
2. Rogue infected plants.
3. Apply tetracycline.

Soybean Bud Proliferation

This may be the same disease as on page 551.

CAUSE: A mycoplasma transmitted by the leafhopper *Scaphytopius acutus* (Say).

DISTRIBUTION: Louisiana.

SYMPTOMS: Symptoms occur at onset of flowering. Affected plants remain green after normal plants have matured. There is proliferation of adventitious buds. Pods are underdeveloped and few. Many contain only one large bean.

CONTROL: No control is reported.

Witches' Broom

CAUSE: A mycoplasmalike organism transmitted by *Orosius orientalis* (Evans) and *O. argentatus* (Evans).

DISTRIBUTION: Brazil, India, Indonesia, Japan, Nigeria, and Tanzania.

SYMPTOMS: Symptoms resemble those of machismo. Symptoms include a phylloid disorder of the floral organs, shoot proliferation, and reduced leaflets.

CONTROL: No control is reported.

CAUSED BY NEMATODES

Lance Nematode

CAUSE: *Hoplolaimus columbus* Sher. has a life cycle similar to *Meloidogyne* spp. that cause root knot. It can survive in dry soil for a long period of time. *Hoplolaimus galeatus* (Cobb) Thorne has been reported to also be pathogenic to soybeans.

DISTRIBUTION: Southeastern United States, particularly in sandy soils.

SYMPTOMS: Infected plants are stunted and chlorotic. Roots are rotted and sparse. In areas of high nematode populations, the root sytem may be almost completely destroyed as the plant nears maturity.

CONTROL:
1. Rotate soybeans with other crops.
2. Plow under infected residue where erosion is of no concern.

3. Clean soil from equipment between fields.
4. Apply nematicides where the expense justifies the return.
5. Subsoil or chiesel to a 36 cm depth where hardpan conditions exist.

Reniform Nematode

CAUSE: *Rotylenchulus reinformis* Linford & Oliveira is a nematode in which only the female invades the roots after feeding on epidermis tissues. It is semiendoparasitic as the posterior portion of the female remains outside the root and swells to a kidney shape. Eggs are laid outside the root. Optimum temperature for root invasion and nematode development is 29°C.

DISTRIBUTION: Africa and southeastern United States.

SYMPTOMS: Plants are stunted and chlorotic. Small galls or root swellings may sometimes occur on damaged roots.

CONTROL:
1. Plant resistant cultivars where feasible.
2. Apply nematicides where expense can be justified.

Root Knot Nematode

CAUSE: *Meloidogyne* spp. including *M. arenaria* (Neal) Chitwood, *M. hapla* (Kofoid & White) Chitwood, *M. incognita* (Kofoid & White) Chitwood, and *M. javanica* (Treub.) Chitwood. Infested soil contains eggs that hatch, producing larvae. The larvae move through the soil and penetrate the root tip. Larvae migrate through the root to the vascular tissues, become sedentary, and commence feeding. Excretions of the larvae stimulate cell proliferation, resulting in the development of syncytia from which larvae derive their food. Females lay eggs that may be pushed out of the root into the soil. Most damage occurs in warm climates in sandy, light-textured soils.

DISTRIBUTION: Generally distributed wherever soybeans are grown.

SYMPTOMS: Plants will be stunted. Leaves turn yellow and wilt during hot, dry weather. Roots will have knots or galls of different sizes that disrupt the flow of water and nutrients to the top of the plant.

CONTROL:
1. Rotate with an alternate crop for three years if the species of nematode is known. Corn can be rotated with soybeans to control

M. hapla. Cotton can be rotated to control all species except *M. incognita*.

2. Plant resistant cultivars; however, different races of *M. incognita* are known to occur.
3. Apply nematicides where the expense can be justified.

Root Lesion Nematodes

CAUSE: *Pratylenchus* spp. are endoparasitic. They enter roots as larvae and attack the root cortex.

DISTRIBUTION: Generally distributed wherever soybeans are grown.

SYMPTOMS: Plants may be yellowed and stunted. Roots at first have small dark lesions. Other soilborne microorganisms may enter the root through wounds caused by *Pratylenchus* spp., resulting in root rot and a browning of the root system.

CONTROL: Plant resistant cultivars.

Soybean Cyst Nematode

CAUSE: *Heterodera glycines* Ichinohe survives in soil as eggs within cysts formed by female bodies. It is spread within a field or over long distances by any agent that is capable of moving infested soil and plant residue. Larvae hatch from eggs and move through soil to enter the plant near the root tip. Nematodes migrate inside the root to the undifferentiated cortical and stelar tissues. Syncytia are formed by salivary excretions. As the female matures, it swells and breaks through the epidermis, with the neck remaining in the root. The female is then inseminated by a male and will produce 200–500 eggs that are retained in the body until a cyst is formed. Eggs are deposited externally from which larvae are hatched to begin a new life cycle. Nematode development is most rapid at 30°C. Different races exist.

DISTRIBUTION: Asia, southeastern and midwestern United States.

SYMPTOMS: Injury is most serious in dry, sandy soils that are low in nutrients and organic matter. Not all plants show aboveground symptoms. Plants are stunted, chlorotic, and may lose their leaves prematurely. A yellowing starts at the leaflet margin and gradually includes all of the blade. Affected leaves are distributed over the plant. Light to moderate infection will stimulate overproduction of lateral roots but severely infected plants will have a small, necrotic root system with few or no Rhizobium nodules. Examination of cleaned roots under magnification will show white to yellow females and brown cysts that are somewhat smaller than the head of a pin clinging to the roots.

CONTROL:
1. Plant resistant cultivars.
2. Rotate soybeans with other crops for two or three years.
3. Practice sanitation; clean soil from equipment between fields.
4. Apply nematicides to soil.

Stem Nematode

CAUSE: *Ditylenchus dipsaci* (Kuhn) Filipjev. is an endoparasite. Injury occurs with high humidity and rainy weather.

DISTRIBUTION: United States.

SYMPTOMS: Stems are stunted, swollen, and distorted. Young leaves are also infected, causing a chlorotic spotting.

CONTROL: Stem nematode is not considered a serious problem on soybeans.

Sting Nematode

CAUSE: *Belonolaimus gracilis* Steiner and *B. longicaudatus* Rau are nematodes that overseason in the soil as eggs and larvae. Feeding by nematodes occurs only on root surfaces but not internally.

DISTRIBUTION: Southeastern United States.

SYMPTOMS: Damage is confined to plants growing on sandy soils. Seedlings may be killed, resulting in a stand loss. Affected plants other than seedlings usually are not killed but appear stunted, chlorotic, or discolored as if suffering from moisture stress.

At first roots have small, dark, sunken lesions that may enlarge to girdle the root tip. Meristematic tissue is destroyed while root proliferation occurs above the point of attachment, resulting in a stubby and dark root system. Eventually roots may become severely rotted.

CONTROL:
1. Use nematicides.
2. Rotate soybeans with other crops.

3. Plow under infected soybean residue where erosion is not a problem.
4. Clean soil from equipment between fields to slow spread of nematodes.

VIRAL CAUSES

Bean Pod Mottle

CAUSE: Bean pod mottle virus (BPMV) is transmitted by several insects, particularly the bean leaf beetle, *Cerotoma trifurcata* (Forster). BPMV is commonly associated with soybean mosaic virus in infected plants.

DISTRIBUTION: United States.

SYMPTOMS: The youngest leaves have a green to yellow mottling particularly during rapid growth and cool weather. As leaves become mature, the mottling disappears. Infected plants under moisture stress are less turgid than normal plants. Such plants may have a yield loss if stress occurs during the reproductive phase.

CONTROL:
1. Grow at least four rows of corn or sorghum between soybean and other legume fields and noncultivated areas.
2. Rogue out infected plants from seed fields.
3. Control weeds, particularly in noncrop areas next to soybean fields.

Brazilian Bud Blight

CAUSE: Tobacco streak virus (TSV) is seedborne. TSV is also sap transmissible but it is not known to be transmitted by insects.

DISTRIBUTION: Brazil and the United States.

SYMPTOMS: Symptoms resemble those of bud blight caused by tobacco ringspot virus and are normally not seen on young plants. Initially irregular yellow spots develop on leaves. Infected plants then develop numerous stunted axillary branches that produce dwarfed leaves. Mosaic symptoms and necrotic streaks develop at nodes. Necrotic blotches appear on pods. Early-infected plants produce fewer pods and seeds.

CONTROL: Follow control measures for bean pod mottle.

Bud Blight

CAUSE: Tobacco ringspot virus (TRSV) is seedborne and overwinters or is reservoired in a large number of legumes. TRSV is sap transmitted. Nematodes, nymphs of *Thrips tabaci* Lindeman, grasshoppers *Melanoplus differentialis* Thomas, and possibly other insects transmit the virus from diseased to healthy plants. TRSV is also transmitted by root grafts and the dagger nematode, *Xiphinema americanum* Cobb.

DISTRIBUTION: Australia, Canada, China, and the United States.

SYMPTOMS: Bud blight occurs sporadically, appearing only in certain years. Losses of 100 percent have been reported. It is frequently more severe in soybean fields that are adjacent to legume-grass pastures or fence rows where insects probably transmit the virus from perennial plants to soybeans. Many strains of the virus exist and some cause severely dwarfed and barren plants.

Symptoms of bud blight vary with the stage of development at which plants become infected. When plants are infected before flowering, the apical bud and shoot turn brown, curve downward markedly to form a crook, and become dry and brittle; hence the name bud blight. The younger leaves often develop a rusty flecking. The plant is stunted and produces little or no seed. Stunting is not evident when plants are grown above 25°C. Plants infected while less than five weeks old are severely stunted because of shortened internodes or fewer nodes. Sometimes the inside of the stem below the blighted terminal bud is discolored, most often at the nodes. Brown streaks occasionally occur on petioles and large leaf veins. Petioles of youngest trifoliate leaves are often thickened, shortened, and occasionally curved, distorting the shoot tips. Leaflets are dwarfed and tend to cup with leaf blades rugose and bronzed. When plants are infected during flowering, small underdeveloped pods are produced. Infection after flowering results in poorly filled pods that have a conspicuous dark blotching. Many of these pods fall to the ground. A good diagnostic symptom is the presence of infected plants that usually remain green after normal plants have matured. These plants are easily found in a field in the autumn.

CONTROL:
1. Do not plant soybeans adjacent to legume fields. If this is not feasible, grow a buffer strip of a few rows of corn or forage sorghum between legumes and soybean field.

2. Control broad leaf weeds in noncrop areas adjacent to soybean fields before soybeans emerge.
3. Rogue infected plants from seed production fields.

Cowpea Chlorotic Mottle

CAUSE: Cowpea chlorotic mottle virus, soybean-infecting strain (CCMV-S). CCMV-S is sap transmissible and is also transmitted by the bean leaf beetle, *Cerotoma trifurcata* (Forster), and the spotted cucumber beetle, *Diabrotica undecimpunctata* Mannerheim.

DISTRIBUTION: United States.

SYMPTOMS: Leaves are mottled, slightly crinkled, and tend to be upright. Mottling is most intense on younger leaves. Plants are stunted and produce fewer seeds of reduced size and quality. Local necrotic lesions may be produced on resistant cultivars and veinal necrosis on cultivars with intermediate resistance.

CONTROL: Plant resistant cultivars.

Cowpea Mild Mottle

CAUSE: Cowpea mild mottle virus (CMMV). The virus is seedborne, transmitted mechanically, and by the whitefly *Bemisia tabaci* (Genn.)

DISTRIBUTION: Ivory Coast and Thailand.

SYMPTOMS: Plants are stunted and show light green mosaic. Inoculated seedlings display vein clearing that evolves into yellow mosiac and occasional crinkling. Leaves may be curled downward or cupped upward. Some cultivars display vein and top necrosis. Vein enations have also been reported as a symptom.

CONTROL: No control is reported.

Cowpea Mosaic

CAUSE: Cowpea mosaic virus (CMV) is transmitted by sap and the bean leaf beatle, *Cerotoma ruficornis*.

DISTRIBUTION: Puerto Rico, Central America, and the United States.

SYMPTOMS: Inoculated primary leaves display chlorotic and necrotic lesions. A systemic light green mosaic occurs on trifoliate leaves together with a very severe leaf malformation and bud blight.

CONTROL: No control is reported.

Cowpea Severe Mosaic

CAUSE: Cowpea severe mosaic virus (CPSMV) is transmitted by the beetle *Cerotoma arcuata* (Oliv.).

DISTRIBUTION: Brazil.

SYMPTOMS: Identical to those of bud blight. Significant reductions occur in plant height, yield, number of pods per plant, and seed germination.

CONTROL: No control is reported.

Delayed Maturity

CAUSE: Possibly a viruslike agent that is seedborne and is possibly spread by insects.

DISTRIBUTION: Louisiana.

SYMPTOMS: Plants remain green with pronounced longitudinal ribs after normal plants have matured. Seeds often germinate inside the pod and rot. Pods often are rough textured with pronounced wrinkles and veins.

CONTROL: No control is reported.

Indonesian Soybean Dwarf

CAUSE: Indonesian soybean dwarf virus (ISDV) is transmitted by the aphid *Aphis glycines* Matsumura in a persistent manner.

DISTRIBUTION: Indonesia.

SYMPTOMS: Infected plants are dwarfed with shortened leaf petioles and internodes and a darker green color than healthy plants. Upper leaves are small and curl upward. Lower leaves are rugose with shortened veins and interveinal white necrosis. Occasionally these leaves will have numerous holes due to their brittleness. Infected plants produce few pods.

CONTROL: No control is reported.

Peanut Mottle

CAUSE: Peanut mottle virus (PMV) is sap transmissible but not seed-borne in soybeans. PMV is transmitted by several species of aphids in a nonpersistent manner. The virus can be spread from nearby peanut fields to soybean fields.

DISTRIBUTION: Southeastern United States.

SYMPTOMS: A general mosaic often occurs that is similar to symptoms caused by other viruses. Small chlorotic areas appear that enlarge but do not become continuous, leaving striking, dark green islands on leaves. Later chlorotic patches, line patterns, or ring patterns occur on third and fourth leaves following infection. Leaves that develop later show a general mosaic without distinct characteristics. Leaves may pucker and curl down at the edges.

CONTROL:
1. Plant resistant cultivars.
2. Do not plant soybeans after peanuts.
3. Separate soybeans from peanut fields by 50 m or more.
4. Rogue volunteer peanut plants from soybean fields.

Peanut Stripe

CAUSE: Peanut stripe virus (PSV). See Peanut Stripe in Chapter 11.

DISTRIBUTION: Southeastern United States.

SYMPTOMS: See Peanut Stripe in Chapter 11.

CONTROL: Resistant lines have been identified.

Peanut Stunt

CAUSE: Peanut stunt virus (PSV) overwinters in several legumes such as lespedeza, crown vetch, peanuts, and others. It is disseminated by aphids.

DISTRIBUTION: Southeastern and central United States.

SYMPTOMS: Symptoms range from diffuse chlorotic lesions to necrotic local lesions on inoculated primary leaves. Vein clearing is evident followed by a general mosaic of trifoliate leaves. Some infected plants may be symptomless.

CONTROL: No control is reported

Soybean Chlorotic Mottle

CAUSE: Soybean chlorotic mottle virus (SCMV) is transmitted by sap but not by aphids or seed.

DISTRIBUTION: Japan.

SYMPTOMS: Plants display mosaic symptoms and are stunted. Mechanically inoculated plants initially show vein clearing, chlorosis, and leaf curling of young leaves, which are reduced in size compared to healthy leaves. Eventually leaves display mottle symptoms. Plants are also stunted with shortened internodes.

CONTROL: No control is reported.

Soybean Crinkle Leaf

CAUSE: Likely a virus that is transmitted by the whitefly *Bemisia tabaci* (Genn.).

DISTRIBUTION: Thailand.

SYMPTOMS: Leaves are twisted or curled with veinal enations on the undersurfaces of the leaves. Foliage of infected plants is darker green than that of healthy plants. In the greenhouse, infected plants show a yellow netting of veins 10–14 days after inoculation.

CONTROL: No control is reported.

Soybean Dwarf

CAUSE: Soybean dwarf virus (SDV) is transmitted by grafting and in a persistent manner by the aphid *Aulacorthum solani* (Kaltenbach). Optimum temperature for transmission is 20°C. SDV is not seed or sap transmitted. Strains are grouped into two types, dwarfing (SDV-D) and yellowing (SDV-Y).

DISTRIBUTION: Japan.

SYMPTOMS: Plants are severely stunted. Leaves are rugose, curl down, and may show interveinal yellowing.

CONTROL: No control is reported.

Soybean Mosaic

This disease is also called soybean crinkle.

CAUSE: Soybean mosaic virus (SMV) is seedborne and overwinters in perennial weeds. It is disseminated by aphids and is readily sap transmissible. Most primary inoculum consists of infected seedlings derived from SMV-infected seed. The distribution of SMV within a field or secondary spread occurs by activity of several aphids that transmit the virus in a nonpersistent manner. Infection before the onset of flowering results in higher levels of seed transmission than later infections. The percentage of SMV-infected seed will gradually decrease when secondary spread is minimal or if spread occurs after the onset of flowering.

DISTRIBUTION: Generally distributed wherever soybeans are grown.

SYMPTOMS: Infected seed may fail to germinate. Diseased seedlings are spindly with crinkled unifoliate leaves that curl downward, eventually becoming prematurely chlorotic. Subsequent trifoliate leaves are severely stunted, mottled, and are more crinkled than the unifoliate leaves. When very susceptible cultivars are infected, the first disease symptom is a yellowish vein clearing that develops in the small, branching veins of the developing leaves. This symptom appears 6–14 days after infections. The best diagnostic symptom is rugose leaves that usually appear on the third leaf formed after infection. If infected plants are grown under cool conditions, the rugose symptoms increase in severity on successive leaves, and the characteristic mosaic pattern of alternate light and dark green patches of leaf tissue may appear. The mosaic symptom is more obvious on some cultivars than others. If infected plants are grown under warm conditions, 31°C and above, symptoms are masked and leaves develop normally. Symptoms may vary from essentially normal-appearing leaves, in which the veins are particularly deep in the lamina, to rugose leaves, in which portions of the upper surface have become invaginated into vesicles. Leaves become leathery to the touch and coarse and brittle as the plant nears maturity. Severely infected leaves may also be quite narrow and have a willow leaf appearance. There may also be seed discoloration.

SMV has been reported to cause depressed black lesions that are irregular in shape on pods. These are similar to those caused by *Cercospora sojina* Hara and *Colletotrichum truncatum* (Schw.) Andrus & W. D. Moore. Yields have been reported to be significantly reduced only when 60 percent or more of plants are infected. The largest decreases in yield occurred in plants inoculated with SMV 20 days after emergence.

CONTROL:
1. Plant seed from disease-free plants.
2. Rogue out infected plants from seed fields.
3. Practice good weed control in and around soybean fields.

Tobacco Necrosis

CAUSE: Tobacco necrosis virus (TNV).

DISTRIBUTION: United States.

SYMPTOMS: Small necrotic spots surrounded by a chlorotic halo occur on secondary leaves. The necrotic areas enlarge toward the veins and spread from small veins to larger veins, the midrib, and the petiole, resulting in systemic infection of the plant. Petioles and stems have necrosis of phloem tissues and adjacent parenchyma.

CONTROL: No control is reported.

Tobacco Streak

CAUSE: Tobacco streak virus.

DISTRIBUTION: Brazil, Iowa, and Oklahoma.

SYMPTOMS: Some plants remain green after maturity with unexpanded, unfilled pods. Plants are stunted with curving and death of stem tips. Necrotic areas occur in stem pith, especially near nodes. Axillary buds develop and leaves are dwarfed. Early infection prevents seed formation and infection at any age delays seed maturation. Symptoms from Iowa include necrotic spots on pods.

CONTROL: No control is reported.

Yellow Mosaic

CAUSE: Bean yellow mosaic virus (BYMV) is disseminated by the insects *Aphis fabae* Scopoli, *Macrosiphum gei* (Kaltenbach), *M. pisi*

(Kaltenbach), *M. solanifolii* (Ashmead), and *Myzus persicae* Sulzer. The virus is not thought to be spread by infected seed but is easily sap transmissible. BYMV has a large host range.

DISTRIBUTION: Canada, Europe, and the United States.

SYMPTOMS: The symptoms caused by BYMV resemble those caused by soybean mosaic virus. Young leaves develop a yellow mottling, scattered in random spots over the leaf or in indefinite bands along the major veins. Rusty, necrotic spots develop in the yellowed areas as the leaf matures. Some strains of BYMV produce severe mottling and crinkling of leaves.

CONTROL:
1. Plant seed from disease-free plants.
2. Rogue infected plants from seed fields.
3. Practice good weed control.
4. Plant cultivars that have some resistance.

DISORDERS OF
UNKNOWN CAUSE

Soybean Bud Proliferation

This may be the same disease as on page 538.

CAUSE: Unknown.

DISTRIBUTION: Arkansas.

SYMPTOMS: Plants have delayed senescence with proliferation of small, undeveloped pods.

CONTROL: No control is reported.

Sudden Death Syndrome

CAUSE: Unknown; however, there are reports that the soybean cyst nematode may be associated with the syndrome.

DISTRIBUTION: United States.

SYMPTOMS: Symptoms are first noticed after flowering, near R3–R4, and continue through maturity. If symptoms develop before R6 there is a strong likelihood that complete defoliation and pod abortion will occur in three to six weeks. External symptoms resemble those of brown stem rot but internal symptoms are different. The most significant difference is lack of pith discoloration. A red-brown vascular discoloration does develop but is quite uniform; no streaking patterns develop.

CONTROL: No control is reported.

———————————BIBLIOGRAPHY———————————

Almeidia, A. M. R., and Kihl, R. 1981. Necrose das vagens um novo sintoma causada pelo virus do mosaico da soja. *Fitopathologia Brasileria* **6**:281–283 (in Portuguese).

Almeidia, A. M. R., and Silveira, J. M. 1983. Efeito da idade de inoculacao de plants de soja com o virus do mosaico comum da soja e da percentagem de plantas infecladas sobre o rendimento e algumas caracteristicas economicas. *Fitopathologia Brasileira* **8**:229-236 (in Portuguese).

Anahosur, K. H., and Fazalnoor, K. 1972. Studies on foliicolar fungi of soybean crop. *Indian Phytopathol.* **25**:504–508.

Anjos, J. R. N., and Lin, M. T. 1984. Bud blight of soybeans caused by cowpea severe mosaic virus in central Brazil. *Plant Disease* **68**:405–407.

Anon. 1986. Sudden death syndrome. Is it the soybean cyst nematode? *Crops and Soils Mag.* **39**:10.

Backman, P. A.; Williams, J. C.; and Crawford, M. A. 1982. Yield losses in soybeans from anthracnose caused by *Colletotrichum truncatum*. *Plant Disease* **66**:1032–1034.

Balducchi, A. J., and McGee, D. C. 1987. Environmental factors influencing infection of soybean seeds by Phomopsis and Diaporthe species during seed maturation. *Plant Disease* **71**:209–212.

Blackmon, C. W., and Musen, H. L. 1974. Control of the Columbia (lance) nematode *Hoplolaimus columbus* on soybeans. *Plant Dis. Rptr.* **58**:641–645.

Boosalis, M. G., and Hamilton, R. I. 1957. Root and stem rot of soybean caused by *Corynespora cassicola* (Berk. & Curt.) Wei. *Plant Dis. Rptr.* **41**:696–698.

Bristow, P. R., and Wyllie, T. D. 1986. *Macrophomina phaseolina*, another cause of the twin-stem abnormality disease of soybean. *Plant Disease* **70**:1152–1153.

Bromfield, K. R.; Melching, J. S.; and Kingsolver, C. H. 1980. Virulence and aggressiveness of *Phakopsora pachyrhizi* isolates causing soybean rust. *Phytopathology* **70**:17–21.

Canady, C. H., and Schmitthenner, A. F. 1979. The effect of nitrogen on Phytophthora root rot of soybeans. *Phytopathology* **69**:539 (abstract).

Cheng, Y. H., and Schenck, N. C. 1978. Effect of soil temperature and moisture on survival of the soybean root rot fungi *Neocosmospora vasinfectum* and *Fusarium solani* in soil. *Plant Dis. Rptr.* **62**:945–949.

Daft, G. C., and Leben, C. 1972. Bacterial blight of soybeans: Epidemiology of blight outbreaks. *Phytopathology* **62**:57–62.

Daft, G. C., and Leben, C. 1973. Bacterial blight of soybeans: Field overwintered *Pseudomonas glycinea* as possible primary inoculum. *Plant Dis. Rptr.* **57**:156–157.

Dale, J. L., and Walters, H. J. 1985. Soybean bud proliferation of unknown etiology in Arkansas. *Plant Disease* **69**:811 (disease notes).

Damsteegt, V. D. 1985. Vector relationships of two strains of soybean dwarf virus. *Phytopathology* **75**:1349 (abstract).

Datnoff, L. E.; Naik, D. M.; and Sinclair, J. B. 1987. Effect of red leaf blotch on soybean yields in Zambia. *Plant Disease* **71**:132–135.

Demski, J. W., and Kuhn, C. W. 1975. Resistant and susceptible reaction of soybeans to peanut mottle virus. *Phytopathology* **65**:95–99.

Derrick, K. S., and Newsom, L. D. 1984. Occurrence of a leafhopper-transmitted disease of soybeans in Louisiana. *Plant Disease* **68**:343–344.

Derrick, K. S.; Newsom, L. D.; and Brlansky, R. H. 1984. Delayed maturity of soybeans. *Phytopathology* **74**:627 (abstract).

Dhingra, O. D., and Muchovej, J. J. 1980. Twin-stem abnormality disease of soybean seedlings caused by *Sclerotium* sp. *Plant Disease* **64**:176–178.

Dianese, J. C.; Riberio, W. R. C.; and Urben, A. F. 1986. Root rot of soybean caused by *Cylindrocladium clavatum* in central Brazil. *Plant Disease* **70**:977–980.

Duncan, D. R., and Paxton, J. D. 1981. Trifluralin enhancement of Phytophthora root rot of soybean. *Plant Disease* **65**:435–436.

Dunleavy, J. M. 1963. A vascular disease of soybeans caused by *Corynebacterium* sp. *Plant Dis. Rptr.* **47**:612–613.

Dunleavy, J. 1966. Factors influencing spread of brown stem rot of soybeans. *Phytopathology* **56**:298–300.

Dunleavy, J. M. 1978. Bacterial tan spot, a new disease of soybeans. *Am. Phytopathol. Soc. 70th Annual Meeting* (abstract).

Dunleavy, J. M. 1980. Yield losses in soybeans induced by powdery mildew. *Plant Disease* **64**:291–292.

Dunleavy, J. M. 1985. Transmission of *Corynebacterium flaccumfaciens* by soybean seed. *Phytopathology* **75**:1295 (abstract).

Dunleavy, J. M.; Chamberlain, D. W.; and Ross, J. P. 1966. Soybean diseases. *USDA Agric. Hdbk. No. 302.*

Ellis, M. A.; Ilyas, M. B.; and Sinclair, J. B. 1974. Effect of cultivar and growing region on internally seedborne fungi and *Aspergillus melleus* pathogenicity in soybeans. *Plant Disease Rptr.* **58**:332–334.

Fagbenle, H. H., and Ford, R. E. 1970. Tobacco streak virus isolated from soybeans, *Glycine max. Phytopathology* **60**:814–820.

Fletcher, J.; Irwin, M. E.; Bradfute, O. E.; and Granada, G. A. 1984. A Machismo-like disease of soybeans in Mexico. *Phytopathology* **74**:857 (abstract).

Fletcher, J.; Irwin, M. E.; Bradfute, O. E.; and Granada, G. A. 1984. Discovery of a mycoplasmalike organism associated with diseased soybeans in Mexico. *Plant Disease* **68**:994–996.

Fortnum, B. A., and Lewis, S. A. 1983. Effects of growth regulators and nematodes on Cylindrocladium black root rot of soybean. *Plant Disease* **67**:282–284.

Gangopadhyay, S.; Wyllie, T. D.; and Luedders, V. D. 1970. Charcoal rot disease of soybean transmitted by seeds. *Plant Dis. Rptr.* **54**:1088–1091.

Garcia-Jimenez, J., and Alfaro, A. 1985. *Colletotrichum gloeosporioides:* A new anthracnose pathogen of soybean in Spain. *Phytopathology* **69**:1007 (disease notes).

Garzonio, D. M., and McGee, D. C. 1981. Pod and stem blight of soybean: The relative importance of seed-borne and soil-borne inoculum. *Phytopathology* **71**:218 (abstract).

Garzonio, D. M., and McGee, D. C. 1983. Comparison of seeds and crop residues as sources of inoculum for pod and stem blight of soybeans. *Plant Disease* **67**:1374–1376.

Gleason, M. L., and Ferriss, R. S. 1985. Influence of soil water potential on performance of soybean seeds infected by *Phomopsis* sp. *Phytopathology* **75**:1236–1241.

Granada, G. A. 1979. Machismo disease of soybeans. I. Symptomatology and transmission. *Plant Dis. Rptr.* **63**:47–50.

Granada, G. A. 1979. Machismo disease of soybeans. II. Suppressive effects of tetracycline on symptom development. *Plant Dis. Rptr.* **63**:309–312.

Grau, C. R., and Radke, V. L. 1984. Effects of cultivars and cultural practices on Sclerotinia stem rot of soybean. *Plant Disease* **68**:56–58.

Grau, C. R.; Radke, V. L.; and Gillespie, F. L. 1982. Resistance of soybean cultivars to *Sclerotinia sclerotiorum*. *Plant Disease* **66**:506–508.

Gray, F. A.; Rodriguez-Kabana, R.; and Adams, J. R. 1980. Neocosmospora stem rot of soybeans in Alabama. *Plant Disease* **64**:321–322.

Gray, L. E. 1971. Variation in pathogenicity of *Cephalosporium gregatum* isolates. *Phytopathology* **61**:1410–1411.

Gray, L. E. 1972. Recovery of *Cephalosporium gregatum* from soybean straw. *Phytopathology* **62**:1362–1364.

Gray, L. E. 1978. *Mycoleptodiscus terrestris* root rot of soybeans. *Plant Dis. Rptr.* **62**:72–73.

Grybauskas, A. P. 1986. First report of soybean naturally infected with *Leptosphaerulina briosiana*. *Plant Disease* **70**:1159 (disease notes).

Hartman, G. L.; Datnoff, L. E.; Levy, C.; Sinclair, J. B.; Cole, D. L.; and Javaheri, F. 1987. Red leaf blotch of soybeans. *Plant Disease* **71**:113–118.

Hartman, G. L.; Manandhar, J. B.; and Sinclair, J. B. 1986. Incidence of *Colletotrichum* spp. on soybeans and weeds in Illinois and pathogenicity of *Colletotrichum truncatum*. *Plant Disease* **70**:780–782.

Heinrichs, E. A.; Lehman, P. S.; and Corss, I. C. 1976. *Nematospora coryli*, yeast-spot disease of soybeans in Brazil. *Plant Dis. Rptr.* **60**:508–509.

Helbig, J. B., and Carroll, R. B. 1982. Weeds as a source of *Fusarium oxysporum* pathogenic on soybean. *Phytopathology* **72**:707 (abstract).

Helbig, J. B., and Carroll, R. B. 1984. Dicotyledonous weeds as a source of *Fusarium oxysporum* pathogenic on soybean. *Plant Disease* **68**:694–696.

Hill, J. B.; Lucas, B. S.; Benner, H. I.; Tachibana, H.; Hammond, R. B.; and Pedigo, L. P. 1980. Factors associated with the epidemiology of soybean mosaic virus in Iowa. *Phytopathology* **70**:536–540.

Hill, J. H.; Bailey, T. B.; Benner, H. I.; Tachibana, H.; and Durand, D. P. 1987. Soybean mosaic virus: Effects of primary disease incidence on yield and seed quality. *Plant Disease* **71**:237–239.

Hirrel, M. C. 1983. Sudden death syndrome of soybean—a disease of unknown etiology. *Phytopathology* **73**:501 (abstract).

Hirrel, M. C. 1984. Influence of overhead irrigation and row width on foliar and stem diseases of soybeans. *Phytopathology* **74**:628 (abstract).

Hobbs, T. W.; Schmitthenner, A. F.; and Ellett, C. W. 1981. Diaporthe dieback of soybean caused by *Diaporthe phaseolorum* var. *caulivora*. *Phytopathology* **71**:226 (abstract).

Hobbs, T. W.; Schmitthenner, A. F.; Ellett, C. W.; and Hite, R. E. 1981. Top dieback of soybean caused by *Diaporthe phaseolorum* var. *caulivora*. *Plant Disease* **65**:618–620.

Iwaki, M.; Isogawa, Y.; Tsuzuki, H.; and Honda, Y. 1984. Soybean chlorotic mottle, a new caulimovirus on soybean. *Plant Disease* **68**:1009–1011.

Iwaki, M.; Roechan, M.; Hibino, H.; Tochihara, H.; and Tantera, D. M. 1980. A peristent aphidborne virus of soybean, Indonesian soybean dwarf virus. *Plant Disease* **64**:1027–1030.

Iwaki, M.; Thongmeearkom, P.; Honda, Y.; and Deema, N. 1983. Soybean crinkle leaf: A new whitefly-borne disease of soybean. *Plant Disease* **67**:546–548.

Iwaki, M.; Thongmeearkom, P.; Prommin, M.; Honda, Y.; and Hibi, T. 1982. Whitefly transmission and some properties of cowpea mild mottle virus on soybean in Thailand. *Plant Disease* **66**:365–368.

Johnson, H. W., and Chamberlain, D. W. 1954. Diseases of soybeans and methods of control. *USDA Cir. No. 931.*

Kennedy, B. W., and Tachibana, H. 1973. Bacterial diseases, in B. E. Caldwell, ed. *Soybeans: Improvement, Production and Uses.* American Society of Agronomy, Madison, WI, pp. 491–504.

Kuhn, C. W.; Demski, J. W.; and Harris, H. B. 1972. Peanut mottle virus in soybeans. *Plant Dis. Rptr.* **56**:146–147.

Kunwar, I. K.; Manandhar, J. B.; and Sinclair, J. B. 1986. Histopathology of soybean seeds infected with *Alternaria alternata*. *Phytopathology* **76**:543–546.

Kunwar, I. K.; Singh, T.; Machado, C. C.; and Sinclair, J. B. 1986. Histopathology of soybean seed and seedling infection by *Macrophomina phaseolina*. *Phytopathology* **76**:532–535.

Lakshminarayana, C. S., and Joshi, L. K. 1978. Myrothecium disease of soybean in India. *Plant Dis. Rptr.* **62**:231–234.

Laviolette, F. A., and Athow, K. L. 1971. Relationship of age of soybean seedlings and inoculum to infection by *Pythium ultimum*. *Phytopathology* **61**:439–440.

Leakey, C. L. A. 1964. Dactuliophora, a new genus of Mycelia Sterilia from tropical Africa. *Br. Mycol. Soc. Trans.* **47**:341–350.

Leath, S., and Carroll, R. B. 1982. Screening for resistance to *Fusarium oxysporum* in soybean. *Plant Disease* **66**:1140–1143.

Lin, M. T., and Hill, J. H. 1983. Bean pod mottle virus: Occurrence in Nebraska and seed transmission in soybeans. *Plant Disease* **67**:230–233.

Lockwood, J. L.; Yoder, D. L.; and Smith, N. A. 1970. *Thielaviopsis basicola* root rot of soybeans in Michigan. *Plant Dis. Rptr.* **54**:849–850.

McGawley, E. C.; Winchell, K. L.; and Berggren, G. T. 1984. Possible involvement of *Hoplolaimus galeatus* in a disease complex of 'Centennial' soybean. *Phytopathology* **74**:831 (abstract).

McGee, D. C., and Biddle, J. 1985. A comparison between isolates of *Diaporthe phaseolorum* var. *caulivora* from soybean seeds in Iowa and stem cankered soybeans in southern states. *Phytopathology* **75**:1332 (abstract).

McGee, D. C., and Biddle, J. 1987. Seedborne *Diaporthe phaseolorum* var. *caulivora* in Iowa and its relationship to soybean stem canker in the southern United States. *Plant Disease* **71**:620–622.

McLaughlin, M. R.; Mignucci, J. S.; and Milbrath, G. M. 1977. *Microsphaera diffusa*, the perfect stage of the soybean powdery mildew pathogen. *Phytopathology* **67**:726–729.

McLaughlin, M. R.; Thongmeearkom, P.; Goodman, R. M.; Milbrath, G. M.; Ries, S. M.; and Royse, D. J. 1978. Isolation and beetle transmission of cowpea mosaic virus (severe subgroup) from *Desmodium canescens* and soybeans in Illinois. *Plant Dis. Rptr.* **62**:1069–1073.

Manandhar, J. B.; Hartman, G. L.; and Sinclair, J. B. 1986. *Colletotrichum destructivum*, the anamorph of *Glomerella glycines*. *Phytopathology* **76**:282–285.

Marchetti, M. A.; Vecker, F. A.; and Bromfield, K. R. 1975. Uredial development of *Phakopsis pachyrhizi* in soybeans. *Phytopathology* **65**:822–823.

Martin, K. F., and Walters, H. J. 1982. Infection of soybean by *Cercospora kikuchii* as affected by dew, temperature and duration of dew periods. *Phytopathology* **72**:974 (abstract).

Mengistu, A., and Grau, C. R. 1986. Variation in morphological, cultural, and pathological characteristics of *Phialophora gregata* and *Acremonium* sp. recovered from soybean in Wisconsin. *Plant Disease* **70**:1005–1009.

Meyer, W. A., and Sinclair, J. B. 1972. Root reduction and stem lesion development on soybeans by *Phytophthora megasperma* var. *sojae*. *Phytopathology* **62**:1414–1416.

Meyer, W. A.; Sinclair, J. B.; and Khare, M. N. 1974. Factors affecting charcoal rot of soybean seedlings. *Phytopathology* **64**:845–849.

Milbrath, G. M., and Tolin, S. E. 1977. Identification, host range and serology of peanut stunt virus isolated from soybean. *Plant Dis. Rptr.* **61**:637–640.

Mishra, B., and Prakash, O. 1975. Alternaria leaf spot of soybean from India. *Indian Jour. Mycol. Plant Pathol.* **5**:95.

Narayanasamy, P., and Durairaj, P. 1971. A new blight disease of soybeans. *Madras Agric. Jour.* **58**:711–712.

Nicholson, J. F.; Dhingra, O. D.; and Sinclair, J. B. 1972. Internal seedborne nature of *Sclerotinia sclerotiorum* and *Phomopsis* sp. and their effects on soybean seed quality. *Phytopathology* **62**:1261–1263.

O'Neill, N. R.; Rush, M. C.; Horn, N. L.; and Carver, R. B. 1977. Aerial blight of soybeans caused by *Rhizoctonia solani*. *Plant Dis. Rptr.* **61**:713–717.

Otazu, V.; Epstein, A. H.; and Tachibana, H. 1981. Water stress effect on the development of brown stem rot of soybeans. *Phytopathology* **71**:247 (abstract).

Park, E. W., and Lim, S. M. 1985. Overwintering of *Pseudomonas syringae* pv. *glycinea* in the field. *Phytopathology* **75**:520–524.

Pataky, J. K., and Lim, S. M. 1981. Effects of septoria brown spot on yield components of soybean. *Phytopathology* **71**:248 (abstract).

Peterson, D. J., and Edwards, H. H. 1982. Effects of temperature and leaf wetness period on brown spot disease of soybeans. *Plant Disease* **66**:995–998.

Phillips, D. V. 1971. Influence of air temperature on brown stem rot of soybean. *Phytopathology* **61**:1205–1208.

Phillips, D. V. 1972. A soybean disease caused by *Neocosmospora vasinfecta*. *Phytopathology* **62**:612–615.

Phillips, D. V.; Vesper, S. J.; and Turner, J. T. Jr. 1983. *Verticillium nigrescens* associated with soybeans. *Phytopathology* **73**:504 (abstract).

Roane, C. W., and Roane, M. K. 1976. Erysiphe and Microsphaera as dual causes of powdery mildew of soybeans. *Plant Dis. Rptr.* **60**:611–612.

Roy, K. W. 1982. Seedling diseases caused in soybean by species of Colletotrichum and Glomerella. *Phytopathology* **72**:1093–1096.

Roy, K. W., and Abney, T. S. 1977. Antagonism between *Cercospora kikuchii* and other seedborne fungi of soybeans. *Phytopathology* **67**:1062–1066.

Roy, K. W., and McLean, K. 1984. Epidemiology of soybean stem canker in Mississippi. *Phytopathology* **74**:632 (abstract).

Roy, K. W., and Miller, W. A. 1983. Soybean stem canker incited by isolates of *Diaporthe* and *Phomopsis* spp. from cotton in Mississippi. *Plant Disease* **67**:135–137.

Rupe, J. C., and Ferriss, R. S. 1984. The effect of moisture on infection of soybean seed by *Phomopsis* sp. *Phytopathology* **74**:632 (abstract).

Saharan, G. S., and Gupta, V. K. 1972. Pod rot and collar rot of soybean caused by *Fusarium semitectum*. *Plant Dis. Rptr.* **56**:693–694.

Schiller, C. T.; Ellis, M. A.; Tenne, F. D.; and Sinclair, J. B. 1977. Effect of *Bacillis subtilis* on soybean seed decay, germination, and stand inhibition. *Plant Dis. Rptr.* **61**:213–217.

Schiller, C. T.; Hepperly, P. R.; and Sinclair, J. B. 1978. Pathogenicity of *Myrothecium roridum* from Illinois soybeans. *Plant Dis. Rptr.* **62**:882–885.

Schlub, R. L., and Lockwood, J. L. 1981. Etiology and epidemiology of seedling rot of soybean by *Pythium ultimum*. *Phytopathology* **71**:134–138.

Schlub, R. L.; Lockwood, J. L.; and Komada, H. 1981. Colonization of soybean seeds and plant tissue by Fusarium species in soil. *Phytopathology* **71**:693–696.

Schneider, R. W.; Dhingra, O. D.; Nicholson, J. F.; and Sinclair, J. B. 1974. *Colletotrichum truncatum* borne within the seedcoat of soybean. *Phytopathology* **64**:154–155.

Schwenk, F. W., and Nickell, C. D. 1980. Soybean green stem caused by bean pod mottle virus. *Plant Disease* **64**:863–865.

Schwenk, F. W., and Paxton, J. D. 1981. Trifluralin enhancement of Phytophthora root rot of soybean. *Plant Disease* **65**:435–436.

Seaman, W. L.; Shoemaker, R. A.; and Peterson, E. A. 1965. Pathogenicity of *Corynespora cassiicola* on soybeans. *Can. Jour. Bot.* **43**:1461–1469.

Sherwood, J. L., and Jackson, K. E. 1985. Tobacco streak virus in soybean in Oklahoma. *Plant Disease* **69**:727 (disease notes).

Short, G. E.; Wyllie, T. D.; and Bristow, P. R. 1980. Survival of *Macrophomina phaseolina* in soil and in residue of soybean. *Phytopathology* **70**:13–17.

Shortt, B. J.; Sinclair, J. B.; and Kogan, M. 1981. Soybean seed quality losses associated with bean leaf beetles and *Alternaria tenuissima*. *Phytopathology* **71**:1117 (abstract).

Sinclair, J. B. (ed.). 1982. *Compendium of Soybean Diseases*, 2nd ed. American Phytopathological Society, St. Paul, MN 104p.

Sinclair, J. B., and Shurtleff, M. C. (ed.). 1975. *Compendium of Soybean Diseases*. American Phytopathological Society, St. Paul, MN.

Sortland, M. E., and MacDonald, D. H. 1986. Development of a population of *Heterodera glycines* race 5 at four soil temperatures in Minnesota. *Plant Disease* **70**:932–935.

Southern Soybean Disease Workers. 1977. *Soybean Diseases Atlas*. Texas Agric. Ext. Serv. MP-1330.

Spilker, D. A.; Schmitthenner, A. F.; and Ellett, C. W. 1981. Effects of humidity, temperature, fertility and cultivar on the reduction of soybean seed quality by *Phomopsis* sp. *Phytopathology* **71**:1027–1029.

Stall, R. E., and Kucharek, T. A. 1982. A new bacterial disease of soybean in Florida. *Phytopathology* **72**:990 (abstract).

Stewart, R. D. 1957. An undescribed species of Pyrenochaeta on soybean. *Mycologia* **49**:115–117.

Stuckey, R. E.; Ghabrial, S. A.; and Reicosky, D. A. 1982. Increased incidence of *Phomopsis* sp. in seeds from soybeans infected with bean pod mottle virus. *Plant Disease* **66**:826–829.

Tachibana, H. 1971. Virulence of *Cephalosporium gregatum* and *Verticillium dahliae* in soybeans. *Phytopathology* **61**:565–568.

Tachibana, H.; Jowett, D. D.; and Fehr, W. R. 1971. Determination of losses in soybeans caused by *Rhizoctonia solani*. *Phytopathology* **61**:1444–1446.

Tachibana, H., and Shih, S. 1965. A leaf-crinkling bacterium of soybean. *Plant Dis. Rptr.* **49**:396–397.

Thompson, T. B.; Athow, K. L.; and Laviolette, F. A. 1971. The effect of temperature on the pathogenicity of *Pythium aphanidermatum, P. debaryanum*, and *P. ultimum* on soybean. *Phytopathology* **61**:933–935.

Thouvenel, J. C.; Monsarrat, A.; and Fauquet, C. 1982. Isolation of cowpea mild mottle virus from diseased soybeans in the Ivory Coast. *Plant Disease* **66**:336–337.

Vakili, N. G., and Bromfield, K. R. 1976. Phakopsora rust on soybean and other legumes in Puerto Rico. *Plant Dis. Rptr.* **60**:995–999.

Vesper, S. J.; Turner, J. T., Jr.; and Phillips, D. V. 1983. Incidence of *Verticillium nigrescens* in soybeans. *Phytopathology* **73**:1338–1341.

Walters, H. J. 1980. Soybean leaf blight caused by *Cercospora kikuchii*. *Plant Disease* **64**:961–962.

Walters, H. J., and Martin, K. F. 1981. *Phyllostica sojaecola* on pods of soybeans in Arkansas. *Plant Disease* **65**:161–162.

Wilcox, J. R., and Abney, T. S. 1973. Effects of *Cercospora kikuchii* on soybeans. *Phytopathology* **63**:796–797.

Wyllie, T. D., and Rosenbroek, S. M. 1985. Crop rotation as a means of controlling populations of *Macrophomina phaseolina* in Missouri soils. *Phytopathology* **75**:1348 (abstract).

19

Diseases of Sugarbeets
(*Beta vulgaris* L.)

_____BACTERIAL CAUSES_____

Bacterial Leaf Spot

CAUSE: *Pseudomonas syringae* van Hall (syn. *P. apatata* (Brown & Jamieson) Stevens). Bacteria survive on living plants, seed, and organic matter in soil. Dissemination is by seed but a wound is necessary for infection. Some research suggests that bacteria penetrate roots and move upward in the plant. Optimum temperatures for bacterial growth are 25–30°C.

DISTRIBUTION: Europe, Japan, and the United States.

SYMPTOMS: A seedling blight may occur. Dark brown to black streaks and spots occur on leaves and infrequently on petioles and seedstalks.

CONTROL:
1. Plant disease-free seed.
2. Apply seed treatment fungicides.

Bacterial Pocket

This disease is also called bacterial canker.

CAUSE: *Xanthomonas beticola* (Smith, Brown & Town) Burkh. survives in soil for long periods of time. The bacteria enter the root through wounds at or near the crown level. Bacteria are disseminated by any means that will move soil such as irrigation water and machinery.

DISTRIBUTION: Colorado, Maryland, Michigan, New Mexico, Utah, Virginia, Wisconsin, and Wyoming.

SYMPTOMS: Normally only scattered plants are affected but occasionally a whole field of plants may show symptoms if a hailstorm has occurred. Galls develop just at the surface of the ground on the crown but may also occur on petioles and lower on the root. The central portion of the galls are water-soaked and yellow due to the presence of the bacteria. As the galls increase in size, the sugarbeet becomes more enlarged at the top of the root and galls become rough and fissured. The surface of the root below the galls usually becomes ridged. In later stages of the disease an abnormal number of leaves may develop.

CONTROL:
1. Plant sugarbeets only every four or five years in the same soil.
2. Provide proper soil fertility.
3. Avoid mechanical injury to plants.

Bacterial Vascular Necrosis and Soft Rot

CAUSE: *Erwinia carotovora* (Jones) Bergey et al. pv. *betavasculorum* Thompson et al. The bacteria survive in unharvested beets and for a time in soil. Weeds may also be a means of overwintering. Bacteria are probably disseminated by soil and are deposited in crowns. Bacteria invade the vascular tissue of the petiole and roots through an injury. Infection usually begins in the crown and progresses toward the root tip. Disease occurs in low, poorly drained areas of fields and in furrow irrigated fields. Temperatures between 25°C and 30°C favor disease development. Young plants are the most susceptible. Other factors contributing to disease are surplus or adequate nitrogen compared to plants with nitrogen deficiency, wide spacing that results in rapid growth, and sprinkler irrigation that results in keeping wounds moist for a long time and disseminating bacteria.

DISTRIBUTION: Arizona, California, Idaho, Texas, and Washington.

SYMPTOMS: Diseased beets are not easily recognized in the field until they are severely rotted. Symptoms include gray to black watery internal rot of the taproot. Vascular bundles in roots are necrotic or discolored. Black longitudinal lesions run up the petioles. A black or foamy white exudate may occur on split crowns of roots and on petiole lesions. Internal infected root tissue surrounding diseased vascular bun dles turns pink to red-brown when exposed to air. Severely infected roots may become hollowed but plants remain alive.

CONTROL:
1. Avoid practices that deposit soil in crowns or cause injury to crowns and petioles.
2. Provide adequate drainage.
3. Plant resistant cultivars.
4. Plant early and maintain optimum stands.
5. Use judicious amounts of nitrogen fertilizer.

Crown Gall

CAUSE: *Agrobacterium tumefaciens* (E. F. Sm. & Town.) Conn. is a bacterium that is a soil inhabitant and pathogen of several hosts. The bacteria enter the root through wounds caused by several agents. The presence of the bacteria causes hypertrophy of sugarbeet tissues. The production of indoleacetic acid by *A. tumefaciens* is associated with gall formation.

DISTRIBUTION: Generally distributed wherever sugarbeets are grown.

SYMPTOMS: At first, small, wartlike growths form on the root and soon develop into galls of various sizes. Seldom are sugarbeets killed but growth may be stunted and sugar content may be lowered. Galls are harvested along with the root and consequently there is little loss in total weight.

CONTROL: No control is necessary since crown gall is seldom serious.

Scab

CAUSE: *Streptomyces scabies* (Thaxt.) Waksman & Henrici (syn. *Actinomycetes scabies* (Thaxt.) Gussow) is an actinomycetes that survives in soil and residue as mycelium and spores. Infection is by hyphae growing into minute wounds or lenticels. Infection occurs more frequently in dry, light, sandy soils that have a neutral or slightly alkaline pH.

DISTRIBUTION: Generally distributed wherever sugarbeets are grown; however, scab is a minor problem.

SYMPTOMS: Small, round spots occur on roots that enlarge, turn brown, with the epidermis rupturing. Eventually raised corky areas occur that vary in size and shape and resemble miniature-raised terraces. The color varies from a gray-white to a dark tan. Sometimes the scabby area may be pitted or pocked; however, the injury is usually superficial and no decomposition occurs. Occasionally, taproots may have brown, corky lesions that can restrict growth and cause roots to be turnip-shaped.

CONTROL: No control is necessary since the appearance rather than the actual value of the beet is affected.

Other Bacteria

Sugarbeet is a symptomless host for *Corynebacterium sepedonicum* (Spieck. & Kotth.) Skapt. & Burkh. (syn. *Clavibacter michiganense* subsp. *sepedonicum* (Spieck. & Kotth.) Davis et al.).

FUNGAL CAUSES

Alternaria Leaf Spot

CAUSE: *Alternaria alternata* (Fr.) Keissler (syn. *A. tenuis* Nees.) is a saprophyte on residue and overwinters as conidia and mycelium. Conidia are windborne and mainly infect plants that are deficient in magnesium, manganese, phosphorous, or potassium. Plants infected with western yellows virus are also likely to be infected. *Alternaria brassicae* (Berk.) Sacc. also infects yellowed tissue but is a foliar pathogen of some hybrids. This fungus causes disease at 7–10°C and high humidity.

DISTRIBUTION: Generally distributed wherever sugarbeets are grown but the disease is more common in Europe.

SYMPTOMS: Symptoms usually only occur late in the growing season. Initially lesions are circular to irregularly shaped areas 2–10 mm in diameter. They are dark brown in color and frequently zonate. Brown necrotic spots may coalesce to form irregularly shaped areas on leaves.

CONTROL: No control is usually necessary since vigorous, healthy plants are ordinarily not infected.

Beet Rust

CAUSE: *Uromyces betae* (Tul.) ex. Kickx is an autoecious rust that produces pycnia, aecia, uredia, and telia only on sugarbeet and other varieties of *Beta vulgaris*. The fungus probably overwinters as uredia and telia on volunteer plants and infected residue. It may also be seedborne. Uredia and telia are produced during damp, cool weather in the summertime. Aecia may be produced in autumn on sugarbeet, or in spring on overwintered, infected plants, usually second-year beets that are being grown to produce seed. Optimum disease development occurs during moist weather at 15–22°C.

DISTRIBUTION: Asia, Europe, Arizona, California, New Mexico, and Oregon; however, it is of minor economic importance.

SYMPTOMS: Initially pustules appear as slightly raised areas with a yellow halo around them. The epidermis ruptures to expose urediospores and teliospores in red-brown pustules that are about 2 mm in diameter on leaves, seedstalks, and petioles. Toward the end of the growing season, pustules become dark brown. Pustules may be randomly dispersed or aggregated in rings, usually on green tissue surrounded by a yellow halo.

CONTROL: Plant resistant cultivars.

Beet Tumor

This disease is also called crown wart.

CAUSE: *Urophlyctis leproides* (Trabut) Magn. survives as resting sporangia in residue in the soil. In Europe the disease is associated with wet soils in irrigated areas.

DISTRIBUTION: Argentina, Europe, North Africa, and California. It is not economically important.

SYMPTOMS: Galls occur on leaf blades, petioles, and crowns of beet. Galls on leaf blades and petioles are green-brown, rough, and less than 1 cm in diameter; however, galls may coalesce to form larger galls. The galls usually extend through the leaf blade. Affected leaves are stunted and malformed. Only older leaves are infected.

Galls on crown are attached by a narrow base and are sometimes 8–10 cm in diameter on mature sugarbeets. Galls on crowns vary from red to green-brown. When galls are cut open, small cavities filled with brown spores are observed. No galls form on taproots.

CONTROL: No control is reported.

Black Root

CAUSE: *Aphanomyces cochlioides* Drechs. survives as oospores in plant residue in soil. Some weeds may aid in the survival and buildup of inoculum in the soil. *Pythium* spp. and *Rhizoctonia* sp. are two fungi that are implicated in the black root complex, causing a damping off of seedlings. However, since *A. cochlioides* infects plants throughout the growing season, this will be the major causal agent discussed here. During warm (22–28°C), wet soil, oospores presumably germinate to form sporangia in which sporangiospores are produced. After three or four hours, each sporangiospore produces a papilla through which a zoospore is released. The zoospore swims to the root and germinates to form a germ tube that penetrates the host. Oospores are eventually formed in diseased tissue.

DISTRIBUTION: Canada, Europe, Japan, and the United States. Black root is most severe in the Great Lakes region of the United States.

SYMPTOMS: Seedlings may be killed but the disease is most severe on older plants. Initially leaves are yellowed and mottled. Plants remain stunted. Most lateral roots are dead and blackened. The main root will have groups of dead, black, lateral roots due to new roots being produced as others die; the new roots are in turn killed. The terminal of the root may have dark lesions, with numerous lateral roots replacing the killed terminal, causing a tasseled appearance.

CONTROL:
1. Do not rotate sugarbeets with legumes.
2. Control weeds.
3. Plant resistant cultivars.
4. Maintain well-drained soil.

Botrytis Storage Rot

This disease is also called clamp rot.

CAUSE: *Botrytis cinerea* Pers. ex Fr. survives as sclerotia in soil or in infected residue. Sclerotia germinate to produce mycelium on which conidiophores and conidia are borne. The conidia are windborne and enter sugarbeet roots through any kind of wound. Conidia are abundantly produced on decayed tissue. Eventually sclerotia develop in mycelium on the outside of roots.

DISTRIBUTION: Generally distributed wherever sugarbeets are grown.

SYMPTOMS: Rotted tissue is normally dark brown or black. In the latter case, *Phoma betae* Frank may also be present. Dark gray fungal growth consisting of mycelium, conidia, and conidiophores will be present on rotting tissue. Eventually dark brown to black, round sclerotia, 2–5 mm in diameter, form in the mycelium on the outside of the root.

CONTROL: No feasible control is presently available. Research is being conducted on breeding for resistance and the use of fungicides in storage piles.

Cercospora Leaf Spot

CAUSE: *Cercospora beticola* Sacc. overwinters as spores and stromata on residue in the soil and on seed when disease is severe. It can also infect several weed hosts and survive in their infected residue. Included as other hosts are pigweed, *Amaranthus retroflexus* L.; lambs quarter, *Chenopodium album* L.; prickly lettuce, *Lactua scariola* L.; common mallow, *Malva neglecta* Wallr.; sweet clover, *Melilotus* spp.; common plaintain, *Plantago major* L.; curly dock, *Rumex crispus* L.; and dandelion, *Taraxacum* spp.

In the spring or early summer, conidia are produced from mycelium and are windborne, splashed by water, or possibly carried by insects to leaves of sugarbeets. Conidia germination and leaf penetration occur in the presence of free water and at optimum temperatures between 25°C and 35°C. Penetration occurs through stomata on the leaf surface. Secondary infection occurs through stomata on the leaf surface from conidia that are produced on leaf spots within 7–21 days.

DISTRIBUTION: Generally distributed wherever sugarbeets are grown. The disease is important only in areas of high relative humidity or abundant moisture.

SYMPTOMS: Spots occur mostly on leaves and occasionally on petioles. Initially small, light brown spots appear on older outer leaves. The circular spots rapidly increase in size to 3–5 mm in diameter and become brownish or purplish in color. Numerous spots may coalesce, forming large necrotic areas. As spots mature, the center becomes gray to black due to production of conidia and conidiophores, and the border turns brown to bright purple.

When spots are numerous, leaves turn yellow and die. In cases of severe disease, the destruction of outer leaves and production of new ones causes the crown to become elongated and cone-shaped. Blighted leaves collapse and fall to the ground but remain attached to the crown. During periods of high humidity, necrotic spots become gray due to production of conidia and conidiophores.

Similar lesions appear on petioles; however, these are normally longer and more elliptical than lesions on leaves.

The upper portion of root that is not covered by soil may also have spots. Such spots are slightly sunken, circular to oval, 2–5 mm in diameter, and initially brown-purple; later spots become gray in the center. Tissue beneath the lesions is brown to a depth of about 1.5 mm.

CONTROL:
1. Plant resistant cultivars.
2. Apply foliar fungicides during the growing season before disease becomes severe.
3. Rotate to nonhosts for two to three years.
4. Plow under infected residue.
5. Locate new fields at least 100 m from old fields.

Charcoal Rot

CAUSE: *Macrophomina phaseolina* (Tassi) Goid. (syn. *M. phaseoli* (Maubl.) Ashby). The anamorph is *Sclerotium bataticola* Taub. The fungus survives as sclerotia in soil and in infected residue. The sclerotia germinate to form a germ tube that penetrates the host. Sclerotia are eventually formed in diseased tissue. Pycnidia may be produced and conidia may be subsequently windborne to plants but this is apparently uncommon. Dissemination of the fungus is most commonly by any means that moves infested soil. The disease is most severe during high soil and air temperatures (optimum 31°C). Only stressed, weakened, or injured plants are infected.

DISTRIBUTION: The hot interior valleys of California, India, and the USSR.

SYMPTOMS: Symptoms first occur on half-grown and mature sugarbeets. Leaves rapidly wilt then gradually turn brown and die, but remain attached to the crowns. Brown-black, irregular-shaped, necrotic areas develop only on the crown or upper part of the root. Diseased areas have a silvery sheen, with the outer portion of the oldest infected area being thin and loosely attached to the underlying tissue. When this outer layer is removed, numerous black sclerotia, resembling charcoal, can be found underneath. When the whole root becomes infected, it becomes brown-black with masses of sclerotia forming in cavities just beneath the outer layer of the root. Roots may shrink and become mummified.

CONTROL: No control is reported.

Dry Rot Canker

Dry rot and Rhizoctonia crown rot are both caused by *Rhizoctonia solani*. However, the symptoms are distinct and the two diseases are discussed separately.

CAUSE: *Rhizoctonia solani* Kuhn survives as bulbils or thickened hyphae in residue. The fungus resumes growth to form mycelium that penetrates the host. Dry rot is caused by a weak strain of *R. solani* that infects under dry conditions and high temperatures (30–35°C). The strain of *R. solani* causing dry rot is apparently different from those causing crown rot and root rot.

DISTRIBUTION: United States.

SYMPTOMS: Initially, leaves wilt during the warmest part of the day but regain turgidity at night. Eventually the leaves wilt permanently and turn brown. Circular brown lesions that have alternate dark and light concentric rings are scattered over the root surface. The lesions develop into cavities filled with brown, pithy remains of the root tissue. There is a sharp brown line between diseased and healthy tissue. In the latter stages of the disease all the vascular tissue becomes brown and darker than normal.

CONTROL:
1. Follow recommended tillage and fertility practices.
2. Rotate with small grains and corn.
3. Avoid hilling up soil around plants.

Fusarium Tip Rot

CAUSE: *Fusarium radicicola* Wr. survives in the soil as chlamydospores or as mycelium in infected residue.

DISTRIBUTION: United States.

SYMPTOMS: The first symptoms appear about midseason when the leaves start to wilt. The tip of the root may be rotted with the white mycelium of the fungus growing on the surface. Eventually the interior of the roots will have a brown rot but the exterior will not show any symptoms.

CONTROL: No effective control is known.

Fusarium Yellows

CAUSE: *Fusarium oxysporum* Schlecht. f. sp. *betae* (Stewart) Snyd. & Hans. survives in soil and residue as chlamydospores, mycelium, and spores. In the spring or summer, chlamydospores germinate in the presence of roots to form a germ tube that penetrates the root and grows into the xylem system. The fungus produces toxic substances that affect the xylem tissue, preventing water from reaching the top of the plant. After plant death or maturity, the fungus grows out of the vascular tissue and sporulates on the plant surface. Chlamydospores are formed either from conidia or mycelium and are able to survive in the soil for several years.

DISTRIBUTION: Europe, India, and the United States.

SYMPTOMS: About midseason, yellowing occurs between the veins of the largest leaves with occasionally only half of the leaf being affected. The large veins and a narrow border of the leaf usually remain green. As the disease progresses, affected leaves wilt during warm weather and touch the soil surface. Eventually parts of these leaves will turn brown and dry out. As the leaves die, the blade pulls upward along the midrib, the margins dry up and turn brown with the leaf tips bending to one side. Leaves utimately become heaped around the crown.

The yellowing and progression of disease symptoms will eventually occur on the younger heart leaves. Roots may not show any external symptoms. In cross sections of the root, the vascular tissue has a gray-brown discoloration. Seed beets have seedstalk wilt and die prematurely.

CONTROL: No effective control exists other than to rotate sugarbeets no less than every four years in the same field.

Downy Mildew

CAUSE: *Peronospora farinora* (Fr.) Fr. f. sp. *betae* Byford (syn. *P. schachtii* Fckl.) overwinters as oospores and mycelium in residue, wild and volunteer *Beta* spp., and in seed. Under cool, moist conditions oospores germinate to produce mycelium on which sporangia are borne. Sporangia are also produced from overwintered mycelium. In either case, sporangia are windborne to foliage where optimum germination occurs between 4°C and 10°C and optimum infection occurs between 7°C and 15°C in the presence of a relative humidity of 80 percent or above. Secondary inoculum is provided by production of sporangia that are windborne to leaves during high relative humidity. Oospores are eventually produced in infected leaves.

DISTRIBUTION: Europe and the United States, primarily California, Oregon, and Washington.

SYMPTOMS: The sugarbeet usually reaches the 4–10 leaf stage before infection occurs. Seedlings may be killed but plants usually have the primary bud destroyed, stopping growth. Symptoms occur on cotyledons, primary leaves, and growing points. Primary leaves in the center of the crown may become infected, causing small, distorted, thickened, light green, and puckered leaves with downward-curled margins. Both leaf surfaces may be covered with a gray downy growth, which is the conidia, and conidiophores of *P. farinora*. Leaves may either appear lighter green than normal or may have a distinct purple cast. Affected leaves may wilt and die. If conditions become unfavorable for downy mildew, secondary heartrot may occur, with chlorosis of older leaves resembling virus yellows infection.

When older leaves become infected, light green, irregular spots that vary in size from 12–25 mm in diameter appear on the upper leaf surfaces. The lower leaf surface opposite the spot is covered with the gray growth consisting of conidia and conidiophores. Infected seed beets may have stunted and distorted young lateral branches. Mycelium and oospores may develop within seed clusters.

CONTROL:
1. In the United States, plant resistant cultivars.
2. In Europe, separate seed crops from root crops by 1.0–1.5 km.

Penicillium Storage Rot

CAUSE: *Penicillium claviforme* Banier. Conidia are windborne and enter roots through wounds and tissue rotted by *Phoma betae* Frank. *Penicillium funiculosum* Thom., *P. variabile* Sopp, *P. stoloniferum* Thom, *P. bordzilowskii* Morotchkovsky, *P. expansum* Lk., *P. duclauxi* Delacroix, *P. rubrum* Stoll, and *P. cyclopium* Westling have also been reported as a storage rot pathogen.

DISTRIBUTION: Generally distributed wherever sugarbeets are grown.

SYMPTOMS: White mycelial growth, resembling tufts, consisting primarily of conidiophores, conidia, and sparse mycelium growing on brown, rotted tissue.

CONTROL: No feasible control is presently available. Research is being conducted on breeding for resistance and the use of fungicide in storage piles.

Phoma Leaf Spot

CAUSE: *Phoma betae* Frank infects sugarbeets during all growth stages. The teleomorph, *Pleospora bjoerlingii* Byford, is found in Europe but apparently is not common in the United States or Canada.

The fungus overwinters as mycelium and conidia on and in seed, and as mycelium and pycnia for up to 26 months in soil and in residue in the soil. In Europe it also overwinters as perithecia within infected residue.

Conidia are produced in pycnidia during moist weather and are exuded in a gelatinous matrix. Dissemination is mainly by splashing rain. Where the teleomorph is present, ascospores are produced in perithecia and are windborne to hosts. Pycnidia develop in spots and conidia are produced that act as secondary inoculum. The disease is most severe during periods of high humidity and temperatures of 15–32°C.

DISTRIBUTION: Generally distributed wherever sugarbeets are grown. Damage is usually small on sugarbeets grown for sugar.

SYMPTOMS: Only the older, outer leaves are infected. Initially, small, light-colored spots develop that soon become light brown. As they enlarge, alternate light and dark brown concentric rings develop in the spots, with at least one dark brown ring being prominant. Spots are round to oval (1–2 cm in diameter), with black pycnidia developing in the dark rings. Spots may coalesce to form large necrotic areas.

CONTROL:
1. Treat seed with a seed-protectant fungicide.
2. Control common lambs quarter (*Chenopodium album*) as this weed is also a host for *P. betae*.
3. Rotate sugarbeets every four years in infested soil.

Phoma Root Rot and Storage Rot

This disease is also called blackleg.

CAUSE: *Phoma beta* Frank. For further information see Phoma Leaf Spot. Presumably seed is the major source of inoculum by conidia. Perithecia of *Pleospora bjoerlingii* Byford overwinter in residue and produce ascospores that also initiate infection. Weather and soil conditions unfavorable to normal growth favor infection by *P. betae*. Only weakened plants are infected. Root rot is favored by temperatures of 5–20°C.

DISTRIBUTION: Generally distributed wherever sugarbeets are grown.

SYMPTOMS: Sugarbeets affected by this disease usually occur randomly over the field. Occasionally many plants in a limited area will be infected. The first visible symptom is a wilting of the leaves, preceded by small brown spots on the upper root near the crown. The outer surface of the root remains unbroken while the rot develops just beneath it. Spots are soft and watery and spread to cover a large area. Often the center of the infected area is dark brown and pithy but the margins are soft and watery. Black vertical growth fissures may occur on the hypocotyl, likely from soilborne inoculum. Tan growth fissures are not a symptom.

Those sugarbeets that survive death in the field and are delivered into storage piles will act as centers of infection. The center of the crown is very susceptible to rot and removal of part of the crown during harvest exposes this tissue to infection by pathogens. About 80 days after roots have been harvested and stored, rot begins. The first indication of rot is in the center of the crown. From here, it spreads in a conelike pattern into the adjacent crown and down into the main taproot. Rotted tissue is black or dark brown, with occasional pockets lined with white mycelium of the fungus.

CONTROL:
1. Treat seed with a seed-protectant fungicide to prevent seedborne infection.
2. Practice cultural methods that promote vigorous growth of host.
3. Rotate sugarbeets every four years in infected soil.

Phymatotrichum Root Rot

This disease is also called Texas root rot.

CAUSE: *Phymatotrichum omnivora* (Dug.) Henneb. (syn. *P. omnivorum* (Shear) Duggar) survives by means of sclerotia in the soil. Infection occurs when temperatures exceed 28°C.

DISTRIBUTION: Southwestern United States and northern Mexico.

SYMPTOMS: Initially leaves are yellowed or bronzed. This is followed by a sudden wilting of plants. A thin, feltlike layer of yellowish mycelium develops on the root system. Eventually roots develop a yellow to tan rot. Occasionally, crustlike spore mats develop on the soil surface.

CONTROL: Do not plant sugarbeets in infected soil.

Phytophthora Root Rot

CAUSE: *Phytophthora drechsleri* Tucker presumably overwinters as oospores in the soil or in infected residue. Oospores usually germinate in the presence of moisture to form sporangia in which zoospores are borne. The zoospores swim to roots where they germinate to form a germ tube that penetrates the plant. The disease normally occurs in wet, poorly drained soils and at high temperatures (28–31°C optimum).
   Phytophthora megasperma Drechs. has been reported to cause a similar root rot in England.

DISTRIBUTION: Iran and the United States. Phytophthora root rot occurs mostly in heavy, poorly drained soils.

SYMPTOMS: The first noticeable symptom is a wilting of the leaves during the warmest part of the day. The leaves may recover their turgidity at night. Necrotic spots appear on the root, usually some distance below the crown, or only the tip of the root may be infected. Later the whole lower part of the root becomes rotted. Adventitious lateral roots often appear above infected points. Frequently the rotted portion breaks down so completely that the lower part of the root is no more than a mass of discolored vascular strands, resembling the end of a frayed rope. The advancing edge of the necrosis is a narrow band of black-brown tissue that is sometimes separated from healthy tissue by a narrow, light, buff band. Small empty cavities frequently occur in the necrotic tissue.

CONTROL: No acceptable control exists except to plant sugarbeets in well-drained soils.

Powdery Mildew

CAUSE: *Erysiphe polygoni* DC. ex Merat (*E. betae* (Vanha) Weltzien) overwinters as mycelium or haustoria in crowns of *Beta* spp. in southwestern sugarbeet-growing areas. In northwestern states the fungus probably survives in axillary bud tissue of seed beets. During the growing season leaves of sugarbeets initially become infected by conidia being windborne northward and eastward. Secondary inoculum is provided by conidia being produced on infected leaves and windborne to healthy leaves. Disease severity is favored by warm (optimum temperature 25°C), dry weather. Conidia germinate at high relative humidity but production and viability are favored by 30–40 percent relative humidity. Disease severity is greater on older plants but this could be due to reduced air circulation, light, and temperature and an increase in humidity under denser canopy of older sugarbeets. Susceptibility is also increased by infection with beet-yellowing viruses. Different races of *E. polygoni* may exist on sugarbeet.

DISTRIBUTION: Europe and the United States.

SYMPTOMS: Symptoms in the United States appear two to six months after planting. The most distinguishing symptom is a white powdery-appearing material that is the mycelium, conidiophores, and conidia of *E. polygoni* on the leaf surface. Leaves may be prematurely killed and turn brown but with the continued presence of the powdery-appearing fungal structures. The fungus growth is mostly superficial on the leaf surface and does not penetrate very far into the plant cells. In Europe, tiny dark specks (cleistothecia) are found in mycelium. These have been found once in the United States.

CONTROL:
1. Apply foliar fungicides during period of rapid root growth.
2. Some cultivars are less susceptible than others to infection.

Pythium Damping Off

CAUSE: *Pythium* spp. including *P. intermedium* deBary, *P. ultimum* Trow., and *P. aphanidermatum* (Edson) Fitzp. The life cycles are discussed under Pythium root rot. Disease is favored by cool (12–20°C), wet soil with the exception of *P. aphanidermatum*, which favors warm, wet soils. Thickly planted seeds are also more susceptible.

DISTRIBUTION: Generally distributed wherever sugarbeets are grown.

SYMPTOMS: Seeds are soft and mushy, light brown, and often overgrown with white mycelium of Pythium or secondary fungi. Seedlings that are killed preemergence have a light brown, soft, water-soaked rot. Seedlings may also be killed postemergence. Plants recover that survive until secondary thickening starts in the hypocotyl.

CONTROL: Treat seed with a seed-protectant fungicide.

Pythium Root Rot

CAUSE: *Pythium aphanidermatum* (Edson) Fitzp. and *P. deliense* Meurs. *Pythium* spp. are fungi that survive as oospores in the soil. Most *Pythium* spp. behave in a manner similar to the following: The oospores germinate in the presence of water and release zoospores that swim to roots. Zoospores germinate to form a germ tube that penetrates the host. Conidia, sporangia, and oospores are formed in rotted tissue. Secondary infection is by further production of zoospores within sporangia that are spread by water to adjacent plants. Sometimes sporangia and conidia germinate directly to form a germ tube that penetrates the host. Root rot caused by *P. aphanidermatum* is more serious under high soil temperature (greater than 26°C at the 10 cm soil depth), excessive moisture, and in heavy, poorly drained soils with a high pH.

DISTRIBUTION: Arizona, California, Colorado, Texas, and Iran.

SYMPTOMS: The disease generally becomes evident later in the growing season. First the older leaves become yellow, wilt, and die while the disease progresses to the younger leaves toward the heart. Leaf petioles have brown, sunken necrotic areas that extend from the petiole base to the blade. Eventually all leaves die and the crown and roots become infected. The diseased portion gradually extends internally from the crown down toward the root tip and from secondary root infections. The diseased root tissue is somewhat firm.

CONTROL: There is no acceptable control other than to ensure a field is properly drained.

Ramularia Leaf Spot

CAUSE: *Ramularia beticola* Fautr. & Lambotte probably overwinters in milder climates as mycelium or conidia in residue in the soil. The fungus may also be seedborne. Conidia are produced during cool (17–20°C), humid conditions and are windborne to foliage.

DISTRIBUTION: British Columbia, Europe, and the United States, primarily northern California, Colorado, Oregon, and Washington.

SYMPTOMS: Older leaves are infected. The spots resemble those caused by *Cercospora beticola* Sacc. but are larger and more angular. Spots are circular, about 4–7 mm in diameter, white to light tan, and surrounded by a brown or purplish border. The spots have tufts of white mycelium growing in the center. Affected leaves turn yellow, then necrotic, and die.

CONTROL:
1. Plant cultivars that are resistant to Cercospora leaf spot. These cultivars are also resistant to Ramularia leaf spot.
2. Apply copper foliage fungicides before disease becomes severe.

Rhizoctonia Crown and Root Rot

This disease is also called brown rot. Dry rot canker and Rhizoctonia crown and root rot are both caused by *Rhizoctonia solani*. However, the symptoms are distinct and the two diseases are discussed separately.

CAUSE: *Rhizoctonia solani* Kuhn survives as bulbils, or thickened hyphae, in residue. The fungus resumes growth to form mycelium that penetrate the host through crowns or roots of leaf petioles. *Rhizoctonia solani* grows between plants along a row by growth of mycelium or dissemination of propagules by water or tillage. The teleomorph is *Thanatephorous cucumeris* (Frank) Donk.

Disease is most severe in heavy, poorly drained soils at a temperature of 25–33°C. Depositing soil in and around crowns (hilling) increases incidence and severity of root rot.

In Ohio, most multinucleate isolates are in either anastomosis group AG2 or AG4. AG2 isolates predominate in fine-texture soils whereas AG4 and AG2 isolates are either equal in number or AG4 predominates in coarser-textured soils. AG4 isolates are more virulent to sugar-beet seedlings whereas AG2 isolates are more virulent to six- to eight-week-old plants.

DISTRIBUTION: Generally distributed wherever sugarbeets are grown.

SYMPTOMS: The first symptoms generally occur on half grown or nearly mature sugarbeet roots. A darkening occurs at the base of the petiole that eventually becomes so weak at the point of infection that leaves fall to the ground, forming a rosette of dead leaves about the crown. The crown beomes rotted and eventually the whole root becomes a mass of brown rotted tissues. On the root surface, infected areas are dark brown to black; internally, tissue is a light to dark brown dry rot. Deep cracks often occur at or near crowns.

Sometimes rot may occur below the crown on the body of the root. Again, the first symptom would be a wilting of the foliage. Just the tip may be rotted off 15–20 cm below the soil surface.

CONTROL:
1. Follow recommended tillage and fertility practices.
2. Rotate with small grains and corn.
3. Avoid hilling up soil around plants.
4. Some breeding lines display levels of resistance.

Rhizoctonia Foliage Blight

CAUSE: *Rhizoctonia solani* Kuhn, teleomorph *Thanatephorous cucumeris* (Frank) Donk., probably survives as resting mycelium and sclerotia in residue and soil. During moist weather in the early part of the growing season, fungal propagules are splashed onto leaves. Disease occurs between 21°C and 25°C and a high relative humidity that approaches 100 percent. Half-grown and mature roots are resistant to infection. About three weeks after infection, basidia and basidiospores of *T. cucumeris* are produced on leaves. The basidiospores function as secondary inoculum and are blown by wind or splashed to surrounding plants. Dry weather in midseason usually prevents further disease development.

DISTRIBUTION: Colorado, Maryland, Michigan, Minnesota, Nebraska, Ohio, and Virginia.

SYMPTOMS: Black spots 6–12 mm in diameter are formed on leaves. The spots become necrotic, dry up, and break away from healthy tissue, giving a leaf a ragged appearance. A circular zone of secondary spots surrounds the original spot or the place where the spot had been. Using magnification, mycelium of Rhizoctonia can be seen growing on the leaf surface. About three weeks after infection, a filmy gray-white growth, which is the mycelium, basidia, and basidiospores of *T. cucumeris*, can be observed on the leaf surfaces. Because of this spread of secondary inoculum, plants may be infected in a 5–10 m diameter circle, with the original infected plant in the center.

CONTROL: No practical control other than to rotate sugarbeets so that they are grown only once every four or five years in the same soil.

Rhizopus Root Rot

CAUSE: *Rhizopus stolonifer* (Ehr. ex Fr.) Vuill and *R. arrhizus* Fisher survive in soil and residue as spores and mycelium. The fungi are generally considered to be weak parasites that normally infect plants through crown wounds or are predisposed by excess soil moisture. Low temperatures (14–16°C) favor infection by *R. stolonifer* and warm temperatures (30–40°C) favor infection by *R. arrhizus*. Plants are infected after thinning. *Rhizopus oryzae* Went & Prin. Geerl. is also reported to be a root rot pathogen.

DISTRIBUTION: Canada, Italy, United States, and the USSR.

SYMPTOMS: Leaves wilt during the warmest part of the day but usually recover their turgidity at night. As the disease progresses, leaves permanently wilt and are transformed into a dry brittle rosette. Superficial gray-brown necrotic areas appear on the base of roots and gradually spread upward toward the crown. Root tissue becomes dark and spongy and eventually the entire root turns black. White mycelium that later turns dark appears on the root surface. Eventually the causal fungi grow deeper into the root, causing internal cavities that become filled with a clear liquid, rich in acetic-like acid.

Symptoms of root rot caused by *R. arrhizus*, have been described as a foamy white exudate often exuding from crowns of dead and dying plants. Diseased root tissue is a soft to spongy texture and varies in color from tan to black.

CONTROL: There is no suitable control other than to maintain proper soil drainage and avoid excessive injury.

Sclerotium Root Rot

This disease is also called southern Sclerotium root rot and southern root rot.

CAUSE: *Sclerotium rolfsii* Sacc. survives as sclerotia in the soil. Sclerotia germinate to produce a germ tube that penetrates the host. Disease development is favored by high temperatures (25–35°C) and moist soil. The fungus is disseminated by anything such as water or machinery that moves soil infested with sclerotia. Sclerotia are formed in mycelium growing over plant surfaces later in the growing season. The teleomorph reported here is *Athelia rolfsii* (Curzi) Tu & Kimbrough; however, *Pellicularia rolfsii* West is the reported teleomorph on other hosts.

DISTRIBUTION: Czechoslovakia, Japan, Korea, Mediterranean countries, Middle East, and southwestern United States.

SYMPTOMS: A general unthrifty appearance occurs to the top part of the plant with the leaves eventually wilting. The root surface is covered with a white, cottony, mycelial growth that extends even into the soil surrounding the root. Scattered throughout the mycelial growth are the small round sclerotia that resemble either white, tan, or brown mustard seeds.

CONTROL:
1. Soils that are high in numbers of sclerotia should be taken out of sugarbeet production and cropped to a nonsusceptible host.
2. Do not put dump dirt on fields that are to be cropped to sugarbeets.
3. Apply heavy amounts of nitrogen to soil.

Seedling Rust

CAUSE: *Puccinia subnitens* Diet. is a dioecious rust. The pycnial and aecial stages occur on sugarbeet; the uredial and telial stages occur on saltgrass, *Distichlis stricta* (Torr.) Rydb.

DISTRIBUTION: USSR and the Rocky Mountain states in the United States. It is of no economic importance.

SYMPTOMS: Generally symptoms occur only on the lower surface of cotyledons and infrequently on the first true leaves. Bright yellow-orange aecial pustules are aggregated in rings on the lower leaf surface. Pycnia may be present on the upper leaf surface.

CONTROL: None is necessary.

Verticillium Wilt

CAUSE: *Verticillium albo-atrum* Reinke & Berth. survives as dark resting mycelium in residue or soil. The fungus enters through lateral roots and grows into the vascular system. Conidia are formed on dead tissue but probably perform a minor role in spread of the pathogen.

DISTRIBUTION: Europe, Colorado, Idaho, Nebraska, and Washington.

SYMPTOMS: Leaves at first become yellow. Eventually the outer leaves wilt and dry up while the inner leaves become twisted and deformed. Roots when cut in cross section may show a slight to sometimes dark browning of the vascular system. Lateral roots through which the fungus entered are usually black and water-soaked; however, little or no root rot occurs.

CONTROL: Rotate sugarbeets for several years with other crops.

Violet Root Rot

CAUSE: *Rhizoctonia crocorum* DC. ex Fr., teleomorph *Helicobasidium purpureum* Pat. overwinters as sclerotia in soil or on roots of weed hosts. Sclerotia germinate late in the growing season to form mycelium that grows over and through the soil to sugarbeet roots. During this time, basidia form on the mycelial mat. When the fungus contacts a root, an infection cushion forms, from which mycelium penetrates the root and eventually grows over it. This mycelial stage is referred to as the *R. crocorum* stage. The fungus spreads from plant to plant by growth of mycelium through soil. Eventually sclerotia are formed within the mycelium to function as survival structures. Reports of the effects of soil moisture and pH are conflicting, with some workers stating disease is more prevalent under wet soil conditions and high soil pH. Others maintain these conditions are not a factor in disease development. Optimum temperature for fungal growth is 13°C.

DISTRIBUTION: Europe and the United States.

SYMPTOMS: Violet root rot is a late-season disease in the United States. Usually all beets in an area will be affected. Initially plants wilt and leaves become a light green. Purplish spots occur on the root surface and a purple fungal growth develops near the root tip that may eventually cover the entire root. The area of the root covered by the fungus is somewhat depressed. The root in cross section shows a sharp line between healthy tissue and that of the surrounding diseased tissue, which is brown. Diseased roots carry excessive amounts of soil.

CONTROL: Most control measures are relatively ineffectual; however, the following may partially aid in controlling violet root rot:
1. Rotate sugarbeets with other crops.
2. Control weeds in fields since these may serve as a host.

_____CAUSED BY NEMATODES_____

Clover Cyst Nematode

CAUSE: _Heterodera trifolii_ (Goff.) Oostenbrink. Females reproduce parthenogenetically. The life cycle is similar to that of _H. schachtii_ Schmidt.

DISTRIBUTION: Europe.

SYMPTOMS: Stands may be lost or have uneven growth. Leaves may wilt with outer leaves first becoming chlorotic. Storage roots are not well formed, have branched root systems, and excessive growth of fibrous root.

CONTROL: Generally the same as those for _H. schachtii._

False Root Knot Nematode

CAUSE: _Nacobbus aberrans_ (Thorne & Schuster) Sher. and _N. dorsalis_ Thorne & Allen.

DISTRIBUTION: _Nacobbus aberrans_ is found in Colorado, Kansas, Montana, Nebraska, South Dakota, and Wyoming; _N. dorsalis_ is found in California.

SYMPTOMS: Necrosis and gall-like swellings occur on roots.

CONTROL: No specific control has been developed. Presumably, measures used to control other nematodes will be of some value.

Nebraska Root Gall Nematode

CAUSE: _Nacobbus batatiformis_ Thorne & Schuster probably overwinters as eggs and adults within galls in the soil. Eggs are deposited in a gelatinous matrix outside the gall. The eggs hatch into larvae that swim in moisture surrounding roots and penetrate roots and rootlets. Galls form at the point of penetration and contain from one to several egg-producing females. Growth and reproduction of nematodes is favored by soil temperatures between 24°C and 35°C. Nematodes are disseminated by any mode that transports soil such as irrigation water and tillage equipment.

DISTRIBUTION: Western Nebraska.

SYMPTOMS: Severely infected sugarbeets are stunted and wilt during warm weather. Numerous galls are formed, particularly on lateral roots that range up to 1 cm in diameter. Many small rootlets are produced on each gall, creating a hairlike or whiskery appearance.

CONTROL:
1. Rotate sugarbeets at least every four years.
2. Plant sugarbeets early because of the inactivity of the nematodes in cool soil.
3. Control weeds, especially lambs quarter and kocia, in sugarbeet fields since they may harbor *N. batatiformis*.
4. Apply nematicides in the form of fumigants to soil.

Needle Nematode

CAUSE: *Longidorus attenuatus* Hooper, *L. caespiticola* Hooper, *L. elongatis* (deMann) Thorne & Swanger, and *L. leptocephalus* Hooper.

DISTRIBUTION: England and the United States; however, they are not a problem in the United States.

SYMPTOMS: Similar to those caused by stubby root nematodes.

CONTROL: Follow general practices used to control nematodes.

Root Knot Nematode

CAUSE: *Meloidogyne arenaria* (Neal) Chitwood, *M. incognita* (Kofoid & White) Chitwood, *M. javanica* (Treub.) Chitwood, *M. hapla* (Kofoid & White) Chitwood, *M. Naasi* Franklin. *Meloidogyne incognita* and *M. javanica* cause the greatest damage in the United States. Nematodes overwinter as eggs or second-stage juveniles in soil, galls, or root tissue. The larvae move through the soil and penetrate the root tip. The larvae migrate through the root to the vascular tissues, become sedentary, and commence feeding. Excretions of the larvae stimulate cell proliferation, resulting in the development of syncythia from which the larvae derive their food. Females lay eggs that may be pushed out of the root into the soil.

DISTRIBUTION: Generally distributed wherever sugarbeets are grown.

SYMPTOMS: Small, light tan galls form on fibrous roots and taproots. Galls may be of different shapes: Rounded galls are caused by *M. hapla*;

elongated or spiral-shaped galls are caused by *M. naasi*; irregular, club-like swellings are caused by *M. incognita*. Later in the season the galls are a darker brown. The galls or swellings along the fine roots resemble a string of beads. Root rot may occur. Aboveground symptom is the presence of small, yellow, stunted plants. Plants under moisture stress may wilt and collapse. Plants may be infected heavily without obvious aboveground symptoms.

CONTROL:
1. Rotate sugarbeets every four or five years with other crops such as small grains. However, this is not considered very effective.
2. Control weeds.
3. Apply nematicides in form of fumigants to soil.

Stem and Bulb Nematode

CAUSE: *Ditylenchus dipsaci* (Kuhn) Filipjev.

DISTRIBUTION: Europe.

SYMPTOMS: Seedlings have swelling of epicotyl, hypocotyl, midribs, and main veins of leaves with the occasional formation of galls. Severely attacked plants may have multiple crowns and be severely stunted. Crown canker and girdling may occur at scars of infected petioles. Root rot may occur later in the growing season.

CONTROL:
1. Rotate sugarbeets with nonhost plants.
2. Control weeds.
3. Remove infected plant debris from fields.

Stubby Root Nematode

CAUSE: *Paratrichodorus anemones* Loof, *P. pachydermus* Seinhorst, *P. teres* Hooper, *Trichodorus cylindricus* Hooper, *T. primitivus* (deMan) Micoletzky and *T. viruliferus* Hooper.

DISTRIBUTION: Europe and the United States; however, they are not considered economically important in the United States.

SYMPTOMS: Tips of seedling taproots may be killed. Lateral roots may have stubby ends that later turn brown and die.

CONTROL: Use general practices recommended to control or manage nematodes.

Sugarbeet Nematode

CAUSE: *Heterodera schachtii* Schmidt survives in soil as cysts; the transformed body of the female nematode is filled with eggs. Eggs hatch over a long period of time, sometimes up to several years. The resulting larvae penetrate the root and take up positions within the cortex. The body of the female breaks through the epidermis with only the head embedded in the root. Eventually eggs develop in the female and largely displace other organs. Upon reaching maturity, the body of the female is transferred into a brown cyst. The nematode grows and multiplies between 13°C and 28°C. Nematode populations increase in soils treated with carbamate herbicides.

DISTRIBUTION: Generally distributed wherever sugarbeets are grown.

SYMPTOMS: At first, injury appears in small, distinct areas where sugarbeets are stunted or killed. Young plants wilt and die shortly after thinning. Small, white cysts that are about the size of a pinhead are attached to feeder roots. Later these cysts turn dark orange-brown and are difficult to see. The taproot may be small and extremely hairy due to proliferation of fibrous roots.

   Aboveground symptom is the presence of small, yellow, stunted plants with sprawling, wilted leaf petioles. In nematode-infested fields, sugarbeets appear to be more susceptible to leaf spot and other diseases.

CONTROL:
1. Rotate sugarbeets every four or five years with crops such as small grains.
2. Equipment should be cleaned between fields.
3. Tare dirt should be returned to nonagricultural soil.
4. Apply nematicides in form of fumigants to soil.
5. Plant as early as feasible.

RICKETTSIAL CAUSES

Beet Latent Rosette

CAUSE: A rickettsia-like organism transmitted by nymphs and adults of the beet lace bug *Piesma quadratum* Fieb., and by dodder *Cuscuta campestris* Yunck. The organism is persistent in the vector and remains infectious for 10–30 days.

DISTRIBUTION: Germany and the United States. It is not considered a serious disease.

SYMPTOMS: Initially there is leaf twisting, downward turning of leaf tips and chlorosis of young leaves. Later in the rosette stage, a cluster of terminal and axillary shoots with straplike leaves is formed, resembling a witches' broom. The rosette remains after the death of mature healthy leaves.

CONTROL: No control measures have been developed.

Yellow Wilt

CAUSE: A rickettsia-like organism transmitted by the leafhopper *Paratamus exitiosus* Beamer, dodder *Cuscuta californica* Choisy, and *C. campestris*, Yunck. and by grafting.

DISTRIBUTION: Argentina and Chile.

SYMPTOMS: Leaves yellow or plants wilt and collapse. New leaves may be dwarfed and often turn downward at the tips. Frequently only portions of leaves or veins of some leaves may turn yellow. Leaves of plants infected for a long time may be necrotic or become straplike. Root tips may become necrotic and result in the formation of tufts of roots. This causes plants to wilt during periods of high temperatures and low soil moisture, often resulting in plant death.

CONTROL:
1. Avoid planting in soils where disease has caused losses.
2. Plant resistant varieties when they become available.

VIRAL CAUSES

Beet Curly Top

This disease is also called curly top, sugarbeet curly top, sugarbeet curly leaf, western yellow blight virus, and tomato yellows.

CAUSE: Beet curly top virus (BCTV) overwinters in a large number of perennial plants and is transmitted by the beet leafhopper *Circulifer tenellus* Baker in a persistent manner. The beet leafhopper can feed for a minute on a diseased plant and acquire the virus. A leafhopper can then carry the virus for a month or more. However, the virus does not multiply in the leafhopper body, and if an insect has fed only once on a diseased plant, its ability to transmit the virus diminishes during successive feedings. The virus is not transmitted through the egg. The vector can disseminate the virus hundreds of kilometers.

DISTRIBUTION: Canada, Mediterranean Basin, and the United States.

SYMPTOMS: Plants infected prior to thinning or immediately after will make little further growth. The most common field symptoms are leaf dwarfing, with leaf edges rolling and curling inward, together with vein clearing, swelling, and spinelike growths. Eventually the swollen veins give rise to small, nipplelike swellings along the veins and veinlets. Sometimes a sticky brown fluid may be exuded that collects in droplets along the petioles and leaves. There may be an increase in number of rootlets. Roots may show concentric black rings in cross section together with phloem tissue that becomes necrotic and cracks.

Infection that occurs late in the growing season produces few symptoms. A slight vein clearing and swelling may occur.

CONTROL:
1. Plant resistant or tolerant cultivars.
2. Control vectors.

Beet Leaf Curl

CAUSE: Beet leaf curl virus (BLCV) is transmitted by the beet lace bug *Piesma quadratum* Fieb. in a persistent manner. The vector remains infective for the rest of its life. The virus overwinters in the vector, which hibernates in protected areas. In spring, vectors move into sugarbeet fields and transmit BLCV to young plants.

DISTRIBUTION: Europe.

SYMPTOMS: Initially vein clearing occurs in youngest leaves. Leaves are crinkled and curl inward, forming a structure similar to a lettuce head. Leaves and roots are badly stunted.

CONTROL:
1. Plant sugarbeets away from areas where disease occurs.
2. Apply insecticides to crop near overwintering areas of vector.

Beet Mild Yellowing

CAUSE: Beet mild yellowing virus (BMYV) probably overwinters in perennial weeds and overwintered sugarbeets. It is transmitted by the peach aphid, *Myzus persicae* Sulzer, which remains infective throughout its life after once acquiring the virus. Isolates of BMYV are serologically indistinct from beet western yellows virus.

DISTRIBUTION: British Isles and, possibly, the western United States.

SYMPTOMS: Affected leaves are an orange-yellow.

CONTROL: Insecticides are used in Great Britain to control the aphid.

Beet Mosaic

CAUSE: Beet mosaic virus (BMV) probably survives in perennial plants and overwintered sugarbeets. BMV is transmitted by several species of aphids in a nonpersistent manner. All stages and forms of the aphid can carry the virus. Aphids only have to feed a few seconds to acquire the virus but rapidly lose it in a few hours during subsequent feeding. Consequently an aphid must frequently feed on diseased plants in order to reacquire the virus. The virus is not transferred to young aphids by the female aphid. Disease is most severe where crops of two growing seasons overlap or where diseased plants have overwintered.

DISTRIBUTION: Generally distributed wherever sugarbeets are grown. Incidence of the disease depends on buildup of the aphid population.

SYMPTOMS: Initially, circular chlorotic spots with sharply defined margins occur on young leaves. Spots often occur as chlorotic rings with green centers. Usually the mosaic pattern consists of irregular patches of various shades of green.

CONTROL:
1. Do not overlap crops of two growing seasons.
2. Destroy wild and escaped beets in vicinity of newly planted fields.

Beet Pseudo Yellows

CAUSE: Beet pseudo yellows virus (BPYV) is transmitted by the common greenhouse whitefly, *Trialeucodes vaporariorum* Westwood. The virus is acquired in one hour of feeding and is also transmitted in a one-hour feeding period. The latent period in the vector is less than six hours. BPYV is retained within the vector for 6 days. Several other plants are also hosts.

DISTRIBUTION: California in the Salinas Valley.

SYMPTOMS: Chlorotic spotting or splotching occurs uniformly on the older and intermediate leaves. Eventually the yellowing becomes more intense and general, with older infected leaves almost entirely chlorotic except for small scattered islands of green tissue. Older leaves also have irregular bright yellow areas 10–15 mm in diameter. Leaves are thickened and brittle.

CONTROL: No control is reported.

Beet Western Yellows

CAUSE: Beet western yellows virus (BWYV) probably overwinters in several weed hosts or overwintered sugarbeets. It is transmitted mainly by the green peach aphid, *Myzus persicae* Sulzer, and infrequently by other species of aphids. An aphid can acquire the virus after a 5-minute feeding on a diseased plant and transmit it after a 10-minute feeding on a healthy plant. An aphid can remain infective for several days. Different strains of BWYV exist.

DISTRIBUTION: Israel, California, Illinois, and Oregon.

SYMPTOMS: It is not possible to differentiate with any certainty between BWYV and beet yellows virus by their symptoms on sugarbeets. At first a light chlorotic spotting occurs on older and middle-age leaves and gradually becomes more intense in color. The chlorotic areas may be sharply delimited by veins. Eventually older leaves become thick, brittle, and yellow except for green areas adjacent to the veins. Such leaves are frequently attacked by *Alternaria* spp.

CONTROL:
1. Plant resistant varieties.
2. Eliminate weeds near and adjacent to sugarbeet fields.

Beet Yellow Net

CAUSE: It is not known for certain the exact nature of the causal agent of beet yellow net (BYN). The causal agent of BYN overwinters in sugarbeets that persist from one season to the next and possibly in weed hosts. It is transmitted mainly by the green peach aphid *Myzus persicae* Sulzer. Once an aphid has acquired the causal agent, it often remains infective for life. Possibly the BYN causal agent cannot be transmitted without beet mild yellowing virus as a carrier virus.

DISTRIBUTION: California and England.

SYMPTOMS: An intense yellow chlorosis of the veins and veinlets of leaves occurs. Infected plants are conspicuous in the field and contrast with the surrounding normal green plants. Eventually the symptoms fade as new leaves appear normal and infected plants are difficult to locate.

Occasionally a severely infected plant will have a complete chlorosis of the older leaves. No stunting or malformation occurs.

CONTROL: No control is necessary.

Beet Yellow Stunt

CAUSE: Beet yellow stunt virus (BYSV) is primarily reservoired in sowthistle, *Sonchus oleraceus* L. It is most efficiently transmitted by the sowthistle aphid *Nasonovia lactucae* (L.), but the aphid apparently does not reproduce on sugarbeet. The green peach aphid *Myzus persicae* Sulzer is found abundantly on sugarbeet but is a relatively inefficient vector. The potato aphid *Macrosiphum euphorbiae* (Thomas) is also an inefficient vector. For this reason, spread of the virus in a field is marginal. Disease incidence is high in rows adjacent to areas where sowthistle is present, but becomes progressively less with increased distance from the virus source.

DISTRIBUTION: California.

SYMPTOMS: Initially there is severe twisting, cupping, and epinasty of one or two intermediate-age leaves. Petioles are shortened and leaves become mottled and yellow. Young leaves are dwarfed, malformed, twisted, and slightly mottled. As leaves age, mottling becomes more intense, with the leaves sometimes becoming completely chlorotic. Plants are severely stunted and sometimes collapse and die.

CONTROL: Control sowthistle adjacent to sugarbeet fields.

Beet Yellow Vein

CAUSE: There is no real evidence that the causal agent of beet yellow vein is a virus. The causal organism can be transmitted by grafting, juice inoculation, and the leafhopper *Aceratagallia calcaris* Oman.

DISTRIBUTION: United States.

SYMPTOMS: Very young leaves are dwarfed and the main veins are yellow. The main veins of all affected leaves are a continuous or broken yellow, with the chlorosis extending 1 mm or more into surrounding tissue. These symptoms are visible on both sides of the leaves. Commonly, half of an infected plant may be stunted severely whereas the other half grows normally.

CONTROL: No adequate control is available.

Beet Yellows

CAUSE: Beet yellows virus (BYV) probably survives in overwintered sugarbeets. The virus is transmitted by the green peach aphid *Myzus persicae* Sulzer, the black bean aphid *Aphis fabae* Scopoli, and infrequently by other aphids. Transmission is of the semipersistent type. The virus is not passed on to the progeny of vectors nor is the virus retained after molting.

DISTRIBUTION: Generally distributed wherever sugarbeets are grown.

SYMPTOMS: Younger leaves have a vein clearing or vein yellowing. The vein clearing may appear bright yellow or necrotic. Secondary and intermediate veins frequently appear sunken and develop an etch symptom. Older leaves are yellow, thickened, and brittle. Thickening and brittleness precedes chlorosis in an infected leaf. Yellowing begins at leaf margins and tips and spreads downward between the veins. Chlorosis may vary from pale green to orange or red, depending on the variety. Small reddish or brown spots develop on many of the older yellowed leaves. Chlorotic areas feel waxy, or in severely infected plants, dry, causing the plants to rustle when shaken or brushed. Such leaves do not wilt easily during dry weather and will splinter when crushed. Early-infected leaves eventually become necrotic starting from where they first became chlorotic. The necrotic spots and yellowing often give leaves a bronze cast.

CONTROL:
1. Insecticides are used in England to control vectors based on a forecasting system.
2. Eliminate escapes and overwintered beets.
3. Separate new beet fields from known sources of infection by 2–3 km.

Cucumber Mosaic

CAUSE: Cucumber mosaic virus (CMV) overwinters in several perennials and has a large host range. It is disseminated by more than 60 species of aphids. Aphids only have to feed a few seconds to acquire the virus but rapidly lose it during subsequent feedings. Consequently an aphid must frequently feed upon diseased plants in order to reacquire the virus. The virus is not transferred to young aphids by the female aphid. CMV is transmitted by seeds of certain plants but not by sugarbeet seed.

DISTRIBUTION: Because of the large host range, CMV is generally distributed wherever sugarbeets are grown; however, it is not considered a serious disease.

SYMPTOMS: Large, irregular patches of bright yellow tissue contrast with dark green of the leaf. Symptoms are of a mosaic type but the pattern is coarser and more contrasting than that found in beet mosaic. Leaf dwarfing and distortion is common with blisterlike areas of green tissue. Necrosis often occurs to large areas of the older leaves.

CONTROL: Control weeds around sugarbeet fields.

Lettuce Infectious Yellows

CAUSE: Lettuce infectious yellows virus (LIYV) survives in a wide range of weeds and commercial crops. LIYV is transmitted by the sweet potato whitefly, *Bemisia tabaci* (Genn.), in a semipersistent manner. Cotton is a major source of high populations of whitefly.

DISTRIBUTION: Imperial Valley of California.

SYMPTOMS: A very mild mottling develops into interveinal yellowing. Necrosis eventually develops in chlorotic areas.

CONTROL:
1. Alter planting of cucurbits to provide a one- to three-week period in July or August when cucurbits are not present.
2. Do not plant sugarbeets near areas of infected plants.
3. Destroy infected residue of cucurbits and lettuce after harvesting.
4. Control weeds in planting field and nearby fields.

Rhizomania Disease

This disease is also called beet necrotic yellow vein.

CAUSE: Beet necrotic yellow vein virus (BNYVV) is transmitted by the soilborne fungus *Polymyxa betae* Keskin and by inoculation of sap. The fungus is an obligate parasite occurring within hair cells as a plasmodium. The plasmodium develops either into a zoosporangium, which releases zoospores at maturity that immediately attack new roots, or into a mass of small resting spores (cystosorus). The resting spores overwinter, and under proper conditions, each spore produces a single zoospore. Resting spores can survive up to 10 years. Disease is severe at an optimum soil temperature of 20°C for an extended time, and with wet soils that are neutral to slightly alkaline.

Different isolates of BNYVV have been found in California. Additionally other virus entities have been found but their relationship to Rhizomania disease is not known.

DISTRIBUTION: France, Italy, Japan, California, and Texas. *Polymyxa betae* has been isolated from sugarbeet roots in Nebraska.

SYMPTOMS: Roots are stunted with a proliferation of lateral rootlets on the main taproot, giving it a bearded look. The root is frequently rotted and constricted below soil level. Vascular system is discolored. Leaves are upright and slightly chlorotic. Leaves proliferate, resulting in excessive crown tissue. Leaves may also lose turgidity and wilt without discoloration. Rarely, distinct veinal yellowing with necrotic lesions is evident.

CONTROL: No control is reported.

Savoy

This disease is also called beet savoy.

CAUSE: Beet savoy virus (BSV) overwinters in the pigweed bug, *Piesma cinerea* Say, which is also the specific insect vector. The insect overwinters in grassy or woody areas. Infection occurs mainly at the edge of fields. BSV is not mechanically transmitted nor is it seedborne. There is a possibility that the causal agent may be a mycoplasma.

DISTRIBUTION: United States, primarily east of the continental divide.

SYMPTOMS: The more pronounced effects are found on the innermost leaves, which are dwarfed, curled downward, and savoyed. This gives the lower leaf surface a netted appearance due to the veinlets first clearing, then thickening. Later roots will be generally discolored and the phloem will appear necrotic. Both plant leaves and roots are markedly stunted.

CONTROL:
1. Sugarbeet fields should be located several hundred meters away from woods or uncultivated land that could serve as an insect source.
2. Grow a barrier crop between the sugarbeet field and potential insect source.

BIBLIOGRAPHY

Altman, J. 1981. Increase in cyst nematode populations in soil treated with cycloate and diallate. *Phytopathology* **71**:199 (abstract).

Bugbee, W. M. 1975. Dispersal of *Phoma betae* in sugarbeet storage yards. *Plant Dis. Rptr.* **59**:396–397.

Bugbee, W. M. 1975. *Penicillium claviforme* and *Penicillium variabile:* Pathogens of stored sugarbeets. *Phytopathology* **65**:926–927.

Bugbee, W. M. 1979. Resistance to sugarbeet storage rot pathogens. *Phytopathology* **69**:1250–1252.

Bugbee, W. M. 1983. Infection and movement of endophytic bacteria in sugarbeet plants. *Phytopathology* **73**:806 (abstract).

Bugbee, W. M., and El-Nashaar, H. M. 1983. A newly recognized symptom of sugarbeet root infection caused by *Phoma betae*. *Plant Disease* **67**:101–102.

Bugbee, W. M.; Gudemestad, N. C.; Secor, G. A.; and Nolte, P. 1985. Sugarbeet: A natural host for *Corynebacterium sepedonicum*. *Phytopathology* **75**:1379 (abstract).

Bugbee, W. M.; Gudemestad, N. C.; Secor, G. A.; and Nolte, P. 1987. Sugarbeet as a symptomless host for *Corynebacterium sepedonicum*. *Phytopathology* **77**:765–770.

Bugbee, W. M., and Nielsen, G. E. 1978. *Penicillium cyclopium* and *Penicillium funiculosum* as sugarbeet storage rot pathogens. *Plant Dis. Rptr.* **62**:953–954.

Bugbee, W. M., and Soine, O. C. 1974. Survival of *Phoma betae* in soil. *Phytopathology* **64**:1258–1260.

Coons, G. H. 1953. Some problems in growing sugarbeets, in *Yearbook of Agriculture*. Washington, D.C., pp. 509–524.

Duffus, J. E. 1965. Beet pseudo-yellows virus, transmitted by the greenhouse whitefly (*Trialeurodes vaporariorum*). *Phytopathology* **55**:450–453.

Duffus, J. E. 1972. Beet yellow stunt, a potentially destructive virus disease of sugarbeet and lettuce. *Phytopathology* **62**:161–165.

Duffus, J. E.; Larsen, R. C.; and Liu, H. Y. 1986. Lettuce infectious yellows virus—a new type of whitefly-transmitted virus. *Phytopathology* **76**:97–100.

Duffus, J. E., and Liu, H. Y. 1987. First report of Rhizomania of sugarbeet from Texas. *Plant Disease* **71**:557 (disease notes).

Giannopolitis, C. N. 1978. Lesions on sugarbeet roots caused by *Cercospora beticola*. *Plant Dis. Rptr.* **62**:424–427.

Giunchedi, L., and Langenberg, W. G. 1982. Beet necrotic yellow vein virus transmission by *Polymyxa betae* Keskin zoospores. *Phytopathol. Medit.* **21**:5–7.

Herr, L. J., and Roberts, D. L. 1980. Characterization of Rhizoctonia populations obtained from sugarbeet fields with differing soil textures. *Phytopathology* **70**:476–480.

Hills, F. J., and Worker, G. F., Jr. 1983. Disease thresholds and increases in fall sucrose yield related to powdery mildew of sugarbeet in California. *Plant Disease* **67**:654–656.

Hine, R. B., and Ruppel, E. G. 1969. Relationship of soil temperature and moisture to sugarbeet root rot caused by *Phythium aphanidermatum* in Arizona. *Plant Dis. Rptr.* **53**:989–991.

Kontaxis, D. G.; Meister, H.; and Sharma, R. K. 1974. Powdery mildew epiphytotic on sugarbeets. *Plant Dis. Rptr.* **58**:904–905.

Langenberg, W. G., and Kerr, E. D. 1982. *Polymyxa betae* in Nebraska. *Plant Disease* **66**:862.

Liu, H. Y., and Duffus, J. E. 1985. The viruses involved in Rhizomania disease of sugarbeet in California. *Phytopathology* **75**:1312 (abstract).

Maas, P. W. T., and Heijbroek, W. 1982. Biology and pathogenicity of the yellow beet cyst nematode, a host race of *Heterodera trifolii* on sugarbeet in the Netherlands. *Nematologica* **28**:77–93.

MacDonald, J. D.; Leach, L. D.; and McFarlane, J. S. 1976. Susceptibility sugarbeet lines to the stalk blight pathogen *Fusarium oxysporum* f. sp. *betae. Plant Dis. Rptr.* **60**:192–196.

Marco, S. 1984. Beet western yellows virus in Israel. *Plant Disease* **68**:162–163.

Ruppel, E. G.; Harrison, M. D.; and Nielson, A. K. 1975. Occurrence and cause of bacterial vascular necrosis and soft rot of sugarbeet in Washington. *Plant Dis. Rptr.* **59**:837–840.

Ruppel, E. G.; Hills, F. J.; and Mumford, D. L. 1975. Epidemiological observations on the sugarbeet powdery mildew epiphytotic in western USA in 1974. *Plant Dis. Rptr.* **59**:283–286.

Ruppel, E. G.; Jenkins, A. D.; and Burtch, L. M. 1980. Persistence of benomyl-tolerant strains of *Cercospora beticola* in the absence of benomyl. *Phytopathology* **70**:25–26.

Ruppel, E. G., and Tomasovic, B. J. 1977. Epidemiological factors of sugarbeet powdery mildew. *Phytopathology* **67**:619–621.

Rush, C. M. 1987. Root rot of sugarbeet caused by *Phythium deliense* in the Texas panhandle. *Plant Disease* **71**:469 (disease notes).

Schneider, C. L., and Robertson, L. S. 1975. Occurrence of diseases on sugarbeet in a crop rotation experiment in Saginaw County, Michigan in 1969–1971. *Plant Dis. Rptr.* **59**:194–197.

Schneider, C. L.; Ruppel, E. G.; Hecker, R. J.; and Hogaboam, G. J. 1982. Effect of soil deposition in crowns on development of Rhizoctonia root rot in sugarbeet. *Plant Disease* **66**:408–410.

Stanghellini, M. E., and Kronland, W. C. 1977. Root rot of mature sugarbeets by *Rhizopus arrhizus. Plant Dis. Rptr.* **61**:255–256.

Stanghellini, M. E.; Von Bretzel, P.; and Kronland, W. C. 1981. Epidemiology of *Pythium aphanidermatum* root rot in sugarbeets. *Phytopathology* **71**:905 (abstract).

Stanghellini, M. E.; Von Bretzel, P.; Olsen, M. W.; and Kronland, W. C. 1982. Root rot of sugar beet caused by *Pythium deliense. Plant Disease* **66**:857–858.

Staples, R.; Jansen, W. P.; and Anderson, L. W. 1970. Biology and relationship of the leafhopper *Aceratagallia calcaris* to yellow vein disease of sugarbeets. *Jour. Econ. Entomol.* **63**:460–463.

Tamada, T. 1975. *Beet Necrotic Yellow Vein Virus*. Commonwealth Mycological Institute. Descriptions Plant Viruses. Set. 9, No. 144.

Thompson, S. V. et al. 1977. Bacterial vascular necrosis and rot of sugarbeet: General description and etiology. *Phytopathology* **67**:1183–1189.

Thompson, S. V.; Hills, F. J.; Whitney, E. D.; and Schroth, M. N. 1981. Sugar and root yield of sugarbeets as affected by bacterial vascular necrosis and rot, nitrogen fertilization, and plant spacing. *Phytopathology* **71**:605–608.

Timmerman, E. L.; D'Arcy, C. J.; and Splittstoesser, W. E. 1984. Beet western yellows virus in Illinois. *Phytopathology* **74**:1271 (abstract).

Western, J. H. 1971. *Diseases of Crop Plants*. The Macmillan Press Ltd. London.

Whitney, E. D. 1971. The first confirmable occurrence of *Urophylyctis seproides* on sugarbeet in North America. *Plant Dis. Rptr.* **55:**30–32.

Whitney, E. D. 1987. Identification and aggressiveness of *Erwinia carotovora* subsp. *betavasculorum* on sugarbeet from Texas. *Plant Disease* **71:**602–603.

Whitney, E. D., and Duffus, J. E. (Ed.). 1986. *Compendium of Beet Diseases and Insects.* American Phytopathological Society, St. Paul, MN, 76p.

20

Diseases of Sugarcane (*Saccharum officinarum* Linn.)

BACTERIAL CAUSES

Bacterial Mottle

This disease is also called leaf stripping and chlorotic mottle.

CAUSE: *Pectobacterium carotovorum* (Jones) Walder var. *graminarium* Dowson & Hayward persists in infected standing cane, cane residue, and grasses; specifically elephant grass, *Pennisetum purpureum* Schumach; para grass, *Brachiaria mutica* Trin.; and guinea grass, *Panicum maximum* Jacq. The bacterium are transmitted by flooding water with most dissemination occurring during the hot, wet season. Bacteria will be exuded through stomata of infected cane or grasses, carried by water to other hosts where they enter through wounds in stem or buds of growing points. Bacteria are not spread to any extent by wind or rain. Seed cane that is infected ordinarily fails to germinate.

DISTRIBUTION: Australia. The disease is most common in low-lying areas subject to flooding.

SYMPTOMS: Initially one to many creamy white stripes, 1–2 mm wide, extend from or near the base of the leaf blade upward and parallel to leaf veins. As the leaf ages, the stripes may enlarge, and orange to brown-red areas develop within them, often involving most of the stripe.

Systemic infection causes a chlorotic mottling of the leaves, often without any distinct striping. Sometimes shoots may become almost completely chlorotic. As leaves become older, abundant, small, brown-red flecks and short, narrow stripes appear, causing a very chlorotic leaf to appear pink.

Infected shoots are stunted and leaf margins wither, causing the leaf to curl inward. Eventually the entire shoot may die. Diseased stalks may tiller excessively due to production of several side shoots at their base. A witches' broom effect may occur several centimeters above soil level when older cane has been infected. Characteristic leaf symptoms are present. During warm, humid weather, small whitish drops of bacterial exudate may occur on the underside of leaves.

CONTROL:
1. Plant the most resistant cultivars.
2. Plant only healthy seed cane.
3. Destroy diseased plants in a field.
4. Control infected grasses in waterways adjacent to cane fields.

Gumming Disease

This disease is also called gummosis, Cobb's disease of sugarcane, and gum disease.

CAUSE: *Xanthomonas campestris* pv. *vasculorum* (Cobb) Dye survives in cuttings from one planting to the next. Bacterium may also infect healthy sets through cutting by a contaminated knife or tool. Secondary infections are primarily by bacteria or gum oozing from wounds of infected plants during wet weather and being disseminated from diseased to healthy plants by wind-blown rain. The bacteria are also disseminated by machinery, the brushing of infected plants against healthy ones, and insects. Bacteria enter into the plant through small wounds caused by leaves brushing against each other. Long-range dissemination is ordinarily through infected cuttings.

DISTRIBUTION: Africa, Australia, South America, West Indies including Puerto Rico.

SYMPTOMS: Symptoms start two to three weeks after wet weather has occurred. Longitudinal streaks occur on mature leaves that have not started to discolor but not on the youngest leaves. The streaks follow the course of the vascular bundles and are uniformly 3–6 mm wide. Young lesions have well-defined margins but the margins of older streaks become diffuse. Most streaks start at the leaf apex or margin and vary in length from a few millimeters to the full length of the leaf; however, streaks do not extend into the sheath. The color of the streaks is yellow to orange flecked with patches of red. Tissue starts to die at the point of infection and progresses along the length of the streak, becoming grayish in color. Symptoms on leaves may seem to disappear during dry weather, particularly if infected leaves are shed as they age.

Systemic infection causes short, narrow, well-defined, dark red streaks to occur on the youngest leaf blades and underside of midribs extending into sheaths. Sometimes creamy white streaks occur that start from the base and are almost half the width of the leaf. Red blotches are lacking at first but may develop as the leaves mature. Additionally, chlorotic areas that are usually speckled with red dots may occur on part of one or many leaves. This symptom may involve the whole top. The growing point usually dies shortly after this symptom occurs but sometimes the shoot recovers. A large percentage of stalks wilt and die.

A yellow to orange gummy mass exudes from the cut ends of badly diseased stalks. Frequently gum pockets may form in the region of the growing point and can be observed when the stalk is cut longitudinally.

CONTROL:
1. Plant resistant cultivars.
2. Plow out diseased crop; however, ratoons usually suffer less loss than plant crops.

Leaf Scald

CAUSE: *Xanthomonas albilineans* (Ashby) Dowson is transmitted primarily by cuttings and tools, particularly cane knives. Tolerant cultivars may transmit the bacterium over considerable distances since symptoms are generally not seen. The bacterium can only be transmitted through wounded buds.

DISTRIBUTION: Worldwide. This disease is not considered of major importance.

SYMPTOMS: Two different types of symptoms have been described: acute and chronic. The chronic or characteristic symptom is white stripes that usually extend the full length of the leaf blade and sheath. Stripes are straight, well-defined, narrow (3 mm wide), and located over and around the vascular bundles.

The acute symptoms are severely diseased plants that may have whole leaves that become chlorotic. Young shoots as they emerge from the soil may also be chlorotic. Frequently there is progressive withering of infected leaves, giving the plant a scalded appearance. Internodes of stalks are shortened, weak, with the production of side shoots. Plants commonly wilt and die.

If a diseased stalk is split longitudinally, the vascular bundles have a light red discoloration. Some vessels are plugged with a reddish gumlike substance; however, there is no bacterial ooze.

Symptoms may disappear as plants grow. However, symptoms may become more severe if plant growth is retarded for any reason as the cane approaches maturity.

CONTROL:
1. Quarantine seed cane into areas where leaf scald does not occur.
2. Plant resistant cultivars.

Mottled Stripe

CAUSE: *Pseudomonas rubrisubalbicans* (Christopher & Edgerton) Krassilnikov likely persists in standing cane or residue. Dissemination is probably by wind and rain.

DISTRIBUTION: Africa, Australia, Caribbean, South America, and the United States. It is not considered a serious disease.

SYMPTOMS: Symptoms occur only on leaves. Stripes are creamy white, on which red areas occur, giving a red and white mottled effect. If the red areas are large, the stripe may have a general red appearance. Stripes are 1–4 mm wide with distinct edges, up to a meter in length, and parallel to leaf veins.

CONTROL: No control is reported. Most cane cultivars are susceptible.

Ratoon Stunting Disease

This disease is also called Q.28 disease.

CAUSE: *Clavibacter xyli* subsp. *xyli* Davis, Gillespie, Vidaver & Harris. The disease affects a wide variety of graminaceous plants including Johnsongrass, sorghum, corn, sweet Sudangrass, and Bermudagrass. The causal bacterium is mechanically transmitted through cutting knives, other tools, and cuttings taken from diseased plants.

DISTRIBUTION: Generally distributed wherever sugarcane is grown.

SYMPTOMS: The only external symptoms are a stunting and unthriftiness of the infected plant. Ratoons display symptoms more than older cane. Ratoons are slow to start, particularly in dry weather, and continue to be retarded in growth. Yield is reduced due to production of thinner, shorter stalks. Stunting is not uniform from stool to stool and a field may have uneven growth. Diseased cane wilts sooner than healthy cane.

A yellow, orange, pink-red or red-brown discoloration occurs with individual vascular bundles in nodes of comparatively mature cane just below region of attachment of the leaf sheath. A pink color occurs in nodes of young cane.

CONTROL:
1. Plant resistant cultivars.
2. Plant only cuttings from healthy plants.
3. Treat infected cuttings with hot water or hot air. Treated cuttings should have a protectant fungicide applied.
4. Destroy all volunteer cane.
5. Sterilize cutting tools.

Red Stripe and Top Rot

CAUSE: *Pseudomonas rubrilineans* (Lee, Purdy, Barnum & Martin) Stapp persists in older, withered leaves and soil for at least 32 days. Bacteria exudate oozes to the leaf surface during warm, wet weather and forms droplets. Bacteria are disseminated mainly by wind and rain but rarely by cuttings or any other mechanical means. Bacteria enter the plant through wounds or stomates. Bacterial exudate may either form galls on the lower portion of the plants or run down the leaf, causing stem infection.

DISTRIBUTION: Generally distributed wherever sugarcane is grown.

SYMPTOMS: All portions of the plant may be infected including the leaves, stalks, and roots. Initially water-soaked stripes surrounded by a chlorotic zone develop on leaves but later the stripes become dark red to maroon. Stripes are 15–40 cm long and 1–4 mm wide, but may extend down into the leaf sheaths where they become wider. Eventually stripes coalesce, forming bands of alternating maroon stripes and chlorotic areas.

The vascular bundles near the growing point of the stalk become slightly red. The discoloration extends down the stalk and shows as a red ring one-fourth to one-half the distance from the rind to the center. The center of the ring is water-soaked and rapidly decomposes but the outside portion changes very little. Eventually the rot extends down to the base of the stalk, leaving a hollow, central cylinder. The terminal bud and spindle leaves die, often resulting in growth of lateral buds that also show a reddening. Stalks in advanced stage of decomposition give off an unpleasant odor that can be detected for a considerable distance.

CONTROL:
1. Plant resistant cultivars.
2. Rogue diseased stools from seedling nurseries.

FUNGAL CAUSES

Alternaria Leaf Spot

CAUSE: *Alternaria alternata* (Fr.) Keissler (syn. *A. tenuis* Nees) is a common saprophyte.

DISTRIBUTION: Cuba, India, and Taiwan.

SYMPTOMS: Initially spots appear as minute water-soaked areas. Eventually spots elongate and become elliptical to irregular with red-brown to dark brown margins. The area between spots becomes necrotic and sometimes the entire leaf dies.

CONTROL: No control is reported.

Banded Sclerotial Disease

CAUSE: *Thanatephorus sasakii* (Shirai) Tu & Kimbrough (syn. *Pellicularia sasakii* (Shirai) Ito). The anamorph is *Rhizoctonia sasakii* Shirai. The fungus survives as sclerotia in soil. Other grasses, particularly Bermudagrass, *Cynodon dactylon* (L.), are also infected. Only the older leaves of cane become infected by coming in contact with a diseased grass leaf or the soil during periods of high humidity.

DISTRIBUTION: Africa, Australia, Asia, Caribbean, Central America, and the United States. It is considered of minor importance.

SYMPTOMS: Older leaves and occasionally leaf sheaths display symptoms that commonly occur as a series of broad bands across the leaf. At first, brown-green, irregularly shaped areas occur that turn brown, then yellow to light tan, with definite red-brown borders. Sclerotia that are irregular to spherical–shaped, 2–5 mm in diameter, and dark brown to black commonly are present on diseased areas.

CONTROL: No control is necessary.

Basal Stem, Root, and Sheath Rot

CAUSE: *Marasmius plicatus* Wakker, *M. sacchari* Wakker, and *M. stenospilus* (Mont.), Singer (syn. *Marasmiellus stenospilus* (Mont.) Singer). *Marasmius* spp. are saprophytic and usually only infect either wounds or plants that are predisposed by other means. Dissemination is by any means that will transport mycelial strands and spores.

DISTRIBUTION: Generally distributed wherever sugarcane is grown. The disease is of minor importance.

SYMPTOMS: White mycelium grows over lower portions of the plant, causing sheaths to adhere tightly to the stalk. White mushrooms develop at the base of infected stalks. Mature leaves of affected plants are covered with red spots that develop during dry periods. The underground portion of the stem and young shoots may also be infected.

CONTROL: Most cultivars are resistant.

Black Rot

CAUSE: *Ceratocystis adiposa* (Butler) C. Moreau (syn. *C. major* (Van Beyma) C. Moreau, *Ceratostomella adiposa* (E. Butler) Sartoris, *C. major* Van Beyma, *Endoconidiophora adiposa* (E. Butler) Davis, *Ophiostoma adiposum* (E. Butler) Nannf., *O. majus* (Van Beyma) Goid, and *Sphaeronema adiposum* E. Butler). The fungus persists as perithecia and mycelium in residue. It is disseminated by windborne conidia, splashing water, and insects. The fungus is also disseminated by any means that will move soil. The disease is more prevalent in loose, cloddy soils that have large air pockets. In the United States the disease is most common in seed cane bedded in the autumn, possibly because of air spaces around cane. Infection occurs through cut ends of seed cuttings.

DISTRIBUTION: Australia, Brazil, China, Dominican Republic, India, Indonesia, Panama, Peru, Taiwan, and the United States.

SYMPTOMS: Seed cuttings, starting at the ends, become soft and watery. The interior of the cuttings is a dark purple at first, then becomes black, with a smell resembling fermenting pineapples. Black mycelial growth of the fungus occurs on the cut ends and is the obvious external symptom.

CONTROL:
1. Plant resistant cultivars.
2. Plant cuttings in the autumn rather than storing them in beds.

Black Stripe

CAUSE: *Cercospora atrofiliformis* Yen, Lo & Chi.

DISTRIBUTION: Taiwan.

SYMPTOMS: Initially small, yellow, round or oval spots occur on leaves. Eventually the spots enlarge to 5–36 mm long, 0.5–1.2 mm wide, and become brown-black.

CONTROL: No control is necessary.

Brown Rot

CAUSE: *Corticium* sp. is a saprophyte or minor pathogen of trees.

DISTRIBUTION: Australia. Brown rot has been found only in sugarcane planted in soil recently cleared of trees. It is not of economic importance.

SYMPTOMS: Stools are unthrifty, leaves wilt and die, then the entire stool dies. A thick layer of brown mycelium extends a few centimeters above the soil line on the stalk and binds the leaf sheaths together. Mycelium also grows over the lower leaf sheaths and belowground plant parts, causing dead roots to look several times thicker than they actually are. Soil is bound to the plant by mycelium. Stalks below soil level are killed and turn a brown color. Killed tissue dries out quickly and is separated from healthy tissue by a narrow, dark brown band.

CONTROL: No control is necessary.

Brown Spot

CAUSE: *Cercospora longipes* Butl. persists in leaf residue or infected leaves of standing cane. Conidia are produced on both leaf surfaces but more abundantly on the lower surface. Dissemination is by wind and rain. Cuttings contaminated with conidia may also transmit the disease.

DISTRIBUTION: Generally distributed wherever sugarcane is grown. However, it has not been reported from Australia or Taiwan. It is ordinarily not considered an important disease.

SYMPTOMS: Numerous spots appear first on both surfaces of older leaves and progress up the plant. Spots are red-brown, surrounded by a narrow yellow halo, oval to linear, and vary in size from specks to 13 mm in length. The centers of older spots become tan and are surrounded by a red zone and yellow halo. Spots may coalesce to form large red-brown patches of irregular shape. Severely infected leaves may die prematurely, giving affected plants and entire fields a fiery appearance.

CONTROL:
1. Plant resistant cultivars.
2. Treat seed cane cuttings with a seed-protectant fungicide.

Brown Stripe

CAUSE: *Cochliobolus stenopilus* Matsumoto & Yamamoto, anamorph *Bipolaris stenospila* (Drechs.) Shoem. (syn. *Drechslera stenospila* (Drechs.) Subram. & Jain and *Helminthosporium stenospilum* Drechs.). It is unknown how the fungus survives in the absence of cane in subtropical areas where cane is not grown year-round. The fungus may either infect other members of the grass family or mycelium may survive in infected residue.

   Conidia develop in lesions on old dead leaves and are airborne a considerable distance. Spores germinate on leaves in presence of free moisture and enter through stomates. The fungus is a weak facultative parasite and is more likely to infect predisposed plants. The disease is most severe in dry weather and when growing conditions for the host are suboptimal.

DISTRIBUTION: Generally distributed wherever sugarcane is grown.

SYMPTOMS: Initially young leaves have small (0.5 mm in diameter), watery spots that soon become elongated and turn red-brown to brown, sometimes surrounded by a narrow chlorotic halo. Sometimes the spots occur in a row across the leaf due to germinating in the moisture of the spindle. Eventually the spots elongate to form definite stripes 2–10 mm in length. As lesions mature, they become even longer, sometimes up to 75 mm in length but only 4–5 mm in width. Stripes have a narrow yellow margin. When disease is severe, stripes coalesce, prematurely killing the leaf. Infrequently a top rot develops.

CONTROL:
1. Plant resistant cultivars.
2. Maintain plants in a vigorous growing condition with proper fertility and moisture.

Collar Rot

CAUSE: *Hendersonina sacchari* Butl.

DISTRIBUTION: Argentina, Bangladesh, India, Indochina, Mauritius, the Philippines, and Sri Lanka. Collar rot is of minor importance.

SYMPTOMS: The top leaves become necrotic and dry from the margin inward, leaving only a green midrib. The interior of the upper nodes is dried up, often with cavities of various sizes that eventually extend throughout the stalk. The tissue of lower internodes is watery and brown with patches of red that become the dominant discoloration in the basal internodes. Roots growing from basal internodes are rotted with dark brown to blackish discoloration. Affected stalks are lighter weight than normal.

CONTROL: No control is necessary.

Common Rust

This disease is also called sugarcane leaf rust.

CAUSE: *Puccinia melanocephala* H. & P. Syd. (syn. *P. erianthi* Padw. & Khan). Urediospores and teliospores are produced on sugarcane. Alternate hosts and other spore stages are not known. Urediospores are disseminated by wind and water. Infection occurs under humid conditions at temperatures of 16–29°C. In Puerto Rico, the optimum rainfall for infection is 13–16 cm per month and the optimum age for infection is two months.

DISTRIBUTION: Widely distributed wherever sugarcane is grown.

SYMPTOMS: Symptoms occur mostly on leaves. Tiny, brownish, elongated spots (uredia) occur on both leaf surfaces, but are more common on the lower surface than the upper surface. Spots increase mostly in length to 2–10 mm, become orange-brown to brown, and are surrounded by a small yellowish halo. Orange urediospores at first are subepidermal but later rupture the epidermis. Uredia eventually darken due to formation of telia either in the uredium or in separate telia, mostly on the lower leaf surface. The surrounding tissue is killed, sometimes resulting in death of young leaves.

CONTROL: Plant resistant cultivars.

Covered Smut

This disease is also called kernel smut.

CAUSE: *Sphacelotheca macrospora* Yen and Wang infects ovules in the spikelet, replacing seed with smut spores.

DISTRIBUTION: Taiwan.

SYMPTOMS: Infected plants appear normal until flowering time when ovaries appear slightly swollen and elliptical or ovoid in shape. At first the structure is covered by a gray-green membrane that later turns brown. The membrane ruptures, exposing a blackish, powdery mass of chlamydospores surrounding a sticklike central column.

CONTROL: No control is necessary.

Other Fungi That Cause Smut

Sphacelotheca consimilis Thirum.
S. cruenta (Kuhn) Potter (syn. *S. chrysopogonis* Clint, *S. holci* Jackson, and *Ustilago cruenta* Kuhn).
S. erianthi (H. & P. Syd.) Mundkur (syn. *U. erianthi* H. & P. Syd.).
S. papuae Zundel.

S. pulverulenta (Cooke & Massee) Ling (syn. *Cintractica pulverulenta* Cooke & Massee, *U. pulverulenta* Cif. and *U. pulverulenta* Boedjin).
S. sacchari (Rabenh.) Cif. (syn. *U. sacchari* Rabehn. and *U. sacchari-ciliaris* (Bref.)).
S. schweinfurthiana (Thum.) Sacc. (syn. *U. schweinfurthiana* (Thum.)).
S. schweinfurthiana (Thum.) Sacc. var. *minor* Zundel.
Sorosporium indicum Mundkur.

Curvularia Leaf Spot and Seedling Blight

CAUSE: *Curvularia lunata* (Wakker) Boed., teleomorph *Cochliobolus lunatus* Nelson & Haasis.

DISTRIBUTION: Africa, Argentina, Asia, Hawaii, and India.

SYMPTOMS: Leaves have circular to oval, scattered, red-brown spots. Spots eventually enlarge into irregular patches that turn dark brown. Leaves may become chlorotic, necrotic, and dry up. Seedings may also become infected and die.

CONTROL: No control is reported.

Cytospora Sheath Rot

CAUSE: *Cytospora sacchari* Butl.

DISTRIBUTION: Widespread; however, it is considered of minor economic importance.

SYMPTOMS: Leaves die from tips. Severely diseased shoots are killed.

CONTROL: No control is reported.

Downy Mildew

This disease is also called leaf stripe and Sclerospora leaf stripe.

CAUSE: *Peronosclerospora sacchari* (T. Miyake) C. G. Shaw (syn. *Sclerospora saachari* T. Miyake). The fungus probably survives as oospores in old shredded leaves. The fungus can infect other plants, particularly corn, *Zea mays* L., and teosinte, *Euchlaena mexicana* Schrad. These plants may be important in the spread of downy mildew but are of little importance in perpetuating the disease from one season to the next. The fungus is probably transmitted mainly by infected seed pieces and conidia.

Conidia are produced on infected plants at an optimum temperature of 2–25°C and a relative humidity above 90 percent. Sporulation also depends on actual sunlight or radiant energy beyond a certain level. Conidia are splashed or airborne up to 400 m from their source. Conidia germinate in the presence of free moisture. Infection probably takes place either on the very young leaf tissues while they are still in the spindle or in the young buds on the stalk. Oospores are produced within mesophyll tissue of the leaf as the cane matures.

DISTRIBUTION: Downy mildew occurs only in the Eastern Hemisphere and has been reported from Australia, Fiji, India, Japan, New Guinea, the Philippines, Taiwan, and Thailand.

SYMPTOMS: Well-defined chlorotic stripes occur on leaves as they emerge from the spindle and become more prominent as leaves elongate. Stripes at first are light green but become yellow as the leaves mature. Stripes are separated by green tissue and are confined between the larger leaf veins. They are about 1–25 mm wide and vary in length from 2.5 to several centimeters, often extending the length of the blade. During high humidity and temperatures, a velvety, whitish growth consisting of conidia and conidiophores, occurs on stripes. As leaves become older, the stripes become dark red.

Following secondary infection, stalks increase in height until some are almost twice as tall as healthy plants. Stalks are light, brittle, and watery. Such a symptom has been called the jump-up stage. Spindle leaves are short, discolored, with yellow stripes; they are fewer in number, fail to unfold, and cling together at the top. These leaves wither, become twisted, and eventually shred.

In contrast, plants that grow from infected seed pieces are severely stunted. Often their leaves are twisted and shredded.

CONTROL:
1. Plant resistant cultivars.
2. Obtain planting material from a disease-free source.
3. Harvest diseased plants early and plow the field.
4. Rogue out diseased stools in fields that have ratooned early.

Downy Mildew

CAUSE: *Peronosclerospora spontanea* (Weston) C. G. Shaw (syn. *Sclerospora spontanei* Weston).

DISTRIBUTION: Philippines and Taiwan.

SYMPTOMS: Similar to downy mildew caused by *P. sacchari*.

CONTROL: Similar to downy mildew caused by *P. sacchari*.

Downy Mildew,
Leaf-Splitting Form

This disease is also called leaf-splitting disease.

CAUSE: *Peronosclerospora miscanthi* (T. Miyake) C. G. Shaw (syn. *Sclerospora miscanthi* Miyake) persists in soil and plant residue as oospores. Infection can be caused by germination of oospores and conidia. Conidia are produced during periods of high humidity and are windborne. Oospores are abundantly produced in leaf tissue as plants mature. Oospores are disseminated by any means that will transport soil. The fungus may also be disseminated by planting diseased seed cane.

DISTRIBUTION: Fiji, Philippines, Papua New Guinea, and Taiwan. The disease is considered of minor importance.

SYMPTOMS: Initially green-yellow stripes are produced that often extend the entire length of the leaf. The color of the stripe progressively changes from yellow to red-brown and finally to dark red. Eventually the leaf tissue disintegrates and splits along the line of the lesion, changing the leaf into a whiplike bundle of fibers. White velvety masses of conidia and conidiophores are produced on the lower surface of leaf lesions.

CONTROL:
1. Plant resistant cultivars.
2. Place cuttings in water at 46°C for 20 minutes and then at 52°C for 20 minutes to eliminate fungus.

Drechslera Leaf Spot
and Seedling Blight

CAUSE: *Drechslera spicifera* (Bainier) von Arx (syn. *Bipolaris spicifera* (Bainier) Subram. and *Helminthosporium spicifera* (Bainier) Nicot.). Teleomorph is *Cochliobolus spicifer* Nelson.

DISTRIBUTION: India.

SYMPTOMS: Symptoms on seedling leaves are small, elongated to elliptic, reddish specks, with a dark reddish margin. Leaf tips die and dry up. Lesions may enlarge and coalesce. In seedling blight phase, plants are stunted with poor root development. Eventually plants wilt and die.

CONTROL: No control is necessary since this is considered an unimportant disease.

Dry Top Rot

CAUSE: *Sorosphaera vascularum* (Matz) M. T. Cook likely survives in residue and soil. A possible life cycle based on what is known about closely related genera is as follows: The fungus persists as resting spores within infected residue. As the residue decomposes, the resting spores are liberated into the soil, where under favorable conditions they germinate to form a zoospore. Zoospores swim to a root hair where they encyst and form an infection hypha that penetrates into the plant. A plasmodium is formed within the plant cells and spreads from cell to cell within the plant. The plasmodia in older-infected cells may have several cell walls form within it, transforming the plasmodia into several zoosporangia. Zoospores may again form within the zoosporangia. The zoosporangia may be liberated into the soil where it will rupture to release the zoospores. The zoospores may again infect a root directly or fuse with a compatible zoospore to form a zygote. The zygote may then infect the root.

Besides infected residue and spores being disseminated by water or by other means, the fungus is also disseminated within seed cane. The resulting disease is most severe in cane grown on wet soils.

DISTRIBUTION: Barbados, Cuba, Puerto Rico, and Venezuela.

SYMPTOMS: At first, leaves lose their green color, then roll and wilt. Growing points may then die, with the entire stalk wilting and becoming stunted. Internodes at the top of the plant are stunted and thinner, giving the entire stalk a tapered appearance.

Vascular bundles in the lower portion of the stalk are discolored orange, yellow, pink, or red, which is continuous through the internode. The discoloration is due to zoosporangia and plasmodia that clog the vascular system, causing the top portion of the plant to die.

CONTROL:
1. Plant disease-free seed cane.
2. If only infected cane is available for seed, take cuttings from the top of young cane.
3. Rotate sugarcane with legumes.

Ergot

CAUSE: *Claviceps purpurea* (Fr.) Tul. in temperate regions and *C. pusilla* Ces. in warmer regions. *Balansia* sp. is also reported to cause ergot. Disease is favored by humid weather. Germination of sclerotia of *C. purpurea* is favored by darkness and a temperature of 35–40°C.

DISTRIBUTION: *Claviceps purpurea* is distributed in the Philippines, India, and Australia. *Claviceps pusilla* is widely distributed in warmer regions of the world.

SYMPTOMS: Initial symptoms are stickiness and drooping of the arrow when spikelet blooming is most active. The inflorescence rapidly turns black due to saprophytic growth. Eventually the typical hard, black, sclerotial bodies replace kernels.

CONTROL: No control is usually necessary. Elimination of susceptible material in and adjacent to sugarcane fields has been suggested as an aid to control.

Eyespot

CAUSE: *Bipolaris sacchari* (Butl.) Shoem. (syn. *Helminthosporium sacchari* (van Breda de Haan) Butler and *Drechslera sacchari* (Butl.) Subram. & Jain). The fungus may persist in residue in areas where cane is not grown during the winter. Conidia are produced on lesions and residue and are windborne a considerable distance. The fungus is also seedborne. The disease is most severe during periods of high moisture and temperatures of 24–31°C. Eyespot is most severe during the winter months in the tropics.

DISTRIBUTION: Generally distributed wherever sugarcane is grown.

SYMPTOMS: Initially the eyespot lesion is a small, watery spot that is darker green than healthy tissue. In three to four days the spot becomes tan; then its center becomes red-brown. A typical spot is oval, 6–10 mm in length, and red-brown with a tan margin. Eventually spots become gray with a less noticeable halo. The red-brown area lengthens to form a streak or runner from the spot to the tip of the leaf. Occasionally a short red-brown streak develops in the opposite direction. Spots and runners may coalesce. As a leaf dies its tip becomes red-brown, causing a field of infected plants to turn a brownish cast. Infection of leaves in the spindle may allow the disease to move into the terminal portion of the stalk and cause a top rot.

Seedling infection is typified by the red-brown spots occurring on the coleoptile a few days after planting. Leaves then become chlorotic and seedlings die. Stalks of highly susceptible cultivars may be infected and have elongated brown lesions.

CONTROL:
1. Plant resistant cultivars.
2. Do not apply large amounts of nitrogen fertilizer before eyespot is likely to occur.

3. A two-year cropping system has been suggested where sugarcane is planted and harvested every three years, immediately
following the eyespot season.

Foliage Blight of Seedlings

CAUSE: *Exserohilum rostratum* (Drechs.) Leonard & Suggs. emend
Leonard (syn. *Helminthosporium rostratum* Drechs., *Drechslera ros-
trata* (Drechs.) Richardson & Frazer, and *Bipolaris rostrata* (Drechs.)
Shoemaker). The teleomorph *Setosphaeria rostrata* Leonard.

DISTRIBUTION: Widespread.

SYMPTOMS: Narrow reddish stripes occur on seedling leaves. Spots
eventually enlarge, coalesce, and become dark brown. Leaf sheaths
become dark brown to olivaceous because of the presence of conidia and
conidiophores. Plants are usually stunted.

CONTROL: No control is reported.

Fusarium Sett or Stem Rot

CAUSE: *Fusarium moniliforme* Sheldon, teleomorph *Gibberella
fujikuroi* (Sawada) Wollenw. (syn. *G. moniliformis* (Sheld.) Wineland),
persists in a wide range of plant residue and is likely as perithecia in
infected cane. Large numbers of conidia are produced on infected residue.
Dissemination is mainly by conidia and ascospores being airborne. The
fungus infects through ends of cane cuttings, young adventitious roots,
and nodal leaf scars of the stalk portion planted in infested soil. The
common sugarcane borer, *Diatraea saccharalis* (Fabricius), and its parasite, *Bassus stigmaterus*, also disseminate and aid infection by wounding the stalk and providing an entrance for insect or airborne spores.
Roots may also be infected without being wounded.

DISTRIBUTION: Generally distributed wherever sugarcane is grown.

SYMPTOMS: Vascular bundles become red starting from the cut ends.
Young roots become purple, then decay. Buds may either swell but
not germinate, or germinate with slight growth occurring. When such
sets are split, the vascular bundles are purple-red and the surrounding
parenchyma tissue is brown.
   In standing cane, nodes and internodes may be infected where infested
by the cane borer. Again vascular bundles are dark red and surrounding
parenchyma tissue is a purplish or brown-red. Leaves turn yellow and
dry up. The top dies even though stalk tissue appears healthy. If harvest
is delayed, the entire stalk will decompose.

CONTROL:
1. Plant resistant cultivars.
2. Dip or spray cuttings with a fungicide.
3. Rotating sugarcane with nonsusceptible crops apparently is an aid in control.

Iliau

CAUSE: *Clypeoporthe iliau* (Lyon) Barr (syn. *Gnomonia iliau* Lyon), anamorph *Melanconium iliau* Lyon. Primarily young plants are attacked in cool, moist weather. Perithecia and pycnidia-like structures in which conidia are produced are commonly found submerged in leaf sheath tissue of mature plants, especially during the autumn. Ascospores are disseminated by wind and water. Conidia are exuded out of pycnidia during moist weather and are disseminated by splashing water and less commonly by wind. The fungi are also disseminated by infected seed cane. Seedlings produced from infected seed cane are infected early and may die before stalk elongation. The fungus spreads rapidly from stalk to stalk when seed cane is stored in covered mats or piles.

DISTRIBUTION: Australia, Brazil, Cuba, Philippines, Hawaii, and Louisiana.

SYMPTOMS: Infected plants remain small and many break over and die. Overlapping leaf sheaths adhere together due to growth of mycelium. Outer leaves die and the white fungal mycelium can be seen after leaf sheaths are stripped. When the fungus grows into the stalk, there is a necrotic area just inside the rind that is reddish while the rest of the stalk interior remains white. The stalk shrivels, becomes weakened, and often breaks. The surface of necrotic tissue becomes rough due to development of perithecia within the tissue whose sharp beaks project above the surface.

CONTROL: Most cultivars are resistant.

Leaf Blast

This disease is also called leaf blast of sugarcane and leaf spot of sugarcane.

CAUSE: *Paraphaeosphaeria michotii* (Westend.) O. Ericks. (syn. *Didymosphaeria taiwanensis* Yen & Chi) likely persists as perithecia in infected residue. Leaves injured by frost apparently are most susceptible to infection.

DISTRIBUTION: Widespread. Leaf blast is considered of minor importance.

SYMPTOMS: Yellow spots (3–25 mm long and 0.5–1.0 mm wide) occur on both leaf surfaces. The spots soon become purple-red and coalesce to form long narrow streaks that turn the entire leaf a purple-red. Leaves may wither and die from the tip back. Perithecia that appear as small black objects develop in the dead portions of the leaves.

CONTROL: No control is necessary.

Leaf Blight

CAUSE: *Leptosphaeria taiwanensis* Yen & Chi., anamorph *Stagonospora taiwanensis* Hsieh. The fungus causes leaf blight throughout the year in high rainfall areas.

DISTRIBUTION: Japan, Okinawa, Philippines, and Taiwan.

SYMPTOMS: Small, narrow, elliptical or elongate, spindle-shaped, yellowish spots with red are produced on both sides of immature leaves in the spindle. The spots become red-brown and elongate into streaks that often coalesce. Severely diseased leaves die and dry up, giving infected plants a red-brown appearance when viewed from a distance. Infection occasionally occurs on a sheath and appears as purple-red lesions, especially in the later stages of the disease. Perithecia that appear as small black specks are produced in the margin of the oldest lesions.

CONTROL: Plant resistant cultivars.

Leaf Scorch (Leptosphaeria)

CAUSE: *Leptosphaeria bicolor* Kaiser, Nimande & Hawksworth, anamorph *Stagnospora* sp.

DISTRIBUTION: Kenya.

SYMPTOMS: Spindle-shaped lesions eventually enlarge and coalesce to give the leaf a scorched appearance.

CONTROL: No control is reported.

Leaf Scorch (Stagnospora)

CAUSE: *Stagnospora sacchari* Lo & Ling persists in infected leaves as mycelium and pycnidia. The fungus apparently does not survive well in the soil. During moist weather, the conidia are exuded out of pycnidia in a gelatinous mass that sticks firmly to the leaf surface when dry.

Subsequent moisture or rain loosens conidia and disseminates them to adjacent plants or leaves. Conidia are rarely airborne. Diseased leaves that adhere to the seed piece also provide a source of inoculum. Disease severity is dependent on the amount of moisture.

DISTRIBUTION: Asia, Africa, Central and South America.

SYMPTOMS: At first, very small red or red-brown spots occur on leaves, especially young leaves. The spots enlarge and become spindle-shaped with a yellow margin. Eventually the spots coalesce and extend along the vascular bundles for a length of 5–17 cm or more and become spindlelike streaks. Initially the streaks are red-brown but later become tan with a dark red border. Numerous pycnidia that appear as small, black, pimplelike objects develop in older lesions. Occasionally lesions extend to the upper part of the leaf sheath but no pycnidia are ordinarily produced there. During dry weather, a severely infected leaf will appear scorched. Initial spots on older leaves remain as small lesions and do not develop into streaks.

CONTROL: Plant resistant cultivars.

Myriogenospora Leaf Binding

This disease is also called leaf binding.

CAUSE: *Myriogenospora aciculisporae* Vizioli likely persists as perithecia in infected residue.

DISTRIBUTION: Argentina, Brazil, and Louisiana. The disease is rare.

SYMPTOMS: Plants are stunted and the tips of unfolded leaves adhere to adjacent older leaves. As the new leaf elongates, its tip is held firmly along the midrib of the previously unfolded leaf by a black stroma of the fungus, giving the plant a whiplike appearance. Although the growing point and the entire shoot may be killed, new shoots commonly develop from buds formed later or basal buds. A few healthy shoots may develop in diseased stools. Perithecia may occur in tips of leaves in the spindle.

CONTROL: No control is necessary because of the rarity of the disease.

Orange Rust

This disease is also called sugarcane leaf rust.

CAUSE: *Puccinia kuehnii* (Kruger) Butl. Urediospores and teliospores are produced on sugarcane. Alternate hosts and other spore stages are not known; however, *P. kuehnii* has been reported on the spontaneums, wild sugarcane, and related genera as collateral hosts. Urediospores are disseminated by wind and water. Infection occurs under humid conditions at 16–29°C.

DISTRIBUTION: Asia, Australia, and Oceania.

SYMPTOMS: Symptoms are similar to those caused by *P. melanocephala* H. & P. Syd. (common rust) except that uredia are cinnamon to yellow-brown.

CONTROL: Plant resistant cultivars.

Philippine Downy Mildew

This disease is also called sleepy disease.

CAUSE: *Peronosclerospora philippinesis* (Weston) C. G. Shaw (syn. *Sclerospora indica* Butl. and *S. philippinesis* Weston).

DISTRIBUTION: India and Philippines.

SYMPTOMS: Seedling leaves have chlorotic streaks. Mature leaves have necrotic streaks and are frequently deformed.

CONTROL: No control is reported.

Phytophthora Seed Piece Rot

CAUSE: *Phytophthora megasperma* Drechs. and *P. erythroseptica* Pethyb. probably persist in soil and infected residue as oospores. Sporangia occasionally germinate directly by producing a germ tube or usually indirectly by producing zoospores. The zoospores swim through soil water to the host where they encyst and germinate to form a germ tube that penetrates the host. Oospores are produced in infected tissue. Infection occurs more frequently through the nodal areas than through the cut ends. The disease is most severe when a long, cool spring follows a cold, wet winter.

DISTRIBUTION: Louisiana.

SYMPTOMS: Seed pieces planted in infested soil do not produce roots and few buds germinate. Initially the interior of an infected seed piece is water-soaked; then pink to orange-red streaks appear. With yellow-stalked cultivars, the streaks appear through the rind. Streaks also may be seen in the rind of cultivars with a dark-colored rind, but the streaks are not as distinct and are darker colored. Eventually the entire seed piece becomes water-soaked and the color darkens to red-brown with an etherlike odor.

CONTROL:
1. Plant resistant cultivars.
2. Improve soil drainage.

Pineapple Disease

CAUSE: *Ceratocystis paradoxa* (Dade) Moreau, anamorph *Thielaviopsis paradoxa* (de Seyn) V. Hohn, persists as a saprophyte on residue in the soil. The fungus infects seed pieces in the soil by conidia growing into the cut ends. Sometimes seed pieces are infected prior to being planted. Standing cane is sometimes infected by airborne spores that enter stalks through injuries caused by rodents, insects, or machinery. The disease is often most severe in poorly drained areas where soil remains wet and relatively cool.

DISTRIBUTION: Generally distributed wherever sugarcane is grown.

SYMPTOMS: Primarily sugarcane cuttings are affected. The disease affects the central core of the stalk. Initially it becomes reddened but stays firm. Later the interior of the stalk breaks down and becomes hollow and black. The vascular bundles remain intact. Cuttings may decay before buds germinate or shoots may die back after growing only a few centimeters. In the early stages of decomposition, an odor may occur that resembles that of pineapple.

Stalks of standing cane may become infected through an injury, causing leaves to wilt, followed by death of the stalk.

CONTROL:
1. A cutting of at least three nodes should be made to protect the center node.
2. Do not plant under conditions of low soil temperatures, or excessive wetness or dryness. Do not plant too deep.
3. Protect seed piece ends with a fungicide.

Pokkah Boeng

This disease is also called Fusarium sett and stem rot, knife cut, and Pokkah boeng chlorosis.

CAUSE: *Fusarium moniliforme* Sheldon, teleomorph *Gibberella fujikuroi* (Sawada) Wollenw. persists as a saprophyte in residue and probably as perithecia in infected cane. Large numbers of conidia are produced on infected residue. The fungus is disseminated mainly by conidia and ascospores being airborne. Infection occurs through the spindle along the margin of a partially unfolded leaf where a small opening is formed during periods of dry weather. Conidia enter the spindle and are carried down to the susceptible region where they germinate during wet weather. Mycelium grows through the soft cuticle to the inner tissues.

DISTRIBUTION: Generally distributed wherever sugarcane is grown.

SYMPTOMS: Pokkah boeng occurs when plants are in a rapid stage of growth, usually three to seven months old. A chlorotic area develops on the basal portion of certain leaves as they emerge from the spindle, mostly the lower leaves. Red spots or stripes often develop on the chlorotic areas. Usually affected areas are confined to only a few leaves on one side of the spindle but sometimes the young stem is penetrated. Infected leaves are frequently deformed and do not unfold normally but remain short. Often infected areas are torn. Sometimes cavities develop in the internode and present a ladderlike appearance. Occasionally during wet weather a soft rot develops in affected areas. Usually affected plants recover and new leaves develop normally.

CONTROL: Plant resistant cultivars.

Purple Spot of Sugarcane

CAUSE: *Pseudocercospora rubropurpurea* (Sun) Yen (syn. *Cercospora rubropurpurea* Sun).

DISTRIBUTION: Taiwan.

SYMPTOMS: Leaf spots are most distinct on the upper surface. They are irregular to circular (2–12 mm diameter) and red-purple to dark purple.

CONTROL: No control is reported.

Pythium Root Rot

CAUSE: *Pythium arrhenomanes* Drechs., *P. aphanidermatum* (Edson) Fitz., *P. catenulatum* Matthews, *P. dissotocum* Drechs., *P. graminicola* Subr., *P. mamillatum* Meurs., and *P. splendens* Braun.

Root rot caused by *P. arrhenomanes* is favored by high early-winter temperatures followed by alternate wet and dry periods in autumn but not in summer plantings. Shoots growing in loam soils are more susceptible than those growing in sand.

DISTRIBUTION: Widely distributed.

SYMPTOMS: Necrotic, dark, reddish lesions on roots. Root tips are flaccid, water-soaked, and necrotic. The fine feeder roots are reduced in growth.

   Shoots exhibit stunting, wilting, and may be killed eventually. Poor ratooning may also occur.

CONTROL: Plant in well-drained soils.

Red Leaf Spot

This disease is also called purple spot.

CAUSE: *Dimeriella sacchari* (van Breda de Haan) Hansford et al. persists as perithecia in infected residue.

DISTRIBUTION: Australia, Bangladesh, Cuba, Fiji, Java, Japan, Nepal, New Guinea, Panama, Philippines, Taiwan, Tanzania, Trinidad, and Tobago. The disease is of minor importance.

SYMPTOMS: Leaf tips are more heavily infected than other parts of the leaf. Spots begin as red dots but later become round or elliptical, 0.5–2.0 mm in diameter, with a yellow margin. Later the color changes to purple-red. Perithecia appear as small black specks in older spots. Premature death of leaves may occur with heavy infections.

CONTROL: No control is necessary.

Red Rot

CAUSE: *Glomerella tucumanensis* Speg. von Arx & Muller (syn. *Physalospora tucumanensis* Speg.), anamorph *Colletotrichum falcatum* Went (syn. *C. graminicola* (Ces.) G. W. Wilson). The fungus persists in infected plants. Cuttings are a major source of infection due to conidia or mycelium being present on or in the seed piece. Incipient infections may be present in bud scales, leaf scars, and other nodal tissues or in stalk borer tunnels. Conidia and ascospores are produced in this infected tissue but are mostly water disseminated; therefore, it takes a relatively long time for leaves higher on plants to become infected. Conidia are produced directly on mycelium or in acervuli in lesions on midribs. This is the principal source of infection for the stalks. Conidia are washed down behind leaf sheaths, germinate, and become established in certain tissues of the node and internodes. Seed pieces taken from this cane are internally infected or have dormant mycelium in bud scales, leaf scars, and other nodal tissues. Conidia are also present on the cutting surface. Injuries caused by insects and machinery predispose sugarcane to infection.

DISTRIBUTION: Generally distributed wherever sugarcane is grown.

SYMPTOMS: All plant parts may be infected but the disease is most important on standing stalks, planted cuttings, and leaf midribs. In the early stage of the disease, stalks display no external symptoms. Tissues inside the stalk become a dull red. These discolored, reddish areas give a mottled appearance to the stalk interior. Red vascular bundles pass through red areas and extend into healthy tissues. The reddish discoloration may eventually occur through the length of the stalk with development of cavities that contain either mycelium or a clear liquid. Eventually the infected internal tissues become darkened, shrink, and dry out. Severely infected plants may have the outside stalk become red-brown, with acervuli present that appear as dark tufts to the eye or black hairlike objects under magnification.

Seed pieces may rot during periods that are adverse to sugarcane growth such as conditions being too wet or too dry. Infected seed pieces are rotted with various shades of red, brown, or gray discoloration. Gray mycelium develops in pith cavities and acervuli develop on the exterior of the cane. Nodal tissue of susceptible cultivars becomes red to almost black.

Initial symptoms on leaves are small red spots on the upper surface of the midrib. Spots enlarge to produce elongated lesions, then may fuse to form long lesions that extend the entire length of the leaf. Spots become tan with dark red margins and are covered with acervuli that look like numerous black tufts. Leaves may wilt and dry.

CONTROL:
1. Plant resistant cultivars.
2. Use only healthy cane for seed pieces.
3. Plant when conditions are proper for rapid germination. Maintain proper soil moisture.
4. Harvest susceptible cultivars before they have passed the peak of maturity.

Red Rot of the Leaf Sheath

CAUSE: *Athelia rolfsii* (Curzi) Tu & Kimbrough (syn. *Botryobasidium rolfsii* (Curzi) Venkatanarayan, and *Pellicularia rolfsii* (West). The anamorph is *Sclerotium rolfsii* Sacc. The fungus persists as sclerotia in soil or residue. The disease is favored by warm, wet weather.

DISTRIBUTION: Generally distributed wherever sugarcane is grown.

SYMPTOMS: Irregular-shaped, orange-red areas, with distinct margins, develop on lower leaf sheaths. White mycelium grows into the inner sheaths and binds them loosely together.

The rind of the underlying stalk becomes light brown and has a distinct margin. As the stalk matures, the affected area dies and dries up, causing an extensive shallow canker. Small, round, brownish sclerotia develop between sheaths, along the edges or over the surface of a diseased area. In severe cases the leaf gradually dries up and dies. Most shoots of a diseased stool are thin, stunted, with many short internodes.

CONTROL: No control is necessary.

Red Spot

CAUSE: *Helminthosporium rostratum* Drechs.

DISTRIBUTION: Mississippi. It is thought to be of little economic importance.

SYMPTOMS: On midribs, lesions are bright red and 15–25 mm by 20–35 mm. Lesions are not sunken.

CONTROL: No control is reported.

Red Spot of Leaf Sheath

This disease is also called red leaf sheath rot of sugarcane.

CAUSE: *Cercospora vaginae* Kruger (syn. *Mycovellosiella vaginae* (Kruger) Deighton). The disease is favored by warm, moist weather.

DISTRIBUTION: Generally distributed wherever sugarcane is grown. The disease is of minor importance.

SYMPTOMS: Small, round, bright red spots with sharp margins occur on the upper leaf sheaths. Spots coalesce to form bright red, irregularly shaped patches that extend through to the inner sheaths. Later dark brown to black conidia and conidiophores that appear like a sooty coating are produced on the affected areas but most abundantly on the inside of the sheaths.

CONTROL: No control is necessary.

Rhizoctonia Sheath and Shoot Rot

CAUSE: *Rhizoctonia solani* Kuhn. Teleomorph is *Thanatephorous cucumeris* (Frank) Donk. (syn. *Hypochnus cucumeris* Frank, *H. filamentosus* Pat., *Pellicularia filamentosa* (Pat.) Rogers, *Ceratobasidium filamentosum* (Pat.) Olive, *H. solani* Prill. & Delacr., *Corticium solani* (Prill. & Delacr.) Bourd. & Galz., *Botryobasidium solani* (Prill. & Delacr.) Donk, *Ceratobasidium solani* (Prill. & Delacr.) Pilat, and *Corticium vagum* Berk. & Curt.).

DISTRIBUTION: Generally distributed wherever sugarcane is grown.

SYMPTOMS: Similar to those of banded sclerotial disease.

CONTROL: None is necessary.

Rind Disease

This disease is also called sour smell and sour rot.

CAUSE: *Phaeocytostroma sacchari* (Ell. & Ev.) Sutton var. *sacchari* (syn. *Melanconium sacchari* Mass. ap. Speg. and *Pleocyta sacchari* (Mass.) Petr. & Syd.). The fungus produces stroma inside infected tissues. Conidia are produced inside a stroma that eventually enlarges and develops pressure that splits the epidermis, forming a pustule. Conidia are then exuded in a long, gelatinous, threadlike structure through the top of the pustule during moist conditions. The conidia are disseminated by wind or rain. Infection normally occurs through wounds and in plants that are not growing vigorously due to drouth or overmaturity.

DISTRIBUTION: Australia, Hawaii, and the West Indies.

SYMPTOMS: Symptoms are most common on stalks but leaf blades and sheaths may also display them. Numerous black, coiled, hairlike fruiting structures of the fungus exude from pustules, breaking through the surface of infected tissue. The internal stalk tissue becomes dark and has a sour smell. Leaves may yellow, dry up, and be covered with the black coiled structures. A symptom from Hawaii is the nodes turning red-brown, and later the internodes become light brown.

CONTROL: Harvest mature cane as soon as possible.

Ring Spot

This disease is also called ring spot of sugarcane leaf.

CAUSE: *Leptosphaeria sacchari* B. de Haan, anamorph *Phyllosticta saccharicola* P. Henn. The fungus persists as a saprophyte on dead leaves.

DISTRIBUTION: Generally distributed wherever sugarcane is grown. The disease is usually of minor importance.

SYMPTOMS: Symptoms occur on leaf blades and leaf sheaths. At first, spots are dark green to brownish but later become red-brown, diamond-shaped, or oval, and are surrounded by a narrow yellowish border. Later the center of the spot becomes tan and is surrounded by a narrow red margin. Often an older spot may still be surrounded by a yellow margin. Older spots often have fruiting structures of various fungi that appear as blackish specks.

CONTROL: Plant resistant cultivars.

Ring Spot of Sugarcane

CAUSE: *Pseudocercospora saccharicola* (Sun) Yen (syn. *Cercospora saccharicola* Sun).

DISTRIBUTION: Taiwan.

SYMPTOMS: Initially, leaf spots are circular to elliptic, dark green to brown, and they are generally surrounded by a narrow red-brown to maroon border. Eventually spots become gray and are surrounded by a purple-brown margin that is up to 7.5 mm wide.

CONTROL: No control is reported.

Sclerophthora Disease

This disease is also called Sclerospora disease and downy mildew.

CAUSE: *Sclerophthora macrospora* (Sacc.) Thirum. Shaw & Naras. (syn. *Sclerospora macrospora* Sacc.). The fungus persists in seed pieces and is transmitted by planting infected seed pieces. A high percentage of infected plants then grow from these infected seed pieces. The fungus also persists in a wide number of grass hosts. Sporangia or conidia are produced most abundantly on the underside of infected leaves during periods of heavy dew or fog. Sporangiophores grow through stomates and produce one to five sporangia on their tips that become windborne. Sporangia germinate indirectly in the presence of free water to produce zoospores. The zoospores swim to host tissue where they encyst and produce a germ tube that penetrates the tissue. The disease is almost entirely confined to low areas in a field that are subject to flooding. Young sugarcane plants have been inoculated by placing sporangia in the spindle. Oospores are abundantly produced in diseased leaf tissue but their role in survival and dissemination has not been determined as they have not been observed to germinate.

DISTRIBUTION: Australia, Mauritius, Peru, South Africa, and the United States. The disease is usually of minor importance.

SYMPTOMS: Stools are severely stunted but often there are some tall plants and a cluster of small dwarfed shoots. Often there is profuse tillering that gives the plant an appearance of a clump of coarse grass. When these stools produce cane, it is not uncommon to find stem galls of various sizes and shapes on them.

Affected leaves are small, coarse, and brittle. They have a yellow-green color due to irregular yellow-white streaks that are a half to several centimeters long and located between the veins. Chlorotic blotches of varying sizes and shapes may also occur on diseased leaves, giving a mosaic appearance.

During rapid growth, the leaves droop and the edges become wavy. However, during poor growing conditions the leaves become stiff and erect. The margins die and dry out, causing shredding and giving leaf edges a ragged appearance. Leaf tips may curl into one or more loops.

Using magnification, numerous white specks, which are the oospores, can be seen alongside the veins. No downy or external mycelial growth can be observed although sporangia are produced on leaf surfaces. Infection of older cane causes the top to develop symptoms while the plant portion below the infection remains healthy. Buds on the upper cane may proliferate to produce clusters of shoots at each node.

CONTROL:
1. Plant only seed pieces from healthy cane.
2. Do not plant in poorly drained areas in a field.
3. Rogue infected stools.
4. Destroy wild grasses adjacent to a cane field.

Sooty Mold

Capnodium sp. and *Fumago* sp. are fungi that grow in insect secretions. A dark brown to black film or crust is the fungal growth that occurs on leaves and sheaths.

Sugarcane Smut

This disease is also called culmicolous smut and inflorescence smut.

CAUSE: *Ustilago scitaminea* Syd. survives as chlamydospores in dry soil or residue; as a latent infection, that is, mycelium in planting material; and as spores on cuttings. Also, chlamydospores and sporidia in soil may be disseminated by water to infect cuttings.

Chlamydospores germinate under moist conditions to form a three or four-celled promycelium. The promycelium produces a hypha that acts either as an infection thread or sporidia bud from each cell. The sporidia, in turn, either germinate to form a hypha or bud off more sporidia.

Ustilago scitaminea is a parasite of meristemic tissues, entering only through the lower part of the bud below the scales. When an infected bud begins to grow, the fungus also grows just behind the growing tip. The apical meristem is stimulated to produce the whiplike appendage in which chlamydospores are produced. The enveloping membrane is ruptured, chlamydospores are exposed, become windborne, and come in contact with standing cane or fall to the ground. Infected buds on standing cane may either give rise to smutted whips the same season or mycelium may remain dormant with the buds. True seedlings may also become infected.

Disease severity is greater during a dry season when plants are predisposed to infection. Additionally, large populations of spores survive under dry soil conditions. Wounding increases infection.

DISTRIBUTION: Generally distributed wherever sugarcane is grown.

SYMPTOMS: When infection occurs early, a clump of spindly canes or small grassy shoots is produced followed by the appearance of smutted whips. At first, infected canes have small, narrow leaves, and slender stalks with widely spaced nodes. The typical symptom is a long whiplike structure that varies in size from a few to several centimeters and is produced from the apex of the affected stalk. Short whips are curved and long ones double back. The whip is narrow, unbranched, with dark brown to blackish chlamydospores surrounded by a thin silvery membrane. Eventually the membrane ruptures, exposing the dark mass of chlamydospores that are soon blown away, leaving only a core parenchyma and vascular elements.

After the production of the smutted top, buds lower on the stalk begin to grow and may terminate in a black whip. Sometimes smutted side shoots occur on an apparently healthy stalk.

Some unusual symptoms have been described from Hawaii. Whips often remained confined under the leaf sheaths and have a characteristic serpentine form. Buds from which lateral bud sori originated are larger than usual infected buds and flattened and elongated.

Callus outgrowths are atypical symptoms and are most abundant at a position on the stalk between normal-appearing tissue and the area showing typical smut symptoms. Galls in this region are generally limited to nodal areas and initiated either in the region of the root band, intercalary meristem, or at a leaf sheath scar. Outgrowths sometimes appear randomly on internodes that are usually positioned near or adjacent to the base of a whip. Initially callus is smooth and lacks chlorophyll. Eventually buds or leaflike structures are differentiated. Tissues form and develop into individual buds. Forty to sixty buds on a gall is not uncommon. Buds continue to grow and terminate by forming a whip. Tissues resembling adventitious leaves and sheaths also differentiate at internodal and nodal positions.

Symptoms on true seedlings are the presence of early-developed, thin, grassy stems, with short, slender internodes and no visible buds or root initials. In the uppermost part of the plant, the leaf sheaths are absent and leaf blades are subtended by long, bristlelike hairs. Most seedlings die from severe lodging, but those that live produce a short, whiplike structure containing chlamydospores.

CONTROL:
1. Rogue out diseased shoots or stools.
2. Select healthy planting material.
3. Treat planting stock with a protectant fungicide.
4. Avoid ratooning of affected cane fields.
5. Rotate sugarcane with a resistant crop.
6. Plant resistant cultivars.

Other Smuts Caused by *Ustilago* spp.

Ustilago consimilis Syd.
U. courtoisii Cif.

Tar Spot

This disease is also called black spot of leaves.

CAUSE: *Phyllachora sacchari* P. Henn. Ascospores in residue may act as a source of primary infection. Spores are disseminated by wind and rain.

DISTRIBUTION: Argentina and Asia.

SYMPTOMS: Black fungal stroma on leaves resembles tar.

CONTROL: No control is reported.

Target Blotch

CAUSE: *Helminthosporium* sp.

DISTRIBUTION: Cuba, South Africa, and Florida. The disease is not economically important.

SYMPTOMS: Symptoms occur only in the wintertime on leaf blades and midribs of mature cane. Initially, small, red-brown spots develop that become tan to brownish necrotic areas with irregular concentric rings, resembling a target. Spots develop first on the rolled leaves in the leaf spindle but the target symptom becomes more pronounced as the leaves unroll. Spots cease to develop when leaves are unrolled from the spindle.

CONTROL: No control is necessary.

Veneer Blotch

CAUSE: *Deightoniella papuana* Shaw.

DISTRIBUTION: Papua New Guinea, New Britain, and British Solomon Islands.

SYMPTOMS: Small, oval, light green to tan spots with a thin redbrown margin occur on the leaves. Next, two more similar spots occur, one on each side of the original spot. Then two more spots occur, each flanking the second pair, and so on until up to 12 spots, each larger than the preceding spot, may be formed on each side of the original spot. Spots may become up to 61 cm long and 1.2 cm wide. The center of each spot finally becomes brown. Conidiophores cause a thick, matlike appearance in older spots on the underside of the leaf. Conidiophores occasionally form on the upper leaf surface. The lesion has a beautiful pattern, like figured veneer of wood.

CONTROL: No control is reported.

White Rash

This disease is also called anthracnose maculata, Elsinoe disease, spotted anthracnose, and white speck.

CAUSE: *Elsinoe sacchari* Lo., anamorph *Sphaceloma sacchari* Lo.

DISTRIBUTION: Brazil, Japan, Malaysia, Philippines, Puerto Rico, Taiwan, and Florida. The disease is considered to be of minor importance.

SYMPTOMS: Symptoms occur on leaf blades, midribs, and occasionally sheaths. At first, lesions are tiny yellowish, round, elliptic or fusiform spots. Some workers have described them as purplish. Later they turn yellow to tan and finally whitish, sometimes with red-brown margins. Spots average 0.2–1.0 mm by 0.1–0.5 mm. Individual spots are slightly elevated with distinct centers. An infected area has the appearance of lacework. Spots may coalesce to form elongate, narrow streaks.

CONTROL: No control is necessary.

Wilt

CAUSE: *Cephalosporium sacchari* Butler. However, some workers have ascribed wilt as being caused by *Gibberella fujikuroi* var. *subglutinans* Edwards. The fungus persists as mycelium in residue. Abundant microconidia are produced on mycelium. The fungus is disseminated by any means that will transport infected residue. Microconidia are also windborne. The fungus is also disseminated in infected seed cane that give rise to infected plants. Disease development is more severe at a soil pH between 7.0 and 8.0, and at an increase in carbon/nitrogen. Infection commonly occurs through insect wounds.

DISTRIBUTION: Generally distributed wherever sugarcane is grown.

SYMPTOMS: Symptoms are not apparent until cane is half grown and they continue until harvest time. The first noticeable symptom is the wilting of a single cane or an entire clum. Leaves at first become yellow, then dry up. Pith in the stem initially has streaks of light purple or reddish discoloration. The stems then become light and hollow, making them worthless for milling. Abundant microconidia can be seen in the mycelium under magnification. A foul odor is noticed after splitting an infected stalk. Cottony mycelial growth can be seen in these pith cavities. An infected seed piece usually fails to produce either eyes or roots. Occasionally an eye may swell or grow slightly before it dies. The interior of such a seed piece is a brown discoloration.

CONTROL:
1. Plant the most resistant cultivars.
2. Plant only healthy seed pieces.

Yellow Spot

CAUSE: *Mycovellosiella koepkei* (Kruger) Deighton (syn. *Cercospora koepkei* Kruger and *Pseudocercospora miscanthi* Katsuki). The fungus persists in infected residue. Conidia are produced on the ends of conidio-phores that usually emerge through stomata in disease spots during wet humid weather. Conidia are more numerous on the lower leaf surface than upper leaf surface and are splashed or blown to new infection sites.

DISTRIBUTION: Africa, Asia, Australia, Brazil, Colombia, and Cuba.

SYMPTOMS: Yellow-green spots that vary in size up to 12 mm occur on the youngest leaves. Spots may coalesce to cover large areas of the leaf and at maturity most of the leaf surface may be affected. Small reddened areas first appear on lower leaf surfaces, then increase on both surfaces, giving the plant and entire field a rusty yellow appearance at normal leaf maturity. A dirty, gray, fungal growth may also develop most abundantly on the lower leaf surface and sparsely on upper leaf surfaces. Consequently leaves shrivel and die prematurely.

CONTROL:
1. Plant resistant cultivars.
2. Apply foliar fungicides during the growing season.

Zonate Leaf Spot

CAUSE: *Gloeocercospora sorghi* D. Bain & Edg.

DISTRIBUTION: Areas where sorghum and sugarcane are grown in the same culture.

SYMPTOMS: See Zonate Leaf Spot of Sorghum, Chapter 17.

CONTROL:
1. Rotate sugarcane with a nonhost.
2. Plow under infected residue.

—————CAUSED BY NEMATODES—————

Lance Nematode

CAUSE: *Hoplolaimus columbus* Sher.

DISTRIBUTION: Unknown.

SYMPTOMS: Root systems are coarse and depleted. Top and root weights are not significantly reduced.

CONTROL: No control is reported.

Root Knot Nematode

CAUSE: *Meloidogyne* spp. Root knot nematodes multiply best in moderately dry, well-drained soils. Weeds in cane fields may also serve as hosts.

DISTRIBUTION: Generally distributed wherever sugarcane is grown.

SYMPTOMS: Knots or galls form at or near the root tips and appear as nodules or elongated thickenings. The thickened roots are often curled. The apical meristem is affected and growth is retarded or stopped with little or no proliferation of lateral roots.

CONTROL: Fumigate the soil with a volatile nematicide.

Root Lesion Nematode

CAUSE: *Pratylenchus* spp. The nematodes feed only in root cortex.

DISTRIBUTION: Australia, Hawaii, Mauritius, and Louisiana.

SYMPTOMS: Roots are thickened with discolored lesions and fewer fine roots.

CONTROL: Fumigate the soil with a volatile nematicide.

Spiral Nematode

CAUSE: *Helicotylenchus* spp., *Scutellanema* spp., and *Rotylenchulus* spp.

DISTRIBUTION: Generally distributed wherever sugarcane is grown.

SYMPTOMS: In pot experiments, roots were blunted, malformed, with a reduction in number of branch rootlets.

CONTROL: Fumigate the soil with a volatile nematicide.

VIRAL CAUSES

Chlorotic Streak

CAUSE: Sugarcane chlorotic streak virus (SCCSV) is not mechanically transmitted. Transmission was accomplished when healthy and diseased plants were grown together in quartz sand and when healthy plants were grown in soil obtained from around diseased plants in the field. This mode of transmission suggests a soilborne vector with infection through the roots. SCCSV is also transmitted in stalk cuttings used for seed and from plant to plant in the field in a manner that suggests an insect vector. The leafhopper *Draeculacephala portola* Ball has been implicated as a possible vector. SCCSV is also disseminated by running water.

The disease is most prevalent in low-lying, poorly drained areas and in plants growing in potassium-deficient soil. Several other grasses are also hosts including *Arundo donax* L., *Brachiaria mutica* Trin., *Panicum maximum* Jacq., and *Pennisetum purpureum* Schumach.

DISTRIBUTION: Australia, Guyana, Java, Puerto Rico, and the United States.

SYMPTOMS: Symptoms are more prevalent in ratoons than in plant cane. One to many yellowish to whitish streaks with wavy, irregular margins occur on both sides of leaves, leaf midribs, and sheaths. Streaks at first may be short, narrow, and very faint. Later, particularly on older foliage, the streaks are well marked, 3–15 mm wide, and may extend the entire length of the leaf blade. The centers of older streaks may become necrotic in sections or along the entire streak length.

Young plants may have stiff, erect leaves, and wilt, even in the presence of abundant moisture. Young shoots may die or become somewhat stunted, eventually causing an entire field to have an uneven appearance. Vascular bundles may be discolored entirely through the node.

Leaf symptoms may disappear when streaked leaves senesce. Symptoms may be most prevalent in early summer.

CONTROL:
1. Plant resistant cultivars.
2. Treat cane in a hot water treatment at 52°C.
3. Obtain disease-free planting material.
4. Rogue out infected plants in a seed field if disease incidence is less than 2 percent.
5. Improve soil drainage.

632 DISEASES OF SUGARCANE (*SACCHARUM OFFICINARUM* LINN.)

Dwarf

CAUSE: Sugarcane dwarf virus (SCDV) is not known to be mechanically transmitted nor are any vectors known. Infection may somehow occur from an alternate host plant or from soil but no evidence presently exists.

DISTRIBUTION: Australia.

SYMPTOMS: Infected plants may be stunted, with short, stiff, erect, distorted, fanlike tops. Numerous, small, white, longitudinal stripes occur most abundantly on younger leaves. The stripes are less than 1 mm wide and vary considerably in length. Wide, diffuse, chlorotic stripes occur mainly toward the margins of leaves. Leaf margins may become necrotic and have a papery texture. Transverse splitting may occur to the midrib.

Stools of susceptible cultivars in advanced stages of disease may appear as a group of grassy shoots. Infected stools that ratoon usually produce diseased plants.

CONTROL:
1. Rogue out diseased plants.
2. Plant resistant cultivars.
3. Plant only disease-free seed cane.

Fiji Disease

CAUSE: Sugarcane Fiji disease virus (SCFDV) is not mechanically transmitted. SCFDV is transmitted by at least two leafhoppers: *Perkinsiella saccharicida* Kirkaldy and *P. vastatrix* Breddin. Both species must evidently acquire the virus as nymphs. Other species such as *P. vitiensis* Breddin may also be vectors. Sugarcane is the only known host for the virus.

DISTRIBUTION: Australia, Fiji, Java, Philippines, and New Guinea.

SYMPTOMS: Elongated swellings or galls occur on the underside of the leaf. Galls extend along the larger veins and vascular bundles and may also be found along vascular bundles of the stalk when it is split open.

The last leaves to unfold from the spindle are shortened or crumpled. Up to that time, a diseased shoot may reach a normal height and produce healthy-looking leaves of the usual length and color until short, deformed, twisted leaves are suddenly produced. Growth then ceases completely. These leaves appear to have had the upper half or two-thirds of the leaf burned or scaled before expanding, leaving short, crumpled stumps.

The shoot may remain alive for several months or it may die in a short time. Galls are produced on both healthy-appearing and deformed leaves. The appearance of galls occurs during the advanced stage of the disease and apical leaf distortion usually occurs just prior to plant death.

CONTROL:
1. Plant resistant cultivars.
2. Plant only disease-free cuttings.
3. Harvest infected plants earlier than normal.
4. Plow under diseased plants.
5. Rogue out infected stools if disease is low.

Grassy Shoot

CAUSE: Sugarcane grassy shoot virus (SCGSV) is transmitted by infected seed pieces, mechanically by cutting knives, and by the aphids *Aphis idiosacchari*, *A. saachari* Zehntner, and *A. maidis* Fitch. Seed pieces can be infected by juice from chlorotic leaves. Besides sugarcane, Jowar appears to be another host.

DISTRIBUTION: India, Taiwan, and Thailand.

SYMPTOMS: Typically, a mass of stunted, crowded shoots arises from diseased seed pieces or ratoon stools. Some shoots may be devoid of chlorophyll, and if enough shoots are affected, the stools may die. Stalks that are produced in a diseased stool usually have long, delicate, scaly buds that may develop into slender chlorotic shoots.

Symptoms of secondary infection show in the leaves at the base of the stool. White to yellow, well-defined stripes that may become diffuse or remain distinct occur on thin and chlorotic leaves. Numerous tillers occur at the base of the stool.

CONTROL: Plant only disease-free cuttings.

Mosaic

CAUSE: Sugarcane mosaic virus (SCMV) has a wide host range and infects a large number of wild and cultivated grasses. The virus is transmitted by numerous aphid species including *Rhopalosiphum maidis* (Fitch), *Carolinaia cyperi* (Thos.), *Aphis gossypii* Glover, *Schizaphis graminum* (Rondani), and *Hysteroneura setariae* (Thomas). SCMV is also mechanically transmitted with difficulty and is rarely seedborne in corn. Several strains of SCMV exist.

DISTRIBUTION: Generally distributed wherever sugarcane is grown.

SYMPTOMS: Symptoms are most obvious in leaves newly unrolled from the spindle. Irregularly oval or oblong, pale green blotches of various sizes that are not delimited by veins occur on leaves. Blotches are of various widths throughout their length and do not resemble stripes. Leaves are not ordinarily distorted but some cultivars have stunted shoots that culminate in twisted and distorted leaves. Some cultivars have a mottling of the stem, causing death of the tissue and causing a cankered area. Small and deformed sticks may occur in other cultivars. Others may have areas of discoloration in internal tissue. Mosaic symptoms sometimes disappear from leaves of growing plants and healthy plants may grow from infected seed cane.

CONTROL:
1. Plant resistant cultivars.
2. Rogue out infected plants if disease incidence is low.

Sereh

CAUSE: Sugarcane sereh disease virus (SCSDV) is transmitted through cuttings. It is not mechanically transmitted nor are other vectors known.

DISTRIBUTION: Australia, Hawaii, India, Indonesia, Sri Lanka, Taiwan, and West Indies.

SYMPTOMS: Severe disease causes the cane stool to be converted into a bunchlike tuft due to shoots produced from an infected cutting, ceasing to grow after reaching a certain height. Usually shoots are of different heights. Secondary shoots arising from basal buds are similarly affected. In some stools, taller stalks may develop with shortened internodes near the top and leaves formed in a fanlike arrangement.

Sometimes hairy, adventitious roots may be produced on some nodes. Abundant side shooting may occur in some varieties.

Less severe disease symptoms are reddish vascular bundles in nodes and sometimes in internodes. Sometimes affected vascular bundles show as narrow, dull, red stripes on the leaf blades.

CONTROL:
1. Plant resistant cultivars.
2. Plant certified disease-free seed cane.
3. Do not grow ratoon crops.

Streak Disease

CAUSE: Sugarcane streak virus (SCSV) cannot be transmitted mechanically but is transmitted by the leafhopper *Cicadulina mibla* Naude, *C. bipunctella zeae* China, and other *Cicadulina* spp. Work done with corn strains shows the leafhoppers can acquire the virus as nymphs or adults in a few minutes of feeding and remain infective throughout life. Apparently several strains of the virus exist; each is virulent to one or a few hosts (mainly grasses) and avirulent or weakly virulent to other hosts. The virus is also transmitted in cuttings from diseased plants.

DISTRIBUTION: Africa.

SYMPTOMS: Symptoms are most obvious on the youngest leaves. A pattern of straight, narrow, uniform in width, translucent stripes that follow the veins occur on leaves. The stripes vary in length from 0.5–10 mm and are about 0.5 mm wide. Streaks may be relatively wide and may coalesce on leaves of young shoots developing from infected cuttings. The leaves may be crinkled and narrower than normal.

CONTROL: Resistant cultivars are being developed.

Striate Mosaic

CAUSE: Sugarcane striate mosaic virus (SCSMV). It is possible that more than one virus causes this disease; one causes the striations and the other causes the stunting. The viruses are sett-transmitted. The disease is associated with the poorer growth areas such as excessive sand or water-logged soils.

DISTRIBUTION: Australia in North Queensland.

SYMPTOMS: Numerous to few, short, 0.5–2.0 mm, fine striations occur on leaves. They are a lighter green, develop first on the youngest exposed leaf, and are difficult to see as the leaf matures. Striations are less numerous around the larger vascular bundles, giving the younger leaves a striping effect. Sometimes there is a yellowing to the entire top of the leaf. Striations can be found on lower midrib but not on the top. Stunting and poor stooling usually occur.

CONTROL:
1. Plant disease-free cuttings.
2. Maintain plants in vigorous growing conditions.

BIBLIOGRAPHY

Abbott, E.V. 1964. Black rot, in *Sugar Cane Diseases of the World*, Vol. II. C. G. Hughes, E. V. Abbott, and C. A. Wismer (eds.). Elsevier Publishing Company, New York.

Abbott, E. V. 1964. Brown spot, in *Sugar Cane Diseases of the World*, Vol. II. C. G. Hughes, E. V. Abbott, and C. A. Wismer (eds.). Elsevier Publishing Company, New York.

Abbott, E. V. 1964. Collar rot, in *Sugar Cane Diseases of the World*, Vol. II. C. G. Hughes, E. V. Abbott, and C. A. Wismer (eds.). Elsevier Publishing Company, New York.

Abbott, E. V. 1964. Dry top rot, in *Sugar Cane Diseases of the World*, Vol. II. C. G. Hughes, E. V. Abbott, and C. A. Wismer (eds.). Elsevier Publishing Company, New York.

Abbott, E. V. 1964. Myriogenospora leaf binding, in *Sugar Cane Diseases of the World*, Vol. II. C. G. Hughes, E. V. Abbott, and C. A. Wismer (eds.). Elsevier Publishing Company, New York.

Abbott, E. V. 1964. Red leaf spot (purple spot), in *Sugar Cane Diseases of the World*, Vol. II. C. G. Hughes, E. V. Abbott, and C. A. Wismer (eds.). Elsevier Publishing Company, New York.

Abbott, E. V. 1964. Red spot of the leaf sheath, in *Sugar Cane Diseases of the World*, Vol. II. C. G. Hughes, E. V. Abbott, and C. A. Wismer (eds.). Elsevier Publishing Company, New York.

Abbott, E. V., and Hughes, C. G. 1961. Red rot, in *Sugar Cane Diseases of the World*, Vol. I. J. P. Martin, E. V. Abbott, and C. G. Hughes (eds.). Elsevier Publishing Company, New York.

Abbott, E. V. and Matsumoto, T. 1964. Banded sclerotial disease, in *Sugar Cane Diseases of the World*, Vol. II. C. G. Hughes, E. V. Abbott, and C. A. Wismer (eds.). Elsevier Publishing Company, New York.

Abbott, E. V., and Tippett, R. L. 1941. Myriogenospora on sugarcane in Louisiana. *Phytopathology* 31:564–566.

Abbott, E. V., Wismer, C. A.; and Martin, J.P. 1964. Rind disease, in *Sugar Cane Diseases of the World*, Vol. II. C. G. Hughes, E. V. Abbott, and C. A. Wismer (eds.). Elsevier Publishing Company, New York.

Ammar, E. D.; Kira, M.T.; and Abul-Ata, A. E. 1980. Natural occurrence of streak and mosaic diseases on sugarcane cultivars at upper Egypt and transmission of sugarcane streak by *Cicadulina bipunctella zeae* China. *Egypt. Jour. Phytopathol.* 12:21–26.

Astudillo, G. E., and Birchfield, W. 1980. Pathology of *Hoplolaimus columbus* on sugarcane. *Phytopathology* 70: 565 (abstract).

Birch, R. G., and Patil, S. S. 1983. The relation of blocked chloroplast differentiation to sugarcane leaf scald disease. *Phytopathology* 73: 1368–1374.

Bourne, B. A. 1961. Fusarium sett or stem rot, in *Sugar Cane Diseases of the World*, Vol. I. J. P. Martin, E. V. Abbott, and C. G. Hughes (eds.). Elsevier Publishing Company, New York.

Breaux, R. D.; Matherne, R. J.; Millhollon, R. W.; and Jackson, R. D. 1972. Culture of sugarcane for sugar production in the Mississippi Delta. *Agric. Res. Serv. USDA Agric. Hdbk. No. 417.*

Byther, R. W., and Steiner, G. W. 1974. Unusual smut symptoms on sugarcane in Hawaii. *Plant Dis. Rptr.* 58:401–405.

Cassalett, C.; Victoria, J. I.; Carrillo, P.; and Ranjel, H. 1983. Resistance to sugarcane smut (*Ustilago scitaminea* Sydow). *Phytopathology* 73:126 (abstract).

Chu, H. T. 1964. Leaf-splitting disease, in *Sugar Cane Diseases of the World*, Vol. II. C. G. Hughes, E. V. Abbott, and C. A. Wismer (eds.). Elsevier Publishing Company, New York.

Comstock, J. C.; Ferreira, S.A.; and Tew, T.L. 1983. Hawaii's approach to control of sugarcane smut. *Plant Disease* **67**:452–457.

Cook, M. T. 1937. The organism causing the dry top rot of sugarcane. *Jour. Dept. Agric. Univ. Puerto Rico* **21**: 85–97.

Damann, K. E., Jr., and Benda, G. T. A. 1983. Evaluation of commercial heat-treatment methods for control of ratoon stunting disease of sugarcane. *Plant Disease* **67**:966–967.

Dean, J. L. 1981. Inoculation of wounded and unwounded sugarcane with *Ustilago scitaminea*. *Phytopathology* **71**:870 (abstract).

Eagan, B. T. 1964. Rust, in *Sugar Cane Diseases of the World*, Vol. II. C. G. Hughes, E. V. Abbott, and C. A. Wismer (eds.). Elsevier Publishing Company, New York.

Edgerton, C. W. 1958. *Sugarcane and Its Diseases*. Louisiana State University Press, Baton Rouge.

Hughes, C. G. 1961. Gumming disease, in *Sugar Cane Diseases of the World*, Vol. I. J. P. Martin, E. V. Abbott, and C. G. Hughes (eds.). Elsevier Publishing Company, New York.

Hughes, C. G., and Ocfemia, G.O. 1961. Yellow spot disease, in *Sugar Cane Diseases of the World*, Vol. I. J. P. Martin, E. V. Abbott, and C. G. Hughes (eds.). Elsevier Publishing Company, New York.

Humbert, R. P. 1968. Control of pests and diseases, in *The Growing of Sugar Cane*. Elsevier Publishing Company, New York.

Kao, J., and Damann, D. E., Jr. 1978. Microcolonies of the bacterium associated with ratoon stunting disease found in sugarcane xylem matrix. *Phytopathology* **68**:545–551.

Liao, C. H., and Chen, T. A. 1981. Isolation, culture and pathogenicity to Sudangrass of a corynebacterium associated with ratoon stunting of sugarcane and with Bermudagrass. *Phytopathology* **71**:1303–1306.

Liu, L. J. 1983. Sugarcane rust: The components of the epidemic and their usefulness in disease forecasting in Puerto Rico. *Phytopathology* **73**:796 (abstract).

Lo, T. T. 1961. Leaf scorch, in *Sugar Cane Diseases of the World*, Vol. I. J. P. Martin, E. V. Abbott, and C. G. Hughes (eds.). Elsevier Publishing Company, New York.

Martin, J. P. 1961. Brown stripe, in *Sugar Cane Diseases of the World*, Vol. I. J. P. Martin, E. V. Abbott, and C. G. Hughes (eds.). Elsevier Publishing Company, New York.

Martin, J. P. 1964. Iliau, in *Sugar Cane Diseases of the World*, Vol. II. C. G. Hughes, E. V. Abbott, and E. A. Wismer (eds.). Elsevier Publishing Company, New York.

Martin, J. P., and Robinson, P. E. 1961. Leaf scald, in *Sugar Cane Diseases of the World*, Vol. I. J. P. Martin, E. V. Abbott, and C. G. Hughes (eds.). Elsevier Publishing Company, New York.

Martin, J. P., and Wismer, C. A. 1961. Red stripe, in *Sugar Cane Diseases of the World*, Vol. I. J. P. Martin, E. V. Abbott, and C. G. Hughes (eds.). Elsevier Publishing Company, New York.

Matsumoto, T., and Abbott, E. V. 1964. Red rot of the leaf sheath, in *Sugar Cane Diseases of the World*, Vol. II. C. G. Hughes, E. V. Abbott, and C. A. Wismer (eds.). Elsevier Publishing Company, New York.

Purdy, L. H.; Liu, L. J.; and Dean J. L. 1983. Sugarcane rust, a newly important disease. *Plant Disease* 67:1292–1296.

Shaw, D. E. 1964. Veneer blotch, in *Sugar Cane Diseases of the World*, Vol. II. C. G. Hughes, E. V. Abbott, and C. A. Wismer (eds.). Elsevier Publishing Company, New York.

Sreeramulu, T., and Vittal, B. P. R. 1970. Incidence and spread of red rot lesions on midribs of sugarcane leaves. *Plant Dis. Rptr.* **54**:226–231.

Steib, R. J. 1961. Phytophthora seed piece rot, in *Sugar Cane Diseases of the World*, Vol. I. J. P. Martin, E. V. Abbott, and C. G. Hughes (eds.). Elsevier Publishing Company, New York.

Steindl, D. R. L. 1964. Bacterial mottle, in *Sugar Cane Diseases of the World*, Vol. II. C. G. Hughes, E. V. Abbott, and C. A. Wismer (eds.). Elsevier Publishing Company, New York.

Stokes, I. E., et al. 1961. Culture of sugarcane for syrup production. *Agric. Res. Serv. USDA Agric. Hdbk. No. 209.*

Subramanium, L. S., and Chona, B. L. 1938. Notes on *Cephalosporium sacchari* Butler (causal organism of sugarcane wilt). *Indian Jour. Agric. Sci.* **8**:189–190.

Taballa, H. A. 1969. Smut on true seedlings of sugarcane. *Plant Dis. Rptr.* **53**:992–993.

Todd, E. H. 1960. Elsinoe disease of sugarcane in Florida. *Plant Dis. Rptr.* **44**:153.

Todd, E. H. 1962. Target blotch of sugarcane in Florida. *Plant Dis. Rptr.* **46**:486.

Van der Zwet, T., and Forbes, I. L. 1961. *Phytophthora megasperma*, the principal cause of seed-piece rot of sugarcane in Louisiana. *Phytopathology* **59**:634–640.

Victoria, J. I.; Carrillo, P.; and Cassalett, C. 1983. Chemical control of sugarcane smut (*Ustilago scitaminea* Sydow) by seedcane treatment. *Phytopathology* **73**:126 (abstract).

Williams, J. R. 1969. Nematodes as pests of sugarcane, in *Pests of Sugar Cane*, Elsevier Publishing Company, New York.

Yen, W. Y. 1964. Leaf blight, in *Sugar Cane Diseases of the World*, Vol. II. C. G. Hughes, E. V. Abbott, and C. A. Wismer (eds.). Elsevier Publishing Company, New York.

Yen, W. Y., and Chi, C. C. 1964. Leaf blast, in *Sugar Cane Diseases of the World*, Vol. II. C. G. Hughes, E. V. Abbott, and C. A. Wismer (eds.). Elsevier Publishing Company, New York.

Zummo, N. 1986. Red spot (*Helminthosporium rostratum*) of sweet sorghum and sugarcane, a new disease resembling anthracnose and red rot. *Plant Disease* **70**:800 (disease notes).

21

Diseases of Sunflowers (*Helianthus annuus* L.)

BACTERIAL CAUSES

Apical Chlorosis

CAUSE: *Pseudomonas syringae* pv. *tagetis* (Hellmers) Young, Dye & Wilkie is seedborne.

DISTRIBUTION: Kansas, Minnesota, North Dakota, and Wisconsin.

SYMPTOMS: Leaf chlorosis without lesions occurs in all vegetative growth stages, although it is more frequent and severe on seedlings. Affected leaves, including veins, are pale yellow to white. Often only a portion of the initially affected leaf is chlorotic but subsequently formed leaves are uniformly chlorotic including veins.

Infected seedlings are frequently stunted and die; however, infection of older plants rarely results in stunting. With seedling infection, systemic chlorosis lasts up to eight weeks and spans 8–10 leaves. Chlorosis on older prebloom plants is frequently limited to a few leaves. Symptoms apparently do not occur on subapical, fully expanded leaves of plants past the bud stage. Thus, subsequent leaves on infected plants are normal in color. Chlorotic leaves do not recover.

Affected plants are usually scattered throughout a field, occurring singly or in small groups within a row.

CONTROL: Differences in resistance exist among sunflower cultivars. Plant the most resistant cultivar.

Bacterial Leaf Blight

CAUSE: *Pseudomonas cichorii* (Swingle) Stapp.

DISTRIBUTION: Brazil.

SYMPTOMS: Leaves have large necrotic lesions surrounded by narrow yellow halos.

CONTROL: No control is reported.

Bacterial Leaf Spot

CAUSE: *Pseudomonas syringae* pv. *helianthi* (Kawamura) Young, Dye & Wilkie.

DISTRIBUTION: Canada, Japan, and Spain.

SYMPTOMS: Angular, brown-green lesions occur on leaves. Light and dark brown lesions occur on stems. The vascular system is discolored and white exudate may be present on leaf axils.

CONTROL: No control is reported.

Bacterial Soft Rot

This disease may be the same as bacterial stalk rot.

CAUSE: *Erwinia carotovora* (Jones) et al. (syn. *Pectobacterium carotovorum* (Jones) Walder). Disease occurred following damage by hailstones.

DISTRIBUTION: Yugoslavia.

SYMPTOMS: Green-black spots occur on stalk. Eventually, these spots coalesce to cover most of the stalk surface, causing the stalk to soften and affected parts to dry up. Flowers and seeds are also discolored.

CONTROL: No control is reported.

Bacterial Stalk Rot

CAUSE: *Erwinia carotovora* pv. *carotovora* (Jones) Bergey et al. and *E. carotovora* pv. *atroseptica* (van Hall) Dye. The disease occurs after extended wet periods late in the growing season, suggesting only stressed or senescing plants become infected. Infection probably occurs at the axil of a petiole. This area often collects water and may serve as a favorable site for bacterial residence and insect activity. Infection is aided by wounding. The sunflower budworm (*Suleima helianthana* Riley) preferentially oviposits in this axil and larvae feed and exit from the same location.

DISTRIBUTION: Europe and the United States.

SYMPTOMS: Stalks are black on the outside and hollow on the inside. Stem blackening is often centered around a petiole axil, suggesting the bacterium enters the plant at this site, probably through wounds caused by insects, hail, or mechanical damage. Normally there is an ink-black, watery breakdown of the pith. This breakdown is odorless unless the stalk is in an advanced state of decomposition. Plants often lodge under weight of maturing heads. In some instances the head will also develop soft rot symptoms.

CONTROL: Do not rotate sunflowers and potatoes.

FUNGAL CAUSES

Alternaria Blight

CAUSE: *Alternaria helianthi* (Hans f.) Tubaki and Nishihara is seed-borne. The causal fungus also overwinters in stem residue on the soil surface and to a lesser extent in residue buried in the soil. Inoculum from infested residue decreases later in the growing season. Disease occurs under warm (25–28°C), humid conditions that promote extended periods of leaf wetness.

DISTRIBUTION: Africa, Argentina, Asia, Australia, Europe, and the United States.

SYMPTOMS: Early-planted seedlings are more severely diseased than later-planted ones. Plants are most susceptible at the anthesis or seed-filling stages of growth. Dark brown, oval, necrotic spots occur on the heads, leaves, petals, petioles, and stems. Often leaves have dark brown necrotic spots surrounded by a chlorotic halo. These spots occur on leaves of younger plants in vegetative or budding stage of growth. Other leaf symptoms are described as dark brown margins, with gray centers ranging from several millimeters to 1.5 cm in diameter. Stem lesions start as black flecks or streaks that later enlarge to cover large areas of stem. The stem may be girdled and the plant may be killed. Severely infected plants are defoliated and frequently lodge.

Head diameter, number of seeds produced per head, and oil content may be reduced.

CONTROL:
1. Apply a foliar fungicide.
2. Plow under residue.
3. Delay planting.
4. Rotate with a nonsusceptible crop for one year.
5. Apply a seed-treatment fungicide.

Alternaria Leaf and Stem Spot

CAUSE: *Alternaria zinniae* Pape is seedborne as conidia contaminating the outside of the seed coat and possibly as mycelium in the seed coat. Other ways of survival are not known but saprophytic growth on residue seems likely. High humidity and warm temperatures are favorable for disease development.

DISTRIBUTION: Canada, Kenya, and the United States.

SYMPTOMS: The disease usually does not become noticeable until after flowering. However, artificially inoculated seeds give rise to seedling blight. Such seedlings either rot completely and fail to emerge or have dark lesions on the hypocotyl and cotyledon, causing seedlings to damp off after emergence.

Leaf spots are circular in shape, uniformly dark colored, with a targetlike appearance. Brown, superficial flecks or streaks occur on stems, petioles, and backs of heads. When numerous, they form large necrotic areas. Severe leaf infection results in defoliation and severe stem infection weakens the plant, causing lodging.

CONTROL: Apply a foliar fungicide.

Botrytis Head Rot

This disease is also called gray mold head rot.

CAUSE: *Botrytis cinerea* Pers. ex Fr. infects during periods of abundant moisture. Disease is common when wet weather prevails in autumn and harvesting is delayed. The fungus may be seedborne.

DISTRIBUTION: Europe.

SYMPTOMS: A brown lesion first occurs on the back of the head that becomes a soft rot. Later lesions become gray with fungal growth. Normally only vascular bundles remain. Seedling blight may occur, thereby reducing stands.

CONTROL: No control is reported.

Cephalosporium Wilt

CAUSE: *Cephalosporium acremonium* Cda. The fungus is weakly pathogenic on young, healthy, growing sunflowers. As plants mature and heads form, they become more susceptible. The fungus likely enters the plant by insect wounds or by mechanical injury from wind, hail, or cultivation.

DISTRIBUTION: Indiana.

SYMPTOMS: Symptoms in a field were caused by a combination of *C. acremonium* and *Erwinia carotovora* pv. *carotovora* (Jones) Bergey et al. Brown lesions are present on the outside of the stalk. The inside of the stalk is hollow, resulting in a bending and collapsing under the weight of the maturing head. Yield is reduced through breaking and lodging of heads.

Plants injected with spores at the onset of senescence developed black necrotic areas on the stem that were covered with mycelium.

CONTROL: No control is reported.

Charcoal Rot

CAUSE: *Macrophomina phaseolina* (Tassi) Goid. (syn. *M. phaseoli* (Maubl.) Ashby) is seedborne as pycnidia and sclerotia. Sclerotia develop on the inner surface of the pericarp and seed coat. Charcoal rot is most severe during high soil temperatures (30°C and above) and low soil moisture. The sunflower stem weevil, *Cylindrocopturus adsperus* (LeConte), can carry *M. phaseolina* as they emerge from overwintered roots and stalks and transmit the pathogen while feeding and ovipositioning in the stalk. However, other fungi are also capable of contributing to the development of black to brown discoloration of stalks infected by *M. phaseolina* and parasitized by larvae of *C. adsperus*.

DISTRIBUTION: Generally distributed wherever sunflowers are grown in warmer climates. Charcoal rot rarely occurs in northern sunflower-growing areas.

SYMPTOMS: Symptoms usually occur after flowering. Premature ripening of stalks occurs together with a poorly filled head. Heads may be small and distorted with a zone of aborted flowers and premature ripening. Stalks have a gray discoloring at the base, with only the vascular bundles or fibers remaining, giving the internal stem a shredded appearance. The vascular fibers become covered with small, black sclerotia, resembling pepper or flecks of charcoal. Stems infected with *M. phaseolina* and parasitized by *C. adspersus* larvae have a brown to black discoloration with or without typical gray areas on the surface of the stem. Portions of the lower stem are hollow. Mixtures of frass and fragments of pith tissue occur in the lower stem and upper taproot.

Infected seed may not germinate, or upon germination, discoloration of roots, hypocotyls, and cotyledons may occur. Pycnidia and sclerotia may eventually develop on infected parts.

CONTROL:
1. Rotate sunflowers with a nonsusceptible crop such as small grains.
2. Plow under infected residue where feasible.
3. Plant cultivars that seem to have some resistance.
4. Irrigate at flowering or at flowering and ripening stages.

Coleosporium Rust

CAUSE: *Coleosporium madiae* Cke. is a heteroecious, long-cycled rust fungus. Monterey pine, *Pinus radiata* D. Don, is the alternate host. The uredial and telial stage occur on sunflower; the pycnial and aecial stage occur on Monterey pine.

DISTRIBUTION: West coast of the United States.

SYMPTOMS: Uredia occur as golden yellow pustules on sunflower leaves.

CONTROL: No control is reported.

Downy Mildew

CAUSE: *Plasmopara halstedii* (Farl.) Berl. et de Toni survives as oospores in the soil and mycelium and zoosporangium in seed. During wet soil conditions in the spring, oospores germinate to produce zoosporangia in which zoospores are formed. Zoospores are released from the zoosporangium and swim to sunflower roots where they encyst and germinate, penetrating the roots. Plants can be infected from germination until flowering but systemic infection usually occurs after emergence, with seedlings becoming increasingly resistant with age. Infection occurs primarily in the zone of elongation of the radicle or adjacent to it. The fungus becomes systemic and grows up the plant. Sporulation occurs on belowground and aboveground plant parts. Zoosporangia are produced and disseminated by wind and rain to infect other plants through apical buds or leaves. Oospores are produced within infected tissues and are returned to soil during harvest. Planting infected seed rarely results in systemically infected seedlings; however, this is a way by which *P. halstedii* may be introduced into soil.

Different races of the fungus exist.

DISTRIBUTION: Generally distributed wherever sunflowers are grown except Australia.

SYMPTOMS: Infected seedlings have a light green or yellowish area that spreads from the midribs of leaves. During wet weather a downy, whitish, fungal growth consisting of zoosporangia and mycelium occurs on the underside of leaves. As plants continue to grow, leaves become wrinkled and distorted and the entire plant is stunted. Infected plants usually produce normal-size heads that remain upright and contain mostly empty seeds.

CONTROL:
1. Plant only clean, healthy seed.
2. Grow sunflowers only once every five years in the same soil. Rotation should include small grains.
3. Control weeds in fields since they may serve as alternate hosts.
4. Treat seed with a systemic fungicide.

Drechslera Leaf Spot

CAUSE: *Drechslera hawaiiensis* (Bugincourt) Subram. & Jain ex M. B. Ellis. Older leaves are more susceptible to infection than younger ones.

DISTRIBUTION: India.

SYMPTOMS: Initially small (pinhead size), brown spots occur on leaves. The spots gradually increase in size (0.1–10 mm in diameter), are round to irregular in shape, with a pink-brown center that becomes gray in older spots. A chlorotic zone usually encircles each spot. Leaves become blighted during severe infections. Plant growth and yield are affected.

CONTROL: No control is reported.

Fusarium Wilt

CAUSE: *Fusarium* sp.

DISTRIBUTION: United States.

SYMPTOMS: Plants wilt with a distortion that affects one side of the leaf. Young infected plants die in a short time. Older infected plants recover.

CONTROL: No control is reported.

Phialophora Yellows

CAUSE: *Phialophora asteria* (Dowson) Burge & Isaac f. sp. *helianthi* Tirilly & Moreau is soilborne, probably overwintering in infected residue. Disease severity is likely increased during moist weather or excess soil moisture.

DISTRIBUTION: Manitoba.

SYMPTOMS: Yellows is an inconspicuous disease. Symptoms may be confused with those caused by nitrogen deficiency, other mineral deficiencies, or excess water. The first symptoms are obvious as sunflowers approach flowering. First the lower leaves turn a dull light green. Then large areas of leaf turn a dull yellow, starting at the apex and margin and extending inward. Angular patches of 5–10 mm long, interveinal leaf tissues and margins become necrotic. Symptoms move progressively up the plant. The bottom leaves may be dry and withered, whereas upper leaves may remain green. The vascular tissue turns brown. Plants are stunted, and heads remain small and sterile.

CONTROL: Some cultivars have varying degrees of resistance.

Phoma Black Stem and Phoma Girdling

This disease is also called black stem. In North Dakota, Phoma girdling is considered as one aspect of premature death.

CAUSE: *Phoma macdonaldii* Boerma, (syn. *P. oleracea* Sacc. var. *helianthi-tuberosi* Sacc.), teleomorph *Leptosphaeria lindquistii* Frezzi. The fungus may possibly overwinter as perithecia, pycnidia, and mycelium in stalk residue. Primary inoculum may result from ascospores and conidia produced during moist weather. Conidia are splashed or windborne and ascospores are likely windborne to hosts. Disease occurs under wet conditions during and after flowering.

DISTRIBUTION: Argentina, Canada, and the United States.

SYMPTOMS: Symptoms are most noticeable after flowering. Lesions normally start at base of leaf petioles and spread along the stem. Numerous spots may join together to form large, black patches. Dark, irregularly shaped spots also occur on leaves and flowers. Leaves infected early may be killed. Pycnidia resembling tiny, black bumps can be observed in mature spots. Infected plants may produce small heads with little seed. Stems may be weakened and lodge.

Symptoms of premature death are difficult to distinguish from normal senescence, especially in early-maturing varieties. Leaves wilt and become necrotic, stalks turn dark brown to black, and plants die. Premature death most often occurs in circular spots in a field, although scattered individual plants may die.

CONTROL:
1. Rotate sunflowers with nonhost plants.
2. Differences in susceptibility exist among sunflower lines.

Phomopsis Stem Canker

CAUSE: *Phomopsis helianthi* Munt.-Cvet. et al., teleomorph *Diaporthe helianthi* Munt.-Cvet. et al. The fungus overwinters as perithecia on sunflower stem residue on the soil. Ascospores produced during wet, cool weather are the principal source of infection throughout the growing season. There may be more than one species or biotype attacking sunflowers.

DISTRIBUTION: United States and Yugoslavia.

SYMPTOMS: Leaf symptoms are necrotic areas surrounded by chlorotic tissue. These areas typically start at the apical end or edge of the leaf. Leaf veins and petioles darken and cankers that are brown to black and later turn ash-gray frequently develop around petiole bases.

Spots on stem are initially light brown and about 10 mm wide. Eventually these spots become darker with a wet appearance, increase in length to 15–20 cm, and increase in width to girdle the stem. With death of epidermal tissue, the color becomes a light brown with dark brown borders of different widths surrounding affected tissue. Pycnidia are abundant in diseased tissue. Leaf symptoms reported from Ohio were characterized by interveinal bronzing. Several internodes may be involved with discoloration that progresses into the interior of the stalk, destroying pith tissue and causing hollow areas. Plants ripen prematurely and oil percentage in seeds decreases.

CONTROL: No control is reported.

Phytophthora Stem Rot

CAUSE: *Phytophthora cryptogea* Pethybr. & Laff. or *P. drechsleri* Tucker. *Phythophthora drechsleri* has been identified as the cause of Phytophthora black stem rot of sunflower in Iran. However, Phytophthora black stem rot is probably the same disease as Phytophthora stem rot.

The fungus overwinters as oospores that germinate to produce a sporangium. Zoospores are then produced within the sporangium, liberated, and swim to the host where they encyst and germinate. Secondary inoculum is by the production of sporangia on infected host tissue. Sporangia are windborne or splashed to healthy host tissue. Eventually oospores are formed in diseased tissue. The disease is most severe under wet soil conditions.

DISTRIBUTION: California and Iran.

SYMPTOMS: Plants are infected where stems contact water or wet soil. A dark brown to black discoloration occurs that partially or totally girdles the stem. Decay spreads to internal tissue and causes plants to fall over.

CONTROL: No control at present but some sunflower hybrids in Iran are more resistant than others.

Powdery Mildew

CAUSE: *Erysiphe cichoracearum* DC. survives as cleistothecia on infected residue. Ascospores are formed in cleistothecia during the summer and are windborne to leaves. Normally the lower leaves become more heavily infected than the upper leaves on a plant. Conidia are produced on diseased leaves and serve as secondary inoculum by being windborne to healthy tissue. Cleistothecia are formed in mycelium on the surface of infected tissue.

DISTRIBUTION: Generally distributed wherever sunflowers are grown.

SYMPTOMS: Symptoms usually occur after full bloom and are more obvious on lower leaves. Initially white powdery areas consisting of mycelium and conidia appear primarily on leaves but all aboveground plant parts may be infected. Later the whitish fungal growth becomes gray and small black objects, which are the cleistothecia, are formed in the fungal growth on infected surfaces. Severely infected leaves may turn yellow and dry up.

CONTROL: No control is necessary since powdery mildew occurs too late in the season to affect yield.

Rhizopus Head Rot

CAUSE: *Rhizopus* spp. occurs mostly as a saprophyte on plant residue. *Rhizopus oryzae* Went & Prin.-Geerl., *R. arrhizus* Fischer, and *R. stolonifer* (Ehr. ex Fr.) Vuill are species that have been implicated. Conidia are windborne to sunflower heads. Wet weather following flowering favors disease development. Heads damaged by birds, hail, or insects, particularly larvae of the sunflower moth *Homoeosoma electellum* (Hulst.), are subject to infection. Larvae of *H. electellum* have been shown to be directly correlated to Rhizopus head rot in Texas.

DISTRIBUTION: Generally distributed wherever sunflowers are grown.

SYMPTOMS: Head tissues turn brown and become soft. In wet weather or in the hollow part of the flower receptacle, the coarse, threadlike strands of the fungus mycelium may be seen. Later, small, black, fruiting structures develop on fungal stands that are the size of a pinhead. Upon dying, heads appear to shred.

CONTROL: No control is available but cultivars with upright heads tend to be infected easier than cultivars with nodding heads.

Rust

CAUSE: *Puccinia helianthi* Schw. is an autoecious, long-cycled rust that produces pycnia, aecia, uredia, and telia only on sunflower. Teliospores germinate in the spring and early summer at 6–28°C to produce basidiospores. Basidiospores germinate and infect leaves; at the point of infection pycnia develop in 10–12 days. Each pycnium produces pycniospores and receptive hyphae. Pycniospores are carried to compatible receptive hyphae of another pycnium by insects or splashing water. The nucleus of the pycniospore fuses with the nucleus of the receptive hypha and migrates down the hypha to form aeciospores in an aecium, usually on the opposite side of the leaf. Aeciospores are windborne to leaves where uredia, in which urediospores are produced, are formed at the point of infection. Urediospores are windborne and further infect healthy tissue. The urediospores are the repeating stage and produce new uredia at each point of infection. Toward maturity or during conditions unfavorable for rust, telia develop in rust lesions. Different physiologic races of the rust exist in nature.

DISTRIBUTION: Generally distributed wherever sunflowers are grown.

SYMPTOMS: Pycnia occur on upper leaf surfaces as small clusters of pale yellow or orange spots. Aecia occur as yellow or orange spots on the lower surface of leaves. Uredia first occur in midsummer as dark brown powdery spots on both surfaces of lower leaves. Sometimes uredia also occur on stems. Severely diseased leaves die and dry up. Telia occur as plants mature and appear as dark brown to black spots.

CONTROL:
1. Plant resistant cultivars.
2. Rotate sunflowers and locate fields as far as possible from fields where sunflowers were grown the previous year.

3. Sunflowers planted early are ordinarily less rusted than those planted late.
4. Destroy volunteer and wild sunflowers in vicinity of commercial fields.

Sclerotinia Head Rot

CAUSE: *Sclerotinia sclerotiorum* (Lib.) deBary (syn. *Whetzelinia sclerotiorum* (Lib.) Korf & Dumont) overwinters mainly as sclerotia in soil or among seeds. It can also survive as mycelium in infected sunflower stalks and in rotted or healthy-appearing seed. The hyperparasite *Coniothyrium minitans* Campbell may parasitize and kill sclerotia on root surfaces, thus potentially reducing inoculum in soil.

Sclerotia on the soil surface or near it germinate two ways: directly, by producing mycelium that infects stalks and roots, or indirectly, by producing apothecia on which ascospores are borne in a layer of asci. The ascospores are easily windborne and are probably the major cause of infection of heads. Circumstantial evidence suggests senescent florets would provide a food base for ascospore infection. Sclerotia are disseminated by any means that transports soil. Disease is most severe if frequent rains and high humidity occur after flowering.

DISTRIBUTION: Generally distributed wherever sunflowers are grown.

SYMPTOMS: Initially, soft, water-soaked spots occur that later dry out, leaving plant tissues with a somewhat pinkish color. A white, cottonlike mold may grow over the diseased area during humid conditions. The heads eventually become partially or entirely rotted, leaving only vascular bundles and fibers that give the head a shredded or brushlike appearance. Seed hulls may be discolored and scurfy. Large black sclerotia develop below the seed layer while others form around the seeds.

CONTROL:
1. Rotate sunflowers with cereals but avoid rotations with field beans, safflower, rapeseed, and mustard since these plants are also hosts for *S. sclerotiorum.*
2. Plant healthy seed that is free of sclerotia.
3. Control weeds in sunflower fields since they may serve as alternate hosts.
4. Deep plow residue since sclerotia that are deeply buried in the soil cannot germinate.
5. Plant spacings of 36 cm or greater and at 26,000–49,000 plants per hectare maximize yield; however, some researchers found plant spacing made no difference.

Sclerotinia Rots

CAUSE: *Sclerotinia minor* Jagger and *S. rolfsii* Sacc. survive in soil as sclerotia. Sclerotia may contaminate seed lots. During moist soil conditions, the sclerotia germinate to produce hyphae that infect the lower stem. Sclerotia are eventually produced in infected area.

DISTRIBUTION: Australia.

SYMPTOMS: A brownish canker forms on stems at about the soil line. Eventually white fungal growth covers the diseased area and black to dark brown sclerotia form within the mycelium. Plants will suddenly wilt.

CONTROL:
1. Plant clean, healthy seed, free of sclerotia.
2. Control weeds in a sunflower field.
3. Deep plow residue since sclerotia cannot germinate that are buried deeply in soil.

Sclerotinia Wilt

This disease is also called base rot, crown rot, Sclerotinia root rot, stalk rot, and stem rot.

CAUSE: *Sclerotinia sclerotiorum* (Lib.) deBary is discussed under Sclerotinia Head Rot. Most root rot originates by direct mycelial germination of sclerotia. The taproot–hypocotyl axis is the primary site of infection. Sclerotia must be adjacent to host tissue, primarily seed, for infection to occur. Sclerotinia root rot spreads through root contact between adjacent plants with *S. sclerotiorum* growing by mycelium from diseased to healthy plants. Plants that are initially infected serve as a primary infection loci from which the pathogen spreads.

DISTRIBUTION: Generally distributed wherever sunflowers are grown.

SYMPTOMS: Wilt symptoms usually appear about flowering time following a period of warm weather. Infected plants occur singly or in groups. Upper leaves droop and within two to three days all leaves dry out and plants die. Young dead plants may turn black but older plants remain tan.

Wilted plants have a canker completely encircling the stem that extends 7–25 cm above the soil line, with some cankers extending up to 125 cm. These cankers are soft, water-soaked, and gray to brown. Dense, white, mycelial growth occurs on the canker surface. Large, black sclerotia that are irregular in shape are produced in mycelium on the canker surface or within the stalk. Stems of infected plants may become shredded and break over at the soil line.

CONTROL: Same as for Sclerotinia Head Rot.

Septoria Leaf Spot

CAUSE: *Septoria helianthi* Ell. & Kell. Little is known of the survival and dissemination of the fungus, but it is likely to overwinter as pycnidia on residue from which conidia are windborne or splashed to host tissue similar to other *Septoria* spp. Humid weather is conducive for disease development.

DISTRIBUTION: Canada, United States, and possibly other sunflower-growing areas in the world.

SYMPTOMS: Septoria leaf spot may occur on plants of any age but symptoms are usually visible after flowering. Spots are normally restricted to leaves. Water-soaked, circular spots, 6–12 mm in diameter, become gray with a dark margin. Other descriptions depict spots as brown, with a narrow yellow ring that fuses gradually with surrounding green tissues. Spots coalesce, producing irregular-shaped dead areas on leaves. If temperatures are moderately high and moisture is abundant, leaves are dropped progressively from the plant bottom upward, until only a few leaves are left on an infected plant.

CONTROL: No control is reported.

Verticillium Wilt

This disease is also called leaf mottle.

CAUSE: *Verticillium dahliae* Kleb. The causal organism was originally identified as *V. albo-atrum* Reinke & Berth. *Verticillium dahliae* survives in soil as microsclerotia that germinate to infect roots. It may also be seedborne as mycelium in the seed, and as microsclerotia on external and internal pericarp tissue and on testa. Eventually mycelium grows into the xylem tissue where microconidia are also produced. Eventually the fungus grows through the xylem tissue of all parts of the plant and into the seed. With death of the plant, microsclerotia are produced from mycelium and returned to soil at harvest time.

DISTRIBUTION: Canada, USSR, South America, and the United States.

SYMPTOMS: Symptoms are most obvious at flowering time when infected plants occur singly or in groups. Symptoms occur on lower leaves first and gradually progress up the plant. Tissue between leaf veins becomes chlorotic, then brown, giving the leaf a mottled appearance. Black areas occur on the stem, particularly near the soil line. In cross section, the vascular system of the stem is a brown to black discoloration. Severely infected plants are stunted and may mature prematurely or die before flowering.

CONTROL:
1. Plant resistant cultivars.
2. Rotate sunflowers every five years.
3. Plant disease-free seed.

White Blister

This disease is also called white rust.

CAUSE: *Albugo tragopogonis* (Pers.) S. F. Gray (sometimes spelled *A. tragopogi*) overwinters as oospores in residue in soil or as mycelium or sporangia on weeds and volunteer sunflowers in milder climates. Little is known of the source and dissemination of the primary inoculum. One possibility is that oospores are disseminated to leaves and then germinate to form zoospores. The zoospores penetrate stomata and encyst. Eventually they germinate and form mycelium. Sporangia or conidia may be windborne from infected overwintered plants. Sporangia germinate to produce zoospores that encyst and eventually germinate.

Secondary inoculum is by formation of sporangia on lower leaf surface. These are windborne to other hosts where they again form zoospores that germinate. Oospores are formed in infected tissue. Disease development is favored by temperatures between 10°C and 20°C and high rainfall.

DISTRIBUTION: Argentina, Australia, USSR, and Uruguay. The disease likely occurs in other countries also.

SYMPTOMS: Creamy white, blisterlike pustules appear most commonly on the underside of leaves near the bottom of plants. Tissue on upper leaf surface, opposite a pustule, is raised and yellow-green. No scorching or shriveling of leaves occurs.

CONTROL: No control is necessary because the disease does little damage.

MYCOPLASMAL CAUSES

Aster Yellows

CAUSE: Aster yellows mycoplasma survives in several dicotyledonous plants and leafhoppers. Transmission is primarily by the aster leafhopper, *Macrosteles fascifrons* Stal.

DISTRIBUTION: Generally distributed wherever sunflowers are grown.

SYMPTOMS: Either the entire head or a pie-shaped sector of it shows symptoms. Flowers remain green instead of normally turning yellow. Structures that resemble small leaves are larger than normal flowers and replace the floral parts in the flower head. Eventually affected portions turn brown and die with the brown discoloration extending downward as a narrow stripe along the stem. Sometimes head symptoms are not evident but a black slimy rot occurs on the stalk below the head.

Infected plants may be stunted or break over. Others may set seed only on the normal portion of their head.

CONTROL: No control is reported.

CAUSED BY NEMATODES

Root Knot Nematode

CAUSE: *Meloidogyne incognita* (Kofoid & White) Chitwood.

DISTRIBUTION: Not known for certain.

SYMPTOMS: Infected plants are reduced in stand and vigor.

CONTROL: No control is reported.

Nematodes Associated
with Sunflowers

Meloidogyne arenaria (Neal) Chitwood
M. hapla (Kofoid & White) Chitwood
M. javanica (Treub.) Chitwood
Helicotylenchus dihystera (Cobb) Sher.
H. pseudorobustus (Steiner) Golden
H. heterodera glycines Ichinoke
Pratylenchus alleni Ferris
P. scribneri Steiner
P. brachyurus (Godfrey) Filip. & Sch. Stek.
P. penetrans (Cobb) Filip & Sch. Stek.
P. zeae Graham
Belonolaimus longicaudatus Rau
Rotylenchulus reniformis Linford & Oliveira
Trichodorus christiei Allen

VIRAL CAUSES

Sunflower Mosaic

CAUSE: Cucumber mosaic virus (CMV). The virus is mechanically transmitted and has a wide host range.

DISTRIBUTION: Maryland.

SYMPTOMS: Symptoms of mosaic and chlorotic rings are more severe on leaves younger than two months than on older leaves. Plants are stunted and sometimes have discrete, narrow, light brown streaks on petioles and stems. Malformed heads are produced with shriveled seed.

CONTROL: No control is reported.

NONPARASITIC CAUSES

Bract Necrosis

CAUSE: Nonparasitic, caused by high temperatures of about 40°C.

DISTRIBUTION: Texas.

SYMPTOMS: Typically a brown discoloration of disk flowers and bracts. The brown discoloration becomes black after a rain. When bract necrosis occurs during the bud stage, buds may remain unopened. Some injured buds may open but produce few or no disk flowers and little pollen.

CONTROL: No control is reported.

BIBLIOGRAPHY

Allen, S. J.; Brown, J. F.; and Kochman, J. K. 1983. Effects of temperatures, dew period and light on the growth and development of *Alternaria helianthi*. *Phytopathology* **73**:893–896.

Allen, S. J.; Brown, J. F.; and Kochman, J. K. 1983. The effects of leaf age, host growth stage, leaf injury and pollen on infection of sunflowers by *Alternaria helianthi*. *Phytopathology* **73**:896–898.

Arsenijevic, J. 1970. A bacterial soft rot of sunflower (*Helianthus annuus* L.) *Acta Phytopathol. Acad. Sci. Hungary* **5**:317–326.

Banihashemi, Z. 1975. Phytophthora black stem rot of sunflower. *Plant Dis. Rptr.* **59**:721–724.

Bernard, E. C., and Keyserling, M. L. 1985. Reproduction of root-knot, lesion, spiral, and soybean cyst nematodes on sunflowers. *Plant Disease* **69**:103–105.

Blanco-Lopez, M. A., and Jimenez-Diaz, R. M. 1983. Effect of irrigation on susceptibility of sunflower to *Macrophomina phaseoli*. *Plant Disease* **67**:1214–1217.

Campos, V. P.; Huang, S. P.; Tanaka, M. A. S.; and Rezende, A. M. 1982. Ocorrencia de *Meloidogyne incognita* em cultural de girassol no Est. Minas Gerais, Brasil. *Fitopathologia Brasileira* **7**:309–310 (in Portuguese).

Carson, M. L. 1985. Epidemiology and yield losses associated with Alternaria blight of sunflowers. *Phytopathology* **75**:1151–1156.

Carson, M. L. 1987. Effects of two foliar pathogens on seed yield of sunflower. *Plant Disease* **71**:549–551.

Cobia, D. W. et. al. 1975. Sunflowers production, pests and marketing. *North Dakota State Univ. Ext. Bull. 25.*

Donald, P. A.; Bugbee, W. M.; and Venette, J. R. 1986. First report of *Leptosphaeria lindquistii* (sexual stage of *Phoma macdonaldii*) on sunflower in North Dakota and Minnesota. *Plant Disease* **70**:352 (disease notes).

Donald, P. A.; Miller, J. F.; and Venette, J. R. 1986. Reaction of sunflower lines to *Phoma macdonaldii*. *Phytopathology* **76**:956 (abstract).

Donald, P. A.; Venette, J. R.; and Gulya, T. J. 1987. Relationship between *Phoma macdonaldii* and premature death of sunflower in North Dakota. *Plant Disease* **71**:466–468.

Dwan, J. M., and Sobrino, E. 1987. First report of *Pseudomonas syringae* pv. *helianthi* on sunflower in Spain. *Plant Disease* **71**:101 (disease notes.)

Fakir, G. A.; Rao, M. H.; and Thirumalachar, J. J. 1976. Seed transmission of *Macrophomina phaseolina* in sunflower. *Plant Dis. Rptr.* **60**:736–737.

Gudmestad, N. C.; Secor, G. A.; Nolte, P.; and Straley, M. L. 1984. *Erwinia carotovora* as a stalk rot pathogen of sunflower in North Dakota. *Plant Disease* **68**:189–192.

Gulya, T. J.; Ooka, J. J.; and Mancl, M. K. 1985. Diseases of cultivated sunflower in Hawaii. *Plant Disease* **69**:542 (disease notes).

Gulya, T. J.; Urs, R. R.; and Banttari, E. E. 1981. Apical chlorosis of sunflower incited by *Pseudomonas tagetis*. *Phytopathology* **71**:221. (abstract).

Gulya, T. J.; Urs, R. R.; and Banttari, E. E. 1982. Apical chlorosis of sunflower caused by *Pseudomonas syringae* pv. *tagetis*. *Plant Disease* **66**:598–600.

Hoes, J. A. 1972. Sunflower yellows, a new disease caused by *Phialophora* sp. *Phytopathology* **62**:1088–1092.

Hoes, J. A. 1975. Sunflower diseases in western Canada, in *Oilseed and Pulse Crops in Western Canada*. Modern Press, Saskatoon, Saskatchewan.

Hoes, J. A., and Huang, H. C. 1985. Effect of between-row and within-row spacings on development of sclerotinia wilt and yield of sunflower. *Can. Jour. Plant Pathol.* **7**:98–102.

Holley, R. C., and Nelson, B. D. 1986. Effect of plant population and inoculum density on incidence of Sclerotinia wilt of sunflower. *Phytopathology* **76**:71–74.

Huang, H. C. 1976. Importance of *Coniothyrium minitans* in survival of *Sclerotinia sclerotiorum* in wilted sunflower. *Can. Jour. Bot.* **55**:289–295.

Huang, H. C., and Dueck, J. 1980. Wilt of sunflower from infection by mycelial-germinating sclerotia of *Sclerotinia sclerotiorum*. *Can. Jour. Plant Pathol.* **2**:47–52.

Huang, H. C., and Hoes, J. A. 1980. Importance of plant spacing and sclerotial position to development of Sclerotinia wilt of sunflower. *Plant Disease* **64**:81–84.

Jeffrey, K. K.; Lipps, P. E.; and Herr, L. J. 1984. Effects of isolate virulence, plant age, and crop residues on seedling blight of sunflower caused by *Alternaria helianthi*. *Phytopathology* **74**:1107–1110.

Jeffrey, K. K.; Lipps, P. E.; and Herr, L. J. 1985. Seed-treatment fungicides for control of seedborne *Alternaria helianthi* on sunflower. *Plant Disease* **69**:124–126.

Jimenez-Diaz, R. M.; Blanco-Lopez, M. A.; and Sackston, W. E. 1983. Incidence and distribution of charcoal rot of sunflower caused by *Macrophomina phaseolina* in Spain. *Plant Disease* **67**:1033–1036.

Kajornchaiyakul, P., and Brown, J. F. 1976. The infection process and factors affecting infection of sunflower by *Albugo tragopogi*. *Br. Mycol. Soc. Trans.* **66**:91–95.

Klisiewicz, J. M., and Beard, B. H. 1976. Diseases of sunflower in California. *Plant Dis. Rptr.* **60**:298–301.

McDonald, W. C. 1964. Phoma black stem of sunflowers. *Phytopathology* **54**:492–493.

McDonald, W. C., and Martens, J. W. 1963. Leaf and stem spot of sunflowers caused by *Alternaria zinniae*. *Phytopathology* **53**:93–96.

Middleton, K. J. 1971. Sunflower diseases in South Queensland. *Queensland Agric. Jour.* **97**:597–600.

Mihaljcevic, M.; Muntanola-Cvetkovic, M.; Vukojevic, J.; and Petrov, M. 1985. Source of infection of sunflower plants by *Diaporthe helianthi* in Yugoslavia. *Phytopathol. Zeitschr.* **113**:334–342.

Morris, J. B.; Yang, S. M.; and Wilson, L. 1983. Reaction of Helianthus species to *Alternaria helianthi*. *Plant Disease* **67**:539–540.

Muntanola-Cvetkovic, M.; Mihaljcevic, M.; and Petrov, M. 1981. On the identity of the causative agent of a serious Phomopsis-Diaporthe disease in sunflower plants. *Nova Hedwigia* **34**:417–435.

Muntanola-Cvetkovic, M.; Mihaljcevic, M.; Vukojevic, J.; and Petrov, M. 1985. Comparison of Phomopsis isolates obtained from sunflower plants and debris in Yugoslavia. *Brit. Mycol. Soc. Trans.* **85**:477–483.

Orellana, R. G. 1971. Fusarium wilt of sunflower, *Helianthus annuus*: First report. *Plant Dis. Rptr.* **55**:1124–1125.

Orellana, R. G., and Quacquarelli, A. 1968. Sunflower mosaic caused by a strain of cucumber mosaic virus. *Phytopathology* **58**:1439–1440.

Parmelee, J. A. 1972. Additions to the autoecious species of Puccinia on Heliantheae in North America. *Can. Jour. Bot.* **50**:1457–1459.

Piening, L. J. 1976. A new bacterial leaf spot of sunflowers in Canada. *Can. Jour. Plant Sci.* **56**:419–422.

Putt, E. D. 1964. Breeding behavior of resistance to leaf mottle or Verticillium in sunflowers. *Crop Sci.* **4**:177–179.

Reddy, P. C., and Siradhana, B. S. 1978. A new leaf spot disease of sunflower in India. *Plant Dis. Rptr.* **62**:508.

Rich, J. R., and Dunn, R. A. 1982. Pathogenicity and control of nematodes affecting sunflower in north central Florida. *Plant Disease* **66**:297–298.

Richeson, M. L. 1981. Etiology of a late season wilt in *Helianthus annuus*. *Plant Disease* **65**:1019–1021.

Rogers, E. C.; Thompson, T. E.; and Zimmer, D. E. 1978. Rhizopus head rot of sunflower: Etiology and severity in the southern plains. *Plant Dis. Rptr.* **62**:769–771.

Sackston, W. E. 1960. *Botrytis cinerea* and *Sclerotinia sclerotiorum* in seed of safflower and sunflower. *Plant Dis. Rptr.* **44**:664–668.

Sackston, W. E. 1981. The sunflower crop and disease: Progress, problems, and prospects. *Plant Disease* **65**:643–648.

Sackston, W. E.; McDonald, W. C.; and Martens, J. W. 1957. Leaf mottle or verticillium wilt of sunflower. *Plant Dis. Rptr.* **41**:337–343.

Stovold, G. E., and Moore, K. J. 1972. Diseases of sunflower. *Agric. Gazette of New South Wales* **83**:262–264.

Styer, D. J., and Durbin, R. D. 1982. Isolation of *Pseudomonas syringae* pv. *tagetis* from sunflower in Wisconsin. *Plant Disease* **66**:601.

Watters, B. L.; Herr, L. J.; and Lipps, P. E. 1983. A new stem canker disease of sunflower. *Phytopathology* **73**:798 (abstract).

Wehtje, G., and Zimmer, D. E. 1978. Downy mildew of sunflower: Biology of systemic infection and the nature of resistance. *Phytopathology* **68**:1568–1571.

Yang, S. M. 1983. Bract necrosis, a nonparasitic disease of sunflower. *Phytopathology* **73**:844 (abstract).

Yang, S. M. 1984. Etiology of atypical symptoms of charcoal rot in sunflower plants parasitized by larvae of *Cylendrocopturus adsperus*. *Phytopathology* **74**:479–481.

Yang, S. M.; Berry, R. W.; Lutterell, E. S.; and Vongkaysone, T. 1984. A new sunflower disease in Texas caused by *Diaporthe helianthi*. *Plant Disease* **68**:254–255.

Yang, S. M., and Gulya, T. 1986. Prevalent races of *Puccinia Helianthi* in cultivated sunflower on the Texas high plains. *Plant Disease* **70**:603 (disease notes).

Yang, S. M.; Morris, J. B.; Unger, P. W.; and Thompson, T. E. 1979. Rhizopus head rot of cultivated sunflower in Texas. *Plant Dis. Rptr.* **63**:833–835.

Yang, S. M., and Owen, D. F. 1982. Symptomatology and detection of *Macrophomina phaseolina* in sunflower plants parasitized by *Cylindrocopturus adspersus* larvae. *Phytopathology* **72**:819–821.

Yang, S. M.; Rogers, C. E.; and Luciani, N. D. 1983. Transmission of *Macrophomina phaseolina* in sunflower by *Cylindrocopturus adspersus*. *Phytopathology* **73**:1467–1469.

Yarwood, C. E. 1969. Sunflower, a new host of *Coleosporium madia*. *Plant Dis. Rptr.* **53**:648.

Zimmer, D. E. 1975. Some biotic and climatic factors influencing sporadic occurrence of sunflower downy mildew. *Phytopathology* **65**:751–754.

22

Diseases of Tobacco (*Nicotiana tabacum* L.)

_____BACTERIAL CAUSES_____

Angular Leaf Spot

This disease is also called blackfire.

CAUSE: *Pseudomonas angulata* (From & Murray) Holland survives in soil and residue. It also persists as a parasite on roots of weeds and winter grains. Leaf tissues must be water-soaked before infection can occur. Bacteria are splashed onto leaves where they are forced into leaf stomates or wounds and then move into the flooded intercellular spaces. Optimal infection conditions occur during rainstorms accompanied by high winds.

Dissemination over a long distance is by contaminated transplants. Wind-blown water is the most important means of dissemination from plant to plant in the field and over short distances.

There is evidence to suggest *P. angulata* arises as a mutant of *P. tabaci* (Wolf & Foster) Stevens.

DISTRIBUTION: Generally distributed wherever tobacco is grown.

SYMPTOMS: Initially symptoms appear as angular to irregular, black to brownish, water-soaked spots that die rapidly, forming dark brown angular spots. These symptoms usually appear suddenly a few days after rainy periods. The centers of these spots often become light tan with dark borders as they age. During rainy weather the dark centers may fall out. A thin yellow band often occurs around the spot. These angular-shaped spots range from about 1–13 mm in diameter. Spots coalesce to involve large areas of a leaf. On rapidly growing plants, infected leaves become puckered and torn. Symptoms in the plant bed are similar to those in the field.

CONTROL:
 1. Plow planting beds in autumn.
 2. Steam or fumigate planting beds plowed in the spring.
 3. Plant only seed from healthy plants.
 4. Treat seed with bactericides to eradicate bacteria.
 5. Apply bactericide sprays to plants in the seed bed and field.
 6. Use only disease-free transplants.
 7. Grow resistant cultivars.
 8. Provide proper drainage for plant bed.
 9. Do not grow tobacco in field for at least two consecutive years.
10. Avoid working in a field when leaves are wet.

Barn Rot

CAUSE: *Erwinia carotovora* pv. *carotovara* (Jones) Bergey et al. (syn. *E. carotovora* (Jones) et al. and *E. aroideae* (Town) Hall), *E. chrysanthemi* Burkholder, McFadden & Dimock, and *Bacillus polymyxa* (Praz.) Migul. The fungi *Botrytis cinerea* Pers. ex Fr., *Sclerotinia sclerotiorum* (Lib.) deBary, *Rhizopus arrhizus* Fischer, *Pythium* spp., and *Alternaria alternata* (Fr.) Keissler are associated with barn rot also.

Overwintering may occur on seed. Infected seed may be an important source of inoculum and possibly the primary source. Dissemination is probably by conveyors.

The disease occurs when wet, mechanically primed leaves are packed tightly in bulk curing barns with poor air circulation. Barn rot also occurs in stick barns with poor circulation when they are overfilled with wet tobacco leaves. Mechanical leaf harvesters injure leaves, creating wounds for bacteria to enter.

DISTRIBUTION: Generally distributed wherever tobacco is grown.

SYMPTOMS: A watery, soft rot occurs during the yellowing stage. A water-soaked brownish discoloration begins in the petiole or lamina. The discoloration darkens to dark brown or black and spreads throughout the leaves, causing them to decompose. Often the leaves break away from the strings and fall to the floor. There is an odor of decaying vegetation. The damage may vary from a few leaves to the entire contents of a barn.

CONTROL:
1. Avoid placing wet tobacco in the barn.
2. Do not crowd leaves on a stick.
3. Do not crowd sticks together in the barn.
4. Use heat and ventilation to dry out excess moisture rapidly if wet tobacco is placed in a barn.
5. Do not injure tobacco.

Blackleg

This is the seedling form of hollow stalk.

CAUSE: *Erwinia carotovora* (Jones) et al. (syn. *E. aroidea* (Town.) Holland and *E. carotovora* pv. *atroseptica* (van Hall) Dye) survives in soil and residue. Cool, heavy, clay soils with a neutral pH favor survival of the bacterium. Disease is most severe during damp, cloudy weather and over a wide range of temperatures. Plants become infected by leaves touching the ground. The bacteria then spread through the petiole into the stem. Bacteria are disseminated to the field on transplants and by any means that moves soil. Bacteria may also be spread within a planting bed by maggot flies; *Hylemyia* spp. *Erwinia carotovora* has a wide host range.

DISTRIBUTION: Generally distributed wherever tobacco is grown. However, black leg is rarely a severe problem.

SYMPTOMS: Black leg occurs in irregular circular areas of about 1 m in diameter in the plant bed. Rotted petioles and stems become black. Eventually rotted areas may split open or rot off.

Some plants may be slightly affected and display no symptoms. When these plants are set in the field, they may grow normally. However, if plants are removed from the bed and are kept overnight for planting the next day, all the plants may become rotted and slimy.

CONTROL: The covering should be removed from the plant bed to reduce humidity as soon as black leg is noticed.

Black Rust

CAUSE: *Bacterium pseudozoogloae* Honing. This may be the same bacterium as *Pseudomonas syringae* pv. *tabaci* (Wolf & Foster) Young, Dye & Wilkie. The bacterium survives in infected crop residue and is disseminated by splashing rain.

DISTRIBUTION: Indonesia and possibly Japan and Italy.

SYMPTOMS: Symptoms occur on lower leaves as plants approach maturity. Initially dark green spots appear that become necrotic and dark brown in the center with zonate concentric rings. A dark green margin continues to surround each spot that eventually reaches 1–2 cm in diameter. The margins remain dark green after infected leaves have dried.

CONTROL: No control other than to remove and burn lower diseased leaves.

Crown Gall

CAUSE: *Agrobacterium tumefaciens* (E. F. Sm. & Town.) Conn. (syn. *Rhizobium radiobacter* Bijerinck & van Delden var. *tumefaciens*). The bacterium is a soil inhabitant that enters a plant through wounds caused by any means. Bacteria then stimulate cells to divide, eventually forming a tumor. The following steps are thought to occur in tumor development: 1. The bacteria enter a wound cell and bind to a specific intracellular wound site. 2. The bacterial DNA is transferred to the plant cell, which causes the plant cell to divide. 3. Finally bacterial DNA increases and is transferred to plant cells free of bacterial cells. These plant cells also divide and contribute to the tumor development.

Tumor development is optimum at 25°C and is inhibited at higher temperatures. The bacteria are disseminated by any means capable of moving soil.

DISTRIBUTION: Crown gall is worldwide in distribution; however, it rarely occurs on tobacco.

SYMPTOMS: More or less round-shaped tumors or overgrowths occur on roots of shoots. The tumors are several times larger than the stem or root. The round tumors have large indentations, making them appear convoluted.

CONTROL: No control is necessary.

Frenching

CAUSE: *Bacillus cereus* Frankland & Frankland is a nonpathogenic bacterium that is common in soil, dust, and on plant surfaces. A diffusate is produced in soil by the bacterium that acts as a toxin to tobacco. The bacterium never enters or infects the plant. The toxin disturbs the nitrogen metabolism of the plant so that an increase in different amino acids occurs, particularly isoleucine. The excess of free amino acids causes the tobacco abnormalities. Frenching is more prevalent on wet soils that sometimes lack available nitrogen or other materials. Frenching does not occur at soil temperature of 21°C or less.

DISTRIBUTION: Generally distributed wherever tobacco is grown. Frenching rarely causes much damage.

SYMPTOMS: Frenching begins to occur on plants from two to three weeks old and can continue until flowering. Initially a chlorosis occurs along the margins of young leaves that eventually involves all interveinal areas. However, veins remain green. Leaves become long and narrow due to only the midrib elongating. Plants are stunted with a large number of small, distorted, and brittle leaves. The axillary buds of severely affected plants are stimulated, causing a rosette of small, narrow, light green leaves that gives a witches' broom effect.

CONTROL:
1. Ensure soils are properly drained and fertilized.
2. Do not plant in alkaline soils and avoid heavy applications of lime.
3. Apply fall applications of elemental sulfur.

Granville Wilt

This disease is also called bacterial wilt and several other names in different countries where the disease occurs.

CAUSE: *Pseudomonas solanacearum* (Smith) Smith is a bacterium that survives in the soil and infects plants through wounds in the root caused by several factors. Bacteria are disseminated by wilt-infected seedlings. During periods of high soil moisture, bacteria are released from infected roots and spread to adjacent healthy roots, particularly where roots intermingle. Disease severity is greatest during high soil temperature (30–35°C) and moisture.

DISTRIBUTION: Generally distributed wherever tobacco is grown in the warm temperate and semitropical areas of the world. *Pseudomonas solanacearum* is rarely found in areas where the mean temperature in the winter falls below 10°C.

SYMPTOMS: The initial symptom on young, succulent plants is the drooping of one or two leaves during the day that may recover their turgidity at night. Often only half a leaf wilts. If the disease progresses slowly, affected leaves become lighter green and may gradually turn yellow; in severe disease conditons, the whole plant wilts and dies within a few days.

Light tan to yellow-brown streaks occur in the vascular tissue. Eventually the streaks darken, and in advanced stages, blackened areas show on the surface of the stalk. A dirty white to brownish slimy ooze in the form of glistening beads can be seen on a cut stem. When the two cut edges of a stem are pulled apart slightly, a bacterial stand will be stretched between the two pieces.

Eventually roots become dark brown to black and rotted. In moist soils, roots become soft and slimy.

CONTROL:
1. Grow resistant cultivars.
2. Do not transplant plants showing wilting or other disease symptoms.

3. Do not plant tobacco in soil containing *P. solanacearum* for at least five years. Crops to include in a rotation are grasses, legumes, small grains, cotton, corn, soybeans, and sorghum.
4. Practice good nematode control.

Hollow Stalk

Blackleg is the seedling form of this disease.

CAUSE: *Erwinia carotovora* (Jones) et al. (syn. *E. aroidea* (Town.) Holland and *E. carotovora* pv. *atroseptica* (van Hall) Dye) survives in soil and in residue composed primarily of decayed root crowns in soil. Soils high in clay content and of neutral pH favor survival of the bacterium. Disease development is most rapid during wet weather, but over a wide temperature range. Bacteria enter plants through wounds made during topping, suckering, and harvesting. Bacteria are disseminated in contaminated soil on workers' hands during topping and suckering. Mineral oil used to control suckering apparently favors the reproduction of *E. carotovora* in the plant and on it.

DISTRIBUTION: Generally distributed wherever tobacco is grown. However, hollow stalk is rarely an important disease.

SYMPTOMS: The disease first appears at topping and suckering time. Discoloration occurs at any stem wound but usually occurs in the pith at the break made at topping. The pith rapidly becomes brown followed by a soft rotting and disintegration. The top leaves wilt first, and as the disease progresses down the stem, successive leaves wilt, hang down next to the stalk, or fall off. The disease may occasionally begin at the bottom of the plant. Eventually the entire stalk may be bare. Black or brown sunken lesions form in leaf axils. Commonly the pith becomes a slimy, foul-smelling pulp that eventually dries up, forming a hollow stem.

Infected leaves may rot in the curing barn. The leaf petiole rots at the point where the string is attached and the leaf falls to the floor. Infected leaves may smell like decaying vegetation.

CONTROL:
1. Do not do topping and suckering during damp or cloudy weather.
2. Reduce humidity in curing barns when wet, succulent tobacco is hung.
3. Avoid manual topping since this favors disease spread.
4. Use sucker control agents other than mineral oils.

Leaf Gall

This disease is also called fascination.

CAUSE: *Corynebacterium fascians* (Tilford) Dows. is commonly found in soil. The bacterium does not need a wound to infect a plant. Symptoms probably occur because of the imbalance of growth hormones in the plant caused by the bacterium. Optimum growth occurs at 25–28°C.

DISTRIBUTION: Not known for certain but leafy gall rarely occurs on tobacco.

SYMPTOMS: Numerous short, fleshy, thick stems or multiple buds with misshaped leaves develop at or below the soil line; however, the main stem usually appears normal but may be stunted. The proliferated growth may reach a diameter of 2–8 cm and resemble a leafy gall.

CONTROL: No control is necessary.

Philippine Bacterial Leaf Disease

CAUSE: *Pseudomonas aeruginosa* (Shroeter) Migula is a common soil saprophyte and is seedborne. Bacteria enter through the wounds and stomates of water-soaked leaves. Dissemination is by splashing rain.

DISTRIBUTION: Philippines.

SYMPTOMS: Bleached white spots may occur on seedling leaves that vary in size from a pinpoint to several millimeters in diameter. Infected areas may develop a wet rot that becomes necrotic and disintegrates. Petioles and stems may also become infected during wet weather, causing the plant to die.

Plants in the field have whitish or opaque lesions that eventually become brown and zonate, bordered by a yellowish zone. The yellow border disappears as the leaves mature. Lesions are most numerous on the bottom leaves and may coalesce to form large, irregular, dead areas. After curing, the brown spots may disappear but the white ones are still evident.

CONTROL: The same controls should be used as for wildfire.

Wildfire

CAUSE: *Pseudomonas syringae* pv. *tabaci* (Wolf & Foster) Young, Dye & Wilkie (syn. *P. tabaci* (Wolf & Foster) Stevens) survives in soil and residue. The bacterium also persists in dry leaves and some types of manufactured tobacco. It also survives as a parasite on the roots of weeds, grasses, and winter grains.

Bacteria are splashed onto leaves where they infect through stomates or wounds. Leaf tissues must be water-soaked before infection can occur. Water-soaked areas develop as a result of flooding of the intercellular spaces. Bacteria are forced into leaf openings and then move into the flooded intercellular spaces. Optimal infection conditions occur during rainstorms accompanied by high winds.

Windblown water is the most important means of dissemination from plant to plant in the field and over long distances but it is also disseminated on contaminated transplants. Optimum temperature for growth in culture is between 24°C and 28°C. The major difference between *P. tabaci* and *P. angulata* is the production of toxin by *P. tabaci*.

DISTRIBUTION: Generally distributed wherever tobacco is grown.

SYMPTOMS: Wildfire may suddenly appear in the plant bed on tender, rapidly growing plants after a cool, rainy period. Diseased areas are in the wettest part of the bed and are about 1–2 m in diameter. During high humidity, small, circular, yellow-green, water-soaked spots develop in leaves. These spots rapidly turn brown and are surrounded by a yellow-green halo, giving the plants a scorched appearance. Severely infected seedlings are located in the middle of a diseased area and may die.

Older plants have spots that are mostly circular, with a brown dead center of about 12–25 mm in diameter surrounded by a yellow-green, water-soaked border or halo. Spots may coalesce, forming large dead areas on the leaf. Leaves affected on one side become twisted and distorted.

CONTROL:
1. Plow planting beds in the autumn to give residue time to decompose.
2. Steam or fumigate spring-plowed planting beds.
3. Plant only seed from healthy plants.
4. Treat seed with a bactericide to eradicate bacteria from infected seed.
5. Apply bactericide to foliage of plants in seed bed and in field.
6. Use only disease-free transplants.
7. Grow resistant cultivars.
8. Provide proper drainage for plant beds.
9. Do not grow tobacco in infested soil for at least two consecutive years.
10. Avoid working in a field when leaves are wet.

Wisconsin Bacterial Leaf Spot

CAUSE: *Bacterium melleum* (Johnson). However, this is probably a strain of *Pseudomonas syringae* pv. *tabaci* (Wolf & Foster) Young, Dye & Wilkie. Some workers suggest *B. melleum* is an intermediate between *P. tabaci* and *P. angulata* (From & Murray) Holland. Therefore, Wisconsin bacterial leaf spot may have been confused with wildfire.

DISTRIBUTION: Europe and Wisconsin. There has been no report of the disease from Wisconsin since 1937.

SYMPTOMS: Initially, circular, pinpoint-sized specks appear that are surrounded by distinct yellow zones. Spots enlarge rapidly during periods of high temperatures and humidity to form spots about 1 cm in diameter. The necrotic tissue in these spots becomes brown. Irregular spots form when smaller spots coalesce. Lesions on veins are brown and sunken.

Seedlings have inconspicuous angular spots surrounded by yellowish margins. As lesions become older, the necrotic tissues become brown. When spots are numerous, foliage will appear to be blighted.

CONTROL: Control is similar to that of angular leaf spot and wildfire.

FUNGAL CAUSES

Anthracnose

CAUSE: *Colletotrichum destructivum* O'Gara overwinters as a saprophyte on various plant residue. It also survives in infected residue and soil, probably as mycelium and acervuli, and is seedborne. Conidia are produced in an acervulus and are airborne or splashed to leaves. Anthracnose develops over a wide temperature range (18–32°C) and high relative humidity.

DISTRIBUTION: Asia, Africa, Australia, Brazil, and the United States.

SYMPTOMS: Young leaves have small, light green, water-soaked spots that are depressed. Spots enlarge up to 3 mm in diameter and give the underside of the leaf a greasy appearance. As the spots dry out they resemble paper and become gray-white. The spots are surrounded by a raised, water-soaked border that becomes brownish. Larger spots are zonate in appearance and have a dark brown center. Affected leaves become wrinkled, distorted, and may die. Small plants may be stunted or killed. Large plants may have cankers up to 12 mm long on the leaf midrib and petiole that are initially water-soaked but later become red-brown.

CONTROL:
1. Apply foliar fungicides to plants in the plant bed and field.
2. Fumigate infested soils.
3. Treat seed with a seed-protectant fungicide.
4. Destroy all weeds in the vicinity of plant beds.
5. Do not grow peppers and tomatoes in seed beds.

Ascochyta Leaf Spot

This disease is also called ragged leaf spot.

CAUSE: *Ascochyta phaseolorum* Sacc. Disease is more severe during cool (22°C), wet weather.

DISTRIBUTION: Asia, Europe, and the United States.

SYMPTOMS: Symptoms occur on plants of all ages. Spots are circular, with gray to brown centers, and are up to 2.5 cm in diameter. Centers of spots may fall out, giving a ragged leaf spot appearance. Scattered black pycnidia may be found in the center of older lesions and on stems of young seedlings. Spots may coalesce to form large dead areas.

CONTROL: Apply foliar fungicides.

Black Mildew

CAUSE: The causal fungus is similar in morphology to *Diporotheca rhizophila* Gordon & Shaw. The fungus has been found only on plants infested with the tobacco cyst nematode.

DISTRIBUTION: Connecticut. Black mildew is of no economic importance.

SYMPTOMS: Not every plant growing in soils infested with the tobacco cyst nematode is infected. Tobacco roots are colonized with a dark mycelium of the fungus.

CONTROL: None is needed.

Black Root Rot

This disease is also called maricume radicale, root rot, and Thielavia root rot.

CAUSE: *Thielaviopsis basicola* (Berk. & Br.) Ferr. survives as chlamydospores and endoconidia in soil. It is also able to grow saprophytically on several different types of plant material and persist indefinitely in the absence of tobacco. Chlamydospores and endoconidia germinate to form hyphae that penetrate roots. Chlamydospores and endoconidia are produced in mycelium on the outside of infected roots and serve as secondary inoculum. The fungus is disseminated on the roots of transplants and by any means that transports soil. Black root rot is most severe at relatively cool temperatures between 17°C and 23°C and in wet soils. Plants grown in soils with a pH of 6.4 and above are more prone to infection by *T. basicola*.

DISTRIBUTION: Generally distributed wherever tobacco is grown.

SYMPTOMS: Seedlings in the plant bed damp off. Roots are partially or entirely black. Smaller roots are rotted off but lesions on larger roots may only appear roughened. Infections that occur later in plant beds produce black lesions on the roots and cause leaves to yellow. Severely infected plants may have a root system that is pruned with a few remaining black stubby roots attached to a stem.

Usually only portions of a field may be affected. Plants may be stunted or uneven in growth. Leaves yellow and wilt during the day. Such plants may flower prematurely. Black lesions form on the main and lateral roots. The fungus present in these lesions can be identified easily microscopically. If the weather becomes warm, many infected plants will produce new roots and assume normal growth.

CONTROL:
1. Grow resistant cultivars.
2. Use a new bed site each year.
3. Treat soil with steam or chemicals if an old bed site is used.
4. Rotate tobacco with nonsusceptible crops such as grass for three years or longer. Soybeans, red clover, alfalfa, and other legumes are also susceptible to *T. basicola*.
5. Plow under cover crops and manure to ensure complete decomposition before transplanting.
6. Do not apply more than 10 tons of manure per acre.
7. Maintain a soil pH of 6.0–6.4.
8. Fertilize according to soil tests. Avoid excessive fertilizer.
9. Avoid transplanting when soils are cold or when the air temperature is low.

Black Shank

CAUSE: *Phytophthora parasitica* Dast. var. *nicotianae* (Breda de Haan) Tucker (syn. *P. nicotianae* var. *nicotianae* (Breda de Haan) Waterhouse) survives as chlamydospores in soil or residue and mycelium in residue. Oospores are formed and overwinter in diseased tissue and probably in soil; however, little is known of their function in the life cycle of the fungus. The chlamydospores germinate to form germ tubes in which either another chlamydospore or sporangia are produced. The sporangia then germinate to form either hyphae or zoospores. Zoospores germinate to produce hyphae that infects the plants along with the hyphae produced directly from sporangia. Sporulation occurs on stems near the soil surface or on residue. Sporangia then may be airborne or waterborne some distance to initiate further infection. Infection is accentuated in the presence of nematodes.

The fungus is disseminated on transplants, water, or by any means that moves soil. Spread of the fungus is greatest between susceptible plants and least between resistant plants. Black shank is most severe in wet soils at temperatures above 21°C. Thus disease incidence is correlated with soil texture and drainage class of the parent soil. In experiments, *Meloidogyne incognita* (Kofoid & White) Chitwood and *Globodera solanacearum* (Miller & Gray) Mulvey & Stone increased the severity of black shank. Different physiologic strains of the fungus exist.

DISTRIBUTION: Africa, Asia, Australia, Caribbean, eastern Europe, North America, and South America.

SYMPTOMS: Symptoms may initially occur in the wetter areas of a field. Plants of all ages may be infected but seedlings are most susceptible. Seedling stems near the soil line become dark brown or black. Roots turn black with the discoloration extending several centimeters above the soil line on the stalk. In the absence of aboveground symptoms, the tops of plants may wilt during the day and not recover at night.

The first symptom on vigorously growing plants is a rapid wilting of the leaves. Soon the leaves turn yellow, then hang down, die, and turn brown. A plant in the early stages of disease will have one or more dead and blackened lateral roots but no discoloration on the stem. As the disease progresses, the entire root system and base of the stalk die and become blackened. The discoloration will also extend several centimeters above the soil line on the stalk.

Older plants that become infected initially have a black discoloration on the stem that extends some distance above the soil line. Soon the leaves turn yellow, then brown, and shrivel, with sometimes a few small, green leaves remaining at the top. Such plants may bloom prematurely.

One of the most characteristic symptoms of black shank is seen when the stem is split lengthwise. The pith is dry, brown to black, and separated into platelike disks.

During rainy weather, large lesions may develop on the lower leaves. At first a lesion is pale green; later it becomes brown and may develop up to 8 cm in diameter.

CONTROL:
Where black shank is present:
1. Plant resistant cultivars.
2. Move plant beds to an area that has not been planted in row crops.
3. Plant an infested field to permanent pasture for at least five years. Other alternatives are two or three years of clean fallow and rye grown for grain. Two or three years of peanuts, soybeans, and cotton are not as satisfactory but still aid in reducing the population of the pathogen.
4. Use good nematode control practices.

5. In tillage work, leave an area or field that is infested with *P. parasitica* var. *nicotianae* until last.
6. Disinfect machinery parts with fungicides after use.
7. Clean shoes after working in infested soil. Use a stomp box, that is, a wood box with a sack in the bottom on which a fungicide is poured.
8. Apply fungicides.

Where black shank is not present:

1. Grow your own plants. This prevents movement of the fungus on transplants.
2. Do not irrigate from water sources if water may originate from infested fields.
3. Do not borrow equipment. Clean off all equipment with a fungicide solution.

Blue Mold

This disease is also called downy mildew.

CAUSE: *Peronospora tabacina* Adam (syn. *P. hyoscyami* deBary f. sp. *tabacina*) survives as mycelium and possibly sporangia within residue in areas with mild winters. Apparently oospores that overwinter in soil for one year or longer are not infectious. Systemic mycelium may overwinter in roots of plants whose aerial portion has been killed. Such plants may produce suckers during favorable weather and become systemically infected. Wild tobacco (*Nicotiana repanda* Willd.) also provides a means of overwintering and a possible significant inoculum source over a long period.

In colder tobacco-growing areas, oospores probably constitute the overwintering means in seedlings. However, it is not known how they germinate to initiate primary infection. Primary inoculum may be airborne from warmer regions.

Conidia are produced from mycelium and are airborne long distances. Conidia germinate in presence of moisture and produce a germ tube that penetrates the leaf. Conidia produced in lesions provide secondary inoculum. Oospores are formed within diseased tissue. The fungus is primarily wind-disseminated. However, any means by which infected plant material containing conidia, mycelium, and oospores can be moved from place to place is capable of disseminating the fungus. Disease severity is favored by temperatures of 15–25°C. Relative humidity in excess of 95 percent together with intermittent rainy weather also favors blue mold development. However, epiphytotics of blue mold have occurred despite temperatures outside the range considered favorable for the disease. Apparently the pathogen population has adapted to higher temperatures in some areas. Availability of moisture on leaves appears to be a predominant factor in disease development. Younger plants are more susceptible to infection than older plants.

DISTRIBUTION: Australia, Canada, Europe, Cuba, Mexico, Mediterranean area, Near East, and the United States.

SYMPTOMS: Small plants in the seed bed have erect leaves. Circular groups of larger plants become yellow with plants in the center having cupped leaves. Some cupped leaves have a bluish, fungal growth on the lower surface consisting of conidia and conidiophores. Plants begin to die and turn light brown. Leaves become twisted and turn the lower surface upward, accentuating the bluish color, especially when moisture is available. The entire plant or all the leaf tissue except the growing tip may be killed.

Older plants infected in seed beds become deformed with partially killed leaves or leaves that are twisted and puckered. Irregular-shaped, necrotic lesions may develop on leaves. Plants become stunted and roots become dark brown. The fungus and resulting disease will spread rapidly in cool, damp weather.

Plants in the field will have circular blotches on leaves that coalesce to form brown, necrotic areas. Leaves become distorted and puckered with large areas that disintegrate. Lesions may also occur on buds, flowers, and capsules. Plants may be stunted and display wilt symptoms. Vascular discoloration may occur as brown streaks. Stems may lodge and roots become dark brown.

CONTROL:
1. Apply foliar fungicides to plants in bed. Treatment must begin before disease appears.
2. Destroy beds after setting is complete.
3. Destroy any live plants remaining after harvest.
4. Locate beds in an area where they will not receive morning shade.

Botryosporium Barn Mold

CAUSE: *Botryosporium longibrachiatum* (Oud.)   Maire and *B. pulchrum* Cda. The fungi probably survive on residue. Plants are apparently infected or have a latent infection at harvest. Secondary infection originates from infected plants brought into the barn. Humid, wet conditions favor disease.

DISTRIBUTION: Widespread.

SYMPTOMS: White mycelium and conidiophores are confined to lesions 1–20 cm long on midribs and on secondary veins of lower leaves. Lesions are initially water-soaked. Sporulation then occurs on laminae adjacent to infected areas on midribs and veins. Lesion development and sporulation occur after tobacco reaches the brown stage of curing. Eventually sporulation spreads further and is evident on all leaf surfaces. Lower leaves drop to the barn floor; those remaining on stalks are severely water-soaked and decayed.

CONTROL: Supply supplemental heating to assist drying.

Brown Spot

CAUSE: *Alternaria alternata* (Fr.) Keissler (syn. *A. tenuis* Nees) survives as mycelium in woody plant residue such as tobacco stems. During periods of wet weather, conidia are produced that are air- and water-disseminated to lower leaves of plants. Secondary inoculum is provided by conidia produced in lesions during moist weather and windborne to plant tissue. The fungus is also disseminated by infected seed and transplants. Infection and disease development occur at temperatures higher than 21°C and are most rapid when higher than 26.5°C.

DISTRIBUTION: Generally distributed wherever tobacco is grown.

SYMPTOMS: Round, brown spots occur on leaves. On inoculated plants, spots are larger on old leaves than on young ones. Frequently the spots are marked by concentric rings. The spots may be small, less than 6 mm in diameter; however, during moist weather they may grow up to 25 mm in diameter. Spots may coalesce until a large area of the leaf is killed.

Spots may occur on stalks late in the growing season. These spots are usually smaller than on leaves but they may be more numerous, giving the stalk a speckled appearance.

CONTROL:
1. Plant only disease-free seed.
2. Fumigate plant beds and remove plant residue from the area.
3. Rotate tobacco in fields where brown spot was present.
4. Plant tolerant cultivars. In general, burley and dark tobacco are less susceptible than flue-cured.

Charcoal Rot

CAUSE: *Macrophomina phaseolina* (Tassi) Goid (syn. *M. phaseoli* (Maubl.) Ashby) survives as sclerotia in soil and residue. Sclerotia germinate to form a germ tube that penetrates the host. Charcoal rot is more severe in dry weather at temperatures of 38°C and above.

DISTRIBUTION: Generally distributed wherever tobacco is grown in tropical or subtropical areas.

SYMPTOMS: Symptoms are associated with injury caused by mineral oil applied for sucker control. A black lesion develops. The fungus may grow inward, causing pith and wood decay, followed by plant death and lodging. Sclerotia may be embedded throughout the bark and wood.

CONTROL:
1. Charcoal rot is most severe on plants to which mineral oil has been applied during temperatures of 32°C and above.
2. Fumigate soil to kill sclerotia.

Collar Rot and Dead Blossom Leaf Spot

CAUSE: *Botrytis cinerea* Pers. ex Fr. and *Sclerotinia sclerotiorum* (Lib.) deBary. *Botrytis cinerea* is discussed under Gray Mold. *Sclerotiorum sclerotiorum* survives as sclerotia in infected stalks and soil. In spring sclerotia germinate to form either mycelium or apothecia. An occasional infection may originate from mycelium but most infections originate from ascospores produced on apothecia that are windborne to plants. Ascospores are produced throughout the summer, infecting flowers that fall onto leaves, causing dead blossom leaf spot. Infection also occurs in wounds where tops or suckers were removed. Sclerotia are then produced in the stalk. Both diseases are more severe during wet weather.

DISTRIBUTION: Generally distributed wherever tobacco is grown.

SYMPTOMS: *Collar rot.* Seedlings may be infected up to transplant size. Initially a small, dark brown lesion with a well-defined margin occurs at the base of a stem. After transplanting, the lesion may circle the stem and kill the plant. The lesion is gray and may be higher up on the stem with less well-defined margins. White, cottony, mycelial growth may grow over an infected area in which small black sclerotia are present.

Dead blossom leaf spot. As infected flowers fall, they adhere to the leaf. Small dark spots develop on the leaf where the flower has fallen. Eventually the spots become larger and brown to gray. During wet weather a spot may enlarge to cover half the leaf. If the midrib is rotted, the leaf becomes yellow and hangs down. Brown lesions several centimeters long may occur on the stem. All leaves above the lesion die.

CONTROL:
1. Control collar rot by steaming or fumigating plant beds. Do not place plant beds in low, wet areas. Provide adequate ventilation.
2. No practical control exists for dead blossom leaf spot except late-flowering cultivars become less infected than early-flowering cultivars.

Corynespora Leaf Spot

CAUSE: *Corynespora cassiicola* (Berk. & Curt.) Wei. survives two or more years as mycelium and chlamydospores in residue and soil. It can also grow saprophytically on many different kinds of plant residue in the soil. Conidia are splashed or blown to leaves and cause infection when free moisture is present or at a relative humidity of 80 percent and above. Several other kinds of plants are hosts.

DISTRIBUTION: Nigeria.

SYMPTOMS: Older leaves are infected first, and younger leaves are infected later in the growing season. Disease incidence increases toward harvest. Initially a dark brown, circular spot occurs on leaves. Spots at first are about 2–3 mm in diameter and grow to 20–30 mm in diameter. Petiole and midribs have dark brown spots, with veins becoming brown starting from the midrib. Spots tend to be zonated with a dark brown center that in turn is surrounded by a light brown zone, a thin dark brown ring, and finally a light brown margin. Spots may coalesce.

CONTROL:
1. Harvest on time since mature leaves are likely to become infected.
2. Exclude any known alternate host, such as cotton, soybeans, cowpeas, cucumbers, lupines, and watermelons, from the vicinity of tobacco fields.

Curvularia Leaf Spot

CAUSE: *Curvularia verruculosa* Tandon & Bilgrami ex. M. B. Ellis.

DISTRIBUTION: India.

SYMPTOMS: Spots appear as round to oval areas with concentric zones. Lesions may enlarge rapidly and affect the entire lamina. Conidia and conidiophores appear on both leaf surfaces. Infected leaves initially are yellow but later turn dark brown.

CONTROL: Do not rotate tobacco with rice.

Damping Off

CAUSE: *Pythium* spp. including *P. aphanidermatum* (Edson) Fitzp., *P. debaryanum* Hesse, and *P. ultimum* Trow. Other *Pythium* spp. may also be involved. The fungi survive in soil as oospores and chlamydospores (sporangia). Both structures germinate by germ tubes or form zoospores, which in turn produce germ tubes that infect stems and roots at or just below the soil line. Oospores are eventually formed in diseased tissue. Damping off may be expected to be more severe in saturated soil at temperatures below 24°C.

Dissemination is by any means that will transport soil but water movement is usually the more important means. The fungi are not seedborne.

DISTRIBUTION: Generally distributed wherever tobacco is grown.

SYMPTOMS: Damping off is primarily a problem in plant beds. Infection usually occurs on the hypocotyl near the soil line. A brown, watery, soft rot develops on the hypocotyl that soon girdles it and causes the seedling to fall over. Roots may also be infected and decayed.

Older seedlings may turn chlorotic. The roots of these plants have a soft, watery rot and the cortex peels away from the woody central cylinder. If plants are infected at the soil line, roots are not rotted but remain white.

Transplants may be infected with brown, watery lesions that cause the stems to become limp, shrivel, and disintegrate. Plants suddenly wilt and die, especially when weather is cool and damp following setting.

CONTROL:
1. Disinfect seed beds.
2. Avoid wet soils as plant bed sites.
3. Avoid dense stands and provide adequate ventilation during wet, humid periods.
4. Apply fungicide.
5. Disk and reset the field if loss of transplanted seedlings is high. New rows should be placed in the middle between the old, diseased ones.

Frogeye

The curing phase of frogeye is called barn spot or green spot.

CAUSE: *Cercospora nicotianae* Ell. & Ev. overwinters as mycelium in infected residue or other plant hosts. In the spring conidia are produced that are windborne or splashed to leaves. Disease development is more severe during warm, wet weather.

DISTRIBUTION: Africa, Central America, southeast Asia, and the United States.

SYMPTOMS: Mature leaves are most susceptible with lesions usually occurring on the lower leaves. Lesions are small (2–15 mm in diameter), brown, gray, or tan with a paperlike center. Dark dots, which are groups of conidia and conidiophores, may be found in lesions. Near harvest time, the upper leaves may develop large necrotic spots, particularly during wet weather. During dry weather, frogeye lesions may be only the size of a pinpoint.

CONTROL:
1. Plant only disease-free seed.
2. Remove residue from plant beds.
3. Apply foliar fungicides before transplanting.
4. Do not overfertilize with nitrogen.
5. Prime tobacco before it becomes overripe.
6. Plow under infected residue in the field.
7. Rotate tobacco with other crops.
8. Cure at 38°C with 100 percent relative humidity.

Fusarium Wilt

CAUSE: *Fusarium oxysporum* Schlecht. f. sp. *nicotianae* (J. Johnson) Snyd. & Hans. survives as chlamydospores in soil and infected residue, and as mycelium in infected residue. The fungus also persists on the roots of symptomless hosts. Chlamydospores germinate to form hyphae or mycelium grows from infected residue and infects roots through wounds. The fungus then grows into the xylem tissue.

Macroconidia and microconidia are produced on the outside of moribund tissue and microconidia are also formed within vessels. Mycelial cells, macroconidia, and microconidia are converted to chlamydospores and returned to the soil. The fungus may be disseminated in tobacco transplants and by any means that transplants soil. Fusarium wilt is most severe at 28–31°C.

DISTRIBUTION: Generally distributed wherever tobacco is grown.

SYMPTOMS: Initially leaves on one side of plants are dwarfed and yellowed. Eventually leaves wilt and turn brown permanently. The plant top usually turns yellow and bends toward the affected side of the plant. This condition may continue for some time before wilting and death occur. When the soft outer bark is removed, the surface of the exposed wood is a dark chocolate-brown.

CONTROL:
1. Grow resistant cultivars.
2. Fumigate soil with a nematicide where nematodes are a problem.
3. Rotate tobacco with a nonsusceptible crop. Flue-cured but not burley and dark tobacco can be grown in rotation with cotton. Do not grow sweet potatoes in the same rotation with cotton.

Gray Mold

CAUSE: *Botrytis cinerea* Pers. ex Fr. survives in soil and residue as sclerotia. The fungus also persists as a saprophyte on a wide range of dead plant material. Sclerotia germinate during wet weather to produce mycelium on which conidia are borne. Spores are splashed or blown to tobacco. Stalk-cut tobacco is infected through wounds and injuries. Sclerotia are formed on surface of infected tissue.

DISTRIBUTION: Generally distributed wherever tobacco is grown.

SYMPTOMS: The first symptoms occur on the lower leaves when seedlings are ready to be transplanted. If weather turns dry after disease has started, spots will turn brown with a yellow margin. During wet weather, a gray mycelial growth covers the spots. This is a good diagnostic characteristic. Eventually the leaf collapses but remains attached to the stem. The fungus grows from the leaf petiole into the stem. The stem lesion may become several centimeters long and covered with gray mycelial growth. Plants may die; however, slightly infected plants may recover during dry weather.

CONTROL:
1. Place seed beds on a well-drained site with adequate ventilation.
2. Apply foliar fungicides.

Olpidium Seedling Blight

CAUSE: *Olpidium brassicae* (Woron.) Dang. survives in soil or residue as resting sporangia. The resting sporangia discharge zoospores in the presence of moisture. Zoospores swim to roots where they encyst. The cyst protoplast moves through the host wall and establishes itself within the host cytoplasm. Resting sporangia eventually form within roots. The disease is most severe in moist and cool soils (10–16°C).

DISTRIBUTION: Generally distributed wherever tobacco is grown.

SYMPTOMS: The plant aboveground will yellow and wither. Roots have a brown decay.

CONTROL: Fumigate seed beds.

Phyllosticta Leaf Spot

CAUSE: *Phyllosticta nicotiana* Ell. & Ev. but three other species of Phyllosticta have been reported on tobacco, *P. tabaci* Pass., *P. capsulicola*

Sacc. and Speg., and *P. nicotianicola* Speg. The fungus probably over-winters as pycnidia in diseased residue. It is also seedborne. During wet weather, pycnidiospores are produced in pycnidia and are splashed or windborne to tobacco leaves that are stressed, dead, or dying. Transplants serve as a means of introducing the fungus into a field.

DISTRIBUTION: Africa, Europe, Caribbean, South America, and the United States.

SYMPTOMS: Leaf lesions are brown, irregular, zonate spots, 1–10 mm in diameter. Lesions are usually dark brown in the center, lighter brown toward the margin, with yellow-green tissue surrounding the necrotic areas. Pycnidia that appear as small, black dots may be embedded in necrotic tissue.

Other symptoms are small, white spots whose centers fall out, giving a shot-hole effect. Eventually large (15 mm in diameter), white blotches may occur. Sometimes all interveinal tissue may fall out, giving a skeletal appearance. Pycnidia develop in white tissue around the holes.

CONTROL: Phyllosticta leaf spot is rarely severe enough to warrant control measures.

Powdery Mildew

CAUSE: *Erysiphe cichoracearum* DC. overwinters as perithecia on residue and as mycelium on tobacco plants left standing in the field. The fungus may also overwinter on weed hosts. Primary inoculum is either ascospores that are released from perithecia or conidia that are produced on mycelium. Both kinds of spores are windborne to leaves. Secondary inoculum is provided by conidia produced on leaves. Conidia are primarily airborne to susceptible tissue. As tobacco matures, perithecia are formed in the fungal growth on leaves and stems.

Erysiphe cichoracearum is tolerant of a wide range of temperatures and humidities. Disease development is more severe under reduced light intensity.

DISTRIBUTION: Africa, Asia, Europe, Central and South America.

SYMPTOMS: Plants in the field do not show symptoms until approximately six weeks after planting. At first, feltlike patches appear on lower leaf surfaces and eventually cover the entire leaf surface. Soon a powdery gray layer consisting of conidia and mycelium covers both leaf surfaces. Eventually brown spots appear on the upper leaf surface.

Affected leaves become thin and papery. Black, round perithecia form on the leaf surface and can be viewed with a hand lens.

CONTROL:
1. Plant resistant cultivars.
2. Properly fertilize soil.

3. Plant tobacco early.
4. Apply foliar fungicide.

Rhizoctonia Leaf Spot

CAUSE: *Rhizoctonia solani* Kuhn, teleomorph *Thanatephorous cucumeris* (Frank) Donk. Disease is favored by cool, wet weather. The fungus belongs to anastomosis group AG2.

DISTRIBUTION: Brazil, Costa Rica, and North Carolina.

SYMPTOMS: Disease occurs on flue-cured tobacco. Initially symptoms are small, circular, water-soaked spots that rapidly expand to light green to tan lesions 2–6 cm in diameter with an irregular margin. Tissue within lesions is almost transparent, often displays a pattern of concentric rings, and frequently drops out, leaving a shot-hole effect on the leaves. Fungal mycelium may be present at margins of lesions on the underside of leaves and occasionally a hymenial layer and basidiospores of *T. cucumeris* is observed.

Lesions are most common on lower leaves but may occur as high up as 85 cm on the stalk.

CONTROL: None reported.

Rhizopus Stem Rot

CAUSE: *Rhizopus arrhizus* Fischer. Disease is apparently favored by high temperatures that occur during summer months.

DISTRIBUTION: Iraq on *Nicotiana glauca* R. C. Grah.

SYMPTOMS: Initially pale green, water-soaked lesions appear on the stem mostly below the inflorescence. Infection extends upward and downward, resulting in a slimy, wet rot of the cortical tissues. Lesions may be confined to one side or may completely encompass the stem. Leaf petioles attached to diseased portions of the stem become flaccid and die. Infected stem tissues appear pale to yellow-brown when dry. Frequently the flower head topples downward, bending at the infected region but remaining attached to the plant, or it may be broken off by strong winds. Severe disease results in plant death. Fluffy mycelial growth of the pathogen is visible in the pith of infected stems.

CONTROL: No control is reported.

Rust

CAUSE: *Uredo nicotianae* Arth. Uredia are produced on tobacco.

DISTRIBUTION: Brazil, Italy, and southern California.

SYMPTOMS: Brown sori are produced on leaves.

CONTROL: No control is necessary.

Rust

CAUSE: *Puccinia substriata* Ell. & Barth. Aecia are produced on tobacco; uredia and telia stages occur on various genera of grasses including *Paspalum* spp., *Digitaria* spp., and *Setaria* spp. The fungus is a weak pathogen that infects during the rainy season.

DISTRIBUTION: Honduras, Nicaragua, and Zimbabwe.

SYMPTOMS: Tobacco leaves are severely spotted. Round, raised lesions up to 1 cm in diameter, composed of thick, hard tissue occur on the upper leaf surface. These spots remain light green after curing. Aecia are formed on the underside of lesions in dense concentric rings, varying in color from cream to orange. Advanced symptoms of what is thought to be the same disease causes tissues to turn black and fall out, leaving a hole.

CONTROL: No control is necessary.

Scab

This disease is also called blotch.

CAUSE: *Fusarium affine* Fautr. & Lamb. The causal pathogen of scab was originally identified as *Septomyxa affinis* (Shub.) Wollenw. Fide Wollenw. The disease is most severe in plant beds during wet years or in damp shaded areas of the bed. Light colored, nitrogen-starved plants and those infected with black root rot are more susceptible to infection.

DISTRIBUTION: The United States and Zimbabwe.

SYMPTOMS: Upper leaf surfaces, petioles, and stems have olive-brown, irregular blotches and streaks. During wet weather, the affected parts develop a soft rot, resulting in leaves having a ragged appearance. The petioles and stems decompose. Slightly infected plants recover when transplanted to the field.

CONTROL: No control is reported.

Sooty Mold

CAUSE: *Fumago vagans* Pers. The fungus grows on honeydew secreted on leaves by aphids but does not infect the plant itself.

DISTRIBUTION: Formosa, Italy, Japan, the United States, and Zimbabwe. Sooty mold rarely occurs on tobacco.

SYMPTOMS: The lower mature leaves are most often affected. Fungus mycelium, chlamydospores, and conidia occur as a superficial black film or sooty layer on the leaf surface that can be scraped easily with a fingernail.

CONTROL: Controlling aphids will control the disease.

Soreshin

This disease is also called canker, collar rot, blackleg, rotten stalk, stem rot, and sore shank.

CAUSE: *Rhizoctonia solani* Kuhn, teleomorph *Thanatephorus cucumeris* (Frank) Donk., persists in soil as a saprophyte on the residues of a wide range of plants; as a parasite and often as a pathogen on roots of other crops used in rotation with tobacco and on weeds. It also survives as mycelium in plant residue and soil, and as sclerotia in soil. Additionally the fungus can survive in soil for a few weeks as basidiospores.

Roots are infected by mycelium from germinating sclerotia and basidiospores or directly from mycelium in soil and plant residue. The fungus is disseminated as sclerotia and mycelium by any means that moves soil or residue. Disease development can occur in either relatively dry or moist soils and over a wide temperature range. Soreshin can occur during temperatures of 20°C and lower in a plant bed and during relatively high temperatures in the field.

DISTRIBUTION: Generally distributed wherever tobacco is grown. Soreshin is generally not a serious disease.

SYMPTOMS: On seedlings and transplants, a dark brown, decayed area occurs at or near the soil line. The decayed area will extend upward and around the stem until the plant topples over. If slightly infected plants are set in a field during cool, wet weather, a poor stand will result.

On older plants in the field, a dark brown lesion occurs at or near the soil line, which eventually girdles the stem. The canker may extend up the stem to the lower leaves, causing them to drop off. The woody part of the stalk becomes hard and brittle. The entire plant may appear stunted, yellow, and wilted.

Infected plants are usually not noticed until the wind or another force topples the plant over. Roots generally remain healthy until the plant dies. The interior of the stalk has a decayed, dried, and brown pith. Light gray patches of fungal growth may also be present in the stalk.

CONTROL: Control measures can be used in the plant bed but not in the field.
1. Grow plants on a well-drained soil.
2. Disinfect seed beds with steam or a fumigant.

Southern Stem and Root Rot

This disease is also called southern blight.

CAUSE: *Sclerotium rolfsii* Sacc., teleomorph *Athelia rolfsii* (Curzi) Tu & Kimbrough, survives in soil as sclerotia and as mycelium in residue. Sclerotia germinate to form infection hyphae that penetrate the stalk. Infection also occurs by mycelium growing from precolonized residue. Sclerotia are formed on the outside of diseased tissue.

The fungus is disseminated as sclerotia in seed lots and by any means that will transport soil. Disease is more severe at high temperatures (30–35°C) and high soil moisture. The fungus is killed by low temperatures.

DISTRIBUTION: Generally distributed wherever tobacco is grown in warm temperate regions and the tropics.

SYMPTOMS: Infected plants are usually scattered throughout a field but may occur in groups. Initially a yellowing and wilting of the lower leaves occurs that gradually progresses up the plant. Eventually leaves die and turn brown. The stem at the soil line has a sunken canker that is brown, dry, and completely girdles the stem. The roots usually do not decay until the entire plant dies. Under moist conditions, a white, cottonlike, fungal growth occurs on the canker. Later, small, white to dark brown sclerotia that are about the size of mustard seeds appear on and in the mycelium.

CONTROL: There are no truly effective controls, but the following may be aids:
1. Bury infected residue at least 8 cm deep.

2. Do not place soil containing slowly decaying organic matter against stems during cultivation.

Tobacco Stunt

CAUSE: Not known for certain but the endogonaceous mycorrhizal fungus *Glomus macrocarpum* (Tul. & Tul.) Gerd. & Trappe is implicated as the primary pathogen.

DISTRIBUTION: Kentucky.

SYMPTOMS: Plants are stunted; flowering is delayed with reduced yield and quality. Affected plants are seldom killed.

CONTROL: Experimentally the disease is controlled by soil fumigation.

Verticillium Wilt

CAUSE: *Verticillium dahliae* Kleb. However, some researchers state that *V. albo-atrum* Reinke & Berth. is the causal organism. The fungus survives as microsclerotia in soil or infected residue. Microsclerotia germinate to produce chlamydospores or conidia, which in turn germinate to produce hyphae that infect roots. The fungus grows into the xylem tissue. Microsclerotia are abundantly formed within infected tissue and are released into soil on residue decay. Microsclerotia and conidia are disseminated by wind, water, and any means that can transport soil. Verticillium wilt is most severe when abundant moisture follows dry weather and at a temperature of 22–28°C.

DISTRIBUTION: Generally distributed wherever tobacco is grown. Verticillium wilt is a severe disease in New Zealand.

SYMPTOMS: Symptoms are not obvious until flowering time. One or more lower leaves wilt, particularly during hot weather. Eventually the interveinal area of lower wilted leaves turns a bright orange. Eventually the tissue yellows, dies, and turns brown, leaving an orange border between living and dead tissue. Wilting sometimes occurs on one side of a leaf or one side of an entire plant. All leaves eventually die. The vascular system of both leaves and stems becomes light brown.

CONTROL:
1. Fumigate plant beds.
2. Control nematodes.
3. Grow resistant cultivars.

MYCOPLASMAL CAUSES

Aster Yellows

CAUSE: Aster yellows mycoplasma infects and persists in more than 175 species of plants. It is disseminated by several species of leafhoppers, including *Hyalesthes obsoletus* Signoret, *Aphrodes bicintus* (Schrank), *Euscelis plebejus* Fn., *Macrosteles laevis* Rib., *M. fascifrons* Stal., and others. The mycoplasma multiplies in the leafhoppers and likely overwinters in the eggs and adults. The optimum temperature for multiplication in plant and insect tissue is 25°C. Leafhoppers may be blown or fly several hundred kilometers.

DISTRIBUTION: Generally distributed wherever tobacco is grown.

SYMPTOMS: Symptoms are the same as those for Bigbud.

CONTROL: Controls are the same as those for Bigbud.

Bigbud

CAUSE: Bigbud mycoplasma infects and persists in several species of plants. It is disseminated only by the leafhopper *Orosius argentatus* (Evans). The mycoplasma multiplies in the leafhopper and likely overwinters in the eggs and adults. The leafhopper does not breed on tobacco, so transmission occurs during periods of migration in late spring and early summer.

DISTRIBUTION: Australia.

SYMPTOMS: Plants are stunted. Apical leaves are small, whitish, and curled. Leaves have interveinal necrosis and hang close to the stem. Numerous small leaves are present on sucker growth. Flowers change to green, leaflike structures together with numerous short, stiff branches. The proliferation of floral parts results in a compact tufted growth habit.

CONTROL:
1. Eliminate weed hosts in tobacco fields and adjacent to them.
2. Plant only disease-free transplants.
3. Avoid growing tobacco near other host plants such as potatoes, tomatoes, and peppers.
4. Plant during periods when leafhopper populations are low.
5. Harvest and cure separately if only a few plants are affected.

Stolbur

CAUSE: Stolbur mycoplasma overwinters in several species of plants but principally the common perennial bindweed, *Convolvulus arvensis* L. *Nicotiana glauca* R. C. Grah is a symptomless host. The mycoplasma is disseminated and multiplies within several species of leafhoppers. It may also overwinter in the eggs and adults of leafhoppers. Stolbur mycoplasma strains are vector specific. Some strains are transmitted principally by one species of leafhopper. Leafhoppers may be blown or fly several hundred kilometers. Leafhoppers inoculate the mycoplasma directly into the phloem.

DISTRIBUTION: Europe.

SYMPTOMS: Younger plants will suffer more damage than will older plants. Symptoms are less severe during cool weather. Apical leaves are small, whitish, and curled. Leaves have interveinal necrosis and hang close to the stem. Flowers change to green, leaflike structures, together with numerous, short, stiff branches. Flower proliferation results in a compact, tufted growth habit. Numerous small leaves are present on sucker growth. Plants in general are stunted. Affected plants do not set seed even though the pollen is normal.

Internal symptoms are a degeneration of the phloem and irregularities in meiotic processes. Cured leaves are more hygroscopic than unaffected ones. Diseased leaves interfere with fermentation when bulked with healthy leaves and decrease quality.

CONTROL: Controls are the same as for Bigbud.

Yellow Dwarf

CAUSE: Yellow dwarf mycoplasma infects and persists in several species of plants. It is disseminated only by the leafhopper *Orosius argentatus* (Evans). The mycoplasma multiplies in the leafhopper and likely overwinters in the eggs and adults. The leafhopper does not breed on tobacco so transmission occurs during periods of migration in late spring and early summer. Conditions favoring leafhopper flight are a 9 PM temperature above 27°C, high humidity, and no wind.

DISTRIBUTION: Australia.

SYMPTOMS: Infection may occur at any stage of growth. Small, rapidly growing plants are highly susceptible. Plants are yellowish, dwarfed, and grow slowly, producing small leaves unsuited for commercial use. Older leaves become thick and wrinkled. Root systems remain undeveloped.

CONTROL: Controls are the same as those for Bigbud.

_____CAUSED BY NEMATODES_____

Brown Root Rot

CAUSE: _Pratylenchus_ spp. particularly the lesion nematodes, _P. pratensis_ (de Man) Filipjev, _P. brachyurus_ (Godfrey) Filip. & Sch. Stek., and _P. zeae_ Graham. The nematodes overwinter in a number of ways: as adults, eggs, or larvae in roots and soil. Eggs are deposited in root tissue and hatch in 6–17 days. The resulting larvae will either continue to feed in the same root or emerge and migrate to another root. Eggs may also be deposited into soil by decaying roots. The larvae penetrates the root and migrates mainly in the cortex.

The different lesion nematodes are favored by different soil temperatures; therefore, disease development can be expected to occur over a wide temperature range. Soil moisture is not ordinarily a factor in disease development although lesion nematodes are sensitive to drying.

DISTRIBUTION: Canada and the United States.

SYMPTOMS: Symptoms usually occur in a definite area within a field. Occasionally an entire field may be affected. Infected plants are stunted and wilt during the day. Plants recover turgidity at night but repeatedly will wilt again. Eventually leaves in the middle and lower portion of the stalk will have brown and necrotic margins.

Roots, at first, will have colorless and water-soaked lesions. Eventually cortical lesions will vary from pale yellow to dark brown or black. Such lesions may result in a girdling of the feeder roots. Lesions break open and cortical tissue sloughs off, leaving only the vascular cylinder.

Infected roots shrivel and die, resulting in extensive root pruning. Numerous adventitious roots will often develop above a lesion, giving a root a bushy appearance. If pruned roots are numerous, the root system will also have a stubby appearance. Severely affected plants may be pulled easily from the soil because of almost complete destruction of the root system. The surviving roots are usually grouped near the soil surface.

CONTROL:
1. There are no resistant cultivars, but some are injured less than others.

2. Crop rotation. Do not plant tobacco after timothy, rye, corn, cotton, bluegrass, and legumes since these plants support large populations of Pratylenchus spp. Some of these plants also release toxic chemicals that may harm tobacco roots, so turn crops over in plenty of time to allow for decomposition before planting tobacco.

Root Knot Nematodes

CAUSE: Root knot nematodes *Meloidogyne arenaria* (Neal) Chitwood, *M. incognita* (Kofoid & White) Chitwood, and *M. javanica* (Treub.) Chitwood survive in soil as eggs and larvae, and in galls and roots as adults and eggs. Eggs in soil hatch to produce larvae that move through soil water to roots. The larva enters the root where it remains all its life. After the female is mature, it produces eggs in a sac that emerges from the female genital opening prior to egg production. Eggs are released into the soil as roots and galls decay.

Nematodes are disseminated by any means that will transport soil. Where soil is subjected to subzero temperatures, the nematodes survive only with difficulty. Disease is ordinarily more severe in lighter sandy soils but may occur in any soil type. Soil moisture is not an important factor in disease development. Different strains exist.

DISTRIBUTION: Generally distributed wherever tobacco is grown.

SYMPTOMS: Infected plants occur at random throughout a field. Plants are stunted and yellowed. Plants may occasionally be killed especially during dry weather. Such plants may wilt during the day but recover turgidity at night. Wilting reoccurs until leaves in the middle and lower part of plants develop necrotic margins and tips.

Galls form on roots and vary in size from a pinhead to several times the thickness of the root. Galls vary in shape from irregular to spherical and are most often found on rootlets resembling beads on a string. Galls may form so close together that they resemble one large gall.

CONTROL:
1. Plant resistant cultivars; however, there is only resistance to southern root knot nematode.
2. Treat soil with a nematicide.
3. Rotate tobacco with nonsusceptible crops such as small grains. Plant tobacco only once every three or four years in the same soil.
4. Plow or disk as soon after harvest as possible. Exposing roots to drying action of the sun and air kills many nematodes. As long as tobacco stalks and root systems are left in the field undisturbed, the plants will live and nematodes will continue to live and reproduce in them.

Stem Break

CAUSE: *Ditylenchus dipsaci* (Kuhn) Filipjev. can survive for several years in dry plant material as the fourth larval stage. Larvae and eggs can also survive in soil. Eggs hatch and the resulting larvae migrate to plant roots where they probably migrate up the plant in a film of water. The larvae enter the stem through stomates, lenticels, or wounds. The larvae move in the stem but are confined to cortical tissue.

Stem break is favored by cool (15–20°C), wet weather. Disease incidence is usually more severe on plants grown in moist loam or clay soils. The nematode has a wide host range.

DISTRIBUTION: France, Germany, Holland, and Switzerland.

SYMPTOMS: Young plants are more severely affected than older ones. Initially, small, yellow swellings or galls occur on stems and may extend 40 cm or more above the soil level. The older galls die, causing the stem to turn black. Infected plants stop growing and the upper leaves turn yellow while the lower ones fall off. Eventually the stem breaks and the plant falls over; the vascular system remains intact; therefore, wilting usually does not occur until just before the plant breaks over.

CONTROL:
1. Remove diseased plants from field.
2. Rotate tobacco for long intervals with nonsusceptible crops.
3. Fumigate soil with nematicides.
4. Apply contact nematicides to plants.

Stubby Root Nematodes

CAUSE: *Trichodorus* spp. are nematodes with a wide host range. Little is known of their life cycle but it is probable that they can survive as eggs, larvae, and adults in the soil. Nematode buildup is favored by soil temperatures between 22°C and 30°C and well-aerated, light sandy soils. *Trichodorus* spp. may be found deeper in the soil than most nematodes at 100 cm.

DISTRIBUTION: Generally distributed throughout the world.

SYMPTOMS: Plant growth is retarded. Infected plants wilt easily even in the presence of moisture. Root systems are small, with fewer and shorter rootlets than normal because the root tip stops growing. Damage is best characterized by the presence of short, stunted, and stubby roots.

CONTROL: Control measures are the same as for root knot; however, the stubby root nematodes are usually more difficult to control.

Stunt Nematode

CAUSE: *Merlinius* spp. and *Tylenchorhynchus* spp. *Tylenchorhynchus claytoni* Steiner is the principal stunt nematode infecting tobacco. It survives for several months in the soil as eggs, larvae, and adults if soil is not dried out. The nematode has a wide host range.

DISTRIBUTION: Canada and southeastern United States.

SYMPTOMS: Plants are stunted and have a small root system. Infected roots are shriveled, flaccid, do not elongate normally, and in general, are poorly developed.

CONTROL: Control measures are the same as for Root Knot Nematodes.

Tobacco Cyst

CAUSE: *Globodera solanacearum* (Miller & Gray) Mulvey & Stone and *G. tabacum* (Lownsbery & Townsbery) Mulvey & Stone. The nematodes survive in soil as eggs and larvae contained in cysts. Larvae emerge from cysts and migrate to roots where they begin to feed with their heads in the stele of the rootlet. Females become spherical and break through root epidermis with only their head and neck in the root. The female is fertilized by males and eggs are produced in the female body. Many eggs hatch and reinvade rootlets. When females die, the bodies become resistant cysts containing hundreds of eggs. The nematodes are disseminated by any means that will transport soil. Injury is most severe under continuous tobacco culture.

DISTRIBUTION: Eastern United States.

SYMPTOMS: Plants are stunted, wilt, and have a small root system. Dark brown oval cysts, about 0.5 mm in diameter, are attached to the roots.

CONTROL:
1. Rotate to a nonhost. Fescue has proven successful in Virginia.
2. Apply nematicides.
3. Grow resistant varieties; however, these varieties are not tolerant.
4. Practice sanitation with equipment and irrigation water.

VIRAL CAUSES

Alfalfa Mosaic

CAUSE: Alfalfa mosaic virus (AMV) overwinters in a wide number of hosts. It is transmitted primarily by aphids. Several strains of the virus exist and may be found in one plant.

DISTRIBUTION: AMV has been reported in tobacco from Europe, Japan, New Zealand, and the United States.

SYMPTOMS: Initially chlorotic spots and blotches occur on leaves. The first leaves infected systemically have vein clearing followed by white rings, arcs, and coalescing line patterns of necrotic tissue. Sometimes bud leaves are distorted and a bright yellow mosaic occurs.

CONTROL: AMV has not been serious on tobacco. However, isolation of tobacco fields from sources of inoculum, weed, and aphid control would be beneficial.

Beet Curly Top

CAUSE: Beet curly top virus (BCTV) overwinters in more than 244 species of host plants. BCTV is transmitted by leafhoppers. *Circulifer tenellus* Baker is the most important vector in North America and *Agallia albidula* Uhler and *Agalliana ensigera* Oman in South America. BCTV can persist up to 85 days in the leafhopper vectors. The acquisition period takes up to 2 days. The latent period within the leafhopper is 4–24 hours but does not multiply within the vector. Infectivity of leafhoppers decreases over 8–10 weeks. Different strains of the virus exist.

DISTRIBUTION: Brazil and the United States.

SYMPTOMS: Young plants are most susceptible to infection. Initially there is a vein clearing with the leaf tips and margins bending downward. The larger veins are restricted in growth with a resulting rolling and crinkling of leaves and veins that are swollen in places. Flowers may also be distorted. Plants in the field are stunted with smaller leaves that are rugose and curl downward.

Other strains in the United States cause similar symptoms. Plants are stunted and have rugose, warty leaves that curl downward. Leaves fully grown at the time of infection will yellow and die. Sometimes plants appear to recover.

CONTROL:
1. Fumigate plant beds.
2. Control weeds and other hosts in plant beds and fields.
3. Exclude insects from plant bed with insect-proof covers.
4. Apply insecticides periodically to plants in plant bed.
5. Plow under residue and any plants left after transplanting.

Bushy Top

CAUSE: A combination of two viruses, the tobacco vein distorting virus (TVDV) and the bushy top virus (BTV). BTV may be a strain of tobacco mottle virus (TMV). BTV can be transmitted by aphids only if TVDV is also present in the same plant.

DISTRIBUTION: Nyasaland and Zimbabwe.

SYMPTOMS: Plants infected early have severe symptoms. Excessive growth of axillary buds produces numerous brittle shoots and crisp leaves. Flowers are small but set seed.

CONTROL: Control measures are the same as for Tobacco Rosette Disease.

Club Root

CAUSE: Club root virus (CRV) is transmitted by grafting.

DISTRIBUTION: Kentucky, Maryland, and Tennessee. Club root is a minor disease.

SYMPTOMS: Symptoms may be confused with those caused by root knot nematode. Root knot galls are found on small and large roots. Club root galls of various sizes are usually only found on older roots. Plants are stunted due to shortening of internodes with curled and otherwise distorted leaves. Infrequently veins enlarge and enations form on the underside of leaves.

CONTROL: No control is necessary.

Cucumber Mosaic

CAUSE: Cucumber mosaic virus (CMV) overwinters in several perennial and annual hosts. Pollen of at least four *Stellaria* spp. can be infected with virus; consequently CMV overwinters in the seed of these plants. Several species of aphids are vectors, with *Myzus persicae* Sulzer and *Aphis gossypii* Glover as two of the principal vectors. CMV is nonpersistent and aphids retain it for less than four hours. Different strains of the CMV exist, with some strains efficiently transmitted by aphids while other strains are not.

DISTRIBUTION: Generally distributed wherever tobacco is grown. The disease is especially severe in Japan and wherever vegetables are grown.

SYMPTOMS: Symptoms are similar to tobacco mosaic virus. Pale green, circular spots appear on leaves, giving a mottled and mosaic pattern. Systemic infection first shows as a slight vein clearing followed by a mild, general mottling. Sometimes leaves are stunted and narrow. Severe strains may cause interveinal discoloration and oak leaf pattern of necrosis on lower leaves. A burn or sun scald sometimes appears on upper leaves. Mild strains cause only a faint mottling of the leaves.

CONTROL: Control measures are the same as for Vein Banding (this chapter).

Leaf Curl

CAUSE: Tobacco leaf curl virus (TLCV) survives in several different species of plants. TLCV is transmitted by the white fly *Bemisia tabaci* (Genn.) and probably other insects. The acquisition period varies between 15 minutes to 2 hours. TLCV persists in the vector for at least six days. TLCV is graft transmitted but not seed or sap transmitted. The disease is more severe during dry, warm (30°C) weather because of increased whitefly activity. Several strains of the TLCV exist.

DISTRIBUTION: The disease is most prevalent in the tropical areas of Africa, Asia, Australia, Central and South America. It has also been found in Switzerland and the USSR.

SYMPTOMS: The most characteristic symptom is the production of leaf outgrowths up to 1 cm wide on the veins of the lower leaf surface. Usually the outgrowths are only a dark green thickening of sections of the veins.

The entire plant is stunted with leaves smaller than normal and twisted and curled. The smaller veins have a vein clearing or chlorotic symptom. In some cases the veins may be greener than normal.

Sometimes the leaf margin is rolled downward. Flower parts are curled and deformed with apical dominance being lost. This gives a broomlike appearance to the plant.

Mild strains cause little stunting. The uppermost leaves of nearly mature plants may be curled and twisted. The other leaves appear normal.

CONTROL:
1. Control weeds in and adjacent to tobacco beds and fields.
2. Do not locate tobacco beds and fields near alternate hosts.
3. Apply insecticides or mulches to control whiteflies in seed beds.
4. Destroy tobacco residue after completion of harvest to eliminate overwintering hosts.
5. Rogue infected plants out of a field.

Peanut Stunt

CAUSE: Peanut stunt virus (PSV) overwinters in a large number of host plants, primarily clover species and other herbaceous perennials. Although PSV can be transmitted mechanically, the most prevalent means of spread is by aphids. *Aphis craccivora* Koch, *A. spiraecola* Patch, and *Myzus persicae* Sulzer transmit PSV nonpersistently.

DISTRIBUTION: Japan, eastern United States, and the state of Washington.

SYMPTOMS: Burley tobacco is stunted. Young leaves have large areas of chlorotic tissue. Older leaves have chlorotic tissue bordering small veins while large veins have spots and rings of chlorotic tissue.

Flue-cured tobacco is stunted. Additionally mosaic symptoms that resemble tobacco mosaic virus and ringspot lesions are present.

CONTROL: Control measures are the same as for vein banding.

Pepper Veinal Mottle

CAUSE: Pepper veinal mottle virus (PVMV) is transmitted by sap, grafting, and the aphid *Myzus persicae* Sulzer. PVMV is closely related to tobacco vein mottle virus.

DISTRIBUTION: Nigeria.

SYMPTOMS: The primary symptom is irregular vein banding of leaves. Additionally a systemic mosaic or chlorotic mottle fades with increasing age of leaves.

CONTROL: None reported.

Rattle

CAUSE: Tobacco rattle virus (TRV) overwinters in at least 100 different plant species. The virus is transmitted by both adult and juvenile nematodes of the genus Trichodorus, including *T. pachydermus* Seinhorst, *T. primitivus* (deMan) Micotelzky, *T. christii* Allen, and *T. alii*. TRV is rarely transmitted by seed of some weed species.

The acquisition and transmission time is one hour. TRV can be retained for 20 weeks in nematodes that have not fed on plants. However, it does not persist through the egg or molt. It enters roots through wounds made by nematode feeding. TRV probably contaminates the stylet. Transmission is favored by temperatures of 25°C and below. Rattle is more prevalent on sandy soils.

DISTRIBUTION: Brazil, Denmark, Germany, Japan, Holland, Scotland, and the United States.

SYMPTOMS: Inverted, spoon-shaped leaves have spots and ring or line patterns of necrotic tissue. Leaves break and rattle when touched. Discontinuous brown to gray stripes of sunken, necrotic tissue develop on stems, petioles, and leaf veins. Leaves become curled with sinuate margins.

Leaves infected with mild strains have numerous tan lesions that form rings. Systemically infected leaves are slightly distorted and mottled with little stem necrosis. Little or no plant stunting occurs. Plants usually recover and produce normal-appearing leaves that are somewhat elongated and pointed at the tip.

CONTROL:
1. Fumigate or use nematicides to control nematodes in the plant beds.
2. Use nematicides in field.
3. Rotate tobacco with nonsusceptible crops.

Ringspot

CAUSE: Tobacco ringspot virus (TRSV) survives in a wide range of host plants including cucumbers and other vegetables, soybeans, horse nettle, ground cherry, pokeweed, sweet clover, lespedeza, and alfalfa. TRSV is mechanically transmitted easily and is seed transmitted in several plants, notably soybeans. There are several vectors that transmit TRSV, notably all stages of the nematode, *Xiphinema americanum* Cobb. Other vectors are thrips, *Thrips tabaci* Lindeman; mites, *Tetranychus* sp.; grasshoppers, *Melanoplus differentialis* Thomas; and the tobacco flea beetle, *Epitrix hirtipennis* (Melsheimer). Different strains of TRSV exist.

DISTRIBUTION: Australia, Canada, Europe, Formosa, Japan, Nyasaland, New Zealand, South Africa, Sumatra, the United States, and the USSR.

SYMPTOMS: Symptoms may appear on plants in the plant bed or shortly after transplanting. Initially numerous or few necrotic rings occur that become blanched or brown after a few days. Frequently there may also be line patterns that parallel the veins or occur as irregular wavy lines between the larger veins. Leaves on only one side of the plant may be infected. Severely infected plants may be dwarfed with small leaves of poor quality. Pollen becomes sterile.

CONTROL:
1. Control nematodes in plant beds and fields.
2. Control weeds in and adjacent to plant beds and fields.
3. Do not grow tobacco after or next to other virus hosts.
4. Control insects in plant beds and fields.

Streak

CAUSE: Tobacco streak virus (TSV) survives in a large number of plants including tomato, sweet clover, cotton, peanut, and several others. The virus is transmitted mechanically and by dodder. No insect vectors are known but some workers have alluded to an insect(s) as a vector that transmits TSV from sweet clover to tobacco.

DISTRIBUTION: Asia, Australia, Canada, Europe, and the United States.

SYMPTOMS: Affected leaves are smaller than normal, narrow, and slightly crinkled. The leaf may be affected on one side only, with midrib curling toward the affected side. Necrotic lesions are surrounded by water-soaked lines or rings that later become brown and necrotic. Lesions spread along the veins with parallel necrotic lines appearing in the tissue surrounding the vein. Sometimes the midrib and petiole collapse.

Systemic symptoms appear as a net pattern, rings, or partial rings in the leaf. These symptoms at first are brown but later become gray-white and are closely associated with the veins and bases of young leaves. Necrotic tissue often falls out of leaves.

CONTROL: No control is known. However, other hosts, particularly sweet clover, should be eliminated from the vicinity of tobacco fields before transplanting occurs.

Tobacco Etch

CAUSE: Tobacco etch virus (TEV) overwinters in several different solanaceous plants including horse nettle, *Solanum carolinense* L., and ground cherry, *Physalis virginiana* Mill. TEV is not seedborne and does not survive in dead plant tissue. TEV can overwinter in tobacco roots and new plants growing from these roots will be infected.

DISTRIBUTION: Canada, Germany, Japan, and the United States.

SYMPTOMS: The disease is most important on burley tobacco. Symptoms in the field are first noticed when plants approach the flowering stage. Younger leaves have mild mosaic symptoms but the tip leaves are not mottled. Vein clearing is the first symptom to occur, followed by line patterns of necrotic tissue, or etching and mottling. On older leaves, chlorotic spots up to 6 mm in diameter usually occur in the interveinal area. Eventually the spots become white and necrotic. Severely infected plants may be stunted and have fired, chlorotic, and tattered leaves at harvest time.

CONTROL:
1. Control weeds, especially perennial ones, in tobacco fields and fields adjacent to them.
2. Use only certified transplants.
3. Destroy live tobacco stalks and roots in which the virus may overwinter.
4. Control aphids.
5. Plant resistant cultivars.

Tobacco Mosaic

CAUSE: Tobacco mosaic virus (TMV) overwinters in several perennial weeds including horse nettle, *Solanum carolinense* L., and ground cherry, *Physalis angulata* L. The virus commonly persists in air-dried tobacco and infected residue in the field in the absence of freezing, desiccation, or complete rotting. TMV is disseminated mechanically. Workers who chew or smoke tobacco, or handle infected residue or weed hosts and then touch tobacco plants, are primary sources of dissemination. Other means of spread occur during tillage, suckering, or other field operations. A species of large grasshopper may mechanically transmit the virus as well as a leaf miner fly; however, these are not efficient vectors.

DISTRIBUTION: Generally distributed wherever tobacco is grown.

SYMPTOMS: Ordinarily local lesions do not form on inoculated leaves, but under high temperature and light conditions, small, circular, faintly chlorotic spots may occur. Systemic infection first causes a vein clearing of the youngest leaf. Eventually the leaf becomes mottled and distorted. The outer leaf edge may turn slightly upward, forming a rim around the leaf. Eventually large blisters of green tissue and raised or sunken yellow areas may develop, together with a marked mottling of dark and light green. Considerable malformation and distortion also occur. Frequently the lamina of the leaf may be so reduced that a shoestring effect is produced.

CONTROL:
1. Fumigate plant bed soils with a chemical that kills weeds and inactivates the virus.
2. Control weeds in and around plant beds.
3. Do not use tobacco while working in the seed bed.
4. Spray plant bed with milk 24 hours before pulling. Laborers should dip hands in milk every 20 minutes while working with seedlings.
5. Rotate tobacco every two years in a field.
6. Rogue out any infected plants prior to the first cultivation.
7. Plant resistant cultivars.

Tobacco Necrosis

CAUSE: Tobacco necrosis virus A (TNV-A) infects and persists in a large number of host plants. A second virus, tobacco necrosis satellite virus (TNSV), is found only in association with TNV-A. So far, TNSV has not been demonstrated to multiply by itself and appears to be incapable of multiplication unless associated with TNV-A.

The viruses are transmitted by the soil fungus, *Olpidium brassicae* (Woron.) Dang. Particles of both viruses leak out of infected roots into soil water where they are absorbed to the flagella and plasmalemmae of *O. brassicae* zoospores. When the zoospores encyst, the flagellae, bearing the virus particles, are pulled into the cytoplasm of the zoospores. Upon germination of the encysted zoospore, the virus particles are carried along with zoospore cytoplasm into the host root. Root infection can also occur when wounded roots are in contact with soil water that has a high content of virus particles. TNV-A is primarily restricted to roots of various plants that appear to be normal. The viruses occasionally move out of the roots into the aerial parts of the plant, but are arrested at a point just above the soil level.

Both viruses are also mechanically transmitted. Any method of moving soil disseminates the viruses. TNV-A persists in infected plant roots but not within resting sporangia of *O. brassicae*. Disease symptoms are more severe during cool temperatures and low light intensity. Different strains of the virus exist.

DISTRIBUTION: Europe, Japan, New Zealand, and the United States.

SYMPTOMS: Tobacco necrosis occurs only in young seedlings. Usually the first symptoms are necrotic spots that often coalesce in the tissue along midribs and veins of the lower leaves. Very young plants may be killed. Only the lower and oldest leaves become necrotic and dry on older plants. The rest of the plant appears normal.

CONTROL:
1. Sterilize soil to kill *O. brassicae*.
2. Remove tobacco residue from plant bed areas.
3. Set only healthy transplants in the field.

Tobacco Rosette Disease

CAUSE: A combination of two viruses: tobacco vein distorting virus (TVDV) and tobacco mottle virus (TMV). TMV is thought to overwinter in several plant species in the Solanaceae and is mechanically transmitted. It is also aphid transmitted, principally by *Myzus persicae* Sulzer, but only if TVDV is also present in the plant.

TVDV also overwinters in several plant species in the Solanaceae. The virus is not mechanically transmitted. Aphids, principally *M. persicae*, are the main means of transmission. TVDV also persists in the aphid body for long periods of time.

DISTRIBUTION: Nyasaland and Zimbabwe.

SYMPTOMS: Rosette may occur in plant beds or fields. Affected plants are stunted. Midribs of young leaves are twisted and distorted, forming a knot in the center of the plant, giving a rosette appearance. Infected leaves curl sharply at the tip.

Infected older plants do not become distorted, but leaves will droop. If plants are affected only on one side, the flower bends over. Splitting of stem and petiole may also occur.

CONTROL:
1. Plant early before aphid populations are too large.
2. Apply systemic insecticides.
3. Control all weeds in tobacco beds and fields adjacent to them.
4. Destroy all plants in beds after transplanting.

Tobacco Stunt

CAUSE: Tobacco stunt virus (TSV) infects and probably persists in a large number of host plants. The principal vector is the soil fungus *Olpidium brassicae* (Woron.) Dang, but it is also mechanically and graft transmitted. The virus is carried into the root by the zoospore cytoplasm. TSV multiplies only after zoospore cytoplasm enters plant roots. It then multiplies both in the cytoplasm of the tobacco cells and in developing zoospores.

TSV is disseminated on infected transplants and by any means that will transport soil. TSV apparently persists within *O. brassicae* in the soil for long periods of time, perhaps several years. Disease severity is increased by cool temperatures and low light intensity.

DISTRIBUTION: Japan.

SYMPTOMS: Symptoms first appear when seedlings have five to eight leaves. The first symptom is a vein clearing. The veins later become necrotic. Eventually small brown to white necrotic spots or necrotic ring-like patterns develop on the bases of lower leaves. Bud leaves become generally yellow, with surface crinkles and tips that curl downward. Internodes fail to develop, causing plants to remain in the rosette stage. A band of necrotic tissue is present on the stem at the soil level. Transplants moved to fields remain yellow and stunted. Frequently severely infected plants die. Plants with mild symptoms have narrow leaves and stems and will flower before a normal plant.

CONTROL: Control measures are the same as for tobacco necrosis virus.

Tobacco Vein Mottle

CAUSE: Tobacco vein mottle virus (TVMV) overwinters principally in dock, horse nettle, and ground cherry. TVMV can also overwinter in living tobacco roots. TVMV is not transmitted mechanically under natural conditions but is transmitted by the green peach aphid and probably other aphid species.

DISTRIBUTION: Kentucky and North Carolina.

SYMPTOMS: Symptoms will vary greatly with the time of infection and cultivar. Initially burley types will have a slight vein clearing. All infected plants will have an irregular dark green area along the veins of some leaves.

Certain cultivars develop a severe systemic necrotic spotting on leaves. This necrosis becomes increasingly more severe during the season.

Plants infected early will be stunted and will not be as green as healthy plants. Tip leaves may have light green spots.

CONTROL:
1. Control overwintering hosts in tobacco fields and adjacent to them.
2. Control insects with insecticide application. However, because an aphid can transmit the virus to a healthy plant a few seconds after it starts feeding, insecticides are often a questionable control. Repeated transmissions to several healthy plants may be reduced.
3. Plant barrier rows of corn around a tobacco field to intercept aphids reaching the field.
4. Plant resistant cultivars. No cultivar is immune but resistance varies greatly between cultivars.

Tomato Spotted Wilt

CAUSE: Tomato spotted wilt virus (TSWV) overwinters in several species of perennial plants in areas that have cold winter temperatures, and in annual plants where winters are not cold enough to kill plants. TSWV is transmitted by several species of thrips, with the onion thrips, *Thrips tabaci* Lindeman, thought to be the principal vector. TSWV must be acquired by the larval form first, with the subsequent adult able to transmit the virus. TSWV persists in insects as long as they live but is not transmitted through the eggs. The disease is more severe during high temperatures and abundant moisture.

DISTRIBUTION: Generally distributed wherever tobacco is grown.

SYMPTOMS: Plants of all ages may be infected. Initially yellow-green spots occur on young leaves. Later the spots turn red-brown and become concentric necrotic rings or zonate necrotic spots. Frequently the spots coalesce to form large, irregular areas. Necrotic streaks develop along the stem and dark necrotic areas or cavities appear in the cortex and pith.

Infected plants are stunted and the apical bud droops or bends over. Young leaves that are infected only on one side may be distorted and puckered. Severely infected plants will not grow for several weeks. The leaves droop and finally die.

CONTROL:
1. Use insecticides to control thrips in the seed bed and for a month after transplanting.
2. Eliminate overwintering hosts and deep plow all tobacco debris in a field.

Vein Banding

CAUSE: Potato virus Y (PV-Y) overwinters in several species of plants including potatoes, tomatoes, and peppers. PV-Y is transmitted from overwintering hosts to tobacco by several species of aphids, principally *Myzus persicae* Sulzer. PV-Y is carried on stylet tips of the insects.

DISTRIBUTION: Generally distributed wherever tobacco is grown. However, the disease tends to be more severe where potatoes are also grown.

SYMPTOMS: Symptoms vary with the tobacco cultivar infected. Most tobacco cultivars develop a faint mottling of the young expanding leaves. A clearing of veins also occurs in younger leaves. Larger leaves have slight interveinal chlorosis, leaving a band of dark green tissue along each side of the vein. Leaf epinasty and stunting may occur.

Severely infected plants have veins that turn dark brown to black. Leaves yellow prematurely and plants may die.

CONTROL:
1. Control weed hosts in tobacco fields and adjacent to them.
2. Plant only disease-free or certified transplants.
3. Grow a nonhost crop to act as a barrier between the source of inoculum and tobacco.
4. Do not grow tobacco near potatoes, tomatoes, and peppers.
5. Plant tobacco when aphid populations are low or before they build up.

DISORDERS CAUSED BY HIGHER PLANTS

Broomrape

CAUSE: Parasitic flowering plant, *Orobanche ramosa* L. It survives as seed in soil for several years. The plant grows on roots of tobacco.

DISTRIBUTION: Africa, Asia, Europe, and North America.

SYMPTOMS: Tobacco plants are not vigorous growing and become yellow. Plants mature earlier than normal. Large masses of blue-flowered broomrape plants can be seen breaking through the soil around the base of the tobacco plant.

CONTROL:
1. Do not plant tobacco in an infested field for several years.
2. Destroy broomrape plants before seed is ripe.

BIBLIOGRAPHY

Anderson, P. J. 1948. Pole rot of tobacco. *Connecticut Agric. Exp. Sta. Bull. 517.*

Anderson, T. R., and Welacky, T. W. 1983. Barn mold of burley tobacco caused by *Botryosporium longibrachiatum. Plant Disease* **67:**1158–1159.

Bower, L. A.; Fox, J. A.; and Miller, L. I. 1980. Influence of *Meloidogyne incognita* and *Globodera solanacearum* on development of black shank of tobacco. *Phytopathology* **70:**688 (abstract).

Clayton, E. E., and McMurtrey, J. E., Jr. 1958. Tobacco diseases and their control. *USDA Farmers Bull. No. 2023 (rev.).*

Davis, J. M., and Main, C. E. 1984. Meteorological aspects of the spread and development of blue mold on tobacco in North Carolina. *Phytopathology* **74:**840 (abstract).

Davis, J. M.; Main, C. E.; and Bruck, R. I. 1981. Analysis of weather and the 1980 blue mold epidemic in the United States and Canada. *Plant Disease* **65:**508–512.

Fajola, A. O., and Alasodura, S. O. 1973. Corynespora leaf spot, a new disease of tobacco (*Nicotiana tabacum*). *Plant Dis. Rptr.* **57:**375–378.

Ferrin, D. M.; and Mitchell, D. J. 1986. Influence of soil water status on the epidemiology of tobacco black shank. *Phytopathology* **76:**1213–1217.

Fortnum, B. A.; Csinos, A. S.; and Dill, T. R. 1982. Metalaxyal controls blue mold in flue-cured tobacco seedbeds. *Plant Disease* **66:**1014–1016.

Fortnum, B. A.; Krausz, J. P.; and Conrad, N. D. 1984. Increasing incidence of *Meloidogyne arenaria* on flue-cured tobacco in South Carolina. *Plant Disease* **68:**244–245.

Fulton, R. W. 1980. Tobacco blackfire disease in Wisconsin. *Plant Disease* **64:**100.

Gayed, S. K. 1978. Tobacco diseases. *Canada Dept. Agric. Pub. 1641.*

Gwynn, G R.; Barker, K. R.; Reilly, J. J.; Komn, D. A.; Burk, L. G.; and Reed, S. M. 1986. Genetic resistance to tobacco mosaic virus, cyst nematodes, root-knot nematodes, and wildfire from *Nicotiana repanda* incorporated into *N. tabacum*. *Plant Disease* **70**:958–962.

Hartill, W. F. T. 1967. A rust of tobacco in Rhodesia. *Rhodesia Zambia Malawi Jour. Agric. Res.* **5**:189.

Hendrix, J. W., and Csinos, A. S. 1985. Tobacco stunt, a disease of burley tobacco controlled by soil fumigants. *Plant Disease* **69**:445–447.

Herr, L. J., and Sutton, P. 1984. Tobacco black shank control with metalaxyl and cultivars. *Phytopathology* **74**:854 (abstract).

Hopkins, J. C. F. 1956. *Tobacco Diseases*. Commonwealth Mycological Institute, Surrey.

Johnson, M. C.; Pirone, T. P.; and Litton, C. C. 1982. Selection of tobacco lines with a high degree of resistance to tobacco etch virus. *Plant Disease* **66**:295–297.

Kelman, A., and Sequeira, L. 1965. Root-to-root spread of *Pseudomonas solanacearum*. *Phytopathology* **55**:304–309.

Komm, D. A.; Reilly, J. J.; and Elliott, A. P. 1983. Epidemiology of a tobacco cyst nematode (*Globodera solanacearum*) in Virginia. *Plant Disease* **67**:1249–1251.

Ladipo, J. L., and Roberts, I. M. 1979. Occurrence of pepper veinal mottle virus in tobacco in Nigeria. *Plant Dis. Rptr.* **63**:161–165.

Latorre, B. A., and Flores, V. 1987. Wilt of tobacco in Chile, caused by *Verticillium dahliae*. *Plant Disease* **71**:101 (disease notes).

Lucas, G. B. 1974. *Diseases of Tobacco*, 3rd ed. Harold E. Parker & Sons Printers, Fuquay-Varina, North Carolina.

McIntyre, J. L.; Sands, D. C.; and Taylor, G. S. 1978. Overwintering, seed disinfestation, and pathogenicity studies of the tobacco hollow stalk pathogen, *Erwinia carotovora* var. *carotovora*. *Phytopathology* **68**:435–440.

Moss, M. A., and Main, C. E. 1985. Temperature tolerance of *Peronospora tabacina* in the U. S. *Phytopathology* **75**:1341 (abstract).

Muller, A. S., and Acedueda, H. A. 1964. Rust on tobacco discovered in Honduras. *Phytopathology* **54**:499 (abstract).

Powell, N. T.; Melendez, P. L.; and Batten, C. K. 1971. Disease complexes in tobacco involving *Meloidogyne incognita* and certain soilborne fungi. *Phytopathology* **61**:1332–1337.

Prasad, S. S., and Acharya, B. 1965. A new leaf-spot disease of tobacco. *Current Sci.* **34**:542.

Ramachandraiah, M.; Venkatarathnam, P.; and Sulochana, C. B. 1979. Tobacco necrosis virus: Occurrence in India. *Plant Dis. Rptr.* **63**:949–951.

Ramachar, P., and Cummins, G. B. 1965. The species of Puccinia on the Paniceae. *Mycopathol. Mycol. Appl.* **25**:7–60.

Reilly, J. J. 1980. Chemical control of black shank of tobacco. *Plant Disease* **64**:274–277.

Reuveni, M.; Nesmith, W. C.; and Siegel, M. R. 1986. Symptom development and disease severity in *Nicotiana tabacum* and *N. repanda* caused by *Peronospora tabacina*. *Plant Disease* **70**:727–729.

Richardson, M. J., and Zillinsky, F. J. 1972. A leaf blight caused by *Fusarium nivale*. *Plant Dis. Rptr.* **56**:803–804.

Rotem, J., and Aylor, D. E. 1984. Development and inoculum potential of *Peronospora tabacina* in the fall season. *Phytopathology* **74**:309–313.

Shew, H. D. 1983. Effect of host resistance level on spread of *Phytophthora parasitica* var. *nicotianae* under field conditions. *Phytopathology* **73**:505 (abstract).

Shew, H. D., and Main, C. E. 1985. Leaf spot of tobacco caused by *Rhizoctonia solani*. *Phytopathology* **75**:1278 (abstract).

Shew, H. D., and Main, C. E. 1985. Rhizoctonia leaf spot of flue-cured tobacco in North Carolina. *Phytopathology* **69**:901–903.

Sidebottom, J. R., and Shew, H. D. 1985. Effects of soil texture and matric potential on sporangium production by *Phytophthora parasitica* var. *nicotianae*. *Phytopathology* **75**:1435–1438.

Sidebottom, J. R., and Shew, H. D. 1985. Effects of soil type and soil matric potential on infection of tobacco by *Phytophthora parasitica* var. *nicotianae*. *Phytopathology* **75**:1439–1443.

Spurr, H. W., Jr.; Echandi, E.; Haning, B. C.; and Todd, F. A. 1980. Bacterial barn rot of flue-cured tobacco in North Carolina. *Plant Disease* **64**:1020–1022.

Stavely, J. R., and Kincaid, R. R. 1969. Occurrence of Phyllosticta leaf spot on Florida cigar-wrapper tobacco. *Plant Dis. Rptr.* **53**:837–839.

Sun, M. K. C.; Gooding, G. V., Jr.; Pirone, T. P.; and Tolin, S. A. 1974. Properties of tobacco vein-mottling virus, a new pathogen of tobacco. *Phytopathology* **64**:1133–1136.

Taylor, G. S. 1969. A black mildew of the genus Diporotheca, found on roots of *Nicotiana tabacum* in Connecticut. *Plant Dis. Rptr.* **53**:85–86.

Tisdale, W. B. 1929. A disease of tobacco seedlings caused by *Septomyxa affinis*. *Phytopathology* **19**:90 (abstract).

Wilson, K. I.; Al-Beldawi, A. S.; and Dwazah, K. 1983. Rhizopus stem rot of *Nicotiana glauca*. *Plant Disease* **67**:526–527.

Wolf, F. A. 1957. *Tobacco Diseases and Decays*. Duke University Press, Durham, North Carolina.

23

Diseases of Wheat
(*Triticum aestivum* L.)

BACTERIAL CAUSES

Bacteria Leaf Blight

This disease is also called bacterial leaf necrosis.

CAUSE: *Pseudomonas syringae* pv. *syringae* van Hall is seedborne and overwinters in residue, on plants in water, and in soil. During cool (15–25°C), wet weather, bacteria may be splashed or blown onto leaves where they enter the leaf through stomates or wounds. Bacteria may also migrate from the seed to the seedling by swimming up the seedling or by being carried along with the growing point. The bacteria can survive as an epiphyte on leaves where under the right conditions they may become pathogenic.

DISTRIBUTION: North central United States.

SYMPTOMS: Symptoms on winter wheat occur during cool, wet weather in the spring and may be confused with spots caused by other pathogens. At first, small (less than 1 mm in diameter), water-soaked spots occur on the top leaves of plants in the boot to early heading stage. Eventually the spots grow larger and become necrotic, changing from a dull gray-green to a light brown color. During continued wet weather, the spots may grow together into ragged streaks or blotches within two to three days, sometimes killing the entire leaf. During wet weather, droplets of bacterial ooze may develop in a spot.

Lesions on spring wheat coalesce into elongated necrotic areas that gradually expand laterally during moist, humid weather until the entire leaf is destroyed. A chlorotic halo may be present around some lesions.

CONTROL: Some cultivars of both spring and winter wheat are resistant.

Bacterial Mosaic

CAUSE: *Clavibacter michiganense* subsp. *tesellarius* (Davis et al.). The bacterium is seedborne.

DISTRIBUTION: Alaska, Iowa, Nebraska, and Canada.

SYMPTOMS: Initially, small, yellow spots are more numerous near the midrib. Eventually the spots grow together into yellow lesions with indefinite margins. Most of the leaf becomes light to dark brown in color. Severely infected leaves turn brown and dry up. Typical water-soaking that is usually associated with bacterial infections is lacking.

CONTROL: None are known for certain. However, there is circumstantial evidence that the bacterium may be seedborne; therefore, planting seed from diseased plants is not recommended.

Basal Glume Rot

CAUSE: *Pseudomonas syringae* pv. *atrofaciens* (McCulloch) Young, Dye & Wilkie (syn. *P. atrofaciens* (McCulloch) Stevens and *Phytomonas atrofaciens* (McCulloch) Bergey et al.) is seedborne and overwinters in residue. Bacteria are disseminated on dust particles that become entrapped in water present in grooves or small spaces of spikelets. Bacteria may also be disseminated by insects and probably splashing water. Bacteria then multiply near glume joints when water is present but remain dormant when moisture is lacking.

DISTRIBUTION: Generally distributed wherever wheat is grown.

SYMPTOMS: Symptoms are most likely to occur during wet weather, particularly at heading time. The main symptom is a brown discolored area at the base of glumes covering a kernel that is more evident on the inside than the outside of the glume. Usually only about the bottom one-third of the glume is discolored, but sometimes the entire glume may be affected. Severely infected spikelets are slightly dwarfed and lighter in color than healthy ones. Sometimes the only sign of disease is a dark line at the attachment of the glume to the spike. A diseased kernel has a faint brown to black discoloration on its base. Infected leaves have small, dark, water-soaked spots that will eventually elongate, turn yellow, then brown as the tissue dies. Seed filling is sometimes limited.

CONTROL:
1. Treat seed with a seed-protectant fungicide.
2. Plant seed that has been thoroughly cleaned.
3. Rotate wheat with resistant crops such as legumes.

Black Chaff and Bacterial Streak

CAUSE: *Xanthomonas campestris* pv. *translucens* (Jones, Johnson & Reddy) Dye (syn. *X. translucens* (Jones, Johnson & Reddy) Dows f. sp. *undulosa* (Smith, Jones & Reddy) Hagb.) is seedborne and is thought to overwinter in residue, on overwintered hosts, and directly in the soil. Primary infection likely occurs by bacteria being disseminated by splashing water and insects, particularly aphids. Primary infection may also occur by seedborne bacteria but seed stored six months or more is not considered an important source of inoculum. Small spaces and grooves that contain water act as a reservoir for bacteria. Secondary spread occurs by plant to plant contact, rain, and insects.

DISTRIBUTION: Generally distributed wherever wheat is grown. It is not considered a serious problem.

SYMPTOMS: Symptoms usually occur after several days of damp or rainy weather. Blighted plants grow slowly and may be stunted, but usually the disease is not noticed until plants are about two-thirds grown. The chaff or glumes have longitudinal, dark, somewhat sunken stripes or spots that are more abundant on the upper glumes where they coalesce to form larger spots or blotches. The inside of diseased glumes have brown or black spots. Beards of bearded cultivars are brown, especially at the base. In moist weather, tiny, yellow beads of bacteria ooze to the surface of dark lesions and dry as small, yellow scales.

Small water-soaked spots occur on tender green leaves and sheaths of older plants and sometimes seedlings. The spots enlarge and coalesce, becoming glossy, olive-green, translucent stripes or streaks of various lengths that later turn yellow-brown. Stripes may extend the length of a sheaf and are usually narrow, limited by leaf veins. Occasionally a spot may become large and blotch-like, causing the leaf to die, shrivel, and turn light brown. Severely diseased leaves die back from the tips.

Under humid conditions, early in the morning, droplets of milky bacterial exudate may be seen on the surface of diseased spots. The droplets dry into hard, yellowish granules that are easily removed from the leaf surface. Brown to black stripes are produced on stems as the crop reaches maturity. These stripes usually occur below the heads and upper joints.

Grain is not destroyed but it may be brown, shrunken, and carry bacteria to infect next year's crop. If a flag leaf is infected, the head may not emerge from the boot but break through the side of the sheath and be distorted and blighted.

CONTROL:
1. Do not plant seed from diseased plants. Seed should be cleaned to remove lightweight infected kernels.
2. Treat seed with a seed-protectant fungicide.
3. Do not rotate wheat with barley.

Pink Seed

CAUSE: *Erwinia rhapontici* (Millard) Burkholder (syn. *E. carotovora* pv. *rhapontici* (Mill.) Dye) infects only damaged kernels, particularly those injured by gall midge, Cecidomyidae, or when grain is harvested too early. It is not known how *E. rhapontici* survives and is disseminated to wheat. It causes a crown rot of rhubarb (*Rheum* spp.), its only other known host.

DISTRIBUTION: Canada and Europe. The disease is not serious.

SYMPTOMS: Seeds are pink and appear as if they were treated with a dye. The endosperm is soft and pink.

CONTROL: None is needed.

Spike Blight

CAUSE: *Clavibacter tritici* (Carlson & Vidaver) Davis et al. (syn. *Corynebacterium michiganense* (Smith) Jensen pv. tritici (Hutchinson) Dye & Kemp, and *C. tritici* (Hutchinson) Burkholder) is seedborne and survives in organic matter only in wet soils, usually in the lower areas in a field. Dissemination within a field is primarily by the seed gall nematode *Anguina tritici* (Steinbuch) Chitwood. Long-distance dissemination is by seed. The nematode larvae become contaminated with bacteria in the soil and carry them up the stem to the plant apex.

DISTRIBUTION: Australia, Canada, China, Egypt, Ethiopia, and India. The disease is not important except in wet areas of a field.

SYMPTOMS: Wet weather or wet low areas in a field favor disease development. Initially leaves may be wrinkled or misshaped as they emerge from a whorl. As the heads eventually emerge, they also will be misshaped and covered with a sticky yellow exudate of the bacteria.

Dried exudate appears as white flecks that distort heads, necks, and upper leaves and inhibit their elongation.

CONTROL:
1. Plant only cleaned seed.
2. Do not grow wheat for at least two years in nematode-infested soil.
3. Drain any wet areas in a field.

Stem Melanosis

CAUSE: *Pseudomonas cichorii* (Swingle) Stapp. The bacterium may overwinter on several weed species found in wheat fields. Disease may occur only with repeated infections over a two- to three-week period that coincides with high temperature and nightly dew formation.

DISTRIBUTION: Canada.

SYMPTOMS: Symptoms initially occur at milky ripe stage of growth. Small, light brown lesions first develop on stems beneath the lower two nodes. During the following two weeks, lesions darken and coalesce on stems, rachis, and peduncles with occasional mottling of glumes. By the soft dough stage, lesions expand and darken. The rachis and upper portions of the peduncles beneath the heads and stems, that are immediately beneath nodes, turn dark brown. Heads become bleached and thin, and kernels are badly shriveled.

The pattern in the field is sharply defined, irregularly shaped, dark patches that range in size from several square meters to several hectares.

CONTROL: No control is reported.

White Blotch

CAUSE: *Bacillus megaterium* deBary pv. *cerealis*. The bacterium is probably spread by water, insects and/or mites, and seed. It may be harbored in soil, plant parts, seeds of susceptible cultivars, and perhaps in insects. Disease is favored by high temperatures and high light intensity.

DISTRIBUTION: North Dakota on hard red spring, hard red winter, and durum wheats.

SYMPTOMS: Initially small yellow or white lesions develop that enlarge into white or very light tan, irregular blotches and streaks on leaf blades, sheaths, and culms. The white blotches are often broader than the streaks symptomatic of bacterial leaf blight (bacterial leaf necrosis) and lack the water-soaking initially associated with the latter.

CONTROL: No control is reported.

FUNGAL CAUSES

Alternaria Leaf Blight

CAUSE: *Alternaria triticina* Pras. & Prab. survives primarily as conidia on seed and mycelium within seed. Plants are infected at about four weeks when leaves, primarily those in contact with the soil, become infected. Secondary inoculum is provided by conidia produced in lesions disseminated by wind. Disease is favored by at least 10 hours of continued leaf wetness and temperatures at 20–25°C.

DISTRIBUTION: India.

SYMPTOMS: Initially small, oval, chlorotic spots occur on the lower leaves. Spots enlarge, become sunken, assume irregular shapes, and turn brown-gray. Spots progress up a plant and may extend to heads and leaf sheaths. During humid conditions spots have a dark powdery appearance due to presence of spores.

CONTROL:
1. Plant resistant cultivars.
2. Apply foliar fungicide where feasible.
3. Apply seed-protectant fungicides; however, these are frequently not effective.

Anthracnose

CAUSE: *Colletotrichum graminicola* (Ces.) G. W. Wilson (syn. *Dicladium graminicola* Ces. and *C. cereale* Manns) is a soil-inhabiting fungus that is a successful saprophytic colonizer of a wide range of plant residues. It also survives as mycelium and conidia on cereals and grasses or their residues. Conidia are produced from mycelium during wet weather at an optimum temperature of 25°C and are wind disseminated or water disseminated to hosts. *Colletotrichum graminicola* may also be seedborne during harvest. Symptoms tend to be more severe during wet, warm (25°C) weather. Anthracnose is most likely to occur when wheat is grown in a rotation with wheat or another cereal on coarse soils that are low in fertility.

DISTRIBUTION: Generally distributed wherever wheat is grown.

SYMPTOMS: Symptoms become apparent toward plant maturity. Infections initially occur aboveground and belowground as lesions 1–2 cm long and elliptical in shape. Lesions at first are water-soaked, then become bleached and necrotic. Plants display gross symptoms of premature ripening or whitening, a general reduction in vigor, dying, lodging, and shriveled grain. Crowns and stem bases become bleached and later turn brown. Later acervuli develop as small, black, raised spots on the surface of lower leaf sheaths and culms. Acervuli may also develop on leaf blades of dead plants when moisture is plentiful. Additionally round to oblong lesions with acervuli may occur on green leaves. Seedling and crown infection may occur under severe disease conditions.

CONTROL:
1. Rotate wheat with a noncereal or grass. Legumes are not susceptible and would be a suitable crop in a rotation.
2. Improve soil fertility; anthracnose is most likely to occur on plants that are not growing vigorously.
3. Plant resistant cultivars. However, most resistant cultivars are not widely grown.
4. Treat seed with a seed-protectant fungicide.

Ascochyta Leaf Spot

CAUSE: *Ascochyta tritici* Hori & Enjoji overwinters as pycnidia and mycelium in infected residue. During hot weather, pycnidiospores are disseminated a short distance to healthy tissue. Other factors that contribute to infection are high humidity caused by dense foliage or weather, and leaves coming in contact with soil. Secondary infection probably occurs from pycnidia produced in older leaf spots. *Ascochyta graminicola* Sacc. and *A. sorghi* Sacc. have also been reported as pathogens.

DISTRIBUTION: Europe, Japan, and North America.

SYMPTOMS: Ascochyta leaf spot is an inconspicuous disease and can be overlooked easily or confused with other diseases, particularly glume blotch. The leaf spots tend to be mostly on lower leaves and initially are yellow, mostly round to elliptical, and vary from 1–5 mm in width. Later the spots merge together and cease to be separate spots but cause large areas of the leaf to be light brown and necrotic. Pycnidia are often formed in necrotic tissue and appear as black dots. A hand lens is useful in seeing pycnidia since most of the structure is submerged in tissue with just the top portion visible.

CONTROL: No control is presently known; however, application of foliar fungicides should be somewhat effective.

Aureobasidium Decay

CAUSE: *Microdochium bolleyi* (Sprague) de Hoog and Herm.-Nijhof (syn. *Aureobasidium bolleyi* (Sprague) von Arx. and *Gloeosporium bolleyi* Sprague).

DISTRIBUTION: Widespread; however, the disease is of little importance. *Microdochium bolleyi* has been reported as part of the common root rot complex in different areas.

SYMPTOMS: Mild necrosis of seedling roots.

CONTROL: No control is reported.

Black Head Molds

This disease is also called sooty head mold.

CAUSE: Several fungi including *Alternaria* spp., *Cladosporium* spp., *Epicoccum* spp., *Sporobolomyces* spp., and *Stemphylium* spp. Black head molds occur when wet weather accompanies maturation especially if harvest is delayed. The disease becomes more severe on heads of plants that are damaged from other causes such as other diseases, nutrient deficiency, or lodging.

DISTRIBUTION: Generally distributed wherever wheat is grown.

SYMPTOMS: Green-black molds develop on heads.

CONTROL: No practical control is known.

Black Point

This disease is also called kernel smudge.

CAUSE: Several fungi including *Aspergillus* spp., *Chaetomium* spp., *Cladosporium* spp., *Curvularia* spp., *Fusarium* spp., *Gloeosporium* spp.,

Helminthosporium spp., *Myrothecium* spp., *Nigrospora* spp., *Penicillium* spp., *Plenodomus* spp., *Rhizopus* spp., and *Stemphylium* spp.

Black point is most severe at relative humidities greater than 90 percent or continuous rainfall during seed maturation. Damage will increase if grain is stored at 95 percent relative humidity and 25 percent moisture (wet-weight basis).

DISTRIBUTION: Generally distributed wherever wheat is grown.

SYMPTOMS: Black point describes the darkened and sometimes shriveled embryo end of the seed. Diseased kernels are discolored, weathered, black-pointed, or smudged. When the embryo is invaded, germination is decreased. Discolored grain is discounted at the market.

CONTROL:
1. Treat seed with seed-treatment fungicide.
2. Store grain at 15 percent moisture (wet-weight basis) or less.

Brown Root Rot

CAUSE: *Plenodomus meliloti* Dearn & Sanford (syn. *Phoma sclerotioides* (Preuss) Sacc.).

DISTRIBUTION: Canada.

SYMPTOMS: Necrosis occurs where pycnidial initials are attached to roots. Pycnidial initials are green-black, globular, or flattened and up to 0.5 mm in diameter. They are found attached to roots and shoot bases in spring by a short attachment process. These stalks develop singly from dished adaxial surfaces of the protopycnidia that resemble sclerotia.

CONTROL: No control is reported.

Cephalosporium Stripe

CAUSE: *Cephalosporium gramineum* Nis. & Ika. (syn. *Hymenula cerealis* Ell. & Ev.) is seedborne and survives as conidia and sporodochia in residue in the top 8 cm of soil for up to five years. Cephalosporium stripe is most severe in no-till and decreases progressively in minimal and conventional tillage plots. This is likely due to a threshold size for plant fragments infested with *C. gramineum*, below which few lesions develop at any inoculum particle concentration, and above which the number of lesions increases in proportion to weight of particles. The size of plant fragments varies between soil types because infection is influenced by total nutrients available from residue and soil. Conidia serve as primary inoculum and infect roots during the winter and early spring through mechanical injuries caused by soil heaving and insects. Stress caused by freezing also predisposes plants to infection.

Infection is most severe in wet, acid soils (pH 5.0) due to increased fungal sporulation and root growth. There is an increase in disease severity with decreasing soil pH, especially with highly susceptible cultivars. High soil temperatures following planting makes plants more susceptible to infection due to an increase in the size of the root system, which makes it more prone to breakage by heaving soil in the spring. Low soil moisture reduces infection and lowers inoculum potential in the soil. The addition of phosphate fertilizer in the drill rows further increases root growth and leads to greater disease incidence at high soil temperatures but not at lower temperatures.

The subcrown internode can also be infected just as the seed is germinating. After infection, conidia enter xylem vessels and are carried upward where they lodge and multiply at nodes and leaves but do not do any futher damage to the roots. The fungus prevents water movement up the plant and also produces metabolites that are harmful to the plant. At harvest time *C. gramineum* is returned to soil in infected residue where it is a successful saprophytic competitor with other soil-borne microorganisms. Disease is more severe in reduced tillage systems than in conventional tillage systems.

DISTRIBUTION: Great Britain, Japan, and most winter wheat-growing areas of North America.

SYMPTOMS: Winter wheat is most severely infected. Spring wheat is susceptible but apparently escapes the disease and usually does not show symptoms. Infected plants are scattered throughout a field but are usually more numerous in the lower and wetter areas. Infected plants are dwarfed. During jointing and heading, distinct yellow stripes, one to four per leaf, but usually one or two, develop on leaf blades, sheaths, and stems, usually occuring the length of the plant. Thin brown lines consisting of infected veins normally occur in the middle of a stripe. The stripes eventually become brown and highly visible on green leaves and are still noticeable on yellow straw. Toward harvest, the culms at infected nodes or below them may become dark. Heads of infected plants are white and do not contain seed, or if seed is present, it is shriveled.

CONTROL:
1. Plant cultivars that have moderate resistance.
2. Rotate wheat for at least two years with a nonhost.
3. Plant later in the autumn or when the soil temperature at 10 cm below the soil surface is below 13°C. Plants under these conditions have limited root growth, thus reducing the number of infection sites.

4. Infected residue should be plowed deeper than 8 cm where erosion is not a problem.

Common Bunt

This disease is also called stinking smut and covered smut.

CAUSE: *Tilletia foetida* (Wall.) Liro (syn. *T. laevis* Kuhn or *T. levis* Kuhn, and *T. foetens* (Berk. & Curt.) Schroet.), and *T. caries* (DC.) Tul. (syn. *T. tritici* (Bjerk.) Wint.) are two closely related fungi with similar life cycles that may occur together in the same diseased plant. Both fungi overwinter as teliospores (chlamydospores) on seed and in soil. When wheat is planted, teliospores that are either on or near seed in the soil germinate if moisture is present. A promycelium (basidium) is formed on which 8–16 basidiospores (sporidia) are formed. Basidiospores fuse in the middle with a compatible basidiospore to form an H-shaped structure that in turn germinates to form yet another structure called a secondary sporidia. Germination is optimum during cool temperatures of 5–15°C. Secondary sporidia germinate to produce mycelium that infect seedlings. Mycelium grows behind the growing point or meristematic tissue of the plant, invades the developing head, and displaces the grain. Eventually teliospores are formed within seed. At harvest time, infected seed is broken and teliospores are windborne to contaminate healthy kernels or soil.

DISTRIBUTION: *Tilletia caries* is generally distributed wherever wheat is grown but is most prevalent in the northwestern United States. *Tilletia foetida* is limited to areas of Europe and North America, being reported from the Midwest and northwestern United States.

SYMPTOMS: Common bunt is more severe during cool soil temperatures after planting. Wheat planted in the spring may be less severely diseased. Infected plants are somewhat stunted but cannot be readily distinguished from healthy ones until heading. Bunted heads are more slender than healthy ones and glumes of spikelets may be spread apart. A bunted head will often stay green for a longer time than normal and have a bluish cast. Bunted kernels are about the same size as healthy seed but are light brown and more round in shape. The pericarp ruptures at harvest and teliospores contaminate healthy seed, giving them a fishy odor; hence the name stinking smut. The smut balls or infected kernels have an oily feeling when crushed.

CONTROL:
1. Treat seed with a systemic seed-treatment fungicide.
2. Plant resistant cultivars.

3. Plant early in the autumn. Seedlings may be far enough advanced and less susceptible to infection by the time secondary sporidia develop.

Common Root Rot

This disease is also called dry land root and foot rot.

CAUSE: *Bipolaris sorokiniana* (Sacc.) Shoemaker (syn. *Helminthosporium sativum* Pammel., King & Bakke and *Drechslera sorokiniana* (Sacc.) Subramanian & Jain), teleomorph *Cochliobolus sativus* (Ito & Kurib.) Drechs. ex Dast. Other causal fungi are *Fusarium culmorum* (W. G. Sm.) Sacc., *F. graminearum* Schwabe, *F. avenaceum* (Fr.) Sacc., and *F. acuminatium* Ell. & Ev. All are widespread fungi that survive in soil as saprophytes on crop residue. Additionally conidia of *B. sorokiniana* and chlamydospores of *Fusarium* spp. survive for several months in soil and may also be seedborne. Inoculum density of *C. sativus* in soil is related to amount of sporulation occurring on crop residues. Primary infections occur on coleoptiles, primary roots, and subcrown internodes. The infected plant does not die as long as it can produce new roots. Conidia are produced when infection progresses aboveground.

Plants put under stress by drouth, freezing, warm temperatures, lack of nutrition, and Hessian fly injury are most subject to infection. Disease severity increases with planting depth. Moisture is required for infection, but once initiated, disease development requires warm temperatures and moisture stress. Disease may be most severe in dry soil when inoculum is seedborne.

Additionally, *Microdochium bolleyi* (Sprague) de Hoog and Herm.-Nijhof (syn. *Aureobasidium bolleyi* (Sprague) von Arx and *Gloeosporium bolleyi* Sprague) is reported to be part of the common root rot complex.

DISTRIBUTION: Wherever dry land wheat is grown.

SYMPTOMS: Seedlings may be killed either preemergence or postemergence. Surviving seedlings have brown lesions on coleoptiles, roots, and culms. Darkening of subcrown internodes is usually caused by *B. sorokiniana*. *Fusarium* spp. infect secondary roots as they emerge from the crown. Infections of crowns and feet usually kill plants. Diseased plants occur in random patches and are stunted and a lighter green than normal plants. Plants mature early, produce few tillers, and have shriveled seed. Heads are bronzed, bleached, or white. There may be a browning of the root system that can be observed only when roots are washed. In particular, *Fusarium* spp. cause roots, culm bases, and lower nodes to dry and darken. Diseased plants then become brittle and break off easily near the soil level. Winter survival of wheat is reduced by root rot pathogens, particularly *B. sorokiniana*.

CONTROL:
1. Treat seed with a seed-protectant fungicide.
2. Soil should have proper fertility to ensure growth of vigorous plants that are able to produce new roots and overcome root rot.
3. Plant winter wheat in late autumn.
4. Some cultivars have a degree of resistance.
5. Leave field fallow for three to four years.

Crater Disease

Crater disease and Rhizoctonia root rot are similar but symptoms differ sufficiently to consider them as two separate diseases.

CAUSE: *Rhizoctonia solani* Kuhn. Crater disease occurs only on very heavy, black clay soil (greater than 40% clay), in which root growth is impeded by soil structure. Disease is also enhanced by dry weather and early infection by *R. solani*. Isolates of *R. solani* that cause crater disease anastomose with each other but not with any other anastomosis group tester strains.

DISTRIBUTION: South Africa.

SYMPTOMS: Severely stunted patches of wheat occur in a field. Under continuous cultivation of wheat, patches enlarge each year and assume an irregular pattern that tends to stretch in the direction of cultivation. Some fields display a rippling effect caused by elongated patches of less obviously stunted plants.
Initially there is a sudden appearance of yellow areas of plants. About three weeks after emergence, plants wilt and lower leaves soon become dry and white. Most plants die as seedlings but some survive until maturity and have reduced tillering and incomplete grain filling. Roots of stunted plants have beadlike swellings composed of hyphae that superficially resemble sclerotia. Xylem vessels in roots are brown and occluded with gel-like material; roots break easily at these points. Frequently the entire root is brown and rotted where the roots are broken.

CONTROL: Stunting is alleviated by thorough and repeated cultivation of soil.

Disease Caused by *Fusarium sporotrichioides*

CAUSE: *Fusarium sporotrichioides* Sherbakoff.

DISTRIBUTION: Minnesota on spring wheat.

SYMPTOMS: Foliar and glume necrosis is severe only under optimal conditions on very young plants. Leaf symptoms are variable and include small tan flecks with dark red borders; tan, oval to elongate lesions of different sizes, occasionally appearing like mechanical abrasion; tan streaking, especially in young leaves; and tip burn. Red or red-tan coloration of lesions also occurs. Leaf sheaths are attacked, occasionally killing the leaf.

Head symptoms are identical to those of scab except kernels are not visibly affected. However, culm infections sometimes kill the head.

CONTROL: No control is reported.

Downy Mildew

CAUSE: *Sclerophthora macrospora* (Sacc.) Thirum., Shaw & Narasimhan (syn. *Sclerospora macrospora* Sacc., and *Phytophthora macrospora* (Sacc.) Ito & Tanaka) survives for years as oospores either embedded in infected leaf and stem tissue or loose in soil when infected tissues decay. It is an obligate parasite and cannot grow saprophytically on dead plant tissue. Oospores are seedborne and disseminated in infected residue and soilborne by wind or water. Oospores germinate in saturated soil to produce sporangia from which zoospores are eventually released. Zoospores swim through soil water, settle on the developing seedling, and produce germ tubes that penetrate plants. In yet another mode of infection, oospores may survive for months in dry soil and germinate either directly to form a germ tube that penetrates a host, or indirectly by producing zoospores. Infection occurs over a wide temperature range of 7–31°C. Following infection, *S. macrospora* develops systemically within the plant, particularly in meristematic tissue.

DISTRIBUTION: Generally distributed wherever wheat is grown. It is not considered a serious disease except in localized wet areas.

SYMPTOMS: Downy mildew occurs only in localized areas of fields where seedlings have been in flooded or water-logged soil for 24 hours or longer. Wheat plants are dwarfed, deformed, and may tiller excessively with several yellow leaves. Infected plants have leathery, stiff, thickened leaves, stems, and heads. Affected parts are twisted and distorted and no seeds are formed in severely diseased plants.

In less severely diseased plants, dwarfing may be slight with one or more of the upper leaves stiff, upright, or variously curled and twisted. Heads and stems are not deformed. Numerous, round, yellow-brown oospores may be found in disease tissue examined under magnification.

CONTROL: Downy mildew is ordinarily not serious enough to warrant special control measures. However, the following are some general measures that will aid in control should a serious problem occur.
1. Where possible, provide proper soil drainage.
2. Control grassy weeds that may serve as hosts.

3. Plant cleaned seed from disease-free plants to ensure no infested residue is disseminated with seed.

Dwarf Bunt

This disease is also called dwarf smut.

CAUSE: *Tilletia controversa* Kuhn survives as teliospores in soil or on the surface of seed. However, disease occurs only when heavily infested seed (equal or greater than 1 g teliospores per kg of seed) is planted in disease-conducive locations. Seedborne teliospores are not believed to serve as the primary inoculum. When wheat is planted in moist soil, teliospores that are either on the seed or close to it will germinate under warm, dry conditions to form a promycelium on which 8–16 basidiospores are borne. Basidiospores fuse in the middle with a compatible basidiospore to form an H-shaped structure that in turn germinates to form yet another structure called a secondary sporidia. The secondary sporidia germinate to produce infective mycelium.

Tilletia controversa is better adapted to survive in soil and has a lower optimum temperature for germination (1–5°C) than the common bunt fungi. Protracted cool (less than 5°C), moist conditions in the spring induce dormancy in teliospores. The infection process takes from 35–105 days depending on temperature. Infection generally requires heavy snow cover over unfrozen ground. After infection, mycelium grows behind the growing point or meristematic tissue of the plant and invades the developing head, displacing the grain. Eventually teliospores are formed within the seed. At harvest time, the infected seed is broken and teliospores are windborne to contaminate healthy kernels or soil.

DISTRIBUTION: In Canada, Europe, and the United States where a heavy snow cover is likely to occur over unfrozen ground.

SYMPTOMS: Dwarf bunt is limited to winter wheat and infrequently winter barley. Infected plants are generally one-fourth to one-half normal size. Smut balls are smaller and rounder than common bunt and the spore mass feels dry. Spores also smell like rotten fish.

CONTROL:
1. Seed treatment with systemic fungicides is usually not effective due to the long period of time that infection occurs.
2. Plant resistant cultivars in adapted areas.

Epicoccum Glume Blotch

CAUSE: *Epicoccum purpurascens* Ehrenb. ex Schlecht.

DISTRIBUTION: India.

SYMPTOMS: Symptoms initially appear in the dough stage. The disease is characterized by purple-brown elliptical spots on the glumes, which later turn gray. Severely infected glumes are almost entirely covered with spots that are occasionally dotted with black spore masses. Infection sometimes extends to awns. In highly susceptible cultivars, the rachis and peduncle are covered with dark brown spots; leaves have irregular spots. Grain is severely shriveled.

CONTROL: No control is reported; however, varying degrees of resistance apparently exist among cultivars.

Eyespot

This disease is also called Cercosporella footrot, culm rot, foot rot, stem break, and strawbreaker.

CAUSE: *Pseudocercosporella herpotrichoides* (Fron.) Deighton (syn. *Cercosporella herpotrichoides* Fron.) survives for several years in absence of grain crops as mycelium in infected residue. Conidia are produced during cool, damp weather either in the autumn or spring on infected straw lying on the soil surface. Dispersal is by splashing rain. Sporulation occurs at optimum temperatures between 8°C and 12°C and when humidity near the soil surface remains near saturation. Crown and basal culm tissue, but not roots, are infected. With early seedling growth, the coleoptile is the most susceptible to infection; however, with decay of the coleoptile, the susceptibility of leaf sheaths increases.

Development of epiphytotics is dependent on production of large amounts of primary inoculum on infected stubble remaining from previous wheat crops. Secondary inoculum is not important in the development of an epiphytotic since no spores occur on developing lesions until later in the growing season. However, late infections add to the amount of inoculum for succeeding crops. Soil or volcanic ash placed around the base of winter wheat in the spring increases disease severity in the state of Washington.

DISTRIBUTION: Generally distributed where wheat is grown in a cool, moist climate.

SYMPTOMS: Winter wheat is more likely to be infected than spring wheat. This disease is most conspicuous near the end of the growing season by lodging of diseased plants. Lodging caused by eyespot tends to cause straw to fall in all directions but lodging caused by wind or rain causes straw to fall primarily in one direction. Whiteheads, similar to symptoms produced in take-all, are also present at maturity.

Eye-shaped or ovate lesions with white to tan centers and brown margins develop first on the basal leaf sheath. Similar spots form on the stem directly beneath those on the sheath and they cause the lodging. Roots are not infected, but a necrosis occurs around the roots in the upper crown nodes. Under moist conditions, the lesions enlarge and a black, stromalike mycelium develops over the surface of the crown and base of the culms, giving the tissues a charred appearance. The stems then shrivel and collapse; otherwise plants are yellowish or pale green with heads reduced in size and numbers. When infection occurs early, individual culms and weaker plants are killed before maturity.

CONTROL:
1. Rotate wheat with noncereals, such as legumes, since *P. herpotrichoides* cannot survive in the absence of infected residue.
2. Spring wheat and late-planted winter wheat are less subject to infection.
3. Apply fungicides to infested residue to decrease primary inoculum.
4. Reduced tillage limits foot rot incidence.

Flag Smut

CAUSE: *Urocystis agropyri* (Preuss) Schroet. (syn. *U. tritici* Koern.) survives as teliospores in soil and on seed. Spores on seeds have remained viable and germinated after four years. Teliospores germinate to produce sporidia that in turn germinate and infect wheat coleoptiles prior to emergence. Infection is optimum at low soil moisture (10–15%) and temperatures between 10°C and 20°C. *Urocystis agropyri* then overwinters as mycelium within seedlings and in the spring grows systemically in the upper portion of the plant, producing sori in which hyphal tips differentiate into teliospores. The sori erupt in a few days or when wheat is harvested to liberate teliospores that contaminate either the soil or seed.

DISTRIBUTION: In localized areas within most wheat-growing areas. Commonly found in the western United States, Australia, and in certain areas of the Midwest.

SYMPTOMS: Symptoms appear in spring and summer on a few tillers or the whole plant. Smut lesions occur on leaves, sheaths, and upper parts of the stem as long, light green stripes that soon become gray-black streaks. These streaks are subepidermal smut sori that have developed between veins, causing leaves to become rolled and twisted. Splits occur along the sori or stripes, liberating gray-black spore masses. Leaves become weakened, frayed, and split longitudinally. Infected plants are dwarfed, tiller excessively, and rarely head out. If they do, glumes and necks will also be striped.

CONTROL:
1. Plant resistant cultivars.
2. Treat seed with a systemic seed-protectant fungicide.

3. Rotate wheat for at least two years and preferably three years with a nonsusceptible host. Teliospores have been known to survive up to three years in soil.
4. Do not plant seed deep since shallow planting partially prevents some infection.

Fusarium Crown and Foot Rot

CAUSE: *Fusarium culmorum* (W. G. Sm.) Sacc. (syn. *F. roseum* (Lk.) emend. Snyd. & Hans. f. sp. *cerealis* (Cke.) Snyd. & Hans. 'Culmorum') and *F. graminearum* Schwabe group 1 (syn. *F. roseum* (Lk.) emend. Snyd. & Hans, f. sp. *cerealis* (Cke.) Snyd. & Hans. 'Graminearum', group 1) are most commonly cited. Additionally, *F. avenaceum* (Fr.) Sacc., *F. acuminatum* Ell. & Ev. and *F. tricinctum* (Cda.) Sacc. have been reported as a minor component of the disease complex. *F. equiseti* (Corda) Sacc. has been reported to cause a crown rot in South Africa. Primary inoculum is soilborne chlamydospores and infected plant debris in the upper 10 cm of soil. Entry into the crown is gained 2–3 cm below the soil surface through openings around crown roots or by infection of newly emerging crown roots. Disease is severe under dry soil conditions (−40 bars midday leaf water potentials).

DISTRIBUTION: Culmorum is widely distributed in central and eastern Washington, north central Oregon, and northern Idaho. Graminearum is limited to south central Washington and equiseti was reported from South Africa.

SYMPTOMS: The crown and basal stem tissues are decayed with a brown discoloration and spongy texture. Infected plants rarely show outward symptoms until after heading. Plants die prematurely, resulting in white heads.

Internodes become chocolate brown; however, leaf sheaths remain symptomless. Pink or burgundy mycelium can be observed within hollow stems.

Symptoms caused by *F. equiseti* are stunted patches of wheat. Roots and crowns have a red-brown discoloration with crowns being severely rotted. Inoculated plants in the greenhouse have lower plant mass and grain yield.

CONTROL:
1. Do not rotate wheat with oats.
2. Till field after harvest to improve water infiltration.
3. Establish a dust and stubble mulch in spring.
4. Apply proper amount of nitrogen.
5. Plant winter wheat in early September in North America as a compromise between early planting that results in larger plants and greater stress and later planting that results in lower yields.

Fusarium Leaf Blight

CAUSE: *Fusarium nivale* (Fr.) Ces.

DISTRIBUTION: Scotland.

SYMPTOMS: Lesions occur on upper leaves at milky ripe growth stage. Lesions are extensive, dull, diffuse, and water-soaked with pustules of *F. nivale* sporulating through the stomata.

CONTROL: No control is reported.

Glume Blotch

This disease is also called Septoria blotch, and in some literature, *Septoria nodorum* blotch of wheat.

CAUSE: *Septoria nodorum* (Berk.) Berk., teleomorph *Leptosphaeria nodorum* Müller, is seedborne and survives in stored seed for more than two years. Additionally *S. nodorum* survives as mycelium in live plants and pycnidia on infected residue for two or three years. During warm (20–27°C), wet weather, in autumn or spring, pycnidiospores are exuded from pycnidia within a gelatinous drop (cirrhi) that protects them from radiation and drying out. Pycnidiospores are disseminated by splashing and blowing rain to lower leaves of healthy plants. Infection requires 6 or more hours of continuous wetness. Coleoptile infection from seedborne inoculum occurs over broad ranges of soil moisture, temperature, and planting depth. Disease severity of seedlings increases until four weeks after planting. The greatest losses occur when rainfall is excessive between flowering and grain harvest. New spores are produced in 10–20 days. Ascospores are produced in perithecia as wheat matures in late summer and early autumn. Several grasses are alternative hosts.

DISTRIBUTION: Generally distributed in all major wheat-growing areas.

SYMPTOMS: Glumes, culms, leaf sheaths, and leaves are infected; however, little damage normally occurs until the crop nears maturity. Small, grayish or brown spots usually occur near the top third of a glume two or three weeks after the head emerges. The spots enlarge and become a chocolate-brown. Later the center of the spot becomes gray, with small, black pycnidia scattered throughout it.

Infected nodes of the stem turn brown, shrivel, and are speckled with black pycnidia. Such infections often cause straw to bend over and lodge just above the nodes.

Infection of leaves causes light brown spots similar in appearance to Septoria leaf blotch. A brown margin may surround leaf spots and pycnidia will be present on both surfaces of a diseased leaf. If a flag leaf is infected, the head may be deformed. Infection of the leaf sheath causes a dark brown lesion that includes most of each leaf sheath. Severely infected plants are stunted.

Symptoms may be confused with black chaff or basal glume rot. Glume blotch spots do not form streaks and are not as sharply defined or as dark brown as those of black chaff. Glume blotch does not have the water-soaked appearance of basal glume rot.

A symptom caused by an atypical form of *S. nodorum* was a brown discoloration at the base of coleoptiles of seedlings grown from infected seed. Inoculated leaves developed brown lesions surrounded by yellowish halos.

CONTROL:
1. Plant certified, disease-free seed that has been cleaned and treated with a seed-protectant fungicide. Seed treatment is effective when primary source of inoculum is from seed but is ineffective when soil is heavily infected.
2. Plant resistant wheat cultivars if they are adapted to the areas.
3. Plow under infected residue where this is feasible.
4. Apply a foliar fungicide if conditions warrant it.
5. Rotate wheat every three or four years with a nonsusceptible crop. However, disease development is not reduced when infected seed is used.
6. Plant winter wheat after Hessian fly-free date.
7. Plant taller cultivars instead of shorter ones. A less favorable microclimate occurs with taller cultivars due to lower canopy densities, less time that leaves are moist, and less dew deposition.

Gray Snow Mold

CAUSE: *Typhula incarnata* Lasch ex. Fr. See Speckled Snow Mold.

DISTRIBUTION: See Speckled Snow Mold.

SYMPTOMS: Plants may be infected either aboveground or belowground. After snow melt, leaves of infected plants are usually overgrown and matted together with gray mycelium. Leaves break freely, releasing sclerotia. Sclerotia are pale on emergence from snow but then turn pink and eventually darken to a red-brown.

CONTROL: See Speckled Snow Mold.

Halo Spot

CAUSE: *Selenophoma donacis* (Pass.) Sprague & Johnson (syn. *Septoria donacis* Pass.) overwinters as pycnidiospores, pycnidia, perithecia, and mycelium in infected residue, seed, and overwintering wheat. During cool, moist weather, pycnidiospores are exuded from pycnidia and are splashed or windborne to hosts.

DISTRIBUTION: In cool, moist climates of Great Britain, northern Europe, and the United States. Halo spot rarely causes much damage.

SYMPTOMS: Numerous spots appear in the spring on leaves and sometimes culms of winter wheat. Spots are less than 4 mm long, and elliptical or diamond-shaped. At first the spots have purple-brown margins that eventually fade as the spots become old. In time, the center of the spots becomes gray with small, black pycnidia that are difficult to see. Sometimes spots may become so numerous that much of the leaf surface is destroyed.

CONTROL: No control is usually necessary; however, the cultivar Gaines is very susceptible.

Kernal Bunt

This disease is also called partial bunt.

CAUSE: *Tilletia indica* Mitra (syn. *Neovossia indica* (Mitra) Mundkur) survives as teliospores in soil and on seed. Most primary inoculum is derived from seedborne inoculum. Teliospores germinate under moist conditions in the spring and produce a promycelia that grows to the soil surface. Germination should presumably occur within 2 mm of the soil surface. Maximum germination of teliospores in vitro has been reported to occur at a temperature of 15–20°C in continuous light. Sporidia are produced on promycelia and are windborne to plants where florets are infected. The pathogen may be spread over long distances by air currents. Kernels are partially or wholly converted into masses of chlamydospores. Spores are disseminated at harvest time to healthy seeds or soil. Disease is favored by cool, wet weather.

DISTRIBUTION: Afghanistan, India, Iraq, Lebanon, Mexico, Nepal, Pakistan, Syria, Sweden, and Turkey.

SYMPTOMS: Normally only a few random kernels per head are infected. An infected kernel is wholly or incompletely converted to smut sori. Infected kernels are fragile, partially darkened, and have a foul odor. Most kernels are either broken or partially eroded at their embryo end.

CONTROL:
1. Plant resistant cultivars. Most durum wheats are immune.
2. Treat seed with a fungicide seed treatment.

Leaf Rust

This disease is also called brown rust, dwarf rust, and orange rust.

CAUSE: *Puccinia recondita* Rob. ex Desm. f. sp. *tritici* (syn. *P. rubigo-vera*) (DC.) Wint. and *P. triticina* Eriks.). The fungus is a heteroecious, long-cycle rust fungus that has species of meadow rue, *Thalictrum* spp., and rarely *Anchusa* spp. and *Isopyrum* spp. as alternate hosts to wheat. However, meadow rue is found primarily in Europe and is not commonly found in most sections of the United States; therefore, the pycnial and aecial stages are usually found only in Europe.

The rust commonly overwinters as urediospores and mycelium on wheat in the southern United States and Mexico. During some years the fungus may survive as mycelium in winter wheat in the Midwest. Circumstantial evidence implies local oversummering and overwintering of certain phenotypes in Pennsylvania. In the spring, urediospores are windborne north and infect if moisture is present. A new generation of urediospores is produced every 7–14 days if favorable conditions of heavy dews, light rains, or high humidity and temperatures at 15–22°C are present. Urediospores are wind disseminated and cause secondary infections until plants mature. As plants near maturity, teliospores are formed.   Plants infected near maturity may not produce teliospores. Teliospores in North America serve no function, but in Europe may serve as a means of overwintering. During late summer and autumn in North America, urediospores are windborne from the northern wheat-growing areas southward to infect winter wheat planted in the southern United States and Mexico.

DISTRIBUTION: Generally distributed wherever wheat is grown and may be the most widely distributed wheat disease. It is usually most prevalent where wheat matures late; however, infection often occurs too late to do severe damage.

SYMPTOMS: Pustules first develop on lower leaves and leaf sheaths, progressing up to the flag leaf under moist conditions. Some may also occur on stems and awns and glumes of the head. Pustules are small, round to oval, raised, and orange-yellow and contain numerous ure-diospores that appear as an orange powder when rubbed on the fingers.

As wheat matures, telia develop as dark gray to black, flattened pustules about the same size as uredia that do not rupture the epidermis.

Leaf rust may be distinguished from stem rust by (1) smaller pustules, (2) orange-yellow color versus brick-red color of stem rust pustules, and (3) lack of jagged fragments of wheat epidermis adhering to sides of pustules.

CONTROL:
1. Plant resistant wheat cultivars.
2. Soils should have balanced fertility. An excess of nitrogen increases susceptibility of wheat to rust.
3. Apply a foliar fungicide if conditions warrant it.
4. Experimentally fungicide seed treatments have proved effective.

Leaf Spot

CAUSE: *Ascochyta sorghi* Sacc. and *A. graminicola* Sacc. have been found in association with *Septoria nodorum*.

DISTRIBUTION: Virginia.

SYMPTOMS: Initially leaf lesions are elliptic to lanceolate and later coalesce into irregular patterns. Lesions are zonate on the upper leaf surface with a narrow outer chlorotic band enclosing a light brown necrotic band and dark brown center. The dark brown center is absent on the lower leaf surface.

Pycnidia of both fungi are present. Those of *A. sorghi* are golden brown and submerged with only the ostiole or neck above the leaf surface. Those of *S. nodorum* are dark brown and superficial.

CONTROL: No control is necessary.

Leaf Spot

CAUSE: *Phaeoseptoria urvilleana* (Speg.) Sprague, teleomorph *Leptosphaeria microscopica* Karst. The fungus likely overwinters as pycnidiospores in pycnidia on infected residue. Free water must be present on leaves for 48 hours or longer for infection to occur.

DISTRIBUTION: The fungus has been found to reproduce on wheat in Great Britain and the United States.

SYMPTOMS: Irregular leaf spots form on leaves of plants in late milk stage in the field. Plants inoculated with *L. microscopica* have irregular, diffuse leaf spots that are first yellow, then tan.

CONTROL: Some cultivars are resistant or partially resistant.

Leptosphaeria Leaf Spot

CAUSE: *Leptosphaeria herpotrichoides* de Notaris overwinters as mycelium in residue and ascospores in asci from pseudothecia on straw. Free water must be present on leaves for 48 hours or longer for infection to occur.

DISTRIBUTION: Canada, Europe, and the United States. It is considered a minor disease.

SYMPTOMS: Irregular, diffuse, yellow to tan spots occur on leaves.

CONTROL: Some cultivars are more resistant than others.

Loose Smut

CAUSE: *Ustilago tritici* (Pers.) Rostr. (syn. *U. nuda* var. *tritici* Schaf. and *U. nuda* (Jens.) Rostr.) survives as dormant mycelium within the embryo of wheat seed. The smutted head emerges from the boot one or two days earlier than heads of healthy plants. Chlamydospores (teliospores) are windborne to flowers of healthy heads. During favorable weather of temperatures 16–22°C and some moisture provided by dews or light showers, chlamydospores germinate with the germ tube penetrating the flower ovary and possibly the stigma. Accessibility to the embryo appears to be a limiting factor in infection. Subsequent mycelial growth occurs in the germ or embryo of the developing seed. As grain matures, mycelium become dormant within the embryo until the following growing season. When infected seed germinates, the mycelium grows systemically within the plant and produces the smutted heads filled with chlamydospores in place of healthy kernels. Different races of *U. tritici* exist.

DISTRIBUTION: Generally distributed wherever wheat is grown.

SYMPTOMS: Infected seed does not show any outward symptoms. Germination is not ordinarily affected and gives rise to smutted heads that emerge from the boot a couple of days earlier than healthy heads. Before heading, infected plants may have dark green, erect leaves with chlorotic streaks. The brown to dark brown spore mass is enclosed within a fragile, gray membrane that soon ruptures, releasing spores that are windborne to healthy flowers. Soon an erect naked rachis that protrudes above reclining heads of healthy plants is all that remains of a smutted head.

CONTROL:
1. Treat seed with a systemic seed-protectant fungicide.
2. Plant resistant cultivars.

3. Plant certified seed known to be smut-free.
4. Hot water treatments have been used to destroy seedborne inoculum.

Phialophora Take-All

CAUSE: *Phialophora graminicola* (Deacon) Walker (syn. *P. radicicola* Cain var. *graminicola* Deacon and *P. radicicola* sensu Scott). Disease occurs only at high soil temperatures (24–29°C).

DISTRIBUTION: Unknown.

SYMPTOMS: Only plants in the seedling stage are susceptible. Seminal roots and coronal roots are rotted. Culms are blackened. Root weight and shoot weight are reduced.

CONTROL: No control is reported.

Phoma Glume Blotch

This disease is also called Phoma spot.

CAUSE: *Phoma insidiosa* Tassi, *P. glomerata* (Cda.) Wr. & Hochapfel, and *Phoma* spp. Disease development is aided by long periods of wetness.

DISTRIBUTION: India, Mexico, and South America.

SYMPTOMS: Symptoms caused by *P. insidiosa* occur primarily on glumes and awns but leaves may occasionally be infected. Spots on glumes are brown, oval, and 5–7 mm by 2–2.5 mm. Pycnidia are present in lesions.

Artificially inoculated plants under humid conditions develop purple-brown lesions usually on tips of glumes. Later lesions enlarge and form a gray center. No yellow halo is present. Grain is discolored and shriveled. Inoculated leaves develop faded green streaks on the upper surface. Pycnidia develop in rows below the epidermis.

Phoma glomerata causes dark brown lesions on leaf sheaths.

CONTROL: No control is reported.

Pink Snow Mold

CAUSE: *Microdochium nivale* (Ces. ex Berl. & Vogl.) Sammuels and Hallett (syn. *Fusarium nivale* (Fr.) Ces., *Microdochium nivalis* (Fr.) Sammuels & Hallett, and *Gerlachia nivalis* (Ces. ex. Berl. & Vogl.) Gams & Muller), teleomorph *Monographella nivalis* (Schaffnit) E. Muller (syn. *Calonectria nivalis* Schaff.). The fungus oversummers as perithecia that develop during cool, humid weather in the spring on lower leaf sheaths. In the autumn, leaf sheaths and blades near the soil line are infected by ascospores and mycelium growing from infected residue. However, plants can also be infected below the soil surface. During cool, wet periods, with or without snow cover, mycelium grows from infected plants to healthy ones. Saprophytic growth on dead tissue may provide inoculum source for increased disease in the spring. Secondary infection also occurs from conidia or ascospores that are windborne to healthy tissue.

DISTRIBUTION: Canada, central and northern Europe, and the United States.

SYMPTOMS: The first infection occurs on leaf blades and sheaths near the soil line during cool, wet weather in the autumn. Damage corresponds to the pattern of snow cover. Pink mycelium and sporodochia are visible on living and dead tissue and on the soil surface at snow melt. When disease severity is light, lesions with straw-colored centers and dark margins are found on leaves above soil level. When disease is severe, leaves of infected plants become chlorotic, then necrotic. Whole leaves, sheaths, and shoots are killed, and necrotic leaves remain intact and do not disintegrate. White or faintly pink mycelium is sometimes present, especially on lower parts of plants. Leaves turn pink on exposure to light. Unless crowns are infected, plants may often recover during warm, dry weather despite extensive leaf infection.

CONTROL:
1. Plant winter wheat later in the autumn.
2. Rotate wheat with a noncereal crop such as a legume.
3. Plant wheat in the spring. Most of the plant growth will be under conditions unfavorable to infection.
4. Do not mow plants late in the autumn. Removal of clippings reduces deleterious effects of mowing.

Platyspora Leaf Spot

CAUSE: *Platyspora pentamera* (Karst.) Wehm. survives as perithecia in infected residue. During wet weather in the spring, ascospores are produced that are windborne to healthy plants. There must be 24–72 hours of continuous moisture for infection to occur. Perithecia are produced in straw as wheat matures.

DISTRIBUTION: North central United States and Canada. The disease is rare on wheat.

SYMPTOMS: While Platyspora leaf spot is rare on wheat, ascospores of *P. pentamera* are abundant in the air of certain wheat-growing areas.

Nondescript yellow-brown spots are randomly produced on wheat straw as plants reach maturity.

CONTROL: No specific controls are necessary at the present time; however, some wheat cultivars are resistant. Control measures for Septoria diseases would also be beneficial.

Powdery Mildew

CAUSE: *Erysiphe graminis* DC. ex Merat f. sp. *tritici* Em. Marchal overwinters as cleistothecia on diseased tissue. In areas where mild winters allow infected leaf tissue to survive, the fungus may survive as mycelium and conidia. Ascospores form within cleistothecia in spring and are windborne to hosts, thus serving as the primary inoculum in northern wheat-growing areas. Conidia are produced almost as soon as the mycelium becomes established on the leaf surface, especially during humid, cool weather but without presence of free water. Conidia account for most of the secondary inoculum and spread of the disease during the growing season. Cleistothecia are formed on the leaf surface as the plant approaches maturity.

Powdery mildew is more severe on tender, rank-growing plants that have been heavily seeded or have had heavy applications of nitrogen fertilizer. Cool, humid, and cloudy weather are conducive to disease severity. Powdery mildew usually ceases to be a problem when the weather becomes dry and warm later in the growing season.

DISTRIBUTION: Generally distributed wherever wheat is grown.

SYMPTOMS: Superficial mycelium and conidia appear as light gray or white spots on the upper surface and infrequently on the lower surface of leaf blades, sheaths, and floral bracts. The fungus is superficial except for haustoria that penetrate epidermal cells. Affected plant parts appear as if they have been dusted with a gray powder that consists of conidia, conidiophores, and mycelium of the fungus. Later the spots enlarge and darken as the plant matures. As the powdery-appearing areas enlarge, the infected portion first yellows, then turns brown and dies. Chlorotic patches appear on the leaf surface opposite gray colonies. In some cultivars, leaf tissue adjacent to mycelium turns brown and becomes necrotic. Numerous, small, round, dark cleistothecia develop on the affected areas. These can be readily observed with a hand lens. Heavily infected plants may be killed.

CONTROL:
1. Plant resistant cultivars.
2. Apply a foliar fungicide where economically feasible.
3. Rotate wheat with noncereal crops.
4. Plow under volunteer wheat where feasible.

Pythium Root Rot

This disease is also called browning.

CAUSE: *Pythium* spp. are fungi that are present in almost all agricultural soils. Although several species of Pythium have been associated with root rot of wheat, *P. aphanidermatum* (Edson) Fitzp., *P. arrhenomanes* Drechs., *P. aristosporum* Vanderpool, *P. graminicola* Subr., *P. heterothallicum* Campbell & Hendrix, *P. myriotylum* Drechs., *P. sylvaticum* Campbell & Hendrix, *P. torulosum* Coker & Patterson, *P. ultimum* Trow. var. *ultimum*, *P. ultimum* Trow. var. *sporangiiferum* Drechs., and *P. volutum* Vant. & Tru. have been demonstrated to be wheat pathogens. *Pythium* spp. survive for five years or more in soil and infected residue as oospores. The oospore germinates to form either a germ tube that infects the host or a sporangium in which usually 10–40 zoospores are produced. The zoospores swim to a host where they rest, eventually producing a germ tube that infects the host plant root. Optimum growth and infection occurs in wet soils at temperatures between 15°C and 20°C in autumn and spring. Oospores are produced within infected tissue and are returned directly to the soil upon host tissue decomposition or remain in tissue while it is intact.

Pythium spp. infect embryos of wheat within 48 hours after planting in soil at −0.3 bar or wetter. Seedlings produced from old seed are more susceptible to infection than seedlings from new seed.

DISTRIBUTION: Generally distributed wherever wheat is grown.

SYMPTOMS: Pythium root rot generally occurs in patches in a field rather than being generally distributed. It is most likely to occur in the wetter areas of a field that are deficient in phosphorous and organic matter. Young seedlings will be partially stunted and pale green. Usually seedlings grow out of this condition, but at low soil temperatures, seeds and seedlings may damp off. The tips of the newest roots of wheat develop soft, wet, and light brown areas. Diseased plants have fewer tillers, poor root growth, and are slow to mature. Leaves may turn yellow, then tan.

CONTROL:
1. Treat seed with a seed-protectant fungicide. This will aid in protecting against seed rot but is not effective against the root rot phase.
2. Maintain adequate phosphate levels in soils. This promotes root growth and allows plants to grow away from disease.
3. Some cultivars are able to tolerate disease due to a better root system than other cultivars.

Rhizoctonia Root Rot

This disease is also called bare patch, purple patch, and Rhizoctonia patch. Crater disease and Rhizoctonia root rot are similar but symptoms differ sufficiently to consider them as two separate diseases.

CAUSE: *Rhizoctonia solani* Kuhn, teleomorph *Thanatephorus cucumeris* (Frank) Donk. Strains of *R. solani* that cause root rot are different from those that cause sharp eyespot. In South Australia the disease occurs in highly calcareous, sandy soils. Disease severity is greater under conservation tillage conditions than under conventional cultivation. This may be due to cultivation fragmenting pieces of residue, which in turn may reduce propagule size and may also affect the pathogenicity of hyphae that have ramified through the soil.

The herbicide chlorsulfuron at 2.5 g/ha significantly increases disease.

DISTRIBUTION: Australia, Canada, England, Scotland, South Africa, Idaho, Oregon, and Washington.

SYMPTOMS: Plants are severely stunted and occur in distinct patches. Stunting persists throughout the growing season. Infected roots have brown sunken lesions in which the cortex of the root is collapsed, leaving only the stele. Lesions may girdle and sever roots, leaving pinched-off, pointed, brown tips. The roots appear severely pruned. Leaves of diseased plants may be stiff and have a purplish cast when viewed from a distance.

CONTROL:
1. Increase soil fertility by application of superphosphate and increase available nitrogen by improving the legume content of pastures and the use of ammonium sulfate fertilizer.
2. Wheat planted in reduced tillage situations becomes more severely diseased than wheat planted under conventional tillage conditions.

Scab

This disease is also known as blight, Fusarium head blight, Fusarium blight, head blight, pink mold, tombstone scab, and white head.

CAUSE: *Fusarium avenaceum* (Fr.) Sacc., teleomorph *Gibberella avenaceum* Cook; *F. culmorum* (W. G. Sm.) Sacc. (syn. *F. roseum* (Lk.) emend. Snyd. & Hans. f. sp. *cerealis* (Cke.) Snyd. & Hans. 'Culmorum'); *F. graminearum* Schwabe (syn. *F. roseum* (Lk.) emend. Snyd. & Hans. f. sp. *cerealis* (Cke.) Snyd. & Hans. 'Graminearum'); *F. nivale* (Fr.) Ces., teleomorph *Calonectria nivalis* Schaff. and *F. tricinctum* (Corda) Sacc. Other Fusaria reported to cause damage to isolated spikelets are *F. acuminatum* Ell. & Ev., *F. equiseti* (Corda) Sacc., *F. poae* (Peck) Wr., and *F. sporotrichioides* Sherbakoff. The fungi overwinter as mycelium and spores on seed, and mycelium and perithecia in residue. Ascospores produced during warm, moist weather are windborne to hosts. Conidia produced from mycelium on residue, root rot, or seedling blight are also windborne and infect hosts. Spores that serve as secondary inoculum are produced on infected heads.

Grain that becomes wet in swaths favors scab development. Disease is more severe when wheat follows corn, when nitrogen and phosphorous fertilization is inadequate, and where weed density is high. Although scab occurs when conditions are warm and humid, disease caused by *C. nivalis* is reported to occur under cool and moist conditions in the Pacific Northwest.

DISTRIBUTION Generally distributed wherever wheat is grown.

SYMPTOMS: A root rot or seedling blight may develop from soilborne inoculum. Infected seed may also cause seedling blight. Scab develops in warm, humid weather during formation and ripening of the kernels. One or more spikelets of an emerged immature head become bleached or white. If the rachis is infected, all spikelets above that point are bleached, sterile, or contain only a partially filled seed.

Infection begins in flowers and spreads to other parts of the head, giving the appearance of premature ripening. The diseased spikelets at first are water-soaked, die, and become light brown beginning at the spikelet base. Eventually the hulls change from a light brown to a dark brown. When infection occurs late in the development of the grain, the hull may show the brown color only at the base, but in severe cases the entire kernel may become shrunken and brown. During humid or wet weather a pink to pink-red mold consisting of mycelium and spores of the causal fungus can be seen on infected spikelets. Later the small, black perithecia can be seen growing in the same diseased area.

If the entire head is infected, it is dwarfed and compressed with infected spikelets closed rather than spreading. Diseased kernels are gray-brown and lightweight. The interior of the kernel becomes floury and discolored. Infected grain may be harmful when fed to hogs, dogs, and humans.

CONTROL:
1. Rotate wheat with other crops. Do not plant wheat after corn, barley, or rye.
2. Plow under infected residue where feasible to hasten decomposition of residue and prevent spores from being windborne.
3. Do not spread manure that contains infected straw or corn stalks on soil where wheat is growing.
4. Early planting may allow wheat to escape much of the warm, moist weather of summer.
5. Treat seed with a seed-protectant fungicide to control external spores and mycelium.
6. Foliar fungicides may offer a means of control.

Septoria Leaf Blotch

This disease is also called speckled leaf blotch, Septoria leaf spot, *Septoria avenae* spot of wheat, and *Septoria tritici* blotch.

CAUSE: *Septoria tritici* Rob. ex Desm., teleomorph *Mycosphaerella graminicola* (Fckl.) Sand, and *S. avenae* Frank f. sp. *triticea* T. Johnson, teleomorph *Leptosphaeria avenaria* Weber f. sp. *triticea* T. Johnson. The fungi survive as mycelium in live wheat plants and as pycnidia on infected residue for two to three years. *Septoria tritici* may be seedborne. During cool (15–25°C), wet weather in the autumn or spring, pycnidiospores are exuded from pycnidia within a gelatinous drop (cirrhi) that protects spores from radiation and drying out. Optimum spore production occurs at 100 percent relative humidity. Pycnidiospores are disseminated to lower leaves of healthy plants by splashing and blowing rain. More pycnidiospores are produced in pycnidia on susceptible cultivars than in pycnidia on resistant cultivars. Infection requires six or more hours of wetness with subsequent disease development favored by relatively high temperatures (18–25°C). However, response to temperature will vary according to cultivar resistance. New spores are produced in 10–20 days. Ascospores are produced in perithecia as wheat matures in late summer or early autumn and are primarily windborne to hosts.

DISTRIBUTION: Generally distributed wherever wheat is produced. Septoria leaf blotch can become a serious disease given the proper environmental conditions.

SYMPTOMS: At first, small, light green to yellow spots occur between veins of lower leaves, especially if they are in contact with the soil. Spots rapidly elongate to form tan to red-brown, irregular-shaped lesions that are often partly surrounded by a yellow margin. Lesions age and become light brown to almost white with small, dark specks, which are the pycnidia, in the center. The presence of pycnidia is a good diagnostic characteristic.

Infection of the stems, especially the nodes, leaf sheaths, and tips of glumes, may also occur. Pycnidia may be produced in all infected areas. Severely infected leaves turn yellow and may die prematurely. Occasionally an entire plant may be killed. Autumn and winter infection causes a reduction in root weight and mass.

CONTROL:
1. Plant certified, disease-free seed that has been cleaned and treated with a seed-protectant fungicide.
2. Plow under infected residue where feasible.
3. Plant resistant wheat cultivars.
4. Apply a foliar fungicide if environmental conditions favor disease development.
5. Rotate wheat every three or four years with a nonsusceptible crop.
6. Plant winter wheat after the Hessian fly-free date.

Sharp Eyespot

CAUSE: *Rhizoctonia cerealis* van der Hoeven CAG-1 and *R. solani* Kuhn AG-4, teleomorph *Thanatephorous cucumeris* (Frank) Donk. Strains of *R. solani* that cause sharp eyespot differ from those that cause Rhizoctonia root rot. The fungi survive as sclerotia in soil or mycelium in infected residue of a large number of plant genera. Wheat may be infected any time during the growing season. Sclerotia germinate to form mycelium or mycelium grows from a precolonized substrate to infect roots and culms, particularly in dry (less than 20% moisture-holding capacity) and cool soils.

DISTRIBUTION: Generally distributed wherever wheat is grown.

SYMPTOMS: Diamond-shaped lesions that resemble those of eyespot occur on lower leaf sheaths. However, sharp eyespot lesions are more superficial than eyespot lesions. Lesions have light tan centers with dark brown margins. Frequently dark mycelium is visible on them. Dark sclerotia may develop in lesions and between culms and leaf sheaths.

Seedlings may be killed but plants may produce new roots to compensate for those rotted off. When roots are infected, plants may lodge and produce white heads. Diseased plants may be stiff, have a grayish cast, and be delayed in maturity.

CONTROL: Vigorous plants growing in well-fertilized soil are not as likely to become severely diseased as unthrifty plants.

Snow Rot

CAUSE: *Pythium aristosporum* Vanterpool, *P. iwayami* Ito, and *P. okanoganense* Lipps. *Pythium ultimatum* Trow has also been reported as a causal fungus. Disease development depends on extended periods of snow cover. In Washington, it was confined to areas where water collects during snow and ice melt. In North Dakota, the absence of ground frost was thought to contribute to the disease syndrome. Near-freezing temperatures are essential for disease development. Survival is likely as oospores in soil and residue. The fungi are probably disseminated by zoospores.

DISTRIBUTION: North Dakota and the state of Washington.

SYMPTOMS: As plants emerge from under a thick snow cover the older leaves have large, dark green, water-soaked areas. Younger leaves are distorted, water-soaked, dark green, and flaccid. Leaves touching the soil are more rotted than those in an upright position. If the fungi invade the growing point, plants die. After snow has melted, the soft tissue becomes brown or tan. Basal leaf sheaths are dark brown and are filled with oospores. Plants cannot be pulled from soil because of rotted crowns.

Root tissues are invaded only when runoff water washes soil from around crowns. Inoculated plants have severe stunting and root rot with stele and cortical tissues extensively rotted. Cortical tissues slough from roots. Few tillers develop and root systems are reduced.

CONTROL: No control is reported.

Snow Scald

This disease is also called Sclerotinia snow mold.

CAUSE: *Myriosclerotinia borealis* (Bub. & Vleug.) Kohn (syn. *Sclerotinia borealis* Bub. & Vleug. and *S. graminearum* Elen.) oversummers as sclerotia. In the autumn, the sclerotia germinate during damp, cool weather to form cup-shaped apothecia. Ascospores are produced in a layer of asci on the upper portion of the apothecium and are windborne to seedlings. Sclerotinia snow scald occurs when cool, damp autumns are followed by a deep snow cover that lasts for five or more months over unfrozen or slightly frozen soil.

DISTRIBUTION: Canada, Europe, Japan, USSR, and Scandinavia.

SYMPTOMS: Snow scald occurs in scattered patches throughout a field. Sparse, gray mycelium covers dead plants at snow melt. Plants are bleached. Leaves wrinkle on exposure to light, eventually turning dark with saprophytic growth of other fungi and crumbling away. Sclerotia are globular, elongate or flakelike, 0.3–7.0 mm long, black when mature, formed in leaves and on them and in leaf sheaths and crowns. Sclerotia are rarely found except on dead plants.

CONTROL: While no control is entirely satisfactory in controlling Sclerotinia snow scald, the following measures will be of some benefit:
1. Rotate winter wheat with spring wheat, another spring cereal, or a legume to help reduce inoculum.
2. Plowing under residue will bury inoculum and prevent sclerotia from germinating, hastening their decomposition.

Speckled Snow Mold

This disease is also called Typhula blight.

CAUSE: *Typhula ishikariensis* Imai, *T. idahoensis* Remsb. (syn. *T. borealis* Ekstr.), and *T. incarnata* Lasch ex Fr. (syn. *T. itoana* Imai and *T. gramineum* Karst.) oversummer as sclerotia-infected residue and soil, or as parasitic growth on live plants. Fungi do not survive well as saprophytes on old infected residue. Infection optimally occurs at 1–5°C. Sclerotia germinate during wet weather in the autumn and form either a basidiocarp or mycelium. Infections occur when basidiospores form on the basidiocarp and are windborne to plants, or mycelium directly infects plants by growing over the surface of some soils. Sclerotia are formed either within necrotic tissue or in mycelium. Further infection occurs by mycelial growth under snow cover. Saprophytic growth on dead tissue may provide inoculum source for increased disease in the spring; however, the fungi are poor saprophytes in nature and depend primarily on parasitism for existence. *Typhula* spp. growth is more closely related to snow cover than *Microdochium nivale*, the cause of pink snow mold.

DISTRIBUTION: Canada, central and northern Europe, Japan, and the northwestern United States.

SYMPTOMS: At snow melt a gray-white mycelium is present on leaves and crowns of plants and soil. The numerous sclerotia appear within tissues or scattered in the mycelium growing over plant surfaces, giving them a speckled appearance. Dead leaves will be common, but unless the crown is infected, plants may recover during warm, dry weather yet will not be as vigorous as healthy plants. Leaves killed by these fungi will crumple easily and be covered with gray to white mycelium.

CONTROL:
1. Rotation with a legume will help reduce the amount of inoculum present in the soil since the fungi are not good saprophytes.
2. Do not mow plants late in autumn. Removal of clippings reduces deleterious effects of mowing.

Spot Blotch

This disease is also called Bipolaris leaf spot and Helminthosporium leaf spot.

CAUSE: *Bipolaris sorokiniana* (Sacc.) Shoemaker. (syn. *Helminthosporium sativum* Pammel King & Bakke and *H. sorokiniaum* Sacc. apud Sorokin), teleomorph *Cochliobolus sativus* (Ito & Kurib.) Drechs. ex Dast. Survival is primarily by conidia and mycelium. Conidia are airborne, with infection occurring during wet weather as plants approach maturity.

DISTRIBUTION: Generally distributed wherever wheat is grown.

SYMPTOMS: Symptoms are most frequent on lower leaves as distinct, elongate, brown-black lesions that are normally less than 1 cm in length. Lesions are evident after senescence.

CONTROL:
1. Ensure proper fertility.
2. Rotate with nonhosts.
3. Plant resistant cultivars.

Stem Rust

This disease is also called black rust and black stem rust.

CAUSE: *Puccinia graminis* Pers. f. sp. *tritici* Eriks. & Henn. is a heteroecious, long-cycle rust that has the barberries, *Berberis vulgaris* L., *B. canadensis* Mill., *B. fendleri* Gray, and *Mahonia* spp. as alternate hosts. *Puccinia graminis* f. sp. *tritici* causes stem rust of wheat by one of two life cycles. Urediospores (repeating spores) are produced in uredia on wheat during spring and summer. Urediospores are windborne to wheat where they infect and produce new urediospores under moist conditions and moderate temperatures (15–25°C). Secondary urediospores are produced in 7–10 days. This cycle will continue indefinitely as long as growing wheat plants are available. Urediospores produced in northern wheat-producing areas are blown south in summer and autumn to infect wheat in southern wheat-growing areas. The urediospores will recycle on wheat until they are wind disseminated north during late winter and spring to reinfect wheat in the north.

The second life cycle is as follows: As wheat ripens, teliospores are formed that survive through the winter. In spring, teliospores germinate to form basidiospores (sporidia) that are windborne and infect young leaves of barberry. A pycnium (spermogonium) is formed on the upper leaf surface. The pycnia are the means by which new races of stem rust arise and function in the exchange of genetic material, thereby creating new races of the fungus. Each pycnium produces pycniospores and special mycelium called receptive hyphae. Pycniospores are exuded out of the pycnium in a thick, sticky, sweet liquid that is attractive to insects. Pycniospores are then either splashed by water or carried by insects from one pycnium to another where they become attached to the receptive hyphae. The pycniospore germinates and the nucleus from the spore enters into the receptive hyphae. This results in the formation of an aecium on the lower leaf surface directly under the pycnium. Aeciospores that differ genetically from either pycniospores or receptive hyphae are produced in the aecia and are windborne to wheat. Aeciospores can infect only wheat; each infection gives rise to a uredium in which urediospores are formed. The urediospores then serve as the repeating stage and are windborne to wheat where new generations of urediospores are formed. Telia are once again produced as the wheat matures, thus completing the life cycle. Epiphytotics develop during moist weather and the disease is not severe during dry weather.

DISTRIBUTION: Generally distributed wherever wheat is grown.

SYMPTOMS: Uredia and telia occur on the stems, leaf sheaths, leaf blades, glumes, and beards. Uredia are red-brown and oblong. The epidermis of the leaves and culms becomes ruptured, pushed back around the pustule, giving it a jagged or ragged appearance and exposing urediospores.

Telia appear just prior to plant maturation, mostly on leaf sheaths and culms. Telia are oblong to linear, dark brown to black. Teliospores are exposed by a rupturing of the epidermis.

Pycnia appear on barberry leaves in the spring as bright orange to yellow spots, with sometimes what appears to be a drop of water in the middle, which is the liquid containing pycniospores. Opposite the pycnial spots on the other side of the leaf are the aecia that resemble raised, orange, bell-shaped clusters.

CONTROL:
1. Plant resistant wheat cultivars.
2. Eliminate the common barberry from wheat-producing areas. The common barberry should not be confused with the Japanese barberry, which is immune to stem rust. There are several characteristics to differentiate between the two. The common barberry has a saw-toothed leaf edge, gray outer bark, bright yellow inner bark, berries borne in bunches, and spines with usually three in a group. The Japanese barberry has a smooth leaf edge, red-brown outer bark, bright yellow inner bark, berries borne in ones or twos, and usually a single spine.
3. Late planting reduces the effectiveness of the slow rusting character.

Stripe Rust

This disease is also called glume rust and yellow rust.

CAUSE: *Puccinia striiformis* West. is a rust fungus that is not known to have an alternate host. It oversummers as urediospores on residual green cereals and grasses during the period between harvest and emergence of wheat planted in the autumn. Mycelium, and infrequently urediospores, remain alive over the winter in or on different cereals such as barley, grasses, and rye. Urediospores are formed during cool, wet weather and are windborne to healthy plants to cause infection. Little infection occurs above 15°C. Warmer than normal winters and cooler April temperatures favor the development of stripe rust epiphytotics. Teliospores are formed but are not known to function as overwintering spores.

DISTRIBUTION: Australia. In North America, stripe rust occurs at the higher elevations and cooler climates along the Pacific Coast and intermountain areas from Canada to Mexico. It also occurs under the same circumstances in South America and in the mountainous areas of Central Europe and Asia.

SYMPTOMS: The most severe symptoms occur during cool, wet weather in the early growth period of wheat. Symptoms occur early in the spring before other rusts, especially in areas with mild winters. Yellow uredia appear on autumn foliage and new foliage early in the spring. Uredia coalesce to produce long stripes between veins of the leaf and sheath. Small, linear lesions occur on floral bracts. Telia develop as narrow, linear, dark brown pustules covered by the epidermis.

CONTROL:
1. Plant resistant wheat cultivars.
2. Make sure that soils are well fertilized; in particular, potassium levels must be adequate.
3. Cool spring temperatures delay temperature-sensitive adult plant resistance in some cultivars.
4. Apply foliar fungicide. In Australia it is recommended to spray when 1 percent of leaf area is affected and predicted yield loss is sufficient to make spraying economical.
5. Experimentally, fungicide treatment of seed has proven effective.

Take All

This disease is also called Ophiobolus patch.

CAUSE: *Gaeumannomyces graminis* (Sacc.) von Arx & Oliver var. *tritici* Walker, *G. graminis* (Sacc.) von Arx & Oliver var. *graminis*, and *G. graminis* (Sacc.) von Arx & Oliver var. *avenae* (E. M. Turner) Dennis (syn. *Ophiobolus graminis* Sacc.). The fungi survive as mycelium in infected plants or as mycelium and perithecia in infected residue. However, ascospores produced in perithecia are not considered important in the dissemination of the disease. In the autumn or spring, seedlings become infected by the roots growing into the vicinity of crop residue colonized by *G. graminis* var. *tritici* and coming in contact with mycelium. Disease occurs under low to moderate temperature. Plant-to-plant spread of take-all occurs by hyphae growing through the soil from an infected plant to a healthy one or by an infected root coming into contact with a healthy one. The disease is more severe under reduced tillage compared to conventional tillage. Ascospores are produced in perithecia during wet weather but apparently are not disseminated a great distance either by splashing water or wind.

Soil phosphorous and mycorrhizae affect disease development by increasing the phosphorous status of the host. This in turn leads to a decrease in net leakage of root exudates, and thereby reduces pathogen activity. Disease becomes more severe as soil is limed and less severe as soil becomes more acid. This may be due to host plant predisposition resulting from inadequate supplies of certain essential plant nutrients at the elevated pH.

DISTRIBUTION: Generally distributed wherever wheat is grown.

SYMPTOMS: Take-all is usually more severe on wheat grown in alkaline soils. The first symptoms are light brown to dark brown necrotic lesions on the roots. By the time an infected plant reaches the jointing stage, most of its roots are brown and dead. At this point many plants die, or if plants are still alive, they are stunted and the leaves are yellow.

The disease becomes most obvious as plants approach heading. The stand is uneven in height and plants appear to be in several stages of maturity. Infected plants at heading have few tillers, ripen prematurely, and their heads are bleached and sterile. Roots are sparse, blackened, and brittle. Plants can be easily broken free of their crown when pulled from the soil. A very dark discoloration of the stem is visible just above the soil line, along with a mat of dark brown fungus mycelium under the lower sheath between the stem and the inner leaf sheaths.

CONTROL:

1. Rotate wheat with a nonhost crop since *G. graminis* var. *tritici* is not a good saprophyte and does not survive long in the soil in the absence of a host.
2. Treat seed with a systemic seed-treatment fungicide.
3. Plant later in the growing season. This extends the period the fungus must survive as a saprophyte. Additionally the lower temperatures associated with late planting favor a more vigorously growing plant.
4. Apply fertilizers containing ammonium nitrogen, phosphorous, and chloride.

Tan Spot

This disease is also called leaf spot, blight, and yellow leaf spot.

CAUSE: *Pyrenophora tritici-repentis* (Died.) Drechs. (syn. *P. trichostoma* (Fr.) Fckl.), anamorph *Drechslera tritici-repentis* (Died.) Shoemaker (syn. *Helminthosporium tritici-repentis* Died. and *H. triticivulgaris* Nisikado) overwinters as pseudothecia on infected residue, primarily on the soil surface. Ascospores, conidia, and hyphae serve as primary inoculum. Conidia produced in older lesions serve as secondary inoculum and are windborne to hosts. In the autumn, pseudothecia are produced on infected culms and leaf sheaths. Disease is more severe in tall stubble than in short stubble. Infections require a 6–48-hour wet period and are most numerous in proximity to host residues.

Several races have been identified on the basis of virulence.

DISTRIBUTION: Generally distributed wherever wheat is grown.

SYMPTOMS: Symptoms first appear in spring on lower leaves and progress upward to upper leaves into early summer. Symptom development is favored by frequent rains and cool, cloudy, humid weather in the early growing season. At first, tan flecks appear on both sides of the leaf. The flecks eventually become diamond-shaped lesions up to 12 mm long, with a yellow border and a dark brown spot in the center that is due to *P. tritici-repentis* sporulating. Lesions may coalesce to cause large areas of the leaf to die, usually from the tip inward. Pseudothecia will eventually develop on straw as dark raised bumps.

CONTROL:
1. Plant resistant cultivars.
2. Apply a foliar fungicide before the disease becomes severe if weather conditions are wet.
3. Plow under infected residue.
4. Certain macronutrients such as phosphorous and micronutrients such as copper, iron, manganese, and zinc have the potential for limiting take-all either by lessening susceptibility of host tissues, promoting formation of new roots, or both.

Tar Spot

CAUSE: *Phyllachora graminis* (Pers.: Fr.) Fckl. overseasons as stromata on residue. Disease is favored by moist, shaded areas.

DISTRIBUTION: Widespread.

SYMPTOMS: Glossy black and somewhat sunken spots occur on leaf blades and sheaths. Spots are 0.1–0.2 mm by 5 mm in size and are the stromata of *P. graminis*.

CONTROL: Control is not necessary since tar spot occurs infrequently on wheat.

Twist

This disease is also called Dilophospora leaf spot.

CAUSE: *Dilophospora alopecuri* (Fr.) Fr. (syn. *D. graminis* Desm. and *Diplodia graminis* Fuckel) survives as mycelium and pycnidia in infected residue and as conidia on seed. Primary infection occurs on seedlings by conidia produced in pycnidia being airborne or splashed to seedlings during wet weather. Secondary infection occurs from conidia produced in pycnidia in spots on seedling leaves during moist weather. The twist phase of the disease is caused by secondary conidia being disseminated into the whorls by larvae of the seed gall nematode, *Anguina tritici* (Steinbuch) Chitwood, as it moves up the plant in a water film.

DISTRIBUTION: Canada, Europe, India, and the United States. Twist has not been observed for several years and is considered to be a rare disease.

SYMPTOMS: At first, small, elongated, yellow spots appear only on leaves. Soon the spots become light brown with black centers consisting of stromata and pycnidia of the fungus. Leaves may be killed if spots become numerous. When the whorl is colonized, leaves do not emerge or they emerge twisted, distorted, and covered with a gray fuzz, which is the mycelium of the fungus. This phase of the disease rarely occurs in the absence of *A. tritici*. When the fungus dries out, it becomes dark with stromata, in which pycnidia are produced, often occurring as dark streaks on leaves.

CONTROL:
1. Rotate wheat with other crops. Most cereals are not susceptible.
2. Plow under infected residue where feasible.
3. Treat seed with a seed-protectant fungicide.
4. Plant certified wheat seed.

Typhula-like Snow Mold

CAUSE: Basidiomyceteous fungus similar to *Typhula* spp. Disease occurred due to a prolonged snow cover. The fungus is considered a weak pathogen and has been identified as OKLA-1 in the literature.

DISTRIBUTION: Oklahoma.

SYMPTOMS: Dead plants occurred under prolonged snow cover especially on north-sloping fields. Small orange to black subepidermal sclerotia were present in basal leaf sheaths of living and dead plants. These are flatter than either *T. ishikariensis* or *T. idahoensis*.

CONTROL: No control is reported.

White Rot

CAUSE: *Sclerotium rolfsii* Sacc.

DISTRIBUTION: Brazil.

SYMPTOMS: White mycelium and sclerotia occur on necrotic areas of roots, crowns, and lower portions of stems. Commonly the plant will turn white and die.

CONTROL: No control is reported.

Winter Crown Rot

This disease is also called Coprinus snow mold, cottony snow mold, LTB (low-temperature basidiomycete) snow mold, and SLTB (sclerotial low-temperature basidiomycete) snow mold.

CAUSE: *Coprinus psychromorbidus* Redhead & Traquair.

DISTRIBUTION: Canada.

SYMPTOMS: After snow melt, patches of dead wheat occur where snow is deepest. Sometimes there is rather sparse mycelium, giving the plants a gray sheen; often little mycelium can be found. Irregularly shaped sclerotia up to 1 mm in length, gray to black, are usually found inside sheaths loosely attached to host tissues and infrequently on roots and subcrown internodes.

CONTROL: No control is reported.

MYCOPLASMAL CAUSES

Aster Yellows

CAUSE: Aster yellows mycoplasma survives in several dicotyledonous plants and leafhoppers. Aster yellows mycoplasma is transmitted primarily by the aster leafhopper, *Macrosteles fascifrons* Stal., and less commonly by *Endria inimica* Say, *Elymana sulphurella* (Zett.), *Athysanus argentarius* Metc., and *M. laevis* Rib. The mycoplasma survives in biennial or perennial weeds. The leafhoppers first acquire the mycoplasma by feeding on infected plants, then transmit the mycoplasma by flying to healthy plants and feeding on them.

DISTRIBUTION: Generally distributed throughout eastern Europe, Japan, and North America. However, aster yellows is rarely severe.

SYMPTOMS: Symptoms become most obvious at 25–30°C. Seedlings will either die two to three weeks after infection, or if they survive, they will be stunted. Leaves are yellow or have yellow blotches, and the heads are sterile with distorted awns.

Infection of older plants causes leaves to become somewhat stiff and discolored from tip or margin inward in shades of yellow, red, or purple. Root systems may not be well developed.

CONTROL: None is available.

CAUSED BY NEMATODES

Cereal Cyst Nematode

Heterodera avenae is also called the oat cyst nematode.

CAUSE: *Heterodera avenae* Woll. (syn. *H. major* (Schmidt) Franklin) is a nematode that is also a pest of barley, oats, rye, and numerous annual and perennial grasses. The entire life cycle of the cereal cyst nematode as it goes from egg, to larva, to adult can be completed in 9–14 weeks. *Heterodera avenae* survive as cysts in soil for a year or more. Larvae emerge in the spring from eggs contained in overwintered cysts. Larvae enter plant roots and begin feeding. Nematodes swell up and break through the roots but remain attached by a thin neck. Males revert to a veriform size but females continue to swell as eggs develop within their bodies. Eventually the female body forms a cyst that detaches from the roots. Only one generation is completed each season. Other *Heterodera* spp. on wheat are *H. bifenestra* Cooper, *H. hordecalis* Andersson, and *H. latipons* Franklin.

DISTRIBUTION: Africa, Australia, Europe, Japan, USSR, southeastern Canada, and Oregon.

SYMPTOMS: The lemon-shaped cysts are at first white, then gradually become dark brown as the cyst hardens. Eggs and larvae are white and microscopic. The cyst itself is barely visible to the naked eye.
   The first symptom is poor growth in one or more spots in a field. Leaf tips of plants that are heavily infested are red or purple. The discolored leaves die off, with plants becoming yellow. The roots are thickened and more branched than normal. Heavy infestations cause wilting, particularly during times of water stress, stunted growth, poor root development, and early plant death.

CONTROL:
1. Rotate wheat with a legume crop.
2. Apply a nematicide to soil before planting; however, this is generally not economically feasible.

Grass Cyst Nematode

CAUSE: *Punctodera punctata* Mul. & Stone (syn. *Heterodera punctata* Thorne). The life cycle is similar to *H. avenae*.

DISTRIBUTION: Canada and the United States.

SYMPTOMS: See Cereal Cyst Nematode.

CONTROL: No control is reported, but would likely be similar to that of cereal cyst nematode.

Grass Root Gall Nematode

This nematode is also called the root gall nematode.

CAUSE: *Subanguina radicicola* (Grf.) Param. (syn. *Ditylenchus radicicola* (Grf.) Filip. and *Anguillulina radicicola* (Grf.) Gdy.) is an endoparasite that survives in host roots. Larvae penetrate roots and develop in cortical tissue, forming a root gall in two weeks. Mature females begin egg production within a gall; eventually the galls weaken and release larvae that establish secondary infections. Each generation is completed in about 60 days.

DISTRIBUTION: Canada and northern Europe.

SYMPTOMS: Seedlings frequently have reduced top growth and chlorosis. Galls on roots tend to be inconspicuous and vary in diameter from 0.5–6 mm. Roots may be bent at the gall site. At the center of larger galls is a cavity filled with nematode larvae.

CONTROL: Rotate wheat with noncereal crops.

Root Knot Nematodes

CAUSE: *Meloidogyne* spp. are endoparasitic nematodes that overwinter as eggs in the soil. Three species are most commonly found on wheat, *M. naasi* Franklin, *M. incognita* (Kofoid & White) Chitwood, and *M. arenaria* (Neal) Chitwood. In the spring, larvae hatch from eggs but may enter roots at any time. By the middle of the summer, females inside the root knot tissue release eggs into the soil. Overwintering is primarily by eggs.

DISTRIBUTION: Generally distributed wherever wheat is grown.

SYMPTOMS: Swellings or thickenings comprised of swollen cortical cells and bodies of nematodes containing egg masses can be found on roots in the spring and summer. When the root knots are cut open, the egg masses will turn dark.

CONTROL:
1. Wheat planted in the autumn does not become as severely diseased as that planted in the spring.
2. Rotate wheat with root crops.

Root Lesion Nematodes

CAUSE: *Pratylenchus* spp., primarily *P. minyus* Sher. & Allen and *P. thornei* Sher. & Allen. Nematodes live free in soil as migratory endoparasites. They overwinter as eggs, larvae, or adults in host tissue or soil. Both larvae and adults penetrate roots where they move through cortical cells, females simultaneously depositing eggs as they migrate. Older roots are abandoned and new roots are sought as sites for penetration and feeding.

DISTRIBUTION: Generally distributed wherever wheat is grown.

SYMPTOMS: Plants in areas of a field will appear yellow and under moisture stress. Roots and crowns will be rotted when *Rhizoctonia solani* infects through nematode wounds. New roots may become dark and stunted with resultant loss in set of grain.

CONTROL:
1. Plant in autumn when soil temperatures are below 13°C.
2. Soil fumigants could be used where the high costs warrant them.

Seed Gall Nematode

CAUSE: *Anguina tritici* (Steinbuch) Chitwood survives as larvae within seed galls for several years. When galls are planted with wheat seed, larvae are released into moist soil. Larvae move up the plant to the apex in a water film, eventually reaching the flower primordia. Mature nematodes copulate and produce eggs. Seed galls develop from undifferentiated flower tissue interacting with nematodes. If gall development is retarded, larvae may be present in healthy-appearing seed. Galls are mixed with normal seed or fall to the ground where nematodes become dormant under dry conditions. Long-range dissemination is by seed galls mixed with grain.

DISTRIBUTION: Eastern Asia, parts of Europe, India, and southeastern United States.

SYMPTOMS: Prior to heading, wheat plants are swollen near soil level and leaves will be twisted, wrinkled, or rolled. After heading, the distortions are not as obvious but plants are stunted and mature slowly. Heads are small and the dark, seedlike gall forces the glumes to spread apart. Galls are dark brown and do not have the brush or embryo markings of normal seed.

CONTROL:
1. Plant clean seed.
2. Seed may be soaked for 10 minutes at 54°C.
3. Rotate wheat for two years with nonhost crops.

Stubby Root Nematode

CAUSE: *Paratrichodorus* spp., principally *P. minor* (Colbran) Siddigi (syn. *P. christiei* (Allen) Siddigi and *Triochodorus christiei* Allen) survive in soil or on roots as eggs, larvae, or adults. These nematodes are migratory ectoparasites that feed only on the outside of wheat roots and move relatively rapidly through soil at 5 cm/hour, especially in fine sandy soils. Feeding is confined to epidermal and external cortical cells.

DISTRIBUTION: Widely distributed in most agricultural soils. Wheat planted early in the autumn in sandy soils is most severely diseased.

SYMPTOMS: Roots are thickened, short, and stubby. Root tips may have brown lesions. Tops of plants may appear to grow poorly and the entire plant may be pulled easily from the soil due to lack of a fibrous root system.

CONTROL: No control is practically available.

Stunt Nematode

CAUSE: *Merlineus brevidens* (Allen) Siddigi survives in soil and in association with host tissue as different morphological stages. It is an ectoparasite feeding on the outside of wheat roots often in association with the fungus *Olpidium brassicae* (Woron.) Dang.

DISTRIBUTION: Indigenous to most soils but rarely is it found at high populations.

SYMPTOMS: Most severe damage occurs on winter wheat growing in wet soils. The lower leaves die and few tillers form on infected plants. Those that do are small and have small seed. Roots of seedlings are short, dark, and shriveled.

CONTROL:
1. Apply nematicides or soil fumigants where economically feasible.
2. Plant cultivars that tolerate stunting.

CAUSED BY VIROIDS

Seedborne Wheat Yellows

CAUSE: Seedborne wheat yellows viroid (SWTV). The viroid is seedborne and is thought to be mechanically transmitted.

DISTRIBUTION: China.

SYMPTOMS: Initially chlorotic spots occur on the upper and middle portions of seedling leaves and subsequently on new leaves as they emerge. Spots coalesce to form large chlorotic areas that become necrotic.

CONTROL: Plant pathogen-free seed.

VIRAL CAUSES

African Cereal Streak

CAUSE: African cereal streak virus (ACSV) is limited to the phloem where it induces a necrosis. It is transmitted by the delphacid leafhopper *Toya catilina* Fennah. The natural virus reservoir is likely native grasses. It is not mechanically or seed transmitted. Disease development is aided by high temperatures (above 20°C).

DISTRIBUTION: East Africa, specifically Kenya and maybe Ethiopia.

SYMPTOMS: Initially faint, broken, chlorotic streaks begin near the leaf base and extend upward. Eventually definite alternate yellow and green streaks develop along the entire leaf blade. Later the leaves become almost completely yellow. New leaves tend to develop a shoe-string habit and die.

Young infected plants become chlorotic, severely stunted, and die. Older infected plants have yellow heads that are distorted. Seed yield is almost completely suppressed. Plants become soft, flaccid, and almost velvety to the touch.

CONTROL: No control is reported.

Agropyron Mosaic

This disease is also called agropyron green mosaic, streak mosaic, and yellow mosaic.

CAUSE: Agropyron mosaic virus (AMV) is transmitted by the mite *Abacarus hystix* Nal. AMV is sap transmissible between gramineous hosts. The virus is reservoired in quackgrass and other grass hosts. Disease tends to occur near these virus sources.

DISTRIBUTION: Canada, northern Europe, and the northern United States. The disease is of little importance.

SYMPTOMS: Disease symptoms usually occur in patches or along grassy borders. Initially pale green or yellow mosaics, streaks, or dashes occur on leaf blades. Moderate stunting may occur. Symptoms with the exception of the stunting tend to become less conspicuous as plants mature.

CONTROL:
1. Plant tolerant cultivars.
2. Eliminate volunteer wheat and grasses.
3. Plant winter wheat later in autumn.

American Wheat Striate Mosaic

CAUSE: American wheat striate mosaic virus (AWStMV) is trans-mitted mainly by the leafhopper *Endria inimica* Say and occasionally by *Elymana virenscens* F. AWStMV overwinters in wheat and grasses in association with *E. inimica*, which overwinters as eggs. Leafhopper adults are migratory, and all stages are dispersed by wind. Symptom development is favored by warm temperatures between 25°C and 33°C. Disease is associated with heavy infestations of leafhoppers.

DISTRIBUTION: North central United States and south central Canada.

SYMPTOMS: Leaves have obvious striations consisting of yellow to white parallel streaks. Older leaves are stunted, chlorotic, then necrotic.

CONTROL: No control is reported.

Australian Wheat Streak Mosaic

This disease is also called chloris striate mosaic.

CAUSE: The causal agent is thought to be a virus that is transmitted by the leafhoppers *Nesoclutha obscura* Evans and *N. pallida* (Evans). It is not sap transmissible. The pathogen is introduced into the mesophyll by the vector. Barley, corn, and oats are also susceptible.

DISTRIBUTION: Australia. The disease is not considered important.

SYMPTOMS: Plants are dwarfed. Leaves display yellow, broken streaks and fine, grayish striping.

CONTROL: Controls are not necessary.

Barley Stripe Mosaic

CAUSE: Barley stripe mosaic virus (BSMV) is transmitted through seed, sap, and pollen. It can remain viable in seed for up to eight years but seed transmission is relatively uncommon in wheat. There are no known vectors. When infected seed germinates, the resulting seedling is infected. The virus is disseminated by leaves rubbing against each other because of wind, hail, and animals. The virus is then transmitted through plant sap. The virus can also be transmitted by infected pollen but not by insects or other means.

DISTRIBUTION: Australia, southern Asia, Europe, Japan, western North America, and the USSR.

SYMPTOMS: Usual symptoms appear as yellow mottling or spots that are either narrow or wide, numerous or few, continuous or broken. Plants may also be dwarfed, have a rosetting, and be excessively tillered. The color of the spots may be light green, tan, yellowish, or bleached white while the rest of the leaf is green. Normally chlorotic stripes develop on leaves that become increasingly yellow or brown. Virulent strains of the virus cause brown stripes that are continuous or broken with irregular margins.

Plants infected by BSMV through sap develop acute and localized symptoms on subsequent leaves. Within a few days symptoms disperse throughout the plant as a mild systemic mosaic. Symptoms are best expressed between 22° and 30°C.

CONTROL: Do not rotate wheat with barley.

Barley Yellow Dwarf

CAUSE: Barley yellow dwarf virus (BYDV) survives in autumn-planted small grains such as barley, oats, and wheat, and in annual and perennial grasses. The virus is transmitted by several species of aphids. An aphid, once it acquires the virus is capable of transmitting it for the rest of its life. Some strains of BYDV are transmitted equally well by all species of aphids; however, some strains display a high degree of vector specificity. Aphid flights may be local or may be assisted by wind and extend for hundreds of kilometers. The virus is not transmitted through eggs, newborn aphids, seed, soil, or by mechanical means. Virus inoculum, at least initially, is brought in by aphids from some distance away. Aphid flights may be windborne for hundreds of miles. The major source of aphids in the spring may be from distant plants; in the fall aphids may be both distant and local. Generally, local grasses, winter wheat, and corn are of little importance as primary virus sources. Epiphytotics occur during cool (10–18°C), moist seasons that favor grass and cereal growth together with aphid multiplication and migration. Infections occur throughout the growing season but are most numerous in spring where aphids overwinter.

DISTRIBUTION: Africa, Asia, Australia, Europe, New Zealand, North America, and parts of South America.

SYMPTOMS: Symptoms are usually not striking and tend to be confused with nutritional disorders or weather related-problems. However, they become more pronounced at cool temperatures (16–20°C) and cloudless days. Single plants or groups of plants are yellow and stunted. Seedling infection slows plant maturity and causes older leaves to turn bright yellow. Frequently, starting from the blade edge, leaves will be various shades of yellow, red, or purple. A later infection causes the flag leaf to turn yellow or red. Leaves tend to be stiff and roots are not as well developed as healthy plants. Phloem tissues are darkened. Frequently the feeding of some vectors will produce tiny brown-black spots on leaves and culms, with adjacent tissues first turning yellow then tan. Cold hardiness is reduced.

CONTROL:
1. Plant tolerant wheat cultivars.
2. Plant winter wheat later in the autumn.

Barley Yellow Striate Mosaic

CAUSE: Barley yellow striate mosaic virus (BYSMV) is transmitted by the planthopper, *Laodelphax striatellus* Fallen. BYSMV is reservoired in volunteer wheat plants.

DISTRIBUTION: Italy; however, the incidence and severity is very low.

SYMPTOMS: A mild yellow mottling occurs on basal leaves and striations occur on upper leaves.

CONTROL: No control is reported.

Barley Yellow Stripe

CAUSE: Causal agent is not known for certain but is transmitted by the leafhopper *Euscelis plebejus* Fn. Disease tends to occur along field borders and near grassy reservoirs of *E. plebejus*.

DISTRIBUTION: Italy.

SYMPTOMS: Fine, continuous stripes occur on leaves sometimes followed by yellowing and death.

CONTROL: No control is reported.

Brome Mosaic

CAUSE: Brome mosaic virus (BMV) survives in cereals and in perennial grasses and frequently in dry leaf tissues. BMV is primarily sap transmitted but in Europe the virus is also inefficiently transmitted by the nematodes *Xiphinema coxi* Tarjan and *X. paraelongatum* Alther. Dissemination probably occurs through plant contact and by any means that transfer sap.

DISTRIBUTION: Northern Europe, USSR, South Africa, United States, and Yugoslavia.

SYMPTOMS: Initially, a streaklike, yellow-green mosaic occurs on leaves and becomes less prominent as plants age. Mild stunting and head deformation may also occur. Some cultivars are symptomless carriers.

CONTROL: Control grassy weeds.

Cereal Chlorotic Mottle

CAUSE: Cereal chlorotic mottle virus (CCMV) is a rhabdovirus transmitted by cicadellids.

DISTRIBUTION: Australia and northern Africa.

SYMPTOMS: Severe necrotic and chlorotic streaks occur on leaves.

CONTROL: No control is reported.

Cereal Tillering

CAUSE: Cereal tillering virus (CTV) is transmitted by the planthoppers *Laodelphax striatellus* Fallen and *Dicranotropis hamata* Boh. CTV is limited to the phloem and is similar to oat sterile dwarf virus, rice black-streaked dwarf virus, and maize rough dwarf virus.

DISTRIBUTION: Italy and Sweden.

SYMPTOMS: Infected plants are excessively tillered and dwarfed with dark green coloration and poor grain yields. Infrequently leaves will become malformed with serrated edges.

CONTROL: No control is reported.

Cocksfoot Mottle

CAUSE: Cocksfoot mottle virus (CFMV) is sap transmitted and is also spread by the cereal leaf beetles *Oulema melanopa* L. and *O. lenchenis* L.

DISTRIBUTION: Great Britain.

SYMPTOMS: Initially leaves are mottled followed by yellowing and death.

CONTROL: No control is reported.

Eastern Wheat Striate

CAUSE: Eastern wheat striate virus is transmitted by the leafhopper *Cicadulina mbila* Naude but not by sap, seed, soil, or aphids. The virus likely overseasons in the perennial Naraenga grass.

DISTRIBUTION: India.

SYMPTOMS: Fine chlorotic stripes that become pronounced with plant age develop on leaves and leaf sheaths. Stripes will often become necrotic. Infected plants are stunted to some degree. Stunting and striate symptoms are more pronounced in barley than in wheat and in plants infected when young. Plants that are infected while very young usually die.

Grain produced by infected plants is shriveled and of poor quality. Infected plants often produce only partially filled heads.

CONTROL: No control is reported.

Enanismo

This disease is also called cereal dwarf.

CAUSE: One or more unidentified viruses and a toxin from the leafhopper *Cicadulina pastusae* Rup. & DeLg. Leafhopper adults and nymphs of both sexes transmit the virus or viruses in a circulative fashion. Females pass the virus(es) transovarially and are more efficient vectors than males. Barley and oats are also hosts.

DISTRIBUTION: Colombia and Ecuador.

SYMPTOMS: Seedlings are killed or stunted. Later infections result in less stunting but leaves display blotches and earlike enations or galls occur. Galls appear one to three weeks after feeding on the newest leaves rather than the ones fed upon. Infection at heading time results in distorted heads and incompletely filled ears.

CONTROL:
1. Plant later in spring after vector activity declines.
2. Plant tolerant or resistant cultivars.

European Wheat Striate Mosaic

This disease is also called oat striate, red disease, wheat striate, and wheat striate mosaic.

CAUSE: The causal agent is unknown. It is transmitted persistently by the planthopper *Javesella pellucida* Fabr. and sometimes by *J. dubia* Fabr. The pathogen is passed through eggs. Nymphs are more efficient vectors than adults. Symptoms occur two to four weeks after feeding.

DISTRIBUTION: Central Europe, Great Britain, and Scandinavia.

SYMPTOMS: Infected seedlings die. Older plants are normally severely stunted and do not head out. Plants infected near heading develop yellow-white leaf stripes, mature early, and develop whiteheads with shrunken seed.

CONTROL: Plant wheat later in the autumn or spring to avoid exposure to vectors.

Maize Dwarf Mosaic

CAUSE: Maize dwarf mosaic virus (MDMV) is transmitted by several aphid species from corn to wheat. Infections occur among plants grown near MDMV-infected corn.

DISTRIBUTION: New York; however, infection of wheat is rare.

SYMPTOMS: Leaves have a mild mottling.

CONTROL: Control is not warranted.

Maize Streak

CAUSE: Maize streak virus (MSV) is transmitted circulatively by five species of Cicadulina leafhoppers, especially *C. mbila* Naude. All stages of leafhoppers transmit MSV. Wheat is most likely to become infected when grown in association with corn.

DISTRIBUTION: Africa and southeast Asia.

SYMPTOMS: Leaves are shortened and curled with fine, linear, chlorotic streaks. Plants are excessively tillered, sterile, with shortened culms and streaks.

CONTROL:
1. Plant resistant cultivars.
2. Plant winter wheat later in the autumn than normal.
3. Do not plant wheat in the vicinity of corn.

Northern Cereal Mosaic

CAUSE: Northern cereal mosaic virus (NCMV) is transmitted by the planthopper *Laodelphax striatellus* Fallen. The insects *Unkanodes sapporonus* (Mats.) and *Delphacodes* spp. have also been reported as vectors. Wheat rosette stunt virus (WRSV) reported from northern China may be the same virus. Barley, oats, rice, and rye are also hosts.

DISTRIBUTION: China (if WRSV and NCMV are the same), Japan, and Korea.

SYMPTOMS: A yellow mosaic and chlorotic leaf streaks. Infected plants are stunted.

CONTROL: Plant resistant cultivars.

Oat Sterile Dwarf

This disease is also called oat base tillering disease and oat dwarf tillering disease.

CAUSE: Oat sterile dwarf virus (OSDV) is transmitted persistently by delphacid leafhoppers especially *Javesella pellucida* Fabr. and *Dicranotropis hamata* Boh. Barley, corn, grasses, millet, and oats are also susceptible.

DISTRIBUTION: Eastern and northern Europe.

SYMPTOMS: Plants are stunted, excessively tillered, sterile, with helical leaf twisting.

CONTROL:
1. Control grassy weeds.
2. Avoid planting wheat adjacent to oats.

Rice Black-streaked Dwarf

CAUSE: Rice black-streaked dwarf virus (RBSDV) is transmitted by the planthoppers *Laodelphax striatellus* Fallen, *Unkanodes sapporonus* (Mats.), and *U. albifascia* Mats. RBSDV is retained from season to season in all stages of the planthoppers but is not transmitted through eggs.

DISTRIBUTION: Japan.

SYMPTOMS: Plants are severely stunted. Leaves are twisted with waxy veinal swellings on their undersides and on culms.

CONTROL: Do not grow wheat adjacent to rice.

Rice Hoja Blanca

This disease is also called white leaf, white spike, and white tip.

CAUSE: Rice hoja blanca virus (RHBV) is transmitted persistently by planthoppers, especially *Sogata cabana* Crawf. and *S. orizicola* Muir. RHBV is circulative and passed through eggs for up to 10 generations.

DISTRIBUTION: Caribbean, Central and South America, and the southern United States.

SYMPTOMS: Upper leaves and spike display a gray-white discoloration. Leaves are also mottled, striped, and chlorotic. Plants become sterile and die.

CONTROL: Do not grow wheat in association with susceptible rice cultivars.

Russian Winter Wheat Mosaic

CAUSE: Winter wheat mosaic virus (WWMV) is disseminated by the leafhoppers *Psammotettix striatus* L. and *Macrosteles laevis* Rib. *Psammotettix striatus* carries WWMV through all life-cycle stages, transovarially passing it to successive generations. Barley, rye, and some grasses are also hosts.

DISTRIBUTION: Eastern Europe and USSR.

SYMPTOMS: Leaves have mosaic and streak mosaic patterns. Plants are mildly retarded and distinct yellow dashes and streaks are oriented parallel with leaf veins.

Severe infections result from seedling infections and include severe stunting, tillering, and necrosis. Surviving seedlings exist as rosettes with typical leaf mosaic symptoms.

CONTROL: Plant resistant cultivars.

Soilborne Wheat Mosaic

This disease is also called eastern wheat mosaic, green mosaic, mosaic rosette, soilborne mosaic, wheat soilborne mosaic, and yellow mosaic.

CAUSES: Soilborne wheat mosaic virus (SBWMV) survives in soil in *Polymyxa graminis* Led., a soilborne plasmodiophoraceous fungus that is an obligate parasite in roots of many higher plants. The fungus enters root hairs and epidermal cells of roots as motile zoospores when soil is water-saturated and at low soil temperatures (10–20°C). Once inside the plant, *P. graminis* replaces plant cell contents with plasmodial bodies that will either segment into additional zoospores or develop in thick-walled resting spores two to four weeks after infection.

SBWMV is spread by any method that will disseminate infested soil. Disease is most common in low-lying areas of fields that tend to be wet.

Fall-planted wheat is most severely diseased because more growth occurs during low temperatures that are most conducive to dissemination and infection by *P. graminis*.

SBWMV in combination with wheat spindle streak mosaic virus (WSSMV) causes a disease in plants resistant to SBWMV. The ratio of SBWMV and WSSMV in infected plants is about 20:1, respectively.

Different strains, labeled the yellow and green strains, have been reported.

DISTRIBUTION: United States. It has also been reported from Argentina, Brazil, Egypt, Italy, and Japan.

SYMPTOMS: Two phases, mosaic and mosaic-rosette, are caused by yellow and green phases respectively. Symptoms are most prominent in the spring on the lower leaves and range from light green to yellow leaf mosaic. Plants may be severely or slightly stunted. The youngest leaves and leaf sheaths are mottled and develop parallel spots or streaks. Warm weather prevents development of disease symptoms.

Symptoms of SBWMV and WSSMV infection are a yellowing and overall bronze appearance to the field with a stunting, mosaic, and reduced tillering of the plants. Symptoms are strongest in lower portions of the field. In late spring, with warm weather, affected plants lose symptoms except for stunting.

CONTROL:
1. Rotate wheat with noncereal crops.
2. Plant wheat later in the autumn.
3. Plant resistant cultivars. Cultivars are resistant to the vector rather than to the virus.

Tobacco Mosaic

CAUSE: Tobacco mosaic virus (TMV). The virus is associated with wheat soilborne mosaic virus.

DISTRIBUTION: Kansas.

SYMPTOMS: Artifically inoculated plants (cultivar Pawnee) incubated at 30°C and in bright sunlight developed faint, chlorotic, local lesions that rapidly disappeared, leaving symptomless plants. A faint, transient mosaic occurred on inoculated leaves of Pawnee, Arthur, and Michigan Amber wheats and Reno barley. No systemic systems developed but virus was recovered from successive leaves, demonstrating systemic movement in the plant.

CONTROL: No control is needed.

Wheat (Cardamom) Mosaic Streak

CAUSE: A virus transmitted through sap and the aphids *Brachycaudus helichrysi* Kalt and *Rhopalosiphum maidis* (Fitch). Infection occurs near reservoirs of the virus in cardamom.

DISTRIBUTION: India.

SYMPTOMS: Plants are mildly stunted and predisposed to infection by *Helminthosporium sativum*. Leaves display chronic yellow-green mosaic symptoms.

CONTROL:
1. Plant resistant cultivars.
2. Grow wheat apart from cardamom.

Wheat Chlorotic Streak

CAUSE: Wheat chlorotic streak virus (WCSV) is presumed to be reservoired in *Agropyron repens* (L.) Beauv., which also maintains the vector, the planthopper *Laodelphax striatellus* Fallen. WCSV is disseminated in a persistent and transovarial manner.

DISTRIBUTION: France; however, it occurs infrequently.

SYMPTOMS: The growth and yield of individual plants is reduced.

CONTROL: No control is reported.

Wheat Dwarf

CAUSE: The cause is thought to be a virus transmitted by the leafhoppers *Psammotettix alienus* Dahlb. and *Macrosteles laevis* Rib. Nymphs are more efficient vectors than adults, but must acquire the virus from diseased plants since it is not passed through eggs. The disease occurs in association with heavy infestations of leafhoppers and the disease results from the combined effects of leafhopper feeding and virus infection.

DISTRIBUTION: Czechoslovakia, USSR, and Sweden.

SYMPTOMS: Plants are severely dwarfed. If infected as seedlings, plants do not head. If plants are infected later, dwarfing is less severe; if heads emerge, seeds are sparse and shriveled.

Leaves develop scattered, fine, light green to yellow-brown spots and blotches. These may coalesce to cause prominent yellowing and necrosis.

CONTROL: Plant resistant cultivars. Soft wheat cultivars are less susceptible than hard wheats.

Wheat Spindle Streak Mosaic

This disease is also called Ontario soilborne wheat mosaic, wheat variegation, and wheat yellow mosaic.

CAUSE: Wheat spindle streak mosaic virus (WSSMV) can survive for six years in soil in association with the soil fungus *Polymyxa graminis* Led. Transmission is similar to soilborne wheat mosaic. Wheat spindle streak mosaic is a cool weather disease. Infection does not occur above 20°C and disease development does not occur above 18°C. The optimum temperature for virus transmission in the soil is 15°C and symptom development is 5–13°C. WSSMV in combination with wheat soilborne mosaic virus (WSBMV) causes a disease of plants resistant to WSBMV. The ratio of WSBMV and WSSMV in infected plants is about 20:1, respectively.

DISTRIBUTION: France, India, Southern Ontario, and the United States.

SYMPTOMS: Symptoms occur on lower, older leaves during cool weather in the spring and tend to be uniformly distributed through a field. Warmer weather above 18°C prevents symptom development on younger leaves. The first leaves produced in spring have yellow-green mottling, dash, or streaks. Streaks are oriented parallel with leaf veins and have tapered ends that resemble spindles. As the leaves mature, brown areas replace the yellow-green areas. A leaf tip or an entire leaf may die. Diseased plants remain slightly stunted and have fewer tillers. As warm weather occurs, new symptomless leaves will hide diseased lower leaves. Cold hardiness is reduced.

For symptoms of WSSMV and WSBMV infection see Soilborne Wheat Mosaic.

CONTROL:
1. Plant resistant cultivars.
2. Plant wheat later in autumn; spring wheat tends to be unaffected.
3. Crop rotation is some help but the virus can remain infective in soil for several years.

Wheat Spot Mosaic

This disease is also called wheat spot and wheat spot chlorosis.

CAUSE: The causal agent is thought to be a virus transmitted by the wheat curl mite *Aceria tulipae* Keifer. It is not sap transmissible.

DISTRIBUTION: Alberta and Ohio.

SYMPTOMS: About three to eight days after mite feeding, light green spots (0.5–1.0 mm in diameter) develop on youngest leaves. Eventually the spots become necrotic, enlarge, and coalesce to form yellowish leaf areas or whole leaves.

Mottling, chlorosis, leaf tip necrosis, stunting, and plant death may occur especially if the wheat streak mosaic virus (WSMV) is present.

CONTROL: Same as for WSMV.

Wheat Streak Mosaic

CAUSE: Wheat streak mosaic virus (WSMV) survives in several annual and perennial grasses. Corn is an oversummering host for WSMV and its vector. The virus is transmitted only by the wheat curl mite, *Aceria tulipae* Keifer. A mite retains the virus for several weeks. As winter wheat plants mature, mites migrate to nearby volunteer wheat, grasses, or corn and will infect them with the virus. However, some grasses are hosts for the mites and not the virus and vice versa; some are susceptible to both, whereas some are resistant to both.

During late summer or early autumn, the virus may be disseminated to volunteer wheat. In time, the virus may be carried by mites from volunteer wheat to early-planted wheat. Mites can be windborne at least 2.4 km.

Only the young mites acquire the virus by feeding 15 or more minutes on infected plants. Neither the mite nor the virus can survive longer than one or two days in the absence of a living plant. No active virus has been detected in dead plants or in seed. Infection depends on three factors: population of wheat curl mites; nearness of virus-infected plants, especially volunteer wheat; and moisture to keep wheat vigorously growing where mites attain maximum reproduction.

DISTRIBUTION: Eastern Europe, western and central North America, particularly the Great Plains and the USSR.

SYMPTOMS: Winter wheat is commonly infected in the autumn but symptoms ordinarily do not appear until the following spring. The greatest losses occur in early-planted, autumn-infected plants.

The first symptoms consist of light green to light yellow blotches, dashes, or streaks in the leaves parallel to the veins. Plants become stunted, show a general yellow mottling, and develop an abnormally large number of tillers that may vary considerably in height. Stunted plants with sterile heads may remain standing after harvest at the same height or shorter than stubble. As infected plants mature, the yellow-striped leaves turn brown and die.

Heads may be sterile or partially sterile with shriveled kernels. In severe cases, plants may die before maturity. Synergistic effects are suspected between wheat streak mosaic and other viruses, such as barley yellow dwarf, making field identification difficult. Feeding mites often cause leaf edges to curl tightly in toward the upper midvein.

CONTROL:
1. Destroy all volunteer wheat and grasses in adjoining fields two weeks before planting and two to four weeks before using the planting field.
2. Plant wheat as late as practical after Hessian fly-free date. If wheat emerges in October or later, it often escapes infection.
3. Resistant cultivars are being developed.

Wheat Yellow Leaf

CAUSE: Wheat yellow leaf virus (WYLV) is transmitted by the aphid *Rhopalosiphum maidis* Fitch in a semipersistent manner. Barley is also a host.

DISTRIBUTION: Japan.

SYMPTOMS: Leaves yellow and blight. Plants often die or ripen prematurely.

CONTROL: No control is reported.

Wheat Yellow Mosaic

CAUSE: Wheat yellow mosaic virus (WYMV) is soilborne and sometimes associated with wheat soilborne mosaic. The virus is sap transmitted.

DISTRIBUTION: Japan.

SYMPTOMS: Yellow mosaic of leaves.

CONTROL: Plant resistant cultivars.

BIBLIOGRAPHY

Adlakha, K. L.; Wilcoxson, R. D.; and Raychaudhuri, S. P. 1984. Resistance of wheat to leaf spot caused by *Bipolaris sorokiniana. Plant Disease* **68:**320–321.

Arneson, E., and Stiers, D. L. 1977. *Cephalosporium gramineum:* A seedborne pathogen. *Plant Dis. Rptr.* **61:** 619–621.

Babadoost, M., and Hebert, T. T. 1984. Factors affecting infection of wheat seedlings by *Septoria nodorum. Phytopathology* **74:**592–595.

Babadoost, M., and Hebert, T. T. 1984. Incidence of *Septoria nodorum* in wheat seed and its effects on plant growth and grain yield. *Plant Disease* **68:**125–129.

Bailey, J. E.; Lockwood, J. L.; and Wiese, M.V. 1982. Infection of wheat by *Cephalosporium gramineum* as influenced by freezing of roots. *Phytopathology* **72:**1324–1328.

Bockus, W. W. 1983. Effects of fall infection by *Gaeumannomyces graminis* var. *tritici* and triadimenol seed treatment on severity of take all in winter wheat. *Phytopathology* **73:**540–543.

Bockus, W. W.; O'Connor, J. P.; and Raymond, P. J. 1983. Effect of residue management method on incidence of Cephalosporium stripe under continuous winter wheat production. *Plant Disease* **67:**1323–1324.

Bonde, M. R. 1987. Possible dissemination of teliospores of *Tilletia indica* by the practice of burning wheat stubble. *Phytopathology* **77:**639 (abstract).

Brakke, M. K.; Estes, A. P.; and Schuster, M. L. 1965. Transmission of soilborne wheat mosaic virus. *Phytopathology* **55:**79–86.

Bretag, T. W. 1985. Control of ergot by a selective herbicide and stubble burning. *Br. Mycol. Soc. Trans.* **85:**341–343.

Broscious, S. C., and Frank, J. A. 1986. Effects of crop management practices on common root rot of winter wheat. *Plant Disease* **70:**857–859.

Brown, J. S., and Holmes, R. J. 1983. Guidelines for use of foliar sprays to control stripe rust of wheat in Australia. *Plant Disease* **67:**485–487.

Brown, W. M. Jr.; Perotti, L. E.; and Hill, J. P. 1985. Wheat take all in Colorado high country irrigated spring wheat. *Phytopathology* **75:**1296 (abstract).

Bruehl, G. W. 1953. Pythium root rot of barley and wheat. *USDA. Tech. Bull. No. 1084.*

Bruehl, G. W., and Cunfer, B. 1971. Physiologic and environmental factors that affect the severity of snow mold of wheat. *Phytopathology* **61**:792–799.

Bruehl, G. W., and Machtmes, R. 1984. Effects of "dirting" on strawbreaker foot rot of winter wheat. *Plant Disease* **68**:868–870.

Bruehl, G. W.; Machtmes, R.; and Murray, T. 1982. Importance of secondary inoculum in strawbreaker foot rot of winter wheat. *Plant Disease* **66**:845–847.

Bruehl, G. W.; Peterson, C. J., Jr.; and Machtmes, R. 1974. Influence of seeding date, resistance and benomyl on Cercosporella foot rot of winter wheat. *Plant Dis. Rptr.* **58**:554–558.

Burns, E. E.; Sharen, A. L.; and Bishop, T. 1975. Septoria leaf blotch and glume blotch of wheat. Montana State Univ. Coop. Ext. Serv. *Plant Diseases. Leaflet No. 202.*

Campbell, W. P. 1958. A cause of pink seeds in wheat. *Plant Dis. Rptr.* **42**:1272.

Carlson, R. R., and Vidaver, A. K. 1981. Bacterial mosaic of wheat: Distribution and hosts. *Phytopathology* **71**:207 (abstract).

Carlson, R. R., and Vidaver, A. K. 1982. Bacterial mosaic, a new corynebacterial disease of wheat. *Plant Disease* **66**:76–79.

Chamswarng, C., and Cook, R. J. 1985. Identification and comparative pathogenicity of Pythium species from wheat roots and wheat field soils in the Pacific Northwest. *Phytopathology* **75**:821–827.

Clement, D. K.; Lister, R. M.; and Foster, J. E. 1986. ELISA-based studies on the ecology and epidemiology of barley yellow dwarf virus in Indiana. *Phytopathology* **76**:86–92.

Conway, K. E., and Williams, E., Jr. 1986. Typhula-like snow mold on wheat in Oklahoma. *Plant Disease* **70**:169–170.

Cook, R. J. 1980. Fusarium foot rot of wheat and its control in the Pacific Northwest. *Plant Disease* **64**:1061–1066.

Cook, R. J., and Hering, T. F. 1986. Infection of wheat embryos by Pythium and seedling response as influenced by age of seed. *Phytopathology* **76**:1061 (abstract).

Cook, R. J.; Slitton, J. W.; and Waldher, J. T. 1980. Evidence for Pythium as a pathogen of direct-drilled wheat in the Pacific Northwest. *Plant Disease* **64**:102–103.

Cook, R. J., and Waldher, J. T. 1977. Influence of stubble-mulch residue management on Cercosporella foot rot and yields of winter wheat. *Plant Dis. Rptr.* **61**:96–100.

Cook, R. J., and Zhang, B. X. 1985. Degrees of sensitivity to metalaxyl within the *Pythium* spp. pathogenic to wheat in the Pacific Northwest. *Plant Disease* **69**:686–688.

Deacon, J. W., and Scott, D. B. 1985. *Rhizoctonia solani* associated with crater disease (stunting) of wheat in South Africa. *Br. Mycol. Soc. Trans.* **85**:319–327.

Dehne, H. W., and Oerke, E. C. 1985. Investigations on the occurrence of *Cochliobolus sativus* on barley and wheat. 1. Influence of pathogen, host plant and environment on infection and damage. *Zeitschr. für Pflanzenkrankheiten und Pflanzenschutz* **92**:270–280.

Diehl, J. A.; Tinline, R. D.; Kochhan, R. A.; Shipton, P. J.; and Rovira, A. D. 1982. The effect of fallow periods on common root rot of wheat in Rio Grande do Sul Brazil. *Phytopathology* **72:**1297–1301.

Eyal, Z.; Scharen, A. L.; Huffman, M. D.; and Prescott, J. M. 1985. Global insights into virulence frequencies of *Mycosphaerella graminicola*. *Phytopathology* **75:**1456–1462.

Fernandez, J. A.; Wofford, D. S.; and Horton, J. L. 1985. Interactive effects of freezing and common root rot fungi on winter wheat. *Phytopathology* **75:**845–847.

Frank, J. A. 1985. Influence of root rot on winter survival and yield of winter barley and winter wheat. *Phytopathology* **75:**1039–1041.

Fryda, S. J., and Otta, J. D. 1978. Epiphytotic movement and survival of *Pseudomonas syringae* on spring wheat. *Phytopathology* **64:**1064–1067.

Gardner, W. S. 1981. Relationship of corn to the spread of wheat streak mosaic virus in winter wheat. *Phytopathology* **71:**217 (abstract).

Goel, R. K., and Gupta, A. K. 1979. A new glume blotch disease of wheat in India. *Plant Dis. Rptr.* **63:**620.

Gough, F. J., and Lee, T. S. 1985. Moisture effects on the discharge and survival of conidia of *Septoria tritici*. *Phytopathology* **75:**180–182.

Gough, F. J., and Merkle, O. G. 1977. The effect of speckled leaf blotch on root and shoot development of wheat. *Plant Dis. Rptr.* **61:**597–599.

Graham, J. H., and Menge, J. A. 1982. Influence of vesicular-arbuscular mycorrhizae and soil phosphorous on take all disease of wheat. *Phytopathology* **72:**95–98,

Grey, W. E.; Mathre, D. E.; Hoffman, J. A.; Powelson, R. L.; and Fernandez, J. A. 1986. Importance of seedborne *Tilletia controversa* for infection of winter wheat and its relationship to international commerce. *Plant Disease* **70:**122–125.

Halfon-Meiri, A., and Kulik, M. M. 1977. *Septoria nodorum* infection of wheat seeds produced in Pennsylvania. *Plant Dis. Rptr.* **61:** 867–869.

Harder, D. E., and Bakker, W. 1973. African cereal streak, a new disease of cereals in East Africa. *Phytopathology* **63:**1407–1411.

Herrman, T., and Weise, M. V. 1985. Influence of cultural practices on incidence of foot rot in winter wheat. *Plant Disease* **69:**948–950.

Hess, D. E., and Shaner, G. 1987. Effect of moisture and temperature on development of *Septoria tritici* blotch in wheat. *Phytopathology* **77:**215–219.

Higgins, S., and Fitt, B. D. L. 1984. Production and pathogenicity to wheat of *Pseudocercosporella herpotrichoides* conidia. *Phytopathol. Zeitschr.* **III:**222–231.

Hoffman, J. A., and Goates, B. J. 1981. Spring dormancy of *Tilletia controversa* teliospores. *Phytopathology* **71:**881 (abstract).

Holmes, R. J., and Dennis, J. I. 1985. Accessory hosts of wheat stripe rust in Victoria, Australia. *Br. Mycol. Soc. Trans.* **85:**159–160.

Hosford, R. M., Jr. 1971. A form of *Pyrenophora trichostoma* pathogenic to wheat and other grasses. *Phytopathology* **61:**28–32.

Hosford, R. M., Jr. 1975. *Platyspora pentamera*, a pathogen of wheat. *Phytopathology* **65:**499–500.

Hosford, R. M., Jr. 1978. Effects of wetting period on resistance to leaf spotting of wheat, barley and rye by *Leptosphaeria herpotrichoides*. *Phytopathology* **68**:591–594.

Hosford, R. M., Jr. 1978. Effects of wetting period on resistance to leaf spotting of wheat by *Leptosphaeria microscopica* with conidial stage *Phaeoseptoria urvilleana*. *Phytopathology* **68**:908–912.

Hosford, R. M., Jr. 1982. White blotch incited in wheat by *Bacillus megaterium* pv. *cerealis*. *Phytopathology* **72**:1453–1459.

Hosford, R. M., Jr.; Hogenson, R. O.; Huguelet, J. E.; and Kiesling, R. L. 1969. Studies of *Leptosphaeria avenaria* f. sp. *triticea* on wheat in North Dakota. *Plant Dis. Rptr.* **53**:378–381.

Huber, D. M., and Hankins, B. J. 1974. Effect of fall mowing on snow mold of winter wheat. *Plant Dis. Rptr.* **58**:432–434.

Inglis, D. A., and Cook, R. J. 1981. *Calonectria nivalis* causes scab in the Pacific Northwest. *Plant Disease* **65**:923–924.

Jacobs, D. L., and Bruehl, G. W. 1986. Saprophytic ability of *Typhula incarnata, T. idahoensis*, and *T. ishikariensis*. *Phytopathology* **76**:695–698

Jacobsen, B. J. 1977. Effect of fungicides on Septoria leaf and glume blotch, Fusarium scab, grain yield, and test weight of winter wheat. *Phytopathology* **67**:1412–1414.

Kane, R. T.; Smiley, R. W.; and Sorrells, M. E. 1987. Relative pathogenicity of selected Fusarium species and *Microdochium bolleyi* to winter wheat in New York. *Plant Disease* **71**:177–181.

Khokhar, L. K., and Pacumbaba, R. P. 1985. Additional alternative grass hosts of *Leptosphaeria nodorum*. *Phytopathology* **75**:1295 (abstract).

Krupinsky, J. M. 1985. Leaf spot diseases of wheat related to stubble height. *Phytopathology* **75**:1295 (abstract).

Lai, P., and Bruehl, G. W. 1966. Survival of *Cephalosporium gramineum* in naturally infected wheat straws in soil in the field and in the laboratory. *Phytopathology* **56**:213–218.

Latin, R. X.; Harder, R. W.; and Wiese, M. V. 1981. Incidence of Cephalosporium stripe as influenced by winter wheat management practices. *Phytopathology* **71**:888 (abstract).

Latin, R. X.; Harder, R. W; and Wiese, M. V. 1982. Incidence of Cephalosporium stripe as influenced by winter wheat management practices. *Plant Disease* **66**:229–230.

LeBeau, J. B. 1968. Pink snow mold in southern Alberta. *Can. Plant Dis. Surv.* **48**:130–131.

Leukel, R. W., and Tapke, V. F. 1954. Cereal smuts and their control. *USDA Farmer's Bull. No. 2069*.

Lipps, P. E., and Bruehl, G. W. 1978. Snow rot of winter wheat in Washington. *Phytopathology* **68**:1120–1127.

Lipps, P. E., and Herr, L. J. 1981. *Rhizoctonia cerealis* causing sharp eyespot of wheat in Ohio. *Phytopathology* **71**:890 (abstract).

Lipps, P. E., and Herr, L. J. 1982. Etiology of *Rhizoctonia cerealis* in sharp eyespot of wheat. *Phytopathology* **72**:1574–1577.

Lockhart, B. E. L. 1986. Occurrence of cereal chlorotic mottle virus in northern Africa. *Plant Disease* **70**:912–915.

Lommell, S. A.; Willis, W. G.; and Kendall, T. L. 1986. Identification of wheat spindle sreak mosaic virus and its role in a new disease of winter wheat in Kansas. *Plant Disease* **70**:964–968.

Loria, R.; Weise, M. V.; and Jones, A. L. 1982. Effect of free moisture, head development and embryo accessibility on infection of wheat by *Ustilago tritici*. *Phytopathology* **72**:1270–1272.

Love, C. S. 1985. Effect of soil pH on infection of wheat by *Cephalosporium gramineum*. *Phytopathology* **75**:1296.

Love, C. S., and Bruehl, G. W. 1987. Effect of soil pH on Cephalosporium stripe in wheat. *Plant Disease* **71**:727–731.

Luke, H. H.; Barnett, R. D.; and Pfahler, P. L. 1985. Influence of soil infestation, seed infection, and seed treatment on *Septoria nodorum* blotch of wheat. *Plant Disease* **69**:74–76.

Luke, H. H.; Pfahler, P. L.; and Barnett, R. D. 1983. Control of *Septoria nodorum* on wheat with crop rotation and seed treatment. *Plant Disease* **67**:949–951.

Luz, W. C. D., and Hosford, R. M., Jr. 1980. Twelve *Pyrenophora trichostoma* races for virulence to wheat in the central plains of North America. *Phytopathology* **70**:1193–1196.

Luzzardi, G. C.; Luz, W. C.; and Pierobom, C. R. 1983. Podridas branca dos cereais causada por *Schlerotium rolfsii* no Brazil. *Fitopatologia Brasileira* **8**:371–375 (in Portuguese).

Maas, E. M. C., and Kotze, J. M. 1985. *Fusarium equiseti* crown rot of wheat in South Africa. *Phytophylactica* **17**:169–170.

McBeath, J. H. 1981. Bacterial mosaic disease on spring wheat and triticale in Alaska. *Phytopathology* **71**:893 (abstract).

McBeath, J. H. 1985. Pink snow mold on winter cereals and lawn grasses in Alaska. *Plant Disease* **69**:722–723.

MacNish, G. C. 1983. Rhizoctonia patch in Western Australian grain belt. *Australia Plant Pathol.* **12**:49–50.

MacNish, G. C. 1985. Methods of reducing Rhizoctonia patch of cereals in Western Australia. *Plant Pathol.* **34**:175–181.

Modawi, R. S.; Heyne, E. G.; Brunetta, D.; and Willis, W. G. 1982. Genetic studies of field reaction to wheat soilborne mosaic virus. *Plant Disease* **66**:1183–1184.

Moore, K. J., and Cook, R. J. 1984. Increased take all of wheat with direct drilling in the Pacific Northwest. *Phytopathology* **74**:1044–1049.

Nagaich, B. B., and Sinha, R. C. 1974. Eastern wheat striate: A new viral disease. *Plant Dis. Rptr.* **58**:968–970.

Nema, K. G.; Dave, G. S.; and Khosla, H. K. 1971. A new glume blotch of wheat. *Plant Dis. Rptr.* **55**:95.

Nguyen, H. T., and Pfeifer, R. P. 1980. Effects of wheat spindle streak mosaic virus on winter wheat. *Plant Disease* **64**:181–184.

Nyvall, R. F., and Kommedahl, T. 1973. Competitive saprophytic ability of *Fusarium roseum* f. sp. *cerealis* 'Culmorum' in soil. *Phytopathology* **63**:590–597.

Otta, J. A. 1974. *Pseudomonas syringae* incites a leaf necrosis on spring and winter wheats in South Dakota. *Plant Dis. Rptr.* **58**:1061–1064.

Paliwal, Y. C. 1982. Role of perennial grasses, winter wheat and aphid vectors in the disease cycle and epidemiology of barley yellow dwarf virus. *Can. Jour. Plant Pathol.* 4:367–374.

Paliwal, Y. C., and Andrews, C. J. 1979. Effects of barley yellow dwarf and wheat spindle streak mosaic viruses on cold hardiness of cereals. *Can. Jour. Plant Pathol.* 1:71–75.

Palmer, L. T., and Brakke, M. K. 1975. Yield reduction in winter wheat infected with soilborne wheat mosaic virus. *Plant Dis. Rptr.* 59:469–471.

Paulsen, A.; Niblett, C. L.; and Willis, W. G. 1975. Natural occurrence of tobacco mosaic virus in wheat. *Plant Dis. Rptr.* 59:747–750.

Pfender, W., and Wootke, S. 1985. Overwintering microbial populations in wheat straw infested with *Pyrenophora tritici-repentis*. *Phytopathology* 75:1350 (abstract).

Pool, R. A. F., and Sharp, E. L. 1969. Some environmental and cultural factors affecting Cephalosporium stripe of winter wheat. *Plant Dis. Rptr.* 53:898–902.

Prabhu, A. S., and Prasada, R. 1966. Pathological and epidemiological studies of leaf blight of wheat caused by *Alternaria triticina*. *Indian Phytopathol.* 19:95–112.

Pumphrey, F. V.; Wilkins, D. E.; Hane, D. C.; and Smiley, R. W. 1987. Influence of tillage and nitrogen fertilizer on Rhizoctonia root rot (bare patch) of winter wheat. *Plant Disease* 71:125–127.

Puning, L. J., and MacPherson, D. J. 1985. Stem melanosis, a disease of spring wheat caused by *Pseudomonas cichorii*. *Can. Jour. Plant Pathol.* 7:168–172.

Raemaekers, R. H., and Tinline, R. D. 1981. Epidemic of diseases caused by *Cochliobolus sativus* on rainfed wheat in Zambia. *Can. Jour. Plant Pathol.* 3:211–214.

Rakotondradona, R., and Line, R. F. 1984. Control of stripe rust and leaf rust of wheat with seed treatments and effects of treatments on the host. *Plant Disease* 68:112–117.

Rao, A. S. 1968. Biology of *Polymyxa graminis* in relation to soilborne wheat mosaic virus. *Phytopathology* 58:1516–1521.

Reis, E. M.; Cook, R. J.; and McNeal, B. L. 1982. Effect of mineral nutrition on take all of wheat. *Phytopathology* 72:224–229.

Reis, E. M.; Cook, R. J.; and McNeal, B. L. 1983. Elevated pH and associated reduced trace-nutrient availability as factors contributing to increased take all of wheat upon soil liming. *Phytopathology* 73:411–413.

Reis, E. M., and Wunsche, W. A. 1984. Sporulation of *Cochliobolus sativus* on residues of winter crops and its relationship to the increase of inoculum density in soil. *Plant Disease* 68:411–412.

Rewal, H. S., and Jhooty, J. S. 1986. Physiologic specialization of loose smut of wheat in the Punjab state of India. *Plant Disease* 70:228–230.

Richardson, M. J., and Noble, M. 1970. Septoria species on cereals—a note to aid their identification. *Plant Pathol.* 19:159–163.

Roane, C. W.; Roane, M. K.; and Starling, T. M. 1974. Ascochyta species on barley and wheat in Virginia. *Plant Dis. Rptr.* 58:455–456.

Rovira, A. D. 1986. Influence of crop rotation and tillage on Rhizoctonia bare patch of wheat. *Phytopathology* **76**:669–673.

Rovira, A. D., and McDonald, H. J. 1986. Effects of the herbicide chlorsulfuron on Rhizoctonia bare patch and take all of barley and wheat. *Plant Disease* **70**:879–882.

Rowe, R. C., and Powelson, R. L. 1973. Epidemiology of Cercosporella foot rot of wheat: Disease spread. *Phytopathology* **63**:984–988.

Rowe, R. C., and Powelson, R. L. 1973. Epidemiology of Cercosporella foot rot of wheat: Spore production. *Phytopathology* **63**:981–984.

Russell, C. C., and Perry, V. G. 1966. Parasitic habit of *Trichodorus christiei* on wheat. *Phytopathology* **56**:357–358.

Scardaci, S. C., and Webster, R. K. 1982. Common root rot of cereals in California. *Plant Disease* **66**:31–34.

Schafer, J. F., and Long, D. L. 1986. Evidence for local source of *Puccinia recondita* on wheat in Pennsylvania. *Plant Disease* **70**:892 (disease notes).

Scharen, A. L., and Krupinsky, J. M. 1971. *Ascochyta tritici* on wheat. *Phytopathology* **61**:675–680.

Scott, D. B.; Visser, C. P. N.; and Rufenacht, E. M. C. 1979. Crater disease of summer wheat in African drylands. *Plant Dis. Rptr.* **63**:836–840.

Scott, P. R.; Benedikz, P. W.; Jones, H. G.; and Ford, M. A. 1985. Some effects of canopy structure and microclimate on infection of tall and short wheats by *Septoria nodorum*. *Plant Pathology* **34**:578–593.

Sellam, M. A., and Wilcoxson, R. D. 1976. Bacterial leaf blight of wheat in Minnesota. *Plant Dis. Rptr.* **60**:242–245.

Shaner, G., and Powelson, R. L. 1973. The oversummering and dispersal of inoculum of *Puccinia striiformis* in Oregon. *Phytopathology* **63**:13–17.

Sinha, R. C., and Benki, R. M. 1972. *American Wheat Striate Mosaic Virus*. Commonwealth Mycological Institute. Descriptions of Plant Viruses. Set 6. No. 99.

Slykhuis, J. T. 1970. Factors determining the development of wheat spindle streak mosaic caused by a soilborne virus in Ontario. *Phytopathology* **60**:319–331.

Slykhuis, J. T., and Barr, D. J. S. 1978. Confirmation of *Polymyxa graminis* as a vector of wheat spindle streak mosaic virus. *Phytopathology* **68**:639–643.

Smilanick, J. L.; Hoffmann, J. A.; and Royer, M. H. 1985. Effect of temperature, pH, light and dessication on teliospore germination of *Tilletia indica*. *Phytopathology* **75**:1428–1431.

Smiley, R. W.; Fowler, M. C.; and Reynolds, K. L. 1986. Temperature effects on take all of cereals caused by *Philaphora graminicola* and *Gaeumannomyces graminis*. *Phytopathology* **76**:923–931.

Smith, J. D. 1981. Snow molds of winter cereals; guide for diagnosis, culture and pathogenicity. *Can. Jour. Plant Pathol.* **3**:15–25.

Southern, J. W., and Wilcoxson, R. D. 1979. The effect of planting date on rusting character of seven spring wheat cultivars. *Phytopathology* **69**:543 (abstract).

Stack, R. W.; Jons, V. L.; and Lamey, H. A. 1979. Snow rot of winter wheat in North Dakota. *Phytopathology* **69:**543 (abstract).

Stack, R. W., and McMullen, M. P. 1985. Head blighting potential of Fusarium species associated with spring wheat heads. *Can. Jour. Plant Pathol.* **7:**79–82.

Taylor, R. G.; Jackson, T. L.; Powelson, R. L.; and Christensen, N. W. 1983. Chloride, nitrogen form, lime, and planting date effects on take all root rot of winter wheat. *Plant Disease* **67:**1116–1120.

Teich, A. H., and Nelson, K. 1984. Survey of fusarium head blight and possible effects of cultural practices in wheat fields in Lambton County in 1983. *Can. Plant Dis. Surv.* **64:**11–13.

Tinline, R. D. 1981. Effect of depth and rate of seeding on common root rot of wheat in Saskatchewan. *Phytopathology* **71:**909 (abstract).

Traquair, J. A., and Smith, J. D. 1982. Sclerotial strains of *Coprinus psychromorbidus*, a snow mold basidiomycete. *Can. Jour. Plant Pathol.* **4:**27–36.

Vargo, R. H., and Baumer, J. S. 1986. *Fusarium sporotrichioides* as a pathogen of spring wheat. *Plant Disease* **70:**629–631.

Vargo, R. H.; Baumer, J. S.; and Wilcoxson, R. D. 1981. *Fusarium tricinctum* as a pathogen of spring wheat. *Phytopathology* **71:**910 (abstract).

Walker, J. 1975. Take all disease of Graminae: A review of recent work. *Rev. Plant Pathol.* **54:**113–144.

Weller, D. M.; Cook, R. J.; MacNish, G.; Bassett, E. N.; Powelson, R. L.; and Petersen, R. R. 1986. Rhizoctonia root rot of small grains favored by reduced tillage in the Pacific Northwest. *Plant Disease* **70:**70–73.

Weste, G. 1972. The process of root infection by *Ophiobolus graminis*. *Br. Mycol. Soc. Trans.* **59:**133–147.

Wiese, M. V. 1987. *Compendium of Wheat Diseases*, 2nd ed. American Phytopathological Society, St. Paul, MN.

Wilkie, J. P. 1973. Basal glume rot of wheat in New Zealand. *New Zealand Jour. Agric. Res.* **16:**155–160.

Wilkinson, H. T.; Cook, R. J.; and Alldredge, J. R. 1985. Relation of inoculum size and concentration to infection of wheat roots by *Gaeumannomyces graminis* var. *tritici*. *Phytopathology* **75:**98–103.

24

Diseases of Wild Rice (*Zizania palustris* L.)

BACTERIAL CAUSES

Bacterial Brown Spot

CAUSE: *Pseudomonas syringae* pv. *syringae* van Hall.

DISTRIBUTION: Idaho and Minnesota.

SYMPTOMS: Initially a small, water-soaked spot or short streak occurs on the leaf. A diffuse green-brown halo with a dark brown margin rapidly develops around the central water-soaked area. Later the area becomes chestnut or tan. Small or narrow lesions appear uniformly dark brown. Typical lesions are roughly elliptical or spindle-shaped, and 1–10 mm wide by 2–12 mm long; however, they may become as long as 100 mm or more. Lesions are often irregularly shaped or diffuse. Rarely, white exudate is present in the center of a lesion. Frequently the original water-soaked areas will become translucent spots or slits in centers of lesions. In older lesions, these translucent areas may fall out, giving leaves a shot-hole appearance.

CONTROL: No control is reported.

Bacterial Leaf Streak

This disease is also called bacterial streak.

CAUSE: *Xanthomonas campestris* (Pammel) Dowson and *Pseudomonas syringae* pv. *zizaniae* (Bowden and Percich).

DISTRIBUTION: Idaho and Minnesota.

SYMPTOMS: Initially lesions on leaves are narrow, 1–2 cm long, dark green, water-soaked, linear, and follow leaf veins. Eventually the lesions become necrotic, dark brown to black, dry, and may be covered with a glistening crust of bacterial exudate.

CONTROL: No control is reported.

FUNGAL CAUSES

Anthracnose

CAUSE: *Colletotrichum sublineolum* P. Henn.

DISTRIBUTION: Minnesota.

SYMPTOMS: Initially leaf lesions are small (1–2 mm in diameter), dark green-brown, and water-soaked. Later lesions become elliptical to fusiform (0.1–1.5 cm by 0.1–0.6 cm) with light tan centers and dark brown margins. Lesions contain numerous, black, setose, vein-limited acervuli covered with orange-pink masses of conidia. Lesions occur on leaves of aerial plants as well as floating leaves of young plants.

CONTROL: No control is reported.

Ergot

CAUSE: *Claviceps zizaniae* (Fyles) Pantidou overwinters as sclerotia. The sclerotia can drop in the water and be washed to shore where they survive unfavorable periods. In the early summer, particularly when wild rice flowers, the sclerotia germinate to produce stalks on which stroma are produced. Perithecia containing ascospores are produced in the stromatic head. The ascospores are windborne to wild rice flowers and infect them. After infection, a sweet, sticky liquid containing conidia is exuded from flowers. Insects are attracted to the liquid and transmit conidia to healthy flowers. Eventually the hard sclerotia are formed in place of the grain.

DISTRIBUTION: Generally distributed wherever wild rice is grown; however, ergot is rare in commercial wild rice paddies.

SYMPTOMS: Large ergot bodies or sclerotia replace kernels. The sclerotia at first are pink to purple when they are growing on the plants, but eventually turn black and very hard when they mature and dry. Sclerotia vary in size from 3–6 mm in diameter by 3–20 mm in length. Prior to formation of sclerotia, drops of the sweet liquid may be seen on the head together with insects.

Yield losses are usually minimal. Ergot bodies may be harmful if consumed, but are larger than kernels and are readily separated from the grain.

CONTROL: No control other than to remove sclerotia from grain during the cleaning process. Chips and pieces of ergot bodies that remain in grain can be floated out with water.

Fungal Brown Spot

This disease is also called brown spot, Helminthosporium brown spot, and Helminthosporium blight.

CAUSE: *Bipolaris oryzae* (Breda de Hann) Shoem. (syn. *Drechslera oryzae* (Breda de Hann) Subramanian & Jain and *Helminthosporium oryzae* Breda de Hann), teleomorph *Cochliobolus miyabeanus* (Ito & Kurbayashi) Drechs. ex Dastur; and *B. sorokiniana* (Sacc.) Shoemaker (syn. *D. sorokiniana* (Sacc.) Subramanian & Jain and *H. sativum* Pammel, King & Bakke). Both fungi overwinter on straw and stubble as mycelium and are seedborne. Conidia are produced in the spring and are windborne to wild rice. Plants are infected as early as the floating leaf stage. Secondary infection is by conidia produced in new lesions and windborne to initiate new infections. The disease is most severe during daytime temperatures of 30–35° C and a relative humidity between 70 percent and 90 percent at night accompanied by dew.

DISTRIBUTION: Idaho and Minnesota; *B. oryzae* is more common in cultivated paddies and *B. sorokiniana* is more common in natural stands.

SYMPTOMS: All parts of the plant may be infected. The first symptoms are tiny brown or purple lesions on leaves. The lesions develop into different shapes. Some lesions remain 1 mm or less in diameter and retain a very dark brown to purple-black color. Usually lesions caused by *B. oryzae* have a brown center with a yellow margin whereas lesions caused by *B. sorokiniana* normally do not have a halo. Some lesions retain a discrete shape and size of 6–10 mm, whereas others lengthen to 3–4 cm. Some lesions are limited by leaf veins whereas others are not. As lesions become more numerous and larger, leaves turn brown to yellow-brown and die.

Stem and sheath lesions develop similarly to those on leaves. Large necrotic areas develop on stems, frequently causing them to weaken and break. The spikes fall into the water. Infected spikes are bleached and florets and caryopses fail to develop. The result is a reduced number of seeds, severe atrophy, or no seeds produced at all. Yield losses vary from slight to 100 percent.

CONTROL:
1. Paddies should be isolated from each other by 1.6–3.2 km.
2. Plant only clean healthy seed.
3. Crop residue should be removed from the paddy.
4. Paddy soil should be plowed.
5. Wild rice should only be grown for two successive years in the same soil with the soil left fallow and an alternative crop grown for at least one year.

Phytophthora Crown and Root Rot

CAUSE: *Phytophthora erythroseptica* sensu lato. Onset of symptoms coincides with hot, windy weather.

DISTRIBUTION: California.

SYMPTOMS: Disease is either evenly distributed or confined to a portion of a field. However, not all plants are affected, with healthy plants compensating for diseased ones, thereby causing severely affected areas to appear normal. Plants ranging in age from early tillering to grain filling stage are killed. Leaves of infected plants initially turn gray-green, then straw colored, becoming dry and brittle. The crown, first internode, portions of leaf sheaths surrounding them, and many adventitious roots become necrotic. Crowns frequently become so rotted that tillers separate from them and float to the surface of the water, leaving roots embedded in the soil.

CONTROL: No control is reported.

Stem Rot

CAUSE: *Sclerotium* sp. and *Helminthosporium sigmoideum* Cav. These two fungi or similar fungi also cause a stem rot on white rice, *Oryzae sativa* L. On white rice, the sclerotial stage is *S. oryzae* (Catt.) Krause & Webster and the conidial stage of the same fungus is *H. sigmoideum*. On wild rice, both *H. sigmoideum* and *Sclerotium* sp. produce sclerotia that are different morphologically but are likely different stages of the same causal fungus, similar to the situation on white rice.

The fungi overwinter as sclerotia in infected residue and soil but survival is probably better in residue on top of the soil. Sclerotia of *Sclerotium* sp. are also produced in lesions of white water lily, *Nymphaea odorata* Ait., and may serve as a site for overwintering and increase of inoculum. Sclerotia germinate and produce conidia that are windborne or sclerotia float to plants where they germinate and produce infective mycelium. Plants are infected at the water level and soil level and crowns at the time water is removed from the paddy. Disease development is slow until water is removed, then increases rapidly, especially if the weather stays dry and temperatures are above 24° C. Sclerotia are formed in culms and leaf sheaths. The fungus may be disseminated from paddy to paddy by sclerotium-infested seed or plant debris that contaminates seed lots.

DISTRIBUTION: Minnesota.

SYMPTOMS: Initially purplish lesions develop on stems at the water level. Black, round sclerotia less than 2 mm in diameter occur in lesions. Later as paddies are drained, brown lesions develop in the crown and lower 15 cm of the stem. The stems become necrotic, dry, and brittle. Stems break easily at the soil line and culms and leaf sheaths are either partially or wholly destroyed. When culms and leaf sheaths are split open, masses of black sclerotia are embedded in white mycelium. Heads are poorly filled, with seed light in color and underdeveloped.

CONTROL:
1. Plant residue should be removed from paddies.
2. Paddy soil should be plowed.
3. Paddies should be left fallow for at least one year.
4. Plant only cleaned seed, free of any plant residue.

Stem Smut

CAUSE: *Entyloma lineatum* (Cke.) Davis.

DISTRIBUTION: Generally distributed wherever wild rice is grown.

SYMPTOMS: Symptoms usually do not occur until plants are almost mature. Glossy black lesions occur on stems, culms, and heads. The black discoloration is due to chlamydospores being produced under the surface of stems and leaves. As plants mature and dry, the black color changes to a lead gray. Lesions can elongate, coalesce, and girdle the stems near the head; however, there is no evidence that lesions kill immature plants. Stem and leaf infections probably do not cause much damage; however, head infection is likely to reduce seed production.

CONTROL: No control but differences in resistance exist among individual breeding lines and selections.

Zonate Eye Spot

CAUSE: *Drechslera gigantea* (Heald & Wolf) Ito. Smooth brome (*Bromus inermis* Leyss), quackgrass (*Agropyron repens* (L.) Beauv, and reed canarygrass (*Phalaris arundinacea* (L.) can serve as alternate hosts. Volunteer wild rice plants may perpetuate the fungus when the paddy is fallow. Disease is associated with dense stands (10–14 plants per 30 cm sq), emergent debris from previous cropping, 72 hours of continuous 90–100 percent relative humidity, and temperatures of at least 30°C.

DISTRIBUTION: Minnesota.

SYMPTOMS: Initial symptoms on leaves are small (approximately 1 mm in diameter), water-soaked, gray-green lesions with a well-defined brown margin. As lesions enlarge, they become tan to brown in the center, with secondary lesions forming around the primary lesions, producing typical zonate eyespot lesions. Lesions are typically 0.8–1.5 cm but may enlarge to 1–2 cm. Lesions may coalesce and cover the entire area of leaves of susceptible cultivars. Under humid conditions white, prostrate mycelial stands may grow outward from lesion margins.

CONTROL: No control is reported.

Miscellaneous Fungi

Diplodia oryzae Mujake on dead culms.
Doassania zizania J. J. Davis causing a stem smut.
Mycosphaerella zizaniae (S). Lindav causing a leaf spot.
Ophiobolus oryzinus Sacc. causing a culm rot.
Phaeoseptoria sp. causing a leaf spot.
Sclerotium oryzae (Catt.) Krause & Webster from stem and foliar lesions.
S. hydrophilum Sacc. from stem and foliar lesions.

——MISCELLANEOUS NEMATODES——

Radopholus gracilis (de Man) Hirschmann from cortex of roots.
Hirschmanniella gracilis (de Man) Luc & Goodey from paddy soil.

VIRAL CAUSES

Wheat Streak Mosaic

CAUSE: Wheat streak mosaic virus wild rice isolate (WSMV-WR). The mite vector *Aceria tulipae* Keifer is commonly found on wild rice, retains WSMV for several days, and is transported long distances by wind. However, the WSMV-WR vector is not known for certain.

DISTRIBUTION: Minnesota.

SYMPTOMS: Infected plants display typical streak symptoms. As disease progresses the chlorotic areas on the lower leaves become necrotic after 10 days. The necrotic areas eventually coalesce, resulting in death of the leaf.

CONTROL: No control is reported.

BIBLIOGRAPHY

Bean, G. A., and Schwartz, R. 1961. A severe epidemic of Helminthosporium brown spot disease on cultivated wild rice in northern Minnesota. *Plant Dis. Rptr.* **45**:901.

Berger, P. H.; Percich, J. A.; and Ransom, J. K. 1981. Wheat streak mosaic virus in wild rice. *Plant Disease* **65**:695–696.

Bowden, R. L., and Eschen, D. J. 1986. First report of wild rice diseases in Idaho. *Plant Disease* **70**:800 (disease notes).

Bowden, R. L.; Kardin, M. K.; Percich, J. A.; and Nickelson, L. J. 1984. Anthracnose of wild rice. *Plant Disease* **68**:68–69.

Bowden, R. L., and Percich, J. A. 1981. Bacterial leaf streak of wild rice. *Phytopathology* **71**:204 (abstract).

Bowden, R. L., and Percich, J. A. 1981. Bacterial leaf streak of wild rice caused by *Xanthomonas campestris* and *Pseudomonas syringae*. *Phytopathology* **71**:862 (abstract).

Bowden, R. L., and Percich, J. A. 1983. Bacterial brown spot of wild rice. *Plant Disease* **67**:941–943.

Bowden, R. L., and Percich, J. A. 1983. Etiology of bacterial leaf streak of wild rice. *Phytopathology* **73**:640–645.

Dore, W. G. 1969. Wild rice. *Canada Dept. Agric. Res. Branch Pub. 1393.*

Gunnell, P. S., and Webster, R. K. 1988. Crown and root rot of cultivated wild rice in California caused by a *Phytophthora erythroseptica* sensu lato. *Plant Disease* **72**:909–910.

Johnson, D. A.; Stewart, E. L.; and King T. H. 1976. A sclerotium species associated with water lilies in Minnesota. *Plant Dis. Rptr.* **60:**807–808.

Kardin, M. K.; Bowden, R. L.; Percich, J. A.; and Nickelson, L. J. 1981. Zonate eyespot of wild rice in Minnesota. *Phytopathology* **71:**885 (abstract).

Kardin, M. K.; Bowden, R. L.; Percich, J. A.; and Nickelson, L. J. 1982. Zonate eyespot on wild rice caused by *Drechslera gigantea*. *Plant Disease* **66:**737–739.

Kernkamp, M. F., and Kroll, R. 1974. Wild rice diseases in Minnesota. *Univ. of Minnesota Agric. Exp. Sta. Misc. Rept. 125.*

Kernkamp, M. F.; Kroll, R.; and Woodruff, W. C. 1976. Diseases of cultivated wild rice in Minnesota. *Plant Dis. Rptr.* **60:**771–775.

Kernkamp, M. F.; Kroll, R.; and Woodruff, W. C. 1977. Wild rice infected by *Sclerotium* sp. isolated from white water lily. *Plant Dis. Rptr.* **61:**187–188.

Kohls, C. L.; Percich, J. A.; and Huot, C. M. 1987. Wild rice yield losses associated with growth-stage-specific fungal brown spot epidemics. *Plant Disease* **71:**419–422.

Morrison, R. H., and King, T. H. 1971. Stem rot of wild rice in Minnesota. *Plant Dis. Reptr.* **55:**498–500.

Pantidou, M. E. 1959. Claviceps from Zizania. *Can. Jour. Bot.* **37:**1233–1236.

Percich, J. A.; Bowden, R. L.; Kardin, M. K.; and Hotchkiss, E. S. 1983. Anthracnose of wild rice caused by a *Colletotrichum* sp. *Phytopathology* **73:**843 (abstract).

Percich, J. A., and Nickelson, L. J. 1982. Evaluation of several fungicides and adjuvant materials for control of brown spot of wild rice. *Plant Disease* **66:**1001–1003.

Glossary

Acervulus (pl. acervuli): Fruiting body of some fungi that is somewhat saucer-shaped and has a layer of conidiophores and conidia, and sometimes setae. Initially it is subepidermal but usually ruptures the epidermis of the host to expose conidia.

Adventitious: Bud arising in a position other than normal.

Aeciospore: Adikaryotic spore stage in the rust fungi that is produced in an aecium.

Aecium (pl. aecia): Bell-shaped structure in the rust fungi in which aeciospores are produced.

Airborne: To be transported from place to place through the air.

Anamorph: Asexual or imperfect stage of a fungus.

Anastomosis: Fusing together of hyphae.

Apothecium(pl. apothecia): Saucer-shaped fruiting structure on which asci are borne.

Ascospore: Asexually produced spore that is usually borne in an ascus.

Ascus (pl. asci): Usually a microscopic, saclike structure in which ascospores are borne.

Autoecious: Referring to all spore stages, usually in the rust fungi, occurring on one host.

Avirulent: Not able to cause disease.

Bacterium (pl. bacteria): Single-celled microorganism that lacks chlorophyll and reproduces by binary fission.

Basidiocarp: Fungal fruiting body on which basidiospores are produced.

Basidiospore: Sexually produced haploid spore stage that is borne externally on a basidium.

Basidium (pl. basidia): Club-shaped structure on which basidiospores are borne.

Binary fission: Splitting in two.

Biotype: Group of individuals having a similar genetic makeup.

Blight: General and rapid killing of a plant part.

Budding: Development of a new cell or spore from the outgrowth of an old one.

Canker: Sunken, necrotic area on a stem or branch that is usually sharply defined, sometimes by callous tissue.

Chlamydospore: Thick-walled asexual spore that develops from a modified fungal cell.

Chlorotic: Symptom characterized by a yellow color caused by a lack or degradation of chlorophyl.

Cirrhus (pl. cirrhi): Thick, mucouslike material, usually in the shape of a tendril, that carries spores out of a structure in which they were formed.

Cleistothecium (pl. cleistothecia): Round or spherical structure in which asci and ascospores are produced.

Coalesce: Merge, grow, or come together.

Colonize: Refers to a microorganism growing on a substrate.

Conidiophore: Specialized hypha on which one or several conidia are borne.

Conidium (pl. conidia): Asexual spore produced on the end of a conidiophore.

Cyst: Hard, resistant body of a female nematode that contains eggs.

Damping off: Killing of seed or seedlings by microorganisms, usually under wet soil conditions.

Decompose: Break down or rotting of colonized plant tissues, usually by microorganisms.

Desiccated: Dried up.

Disease: Continuous harmful process in plants that is usually caused by a microorganism and manifested by visible morphological changes called symptoms.

Disseminated: To be moved from place to place.

Enations: Small swellings or growths.

Epiphytotic: Rapid increase in a plant disease over time. Similar to a disease epidemic in animals and humans.

Ergot body: The hard, fungal mass or sclerotium that replaces a grain in the disease ergot.

Exudate: Material that has been oozed or extruded out.

Flag leaf: Normally the last leaf to be formed or the uppermost leaf on grains and grasses.

Fungal propagule: Any portion of a fungus that is capable of germinating and infecting a host.

Fungus (pl. fungi): Microscopic plant that lacks chlorophyll and reproduces by means of spores.

Fungicide: Material that kills fungi, usually a chemical.

Gall: Swelling or growth caused by infection or a microorganism.

Germinate: To begin growth of a spore or propagule.

Germ tube: The first mycelial growth during germination.

Halo: Symptom normally characterized by an area of diseased tissue that surrounds a lesion.

Heteroecious: Requiring more than one plant host in order for a rust fungus to complete its life cycle.

Honeydew: Sweet liquid produced by some fungi that is an attractant to insects.

Host: Plant from which a parasite derives its sustenance.

Hypha (pl. hyphae): Single mycelial filament that is usually composed of two or more cells.

Hyphal fragments: Bits or portions of hyphae.

Imperfect stage: Asexual formation of spores.

Incubation: Time in which viruses multiply within a vector. Time between inoculation and symptom development.

Infection: Establishment of a parasite within a host that may or may not be accompanied by symptoms.

Inoculate: To place a microorganism in a substrate or on one.

Inoculum: Portion of a microorganism that is capable of causing infection.

Isolate: Process of obtaining a microorganism from a substrate and getting it into pure culture.

Larva (pl. larvae): Juvenile growth stage of nematodes between embryo and adults.

Lesion: Limited area of diseased tissue.

Macroconidium: Large or long conidium.

Microconidium: Small or short conidium.

Microorganism: Small organism that can be seen only with the aid of magnification.

Microsclerotium (pl. microsclerotia): Small sclerotia.

Mosaic: Symptom of mixed light green, green, or yellow patches.

Mycelium: Mass of hyphae that comprises the vegetative body of a fungus.

Mycoplasma: Procaryotic microorganisms that lack a rigid cell wall, are variable in shape, and are usually smaller in size than bacteria.

Necrosis: Death of plant tissue.

Nematodes: Small, nonsegmented worms that are usually found in the soil.

Obligate parasite: Parasite that must grow and survive on a live host. Special survival structures may allow a means of survival during winter in temperate climates.

Oospore: Thick-walled survival spore of some fungi that may be of sexual or asexual derivation.

Overwintering: To survive by some means during period of plant dormancy or absence.

Parasite: Organism that gets its sustenance from another organism.

Pathogen: Parasite that causes disease.

Perfect stage: Sexual formation of spores.

Perithecium (pl. perithecia): Flask-shaped fruiting structure with an ostiole in which ascospores are formed.

Plasmodium: Mass of protoplasm without cell walls that is multinucleate.

Postemergence: After a seedling has germinated and emerged through the soil surface.

Predisposed: To make a host more susceptible to infection.

Preemergence: After a seedling has germinated but before the seedling has emerged through the soil surface.

Primary inoculum: The first inoculum to parasitize a host in a growing season.

Promycelium: The first structure formed when a teliospore germinates.

Propagule: Part or portion of a microorganism that is usually capable of infecting a host.

Protectant: Fungicide, normally of short efficacy, that prevents fungi from entering a plant.

Pseudostromata: Stromatalike structure.

Pseudothecium (pl. pseudothecia): Peritheciumlike structure.

Pustule: Blister like fruiting structure comprised of host epidermis pushed up by spores.

Pycnidiospore: Conidium produced in a pycnidium.

Pycnidium (pl. pycnidia): Asexual flask-shaped structure in which conidiophores and conidia are produced.

Pycniospore: Haploid sexually derived spore produced in a pycnium. Usually a stage in the life cycle of a rust fungus.

Pycnium (pl. pycnia): Flask-shaped structure in which pycniospores are formed.

Receptive hypha: Specialized hypha that protrudes from the top of pycnium and functions as female gametes.

Residue: Dead plant debris.

Resistance: Property of a host to resist disease development by a parasite.

Rickettsialike bacteria: Microorganism similar to bacteria but capable of only multiplying in living plant cells.

Rot: Breakdown and decomposition of plant tissue by microorganisms.

Rust: Symptom caused by rust fungi.

Saprophyte: Microorganism that lives on dead tissue.

Sclerotium (pl. sclerotia): Hard, compact mass of mycelium that is capable of surviving adverse conditions.

Secondary inoculum: Inoculum produced in infection first caused by primary inoculum.

Seedborne: Carrying or overwintering of inoculum in or on seed.

Seed treatment: The application of a material, usually a fungicide, to control seedborne inoculum.

Seta (pl. setae): Hairlike structures found in certain fruiting bodies.

Smut: Disease caused by smut fungi characterized by masses of dark powdery spores.

Soilborne: Property of a microorganism living and surviving in the soil.

Sorus (pl. sori): Compact mass of spores usually found in smut or rust fungi.

Spermatium: A uninucleate cell that functions in fertilization.

Spiroplasma: Helical-shaped mycoplasma.

Sporangium (pl. sporangia): Structure whose contents differentiate into asexual spores.

Spore: Reproductive unit of bacteria and fungi, similar to seeds of higher plants.

Sporidium (pl. sporidia): Basidiospore.

Sporulation: Act of spore production.

Stroma (pl. stromata): Compact mass of mycelium on which fruiting bodies are formed.

Stylet: Hollow feeding organ of plant parasitic nematodes.

Substrate: Substance on which microorganisms grow.

Survival: Property of microorganisms to exist in some way from growing season to season.

Susceptible: Property of a host to become diseased.

Symptom: Manifestation of a disease by a host.

Syndrome: Symptoms that typify a disease.

Teleomorph: Sexual or perfect stage of a fungus.

Teliospore: Thick-walled resting spore of rust or smut fungi.

Telium (pl. telia): Structure of a rust fungus in which teliospores are formed.

Tolerance: Property of a host that becomes infected to sustain disease without serious damage.

Transmission: Movement of a pathogen to a plant; normally with the aid of another agent.

Urediospore: Asexual spore stage in rust fungi that functions as the summer or repeating stage.

Uredium (pl. uredia): Structure in which urediospores are formed.

Vector: An agent that transmits inoculum.

Viroid: Small, low molecular weight ribonucleic acid.

Virulent: Ability to cause disease.

Virus: Strand of nucleic acid surrounded by a protein coat.

Water-soaked: A disease symptom where plant tissue appears wet.

Wilt: Symptom characterized by lack of turgor.

Windborne: Inoculum transported from place to place by wind.

Zoospore: Fungal spore with flagella that is capable of moving.

Index